PRINCIPLES
of
CELL ADHESION

Edited by

Peter D. Richardson, Ph.D., D.Sc.(Eng.), D.Sc., F.R.S.

School of Engineering
Brown University
Providence, Rhode Island

Manfred Steiner, M.D., Ph.D.

Division of Hematologic Research
Memorial Hospital of Rhode Island
Pawtucket, Rhode Island

CRC Press
Taylor & Francis Group
Boca Raton London New York

CRC Press is an imprint of the
Taylor & Francis Group, an **informa** business

First published 1995 by CRC Press
Taylor & Francis Group
6000 Broken Sound Parkway NW, Suite 300
Boca Raton, FL 33487-2742

Reissued 2018 by CRC Press

A Library of Congress record exists under LC control number: 94011521

Publisher's Note
The publisher has gone to great lengths to ensure the quality of this reprint but points out that some imperfections in the original copies may be apparent.

Disclaimer
The publisher has made every effort to trace copyright holders and welcomes correspondence from those they have been unable to contact.

ISBN 13: 978-1-138-50580-3 (hbk)
ISBN 13: 978-0-203-71085-2 (ebk)

Visit the Taylor & Francis Web site at http://www.taylorandfrancis.com and the CRC Press Web site at http://www.crcpress.com

THE EDITORS

Peter D. Richardson, Ph.D., D.Sc.(Eng.), D.Sc., F.R.S., received his B.Sc. in engineering from the Imperial College of Science, Technology and Medicine of the University of London in 1955, and his Ph.D. from the same institution in 1958. Immediately afterward he joined Brown University, where he has been professor of engineering since 1968 and professor of engineering and physiology since 1984. He has spent periods of leave at various institutions, including the Université de Paris, Orta Dogu Universitesi in Ankara, Abteilung Physiologie RWTH Aachen, and Centro Piaggio of the University of Pisa. He is the author or co-author of more than 180 scientific and engineering publications. Honors and awards include the following: Unwin Scholar (1958), D.Sc.(Eng.) in mechanical engineering of the University of London (1974), Humboldt-Preis (Senior Scientist Award) (1976), D.Sc. in applied physiology of the University of London (1983), election as Fellow of the Royal Society of London (1986), Laureate in Medicine, Ernst Jung Foundation (1987), and Founding Fellow, American Institute for Medical and Biological Engineering (1992).

Dr. Richardson has collaborated with Dr. Steiner, previous to preparing this book, in research on cell adhesion and in teaching a graduate course on thrombosis and hemostasis.

Manfred Steiner, M.D., Ph.D., received his M.D. from the University of Vienna in 1955 and his Ph.D. in biochemistry from the Massachusetts Institute of Technology in 1967. He joined the faculty of Brown University in 1968 as an assistant professor, became an associate professor in 1972 and professor of medicine in 1978. In 1982 he was appointed section leader of the Division of Hematology, Brown University, and Director, Division of Hematology/Oncology at Memorial Hospital of Rhode Island. Honors include Sigma Xi, Fellowship Award Hospital of Boston Medical Foundation, and Visiting Scientist of the Roche Research Foundation, Theodor Kocher Institute, University of Berne, Switzerland. He is a member of the American Society of Hematology, American Physiological Society, American Society for Biochemistry and Molecular Biology, American Association of Pathologists, American Association for the Advancement of Science, and the Federation of American Scientists. He is an associate editor of the *Journal of Atherosclerosis and Thrombosis* published by the Japan Atherosclerosis Society and is a member of the editorial board of the *Journal of Optimal Nutrition.* Dr. Steiner is author or co-author of more than 150 research papers and book chapters, most regarding the subject of platelet function.

PREFACE

Cell adhesion is a process crucially important for virtually all forms of life. As a subject of investigation it has interested researchers of many different scientific disciplines. The last two decades have seen an incredible growth in basic understanding of cell adhesion. As a biological phenomenon, adhesion is a process of considerable complexity. It can, however, be broken down into more easily analyzable components. Over the recent past, many publications have appeared that provide up to date reviews and analyses of various parts of the adhesion process. In the 1960s A. S. G. Curtis wrote a book overviewing physicochemical as well as biological aspects of cell adhesion (*The Cell Surface: Its Molecular Role in Morphogenesis,* Academic Press, 1967, Chapters 3 and 4), and since then monographs have emphasized special aspects (such as J. M. Lackie's fine "Cell Movement and Cell Behaviour," Allen & Unwin, 1986, with its emphasis on motion of adhered cells) without seeking to cover such a wide range of aspects.

We provide the reader with three different aspects of a picture of cell adhesion. These aspects are the underlying themes in the three parts of the book. In the first part of this book, an update is provided on surfaces in general and on theoretical and experimental tools to investigate adhesion to them. This section starts from an illustration of diverse viewpoints of adhesion that are approaching a mutual understanding of the physical and molecular biochemical aspects of cell adhesion and then progresses to allow the reader to gain insight into the thermodynamics and biodynamics of cell adhesion and to understand the surface chemical factors necessary for selective cellular retention. The biological significance of adhesion for unicellular organisms is best documented by the example of *Dictyostelium discoideum* and is continued by illustrating the importance and mechanism of adhesion in bacteria. Bioadhesives provide a different slant on cell adhesion, and leukocyte adhesion is examined as an example of how flow and a surface that is cellular — the vascular endothelium — impact on the adhesion process for single cells supported in suspension.

In the second part of this book, the basic biochemistry and molecular biology of membrane proteins important for adhesion are examined. Recent years have been filled with reports on cell adhesion molecules, thanks especially to modern molecular biology techniques. Some of the membrane proteins of cells provide specific receptors for these adhesion molecules. The interaction of membranes and their constituents with the cellular cytoskeleton as well as changes that adhesive phenomena confer upon these structures are also addressed.

In the third part, the physiology of adhesion is highlighted. The platelet, a cell singularly destined to undergo adhesive processes, is probably better suited than any other cell to studies of adhesion-related phenomena. Certainly, the platelet is the cell whose adhesive physiology and pharmacology have been spotlighted particularly intensively in the past two decades. From the standpoint of basic understanding of both cell adhesion and of the clinical significance of this process for normal physiological processes (e.g., hemostasis) as well as for pathological events (e.g., thrombosis), investigation of the platelet can give answers unequaled by studies of any other cell type. This third part also addresses an aspect of adhesion not readily recognized as such, i.e., homing of cells, using the hemopoietic stem cell as a model. Much material is new.

We expect that this book will be of value, not only to the novice, but also to the expert who may wish to broaden his or her view of this subject.

CONTRIBUTORS

Kailash C. Agarwal, Ph.D.
Division of Biology and Medicine
Brown University
Providence, Rhode Island

Steven M. Albelda, M.D.
Department of Pulmonary and Critical Care
Hospital of the University of Pennsylvania
Philadelphia, Pennsylvania

Robert E. Baier, Ph.D.
Department of Biomaterials
State University of New York at Buffalo
Buffalo, New York

Konrad Beck, Ph.D.
Research Unit
Shriners Hospital for Crippled Children
Portland, Oregon

Josef Beuth, Ph.D., M.D.
Institute of Medical Micrbiology and Hygiene
University of Cologne
Cologne, Germany

Yong Q. Chen, Ph.D.
Department of Radiation Oncology
Wayne State University
Detroit, Michigan

Doloretta D. Dawicki, Ph.D.
Department of Medicine
VA Medical Center
Providence, Rhode Island

Paul A. DiMilla, Ph.D.
Department of Chemical Engineering
University of Pennsylvania
Philadelphia, Pennsylvania

Jonathan M. Edelman, M.D.
Department of Clinical Development
Merck & Co., Inc.
West Point, Pennsylvania

Donna R. Fontana, Ph.D.
Department of Microbiology
University of Minnesota
Minneapolis, Minnesota

Joan E. B. Fox, Ph.D.
Childrens Hospital Oakland Research Institute
Oakland, California

Tanja Gruber, Dipl. Ing.
Department of Biochemistry and Molecular Biology
The Pennsylvania State University
University Park, Pennsylvania

Karolyn M. Hansen, Ph.D.
College of Marine Studies
Cannon Laboratory
Lewes, Delaware

John J. Hemperly, Ph.D.
Department of Cell Biology
Becton Dickinson Research Center
Research Triangle Park, North Carolina

Kenneth V. Honn, Ph.D.
Cancer Biology Division
Department of Radiation Oncology
Wayne State University
Detroit, Michigan

David A. Jones, Ph.D.
Cox Laboratory for Biomedical Engineering
Rice University
Houston, Texas

Eric Larsen, M.D.
Department of Pediatric Hematology and Oncology
Dartmouth Hitchcock Medical Center
Lebanon, New Hampshire

Larry V. McIntire, Ph.D.
Institute of Biosciences and Bioengineering
Rice University
Houston, Texas

Anne E. Meyer, Ph.D.
Industry/University Center for Biosurfaces
State University of New York at Buffalo
Buffalo, New York

S. Fazal Mohammad, Ph.D.
Artificial Heart Research Laboratory
University of Utah
Salt Lake City, Utah

Peter D. Richardson, Ph.D. D.Sc.(Eng.), D.Sc., F.R.S.
School of Engineering
Brown University
Providence, Rhode Island

Karen H. Rittle, Ph.D.
Medical Department
Abbott Laboratories
Abbott Park, Illinois

Leszek M. Rzepecki, Ph.D.
Department of Immunological Diseases
Boehringer Ingelheim Pharmaceuticals, Inc.
Ridgefield, Connecticut

C. Wayne Smith, M.D.
Department of Pediatrics and Microbiology and Immunology
Baylor College of Medicine
and
Leukocyte Biology Section
Texas Children's Hospital
Clinical Care Center
Houston, Texas

Manfred Steiner, M.D., Ph.D.
Division of Hematologic Research
Memorial Hospital of Rhode Island
Pawtucket, Rhode Island

Dean G. Tang, Ph.D., M.D.
Department of Radiation Oncology
Wayne State University
Detroit, Michigan

Mehdi Tavassoli, Ph.D., (Posthumous)
Research Department
Department of Veterans Affairs Medical Center
Jackson, Mississippi

Gerhard Uhlenbruck, M.D.
Institute of Immunobiology
University of Cologne
Cologne, Germany

J. Herbert Waite, Ph.D.
Department of Marine Biology/Biochemistry
College of Marine Studies
University of Delaware
Newark, Delaware

Cheng Zu, Ph.D.
School of Mechanical Engineering
Georgia Institute of Technology
Atlanta, Georgia

CONTENTS

SECTION I: Fundamental Facets of Adhesion

Chapter 1
Physical and Molecular Biochemical Aspects of Cell Adhesion: Viewpoints Approaching a Mutual Understanding 3
Peter D. Richardson

Chapter 2
Biomechanics and Thermodynamics of Cell Adhesion 23
Cheng Zhu

Chapter 3
Influences of Surface Chemical Factors on Selective Cellular Retention 41
Robert E. Baier, Karen H. Rittle, and Anne E. Meyer

Chapter 4
Dictyostelium discoideum Cohesion and Adhesion 63
Donna R. Fontana

Chapter 5
Adhesive Properties of Bacteria 87
Josef Beuth and Gerhard Uhlenbruck

Chapter 6
Bioadhesives: DOPA and Phenolic Proteins as Components of Organic Composite Materials 107
Leszek M. Rzepecki, Karolyn M. Hansen, and J. Herbert Waite

Chapter 7
Flow Effects on Leukocyte Adhesion to Vascular Endothelium 143
David A. Jones, C. Wayne Smith, and Larry V. McIntire

SECTION II: The Molecular Biochemistry of Adhesion

Chapter 8
The Integrin Cell Adhesion Molecules 163
Jonathan M. Edelman, Paul A. DiMilla, and Steven M. Albelda

Chapter 9
The Structure and Biological Function of the Neural Cell Adhesion Molecule N-CAM 187
John J. Hemperly

Chapter 10
The Selectin Family of Adhesion Molecules 203
Eric Larsen

Chapter 11
Structure and Assembly of Basement Membrane and Related Extracellular Matrix Proteins 219
Konrad Beck and Tanja Gruber

Chapter 12
The Interaction of the Cytoskeleton with Adhesive Receptors: Effects on Cell Adhesion 253
Joan E. B. Fox

SECTION III: Physiology of Adhesion

Chapter 13
Homing of Cells: Stem Cell as a Model 271
Mehdi Tavassoli (Posthumous)

Chapter 14
Platelet Aggregation and Inhibitors 287
Kailash C. Agarwal

Chapter 15
Platelet Adhesion 307
Manfred Steiner

Chapter 16
Interactions Among Platelets, Tumor Cells, and the Vessel Wall 317
Kenneth V. Honn, Yong Q. Chen, and Dean G. Tang

Chapter 17
Pharmacological Intervention of Platelet Adhesion 337
Doloretta Dawicki

Chapter 18
Platelet Adhesion to Damaged Vessel Wall 349
S. Fazal Mohammad

Index 369

Section I

Fundamental Facets of Adhesion

Chapter 1

Physical and Molecular Biochemical Aspects of Cell Adhesion: Viewpoints Approaching a Mutual Understanding

Peter D. Richardson

CONTENTS

I. Introduction: Three Complementary Viewpoints ... 3
II. Macroscopic Viewpoint ... 4
III. Atomic/Molecular Viewpoint ... 5
IV. Cellular and Subcellular Viewpoint ... 6
V. Mutual Reconciliation Between the Three Viewpoints ... 7
VI. Some Methods Used in Quantifying Cell Adhesion ... 7
VII. Investigation Stimulated by Prospective Applications ... 9
A. Adhesion and Aggregation of Red Blood Cells ... 9
B. Adhesion and Aggregation of Blood Platelets ... 10
C. Adhesion of Platelets in the Presence of a Flow Over a Surface ... 10
D. Adhesion of Endothelial Cells in the Presence of Flow ... 18
E. Adhesion of Fibroblasts in the Presence of Flow ... 18
VIII. Concluding Remarks ... 19
References ... 19

There is an old legend that solids are the work of God and that surfaces are the work of the Devil. Solids, especially crystalline solids, have a structure that reflects the harmony of creation, while surfaces often have a form that represents disharmony, fracture, and the end of order. People who investigate surface phenomena (including, presumably, cell adhesion) are studying the works of the Devil; and in the past, societies used to punish those attracted to study the works of the Devil. In more modern and enlightened times, societies no longer punish those who study such matters, as it is now recognized that such studies are punishment enough themselves.

I. INTRODUCTION: THREE COMPLEMENTARY VIEWPOINTS

Adhesion, like other surface phenomena, can be viewed from at least three positions. Each conceptual position prevails as being particularly useful at a specific range of physical scale relevant in some phase of the adhesion process and sometimes is illuminating at other scales, too. One is the macroscopic viewpoint, which (like classical mechanics or thermodynamics) does not invoke any specific theory about the nature of matter. Instead, a macroscopic theory makes use of certain laws and defines coefficients that can be measured, such as (in classical, macroscopic mechanics) the coefficient of restitution of a ball when it bounces. A second position is the atomic viewpoint. For this there are again certain laws of interactions. Electric fields associated with the charges on protons and electrons are very important at this scale. Some ideas from macroscopic theories are borrowed and used in the atomic viewpoint. For example, for two atoms that collide and then separate, the coefficient of restitution is 1.0 exactly. A third viewpoint is intermediate in scale between these two. It considers structures that are microscopic yet already complex from the atomic viewpoint. It is the viewpoint that, for example, accepts membranes of cells as having specialized receptors that are distinguishable, yet for which detailed structures may not be known. This viewpoint looks especially at biochemical aspects and is the viewpoint taken by many of the authors of other chapters in this review.

Studies of cell adhesion based on the purely macroscopic viewpoint have not been sufficient in resolving questions, although they are useful. Studies based on the purely atomic-level microscopic viewpoint have also not been sufficient. And studies from the third viewpoint, which is based at the

0-8493-4559-6/95/$0.00+$.50

cellular and membrane scales, have been hampered by inadequate knowledge of cell structure, biochemistry, and physiology, as well as inadequate knowledge of surfaces. Great strides, especially from the third viewpoint, have been made possible by improvements in labeling (e.g., monoclonal antibodies) and in microscopy of events as they occur *in vivo* and *in vitro*. New insight has come from the capability for cloning receptors themselves and sequencing their primary structure. Developments have also come from studies of receptors and ion channels by various methods. These methods have allowed advances, without yet giving enough information that a predictive science of cell adhesion is established.

A basic problem in studying cell adhesion is that experimental methodologies available do not allow as close and direct an examination of what occurs as we would like, for all the cells of interest. If we could watch what occurs, in real time, with a spatial resolution down to a few angstroms, *in vivo,* we would have direct kinematic and structural information about the adhesion process of cells. High resolution currently requires electron microscopy, and the physical and chemical conditions for it (vacuum, electron-dense specific stains) are incompatible with real-time use *in vivo.* Optical microscopy affords much less resolution, and even with optical contrast enhancement (phase or interference methods) the real-time methodology affords little more than a two-dimensional view; the methods for three-dimensional optical microscopy, such as confocal scanning, are too slow for the time and space scales involved in cell adhesion. We are put in a position somewhat like that in astronomy, needing combinations of techniques and viewpoints for inference about processes we cannot fully reach to see as we would like. Macroscopic modeling of planetary orbital mechanics and postulation of a law of gravitational attraction from a distance were enough to represent solar system structure. Spectroscopy of elements, basically a microscopic (quantum) viewpoint, provided rich information about stellar structure. The development of synthetic membranes as simplified models of real membranes and the cloning of receptors and expressing them then on other cells (e.g., CHO cells) have allowed investigation by separating functional components of cell membranes for study. The development of antibody techniques has given biomolecular-scale probe opportunities to cell adhesion study as richly as spectroscopy did to astronomy, and the harvest of information from use of these techniques is not yet all at hand.

II. MACROSCOPIC VIEWPOINT

Even within the framework of the purely macroscopic viewpoint, there are two mainstreams of thought and representation, complementary to each other, for considering the mechanics. One springs from classical mechanics as it is applied to particles; adhesion is viewed as a process executed by particles that follow various rules. For particles that can be treated as points, or as simple spheres, that move within a fluid (liquid or gas), the study of adhesion processes is part of physicochemical hydrodynamics. The other macroscopic viewpoint springs from classical mechanics as it is applied to continua. In this, the adhering entities themselves have continuum-like aspects, rather than being represented as points or spheres, such as tapes, layers of conforming polymeric substances, and so forth. A now classical example, which bears upon cells that have been studied for certain adhesion behavior, is the use of continuum-mechanical representation of red cell membranes as material capable of a specific elastic response to bending forces.[1,2] This type of representation is not limited to just one cell type but can be applied to membranes widely.[3] After some years of using simple forms of representing the elastic response, we can note questions arising as to the need in some cases to take a more complex view.[4]

This need has also contributed to studies of cells of which the adhesion and locomotion may depend on concentrations and gradients of chemical agents diffusing in their surroundings, and correspondingly the *in vitro* techniques may be designed to facilitate calculation of the concentration distribution of such agents in the immediate surroundings of the cells.[5] Results of this study, cast in the macroscopic framework, show that the front face of a cell, locomoting in the micropipette used in the Usami technique,[5] experiences increasing concentration of the chemoattractant during cell motion. The sequence of events includes receptor binding of the chemoattractant, signal transduction, polymerization of the cytoskeleton at the membrane of the front (advancing) face of the cell, spatially-varied adhesion to the micropipette wall, and localized contraction of the cytoskeleton. In performing such studies, it is typically necessary to make some assumptions regarding processes at the submacroscopic scale, e.g., (as in Reference 5) that the reaction kinetics (for association and dissociation) of the ligand–receptor complexes is not much affected by the movement of the receptors in the plane of the cell membrane.

A frequent partner in the analysis of topics by classical mechanics is thermodynamics. A recent analysis in this framework[6] assumes that adhesion cell-to-cell and cell-to-surface is mediated by binding of specific membrane receptors to form crossbridges, and the cross-linking molecules are assumed to be laterally mobile on the plane of cell surfaces by diffusion. Nonspecific and specific binding effects can be treated.[7] (See especially Chapter 2 of this volume.)

Mathematical modeling consists of proposing specific rules and working out consequences in circumstances of interest. Where these circumstances match those occurring in experimental situations, model results may be compared with experimental results. An example is provided by the analysis of Skierczynski et al. cited above.[5] As another example, if one starts with an initially uniform dispersion of small spherical particles that can adhere together (making "dimers"), provided they achieve a certain minimum distance of mutual approach and if the particles move with Brownian motion, it is possible to assess the rate of formation of dimers. If the same circumstances apply but the rate of collision due to Brownian motion is small compared with that due to encounters in a uniform shear flow, the rate of formation of dimers can also be predicted. If the rate of formation observed is smaller by some regular amount, the theory can be adapted by ascribing a collision efficiency (less than unity) to characterize the process. This can give a consistent description but leaves unanswered the question of why, at the level of structural detail of the particles, there is a collision efficiency of less than unity. Are there specific, local "sticky" spots which, if they fail to be close enough during a collision, do not exert enough mutual force to prevent separation? Or are there local sticky spots in different stages of deployment, which (in the case of cells) can be manipulated pharmacologically? A macroscopic theory does not answer these questions, but experiments analyzed in its terms allow coefficients to be measured. Further, experiments carried out and analyzed in macroscopic terms can be enormously helpful in screening candidate pharmacological agents for possible effects on collision efficiency and therefore on potential physiological significance.

An aspect of the use of continuum models for detachment is the model where the membrane peels away from the surface to which it has attached, like a tape. One example of this was illustrated recently using a red blood cell doublet (each of the common biconcave form), of which one cell was adherent to the floor of a flow channel in which an oscillating flow could be set up.[8] By observing a particle adherent to a spot on the upper surface of the membrane of the upper cell of the pair, it was determined that the doublet separates by rolling of the membrane rather than by sliding of the sheared cell. A recent example of the use of the peeling model[9] is applied to the adhesion of glioma cells to fibronectin, with a match around 1.4 to 3.4×10^5 receptors per cell.

Adhesion as that which occurs with coated tapes and other surfaces has long been of commercial significance. Standardized methods for testing have been developed, cast within the macroscopic viewpoint. For example, the American Society for Testing and Materials (ASTM) has been issuing descriptions of standardized methods for testing, developed by committees that seek practical consensus with appropriate reflection of scientific advances that have shown themselves reliable in the hands of different investigators. The methods are basically macroscopic, to allow testing of materials produced by different manufacturers, and with sufficient specification of controls needed of associated factors that there is reproducibility and comparability of results.[10]

A physicochemical interaction that has drawn attention involves the wettability of the surfaces that are in contact. (An author of chapter 3 in this volume, Dr. R. Baier, has published extensively on this topic.) By the nature of the tests for investigating wetting, this is an assessment associated with midrange forces. It is not capable of representing the effects of specific cell receptors and short-range interactions. A recent study of cell adhesion[11] illustrates well the combination of optical measurements and interpretation through a wetting model.

III. ATOMIC/MOLECULAR VIEWPOINT

A seminal study from the microscopic, biophysics viewpoint was made by Bangham and Pethica.[12] One of the important concepts that they brought forward is that cell membranes carrying fixed charges of like type (i.e., negative) would tend to repel each other and thereby make it difficult to achieve small distances of approach when their radii of curvature were large. Correspondingly, some cells undergo morphological changes and thereby provide parts that have small radii of curvature. With these parts, the minimum distance of approach allowed by fixed, like-surface charges is reduced. Thus, long thin processes, or pseudopodia, would (for a given surface charge density) allow a smaller distance of approach to be achieved, and especially so if two thin processes crossed each other at some mutual angle, limiting the region of mutual charge-based repulsion to a very small area. This would give an intrinsic advantage to cells that can develop ruffles, and cells that can develop pseudopodia, in achieving adhesion.

Matching the apparently selective ignoring of masses of theory, as noted by Pethica[26] in a review in 1980, the many works describing identification of specific receptors and ligands have continued the diminished interest in the less specific attraction-repulsion analysis based on surface charge density, although the issue of cell fusion remains a driving force for continued studies.[13-15] However, the modeling of cell membrane shapes (with the aid of computer-based structural modeling techniques, as are discussed in Reference 16, also unavailable at the time of Reference 12) has maintained an interest in the microscopic model of membranes.

Even within the microscopic viewpoint, there are different hypotheses about the submolecular-scale mechanisms of adhesion. One is the diffusion hypothesis: mutual diffusion of polymers at the interface leads to intertwining and molecular cross-linking. Where ligands are involved, the diffusion of ligand molecules into positions where they can perform their function is a related issue. This may be affected by the distance of separation of the membranes. Another hypothesis about the mechanism of adhesion is the adsorption hypothesis: ionic and covalent bonds may form across the interface without requiring diffusion of ligands. There may be movement of receptors in the plane of the membrane, especially when ligands are tied in. These hypotheses lead to different predictions of the mechanical strength of adhesion as a function of time. They are not mutually exclusive.

Adhesion dominated by physicochemical forces has received attention as the basis for the theory of flocculation. Flocculation is the process that occurs when colloidal particles are drawn together, with forces of attraction overcoming forces of repulsion. The analogy with cells arises from the existence of like charges on the surfaces of flocculating particles and on the surfaces of cells. Cells typically have a net negative charge because of charge-bearing residues on the membrane, e.g., sialic acid residues. The cells differ from the flocculating particles in that the charges are clearly localized. However, the number of charged sites is large (e.g., about 50,000 for a platelet) and the electrostatic interaction should be found.

Yet another mechanism for adhesion is the introduction of specific connecting bridges, or ligands, between cells or between a cell and surfaces. There are variations on this concept that allow for specificity on the one hand and variability of the adhesion spot on a noncell surface. And, of course, virtually all glycoproteins have electric charges. Some ligands are known to have fixed charges at specific locations in their molecular structure (e.g., fibrinogen), and knowledge of these assists in stereochemistry studies of adhesion.

The Debye length is the distance over which a local fixed charge on a surface causes an imbalance in local ion concentrations in a solution in contact with the surface, compared with that in the bulk. In physiological saline solution, the Debye length is about 8 Å.[17] This is small, even in comparison with many ligands.

IV. CELLULAR AND SUBCELLULAR VIEWPOINT

It was recognized quite early in the study of cell behavior that cells must be able to attach themselves to their typical environment. Many cells originate in a coherent tissue structure, already rich in mutual adhesion. Such cells do not have to cope with being free of attachment, carried in a flow, and then becoming attached. Some cell types (e.g., platelets, white blood cells, lymphocytes) are adapted to this, however. The attachment could be by some mechanical anchorage, for example, small processes that are able to protrude locally from the cell membrane and anchor like fingers or feet if the typical environment is textured or structured on an appropriate scale, or have particular properties, such as Dembo's "catch bonds".[18] When the existence of the cytoskeleton is considered, and it is recognized that a cell is able to exert appropriate forces on the reverse (internal) side of its membrane to match local forces applied from outside, it is possible that cells can find local sites that they can grip even if the environment is smooth on the scale of the cell size. The local secretion of a very viscous fluid could provide a degree of local attachment sufficient for a cell to be anchored effectively for a while, even if it slipped slowly over the viscous fluid over a period of time. Some cells could alter the environment itself (as indeed monocytes appear to do, for example) and through this modification achieve a more favorable setting for anchorage even by mechanical means. Such speculations have yielded to identification of specific adhesion receptors, such as platelet glycoprotein IIb-IIIa.[19]

An investigative strategy followed with many complex cells is to try to isolate components to study them individually (as with patch-clamp techniques) or to examine behavior in the absence of a feature in a genetically-different strain, e.g., examining red cell membranes in spectrin-free mice, or with overexpression of some feature in transgenic animal models. Specialized receptors for adhesion have been cloned and expressed on other cells to facilitate study. Techniques have been developed, moreover, to produce and experiment with lipid model membranes, which might be regarded as receptor-free models for cell membranes. Where these are physically and chemically well-defined, they have been used to assess adhesion and fusion phenomena.[13,14] Where the lipid membrane is made of a single component, correlation between the threshold of vesicle fusion and increased surface tension or decreased surface dielectric constant holds well. The increased interfacial tension or decreased surface dielectric constant corresponds to the increased hydrophobicity of the membrane surface.[15] The hydration pressure forms a major barrier to the close approach of two polar or charged surfaces for small separations, e.g., 5 to 20 Å. For a variety of lipid bilayer membranes, it has been found that the hydration pressure decays exponentially with increasing fluid separation between the layers.[20] The decay length for hydration pressure measured by McIntosh et al. is about 2 Å.

V. MUTUAL RECONCILIATION BETWEEN THE THREE VIEWPOINTS

In principle, one expects hypotheses and results to provide mutual reconciliation between the three viewpoints. The magnitudes of forces understood at the microscopic level should reconcile with the magnitudes of forces experienced at cell-membrane receptors when the chemical structures of the receptors and ligands are known. Angarska et al.[11] have proposed that there are three principal parameters associated with cell adhesion, specifically F, the minimum force required for detachment, A, the area of contact between two adhering surfaces, and H, the typical distance between two adhering surfaces. The distance H is particularly important because it may enable one to calculate the forces of interaction. Broadly, for H >1000 Å, the forces are determined predominantly by the flow field. The drag on particles approaching each other should be predicted from classical Stokes flow theory, or such special-case analyses as Poe and Acrivos[21] for particles in shear flows. For 1000 A > H > 200 Å, for the cells of sizes of interest here, there should be long-range interactive (van der Waals and electrostatic) forces as well as hydrodynamic forces. For H <200 Å, the forces should be dominated by short-range intermolecular forces (hydration, steric, etc.) and forces from chemical bonds. These correspond roughly to the concept of the staging of adhesion through adsorption, contact, and attachment.

Thus, macroscopic theories, which include aspects such as continuum fluid mechanics, may function on scales of closest approach down to a fair fraction of the dimensions of a red blood cell, or platelet, and the force fields associated with details of membrane structure function at length scales that are less than a tenth of the size of the cells involved. A ligand, fibrinogen, for red cell to red cell adhesion is about 465 Å in length overall.[22] The extremes remind one of the difference in golf between driving from the tee to the green (the phase in which the rolling of the ball on the ground has a very minor role) and putting (where the behavior of the ball in rolling on a slightly curved surface is very important).

A good example of reconciliation of the different approaches has been provided very recently,[23] where the variability in cell detachment kinetics[24,25] has led to seeking an understanding of the possible heterogeneities of specific receptor binding. As the authors remark, "... a healthy tension always exists between rigor and simplicity, which can only be settled by utility to the problem at hand," and the fact that macroscopic approaches typically allow predictive formulations that can be used in all sorts of different circumstances, once some coefficients are known well enough, additionally stimulates reconciliation between the different viewpoints.

Historically speaking, the microscopic viewpoint is identified as the approach of physical chemists to adhesion, the macroscopic viewpoint is the approach of fluid dynamicists and engineers to adhesion and aggregation, and the cellular and subcellular viewpoint is the original direction of approach of biochemists and cell physiologists. In 1980 Pethica wrote: "It is a pleasure to note that the biochemists and cell physiologists have meanwhile selectively ignored the mass of theory in favour of the experimental pursuit of the facts of the case, and the weight of evidence for specific effects (structural and chemical) is rapidly growing. This in no way diminishes the value of long-range force arguments of the DLVO approach, or of broad phenomenological descriptions of the surface energy type, but more attention will be required to the stereochemical subtleties that biological intuition would suggest in cell contact processes...."[26] In view of the great and fruitful efforts on ligands and receptors in the 1980s, these words were timely and apt.

VI. SOME METHODS USED IN QUANTIFYING CELL ADHESION

Methodology of investigation clearly has an effect on what may be determined; and with cell adhesion, where there are details of the process that are not all well understood, there is a particular prospect that specific protocols detect and measure only some of the factors involved. The fruitfulness of applying a specific viewpoint, e.g., for macroscopic, microscopic, or cell/membrane levels, is not immediately obvious. Thus, it is useful to review some of the methods that are used and to consider what aspects of the cell adhesion process they will and will not determine. In particular, there is the issue that measuring forces at small physical scales is not easy, and there is, therefore, an effort to obtain information about forces indirectly from observations of motion. This approach was so successful in observing the motion of the planets of the solar system and inferring from the detailed motion the forces mutually experienced by the bodies in the system, that it has been used widely outside astronomy. Some of the experimentation on cell adhesion is very simple in its physics aspects and can tell little or nothing about forces of adhesion between a cell and a surface. Some of the experimentation — usually with a protocol that makes such experiments distinctly more expensive to perform — is more telling quantitatively about the physical forces of adhesion.

The development of manufactured devices for use in the human body, such as synthetic vascular grafts to replace diseased or damaged segments of arteries, cardiac valves, fully artificial hearts, and so forth (all of which are classed as artificial organs), requires surfaces that can handle flowing blood for extensive periods of time without promoting reactions that defeat their purpose, such as thrombotic occlusion. This has led in particular to studies of cell adhesion that include well-described flows. These have provided both need and better opportunity for determination of forces present in cell adhesion.

In vitro testing of cell adhesion at its simplest is carried out using cell suspensions placed into wells. The suspensions are not stirred mechanically, nor is flow of the suspension caused by other means. After a predetermined period of time has passed, the fraction of cells present in the initial inoculum that have adhered to the well surface is determined. The well surface may be as-supplied by the manufacturer or modified through a protocol that causes a coating to be made on the surface, e.g., of fibronectin. The adhesion is not generally "challenged" by imposition of a defined and maintained force field. However, the protocol usually assures that cells counted have not simply sedimented onto the surface. The forms assumed by the adherent cells are often examined by microscopy and an assessment made whether the adherent cells form a confluent layer or not. The shapes taken by the adherent cells are often indicative of their adhesion, and adhesion is "proven" by their shape rather than by showing that they cannot be dislodged by a specific force applied. For many cell suspensions, a confluent monolayer on a surface is sufficient to inhibit further adhesion, and this observation at the cellular level is sometimes taken over to the macroscopic viewpoint by asserting there that the opportunity for adhesion is proportional to the fraction of the surface still free of adherent cells.[27] A useful review of measurement of adhesion in simpler systems was published over a decade ago.[28]

Blood platelets have a considerable tendency to adhere together (aggregate) in clumps, shaped roughly like beehives when attached to a surface over which blood flows, even when only a few of the possible sites for their adhesion are occupied, so platelet attachment as measured by, say, radioactive labeling, is not definitive for adhesion of the labeled platelets to a surface. Adhesion and aggregation are somewhat different processes. Adhesion involves an attachment between a cell and a surface that is not that of a like cell, which can be maintained in the presence of a finite disturbing force. Aggregation involves attachment between like cells, initially free and separate from each other, which can be maintained in the presence of a finite disturbing force. With blood platelets and normal blood red cells, for example, adhesion and aggregation are distinct phenomena.

An approach rooted in the thermodynamic and continuum approach to cell membrane properties was applied to a study of adhesion of human red cell membranes to the surface of a specially designed experimental chamber (Angarska et al.[11]). With RBC, they found that there seemed to be three types of adhesion "films" between the RBC and the glass surface, based on the approach of Ivanov.[29] They speculated that the differences in area were caused by different ages and membrane properties of the individual RBC.

It was proposed in the 1970s that adhesion of cells occurs in four distinct phases: adsorption, contact, attachment, and spreading.[30] It was also supposed in the 1970s that adhesion was a one-time phenomenon for a site under most circumstances. The force of detachment (or, more precisely, the minimum force required to achieve detachment), F, might be assessed by applying a shearing flow over a flat wall on which a cell was attached, but unfortunately it is not easy at all to assess the force applied to a cell in this manner. It may be more accurate to apply this approach when examining the detachment of thin "rafts" of tissue, measuring in the order of 200 μm across and up to about 10 μm thick. For example, in some experiments in our laboratories on the fluid shear required to remove dental plaque from glass substrates on which it had been cultured, such rafts were observed to become detached and included not only cells but also segments of a highly fibrous and largely planar network laid down by the cells as constituting each "raft". The shear rates of water flow over the surfaces required for detachment were of the order of

20,000 s^{-1} and varied between different cell lines of plaque in culture. The force applied had to tear apart the filaments of the fibrous network, at least at the perimeter of the "raft", as well as detach cells within its framework.

A different approach to assessing the force required for detaching a cell has involved direct pulling on individual members of cell doublets.[31] An estimate for F of 10 to 20 pN was obtained, and this was interpreted as the force required to extract the receptor from the cell membrane. Another method has involved putting doublets into shear flows; these do not involve directly touching parts of the cells, but because the cell doublets rotate, the strength of the force applied varies over time.[32] For this study, the cells were fixed and sphered before being cross-linked by antibodies. This method gave an estimate of the hydrodynamic normal force over the range of 2 to 200 pN. In this latter study, it was noted that such doublets break up under shear, but with times-to-breakup ranging over a factor of ten. The average time to breakup decreased with increasing force, and the fraction of a cohort of doublets that broke up decreased with increasing levels of IgM. For breakup under a slow acceleration, the similarity of the force required for breakup with either IgM or IgA was considered compatible with Evans' conclusion that the receptor is extracted from the cell membrane.

Several years ago, the relative importance in cell suspensions of cell–cell encounters by Brownian motion and by shear-enhanced, orderly laminar motion was assessed.[33] This study also provided an estimate of cell–cell adhesion efficiency for the cell type investigated (blood platelets), and showed that efficiency was much increased when the experiment was performed after ADP was introduced. A recent study[34-36] has taken this approach further. A population-balance method[34] predicts the evolution over time of the spectrum of particle sizes (including aggregates, and allowing disaggregation), and is compared with experiments at high shear rates (up to 8000 s^{-1}) in a plate-cone viscometer,[35] in which the collision efficiency is found to be very small (0.01 or less), and with experiments at a more modest shear rate (450 s^{-1}) but in which ADP is added, so that aggregation is stimulated but also disaggregation can occur, a breakage frequency, 9/s, being assumed as a function of the particle volume.[36]

VII. INVESTIGATION STIMULATED BY PROSPECTIVE APPLICATIONS

As in many sciences, there has been a stimulus for investigation into cellular adhesion to nonbiological surfaces arising from prospective applications. Marine cellular adhesion is discussed in Chapter 6 of this volume, and the examples cited here come mainly from interests driven by the desires to create artificial internal organs, which can even be hybrid: natural tissues supported on a fabricated system, the system being bathed in extravascular fluids or carrying blood channels. Another field of application has been the interpretation of membrane fusion, itself important in fertilization, virus infection, cell membrane assembly, and endo- and exocytosis.

The efforts to develop, for clinical use, implantable devices in contact with flowing blood began a few decades ago. It was clear even before these efforts began that tissue reactions would occur. Blood held *in vitro,* even if continuously moving, clotted quite quickly. Anticoagulants used in blood banking were not considered suitable for systemic anticoagulation. Biologically derived anticoagulants, such as heparin, which could be neutralized at will following surgery, or even following bypass carried out for a few days (as in prolonged extracorporeal oxygenation), made it possible to reduce the aggregative phases of blood reaction to nonbiological surfaces and thereby to derive some clinical benefits. Over the past 20 years or so, there have been several attempts to bind heparin to fabricated surfaces, as a way of achieving a permanent local anticoagulation effect, or, failing this, to allow it to leach out slowly. Reports on use of such an approach in extracorporeal gas exchange began to appear in the mid-1980s.[37] This goal of surface binding is seen as a superior prospect because the effect is local, not systemic, and the heparin would not suffer the fate of circulating heparin — that it is metabolized and needs to be supplemented from time to time to achieve its effect over the time course of use of the device.

A. ADHESION AND AGGREGATION OF RED BLOOD CELLS

Adhesion of red cells to each other — aggregation into rouleaux — requires the presence of long molecules that bridge modest gaps between the cell membranes. In normal physiological circumstances, these molecules are fibrinogen, with a divalent cation (calcium) being an essential cofactor. Other long molecules, such as dextrans, can also provide the surface-connecting function. The cell–cell adhesion appears to be completely reversible by shear, when it is applied to the plasma or suspending fluid in which the rouleaux are. A kinetic theory for the aggregation process has been developed,[38] in which the kinematics of bringing red cells to collide with each other (to form dimers), with dimers (to form trimers), and so forth, are "swallowed" into reaction rate constants. This approach does not predict the rate constants *a priori* from the flow field but does allow consistent estimation of the distributions of aggregate sizes achieved concurrently as the process evolves. Limits to the sizes of rouleaux generated can be predicted by using a statistical method developed in polymer chemistry to find the distribution of molecular weights of the products of a condensation polymerization reaction.[39] It may be possible to make assessments of red blood cell aggregation using ultrasound techniques.[40] Indeed, this was demonstrated with blood in vessels[41] and was used intraoperatively to assess the quality of vascular anastomoses following surgery. While it is well recognized that rouleaux are rapidly broken up in shear flows and the red cells become monodisperse at sufficiently high shear rates, there is no corresponding theory of disaggregation for red blood cell rouleaux that accounts for fluid-mechanical effects.

Formed elements of blood can be deposited on walls without necessarily requiring an adhesive process: they can be deposited as a consequence of ultrafiltration. Forstrom et al.[42] considered the balance between the drag forces on a particle due to ultrafiltration and the repulsive force to fluid shear flow past the surface. They concluded from their analysis and corresponding experiments using dialysis membranes that particles are expected to be deposited passively onto filtering surfaces, provided the nondimensional parameter

$$\nu^{1/2}U\lambda / R^2S^{3/2}$$

exceeds a certain value, where ν is the kinematic viscosity, U is the ultrafiltration velocity normal to the wall, R is the cell or particle size, and S is the wall shear rate. For erythrocytes, this was measured as 0.15, and for platelets the value ranged from 0.01 to 0.15. The factor λ depends on particle volume concentration, H, as given by

$$\lambda = \frac{4 + 3H + 3(8H - H^2)^{1/2}}{(2 - 3H)^2}$$

and has a limiting value of 1.0 as $H \to 0$.

Red cell membranes seem quite robust. Portions have been sucked into micropipettes and yet retained their integrity. RBC membranes have been seen to elongate considerably when some local point is adherent to a surface and a flow passes over it, applying a drag force to the cell.[43] The robustness is not infinite, of course, because red cell membranes can be disrupted and small fragments formed even without adhesion to a surface but just in the presence of sufficiently strong shear flow.

B. ADHESION AND AGGREGATION OF BLOOD PLATELETS

Platelets, suspended in plasma, sediment very slowly, as anyone who has carried out platelet counting by the early method of observing sedimented platelets on a gridded slide will recall. Adhesion of platelets to surfaces in the absence of controlled shear flow would be at least as lengthy to observe unless centrifugation is used to increase the sedimentation rate. Centrifugation is used in the George Test.[44] An aliquot of diluted plasma is placed in a small chamber assembled from two circular cover slips, held parallel and coaxial with each other by a circular gasket separating them at their rim. This chamber is spun in a centrifuge, with one of the cover slips radially outermost, for a time sufficient to make all platelets in suspension in the plasma come into contact with that outermost coverslip. The centrifuge is stopped, the chamber reversed in its orientation so that the opposite coverslip becomes outermost, and the chamber is spun for a similar time. The number of platelets that remain on the initially outermost coverslip is counted — these must all be adherent — and the number in contact with the initially innermost coverslip are counted as well, so the fraction of the total number of platelets present that adhered to the initially outermost coverslip can be determined. This fraction has proven to be closely reproducible on replications of the test. Another test that became well-known was one from Mustard's group.[45] In this the platelets are exposed to a fibrinogen-coated tube, but the episodic rotation of the tube does not give a well-defined flow over the tube wall surface.

C. ADHESION OF PLATELETS IN THE PRESENCE OF A FLOW OVER A SURFACE

The early, pragmatic test of platelet adhesion — and subsequent thrombus build-up — on surfaces of materials that were considered candidates for use in cardiac bypass and other settings used a ring of the material implanted in the vena cava of a dog. The experiments required rings of a suitable diameter, formed so that they could be surgically implanted. At first, the choices of materials were so poor that gross thrombosis was observed, with a significant degree of occlusion. As the choices of materials improved, the extent of thrombosis diminished, and it even became necessary to make the rings with a stepped interior shape (instead of having the smooth bores used initially) in order to provoke an observable reaction. This test was known as the Gott Ring Test.[46,47]

It was recognized quite early in the years of pragmatic testing of such materials for implants that there was loss of adherent material from surfaces over a period of time. Another pragmatic test, in this case for embolization, was developed, in which a ring is introduced surgically into the aorta of a test animal, in the region that is just proximal to the kidneys, and the infrarenal aorta is constricted by a ligature. This configuration ensures that a large fraction of the abdominal aortic blood flow passes into the kidneys. The kidneys are fairly easy to examine for accumulation of embolized thrombi.

These semiquantitative empirical tests for platelet adhesion, thrombus formation, and embolization fell into disfavor because

1. there was a large variation between replicate tests, requiring a fair number of replications to obtain statistically significant differences for different materials and surface treatments,
2. the tests took a long time between implantation and explantation to "read" the results,
3. the tests distinguished largely the degree of badness of poor choices of materials for use, and did not discriminate clearly between the better choices of materials once these had been found,
4. the tests did not give helpful information on the temporal variation of adhesion, thrombus formation, or embolization, or support a "fundamental approach" to understanding the phenomenology.

The role of surface roughness was questioned, in part because people familiar with fluid dynamics could imagine the complex local flows around small projections or in small cavities, providing relative stasis and thereby opportunity for the normal coagulation pathway to run its course; and the presence of gas nuclei, stabilized in fine surface roughness, was noted to promote adhesion.[48]

An *ex vivo* test device that provided contact of blood with test surfaces, with only brief prior contact with nonbiological materials, was developed, and it attracted attention for a number of reasons (49; see Reference 23). It used a small plastic tube filled with glass beads; the beads could be coated with materials to be tested. A typical test, where blood was run from a venipuncture directly through the column without anticoagulation, took only 12 min or so, and the amount of blood taken was small enough that it was practical to perform experiments with human blood instead of with animals. Because of platelet adhesion to the large exposed surface area of the beads, the number-concentration of platelets in the effluent blood was much smaller than that of the source. A sound analysis of platelet adsorption on such a column was developed,[27] and the analysis indicated that by making measurements of platelet count in the effluent blood repeatedly during the test, rather than making one measurement at the end, one could determine a well-founded coefficient for adhesion (adsorption rate constant). Hematologists did not turn to using the protocol for data gathering needed to assess this coefficient of adhesion, however, and the main consequence was that the device proved useful mainly to detect subjects' deficiencies in the von Willebrand factor.

While the glass-bead column provided a test vehicle for short-term tests, the local flow distribution around the beads is very nonuniform, and it was not expected that the bead column could give clear and reliable answers for circumstances where the flow field itself could be dominant, rather than the platelet–surface interaction. In some devices, blood is introduced by displacing a priming liquid, as in routine start-up with artificial kidneys, membrane artificial lungs, and so on. In such circumstances, it is expected that within the first few seconds and minutes there is adhesion of plasma proteins to the artificial surface, turnover of some of these proteins, and episodes of platelet adhesion as well. In priming an artificial organ, the first step has often been to put in a buffered salt solution and subsequently displace this solution with whole blood. If the device is supplied dry, filling it with gaseous carbon dioxide before introducing the buffered salt solution diminishes the risk of having undissolved gas nuclei in blood passages. Sometimes, albumin has been included with the salt solution. Of the constituents of blood not already present in the salt solution, the plasma proteins have higher effective diffusivities than the formed elements, so it is expected that they will reach and react with the blood-passage surfaces of the artificial organ before the formed elements do so.

It is important to remember, when discussing shear flows quantitatively, that two related aspects are used by different authors to describe them. One aspect is the local velocity gradient of the shear flow, usually expressed in "inverse seconds" because the spatial gradient of velocity has the dimensions of

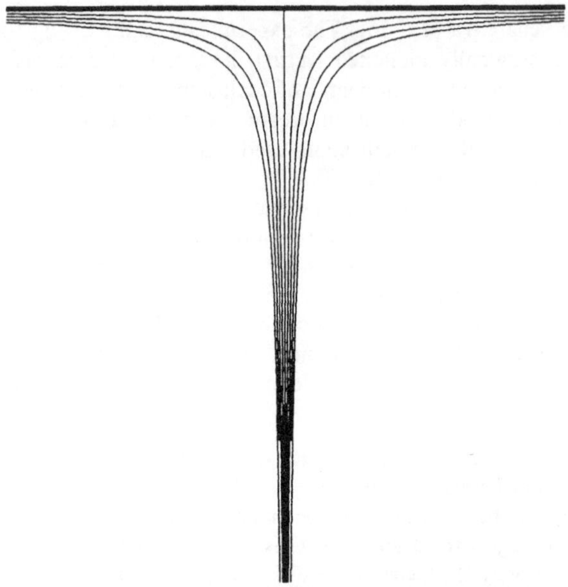

Figure 1 Streamlines of flow approaching a forward stagnation point on a flat surface.

velocity (i.e., length/time) divided by length. The other aspect is the shear stress, usually expressed as (shear) force per unit area, such as dynes per square centimeter. The viscosity of the fluid is the coefficient which, when multiplied by the velocity gradient, equals the shear stress. With blood flows, the effective viscosity depends on the hematocrit and on the shear rate, so the relation between shear rate and shear stress is more complex than it might appear at first.

In an attempt to define and control the flow more precisely, and to make the first contact of blood *ex vivo* be with a test surface, Petschek et al.[50] everted the carotid artery of a dog over a length of plastic tubing and brought the effluent blood with perpendicular symmetry to a microscope cover slip that could be observed continuously from its opposite side using a microscope. From the fluid mechanics viewpoint, an attractive attribute of this chamber design is that the flow near the solid surface is well known and is called the axisymmetric forward stagnation point flow.[51] Strictly speaking, the flow for this geometric configuration is known best for Newtonian fluids. The flow pattern is illustrated in Figure 1. The thick line illustrates the flat plate at which the flow is directed, such as the cover slip. Whole blood coming from a carotid artery is likely to have RBC rouleaux in its central region, which will experience progressively stronger disaggregating forces as the blood flows radially outward over the test surface. How different this may make the flow from the ideal, Newtonian fluid case is not well known. Petschek et al. reported observing platelet adhesion occurring after a "conditioning" time for the surface (glass or a polyurethane coating) of about 1 min, and large-scale thrombus formation (away from the stagnation point itself) after about 8 min of blood flow. The stagnation-point flow continues to attract users for adhesion studies.[52]

The flow streamlines (shown on Figure 1) associated with stagnation point flow should be created by the shape of the walls of the flow chamber. It is obvious from Figure 1 that there is a progressive divergence of streamlines as the flow approaches the flat plate. The lack of divergence that might occur in an everted artery held near a plate is less troublesome in altering the flow from that expected than is failure to use chamber walls that provide a tapering gap in the region where the flow is more conspicuously radial.

Convective transport theory, a component especially in the macroscopic approach, was applied speculatively to predict how rapidly plasma proteins such as albumin and fibrinogen would be transported at a freshly-exposed surface, and how rapidly platelets might be transported through the concentration gradient of platelets perpendicular to the surface (baseline value in the entering blood, zero at the surface initially covered with the priming solution). This involved adaptation of a macroscopic viewpoint already established in chemical engineering circles, the only adaptation required being that platelets were regarded like a chemical substance in the flow, described in terms of their local concentration (number per unit volume) and moving like any diffusible substance. A consequence of this would be that there are no patterns discernible in adherent platelets, because the deposition of each platelet would be the result of the individual random walks that represent at the particulate level the phenomenon of diffusion of the overall population of particles.[53] Butruille et al.[54] made studies of platelet adhesion that indicated, for platelet deposits getting close to surface saturation, that the platelet locations were not fully random, and proposed (as a consequence of clustering in excess of what might appear randomly) that adherent platelets secrete substances that enhance adhesion of other platelets nearby (especially close downstream).

Various flow chambers were developed, some for mounting in an *ex vivo* shunt and others for entirely *ex vivo* use. The flow channels were either tubular or flat. The tubular channels generally have less surface area and require less flow to achieve a particular velocity gradient near the wall. One type of tubular flow chamber uses the unusual geometry of the annulus; i.e., the chamber space is between two circular cylinders, mounted coaxially. This seemingly curious geometry serves the very practical purpose of allowing an everted vessel, such as a rabbit aorta, to be mounted around the inner cylinder.[55] A flat-walled chamber could be used to imitate the flow environment of flat-sheet dialysers with ports in diagonally opposite corners.[56] The flat-walled chambers allow a greater wall surface area to be kept in focus in a microscope. An outcome of the studies of inferred platelet transport rates from whole blood was that the rate of transport to the wall was roughly 100 times what had been expected, based on the equivalent diffusivity of the platelets assessed from their mol wt. This implied that the initial analysis had lacked representation of some essential feature, and this missing feature was identified as the motion of the red cells in the shear flow. The motion of the red cells, even at small concentrations of RBC, induces internal microflows in a blood flow that enhance cross-stream transport,[57] and there are direct cell-collision effects. The RBC concentration is raised toward physiological levels, leading to "enhanced" or augmented diffusivity of platelets.

Flow chamber designs of the axisymmetric forward stagnation point type have a flow in which the velocity gradient at the surface increases linearly with the distance from the stagnation point itself when using a Newtonian fluid.[51] A consequence is that if a sufficient shear force is required to activate platelets for adhesion, no platelets would be seen adherent at small distances from the stagnation point. Reports of observations with such chambers do not indicate any such exclusion zone. Moreover, Petschek et al.[50] reported the formation of thrombus at a range of radii that are within the range of shear rates that give the largest thrombus growth rates *in vivo*,[58] albeit in a different species.

Richardson et al.[59] have described and used a flow chamber in which it is possible to create known variations of local shear rates and shear stress along the streamlines, i.e., along the paths of mean motion followed by any particles (including platelets) that may be carried through the flow chamber. In this flow chamber, it is possible to provide flow paths along which the shear experienced by a particle rises to a maximum and then decreases again as it progresses along a streamline, while along another streamline in the same flow chamber another particle can experience an initial shear and then a progressively diminishing shear, falling to a finite minimum value, after which the particle experiences a progressively rising value to a maximum, after which the opposite sequence is experienced before the particle leaves the chamber. This capability is provided by the chamber being, fluid dynamically, in the class of Hele-Shaw chambers, in which flow around a shape that spans the width of the chamber and that is inserted during assembly can be computed on the basis of potential flow theory.[60] Another feature of the design of this chamber, not related to the determination of the fluid motion but a convenience in working with the types of surface-coated slides used in adhesion studies, is the use of a perimeter vacuum channel to secure the test slide to the rest of the chamber, minimizing manipulation during assembly and facilitating disassembly after use so that the adherent material can be examined. Figure 2 illustrates a chamber of this type, in which a circular obstacle is introduced. The flow has to make its way around the obstacle and in doing so provides a range of different shear rates for particles passing through, depending on the streamline individually followed. The streamlines for the flow are computed using potential flow theory,[60] and an infinite series of "images", ranged on each side of the channel of interest, is used to determine the flow with straight line portions to correspond to the side walls of the actual chamber, as illustrated in Figure 3. The real flow is viscous, and the spatial distribution of wall shear rate (which matches the distribution of wall shear stress when the fluid is Newtonian: platelet-rich plasma, for example) can be calculated readily. Figure 4 shows contours of constant wall shear rate (as solid lines) mapped onto the wall space available for flow.

With a number of chamber designs, the investigations remained one-time studies; i.e., the variation of adhesion over a continuous interval of time was not studied, and only the "snapshot" of the adhesion situation at one time was obtained in each run. In the mid 1970s, two laboratories independently used methods to watch flow chamber walls continuously through periods of exposure to flow of blood or platelet-rich plasma and observed that many platelets that adhere to surfaces in flow chambers leave their

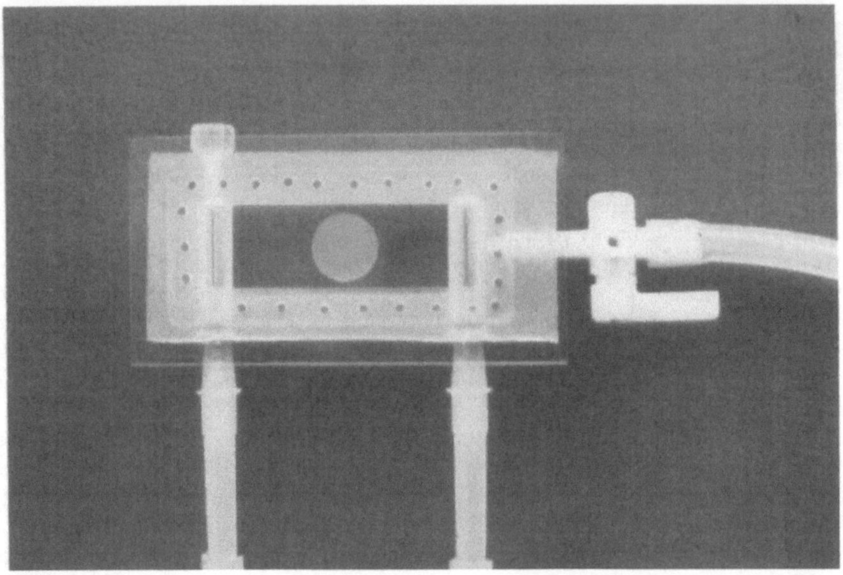

Figure 2 Flow chamber with flat, parallel walls. Inflow to the channel is through a slot fed from a transverse cylindrical hole, at one end of which is the inflow port and at the other end is a plug (which can be replaced by a slotted needle). Exit from the channel is through a similar slot into a cylindrical hole, blind at one end and having the exit port at the other. At the middle of a short side there is a connector for a vacuum line; this connects to a perimeter channel. Holes in the template-cut gasket communicate the vacuum to matching parts of the face of the test slide, which is thereby held integral with the other parts of the flow chamber without other clamping. Assembled into the flat channel is a circular disk of silicone rubber, which forms an obstacle that partly occludes the channel and causes streamlines to flow around it, as illustrated in Figure 3.

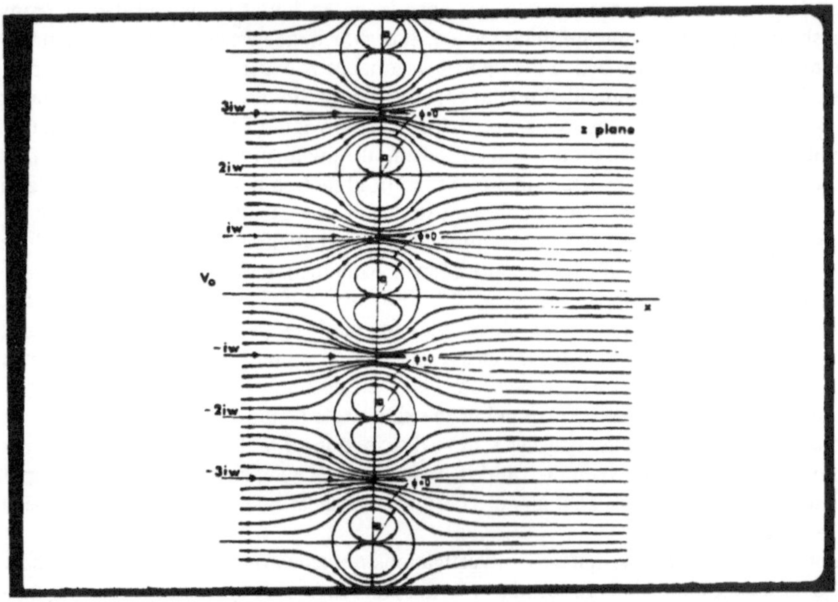

Figure 3 Computation of streamlines for flow in a flat channel having a circular obstacle placed symmetrically in it (see Figure 2). The series of mirror images of the basic shape in parallel with each other is needed to make the streamlines for the sidewalls of the chamber truly straight lines.

Figure 4 Map of contours of constant wall shear rate in a flat channel having a circular obstacle placed in it (see Figures 2 and 3). The shear rate diminishes as the flow approaches the front stagnation point of the circular obstacle. The shear rate increases as the flow accelerates to pass through the narrow spaces between the side walls and the sides of the disk.

adhesion sites after some seconds of adhesion.[61-63] It was also noted that the sites that had been occupied by platelets and then vacated were very highly preferred as adhesion sites by platelets that passed over them subsequently. The experimental observation accounting method kept track of all adhesion sites, and the reuse of sites was readily accounted. Indeed, the phenomenon occurred with sufficient regularity that both the primary rate of adhesion and the reuse of sites provide indices of the platelet adhesion process. Each was found susceptible to alteration by platelet-active pharmacological agents or specific dietary components,[61,64,65] but the effects on platelet adhesion did not simply mirror effects found with the same agents on platelet aggregation in the Born aggregometer.

The existence of a footprint left by a departing platelet presented some ambiguity: did the platelet depart intact and take with it some of the protein layer coating over the underlying glass surface (the latter known to be more adhesive for platelets), or did the platelet leave part of its own membrane behind? Experimental statistics on how long the enhanced adhesiveness of the footprints endured showed the enhancement appeared to diminish over time for about 6 to 7 min, as illustrated in Figure 5. This made the torn-protein-coating hypothesis doubtful, because even a virgin glass slide exposed to blood may be covered in less time. Some preliminary experiments were made with fluorescent dyes that are readily taken up by platelet membranes. Slides from flow chambers were carefully examined following use for adhesion measurements for match of dye-stained spots with adhesion sites observed during flow. Matching was observed, but also some spots showed fluorescence that had not been sites of observed adhesion. This favored the hypothesis that platelets can leave part of their membranes behind and that some such parts (too small to distinguish in optical phase microscopy) might subsequently detach and reattach on their own at different locations.

A possible resolution of the question of platelet damage came years later, although whether it is associated with the "footprint" observations needs definitive determination. Recent studies have shown that platelet microparticles, considerably smaller than platelets themselves, are markers of an agonist-induced platelet activation event.[66,67] Fluorescent-activated flow cytometry has allowed specific analysis of some of the sources of platelet microparticles, e.g., with the use of antibody for IIb-IIIa receptors. There is some indication that complement activation is involved in the platelet activation that allows microparticle formation. It is possible that platelet microparticle formation may arise not only from platelet adhesion and departure from adhesion sites but also in embolization from thrombi.

It is possible to use a mathematical modeling approach to show that the reuse of sites index, developed in representation of experimental observations of platelet adhesion in a flow chamber, is consistent with

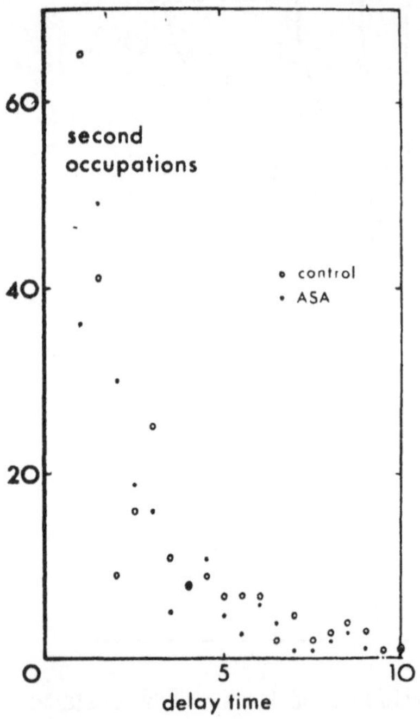

Figure 5 Number of vacated sites reoccupied by platelet "tenants" as a function of the waiting time following vacation. Note that the frequency of reoccupation diminishes as elapsed time increases, falling almost to background level in 6 min.

the existence of platelet microparticles being left on a site when the main body of a platelet departs and the platelet microparticles being active at the sites for a finite period of time. Thus, we can define site adhesiveness as the probability of an adhesive event (occupation by a platelet) in a specific time interval, say 30 s, at any unoccupied site, whether or not the adhesion is followed by detachment (vacation of the site). There appears to be a base level of site adhesiveness appropriate for sites in any zone in a flow chamber, such as in Reference 59, in which the flow is essentially uniform over the field in which adhesive events are viewed, and which depends on the platelet concentration, on the flow rate, and on the donor. For sites which have been occupied and then vacated, there seems to be an enhanced adhesiveness. Let us suppose this is because departing platelets leave a platelet microparticle on the site, and this microparticle enhances the local probability of adhesion of some platelet passing subsequently over the site. It is not clear why there should be an enhanced probability of adhesion, but it might be associated with the particle protruding from the surface and being accessible to a larger number of platelets passing than if there were no protrusion.

The probability of a previously occupied site being reoccupied appears, from experimental data, to diminish as the time since site-vacation occurred, falling roughly linearly to the base level in about 6 min, as seen in Figure 5. (Note that the time scale involved here is not measured from the beginning of the flow through the chamber; the clock measuring time for this process starts when an occupied site is vacated, which might be 2 min after the start of flow, or 10 min, or whatever.) It is not clear whether this finite time of increased adhesiveness is due to a corresponding distribution of times taken for platelet microparticles that had been left on sites to embolize themselves, or for their features that enhance adhesiveness to be masked while they remain on site. From the macroscopic modeling point of view, it does not matter which, nor will analysis of results using the model be able to resolve which.

Indeed, the model can even be simplified to a situation where the enhanced adhesiveness exists at a single value, E times the base probability, which lasts for elapsed time T, after which the adhesiveness falls back sharply to its base value. It is useful to consider the case, typical of the early phase in platelet adhesion experiments, where the primary adhesion events, which usually provide vacated sites within 30 s or so, number up to 100 to 500. With a total field of view of typically 10,000 sites (symbolically K), the chance of there being a reoccupation of a vacated site through the baseline probability effect is small compared with the reoccupation rate found experimentally; so in a simplified analysis, it can be ignored. The rate at which once occupied and newly vacated sites occur, in the field of view K, is A, and the cumulative count of such sites is At, i.e., linearly increasing with time (the time scale for this having a slight lag from the time of start of flow through the chamber), and $At << K$. Following a start-up period, there will be a basic availability of sites that have enhanced adhesion (from the rate of site generation minus the rate at which sites lose their enhanced adhesiveness) of AT. Let the average time delay for reoccupation of a vacated site, while the adhesiveness is enhanced, be qT, where $q < 1$. Then the rate of adhesion at reoccupation sites, vacated once, is EA times the opportunity ratio AT/K of the sites in the field of view, i.e., $EA(AT/K)$, also to be denoted here as B, and the cumulative count at time t is $EA\ (t-qT)\ (AT/K)$. Clearly there is an upper limit on E, in that $EAT/K < 1$. If $EAT/K = 1$, it would mean that every vacated site was reoccupied during the enhanced adhesiveness phase.

If the data in Figure 1A of Reference 68 are examined, for example, A is about 5 vacation events per minute, qT is about 1 min, and cumulative count of once reoccupied sites is linear with respect to $(t-qT)$. B is about 3. (There is clearly an increasing number of sites that have been occupied and vacated once, as time t progresses, and if T were long, the start-up time would be long and the graph of reoccupation would be less linear with respect to time and would rise more with the square of time instead. The data seem appropriate for applying the model.) Thus, from

$$B = EA(AT/K)$$

we can estimate $3 = 25\ ET/10{,}000$, i.e., $ET = 30{,}000/25 = 1200$. If the effective time T of adhesion enhancement is, say, 3 min, $E = 400$, which is a very significant enhancement. The data in Reference 62, Figure 3, are even more clearly linear in the first vacation and first reoccupancy lines. The reuse of sites index, *RUSI*, is the slope of the logarithm of the sites occupied vs. the number of occupations per site,[61,62] so it can be estimated in relation to the model as

$$\ln \frac{EA(t - qT)(AT/K)}{At}$$

or, if $qT \ll t$, it can be estimated as

$$RUSI = \ln (EAT/K)$$

This is, of course, negative, because $EAT/K < 1$. It is possible to estimate this with modest numbers of primary platelet adhesion events, say 50 to 100. The Bodziak and Richardson[62] study has added interest because one can reexamine the data to estimate whether the product ET was altered when platelets had been incubated with ASA, which led to changes in A and *RUSI*. Now, from the approximate model above,

$$\exp (RUSI) = EAT/K$$

so

$$ET = (K/A) \exp (RUSI)$$

and for the baseline data at 25 to 26 s^{-1}, with $K = 10{,}000$,

$$ET = (10{,}000/30.8) \exp (-2.65) = 22.9$$

and after incubation with ASA

$$ET = (10{,}000/13.5) \exp (-4.14) = 11.8$$

In this case it would appear that the product of the enhancement factor E and the characteristic time of enhanced adhesiveness T have been reduced following incubation with ASA. The data shown in Figure 5 do not suggest a change in the characteristic time of enhanced adhesiveness. This suggests that E itself was reduced following platelet incubation with ASA. This illustrates how a simple model of the site occupation (adhesion)-vacation site reoccupation can be used to provide a small number of parameters that can be estimated and compared.

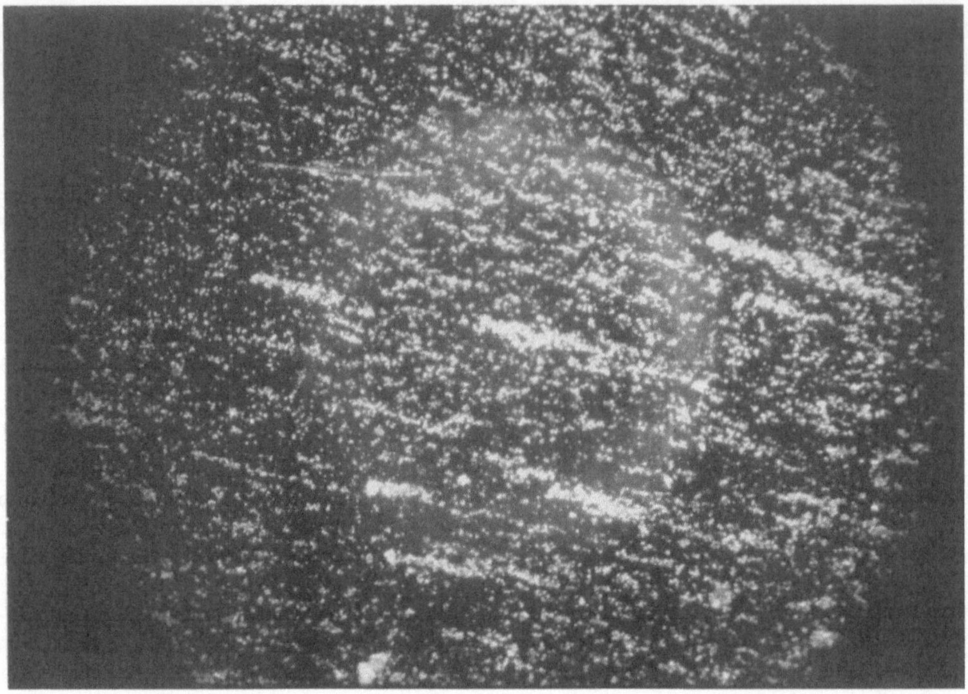

Figure 6 View of platelet adhesion and platelet aggregation (thrombus formation) on a flow chamber wall.

The interactions of the processes of platelet adhesion and platelet aggregation on a surface are suggested by photographs taken in flow chambers. Figure 6 was taken in a flow chamber previously mentioned.[59] It is a local, enlarged view of a small region of the flow chamber wall. Visible features include (1) platelets adherent (small white dots in the figure), (2) thrombi — roughly circular — and most with "tails" on their downstream sides, (3) minithrombi, larger than the widespread and seemingly randomly located single platelets, frequently located in streaks and more tightly packed per unit area. Even with a simple geometry (flat surface) and steady flow, the pattern is complex. The fact that the streaks are finite in length suggests that whatever the influence of thrombi may be in enhancing adhesion close downstream, that influence diminishes with time or distance.

D. ADHESION OF ENDOTHELIAL CELLS IN THE PRESENCE OF FLOW

The role of endothelial cells in countering thrombus formation was recognized in particular once their production of prostacyclin, a potent inhibitor of platelet aggregation, was demonstrated. Techniques for sustaining cultures of endothelial cells in confluence, *in vitro,* allowed studies of their ability to orient their major axes in response to the direction of flow over them.[69] The changes were accompanied by shifts in cytoskeleton organization.[70] What came as a surprise was the ability to endothelial cells to respond in their production rate of prostacyclin to fluid shear stress[71] and also to changes in fluid stress.[72] This showed that a cell type could be a mechanoreceptor for fluid stress and influence the aggregation capability of another cell type passing by. Adhesiveness could, in some measure, be influenced by the flow over mechanoreceptive cells in physiological situations.

E. ADHESION OF FIBROBLASTS IN THE PRESENCE OF FLOW

The interest in cell attachment and conformation on surfaces under flow is not limited to platelets or endothelial cells. Fibroblasts adherent to glass have been observed to change their shape in response to pulsatile flow over them.[73] We may speculate that they, too, may prove to have mechanoreceptor aspects in their behavior. Detachment of normal and transformed fibroblasts from a confluent layer of normal fibroblasts has been studied[74] using a parallel plate flow chamber of a type initially used for platelet adhesion.[59] It was found that detachment of normal, transformed, and reverted rat fibroblasts increased as the shear stress imposed was increased (range 2.5 to 20 dyn/cm^2), and the detachment of the highly metastatic cells was higher than that of the normal or reverted fibroblasts. While it is clear that success of a cell type in metastasis may be increased by reduction in adhesion, the process by which the fraction of released cells subsequently adheres and forms a tumor somewhere else in the circulation is an interesting issue.

VIII. CONCLUDING REMARKS

Cell adhesion is not simply a biological phenomenon; it depends in practice upon physical phenomena, too. In many circumstances where cell adhesion is important and where a sufficient understanding of cell adhesion to allow a predictive approach would be useful, such as in designing better implants for medical use, we lack sufficient knowledge of cell adhesion, despite an enormous amount of research already performed.[75] We know that cells, before they adhere to a surface, may well be in a flow, and that aspects of the flow have effects on the cell adhesion. This is particularly complicated *in vivo,* where the cells present on a vessel wall can respond biochemically to aspects of the flow and then have effects on the adhesion process.

The approach of a cell, with a separation greater than 1000 Å from a potential adhesion site, is dominantly determined by the flow field that carries the cell. In the range 1000 to 200 Å there should be, in addition, long-range interactive forces (van der Waals and electrostatic), and ligands may come into effective action. The range below 200 Å should be dominated by short-range intermolecular forces and chemical bonds. The midrange of 1000 to 200 Å may be the range of closest approach at which the collision efficiency is often determined. Detachment, when a specific receptor is involved, may involve pulling the receptor out of the membrane. The heterogeneity of binding is also a consideration.[76] Much remains to be learned here before a full quantitative analysis of cell adhesion can be constructed.

Theoretical models can be useful in many ways. They can focus attention on what aspects of phenomena are important to know, in what detail, to provide a representation of adhesion that is consistent with all reliable observations and measurements. They can provide parameters that can be inferred from experiments and compared in different geometries. They can inspire experimental designs that are directed particularly at evaluating specific aspects. These experimental designs are often superior for testing and evaluating models and may be more specific in determining what factors influence adhesion. The pursuit of an understanding of how the locomotion of human leukocytes is controlled by chemotactic factors is an illustration of the refinement of experimental methods. These methods range from the simple Boyden chamber through the Zigmond chamber to the Usami method, with concurrent improvements in direct observation of the cells and assessment of the relative significance of chemotactic agent concentration, concentration gradient around the cell, and concentration as a function of time. The use of multiple test geometries that separately favor discovering different aspects of the cellular response is encouraged when there are models to be tested. To use the leukocyte example again, the large increase in cell equivalent viscosity (from 45 to 668 p) on exposure to the attractant FMLP,[77] inferred from measurements of cell dynamics under a flow, will need to be checked against quantitative assumptions required to account for movement of such a cell into a narrow capillary in the presence of the same attractant.

In the science of cell adhesion, there is room and need for consideration of intrinsically different yet mutually complementary viewpoints and reconciliation of results achieved in studies started from these viewpoints. Many of the recent advances have occurred at the scale of the cell, the membrane, and the adhesion receptors, as shown by their emphasis in this volume. Integration into an overall science of cell adhesion calls for advances in understanding at larger physical scales as well. Analyses for these larger physical scales have typically come from persons with backgrounds in biophysics and bioengineering, while receptors have been studied especially by persons with pure or applied biochemistry backgrounds; the future looks interdisciplinary.

REFERENCES

1. **Rand, R. P. and Burton, A. C.,** Mechanical properties of red cell membrane. I. Membrane stiffness and intracellular pressure, *Biophys. J.,* 4, 115, 1964.
2. **Rand, R. P. and Burton, A. C.,** Mechanical properties of red cell membrane. II. Viscoelastic breakdown of the membrane, *Biophys. J.,* 4, 303, 1964.
3. **Evans, E. A.,** Bending resistance and chemically induced moments in membrane bilayers, *Biophys. J.,* 14, 923, 1974.
4. **Fischer, T. M.,** Bending stiffness of lipid bilayers. IV. Interpretation of red cell shape change, *Biophys. J.,* 65, 687, 1993.

5. **Skierczynski, B. A., Usami, S., Chien, S., and Skalak, R.,** Active motion of polymorphonuclear leukocytes in response to chemoattractant in a micropipette, *Trans. Am. Soc. Mech. Eng. J. Biomech. Eng.*, 115, 503, 1993.
6. **Zhu, C.,** A thermodynamic and biomechanical theory of cell adhesion. Part I: General formalism, *J. Theor. Biol.*, 150, 27, 1991.
7. **Bell, G., Dembo, M., and Bongrand, P.,** Cell adhesion. Competition between nonspecific repulsion and specific binding, *Biophys. J.*, 45, 1051, 1984.
8. **Chien, S., Feng, S.-S., Vayo, M., Sung, L. A., Usami, S., and Skalak, R.,** The dynamics of shear disaggregation of red blood cells in a flow channel, *Biorheology*, 27, 135, 1990.
9. **Ward, M. D. and Hammer, D. A.,** A theoretical analysis for the effect of focal contact formation on cell-substrate attachment strength, *Biophys. J.*, 64, 936, 1993.
10. **Anon.,** ASTM Test Method D903-49: Peel or stripping strength of adhesive bonds. ASTM Test Method D773-47: Adhesiveness of gummed tape. American Standards for Testing and Materials, Part 16 (issued annually).
11. **Angarska, J. K., Tachev, K. D., Ivanov, I. B., Kralchevsky, P. A., and Leonard, E. F.,** Red blood cell interaction with a glass surface, in *Cell and Model Membrane Interactions*, Ohki, S., Ed., Plenum Press, New York, 1991, 199.
12. **Bangham, A. D. and Pethica, B. A.,** The adhesiveness of cells and the nature of chemical groups at their surfaces, *Proc. Phys. Soc. Edinburgh*, 28, 43, 1960.
13. **Ohki, S., Doyle, D., Flanagan, T. D., Hui, S. W., and Mayhew, E., Eds.,** *Molecular Mechanisms of Cell Fusion*, Plenum Press, New York, 1988.
14. **Sower, A. E., Ed.,** *Cell Fusion*, Plenum Press, New York, 1987.
15. **Ohki, S.,** Physicochemical factors underlying membrane adhesion and fusion, in *Cell and Model Membrane Interactions*, Ohki S., Ed., Plenum Press, New York, 1991, 267.
16. **Grebe, R. and Zuckerman, M. J.,** Erythrocyte shape simulation by numerical optimization, *Biorheology*, 27, 735, 1990.
17. **Nossal, R. and Lecar, H.,** *Molecular and Cell Biophysics*, Addison-Wesley, Redwood City, CA, 1991.
18. **Dembo, M., Torney, D. C., Saxman, K., and Hammer, D.,** The reaction-limited kinetics of membrane-to-surface adhesion and detachment, *Proc. R. Soc. B.*, 234, 55, 1988.
19. **Plow, E. F. and Ginsberg, M. H.,** Cellular adhesion: GPIIb-IIIa as a protypic adhesion receptor, in *Progress in Hemostasis and Thrombosis*, Vol. 9, Coller, B.S., Ed., W. B. Saunders Co., Philadelphia, 1989, 117.
20. **McIntosh, T. J., Magid, A. D., and Simon, S. A.,** Short-range repulsive interactions between the surfaces of lipid membranes, in *Cell and Model Membrane Interactions*, Ohki, S., Ed., Plenum Press, New York, 1991, 249.
21. **Poe, G. G. and Acrivos, A.,** Closed-streamline flows past rotating single cylinders and spheres: inertia effects, *J. Fluid Mech.*, 72, 605, 1975.
22. **Davie, E. W., Huang, S., Farrell, D. H., and Chung, D. W.,** The structure and function of fibrinogen, in *New Horizons in Coronary Heart Disease*, Born, G. V. R. and Schwartz, C. J., Eds., Current Science, London, 1993, chap. 18.
23. **Salzman, E. W.,** Measurement of platelet adhesiveness. A simple in vitro technique demonstrating abnormality in von Willebrand's disease, *J. Lab. Clin. Med.*, 62, 724, 1963.
24. **Cozens-Roberts, C. D., Lauffenburger, D. A., and Quinn, J. A.,** Receptor-mediated cell attachment and detachment kinetics. I. Probabilistic model and analysis, *Biophys. J.*, 58, 841, 1990.
25. **Cozens-Roberts, C. D., Lauffenburger, D. A., and Quinn, J. A.,** Receptor-mediated cell attachment and detachment kinetics. II. Experimental model studies with the Radial Flow Detachment Assay, *Biophys. J.*, 58, 857, 1990.
26. **Pethica, B. A.,** Microbial and cell adhesion, *Microbial Adhesion to Surfaces*, Berkeley, R. C. W., Lynch, J. M., Melling, J., Rutter, P. R., and Vincent, B., Eds., Ellis Horwood, Chichester, 1980, chap. 1, p. 19.
27. **Robertson, C. R. and Chang, H. N.,** Platelet adhesion in columns packed with glass beads, *Ann. Biomed. Eng.*, 2, 361, 1974.
28. **Fowler, H. F. and McKay, A. J.,** The measurement of microbial adhesion, in *Microbial Adhesion to Surfaces*, Berkeley, R. C. W., Lynch, J. M., Melling, J., Rutter, P. R., and Vincent, B., Eds., Ellis Horwood, Chichester, 1980, chap. 7, p. 143.
29. **Ivanov, I. B., Ed.,** *Thin Liquid Films*, Marcel Dekker, New York, 1988.
30. **Grinell, F.,** Cellular adhesiveness and extracellular substrata, *Int. Rev. Cytol.*, 53, 65, 1978.

31. **Evans, E., Berk, D., and Leung, A.** Detachment of agglutinin-bonded red blood cells. I. Forces to rupture molecular-point attachments, *Biophys. J.*, 59, 838, 1991.
32. **Tees, D. F. J., Coenen, O., and Goldsmith, H. L.**, Interaction force between red cells agglutinated by antibody. IV. Time and force dependence of break-up, *Biophys. J.*, 65, 1318, 1993.
33. **Chang, H. N. and Robertson, C. R.**, Platelet aggregation by laminar shear and Brownian motion, *Ann. Biomed. Eng.* 4, 151, 1976.
34. **Huang, P. Y. and Hellums, J. D.**, Aggregation and disaggregation kinetics of human blood platelets. I. Development and validation of a population-balance method, *Biophys. J.*, 65, 334, 1993.
35. **Huang, P. Y. and Hellums, J. D.**, Aggregation and disaggregation kinetics of human blood platelets. II. Shear-induced platelet aggregation, *Biophys. J.*, 65, 344, 1993.
36. **Huang, P. Y. and Hellums, J. D.**, Aggregation and disaggregation kinetics of human blood platelets. III. The disaggregation under shear stress of platelet aggregates, *Biophys. J.*, 65, 354, 1993.
37. **Bindslev, I., Gouda, I., Inacio, J., Kodama, K., Lagergren, H., Larm, O., Nilsson, E., and Olsson, P.**, Extracorporeal elimination of carbon dioxide using a surface-heparinized veno-venous bypass system, *Trans. Am. Soc. Artif. Intern. Organs*, 32, 530, 1986.
38. **Samsel, R. W. and Perelson, A. S.**, Kinetics of rouleau formation. I. A mass action approach with geometric features, *Biophys. J.*, 37, 493, 1982.
39. **Samsel, R. W. and Perelson, A. S.**, Kinetics of rouleau formation. II. Reversible reactions, *Biophys. J.*, 45, 805, 1984.
40. **Razavian, S. M., Guillemin, M.-Th., Guillet, R., Beuzard, Y., and Boynard, M.**, Assessment of red blood cell aggregation with dextran by ultrasonic interferometry, *Biorheology*, 28, 89, 1991.
41. **Sigel, B., Machi, J., Beitler, J. C., Justin, J. R., and Coelho, J. C. U.**, Variable ultrasound echogenicity in flowing blood, *Science*, 218, 1321, 1982.
42. **Forstrom, R. J., Bartelt, K., Blackshear, P. J., Jr., and Wood, T.**, Formed element deposition onto filtering walls, *Trans. Am. Soc. Artif. Intern. Organs*, 21, 602, 1975.
43. **Hochmuth, R. M. and Mohandas, N.**, Uniaxial loading of the red cell membrane, *J Biomech.*, 5, 501, 1972.
44. **George, J. N.**, Direct assessment of platelet adhesion to glass: a study of the forces of interaction and the effects of plasma and serum factors, platelet function, and modification of the glass surface, *Blood*, 40, 862, 1972.
45. **Cazenave, J. P., Packham, M. A., and Mustard, J. F.**, Adherence of platelets to a collagen-coated surface: development of a quantitative method, *J. Lab. Clin. Med.*, 82, 978, 1973.
46. **Gott, V. L., Ramos, M. D., Allen, J. L., and Becker, K. E.**, The causes and prevention of thrombosis on prosthetic materials, *J. Surg. Res.*, 6, 274, 1966.
47. **Gott, V. L. and Furose, A.**, Anti-thrombogenic surfaces classification and in vivo evaluation, Conference on Moving Blood, *Fed. Proc.*, 30, 1679, 1971.
48. **Ward, C. A. and Forest, T. W.**, On the relation between platelet adhesion and the roughness of a synthetic biomaterial, *Ann. Biomed. Eng.*, 4, 184, 1976.
49. **Hellem, A. J.**, The adhesiveness of human blood platelets in vitro, *Scand. J. Clin. Lab. Invest.*, 12 (Suppl. 51), 1960.
50. **Petschek, H., Adams, D., and Kantrowitz, A. R.**, Stagnation flow thrombus formation, *Trans. Am. Soc. Artif. Intern. Organs*, 14, 256, 1968.
51. **Rosenhead, L., Ed.**, *Laminar Boundary Layers*, Clarendon Press, Oxford, 1963, 419.
52. **Xia, Z., Goldsmith, H. L., and van de Ven, T. G. M.**, Kinetics of specific and nonspecific adhesion of red blood cells on glass, *Biophys. J.*, 65, 1073, 1993.
53. **Berg, H. C.**, *Random Walks in Biology*, Expanded ed., Princeton University Press, 1993.
54. **Butruille, Y. A., Leonard, E. F., and Litwak, R. S.**, Platelet-platelet interactions and non-adhesive encounters on biomaterials, *Trans. Am. Soc. Artif. Intern. Organs*, 21, 609, 1975.
55. **Baumgartner, H. R.**, The role of blood flow in platelet adhesion, fibrin deposition and formation of mural thrombi, *Microvasc. Res.*, 5, 167, 1973.
56. **Lindsay, R. M.**, Platelets, foreign surfaces, and heparin, in *Physiological and Clinical Aspects of Oxygenator Design*, Dawids, S. G. and Engell, H. C., Eds., Elsevier, Amsterdam, 1976.
57. **Fischer, T. and Richardson, P. D.**, Blood cell interactions in shear flow and consequences for diffusion and aggregation, *American Society of Mechanical. Engineers. 1980 Advances in Bioengineering*, Mow, V. C., Ed., American Society of Mechanical Engineers, New York, 1980, 305.
58. **Begent, N. and Born, G. V. R.**, Growth rate in vivo of platelet thrombi, produced by iontophoresis of ADP, as a function of mean blood flow velocity, *Nature*, 227, 926, 1970.

59. **Richardson, P. D., Mohammed, S. F., and Mason, R. G.,** Flow chamber studies of platelet adhesion at controlled, spatially varied rates, *Proc. Eur. Soc. Artif. Organs,* 4, 175, 1977.
60. **Batchelor, G. K.,** *An Introduction to Fluid Dynamics,* Cambridge University Press, 1967, 222.
61. **Richardson, P. D., Kane, R., and Agarwal, K.,** Effects of drugs on adhesion of human platelets, *Trans. Am. Soc. Artif. Intern. Organs,* 27, 203, 1981.
62. **Bodziak, K. and Richardson, P. D.,** Turnover of adherent platelets: some effects of shear rate, ASA, and reduced anticoagulation, *Trans. Am. Soc. Artif. Intern. Organs,* 28, 426, 1982.
63. **Feuerstein, I. A., and Kush, J.,** Measurements of platelet collision efficiency for virgin and platelet-primed (foot-printed) fibrinogen with fluorescent video-microscopy, *Trans. Am. Soc. Artif. Intern. Organs,* 29, 430, 1983.
64. **Li, X. and Steiner, M.,** Fish oil: a potent inhibitor of platelet adhesiveness, *Blood,* 76, 938, 1990.
65. **Li, X. and Steiner, M.,** Dose response of dietary fish oil supplementations on platelet adhesion, *Arteriosclerosis Thrombosis,* 11, 39, 1991.
66. **Abrams, C. S., Ellison, N., Budzynski, A. Z., and Shattil, S. J.,** Direct detection of activated platelets and platelet-derived microparticles in humans, *Blood,* 75, 128, 1990.
67. **Gemmell, C. H., Sefton, M. V., and Yeo, E. L.,** Platelet-derived microparticle formation involves glycoprotein IIb-IIIa, *J. Biol. Chem.,* 268, 14586, 1993.
68. **Richardson, P. D.,** Interaction of transport phenomena and surface reactions, *Ann. N.Y. Acad. Sci.,* 516, 492, 1987.
69. **Dewey, C. F., Bussolari, S. R., Gimbrone, M. A., and Davies, P. F.,** The dynamic response of vascular endothelial cells to fluid shear stress, *Trans. Am. Soc. Mech. Eng. J. Biomech. Eng.,* 103, 177, 1981.
70. **White, G. E., Fujiwara, K., Shelton, E. J., Dewey, C. F., and Gimbrone, M. A.,** Fluid shear stress influences cell shape and cytoskeletal organization in cultured vascular endothelium, *Fed. Proc.,* (Abstr.) 41, 321, 1982.
71. **Frangos, J. A., McIntire, L. V., Eskin, S. G., and Ives, C. L..,** Flow effects in prostacyclin production by cultured human endothelial cells, *Science,* 227, 1477, 1985.
72. **Grabowski, E. F., Jaffe, E. A., and Weksler, B. B.,** Prostacyclin production by cultured endothelial cells exposed to step increases in shear stress, *J. Lab. Clin. Med.,* 105, 36, 1985.
73. **van Kooten, T. G., Schakenrad, J. M., van der Mei, H. C., and Busscher, H. J.,** Influence of pulsatile flow on the behaviour of human fibroblasts adhered to glass, *J. Biomat. Sci. Polymer Edn.,* 4, 601, 1993.
74. **Cezeaux, J. L., Austin, V., Hosseinipour, M. C., Ward, K. A., and Zimmer, S.,** The effects of shear stress and metastatic phenotype on the detachment of transformed cells, *Biorheology,* 28, 195, 1991.
75. **Anderson, J. M. and Kottke-Marchant, K.,** Platelet interactions with biomaterials and artificial surfaces, *CRC Rev. Biocompatibility,* 1, 111, 1985.
76. **Saterbak, A., Kuo, S. C., and Lauffenburger, D. A.,** Heterogeneity and probabilistic binding contributions to receptor-mediated cell detachment kinetics, *Biophys. J.,* 65, 243, 1993.
77. **Lipowsky, H. H., Riedel, D., and Shi, G. S.,** In vivo mechanical properties of leukocytes during adhesion to vascular endothelium, *Biorheology,* 28, 53, 1991.

Chapter 2

Biomechanics and Thermodynamics of Cell Adhesion

Cheng Zhu

CONTENTS

I. Introduction 23
II. Continuum Mechanics Analysis 24
III. Molecular Considerations 28
IV. Thermodynamics Formulation 32
V. Qualitative Discussion 35
VI. Conclusion and Future Directions 36
References 38

I. INTRODUCTION

Recent advances in the biology of cell adhesion have focused on the identification and characterization of cell adhesion molecules, which mediate the adhesive interactions. The biological relevance of cell adhesion molecules has been defined using functional assays, which have two basic aspects: biophysical definition of adherence and biochemical alterations of cell adhesion molecules. Although biochemical studies, such as the development of monoclonal antibodies against cell adhesion molecules and the use of biological response modifiers to treat cells, have attracted much attention, the biophysical basis for any functional assay is the mechanical strength of adhesiveness. Only when a scale of adhesiveness is defined in a controlled experiment can the effects of biochemical manipulations be compared quantitatively.

From a mechanics standpoint, the mechanical strength of adhesiveness can be defined by either the force or the energy of the interaction. The adhesive strength can be calculated either at the level of a single pair of cells or at the level of a per molecular bond. At the cellular level (avidity), fracture stress is defined as contact stress (force per unit area of adhesion) at the point of detachment. The surface adhesion energy density is defined as the mechanical work done to separate a unit contact area. At the molecular level (affinity), bond strength is defined as the maximum force a single molecular crossbridge can sustain. Bonding energy is defined as the energy required to break a single crossbridge.

The direct measurements from experiments, however, are generally not the force and the energy of the adhesive interactions per se, but the indices thereof. An adherence index is one that reflects the adhesive strength and is convenient to measure directly from an experiment, but it also depends upon other factors and experimental conditions. For instance, the number of adherent cells from the flowing suspension to the cell monolayer in a flow chamber and the number of adherent cells to the cell monolayer in a 96-well plate after centrifugation are two adherence indices. They both reflect the fraction of cells in a population whose adhesion is strong enough to sustain the applied forces. Under controlled experimental conditions, each index can be used as a scale to compare the effects of biochemical manipulations, such as cytokine stimulation of cells. However, these two indices are not directly comparable because they depend on different sets of factors. In the centrifugation experiment, the applied force is perpendicular to the surface of adhesion. The contact area is finite after the incubation period. In the flow chamber experiment, by contrast, the hydrodynamic force is neither perpendicular nor tangential to the adhesion surface. The adhesion area is so small that it can be regarded as a "point contact". Although the external forces (shear stress and centrifugal force, respectively) are measured directly, they differ from the local stress responsible for the separation of the contact. Thus, in order to compare experimental data obtained from different types of experiments, the directly measured adherence indices need to be converted through mechanics analysis to the objective and absolute scales of adhesive strength.

Furthermore, changes of the measured adherence indices in response to biochemical manipulations are the results of changes of molecular and cellular properties. For example, two different pathways of avidity modulation are, respectively, the regulation of the expression and binding affinity of the cell adhesion molecules. To extract information regarding changing molecular and cellular properties from changing

0-8493-4559-6/95/$0.00+$.50

adherence indices requires micromechanics analysis, including molecular considerations of receptor/ligand binding.

In this chapter, selected methods and results for mathematical modeling of cell adhesion will be outlined. These models are based on abstract physical descriptions of the cell (e.g., mechanical properties of the cell and of the cell adhesion molecules) and mechanics and thermodynamics principles (e.g., mass, force, and energy balances). General equations will be presented to indicate the kinds of variables and relations involved, but detailed derivations will be omitted. Wherever possible and appropriate, numerical values will be given to indicate the order of magnitude of the important quantities. It is intended to provide an overview of the field rather than a comprehensive review of the existing literature.

II. CONTINUUM MECHANICS ANALYSIS

A starting point of continuum mechanics analysis is the following equation of conservation of energy

$$\frac{dW}{dt} = \frac{dE_k}{dt} + \frac{dE_s}{dt} + \frac{d\Phi}{dt} - \frac{dE_a}{dt} \tag{1}$$

which states that the rate of external work done by applied forces dW/dt is equal to the sum of the rate of change of kinetic energy dE_k/dt, the rate of change of elastic strain energy dE_s/dt, and the rate of viscous dissipation $d\Phi/dt$ minus the rate of change of adhesion energy dE_a/dt. The last term in Equation 1 is unique to the adhesion problem. The minus sign indicates the fact that energy is released during the formation of adhesion (when $dE_a/dt > 0$), and work must be done during the separation of adhesion (when $dE_a/dt < 0$).

Adhesion is a surface phenomenon. In general, the larger the adhesion area, the greater the adhesion energy. Therefore, the extensive property E_a can be expressed as an integral of an intensive property γ,* called surface adhesion energy density (energy per unit area, J/m²), over the adhesion region A_c, i.e., $E_a = \iint_{A_c} \gamma dA$. Alternatively, the surface adhesion energy density γ can also be directly defined as the work done by the adhesive stress σ (force per unit area, N/m²) in bringing together a unit adhesion area from infinity to an equilibrium distance of separation z,[1] i.e.,

$$\gamma = \int_z^\infty \sigma dz \tag{2}$$

For planar contact surface, the integral of σ over the adhesion region A_c yields the resultant of total adhesion force, $F = \iint_{A_c} \sigma dA$.

Adhesion energy may change because its density changes or because the area of adhesion changes. Direct differentiation of the integral expression of E_a yields

$$\frac{dE_a}{dt} = -\iint_{A_c} \sigma \frac{\partial z}{\partial t} dA + \oint_S (\boldsymbol{n} \cdot \boldsymbol{v}) \gamma ds \tag{3}$$

where S is the moving boundary contour of the changing adhesion area whose outward normal velocity is $(\boldsymbol{n} \cdot \boldsymbol{v})$. Equation 3 can be interpreted as stating that the rate of change of the total adhesion energy is equal to the sum of the rate of work done by the contact stress to bring the two surfaces closer together inside the adhesion region (the minus sign indicates the fact that $\partial z/\partial t < 0$ if the two surfaces move closer to each other due to the adhesive stress) and the rate of energy released from formation of new adhesion area at the boundary.

The complete specification of the surface adhesion energy density γ and the adhesive stress, including the use of Equation 2, requires a detailed description of the molecular interactions within the adhesive layer of a very small thickness z, as will be discussed in Section III. Within the framework of continuum mechanics at the cellular level, however, such details are omitted so that only macroscopic variables will

* In the physical chemistry of surfaces, γ is further broken down into $\gamma_1 + \gamma_2 - \gamma_{12}$ for contributions from each adhering surface and the interface, respectively.

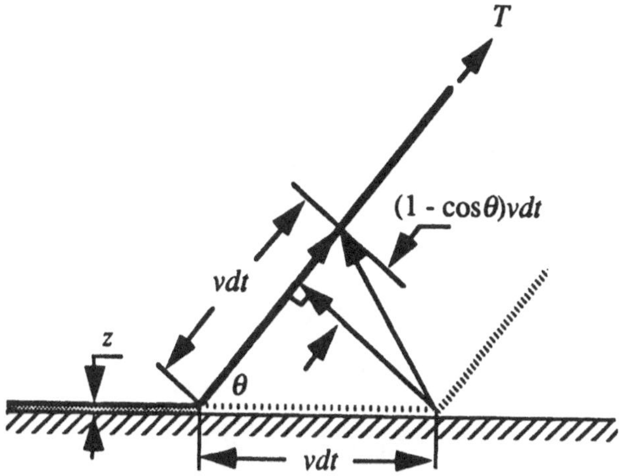

Figure 1 Peeling of a light, inextensible, flexible tape from a rigid surface to which the tape adheres. In an infinitesimal time interval *dt*, the distance that the edge of the adhesion region moves is *vdt*, the area of separation is *Svdt*, the displacement of a membrane point tangent to the tension *T* is (1 – cosθ)*vdt*, the amount of work done by the applied tension is *T*(1 – cosθ)*Svdt*, and the work done to separate a unit area of adherent tape is *T*(1 – cosθ).

be used. The justification for this approximation is the fact that the optically invisible separation distance z between the two adherent cell membranes (tens of nanometers, see Section III) is usually orders of magnitude smaller than the characteristic length scale at the cellular level. Physically, the omission of z is valid if the change of adhesion energy occurs primarily along the contact boundary so that the line integral is dominant in Equation 3.

With the omission of the first term on the right-hand side of Equation 3, the conservation of energy of Equation 1 then provides a practical way to compute γ without using its definition of Equation 2. However, a price has to be paid in ignoring z, namely, that the surface adhesion energy density obtained from continuum mechanics analysis (Equations 4, 5, 10, and 11 below) is only the γ value evaluated at the edge of the adhesion area. The γ value inside the adhesion area (which should be a function of z) cannot be obtained in this macroscopic approach.

To illustrate the use of Equations 1 and 3, let us consider a simple example of the peeling of a light, inextensible, flexible tape of width S from a surface to which the tape adheres. As shown in Figure 1, the rate of work done by the applied membrane tension T (force per unit length, N/m) is $dW/dt = T(1 - \cos\theta)Sv$, where θ is the angle between the tape and the planar surface. The line integral in Equation 3 becomes $\oint_s (n \cdot v)\gamma ds = -\gamma Sv$, where the minus sign on the right-hand side comes about because, in a peeling process, the velocity of the moving boundary is in the opposite direction to the outward normal of the boundary. The kinetic energy, the elastic strain energy, and the viscous dissipation are zero due to the assumptions of light, inextensible, and flexible tape. Therefore, Equations 1 and 3 are reduced to

$$\gamma = T(1 - \cos\theta) \tag{4}$$

which is the classic Young-Dupre equation.[2]

For axisymmetric problems, γ is a constant along the boundary of the circular adhesion area and can therefore be taken out from the integral sign of the contour integral on the right-hand side of Equation 3 to yield $\gamma\oint_s (n \cdot v)ds = \gamma dA_c/dt$. For quasi-static processes in which the kinetic energy and the viscous dissipation are negligible, Equations 1 and 3 can be written in the form of a variational principle, $\delta W = \delta E_s - \gamma\delta A_c$.[3,4] For the case in which adherent cells are in static medium with no external work done to them, the variational principle reduces to $\gamma = \partial E_s/\partial A_c$,[3] which states that, in the absence of other energy sources and sinks, the energy released from formation of a unit area of adhesion will be converted into the change of elastic strain energy per unit change of adhesion area. This equation has been applied to compute γ from the known mechanical properties and the observed shape changes of red blood cells in rouleau formation[3,5] and in particle encapsulation.[6] In these computations, the elastic strain energy E_s of the red blood

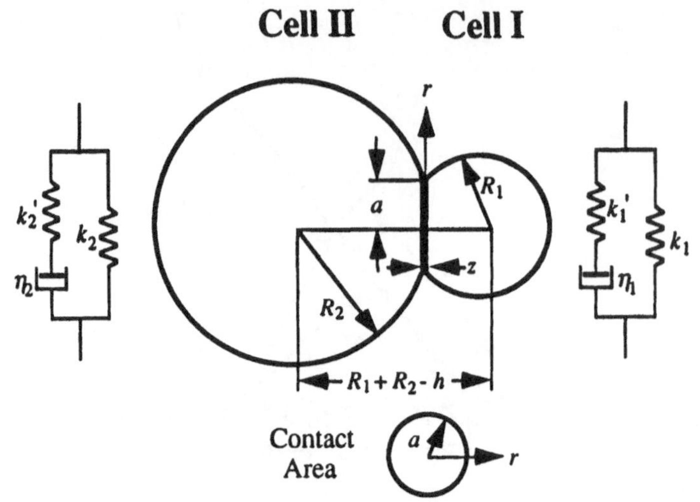

Figure 2 Schematic of the adhesion of two cells from a continuum viewpoint. The mechanical properties of the cells are modeled as incompressible, viscoelastic standard solids as shown by the spring and dashpot elements. Upon adhesion, the cell pair has the shape of two indented spheres with radii of R_1 and R_2, respectively, and forms a circular contact area of radius *a*. The centroids of the two cells move toward each other by a distance *h* after the cells deform from a point contact to a finite contact. The separation distance *z* between the two cell membranes within the contact area is negligible compared with other cellular dimensions such as *a* or *h*.

cells is due to the shear and bending deformations of the cell membrane from the natural biconcave shape to the shape of a rouleau and is therefore a function of the adhesion area A_c.

Let us consider a more involved example of the adhesion of a pair of incompressible, elastic spherical cells, as shown in Figure 2. The mechanical properties of neutrophils,[7] lymphocytes,[8] and endothelial cells[9] have been measured using an incompressible, viscoelastic standard solid model, which reduces to the elastic model in a quasi-static process when the viscous component is completely relaxed. The contact stress distribution within the adhesion area is[10]

$$\sigma = \left\{ \frac{8G_\infty}{\pi R} \left(\frac{a^2 - Rh}{2\sqrt{a^2 - r^2}} - \sqrt{a^2 - r^2} \right) \right. \tag{5}$$

where *h* is the relative displacement of the centroids of the two cells, *a* is the radius of the contact circle, and *r* is the polar coordinate in the adhesion plane. $G_\infty = 0.5k_1k_2/(k_1 + k_2)$, in which k_1 and k_2 are, respectively, the elastic coefficient of the two cells. $R = R_1R_2/(R_1 + R_2)$, in which R_1 and R_2 are, respectively, the radii of the two cells. Figure 3 shows graphically the contact stress distribution, along with the cell shapes before and after the adhesion occurs. The stress is compressive near the center of the adhesion area in order to deform the elastic cells into indented spheres. Along the periphery of the contact area, the stress is tensile in order to hold the two cells together. The total force resultant, $F = \iint_{A_c} \sigma dA = 8G_\infty a(a^2/3R - h)$, can be either compressive ($F < 0$) or tensile ($F > 0$) depending on whether the externally applied forces tend to push the two cells together or pull them apart. It should be pointed out that the σ given by Equation 5 is an approximation of the true contact stress because of the omission of the adhesive layer, as evidenced by the singularity of the stress at the boundary, which approaches infinity. It cannot be used in Equation 2 for a direct calculation of γ because the dependency of σ on *z* is missing. However, γ can be computed from the energy balance of Equation 1, as discussed below.

The rate of change of elastic strain energy can be calculated from the integral of the contact stress times the surface velocity over the adhesion area. This consists of two contributions. One contribution is from the regular integral inside the adhesion area, which is equal to the force resultant, *F*, times the relative velocity of the two centroids, $-dh/dt$ (the minus sign is due to the fact that $dh/dt < 0$ if the two spheres are approaching each other). It can be readily identified that this term is equal to the rate of external work

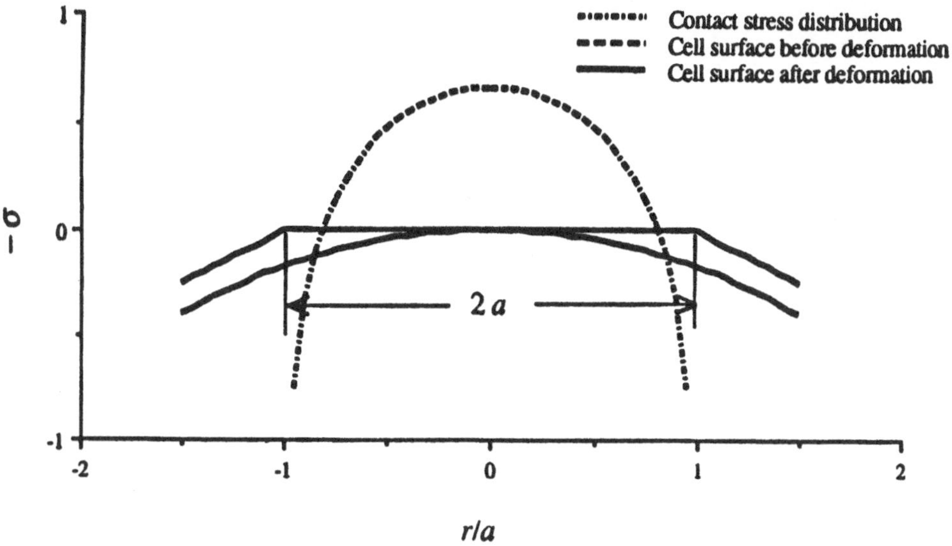

Figure 3 Contact stress distribution and shapes of one cell near the contact area before and after the adhesion. The contract stress is compressive (negative) near the center and is tensile (positive) along the periphery.

done. The other contribution comes from the integral of the product of the concentrated stress and the singular velocity over an infinitesimal thin boundary "ring". From Equations 1 and 3, this term is due to the change of adhesion energy. The result is[10]

$$\gamma = \frac{1}{16\pi G_\infty a^3}\left(F + \frac{16G_\infty a^3}{3R}\right)^2 \tag{6}$$

This calculation directly demonstrates the notion that the change of adhesion energy comes primarily from the contribution at the boundary of the adhesion region. It is worth pointing out that the stress concentration at the adhesion boundary resembles that at a Griffith crack tip in linear elastic fracture mechanics. Therefore, Equation 6 can also be obtained using the method from fracture mechanics, i.e., $\gamma = \pi N^2/2G_\infty$, where $N = \lim_{r\to a}\sqrt{a-r}\,\sigma$ is the stress intensity factor.[11]

Setting $F = 0$ in Equation 6 provides us with an expression for the adhesion energy density for the case in which the adhesion area increases in the absence of externally applied forces. Namely, γ is proportional to the cube of the radius a of the contact area, is proportional to the rigidity G_∞ of the cell pair, and is inversely proportional to the square of the size R of the cell pair. From the measured geometry and viscoelastic coefficients (R_1 = 4.0 μm, R_2 = 7.5 μm, k_1 = 439 dyn/cm^2, k_2 = 641 dyn/cm^2),[8] it can be calculated that the surface adhesion energy density for the conjugation of a cytotoxic T-lymphocyte (CTL, designated as cell I) to a target cell (TC, designated as cell II) is $\gamma_\infty = 3.2 \times 10^{-3}$ dyn/cm after the adhesion has spontaneously grown to the maximum conjugation area (with a radius of a_∞ = 1.43 μm).[12]

The adhesion energy densities of inert adhesives used in engineering are usually constants in a peeling test, which can thereby be characterized as a property of a given adhesive material. Assuming γ = constant in Equation 6 provides a relationship between the applied force F and the contact radius a in a process of separation of the two adherent solid spheres. The maximum force that the conjugation can sustain (critical force) can then be calculated by setting $dF/da = 0$ to yield $F_c = 0.75\pi R\gamma$. This theoretical prediction compared well with experimental measurements on gelatine spheres in contact with perspex.[10]

The adhesion between biological cells, by contrast, often exhibits a variable γ. As given by Equation 6, γ increases with increasing adhesion area in the process of spontaneous spreading of adhesion between two cells in the absence of applied force. In the process of forcible separation, γ often increases with decreasing adhesion area. For CTL-TC adhesion, the critical force predicted under the assumption of a constant $\gamma(=\gamma_\infty = 3.2 \times 10^{-3}$ dyn/cm) is $F_c = 4.5 \times 10^{-6}$ dyn, which is two orders of magnitude smaller than the experimentally measured minimum force required to separate the cell pair ($F_c = 4.8 \times 10^{-4}$ dyn).[12] In other words, the γ value at the point of separation by force should be much greater than that at maximum conjugation area without applied force. Tözeren et al.[13] have applied the Young-Dupre Equation 4 to

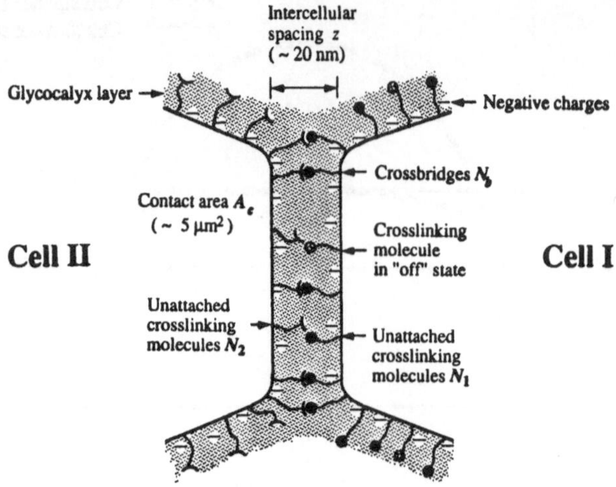

Figure 4 Schematic of the adhesion of two cells from a molecular viewpoint. The cell surfaces are negatively charged and coated with a glycocalyx layer, which gives rise to the repulsive forces as a function of the intercellular spacing *z*. The adhesive forces are mediated by specific adhesion molecules, which are expressed on the cell surface, bind to their ligands on the other cell, and may move laterally by diffusion.

analyze the experimental data of the peeling of an adherent CTL-TC pair by micropipette manipulation.[12] The $1/\gamma$ value calculated from the measured inclined angle θ and the membrane tension T is nearly a linear function of the conjugation area, and γ can be as high as 1.4 dyn/cm at the point of separation.[13]

Continuum mechanics analysis alone cannot explain why the boundary value of γ should change with the applied force and with the adhesion area, nor can it tell how the adhesion energy density will change. Moreover, computation of the boundary value of γ from Equation 1 requires that γ be a constant along the boundary, and is therefore limited to one-dimensional or axisymmetrical problems. In the case in which γ is a variable along the boundary, such as in the case of a leukocyte rolling on an endothelial cell with formation of adhesion at the leading edge and separation of adhesion at the rear edge of the contact area, the distribution of γ must be assumed in order for Equation 1 to be used.[14] These limitations call for an argument that the continuum mechanics approach include molecular considerations.

III. MOLECULAR CONSIDERATIONS

To obtain a more detailed description of adhesion energy density γ than its boundary value computed from continuum mechanics analysis, the molecular interactions within the thin layer between the two cell membranes that mediate the adhesion need to be considered. These molecular forces can be classified into two categories: specific binding via cell adhesion molecules and nonspecific interactions. Figure 4 is a schematic of two interacting cell surfaces, showing the physical and chemical factors involved at the molecular level. It should be pointed out that the smooth membrane shown in Figure 4 is only an idealization of a cell surface. This probably is a good model for red blood cells whose surfaces are smooth, but may be a poor model for white blood cells, whose surfaces have many membrane folds. The initiation of adhesion between smooth surfaces and between rough surfaces may be quite different. In a recent paper, the irregular microvilli were described statistically in a simulation of cell rolling and adhesion on surfaces in shear flow.[15] However, after adhesion has been initiated, the subsequent progression of the adhesion zone no longer appears to be mediated by the microvilli alone.[16] We shall limit the scope of our discussion to the physical model as sketched in Figure 4.

In the nanometer range, the nonspecific intermolecular and surface forces that are believed to be relevant to cell adhesion include electrostatic repulsion, van der Waals force, and steric stabilizing force.[16,17] Electrostatic repulsion is due to the negative charge groups of the glycocalyx layer covering the cell surface, which attract counterions from the suspending medium.[18] Repulsive pressure can be calculated assuming two parallel double-layers of ions.[19,20] Roughly speaking, the repulsive pressure is proportional to z^{-2} when z is smaller than the thickness of the glycocalyx layer (~10 nm), it is nearly a constant (~10^5 dyn/cm^2) when z is between one and two times the thickness of the glycocalyx layer and it drops off exponentially (with a decay length scale as the Debye length of about 0.8 nm) when z is larger than twice of the thickness of the glycocalyx layer.[16] Van der Waals force originates from the attraction among the freely rotating dipoles and polarizable structures.[17] The interaction energy between two semi-infinite media can be calculated by summation over all the pair interactions, which yields an attractive stress that is proportional to z^{-3}, with a coefficient of proportionality of about 10^{-15} erg (Hamaker constant divided by 6π). Hence, in the nanometer range, van der Waals force is about two orders of magnitude smaller than electrostatic force. Steric stabilizing force is due to the interaction between the polymers of the glycocalyx layers of the two cells.[16,21] It is essentially zero until the two layers overlap (i.e., the polymers of one layer interpenetrate the other layer). After this point, the steric interaction has the same functional dependency on z as, but is about an order of magnitude larger than, the electrostatic repulsion. Other forces that may be relevant to cell adhesion include hydration force,[17] hydrogen bonding,[22] macromolecular depletion force, and multivalent ion correlation force.[23]

A simple phenomenological function has been proposed by Bell et al.[24] to represent overall nonspecific repulsive stress:

$$\sigma_r = -\left(\frac{1}{z} + \frac{1}{d}\right)\frac{e}{z}\exp\left(-\frac{z}{d}\right) \tag{7}$$

where d is the Debye length and e is a compressibility coefficient (in Newtons). It has the z^{-2} repulsive core as z approaches zero and drops off exponentially as z becomes large. e Is proportional to the concentration of the polymers in the cell's "fuzzy coat" if these "repellers" are modeled as freely mobile.[25]

Since the nonspecific interactions are repulsive, adhesion of cells will not occur unless an adhesive force counterbalances the repulsive force. In the absence of macromolecular depletion forces and multivalent ion correlation forces,[23] this adhesive force is mediated by the binding of specific cell adhesion molecules. The binding of v_{1j} receptors N_{1j} expressed on the surface of cell I to v_{2j} counterreceptors N_{2j} expressed on the surface of cell II to form v_{bj} crossbridges N_{bj} resembles a chemical reaction, as symbolized by the following chemical equation ($j = 1, \ldots, \ldots J$)

$$v_{1j}N_{1j} + v_{2j}N_{2j} \underset{k_{-j}}{\overset{k_{+j}}{=}} v_{bj}N_{bj} \tag{8}$$

where the subscript j denotes a particular type of binding reaction and the total number of the binding reactions is J. Except in some experimental systems where purified molecules were used,[26,27] the adhesion of a particular cell pair is in general mediated by more than one type of cell adhesion molecule. For example, at least four receptor–ligand pairs have been identified to be involved in the adhesion of a cytotoxic T-lymphocyte and a target cell: LFA-1/ICAM-1, LFA-1/ICAM-2, CD2/LFA-3, and TcR-CD8/MHC-I.[28] These are schematically shown in Figure 5, together with the numerical values of the stoichiometric coefficients v_{ij}. Different types of cell adhesion molecules may play different roles in different processes, and they may be cooperative in a particular process via subtle interactions.[26]

Detailed information regarding the kinetics of receptor–ligand binding is very limited. Models published to date all assume an elementary process with a simple kinetics law for the binding rate

$$q_j = k_{+j} n_{1j}^{v_{1j}} n_{2j}^{v_{2j}} - k_{-j} n_{bj}^{v_{bj}} \tag{9}$$

where k_{+j} and k_{-j} are, respectively, the "on" and "off" rate coefficients. The binding affinity, or association parameter, K_j, is:

$$K_j = \frac{k_{+j}}{k_{-j}} = \left.\frac{n_{bj}^{v_{bj}}}{n_{1j}^{v_{1j}} n_{2j}^{v_{2j}}}\right|_{eq} \tag{10}$$

where n_{ij} denotes the surface number of density (molecules per unit area of cell surface) of species N_{ij}. The second half of Equation 10 is obtained at chemical equilibrium when q_j vanishes.

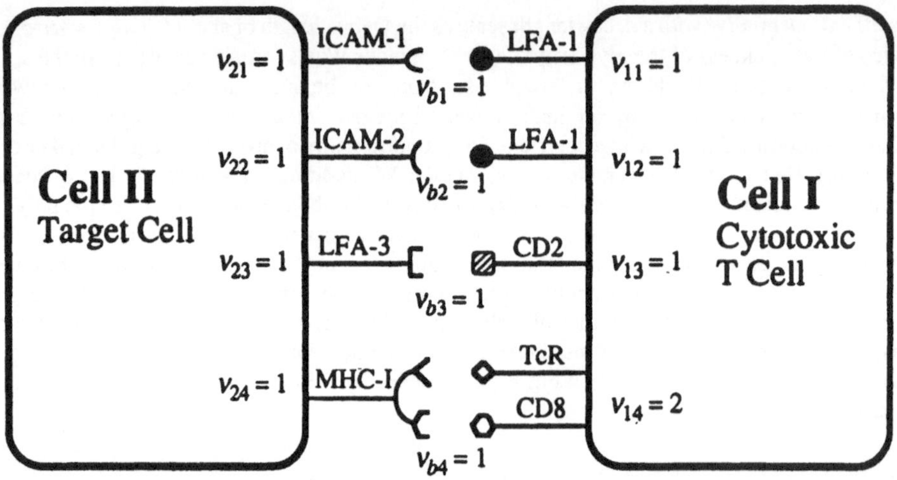

Figure 5 Schematic of various pairs of receptors and ligands known to be involved in the adhesion of a cytotoxic T lymphocyte to a target cell. The stoichiometric coefficients v_{ij} for each receptor–ligand pair are shown, in which i (= 1, 2, or b) denotes the ith cell or bond, and j (=1, 2, 3, or 4) denotes the jth receptor–ligand pair. Abbreviations: ICAM-1(2) = intercellular adhesion molecule-1(2), LFA-1(3) = lymphocyte function-related antigen-1(3), CD2(8) = cluster of differentiation 2(8), MHC-I = major histocompatibility complex class I, TcR = T cell receptor.

The crossbridges are normally subject to forces, since the cellular binding is mediated by these molecular bonds. A mechanical model specifying how a crossbridge responds to forces is required to describe the influences of the stress state of bonds on their binding kinetics, affinity, lateral mobility, and chemical potential. Because of the lack of experimental data, only the simplest model is assumed in the literature published to date, namely, the bond force f_j is proportional to the bond stretch, $f_j = \kappa_j(z - z_{0j})$. This linear elastic spring model should be a good approximation if the intercellular spacing z is close to the natural length z_{0j} of the crossbridge. The value of the elastic modulus κ_j has been estimated to range from 10^{-2} to 10^3 dyn/cm.[24] For a spring constant $\kappa_j = 0.1$ dyn/cm, a bond force $f_j = 10^{-8}$ dyn[29] and a natural length $z_{0j} = 20$ nm, the bond stretch is $z - z_{0j} = 1$ nm, and the bond strain is $(z - z_{0j})/z_{0j} = 0.05$.

Thus, from molecular considerations, the total contact stress σ is the sum of the binding stress σ_b and the repulsive stress σ_r,

$$\sigma = \sum_{j=1}^{J} n_{bj}\kappa_j(z - z_{0j}) - \left(\frac{1}{z} + \frac{1}{d}\right)\frac{e}{z}\exp\left(-\frac{z}{d}\right) \tag{11}$$

which, upon integration, should give the surface adhesion energy density γ. If the polymers of the glycocalyx layers are modeled as fixed,* direct integration on the repulsive stress indeed yields a repulsive energy density: $\gamma_r = \int_z^\infty \sigma_r dz = -(e/z)\exp(-z/d)$. However, direct integration of the binding stress σ_b (the first term on the right-hand side of Equation 11) is not trivial because the bond density n_{bj} also depends on the separation distance z via several mechanisms.

First, from thermodynamics (see Section IV), the association parameter K_j is an exponential decaying function of the bond strain energy (scaled by the thermal energy kT), i.e.,

$$K_j = K_{j0} \exp\left\{-v_{bj}\kappa_j(z - z_{0j})^2/2kT\right\} \tag{12}$$

where k is the Boltzmann constant, T is the absolute temperature, and the constant K_{j0} is the intrinsic association constant (binding affinity). Second, Equations 10 and 12 provide a restriction on the association and disasocciation rate coefficients which also depend on z. For example,[30]

*For the case of freely mobile repellers, see Torney et al.[13]

$$k_{+j} = k_{+j}^0 \exp\left\{-v_{bj}\zeta_j\kappa_j(z - z_{0j})^2/2kT\right\}$$
$$k_{-j} = \frac{k_{+j}^0}{K_{j0}} \exp\left\{-v_{bj}(\zeta_j - 1)\kappa_j(z - z_{0j})^2/2kT\right\} \tag{13}$$

The splitting parameter ζ_j is an important property of a bond. If $\zeta_j > 1$, the bond is said to be a catch-bond, if $\zeta_j = 1$, it is said to be an ideal bond, and if $\zeta_j < 1$, it is said to be a slip-bond.[30] The stretch of the bonds will increase the disassociation rate for slip-bonds, but not for ideal and catch-bonds. The implications of the catch-bond property will be discussed in Section V.

Third, due to the spatial variation of the separation distance z, the bond force has a component tangent to the inclined surface, which tends to drag the crossbridge toward the area of lesser bond strain. This has been modeled by a flux $\boldsymbol{J}_j$ of crossbridges in addition to the density gradient, i.e.,[13]

$$\boldsymbol{J}_j = -n_{bj}\beta_j\kappa_j(z - z_{0j})\nabla(z - z_{0j}) \tag{14}$$

where the drag coefficient β_j can be related to the diffusion coefficient D_{bj} by Einstein's relation, $\beta_j = D_{bj}/kT$. Thus, direct integration of Equation 11 is not possible until the dependency of n_{bj} on z is solved from the general equations governing the densities n_{ij} on the cell surface (see Section IV).

The governing equations for n_{ij} are a set of diffusion-reaction equations derived from the species conservation of the cell adhesion molecules.[31]

$$\frac{\partial n_{1j}}{\partial t} = D_{1j}\nabla^2 n_{1j} + r_{1j} - v_{1j}q_j$$
$$\frac{\partial n_{2j}}{\partial t} = D_{2j}\nabla^2 n_{2j} + r_{2j} - v_{2j}q_j \tag{15}$$
$$\frac{\partial n_{bj}}{\partial t} = D_{bj}\nabla^2 n_{bj} + \frac{D_{bj}\kappa_j}{2kT}\nabla\cdot\left\{n_{bj}\nabla(z - z_{0j})^2\right\} + v_{bj}q_j$$

which states that the rates of changes of the densities n_{ij} are due to diffusion, binding, upregulation (for the cases of n_{1j} and n_{2j}), and forcible migration (for the case of n_{bj}). The equation for attached molecules n_{bj} applies to the contact area A_c only. The equations for the unattached molecules n_{1j} and n_{2j} apply to the entire cell surfaces, A_1 and A_2, respectively, keeping in mind that $q_j = 0$ outside of A_c. The diffusion coefficients D_{ij} range from 10^{-9} to 10^{-12} cm^2/s, depending on the degree of anchorage of the cell adhesion molecules to the cytoskeletal structures underneath the plasma membrane.[32,33] The upregulation rates r_{ij} are negligible if it is due to the synthesis of new adhesion proteins, because protein synthesis requires a time period of several hours, which is much longer than the time scale of the adhesion process under consideration (minutes). However, r_{ij} may be significant if the upregulation is due to the active expression of the adhesion molecules from internal storage or the activation of adhesion molecules already present on the membrane via conformational changes, which may occur in a very short time scale.[28] In such a case, additional kinetics equations describing the rates r_{ij} are needed.

The boundary conditions for n_{ij} at the edge of the adhesion region can also be derived from the species conservation of the cell adhesion molecules. These are a set of jump conditions.[31]

$$\boldsymbol{n}\cdot\left\{\boldsymbol{v}\left(\frac{v_{1j}}{v_{bj}}n_{bj} - [n_{1j}]\right) + \frac{v_{1j}}{v_{bj}}D_{bj}\left(\nabla n_{bj} + \frac{n_{bj}\kappa_j}{2kT}\nabla(z - z_{0j})^2\right) - D_{1j}[\nabla n_{1j}]\right\} = 0$$
$$\boldsymbol{n}\cdot\left\{\boldsymbol{v}\left(\frac{v_{2j}}{v_{bj}}n_{bj} - [n_{2j}]\right) + \frac{v_{2j}}{v_{bj}}D_{bj}\left(\nabla n_{bj} + \frac{n_{bj}\kappa_j}{2kT}\nabla(z - z_{0j})^2\right) - D_{2j}[\nabla n_{2j}]\right\} = 0 \tag{16}$$

where the square brackets denote jumps (difference between the outer and inner limits of the quantities across the boundary). Equation 16 omits the details within the narrow edge zone (where spatial variations of n_{ij} may be large) by treating the boundary as a singular curve of infinitesimal thickness. The binding rate of Equation 9 has been integrated into the jumps in densities n_{ij}. This is analogous to the treatment of shock waves as discontinuous surfaces in aerodynamics.

If the adhesion is a fixed domain, Equations 15 and 16, plus the global mass conservation equation and the initial conditions, form the boundary value problem for the determination of the spatial and temporal distribution of the cell adhesion molecules on the surface of the interacting cells. However, in the case in which the time course of growth of the adhesion area is considered, the domain of definition of Equation 15 is itself a variable and must be obtained as part of the solution. Mathematically, this is a moving boundary problem. In this case, an additional boundary condition is needed, which can be obtained from the energy balance along the boundary, as will be discussed in Section IV.

IV. THERMODYNAMICS FORMULATION

As in the continuum mechanics approach, a practical way to compute γ without directly using its definition, Equation 2, is to use the equation of conservation of energy. Since we would like to obtain an expression for γ in terms of the molecular variables instead of continuum mechanics variables, however, Equation 1 needs to be augmented to include thermal and internal energies, i.e., the first law of thermodynamics

$$\frac{dW}{dt} + \frac{dQ}{dt} = \frac{dE_k}{dt} + \frac{d}{dt}(H + TS) \tag{17}$$

which states that the rate of external work done plus the rate of heat added *(dQ/dt)* is equal to the rate of changes of kinetic and internal energies. Here, the internal energy is written as the sum of Helmholtz free energy *H* and the temperature *T* times the entropy *S* according to the thermodynamic relation, because temperature is chosen as a primary state variable instead of entropy (which is to be determined as a function of state variables).

The molecular variables enter the thermodynamics formulation through the constitutive hypothesis of Helmholtz free energy:*[31]

$$H = E_s - \iint_{A_c} \gamma_r dA + \sum_{j=1}^{J} \left\{ \oiint_{A_1} \mu_{1j} n_{1j} dA + \oiint_{A_2} \mu_{2j} n_{2j} dA + \iint_{A_c} \mu_{bj} n_{bj} dA \right\} \tag{18}$$

which consists of, in order of appearance, the elastic strain energy of the cells, the repulsive potential for the nonspecific interactions, and the Gibbs free energy for each molecular species N_{ij}. The chemical potentials are[24]

$$\begin{aligned} \mu_{1j} &= \mu_{1j}^0 + kT \ln n_{1j} \\ \mu_{2j} &= \mu_{2j}^0 + kT \ln n_{2j} \\ \mu_{bj} &= \mu_{bj}^0 + kT \ln n_{bj} + \frac{\kappa_j}{2}(z - z_{0j})^2 \end{aligned} \tag{19}$$

which consists of a standard constant and a contribution from the variable density n_{ij}. For the case of attached crossbridges, an additional term for the strain energy stored in the stretched bond has been included to account for the influence of bond force on the chemical potential μ_{bj}. The adhesion energy E_a introduced in Equation 1 has already been implicitly included in the Helmholtz free energy, and therefore allows derivation of γ upon substitution of Equation 18 into Equation 17.

Bell et al.[24,25,34] have introduced the thermodynamics approach to calculate equilibrium properties of an adherent cell system. In thermodynamic equilibrium, the use of Equation 17 is equivalent to a

*The nonspecific interactions are assumed to be mediated by immobile repellers. For the case of freely mobile repellers, see Torney et al.[13]

variational principle (similar to that discussed in Section II) that minimizes the Helmholtz free energy. However, this time, the Helmholtz free energy is minimized in terms of not only the continuum mechanics variables, but also the molecular variables. The minimization of H with respect to z, $\partial H/\partial z = 0$, leads to the force balance of Equation 11. The minimization of H with respect to n_{bj}, $\partial H/\partial n_{bj} = 0$, leads to the condition for chemical equilibrium, Equations 12 and 13. The minimization of H with respect to A_c can be written as $\partial(E_s - H)/\partial A_c = \partial E_s/\partial A_c$. The right-hand side of this equation is the macroscopic expression for the adhesion energy density γ (see Section II). The left-hand side, therefore, is the microscopic expression for γ in terms of the molecular variables.* For the case in which only one type of receptor–ligand pair exists ($J = 1$, the subscript j can therefore be omitted), the surface adhesion energy density obtained by these authors is $\gamma = \gamma_b + \gamma_r$,* where γ_r is the contribution from the nonspecific repulsion (see discussion following Equation 11 in Section III). The contribution from specific molecular binding is

$$\gamma_b = n_b kT \tag{20}$$

which states that each molecular crossbridge contributes an equilibrium binding energy of the amount of kT ($= 4.1 \times 10^{-14}$ erg at room temperature of 25°C). For a bond density of 10^5 molecules per square micron, this calculates to a γ_b of 0.41 dyn/cm.

Equation 20 resembles the state equation for ideal gas, where the binding adhesion energy density γ_b is analogous to the pressure p, the adhesion area A_c is analogous to the volume V, and kN_b($N_b = n_b A_c$ is the total number of bonds) is the "gas constant" for molecules moving in a two-dimensional surface. This result is not surprising because of the ideal solution form of the chemical potentials (Equation 19) used and the assumption of thermodynamic equilibrium.

If the cross-linking molecules are assumed to be immobile, Equation 20 needs to be modified as[36]

$$\gamma_b = \left\{ n_b - (n_1^+ - n_1^-) - (n_2^+ - n_2^-) + n_1^+ \ln\left(\frac{n_1^+}{n_1^-}\right) + n_2^+ \ln\left(\frac{n_2^+}{n_2^-}\right) \right\} kT \tag{21}$$

where the superscripts + and – denote, respectively, the constant density values outside and inside the adhesion area. In the case of excessive ligands ($v_2 = 0$), Equation 21 reduces to $\gamma_b = n_1^+ kT\ln(1 + K)$,[30,34,35] which calculates to a γ_b of 0.18 dyn/cm for a cell of surface area of 154 μm, cell adhesion molecules of 10^6 receptors per cell, and an association parameter of $K = 10^3$.

The different results given by Equations 20 and 21 are due to the different mass conservation equations used (as restrictions in the minimization of H): for the case of mobile molecules, this is

$$\begin{aligned} n_1 A_1 + \frac{v_1}{v_b} n_b A_c &= n_1^0 A_1 \\ n_2 A_2 + \frac{v_2}{v_b} n_b A_c &= n_2^0 A_2 \end{aligned} \tag{22}$$

and for the case of immobile molecules, it is

$$\begin{aligned} n_1^- + \frac{v_1}{v_b} n_b &= n_1^+ \\ n_2^- + \frac{v_2}{v_b} n_b &= n_2^+ \end{aligned} \tag{23}$$

Equation 22 states that, in the absence of upregulation, the total cell adhesion molecules (unattached plus attached) on the surface of each cell remains a constant. Since the molecules are freely mobile, the

* E_s was neglected in the paper of Bell et al.[9] and therefore γ had to be set to equal zero for equilibrium.

unattached molecules are uniformly distributed over the entire cell surface. For immobile molecules, the conservation statement becomes a local one instead of a global one. Equation 23 is valid at every point inside the adhesion region. However, the difference in the densities of the unattached molecules inside and outside the adhesion region cannot be smoothed out because the molecules are fixed. Thus, Equation 23 recovers the jump conditions of Equation 16.*

The reader can verify that γ_b given by Equations 20 or 21 is related to the binding stress σ_b (the first term on the right-hand side of Equation 11, for the case of $J = 1$) by Equation 2. The bond density n_b in Equation 20 as a function of the intercellular spacing z can be obtained as an equilibrium solution to Equations 15 and 16 when all the time derivatives and fluxes vanish. The result is that the unattached molecules are uniformly distributed, and the bond density obeys a Gaussian distribution, $n_b(z) = n_b(0)\exp\{-\kappa(z - z_0)^2/2kT\}$. Thus, Equation 2 can readily be verified by direct differentiation of Equation 20 and by comparison of the result with the binding stress σ_b. In a reversed approach, Equation 20 was obtained by direct integration of σ_b from the consideration of force balance using the Gaussian distribution for the bond density.[13] The fact that γ_b given by Equation 21 and σ_b are related by Equation 2 can be shown by differentiating Equation 21 and using Equations 10, 12, and 23 to eliminate $\partial n_b/\partial t$ and $\partial n_i/\partial t$.

The use of Equation 17 alone is subject to the same limitation as the use of Equation 1 in the sense that γ must be a constant along the boundary of the adhesion area and therefore is limited to equilibrium problems. To remove this deficiency, the second law of thermodynamics has been used in connection with the first law to obtain the Clausius-Duhem inequality:[31]

$$\frac{d_iS}{dt} = \frac{1}{T}\left\{\sum_{j=1}^{J}\sum_{i=1}^{2}\oint\!\!\oint_{A_i}(\mu_{ij} + kT)r_{ij}dA + \frac{dW}{dt} - \frac{dE_k}{dt} - \frac{dH}{dt} - S\frac{dT}{dt}\right\} \geq 0 \tag{24}$$

which states that the rate of spontaneous entropy production due to irreversible processes, d_iS/dt, is non-negative. The right-hand side of Equation 24 can be expressed in terms of the system variables upon substitution of Equation 1 for dW/dt and Equations 18 and 19 for H. To allow the densities n_{ij} to be nonuniform for the general nonequilibrium case, Equations 15 and 16 are required to carry out the derivative of H. After extensive mathematical manipulations, the same equations can be obtained for the force balance (Equation 11) and association equilibrium parameter (Equation 12). The equation for the reversible part† of the surface adhesion energy density γ becomes[31]

$$\gamma = kT\sum_{j=1}^{J}\left(n_{bj} - \sum_{i=1}^{2}\left\{[n_{ij}] - \left(n_{ij}^{+} + D_{ij}\frac{\boldsymbol{n}\cdot\nabla n_{ij}^{+}}{\boldsymbol{n}\cdot\boldsymbol{v}}\right)[\ln n_{ij}]\right\}\right) - \frac{e}{z}\exp\left(-\frac{z}{d}\right) \tag{25}$$

Several points regarding Equation 25 are worth noting. First, Equation 25 is valid for any degree of mobility for the cross-linking molecules. This can be checked by the two extreme cases: for the case of freely mobile molecules (in which $[n_{ij}] = [\ln n_{ij}] = 0$) and for the case of immobile molecules (in which $D_{ij} = 0$), the first term on the right-hand side of Equation 25 recovers, respectively, Equations 20 and 21. Second, Equation 25 is not limited for equilibrium cases because Equation 24 is valid for nonequilibrium as well as equilibrium processes. However, in the nonequilibrium case, Equation 25 holds only along the boundary of the adhesion area, as opposed to the equilibrium case in which Equations 20 or 21 are valid on the entire adhesion area. The reason is that the jumps in Equation 25 are not defined inside the adhesion area when the densities are nonuniform. Third, Equation 25 holds at every point along the contact boundary and can therefore be used as an additional boundary condition, in connection with the jump conditions of Equation 16, to determine the shape and position of the moving boundary of the contact area as a function of time for the case in which the dynamic evolution of the adhesion region is considered.

So far, the formulation for the case of multiple species binding has been presented for the sake of generality. However, examination of Equations 11 and 25 reveals that, as a result of the idealizing

---

* Note that the diffusion coefficients are zero for immobile molecules!

† The reversible part of γ is the part that is an odd function of the outward normal velocity of the changing adhesion boundary $(\boldsymbol{n}\cdot\boldsymbol{v})$, while the irreversible part is an even function of $(\boldsymbol{n}\cdot\boldsymbol{v})$.

assumptions of the present theory, the contributions from the individual molecular species N_{ij} to the overall adhesive stress σ_b and specific adhesion energy density γ_b are additive. In other words, different types of cell adhesion molecules can all be lumped into a single type of composite surrogate in a similar manner, as different kinds of gases can be treated collectively as air in aerodynamics.

In this chapter, the continuum mechanics analyses and molecular considerations are decoupled and carried out at two separate levels. This decoupling is made possible by omitting the intercellular spacing z in the analysis at the cellular level and by omitting the thickness of the edge zone of the adhesion area in the analysis at the molecular level. As has already been pointed out, however, the stress computed via continuum mechanics is approximate when z is omitted. This is evidenced by the presence of a stress singularity at the contact boundary where z becomes comparable with the cellular dimension and thus causes the approximation to break down (Equation 5). An alternative approach is to solve simultaneously the coupled problem for the distributions of both stress and bonds.[13,30] This approach is necessary to obtain detailed information within the edge zone. The fact that the results so obtained match asymptotically the results derived from the separate microscopic and macroscopic approaches validates the decoupling of continuum mechanics and molecular equations.

V. QUALITATIVE DISCUSSION

The equations previously presented are general and flexible enough to account for a wide variety of cell adhesion phenomena. They help us to identify the basic physicochemical processes involved, the important variables used to describe these processes, and the fundamental relationships governed by the laws of mechanics and thermodynamics. No general theoretical solutions have been found to date, due to the complexity of the general equations. A major difficulty is the lack of specific experimental information. In some cases, it is not known among several plausible processes which specific one actually occurs or dominates. In other cases, the numerical values of important coefficients are not available. Consequently, applications of the general formulation to particular problems require specific assumptions be made to simplify the equations to allow a tractable analysis and to fill in the missing links. In the following paragraphs, we will discuss the physical bases of some simplified models and their limitations.

Equilibrium models are concerned with the final outcome after adhesion processes cease to change with time, or quasi-static processes whose time rates of change are negligible. These models calculate the surface adhesion energy density and relate the experimentally measured adhesion area and separation force to the receptor number and binding affinity. An important question is the reversibility of a given adhesion process. From thermodynamics, the adhesion energy density calculated from equilibrium models is reversible and should be the lower bond of that calculated from dynamic models. It has been observed experimentally that some adhesion processes appear to be reversible, while other are not. Irreversibility may manifest as dissipation in a cycle of adhesion and separation,[14] or as higher tension required to separate an adhered area than that required to prevent spread of an adhering area,[36-38] or as lack of the ability to readhere once the surfaces are separated.[39,40]

Several mechanisms may be responsible for the observed irreversibility. Tözeren[41] has shown that, if the bonds are allowed to be drawn from the contact boundary toward the adhesion region by the peeling force (see Equation 14), a concentration of bonds at the edge of the contact area can be produced during separation of adhesion, but not during formation of adhesion. Such a redistribution of bonds may account for the difference between the energy released in the formation of adhesion and work done in the separation of adhesion. An alternative explanation was offered by Evans,[36] who showed that the different membrane tensions required to separate adhesion and to prevent further adhesion could be produced if the bonds were treated as discretely distributed. Another plausible mechanism may be the different responses of the rate coefficients to bond force. The membrane tension required to prevent formation of adhesion would be less than that required to separate adhesion if the bonds have the property of the so-called "catch-bond" ($\zeta_j > 1$ in Equation 13). The applied force would prevent the breakage of existing bonds, if separation proceeds at a rate faster than the equilibrium rate for dissociation of bonds (which is infinitely slow).[30] Yet another mechanism may be the rupture of bonds or extraction of bonds from the cell membrane. This mechanism can account for the inability of surfaces to readhere once bonds have separated.[40] There have been several experimental observations that indicate some redistribution of bonds[42] and extraction of intact bonds from the plasmic membrane.[40] At the present time, however, available data are not sufficiently extensive to allow definite selection among plausible models.

Dynamic models are concerned with adhesion processes whose time rates of change are important. Various dynamic adhesion phenomena can be understood based on the time scale of the external control factors in relation to the intrinsic time scales of the molecular events that mediate cell adhesion. In the absence of upregulation, these are the time scales of binding kinetics and of lateral mobility.[43] The binding time scale t_b is determined by the association rate constant of the binding reaction, k_+. If the binding is energetically favorable, t_b is the time a receptor takes to find a counter-receptor in its vicinity, which is limited by the short-range diffusion of the receptor and the local density of the counter-receptor. Using typical values of $k_+ \sim 10^{-12} - 10^{-7}$ cm²/s,[26,44] $D_1 = 10^{-12} - 10^{-9}$ cm²/s,[32,33] and $n_1^0 = 500 - 5000$ molecules/μm², it can be estimated that $t_b \sim 1/(k_+ n_1^0) \sim 1/(D_1 n_1^0)$ ranges from a fraction of a second to seconds. The diffusion time scale t_d is determined by the long-range diffusion of cross-linking molecules on the surface of a cell and can be estimated from $t_b \sim R_1^2 / D_b$ to range from a few to tens of minutes for a cell radius of $R_1 = 5$ μm.

The rolling and transient adhesion of flowing blood cells to endothelial cells in small venules operates in a time scale of $t_r \sim R_1/V$, which is comparable to the binding time scale for a rolling velocity V of a few microns per second. Within the binding time scale after the first contact, only those cross-linking molecules that are originally present in the contact "point" can participate in binding, because there is not enough time for molecules outside the contact area to diffuse into the contact area. If the binding energy released from the bond formation of the "on-site" cross-linking molecules is insufficient to deform the cell to form a "finite" conjugation area, the adhesion will remain a "point contact" with a small number of bonds. The key issue here is whether a few bonds can be formed to stop the flowing cell during the short time of "encounter duration". The governing process is the binding reaction, which has been modeled by Hammer and Lauffenburger using a simple kinetics equation.[15,44,45] The mechanics enters the problem in several ways. The frequency of collision and the duration of encounter between a flowing cell and an endothelial cell are determined by the hydrodynamic interactions among the fluid, the particle, and the solid wall. So are the force and torque exerted on the blood cell. The applied force and torque act on the molecular bonds and stretch them, thereby affecting the rate coefficients. These notions are also applicable to a related problem of the aggregation and disaggregation of blood cells under shear flow, which is determined by the binding time scale in relation to the shear rate (of unit 1/s).[46-48]

The binding reaction achieves equilibrium at the end of the binding time scale when all available molecules originally present in the point contact have been "used up". Further growth of the contact area then requires additional cross-linking molecules to be brought into the adhesion region. If there is no up-regulation and active transport of the receptors, this can only be done through long-range diffusion. The strengthening of adhesion via spreading of the adhesion area often takes 10 to 20 min before a plateau is reached.[12,26,27,29] It is probably not a coincidence that this is the same time scale as that of the lateral mobility of cross-linking molecules. The dynamic process of spontaneous growth of the adhesion area has been modeled as a diffusion-moving boundary problem. The binding reaction can be treated as in chemical equilibrium, except at the boundary. The rate of growth of the adhesion area depends on the rate of diffusion into the contact area. The final equilibrium contact area is limited by the deformability of the cell.[49]

The rouleau formation in red blood cells is an example that is different from the two cases discussed before. The mechanical properties of a red blood cell are primarily determined by its membrane, which is essentially elastic. In the absence of external hemodynamic forces, the mechanical response time of a red blood cell as represented by the ratio of the surface viscosity and membrane tension is less than a second.[43] Moreover, erythrocytes are much more deformable than leukocytes. The binding energy released from the bond formation of the cross-linking molecules originally present in the contact region is often sufficient to deform the cell to form a finite conjugation area without additional supply of molecules from long-range diffusion. Thus, the growth of the adhesion area in erythrocytes can proceed in the time scale of the binding reaction in a fashion similar to zipping up a zipper. Experimentally, the rouleau formation of red blood cells induced by dextran has been found to proceed through at least three different modes involving different time rates: sliding, rolling, and clapping, depending upon the concentration and molecular weight of the dextran.[50] This interesting phenomenon has not yet been modeled but warrants further study.

VI. CONCLUSION AND FUTURE DIRECTIONS

At the cellular level, the basic equations for continuum mechanics analysis of cell adhesion problems have been well established. However, the analysis critically depends on the cell surface morphology and the mechanical property model used for the cells. In a recent paper, the irregular microvilli on the cell

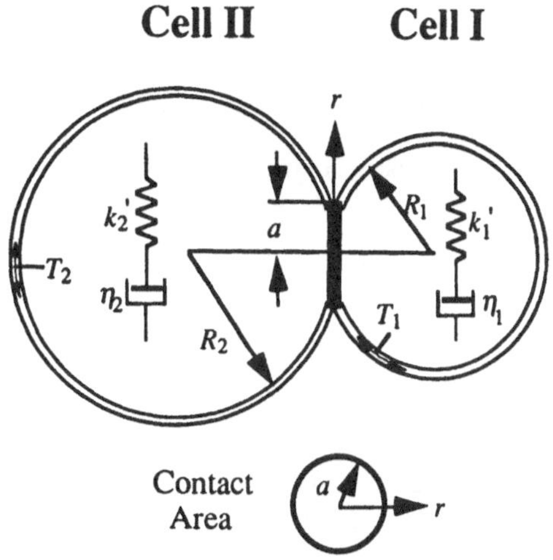

Figure 6 An alterative rheological model for the adhering cells. The cells maintain their spherical shapes (except the adhesion region) through membrane tensions T_1 and T_2, respectively. The cell interiors underneath the cortical layer are modeled by incompressible Maxwell fluid, as shown by the spring and dashpot elements.

membrane have been described statistically in a simulation of cell rolling and adhesion on surfaces in shear flow.[15] More work in this direction is needed to establish the influence of surface roughness on adhesion, especially in the presence of flow.

To illustrate the importance of the mechanical property model, let us reconsider the adhesion of two spherical cells. Instead of viscoelastic solids as used in Section II, this time, the cells are modeled as liquid drops, i.e., a liquid interior enclosed by a constant cortical tension T (Figure 6). Recent studies have suggested that the liquid drop model may better describe the slow deformation behavior (in a time scale of tens to hundreds of seconds) for leukocytes than the standard solid model.[51-53] Using the Young-Dupre Equation 4, it can be shown readily that, in the absence of externally applied forces,

$$\gamma = \left(\frac{T_1}{R_1^2} + \frac{T_2}{R_2^2} \right) \frac{a^2}{2} \tag{26}$$

Here γ depends on the cellular rigidity (manifest as the cortical tensions T_1 and T_2) and size (R_1 and R_2) in a similar manner as it does in Equation 6 (the case of $F = 0$) for the solid sphere model (see discussion following Equation 6). But, γ is proportional to the square of the contact radius a for the liquid drop model, as opposed to proportional to the cube of a for the solid sphere model. For a neutrophil adhesion to a flat, rigid surface ($R_2 = \infty$, $k_2 = \infty$) with a contact radius of $a = 1$ μm, the γ values calculated from Equations 6 and 26 are, respectively, 6.35×10^{-4} dyn/cm and 1.27×10^{-3} dyn/cm using typical values of $R_1 = 3.5$ μm, $k_1 = 275$ dyn/cm^2,[54] and $T_1 = 0.031$ dyn/cm.[51] This difference in the functional dependencies of γ on a can affect qualitatively the result in the analysis of the time course of growth of the adhesion area.

At the molecular level, important steps have been taken to describe the molecular events that mediate cell adhesion using physical and chemical variables and mathematical equations. The present theoretical framework allows modeling of many cell adhesion phenomena. There are, however, still many uncertainties about the basic equations and parameter values. Some critical relationships were proposed based on physical arguments that were only or even purely speculations because of the lack of experimental data. Only little or no data are available for such fundamental properties as the binding rate coefficients, diffusion constant, and elastic modulus of a crossbridge. It has not been tested experimentally whether a bond behaves as a catch-bond (Equation 13 with $\zeta_j > 1$) or as a fracturable material (the disassociation rate coefficient $k_- = k_-^0 \exp\{\alpha f/kT\}$, where f is the bond force and α is an effective working distance of the bond[55]). The simple linear spring model for a crossbridge needs to be augmented to include the description for the failure of a bond and the extraction of a bond from the cell membrane. The modeling of receptor upregulation has barely begun, despite the fact that it is a fundamental property and is of much interest to biologists. Thus, a large number of physical issues and theoretical possibilities remain to be elucidated. Further progress in this area will depend on more detailed and quantitative information on fundamental relations and properties of the cell adhesion molecules.

REFERENCES

1. **Skalak, R. and Zhu, C.,** Rheological aspects of red blood cell aggregation, *Biorheology,* 27, 309, 1990.
2. **Adamson, A. W.,** *Physical Chemistry of Surfaces,* John Wiley & Sons, New York, 340, 1976.
3. **Skalak, R., Zarda, P. R., Jan, K.-M., and Chien, S.,** Mechanics of rouleau formation, *Biophys. J.,* 35, 771, 1981.
4. **Evans, E. A.,** Minimum energy analysis of membrane deformation applied to pipet aspiration and surface adhesion of red blood cells, *Biophys. J.,* 30, 265, 1980.
5. **Skalak, R. and Chien, S.,** Theoretical models of rouleau formation and disaggregation, *Ann. N.Y. Acad. Sci.,* 416, 138, 1983.
6. **Evans, E. A. and Buxbaum, K.,** Affinity of red blood cell membrane for particle surfaces measured by the extent of particle encapsulation, *Biophys. J.,* 34, 1, 1981.
7. **Schmid-Schönbein, G. W., Sung, K.-L. P., Tözeren, H., Skalak, R., and Chien, S.,** Passive mechanical properties of human leukocytes, *Biophys. J.,* 36, 243, 1981.
8. **Sung, K.-L. P., Sung, L. A., Crimmins, M., Burakoff, S. J., and Chien, S.,** Dynamic changes in viscoelastic properties in cytotoxic T-lymphocyte-mediated killing, *J. Cell Sci.,* 91, 179, 1988.
9. **Sato, M., Theret, D. P., Wheeler, L. T., Ohshima, N., and Nerem, R. M.,** Application of the micropipette technique to the measurement of cultured porcine aortic endothelial cell viscoelastic properties, *J. Biomech. Eng.,* 112, 263, 1990.
10. **Johnson, K. L.,** *Contact Mechanics,* Cambridge University Press, Cambridge, 1985.
11. **Greenwood, J. A. and Johnson, K. L.,** The mechanics of adhesion of viscoelastic solids, *Philos. Mag.,* 43, 697, 1981.
12. **Sung, K.-L. P., Sung, L. A., Crimmins, M., Burakoff, S. J., and Chien, S.,** Determination of junction avidity of cytolytic T cell and target cell, *Science,* 234, 1405, 1986.
13. **Tözeren, A., Sung, K.-L. P., and Chien, S.,** Theoretical and experimental studies on cross-bridge migration during cell disaggregation, *Biophys. J.,* 55, 479, 1989.
14. **Schmid-Schönbein, G. W., Skalak, R., Simon, S., and Engler, R. L.,** The interaction between leukocytes and endothelium *in vivo, Ann. N.Y. Acad. Sci.,* 516, 348, 1987.
15. **Hammer, D. A.,** Simulation of cell rolling and adhesion on surfaces in shear flow: microvilli-coated hard spheres with adhesive springs, *Cell Biophys.,* 1992 (submitted).
16. **Bongrand, P. and Bell, G. I.,** Cell–cell adhesion: parameters and possible mechanisms, in *Cell Surface Dynamics: Concept and Models,* Perelson, A. S., DeLisi, C., and Wiegel, F. W., Marcel Dekker, New York, 1984, 459.
17. **Bongrand, P., Capo, C., and Depieds, R.,** Physics of cell adhesion, *Prog. Surface Sci.,* 12, 217, 1982.
18. **Jan, K.-M. and Chien, S.,** Role of surface electric charge in red blood cell interactions, *J. Gen. Physiol.,* 61, 638, 1973.
19. **Israelachvili, J. N.,** *Intermolecular and Surface Forces,* Academic Press, San Diego, 1985.
20. **Ohsima, H., Makino, K., and Kondo, T.,** Electrostatic interaction of two parallel plates with surface charge layers, *J. Colloid Interface Sci.,* 116, 196, 1987.
21. **Napper, D. H.,** *Polymeric Stabilization of Colloidal Dispersions,* Academic Press, San Diego, 1983.
22. **Jan, K.-M.,** Role of hydrogen bonding in red cell aggregation, *J. Cell. Physiol.,* 101, 49, 1979.
23. **Evans, E. A.,** Force between surfaces that confine a polymer solution: derivation from self-consistent field theories, *Macromolecules,* 22, 2277, 1989.
24. **Bell, G. I., Dembo, M., and Bongrand, P.,** Cell adhesion. Competition between nonspecific repulsion and specific bonding, *Biophys. J.,* 45, 1051, 1984.
25. **Torney, D. C., Dembo, M., and Bell, G. I.,** Thermodynamics of cell adhesion. II. Freely mobile repellers, *Biophys. J.,* 49, 501, 1986.
26. **Lawrence, M. B. and Springer, T. A.,** Leukocytes roll on a selectin at physiologic flow rates: distinction from and prerequisite for adhesion through integrins, *Cell,* 65, 859, 1991.
27. **Tözeren, A., Sung, K.-L. P., Sung, L. A., Dustin, M. L., Chan, P.-Y., Springer, T. A., and Chien, S.,** Micromanipulation of adhesion of a jurkat cell to a planar membrane containing LFA-3 molecules, *J. Cell Biol.,* 1991.
28. **Springer, T. A.,** Adhesion receptors of the immune system, *Nature,* 346, 425, 1990.
29. **Sung, K.-L. P.,** LFA-1/ICAM-1 Mediated Cell Adhesion, The American Society of Mechanical Engineers, Atlanta, GA, 1991, 69.

30. **Dembo, M., Torney, D. C., Saxman, K., and Hammer, D. A.,** The reaction-limited kinetics of membrane-to-surface adhesion and detachment, *Proc. R. Soc. Lond. B,* 234, 55, 1988.
31. **Zhu, C.,** A thermodynamic and biomechanical theory of cell adhesion. Part I: general formulism, *J. Theor. Biol.,* 150, 27, 1991.
32. **Axelrod, D.,** Lateral motion of membrane proteins and biological function, *J. Membr. Biol.,* 75, 1, 1983.
33. **Chan, P. Y., Lawrence, M. B., Dustin, M. L., Ferguson, L., Golan, D., and Springer, T. A.,** The influence of receptor lateral mobility on adhesion strengthening between membranes containing LFA-3 and CD2, *J. Cell Biol.,* 115, 245, 1991.
34. **Dembo, M. and Bell, G. I.,** The thermodynamics of cell adhesion, *Curr. Top. Membr. Trans.,* 29, 71, 1987.
35. **Tözeren, A.,** Cell–cell, cell–substrate adhesion: Theoretical and experimental considerations, *J. Biomech. Eng.,* 112, 311, 1990.
36. **Evans, E. A.,** Detailed mechanics of membrane–membrane adhesion and separation. II. Discrete kinetically trapped molecular cross-bridges, *Biophys. J.,* 48, 185, 1985.
37. **Berk, D. and Evans, E.,** Detachment of agglutinin-bonded red blood cells. III. Mechanical analysis for large contact areas, *Biophys. J.,* 59, 861, 1991.
38. **Evans, E., Berk, D., Leung, A., and Mohandas, N.,** Detachment of agglutinin-bonded red blood cells. II. Mechanical energies to separate large contact areas, *Biophys. J.,* 59, 849, 1991.
39. **Evans, E. A. and Leung, A.,** Adhesivity and rigidity of erythrocyte membrane in relation to wheat germ agglutinin binding, *J. Cell Biol.,* 98, 1201, 1984.
40. **Evans, E., Berk, D., and Leung, A.,** Detachment of agglutinin-bonded red blood cells. I. Forces to rupture molecular-point attachments, *Biophys. J.,* 59, 838, 1991.
41. **Tözeren, A.,** Adhesion induced by mobile cross-bridges: steady state peeling of conjugated cell pairs, *J. Theor. Biol.,* 140, 1, 1989.
42. **Vayo, M. M., Skalak, R., Brunn, P., Usami, S., and Chien, S.,** Role of the surface adhesive energy in the disaggregation of red cell rouleaux by fluid shear. II. Experiment, *Biorheology,* 1990.
43. **Evans, E. A.,** Detailed mechanics of membrane–membrane adhesion and separation. I. Continuum of molecular cross-bridges, *Biophys. J.,* 48, 175, 1985.
44. **Hammer, D. A. and Lauffenburger, D. A.,** A dynamical model for receptor-mediated cell adhesion to surfaces, *Biophys. J.,* 52, 475, 1987.
45. **Cozens-Roberts, C., Lauffenburger, D. A., and Quinn, J. A.,** Receptor-mediated cell attachment and detachment kinetics. I. Probabilistic model and analysis, *Biophys. J.,* 58, 841, 1990.
46. **Bell, G.,** Estimate of the sticking probability for cells in uniform shear flow with adhesion caused by specific bonds, *Cell Biophys.,* 3, 289, 1981.
47. **Tha, S. P. and Goldsmith, H. L.,** Interaction forces between red cells agglutinated by antibody. I. Theoretical, *Biophys. J.,* 50, 1109, 1986.
48. **Tha, S. P., Shuster, J., and Goldsmith, H. L.,** Interaction forces between red cells agglutinated by antibody. II. Measurement of hydrodynamic force of breakup, *Biophys. J.,* 50, 1117, 1986.
49. **Zhu, C.,** A model of transient processes in cell adhesion, *FASEB J.,* 5, A652, 1991.
50. **Sung, K.-L. P. and Chien, S.,** Rouleau Formation in Red Blood Cells Induced by Dextran, The American Society of Mechanical Engineers, Washington, D.C., 1990, 69.
51. **Dong, C., Skalak, R., Sung, K.-L. P., Schmid-Schönbein, G. W., and Chien, S.,** Passive deformation analysis of human leukocytes, *J. Biochem. Eng.,* 110, 27, 1988.
52. **Yeung, A. and Evans, E. A.,** Cortical shell-liquid core model for passive flow of liquid-like spherical cells into micropipette, *Biophys. J.,* 56, 139, 1989.
53. **Needham, D. and Hochmuth, R. M.,** Rapid flow of passive neutrophils into a 4 μm pipet and measurement of cytoplasmic viscosity, *J. Biomech. Eng.,* 112, 269, 1990.
54. **Schmid-Schönbein, G. W.,** Rheology of leukocytes, in *Handbook of Bioengineering,* Skalak, R. and Chien, S., McGraw-Hill, New York, 1987, 13.
55. **Bell, G. I.,** Models for the specific adhesion of cells to cells, *Science,* 200, 618, 1978.

Chapter 3

Influences of Surface Chemical Factors on Selective Cellular Retention

Robert E. Baier, Karen H. Rittle, and Anne E. Meyer

CONTENTS

I. Summary 41
II. Introduction 42
III. Materials and Methods 43
A. Materials 43
B. Analytical Methods 44
C. Cell Assays 45
IV. Results 45
A. Gas-Plasma-Treated PS and PE 45
B. UV-Treated PS and PE 49
C. Cast PS and PE Films 49
D. The "Conditioning Film" 49
E. Cell Attachment Studies 50
F. Growth Rate Studies 52
G. Strength of CHO Cell Attachment on Test Substrata 52
H. CHO Cell Behavior on UV-Treated PS 56
I. Patterning Through Masks using RFGPT 56
V. Correlations 56
A. XPS Analysis of Surfaces Correlated with Cell Behavior 56
B. Surface (Contact) Potential Data Correlated with Cell Adhesion 57
C. Cellular Processing of "Conditioning Films" 57
D. Cell Attachment/Retention and Morphology 58
VI. Discussion 59
VII. Conclusion: 20 Years of Education of a Biosurface Scientist 60
References 60

I. SUMMARY

For the past 20 years, biosurface scientists have developed and refined their understanding of the physical/chemical aspects of "bioadhesion". Living cells attach to solid synthetic substrata with varying efficiencies, as a function of cell type, suspending medium constituents, and substratum treatment. There is a significant need for a better understanding of the "selection rules" that can qualitatively and, perhaps, quantitatively relate the degree of cell retention on synthetic solid supports to controllable surface properties of those substrata. As an example, a detailed physical/chemical study is described, wherein Chinese Hamster Ovary (CHO-K1) cells were incubated with polystyrene (PS) and standard reference polyethylene (PE) modified to produce variations in surface chemical composition by radio-frequency-gas-plasma treatment (RFGPT). Air plasmas were applied over the substratum faces uniformly or in patterned arrays. All substrata were characterized by several surface analytical techniques. The retentive strengths of initially attached cells were assessed. CHO cells attached but remained round on untreated PS and control PE. Upon RFGPT, oxygen in the form of alcoholic (C–OH), carbonyl (C=O), and carboxyl (O–C=O) functional groups was introduced to the PS surfaces, resulting in increases in surface energy; CHO cell attachment was followed by full spreading; CHO cell growth rate was increased over control values. The ultraviolet component of RFGPT did not contribute to the observed surface modification of either polymer. Shear challenges of attached cells resulted in easy removal of all round cells (at rates of 91 s^{-1}), whereas larger shear rates were required to remove spread cells. Cell removal occurred due to cohesive failures, leaving substratum-attached materials on the test surfaces.

0-8493-4559-6/95/$0.00+$.50

As a result of attempts to mask specific regions, to provide RFGPT modifications only in discrete zones on a scale of tens of microns, it was discovered that the plasma also was generated within covered petri dishes and within any cracks and crevices existing between masks and substrata. Patterns were created successfully with RFGP treatment only when close apposition of mask to substratum was achieved. Reliable production of patterned surfaces that selectively retain cell populations in prescribed areas allows for valuable new applications in biotechnology and medical device development.

II. INTRODUCTION

In bioreactor systems under optimal conditions, immobilized cells (both procaryotes and eucaryotes) would produce large quantities of daughter cells and/or metabolites that could be readily collected for biotechnological applications. Such systems currently have limited efficiencies and/or operational lives due to either poor cell attachment or, alternatively, overgrowth. If cells retained on solid supports could be maintained in specific arrays such that growth with its concomitant production and elution of inherent or bioengineered macromolecules (i.e., growth factors, antibiotics, insulin, etc.) could be sustained, many of the problems facing the biotechnology industry could be solved. This task will require more careful selection of support materials with controllable surface properties.

Better understanding is needed regarding cell attachment, retention, spreading, and growth, from both biological and physical perspectives. This work sets out to provide that understanding by evaluating the solid supports, on which mammalian cells are immobilized, by chemical compositional and surface energetic criteria related to the biological/physical aspects of the growing cells. In particular, the contributions from growth medium constituents are identified with respect to amounts and physical orientations of glycoproteinaceous molecules, as they are spontaneously deposited from the medium to the solid support prior to first cell attachment.

Considerable attention has been given to the need for "special" surface characteristics of the substrata upon which anchorage-dependent mammalian cells are cultured. A substratum used widely today is PS. In its virgin state, PS has surface properties that do not prompt good cell coverage. When treated by processes of gas-plasma or glow-discharge, or when exposed to acids, PS acquires oxygen-containing functional groups on the surface that do promote effective cell attachment and growth.[1-4]

There is increasing attention being paid to the glycoproteinaceous interfacial layers deposited upon the substratum spontaneously from culture medium before inoculated cells settle and attach. Culture media supplemented with serum (fetal bovine, calf, horse, human) obviously comprise varying macromolecules of varying concentrations.

A system developed with *E. coli* has provided some understanding of cell division and growth kinetics. This system of retention of *E. coli* B/r on membrane filters, their growth, and subsequent elution of daughter cells has been utilized for over 20 years to produce synchronized cell populations from a device called the "baby machine".[5] Unfortunately, the membrane supporting *E. coli* "fouls" after four to five doublings, an increasing fraction of daughter cells remains on the filter after each doubling, and crowding occurs, preventing reliable elution of daughter cells of the same age.

The synchronization of *E. coli* B/r, even for only several doublings, was a boon for research since only then was the rate of DNA synthesis determined[5,6] and the sequence of events involved in DNA replication opened to discovery. Equally exciting would be the development of ways to synchronize large populations of mammalian cells.

Mammalian cells have been shown, as a generalization, to be retained better on substrata of greater water wettability. S.B. Carter[7-9] was among the first to put forth a "passive" theory for cell attachment and movement, making an analogy between cells and liquids; liquids and cells spontaneously spread upon contact with surfaces that are highly wettable. In his experiments, settled cells attached and spread well on palladium layers sputtered in varying thicknesses over cellulose acetate (which generally does not support firm cell attachment). The cells migrated toward the more dense (thickest) palladium-coated zones, the regions of highest wettability and potential adhesivity. Cells first attached to these thick palladium film regions moved randomly about the regions but rarely departed from them. In fact, the cells often were seen to be confined to surface "islands" of the higher wettability states.[8,10] Once the islands were full, cells did move to the peripheries (even from glass to cellulose acetate),[11] diminishing notions of a very strict dependence on water-wettability alone.

When more robust but complex substratum surface considerations also were explored with respect to the attachment and retention of bacteria under calibrated laminar flows,[12,13] more cells were retained on the higher energy surfaces. It was possible routinely to confirm prior observations that surface energies of various substrata correlate well with a variety of cell attachment phenomena: the extent of biological fouling of medical devices used in blood,[14,15] the extent of bacterial retention,[16,17] and the extent of spreading of fibroblasts[18] and platelets.[4] When Chinese Hamster Ovary cells (CHO-K1 cells), a typical cell culture line, settled onto polyHEMA [poly(hydroxyethylmethacrylate)] films, little or no attachment was observed. CHO cells did, however, attach to 10-μm "spots" deliberately created by RFGP-treatment through a Nucleopore filter mask.[19] CHO cells attached and divided only in the modified regions.

With this encouraging background, it is useful now to report much more comprehensive studies with PE and PS materials that underwent surface modification via exposure in two radio-frequency-gas-plasma treatment (RFGPT) devices. The surface characters of all the test substrata were assessed, and special attention was given to results attributable to the UV light component of RFGPT alone. Short-term (at and just before cell attachment) and longer-term bioadhesion consequences were revealed by the attachment morphology and growth of CHO cells from serum-containing culture medium. Control, untreated, and RFGP-treated surfaces were evaluated, and the physical/chemical data were correlated for both the starting and deposited-protein-film-modified substrata. Important and novel features of this work were the full surface characterization of the materials both before *and after* cell attachment/retention, through 24 h of incubation, and determination of the shear strengths of the variously attached cells under controlled laminar flow conditions.

III. MATERIALS AND METHODS

A. MATERIALS

The generic materials chosen for investigation were free-standing PS and PE films and the same polymers cast as thin films on germanium test plates. PS is used most widely in research laboratories in both its native and corona- or gas-plasma-treated forms.[2] It is optically clear and, untreated, possesses a regular chemical composition, based on a repeating carbon backbone with a phenyl ring attached to alternate carbons. PE has an even simpler repeating aliphatic hydrocarbon chemistry. It is an important "basic reference material" in its low-density form [LDPE free-standing film, available as a National Institutes of Health (NIH) standard]. Although it is not optically clear, it is sufficiently transparent to allow conventional transmitted light microscopy to be used routinely.

Prior to use, all free-standing PS samples were cleaned with a Sparkleen (Fisher Scientific) detergent slurry in tap water and sonicated for 15 min. The surfaces then were rinsed thoroughly with distilled water and sonicated in distilled water for another 15 min. Finally, the samples were rinsed with distilled water and allowed to dry on clean, absorbent paper under ambient laboratory conditions. The PE samples were used "as received" from the recommended NIH supplier, in clean and sterile packaging.

PS samples were exposed to RFGPT utilizing a discharge plasma created by an AC radio-frequency circuit.[20] In this treatment, partially ionized gas is generated through inelastic collisions between high-energy electrons (from the RF) and residual gas molecules remaining in a low pressure environment (500 to 1000 mtorr). The gas contains ions, electrons, and photons, but mostly neutral species. These reactive species collide with the test surface to remove surface contaminants and even intrinsic components (at approximately 0.2 to 1 nm/min in argon),[21] leaving a clean, disinfected, higher-energy surface.[22] The gas molecules in the RFGPT chamber can be residual air or any infused noble gas (e.g., argon, helium, or xenon) or any vapor (e.g., H_2O_2, H_2O) or gas. Many gases have been used for RFGPT surface modification, by eliciting many additional chemical reactions at the surface. Oxygen, hydrogen, and nitrogen are examples of such reactive gases.[21]

Two RFGP devices were used in this study, (1) a commercially available unit (Harrick Scientific Co., Ossining, NY, model PDC-32G, 100 W, 35 MHz) and (2) an increased power, laboratory-assembled unit ("Hiditron"). Both units have inductively coupled chambers[21] and were chosen for this study based on a preliminary survey of the surface modifications produced by numerous alternate devices.[23]

PS samples were subjected to routine RFGPT for 3 min (unless otherwise stated) at a chamber pressure of 500 mtorr in residual air. The 3-min exposure period was chosen after demonstrating its sufficiency to remove multilayer organic contaminants and increase the surface energy of germanium prisms without risking deposition of silicon dioxide sputtered from the quartz chamber.[20] Polymeric surfaces could be significantly modified, however, at even very short exposure times of a few seconds, as easily determined by observations of increased water wettability after RFGPT.

PS specimens were exposed to RFGPT in two ways, (1) uniformly over all faces and (2) in patterns by using a mask. The use of masks allowed for comparison of treated and untreated portions of the same optically clear materials and were chosen to determine the effects of exposure of the test surfaces to gas-plasma-discharge-generated photons alone.

In addition, the effects of predominantly UV light from two different UV sources, one of specific UV wavelength (254 nm — Eprom Eraser, Memorase Model C-25, UVP, Inc., San Gabriel, CA) and one of a broad spectrum of UV (250 to 400 nm — Kratos 2500 W xenon arc lamp solar simulator; 7.74 J/cm^2 delivered in 3 min), were investigated for comparison with the UV-mediated effects possible within the RFGPT devices. The effects of UV treatment of samples were followed by comprehensive contact angle analysis (described below) and by subsequent observation of CHO cell attachment and degree of spreading.

PE and PS films also were solvent cast from xylene onto germanium prisms. The PE was supplied in powder form (Scientific Polymer Products, Ontario, NY, Cat #560), the PS in flake form (Scientific Polymer Products, Cat #400). Prior to solvent-casting, detergent-washed germanium prisms were RFGP treated in the Harrick unit for 3 min (500 mtorr) in air to obtain a high-energy surface that would "hold" the cast film. The cast films were analyzed "as prepared" and after RFGPT. Since the films were relatively thin, the glow-discharge treatment was reduced to 30 s for the PE films and 1 min for the PS films. (In some cases, 3 min of RFGPT would remove the entire polymer film.)

B. ANALYTICAL METHODS

Stylus profilometry was coupled with results from scanning electron microscopic analysis to provide information on the surface texture of the polymers before and after RFGPT modification.[24] This information was applicable to the current studies as an aid to determine the possible "roughness" contributions to observed cell retention differences.

X-ray photoelectron spectroscopy (XPS or ESCA) was used to probe valence and core electrons of atoms within the outermost 10 to 100 Å of the surfaces, using the photoelectric process.[25] By changing the angle of the electron detector, the "sampling depth" of the technique was controlled. The X-ray Photoelectron Spectrometer [Surface Science Instruments (SSI), model SSX-100] has a monochromatized aluminum K-alpha excitation source (operating pressure = 3.8 × 10 E-9 torr). The spot size used was either 1000 or 600 μm. Curve-fitting of resolved lines was done with the SSI ESCA software biased by the operator.

Comprehensive contact angle analyses[26] were used to characterize untreated controls and surfaces immediately after treatment, after 3 h, 24 h, 41 d, and 165 d to determine the stability of changes induced by RFGPT. Interfacial changes due to adsorption of the proteinaceous "conditioning film" (prior to cell attachment) also were tracked by comprehensive contact angle analyses.

Ellipsometry[27] was used to determine the thicknesses of the polymer films cast on the germanium prisms and the amounts of polymer ablated by gas-plasma treatment.

Surface (contact) potential measurements were made to reflect the changed structural arrangements of the cast polymers and any serum components they later adsorbed.[1] Correlation was sought between the contact potential differences of polymers with adsorbed protein films (those adsorbed from the culture medium) and subsequent cell behavior.

Infrared spectroscopy, in the mode of multiple-attenuated-internal-reflection (MAIR),[28] was used to confirm the chemical composition (covalent-bound moieties) of the polymers and to determine the nature and relative amounts of adsorbed protein in the conditioning film experiments with polymers cast on germanium internal reflection test plates.

The excellent natural transparency of the PS and adequate transparency of the thin PE made it possible to routinely observe and photograph the attached CHO cells using phase contrast microscopy. Five preselected fields on each polymer sample were photographed and analyzed to give a total of 15 fields for each polymer/treatment condition. In addition to phase contrast microscopy and photography of all samples, at least one sample from each substratum/treatment group subsequently was analyzed by contact angle analysis; another was used in cell retention studies (described below). Cell viability was confirmed throughout by monitoring cell growth during incubation.

Image analysis utilized tracings of the peripheries of the cells from the photographic negatives. While not the most efficient approach, this technique allowed the use of older image analysis equipment. Round cells were differentiated from one another as well as from spread cells. Rounded cells noted to be just ready to divide were counted as one cell. When a division plane was present, the count was as two round cells. The cells retained on the surfaces were quantitated by number present, the percent of total cells present that were round, and total cell spread area. The data were compared using the paired t-test (Statgraphics, STSC, Inc.).

A fine wire of known diameter also was photographed at each experimental occasion. The diameter was traced in two perpendicular directions to allow determination of the area of the resulting square. The square was marked with the image analyzer, and the number of pixels in the square was recorded; the area/pixel was calculated. The equipment used was Image-Pro II, Image Processing System 2.0 (Media Cybernetics, Silver Spring, MD). An average number of pixels/cell was determined for 15 photographs for each substratum/treatment condition. The average pixels/cell was multiplied by the area/pixel to get the average area/cell in the field. Error estimation was done by using single photonegatives of cells traced five times and separately analyzed.

After *in situ* phase contrast photography, representative samples were inspected by scanning electron microscopy to observe details of CHO cell morphology. Energy-dispersive X-ray analysis (EDX) also was utilized for samples in the SEM vacuum chamber, to probe the composition of CHO cell detachment sites and cellular debris on the filters. The extent of shrinkage during sample processing was assessed from several samples observed with phase contrast microscopy (×40) both before and after fixation/dehydration.

C. CELL ASSAYS

Exposure to CHO cells was in serum- and antibiotic-supplemented medium at 37°C. Triplicates from each substratum/treatment group (untreated PE control, untreated PS, Harrick-treated PS, and Hiditron-treated PS) were placed in petri dishes inoculated with 15 ml of Ham's F-12 medium (with 1% penicillin/streptomycin and 10% fetal bovine serum) and CHO cells. Each experiment was done twice using two different concentrations of CHO cells: 1.5×10^5 and 8×10^5 in 10 ml of medium. After the cells were allowed to gravitationally settle, attach, and grow for 24 h, reference photographs were taken of the cell arrays *in situ* for later image analysis. Preliminary studies had shown that most CHO cells were well spread on commercial "tissue-culture" PS after 20 h of incubation.

CHO cell *growth rate studies* on the test surfaces were performed by photographing five preselected areas on each sample at set periods of time following inoculation. Two different studies were done to determine the relative growth of the cells: (1) in combination with cell arrival (or recruitment) at the surface and (2) in the absence of recruitment. For the latter, CHO cells were inoculated and allowed to attach and incubate for 1 h, after which time the unattached cells were removed with the medium and new medium added. The relative growth rates of CHO cells on the different test surfaces were determined by first-order least-squares approximation through the data plots. In addition, second-order curves were fitted to the cell growth + cell arrival (recruitment) data to extract an estimate of cell growth rate vs. new cell attachment rate on each test surface.

Cell *retention studies* used special flow cells[29,30] as follows: two samples from each substratum/treatment group were removed from their incubating medium after 1 d of growth and placed directly into silicone-rubber flow cell bodies, with the cell-covered plates comprising a rectangular capillary when separated at their extreme edges by standard silicone-rubber shims. A recirculating physiological saline (0.85%) rinse at shears up to 0.3 N/m^2 was conducted for 10 min. Any removed matter was collected on filters (0.45 μm) within the rinse exit path. After the rinse, the retained cells and/or their detachment sites were observed by phase contrast microscopy, and then fixed, dehydrated, and critical-point dried for SEM analysis. The detached cells and cell fragments (now on a filter) also were prepared for SEM to determine the fraction of cells removed intact vs. those that were ruptured.

Duplicate studies run under continuous observation by phase contrast microscopy obtained photographs before, during (between incremental increases in flow rate up to and including the maximum flow rate), and after the rinse. Several controls were run to support observations from the rinse studies. Samples from each substratum/treatment group were rinsed without prior incubation with medium/CHO cells as described above and prepared for SEM. These confirmed that no debris or microbes were deposited on the surfaces from the saline rinse solution. Also, the solution effects of the saline vs. those by the shear forces were observed on the morphology of the cells after 1 d of incubation on the test substrata.

IV. RESULTS

A. GAS-PLASMA-TREATED PS AND PE

MAIR-IR spectroscopy confirmed the "bulk" composition of the test polymers (see Figure 1). Identical spectra were obtained for PS before and after cleaning.

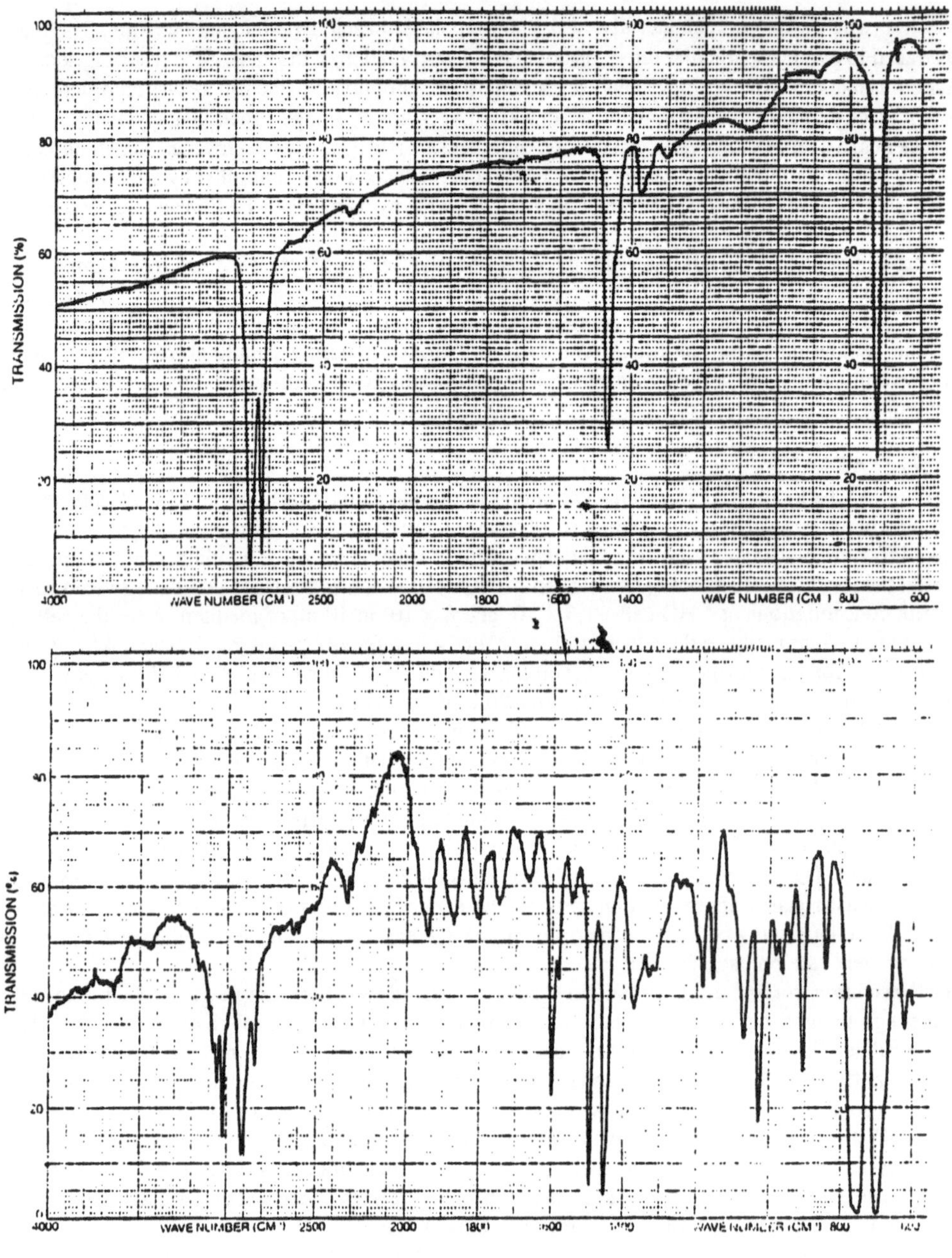

Figure 1 MAIR-IR spectra of PE (top) and PS (bottom). "KRS-5" (thallium bromide/thallium iodide blend) internal reflection test plates were used.

Contact angle analyses of untreated samples resulted in typical "Zisman plots" for PE and PS. Figure 2 gives the Zisman plots for untreated and RFGPT samples immediately after treatment. The plots in Figure 3 show data taken from PE and PS samples 3 h (PE only), 24 h (PE only), 41 d, and 165 d after treatment in the Harrick gas plasma device. Table 1 lists the critical surface tension (γ_c), the dispersive (γ_d), and polar components (γ_p) of the sample's surface free energy ($\gamma_s = \gamma_d + \gamma_p$), and the $\%\gamma_p$ in γ_s for each test substratum. Gamma (γ) values could not be calculated for the RFGPT PS immediately after treatment due to limited or no nonzero angle measurements (identified by two asterisks, **), indicating only that the apparent surface free energy had been elevated (at least temporarily) to greater than 72 mN/m. RFGPT clearly increased the surface energies of both polymers, but the treatment effects slowly decayed with time to an apparent plateau. On RFGP-treated PE (both units), γ_p was restored to its original value after 41 d. However, γ_c and γ_d remained higher than that for the untreated PE for at least 165 d. The γ_c, γ_d, and

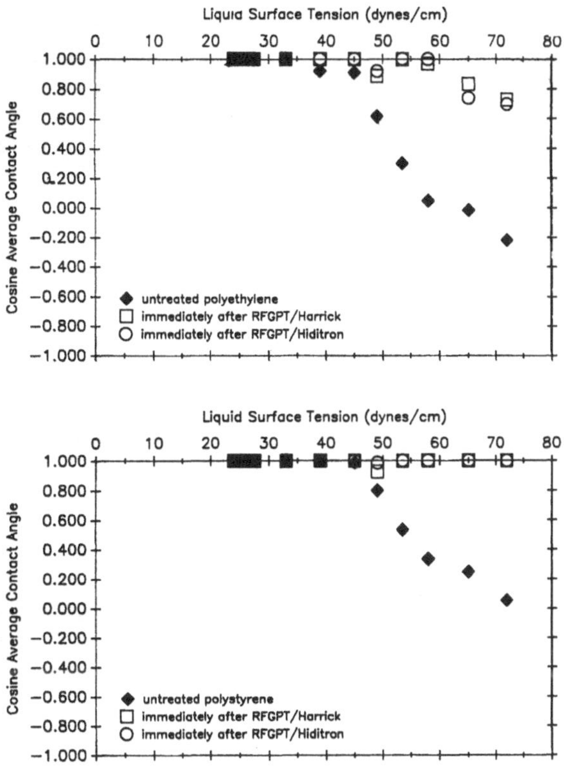

Figure 2 Contact angle data plots ("Zisman plots") for untreated and gas-plasma-treated (3 min) free-standing PE and PS films. Post-treatment data were taken immediately after removal of each sample from the device chamber.

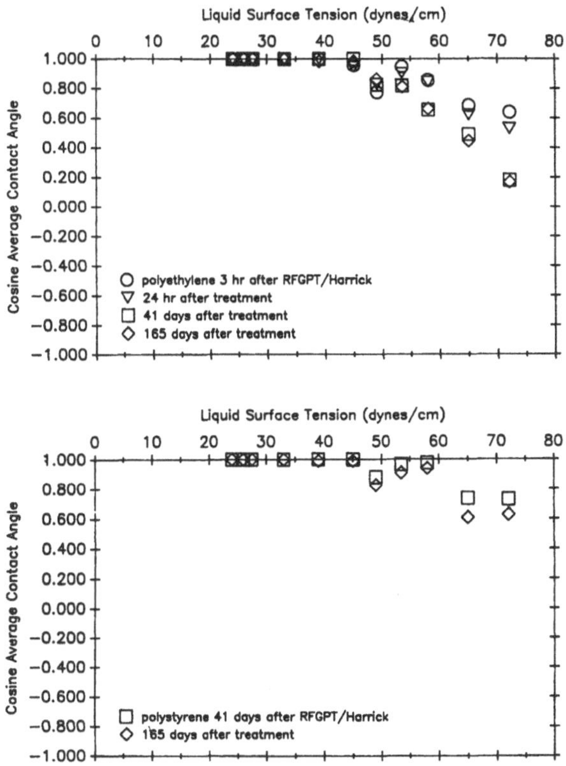

Figure 3 Contact angle data plots for gas-plasma-treated (3 min/Harrick device) PE and PS films after storage for up to 165 d.

Table 1 **The calculated critical surface tensions (γ_c) and the dispersion (γ_d) and polar (γ_p) components of the composite surface free energies (γ_s) for control and RFGPT PE and PS. The percent of γ_p in γ_s is also shown for each sample**

Sample	γ_c	γ_d	γ_p (dyn/cm)	γ_s	$\%\gamma_p$
PE	35	35	5	40	13
RFGPT PE (Harrick)	47	31	25	56	45
after 3 h	40	30	21	51	41
after 24 h	40	30	19	49	39
after 41 d	45	43	4	47	9
after 165 d	39	47	4	51	8
RFGPT PE (Hiditron)	43	27	25	52	48
after 3 h	45	35	20	55	36
after 24 h	45	33	20	53	38
after 41 d	48	46	6	52	12
after 165 d	45	38	5	43	12
PS	38	37	4	41	10
RFGPT PS (Harrick)	72**	32	22	54	41
after 41 d	47	29	26	55	47
after 165 d	38	28	22	50	44
RFGPT PS (Hiditron)	72**	—	—	—	—
after 41 d	48	39	13	52	25
after 165 d	47	36	16	52	31

Note: **Indicates little or no nonzero contact angle data.

γ_p values all remained higher throughout the 165-d test period on RFGP-treated PS (from both gas-plasma treatment units) when compared to values for untreated PS.

The XPS results for untreated and RFGPT PE and PS included high resolution scans of the C1s binding region for the starting materials that confirmed the nominal composition of the reference grade PE. No contamination was detected; the sample was 100% PE. Carbon exists in only one functionality in pure PE, as aliphatic carbon (CH_x). The ratios of the percent atomic composition of carbon to oxygen (C/O) for both polymers are listed in Table 2. For PE, the amount of oxygen imparted to the surface was greater due to the Hiditron treatment. Oxygen was detected on untreated PS, indicating slight oxidation of the otherwise entirely hydrocarbon-based material. Upon RFGPT, more oxygen was detected; little difference was seen between the Hiditron-treated and Harrick-treated PS. Oxygen was added to the PS surface predominantly in the form of carbonyl and carboxyl groups.

When the polymers' surface morphologies and roughnesses were probed with SEM, NHLBI reference grade PE exhibited fine structure (<1 μm) oriented in the apparent film extrusion direction. At high

Table 2 **Ratios of percent atomic composition of carbon to oxygen for untreated and RFGPT PE and PS (XPS data)**

Sample/treatment	C/O ratio
PE/untreated	NOD
/RFGPT Harrick	7.1
/RFGPT Hiditron	5.9
PS/untreated	58.2, 53.6
/RFGPT Harrick	4.6
/RFGPT Hiditron	4.2

Note: NOD, no oxygen detected.

magnification (×2000), cracks (still <1 μm) were seen across the oriented grains. In comparison, the PS was relatively smooth. After RFGPT with the Harrick unit, pits were seen in the PE surface, but no change in the PS surface was noted. After treatment with the Hiditron, the apparent superficial grains and some cracks in the PE were partially obliterated by an amorphous overlayer. The PS surface morphology remained unchanged by Hiditron treatment. Profilometry results confirmed the SEM observations for the RFGPT free-standing films.

B. UV-TREATED PS AND PE

Contact angle data collected immediately after a 30-min UV treatment of PE were nearly identical to those for untreated PE. When PS was analyzed immediately after UV treatment (exposure time ≤30 min), an increase of the polar component, γ_p, of the surface free energy was observed as a function of increasing UV exposure time. Some yellowing of the PS was observed after 30-min exposures to UV irradiation.

To assess the effects of only the UV radiation components of RFGPT, PS samples were shielded from the gas-plasma itself by variously transparent masks. Contact angle data for PS treated in both units, while tightly masked by UV-transparent film (polyvinylidene chloride "kitchen wrap"), showed little change from those for untreated PS.

C. CAST PS AND PE FILMS

Ellipsometry and infrared data were collected on each germanium prism used for the cast films. The thicknesses of every PE and PS film prepared and RFGP-treated (30 s) in this study were compared to the germanium prism baselines (defined as zero thickness).

Contact angle data for unmodified "as prepared" PE and PS films and replicates RFGPT in the Harrick and Hiditron units were collected 30 min after treatment of the same films for which ellipsometry and MAIR-IR spectral data were gathered. RFGPT (in both units) left the polymer films with increased surface energies (γ_s); increased γ_p components were especially noted. Contact angle data for untreated and treated PE and PS thin films were essentially identical to those already presented for the bulk samples of those same materials. The XPS data did show addition of a small amount of silicon to both films, as well as their significant oxidation.

D. THE "CONDITIONING FILM"

CHO cells in medium were incubated with untreated and RFGPT cast polymer films and clean control germanium prisms for 30 min, a time just less than that at which most of the settling cells were able to become sufficiently attached to resist movement when the fluid medium was swirled. The "conditioned" samples were removed and analyzed (a) as-taken, (b) after distilled water leaching, and (c) after vigorous rinsing.

The presence of bound protein was clearly indicated in all MAIR-IR spectra of the test samples. Protein deposits decreased as a result of leaching and rinsing, as loosely associated portions of the initial protein deposit were removed. The sum of the amide I and II infrared absorption bands was taken as an indication of the mass of protein present and, in combination with the independently measured thickness (ellipsometry) of the same film, allowed an estimation of the remaining protein films' densities. The protein films on the RFGPT germanium and PE samples, after leaching, were higher in apparent density than those retained on the germanium and PE controls. This cannot be said with the same certainty for the PS films, since higher initial film thicknesses obscured ellipsometric readings, and PS layers loosened and were removed from some prism areas upon incubation in medium. Contact potential data for PE films, PS films, and clean germanium also were collected after exposure to medium/CHO cells. The contact potential readings increased significantly over those for the pre-exposure specimens. As water-leaching removed much of the excess medium, the contact potentials also decreased with the final rinsing step, causing a further decrease in both contact potentials and film thicknesses.

Contact angle analyses also were performed after leaching and, on replicate samples, after rinsing. On several medium-exposed substrata (all germanium samples; untreated PS), the data taken after leaching deviated from a straight line fit. The distribution of the data indicated that the protein remaining on the surfaces interacted preferentially with some of the diagnostic contact angle liquids (particularly water and formamide). The calculated surface energy values for each substratum before exposure to medium and after leaching and rinsing are shown in Tables 3 to 5. The critical surface tension values for all substrata tested fell between 31 to 38 dyn/cm; upon rinsing, γ_p of the protein films consistently dropped.

"Conditioning films" formed after 30 min on free-standing polymer films also were characterized, after leaching with distilled water, with comprehensive contact angle measurements. Table 6 presents the

Table 3 **The calculated critical surface tensions (γ_c) and the dispersion (γ_d) and polar (γ_p) components of the composite surface free energies (γ_s) for control and RFGPT PE exposed to Ham's F-12 medium for 30 min, leached, and rinsed (see text)**

Sample	γ_c	γ_d	γ_p (dyn/cm)	γ_s	$\%\gamma_p$
Controls (No Medium)					
PE on Ge	36	34	4	38	11
RFGPT 30s/Harrick	43	20	40	60	67
RFGPT 30s/Hiditron	46	21	44	65	68
After Medium Exposure					
PE on Ge/leached	31	27	15	42	36
/rinsed	34	32	8	40	20
RFGPT/Har/leached	37	42	8	50	16
/rinsed	36	38	8	46	17
RFGPT/Hidi/leached	36	30	15	45	33
/rinsed	35	32	13	45	29

resulting surface energy values; the data are similar to those for water-leached "conditioning films" on the cast thin films. It is interesting to note that the expressed polar component (γ_p) of the surface free energy is higher for the films on untreated PS and control PE than on RFGPT PS; the polar components of RFGPT samples without proteinaceous "conditioning films" were considerably greater than control samples. The composite surface free energy, (γ_s), was similar for "conditioning films" on all of the four test substrata.

E. CELL ATTACHMENT STUDIES

CHO cells began to change in shape from rounded to more flattened forms at 90 min of incubation, and some cells had spread completely by 3.5 h on commercial "tissue culture" PS. At longer incubation periods of 5, 6.5, 9, 16, and 20 h, more and more cells had spread. At the 20-h exposure, >95% of the cells were maximally spread. After these observations, it was decided to allow subsequent cultures to incubate for 24 h before quantification of maximum cell spread areas. After 24 h, most cells would be in their early exponential growth phase, and plenty of substratum surface area would still be available to

Table 4 **The calculated critical surface tensions (γ_c) and the dispersion (γ_d) and polar (γ_p) components of the composite surface free energies (γ_s) for control and RFGPT PS exposed to Ham's F-12 medium for 30 min, leached, and rinsed (see text)**

Sample	γ_c	γ_d	γ_p (dyn/cm)	γ_s	$\%\gamma_p$
Controls (No Medium)					
PS on Ge	39	49	4	53	8
RFGPT 60s/Harrick	72	—	—	—	—
RFGPT 60s/Hiditron	72	30	28	58	48
After Medium Exposure					
PS on GE/leached	37	29	18	47	38
/rinsed	37	38	8	46	17
RFGPT/Har/leached	36	35	8	43	19
/rinsed	34	37	8	45	18
RFGPT/Hidi/leached	36	31	14	45	31
/rinsed	37	36	8	44	18

Table 5 **The calculated critical surface tensions (γ_c) and the dispersion (γ_d) and polar (γ_p) components of the composite surface free energies (γ_s) for control and RFGPT germanium exposed to Ham's F-12 medium for 30 min, leached, and rinsed (see text)**

Sample	γ_c	γ_d	γ_p (dyn/cm)	γ_s	$\%\gamma_p$
Controls (No Medium)					
Germanium	33	28	29	57	51
RFGPT/Harrick	72	23	40	63	63
RFGPT/Hiditron	72	27	34	61	56
After Medium Exposure					
Germanium/leached	37	30	25	55	45
/rinsed	38	29	16	45	35
RFGPT/Har/leached	37	31	22	53	42
/rinsed	38	31	16	47	34
RFGPT/Hidi/leached	37	31	23	54	43
/rinsed	38	35	11	45	24

allow for maximum spreading (cell spread area decreases as cell density increases). Contact angles were measured to assess the natures of the possibly transformed "conditioning film" surfaces after 24 h. Table 7 presents the calculated surface energy values for each test substratum. In contrast to the contact angle data for the "conditioning films" at 30 min, the $\%\gamma_p$ at 24 h now was higher on the RFGPT PS than on the untreated PS or control PE.

Scanning electron micrographs of CHO cells on the test surfaces also revealed substratum-related differing morphologies. Cells on control PE were round (diameter approximately 10 µm) and attached by only small portions of their membranes. Similarly, only relatively round cells were seen on untreated PS. On the RFGPT PS, many CHO cells were of spread form: raised dense bodies identified as the cells' nuclei were surrounded by flattened cytoplasmic regions and extended cell membranes (of diameter approximately 20 to 25 µm). Extensions of cell membranes in the form of pseudopodia were evident toward the peripheries of many cells, appearing to provide anchorage. Some cells in groups appeared to be fused together, making it difficult to discern where one cell ended and another began. Figure 4 presents SEM views of CHO cells on control PE and PS after another day in culture; the morphologies observed after only 24 h were maintained. Figure 4 also demonstrates that the shape and size of CHO cells on clean germanium test plates, which have surface properties similar to RFGPT PS, were like those observed after 24 h on the gas-plasma-treated PS.

Controls run to test the effects of the fixation/dehydration processes (prior to SEM) on CHO cell morphology and retention proved the absence of significant preparation artifacts. Some shrinkage occurred during sample processing, but only a few loosely associated cells or cell aggregates were lost.

Table 6 **The calculated critical surface tensions (γ_c) and the dispersion (γ_d) and polar (γ_p) components of the composite surface free energies (γ_s) for control and RFGPT free-standing polymers exposed to CHO cells and medium for 30 min, leached, and rinsed (see text)**

Sample	γ_c	γ_d	γ_p (dyn/cm)	γ_s	$\%\gamma_p$
PE	36	32	13	45	29
PS	35	29	19	48	40
PS/RFGPT/Harrick	37	38	8	46	17
PS/RFGPT/Hiditron	33	35	8	43	19

Table 7 **The calculated critical surface tensions (γ_c) and the dispersion (γ_d) and polar (γ_p) components of the composite surface free energies (γ_s) for control and RFGPT free-standing polymers exposed to CHO cells and medium for 24 h and leached (see text)**

Sample	γ_c	γ_d	γ_p (dyn/cm)	γ_s	$\%\gamma_p$
PE	33	34	11	45	24
PS	36	33	11	44	25
PS/RFGPT/Harrick	34	28	20	48	42
PS/RFGPT/Hiditron	32	30	15	45	33

The results of the image analysis study are reported in Tables 8 and 9. Mean spread cell areas (μm^2/cell) with standard deviations (SD) and standard errors (SE) are reported, with the total numbers of cells analyzed in n fields and the percent of round cells. The percent surface coverage by attached cells was determined by dividing the total spread area by the total surface area available. The data in the two tables can be compared absolutely within each table but only relatively between tables, since photographic magnification changed slightly in separate experiments.

In general, the relative values within each table are similar. Only round cells were seen on control PE and untreated PS. Both round and spread cells were seen on the RFGPT PS specimens. The mean total surface area per round cell remained nearly the same for all samples, treated or not, within each experiment. However, the spread cells, which were seen only on the treated PS specimens, occupied more area than round cells when present on the same substratum. The mean total spread cell area was greater on RFGPT PS than on either untreated PS or control PE. Percent coverage of the available surface area by the cells was greater in Experiment 1 (8×10^5 cells inoculated) than in Experiment 2 (1.5×10^5 cells inoculated).

F. GROWTH RATE STUDIES

The growth rate results are shown as graphs in Figure 5. Those for a growth + recruitment rate study are shown in Figure 6. First order and second order equations were fitted to each data set; first order constants are listed in Table 10. The growth rate is the change in the number of cells per change in unit time, which is the slope, m, of a first order equation. The growth rates for CHO cells on control PE and untreated PS were similar; the growth rates on RFGDT PS were higher in comparison. Little difference was seen between the growth rates of CHO cells on the two types of RFGPT PS surface. In the growth + cell recruitment rate study, the slopes increased in the following order: PE < PS (Hiditron) < untreated PS < PS (Harrick).

The growth + cell recruitment rate data were analyzed further as a composite of two processes, one reflecting only recruitment and the other including growth with recruitment. An example of how these rates were determined is shown in Figure 7. The difference between the two rates was taken to equal the growth rate. The time corresponding to the intersection of the two lines drawn gives an indication of the lag time before cell division occurs. The lag time was 14 to 15 h for the four substrata tested. Table 11 lists the growth + recruitment rate, the recruitment rate, their difference (i.e., the growth rate), and the vertical distance between the two tangents at 29 h (δ). CHO cell growth rate increased in the following order, when judged by this analysis: untreated PS < PE < Hiditron-treated PS < Harrick-treated PS.

G. STRENGTH OF CHO CELL ATTACHMENT ON TEST SUBSTRATA

Test substrata were incubated with CHO cells for 24 h and subsequently rinsed with saline in a flow cell for 10 min. After flow challenges between 1.5 and 3 dyn/cm^2, the substrata were observed in saline by phase contrast microscopy; no cells could be seen on any surface. The shear forces removed all cells, leaving only portions of cell membranes behind (determined by scanning electron microscopic inspection). Strands of additional surface-associated material and scattered debris also were present from the shear-challenged cells attached to RFGPT PS specimens. Filters in-line with the flow cell were positioned to "catch" the detached cells. Many round bodies of 7 to 8 μm in diameter were captured as detached from the PE control surface. These bodies were apparently whole, minimally damaged CHO cells. Parts of disrupted cells also were observed. Whole cells were seen on the filters in-line with the rinsing of the

a

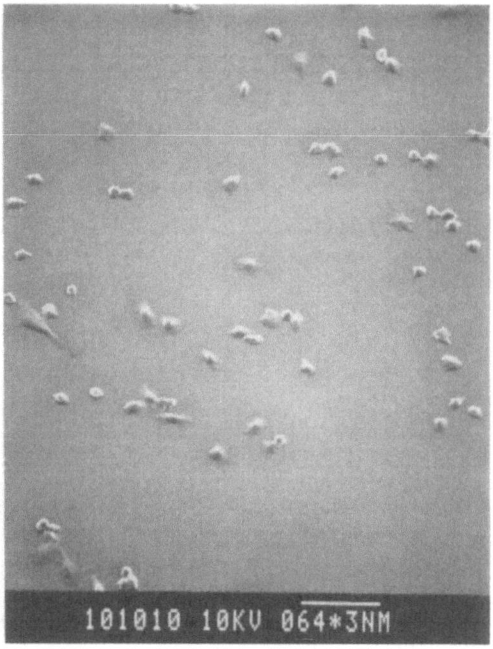

b

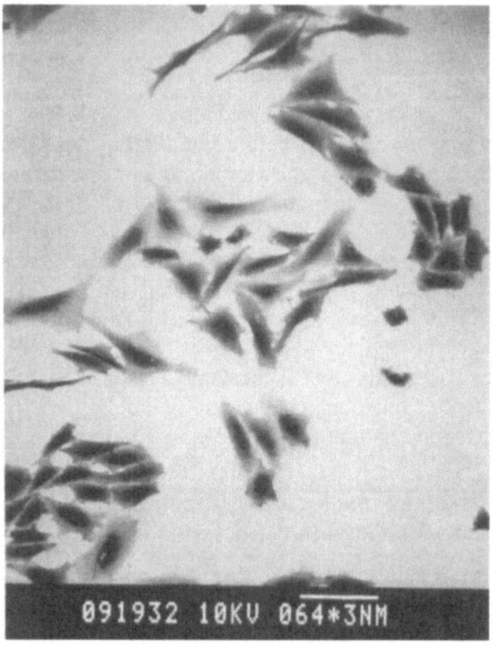

c

Figure 4 SEM photos of CHO cells on three different substrata after 2 d in culture: (a) low density PE, (b) bacterial grade PS ("untreated PS" in this study), and (c) germanium. The morphologies observed at 2 d were the same as those at 1 d. Note the good contrast for the germanium sample; because the substratum is conductive, it is not coated with evaporated metal or carbon prior to SEM analyses. (Magnification ×250.)

Table 8 **Image analysis results for CHO cell Experiment #1 (8×10^5 cells inoculated in 10 ml medium; 1 d exposure)**

Sample	Mean spread area (μm²/cell) ± sd [SE]	n	Total cells	% Round cells	Percent coverage
PE	290 ± 50 [10]	14	2823	100	3.5
PS	300 ± 20 [<6]	15	2907	100	3.5
PS/Harrick					
Round	320 ± 40 [10]	14	416		0.6
Spread	420 ± 40 [10]	14	2619		4.8
All cells	400 ± 50 [10]	14	3035	14	5.3
PS/Hiditron					
Round	290 ± 30 [10]	15	591		0.7
Spread	380 ± 20 [10]	15	4810		7.4
All cells	370 ± 20 [10]	15	5401	11	8.1

Harrick- and Hiditron-treated PS specimens, but the cells were of a flatter morphology corresponding to the spread appearance of the attached cells on the prerinse substrata. Since cells were observed on the filters and detachment remnants also were seen on the gas-plasma-discharge-treated substrata, the shear challenge was apparently strong enough to remove the cells by rupture in or near their attachment plaques, but not directly at the substratum interface. Microscopic observation of the shear challenge utilized a flow cell modified so that the attached cells could be photographed at several stages in each experiment. Statically attached (for 24 h) cells first were observed on unmodified PS before saline flow was started. Many aggregates were present. Even before continuous flow had begun, most aggregates were easily removed by the saline flush, and only single cells remained. As higher shear rates were produced, proportionately more cells were detached. After increasing the flow to produce shear stress of approximately 1 dyn/cm², no remaining CHO cells were seen. In experiments with various substrata *not* supporting spread cells, increased incubation times (up to 5 d) did not change this susceptibility to complete detachment at very modest shear rates.

When CHO cells were similarly statically incubated on gas-plasma-discharge-treated PS, aggregate-free confluent CHO cell layers were observed before initiation of the shear challenge. As saline was introduced into the flow cell, little change in the morphology or distribution of the cells was noted at shears up to 1 dyn/cm². At higher shears, up to 3.5 dyn/cm², the cells were detached from substrata upon which they had been previously incubated for only 24 h. In contrast, when the preshear incubation time was extended to 48 h, only small spaces developed between the cells as they rounded up at the higher shear rates; few cells were removed. These observations were reproducible on gas-plasma-discharge-treated PS, even when the initial cell layers were not confluent. Few differences, beyond occasional cell rounding, were seen in cell morphology due to the increasing shear rate.

The possible effects of the rinsing fluid alone (saline, under static conditions) on cell morphology were evaluated with CHO cells cultivated on PS, untreated and RFGPT. Saline exposure did slightly shrink the CHO cells and cause some to round up, but no CHO cells were detached.

Table 9 **Image analysis results for CHO cell Experiment #2 (1.5×10^5 cells inoculated in 10 ml medium; 1 d exposure)**

Sample	Mean spread area (μm²/cell) ± sd [SE]	n	Total cells	% Round cells	Percent coverage
PE	220 ± 20 [<10]	15	524	100	0.5
PS	190 ± 20 [<10]	15	1119	100	0.8
PS/Harrick					
Round	230 ± 20 [<10]	15	193		0.2
Spread	370 ± 60 [20]	15	1870		2.7
All cells	350 ± 50 [10]	15	2063	9	2.8
PS/Hiditron					
Round	290 ± 90 [20]	15	88		0.1
Spread	380 ± 50 [10]	15	1240		1.9
All cells	370 ± 50 [10]	15	1328	7	1.9

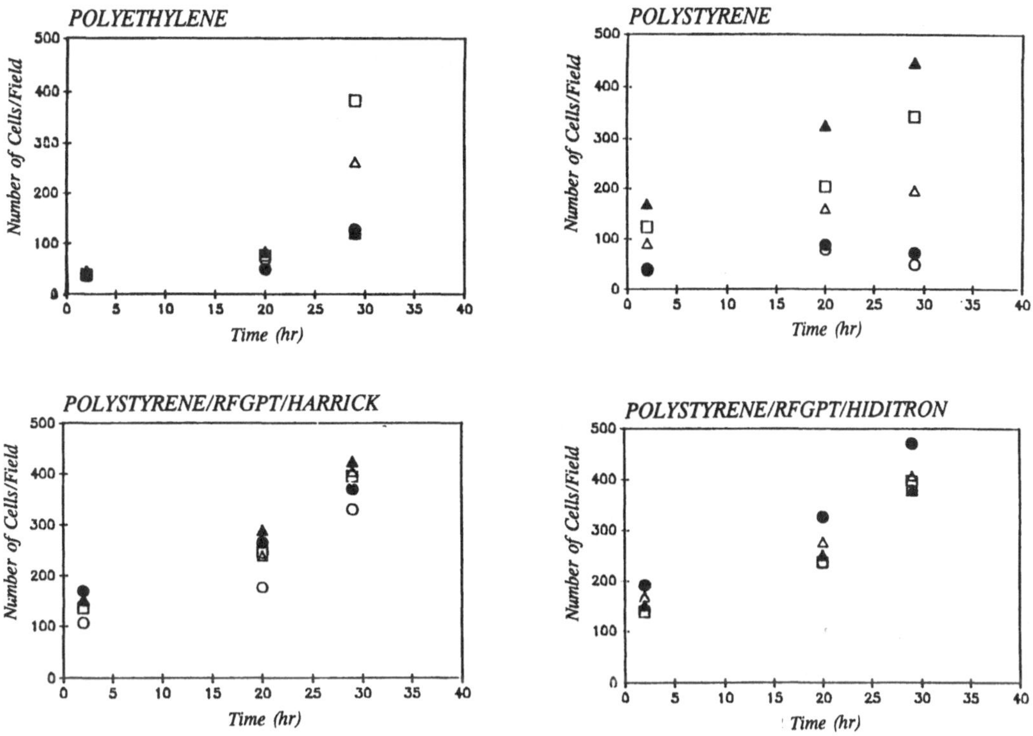

Figure 5 Graphs used to determine growth rates (see Table 10). CHO cell number/field (n = 5; 5 symbols) vs. time for PE, untreated PS, and RFGPT PS (both units).

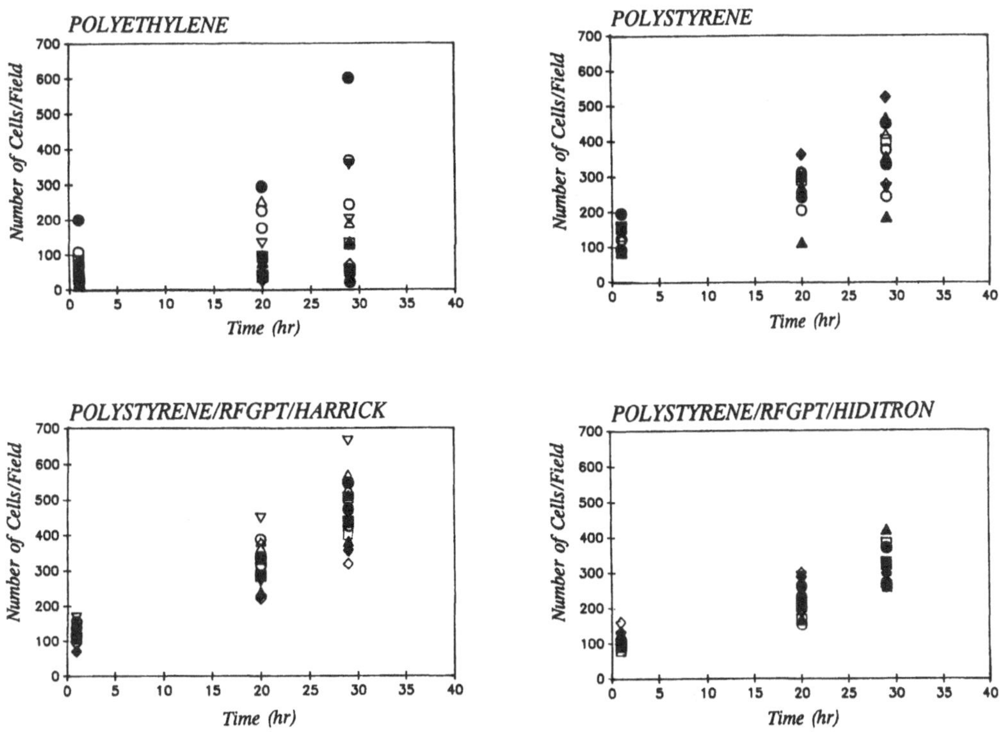

Figure 6 Graphs used to determine growth rates and recruitment rates (see Table 10). CHO cell number/field (n = 15; 15 points at each time) vs. time.

Table 10 **Constants determined from first order fits of data for CHO cell growth rate alone and growth rate + recruitment**

Sample	m	b
Growth Rate Study (n = 5)		
PE control	5.40	13.5
PS control	4.71	81.7
PS/RFGPT/Harrick	8.50	112.9
PS/RFGPT/Hiditron	8.77	127.6
Growth + Cell Recruitment Rate Study (n = 15)		
PE control	3.67	55.6
PS control	8.08	116.0
PS/RFGPT/Harrick	12.20	103.4
PS/RFGPT/Hiditron	7.27	98.6

H. CHO CELL BEHAVIOR ON UV-TREATED PS

When CHO cells were incubated with PS earlier exposed to only UV light, either under intense UV lamps or masked with UV-transparent film during RFGPT, round CHO cell morphology was seen on all surfaces.

I. PATTERNING THROUGH MASKS USING RFGPT

Several masking schemes were employed to pattern surfaces so that different cell morphologies would be elicited. Flattened CHO cells that were cultured on PS grew right up to the edge of where a piece of tape was placed prior to RFGPT (removed before adding round cells). Round cells from the same seed population were seen where the tape had been. Controls showed that the tape did not leave residues that inhibited spreading.

V. CORRELATIONS

A. XPS ANALYSIS OF SURFACES CORRELATED WITH CELL BEHAVIOR

XPS analysis did not detect oxygen on the surface of the untreated PE specimens; these substrata did not support cell spreading. Very small amounts of oxygen were detected on untreated PS, although PS has no oxygen in its nominal structure. After RFGPT, more oxygen was found on the PS surface; C–O, C=O,

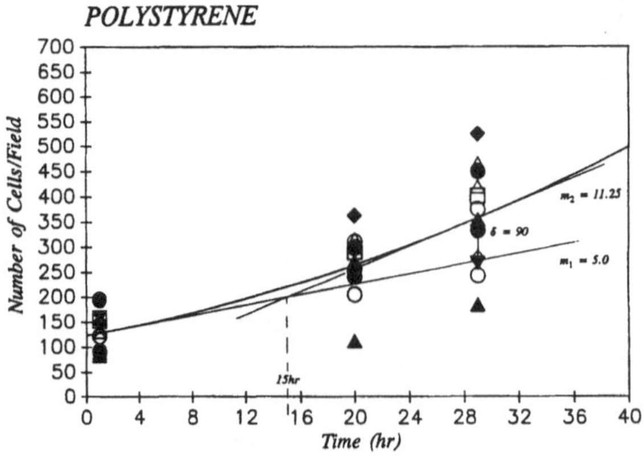

Figure 7 Example (untreated PS) of a second order fit to the recruitment + growth data. The slopes of the tangents drawn give an estimate of the recruitment rate (m_1) and the recruitment + growth rate (m_2) (see Table 11). The intersection of the two tangents gives an estimate of the lag time prior to division.

Table 11 **CHO cell study: recruitment rate (m_1), growth + recruitment rate (m_2), growth rate ($m_3 = m_2 - m_1$), and δ for the four test substrata**

Substratum type	m_1	m_2	m_3	δ
PE control	0.0	7.5	7.5	110
PS control	5.0	11.3	6.3	90
PS/RFGPT/Harrick	5.0	13.8	8.8	210
PS/RFGPT/Hiditron	3.8	11.3	7.5	125

and O–C=O (carboxyl) functionalities were detected. Alcoholic oxygen was the most abundant oxygen-containing functionality present and correlated best with the excellent cell spreading results.

Previous studies of plasma-treated PE and PS corroborate these results. XPS studies of PE plasma-treated via corona discharge in air confirmed surface incorporation of alcoholic (C–O), carbonyl (C=O), and carboxyl (O–C=O) groups.[31,32] In the autoadhesion of PE, it was postulated that corona discharge treatment forms keto groups (R–CO–CH_2–R′) and carbonyl groups on PE molecules. The ketone groups enolize (R–COH=CH–R′); the enolic hydrogen then bonds with carbonyl groups on the adjacent PE sheet to which self-adhesion is desired.[32] Derivatization techniques applied to PE surfaces after corona-discharge treatment resulted in the detection of hydroxyl (–C–OH), carbonyl (–C=O), and carboxyl (–COOH) groups.[33] Another study employing derivatization techniques isolated the aforementioned functional groups as well as hydroperoxy and epoxy groups.[34]

RFGP treatment first acts to remove contaminants from the surface of test substrata and then intrinsic components of the surface itself, leaving unsaturated chemical bonds that spontaneously adsorb atmospheric molecules.[35] Angle-resolved XPS studies of PE showed the primary formation of ether linkages after RFGP treatment in pure oxygen in addition to C–O (alcoholic), O–C–O, C=O, and O–C=O.[36] Carbonyls were detected on PE after RFGP treatment in argon.[37]

XPS results for RFGPT PS reported previously in the literature have indicated the addition of carbonyl from an argon plasma[37] and the addition of carbonate (O=COO), alcohol (C–O), carbonyl (C=O), and carboxyl (O–C=O) from pure oxygen plasma.[36] PS in cast thin films demonstrated significant oxygen addition upon RFGPT, with alcoholic groups (C–O) being dominant. UV radiation of PS, in contrast to RFGPT, results in oxidation mainly in the form of carbonyl groups (i.e., chain scission occurs), and this process intensifies with increasing exposure time.[38] Contact angles of water on UV-treated PS are lower than for the untreated controls; however, extraction in methanol or aqueous detergent solution removes some of the low mol wt fragments generated via chain scission, and the water contact angles increase again toward control values.[38]

B. SURFACE (CONTACT) POTENTIAL DATA CORRELATED WITH CELL ADHESION

Contact potential readings were taken, first, of protein films (>1000 Å) present after incubation in media/CHO cells for 30 min. These values, three- to five-fold greater than those for the test substrata on which they deposited, persisted after leaching away the loosely associated portions of the "conditioning film". At least 70% of the protein-contributed contact potential readings remained, and even after vigorous rinsing, at least 42% (and as much as 76%) of protein contributions to the surface potentials remained. Interestingly, all RFGPT surfaces retained greater percentages of their initially deposited "conditioning films" than did their companion, untreated substrata.

These data indicate that, within the protein films remaining on the RFGPT substrata after rinsing, there was considerable orientation of the molecules, with a large portion of the negatively charged groups present in the film "standing" upright. Since RFGP treatment increases the polar and composite surface energies of all test substrata, these contact potential results illustrate that deposited proteins bind in an organized manner with most of their charged groups (predominantly negative) facing the polar substratum. That is, the predominant binding force is polar to polar, even though hydrophobic bonding almost certainly plays a major role in establishing protein-to-surface contact initially.

C. CELLULAR PROCESSING OF "CONDITIONING FILMS"

Contact angle results for PS incubated with medium/CHO cells for 24 h showed that changes took place in the "conditioning films" on both the untreated and RFGPT samples, compared to the films present at the time of first attachment after only 30 min of incubation. The $\%\gamma_p$ characterizing the outermost chemical groups probed by contact angles changed by 14 to 15% in a substratum-dependent manner. The polarity of the protein layers remaining on untreated PS diminished by 24 h; the polarity of the protein layers remaining after 24 h of cell incubation on RFGPT PS increased. These data indicate that, upon initial contact with untreated PS, proteins deposit with abundant polar groups outermost, as might be expected given the apolar quality of the unmodified PS surface. Over the 24-h incubation period, protein exchanges and/or reorientations occur, such that the polar content of the outermost layer is reduced. Similarly, protein rearrangements that occur on the initially quite polar RFGPT PS result in more polar surfaces after 24 h of incubation.

Proteins can deposit spontaneously from biological solutions onto solid substrata and take up varying conformations.[3,39-41] Fibrinogen, as an example, denatures to different extents on surfaces of low vs. high surface energy, allowing some biomaterials with low surface energy to remain thrombus-free, due to minimal denaturation of their predominantly fibrinogen "conditioning films".[3] Studies of albumin, fibrinogen, and γ-globulin adsorbed to amino acid copolymers revealed that the extent of their conformational modification increased with time.[41] Upon desorption, the α-helical contents of albumin and fibrinogen and the β-sheet content of γ-globulin were lower in comparison to the same secondary structures of these proteins in solution. Antifibronectin was bound to adsorbed fibronectin on bacteriological (untreated) PS plates in greater quantities than to adsorbed fibronectin on (glow-discharge-treated) tissue-culture PS, indicating that fibronectin (like fibrinogen) also binds to the lower energy and less polar surfaces in more native conformation.[39]

D. CELL ATTACHMENT/RETENTION AND MORPHOLOGY

CHO cell morphology correlated with RFGP treatment, since better spreading was observed on all treated polymer samples. No effects of surface microtopography were observed, since spread cells were oriented in a random fashion on all RFGPT surfaces. On untreated and UV-treated PE and PS, only round CHO cells were present. The spread cell areas on the RFGPT surfaces were significantly greater than the round cell areas on the untreated and control surfaces. Occasional high standard deviations for the round cells on the gas-plasma-treated PS resulted from larger portions of the round cells being present from retraction of spread cells just ready to divide. Their cell masses and, therefore, cell areas were much greater than for the individual daughter cells produced just after division. Cells cultured on gas-plasma-treated PS at subconfluent densities were organized in colonies touching; cells at higher seeding densities on gas-plasma-treated PS were well distributed across the available surface. The distribution of cell spread areas was larger on the latter specimens, as indicated by larger standard deviations, perhaps reflecting prior observations that cell spread area usually decreases with increases in cell density.[42]

Growth rate data showed that CHO cells grew at accelerated rates on all RFGPT surfaces, 40% faster than cells on untreated PS and control PE. Cells on control PE and untreated PS were observed to be mobile, moving in and out of the counting field in response to shear forces occurring upon transport of the petri dish from the incubator to the microscope. Cells on RFGPT PS, once attached, remained attached, spread, and grew; few to no cells were observed in suspension.

The combined image analysis and cell growth data indicate that RFGPT elicits significant metabolic cell responses via changes in cell morphology (i.e., average cell spread areas). Cell shape has been correlated with cell growth in previous studies.[42-44] Spread cells synthesize more DNA than round cells.[44]

Fibroblasts in suspension culture for 3 d ceased all protein, mRNA, rRNA, and DNA synthesis. Upon attachment to solid substrata (but where spreading was prevented either by choice of substrata or with drugs), protein synthesis fully recovered within 6 h. Only upon physical spreading did mRNA, rRNA, and DNA synthesis resume.[43]

Actively growing anchorage-dependent cells (spread cells) round upon an increase in cell density and cease DNA production upon confluence; cell division stops. Prior to studies that determined a relationship between cell geometry and protein/DNA synthesis, it was believed that physical contact between cells was responsible for slowing and stopping growth, a phenomenon known as contact inhibition. Less cell membrane is available for nutrient transport in confluent cultures, and accumulated toxins from large numbers of cells must be considered also. A spread cell, on the other hand, may have a more permeable membrane.[42] All of these factors contribute to the control of population density in cell cultures.

Studies of the forces required to remove attached CHO cells on the test substrata support the assertion that a spread cell is a more firmly attached cell. Round cells on control PE and PS were easily removed. Little or no cellular mass remained on PE surfaces after controlled-shear (flow cell) rinsing at 3 dyn/cm^2. Discrete areas of debris left behind after controlled-shear challenges of CHO cells attached to untreated PS indicated that cohesive failure at the level of the cell membrane occurred at less than 2 dyn/cm^2. The number of discrete areas of debris remaining on the gas-plasma-treated PS after controlled-shear challenges was always larger in comparison to the number on untreated PS. In addition, other stray strings of debris were seen on some gas-plasma-treated surfaces in regions where detachment occurred.

Critical shear stress has been defined in the literature as the stress at which cell removal begins to occur.[45,46] The critical shear stress required to remove red blood cells from glass (after 10 min of gravitational settling) was 11 dyn/cm^2. Critical shear stress increased with an increase in cell surface contact time.[46] The critical shear stress required to remove 3T3 cells from copolymers of polyHEMA/polyEMA decreased with increasing polyHEMA content; 18 dyn/cm^2 was needed to remove cells from 100% polyEMA.[45]

The fact that disrupted cellular material was found on both filter and test substrata after the rinse indicates that portions of the cell were left on the test substrata. Energy-dispersive X-ray analysis confirmed that the relative elemental compositions of the discrete deposits observed on the rinsed substrata were identical to those for whole spread cells. Results from control studies confirmed that the discrete areas were not deposited from the rinsing solution itself.

Cellular debris of identical morphological features to those observed in this study has been called substrate-attached material (SAM), and reported in the literature to be cell membranes' focal contact areas,[45,47,48] including hyaluronate proteoglycans, microfilament-associated proteins, large-external-transformation-sensitive (LETS) glycoproteins, and other proteins. The "strings" of debris are said to be pseudopodia, which extend from the cell to link the focal contact area in direct apposition to the substratum.

Cells removed by controlled-shear rinsing from gas-plasma-treated PS (and "caught" on a filter) were deformed and flat. Only round cells were observed on the filter after controlled-shear removed them from untreated PS and PE.

Observations of CHO cells on the test substrata *during* flow confirmed the easy removal of round cells from untreated PS and PE. Confluent cells on the gas-plasma-treated PS specimens actually flattened upon increasing flow rates so that more of the cell bodies came into focus.[47] Although some CHO cells on gas-plasma-treated PS appeared to round up slightly during the controlled-shear challenge, only a few cells were removed. Well-distributed spread cells on gas-plasma-treated PS did not round up during the controlled-shear challenge, and no cells were removed. Photographs of both types of RFGPT samples after the shear challenge showed retained basal membranes of cells whose remaining bodies had been sheared away and carried to the downstream filters.

VI. DISCUSSION

The functional groups added to PE and PS surfaces during RFGP treatment in this study were essentially alcoholic (C–OH), carbonyl (C=O), and carboxyl (O–C=O). Can cell attachment and/or spreading be attributed to specific functional groups present on the substratum to which it attaches? An earlier study of RFGPT PE and PS reported that carboxyl groups were responsible for cell adhesiveness and growth.[49] (No correlation was found between total surface oxygen content and cell adhesion.) However, a complete XPS analysis was not reported. While mention was made of the possible formation of alcohol and ether groups, data were not presented to allow the reader to assess the significance of these functionalities.

Hydroxyl and carboxyl functionalities introduced to PS plates, via chloric acid treatment, were systematically blocked by derivitization reactions (diazomethane to methylate carboxyl groups, acetic anhydride to esterify hydroxyl groups) and their contribution assessed regarding the cell adhesion process.[50,51] Hydroxyl groups were required for cell adhesion. At very high –OH densities, however, no cell adhesion occurred. Carboxyl groups either decreased cell adhesion (because the blocking of these groups led to increased adhesion)[50] or did not affect adhesion at all. [51] We conclude from our data and these earlier studies that H-bonding plays a role in cell adhesion.

PolyHEMA treatment with sulfuric acid introduces carboxyl groups to a surface that is hydroxyl rich; growth of vascular endothelial cells is then supported.[52] These results, together with our data, suggest that the partial conversion of polyHEMA to polyMMA alters the degree of hydration to allow more prompt — if not more secure — cell attachment.

Red cells attach to Langmuir-Blodgett films of docosonol ($C_{22}H_{45}OH$) where OH groups are outermost (the films being kept under water).[53] Shear forces beyond 10 dyn/cm^2 just begin to remove the red cells, leaving broken cellular attachment strands behind; nearly all are removed at shears of 30 dyn/cm^2. Thus, the contribution of specific surface functional groups in determining the "adhesiveness" of a surface is now clearer. The highest significance of hydroxyl groups is strongly implicated.

Cells do attach but remain round, on surfaces of critical surface tension between 20 to 30 dyn/cm. Such surfaces can be predominantly populated either with hydrophobic, closely packed methyl (CH_3) groups (e.g., from closely packed aliphatic compounds or polydimethyl siloxane) or with hydrophilic, hydroxyl (OH) groups (e.g., organized water on polyacrylamide or agarose).[10,54] BHK fibroblasts incubated with copolymers of ethyl methacrylate (EMA, with terminal CH_3 on its side chain) and hydroxyethyl methacrylate (HEMA, with a terminal OH on its side chain) did not attach well when only 1 to 8% EMA was used.[55] The same cells quickly attached and spread completely on copolymers of 9 to 100% EMA. Interestingly, the critical shear stress needed to remove cells from identical HEMA/EMA surfaces increased with increasing EMA content.[45] Some cells, therefore, and possibly all cells can eventually attach to all surfaces, at different rates and with varying attachment strengths. In the same study, copolymers of HEMA and PS were tested; attachment and full cell spreading were seen on all surfaces of 4 to 100% PS. "There was no continuous modulation of cell response to hydroxylation levels" in either copolymer series.[55] Interestingly, different hydroxylation concentrations were needed for prompt cell attachment vs. cell spreading. High densities of hydroxyl groups, as in 100% polyHEMA, inhibited cell adhesion;[50,51] the equilibrium water content (EWC) and fractional polarity correlated with cell spreading.[55] The organization of water at the substratum/medium interface may be related to cellular interaction. Many polymers with a wide range of surface properties were considered in this assessment. No polymer of 35 to 60% EWC supported cell attachment over the time frame of the experiment. However, those polymers with 2 to 34% EWC did support rapid cell attachment.

VII. CONCLUSION: 20 YEARS OF EDUCATION OF A BIOSURFACE SCIENTIST

In closing, the development and refinement of a "bioadhesion" concept has taken many turns and paths over the past two decades. An "Eight-Fold Way" might be used to demonstrate the concept's evolution:

1. Recognition of differential adhesion as a function of substratum energy,[15]
2. Recognition of the universality of the presence of preadsorbed macromolecular films before any cell (or particle) adhesion,[14]
3. Recognition of nonlinearity of surface energy vs. bioadhesion relationship, with 20 to 30 dyn/cm being the minimum,[56]
4. Recognition of universality (within a given system) of protein deposition and protein identity on all substrata, so that differences must be in conformations rather than compositions,[57]
5. Recognition of universality (within a given system) of deposition of specific particles (cells) on all substrata, so that differences must be in patterns and degrees of (spreading) interactions,[58]
6. Recognition that differences among solid substrata–biosystem interactions are based in differences in postdeposition retention of films and particles, rather than in initial associations,[59]
7. Recognition of the important role of mechanical work (shear rate) in revealing substratum-dependent differences by overcoming variable adhesive (retentive) strengths of film/particle deposits on different substrata,[30]
8. Recognition that differential "processibility" by both shear and adjacent biological/biochemical phases (including "Vroman Effect" molecular replacements) is basis for longer term bioadhesive effects.[60]

REFERENCES

1. **Adamson, A. W.,** *Physical Chemistry of Surfaces,* 2nd ed., Interscience Publishers, New York, 1967.
2. **Amstein, C. F. and Hartman, P. A.,** Adaptation of plastic surface for tissue culture by glow discharge, *J. Clin. Microbiol.,* 2, 46, 1975.
3. **Baier, R. E.,** Modification of surfaces to meet bioadhesive design goals: a review, *J. Adhesion,* 20, 171, 1985.
4. **Baier, R. E., DePalma, V. A., Goupil, D. W., and Cohen, E.,** Human platelet spreading on substrata of known surface chemistry, *J. Biomed. Mater. Res.,* 19, 1157, 1985.
5. **Helmstetter, C. E.,** Rate of DNA synthesis during the division cycle of *Escherichia coli* B/r, *J. Mol. Biol.,* 24, 417, 1967.
6. **Helmstetter, C. E.,** DNA synthesis during the division cycle of rapidly growing *Escherichia coli* B/r, *J. Mol. Biol.,* 31, 507, 1968.

7. **Carter, S. B.,** Principles of cell motility: the direction of cell movement and cancer invasion, *Nature*, 208, 1183, 1965.
8. **Carter, S. B.,** Haptotactic islands: a method of confining single cells to study individual cell reactions and clone formation, *Exp. Cell Res.*, 48, 189, 1967.
9. **Carter, S. B.,** Haptotaxis and the mechanism of cell motility, *Nature*, 213, 256, 1967.
10. **Maroudas, N. G.,** Chemical and mechanical requirements for fibroblast adhesion, *Nature*, 244, 363, 1973.
11. **Harris, A.,** Behavior of cultured cells on substrata of variable adhesiveness, *Exp. Cell Res.*, 77, 285, 1973.
12. **Rittle, K. H., Helmstetter, C. E., Meyer, A. E., and Baier, R. E.,** *Escherichia coli* retention on solid surfaces as functions of substratum surface energy and cell growth phase, *Biofouling*, 2, 121, 1990.
13. **Rittle, K. H., Baier, R. E., and Helmstetter, C. E.,** Substrata influences on patterns of bacterial attachment from laminar flows, *Trans. Soc. Biomater.*, XII, 162, 1989.
14. **Baier, R. E. and Dutton, R. C.,** Initial events in interactions of blood with a foreign surface, *J. Biomed. Mater. Res.*, 3, 191, 1969.
15. **Baier, R. E., Shafrin, E. G., and Zisman, W. A.,** Adhesion: mechanisms that assist or impede it, *Science*, 162, 1360, 1968.
16. **Dexter, S. C., Sullivan, J. D., Williams, J., and Watson, S. W.,** Influence of substrate wettability on the attachment of marine bacteria to various surfaces, *Appl. Microbiol.*, 30, 298, 1975.
17. **Dexter, S. C.,** Influence of substratum critical surface tension on bacterial adhesion — *in situ* studies, *J. Colloid Interface Sci.*, 70, 346, 1979.
18. **van der Valk, P., van Pelt, A. W. J., Busscher, H. J., de Jong, H. P., Wildevuur, C. R. H., and Arends, J.,** Interaction of fibroblasts and polymer surfaces: relationship between surface free energy and fibroblast spreading, *J. Biomed. Mater. Res.*, 17, 807, 1983.
19. **Rittle, K. H., Baier, R. E., and Helmstetter, C. E.,** Attachment and growth of CHO cells on radio-frequency-glow-discharge-treated PHEMA, *Trans. Soc. Biomater*, XII, 41, 1989.
20. **Baier, R. E. and DePalma, V. A.,** Electrodeless Glow Discharge Cleaning and Activation of High-Energy Substrates to Ensure Their Freedom from Organic Contamination and Their Receptivity for Adhesives and Coatings, Report #176, Calspan Corporation, Buffalo, NY, 1970.
21. **Hess, D. W.,** Plasma-surface interactions in plasma-enhanced chemical vapor deposition, *Annu. Rev. Mater. Sci.*, 16, 163, 1986.
22. **Baier, R. E., Carter, J. M., Sorenson, S. E., Meyer, A. E., McGowan, B. D., and Kasprzak, S. A.,** Radiofrequency gas plasma (glow discharge) disinfection of dental operative instruments, including handpieces, *J. Oral Implantol.* XVIII, 236, 1992.
23. **Mondi, P. G.,** Influence of Substratum Surface Properties on Cell Attachment and Growth, Master's project, Roswell Park Division of the Graduate School of SUNY at Buffalo, 1990.
24. **Meyer, A. E., Baier, R. E., DePalma, V. A., and Goupil, D. W.,** Investigation of surface roughness as a parameter for blood/biomaterial biocompatibility, *Proceedings of the 40th Annual Conference on Engineering in Medicine and Biology*, Sept. 10–13, 1987, Niagara Falls, NY, The Alliance for Engineering in Medicine and Biology, Washington, D.C., 196.
25. **Ratner, B. D.,** Surface characterization of biomaterials by electron spectroscopy for chemical analysis, *Ann. Biomed. Eng.*, 11, 313, 1983.
26. **Baier, R. E. and Meyer, A. E.,** Surface analysis, in *Handbook of Biomaterials Evaluation*, von Recum, A., Ed., Macmillan, New York, 1986, 97.
27. **Azzam, R. M. A. and Bashara, N. M.,** *Ellipsometry and Polarized Light*, North-Holland, Amsterdam, 1977.
28. **Harrick, N. J.,** *Internal Reflection Spectroscopy*, Interscience, New York, 1967.
29. **Baier, R. E. and DePalma, V. A.,** Flow Cell and Method for Continuously Monitoring Deposits on Flow Surfaces, U.S. Patent 4,175,233, 1979.
30. **Baier, R. E., Meyer, A. E., DePalma, V. A., King, R. W., and Fornalik, M. S.,** Surface microfouling during the induction period, *J. Heat Transfer*, 105, 618, 1983.
31. **Blythe, A. R., Briggs, D., Kendall, C. R., Rance, D. G., and Zichy, V. J. I.,** Surface modification of polyethylene by electrical discharge treatment and the mechanism of autoadhesion, *Polymer*, 19, 1273, 1978.
32. **Owens, N. F.,** Mechanism of corona-induced self-adhesion of polyethylene film, *J. Appl. Polymer Sci.*, 19, 265, 1975.

33. **Briggs, D. and Seah, M. P.,** Eds., *Practical Surface Analysis by Auger and X-Ray Photoelectron Spectroscopy,* John Wiley & Sons, New York, 1983.
34. **Gerenser, L. J., Elman, J. F., Mason, M. G., and Pochan, J. M.,** E.s.c.a. studies of corona-discharge-treated polyethylene surfaces by use of gas-phase derivatization, *Polymer,* 26, 1162, 1985.
35. **Kasemo, B. and Lausmaa, J.,** Biomaterial and implant surfaces: on the role of cleanliness, contamination, and preparation procedures, *J. Biomed. Mater. Res.,* 22, 145, 1988.
36. **Clark, D. T. and Dilks, A.,** ESCA applied to polymers. XXIII. RF glow discharge modification of polymers in pure oxygen and helium-oxygen mixtures, *J. Polymer Sci. Polymer Chem. Ed.,* 17, 957, 1979.
37. **Yasuda, H., Marsh, H. C., Brandt, S., and Reilley, C. N.,** ESCA study of polymer surfaces treated by plasma, *J. Polymer Sci.,* 15, 1991, 1977.
38. **Price, T. R. and Fox, R. B.,** Surface effects in the photodegradation of polymer films, Report #6563, Office of Naval Research, Washington, D.C., 1967.
39. **Grinnell, F. and Feld, M. K.,** Adsorption characteristic of plasma fibronectin in relationship to biological activity, *J. Biomed. Mater. Res.,* 15, 363, 1981.
40. **Morrisey, B. W.,** The adsorption and conformation of plasma proteins: a physical approach, *Ann. N.Y. Acad. Sci.,* 283, 50, 1977.
41. **Soderquist, M. E. and Walton, A. G.,** Structural changes in proteins adsorbed on polymer surfaces, *J. Colloid Interface Sci.,* 75, 386, 1980.
42. **Folkman, J. and Greenspan, H. P.,** Influence of geometry on cell growth, *Biochim. Biophys. Acta,* 417, 211, 1975.
43. **Ben-Ze'ev, A., Farmer, S. R., and Penman, S.,** Protein synthesis requires cell-surface contact while nuclear events respond to cell shape in anchorage-dependent fibroblasts, *Cell,* 21, 365, 1980.
44. **Folkman, J. and Moscona, A.,** Role of cell shape in growth control, *Nature,* 273, 345, 1978.
45. **Horbett, T. A., Waldburger, J. J., Ratner, B. D., and Hoffman, A. S.,** Cell adhesion to a series of hydrophilic-hydrophobic copolymers studied with a spinning disc apparatus, *J. Biomed. Mater. Res.,* 22, 383, 1988.
46. **Mohandas, N., Hochmuth, R. M., and Spaeth, E. E.,** Adhesion of red cells to foreign surfaces in the presence of flow, *J. Biomed. Mater. Res.,* 8, 119, 1974.
47. **Hertl, W., Ramsey, W. S., and Nowlan, E. D.,** Assessment of cell-substrate adhesion by a centrifugal method, *In Vitro,* 20, 796, 1984.
48. **Rosen, J. J. and Culp, L. A.,** Morphology and cellular origins of substrate-attached materials from mouse fibroblasts, *Exp. Cell Res.,* 107, 139, 1977.
49. **Ramsey, W. S., Hertl, W., Nowlan, E. D., and Binkowski, N. J.,** Surface treatments and cell attachment, *In Vitro,* 20, 802, 1984.
50. **Curtis, A. S. G., Forrester, J. V., and Clark, P.,** Substrate hydroxylation and cell adhesion, *J. Cell Sci.,* 86, 9, 1986.
51. **Curtis, A. S. G., Forrester, J. V., McInnes, C., and Lawrie, F.,** Adhesion of cells to polystyrene surfaces, *J. Cell Biol.,* 97, 1500, 1983.
52. **McAuslan, B. R. and Johnson, G.,** Cell responses to biomaterials. I: Adhesion and growth of vascular endothelial cells on poly (hydroxyethyl methacrylate) following surface modification by hydrolytic etching, *J. Biomed. Mater. Res.,* 21, 921, 1987.
53. **Owens, N. F., Gingell, D., and Trommler, A.,** Cell adhesion to hydroxyl groups of a monolayer film, *J. Cell Sci.,* 91, 269, 1988.
54. **Baier, R. E., Gott, V. L., and Feruse, A.,** Surface chemical evaluation of thromboresistant materials before and after venous implantation, *Trans. Am. Soc. Artif. Intern. Organs,* 16, 50, 1970.
55. **Lydon, M. J., Minnett, T. W., and McTighe, B. J.,** Cellular interactions with synthetic polymer surfaces in culture, *Biomaterials,* 6, 396, 1985.
56. **Baier, R. E.,** Surface properties influencing biological adhesion, in *Adhesion in Biological Systems,* Manly, R. S., Ed., Academic Press, New York, 1970, 15.
57. **Baier, R. E., Loeb, G. I., and Wallace, G. T.,** Role of an artificial boundary in modifying blood proteins, *Fed. Proc.,* 30, 1523, 1971.
58. **Baier, R. E.,** The organization of blood components near interfaces, *Ann. N.Y. Acad. Sci.,* 283, 17, 1977.
59. **Baier, R. E.,** Conditioning surfaces to suit the biomedical environment: recent progress, *J. Biomech. Eng.,* 104, 257, 1982.
60. **Baier, R. E., Meyer, A. E., Natiella, J. R., Natiella, R. R., and Carter, J. M.,** Surface properties determine bioadhesive outcomes: methods and results, *J. Biomed. Mater. Res.,* 18, 337, 1984.

Chapter 4

Dictyostelium discoideum Cohesion and Adhesion

Donna R. Fontana

CONTENTS

I. Introduction to *Dictyostelium discoideum* 63
A. Life Cycle 63
B. cAMP Relay Regulates Cellular Movements During Development 64
II. Adherence to Bacteria and the Substratum 65
A. Function and Mechanisms of Adherence During Growth 65
B. Role of Adhesion in Development 65
III. Cohesion Prior to Aggregation 66
A. Definition of Contact Sites A and B 66
B. Identification of Contact Sites A 67
C. gp80-Mediated Cohesion is Homophilic 67
D. Structure–Function Analysis of gp80 67
1. Predicted Structure of gp80 67
2. Localization of the Cell-Binding Site 68
3. Posttranslational Modifications of gp80 69
E. Developmental Relevance of gp80-Mediated Cohesion 70
1. Reduction or Elimination of gp80-Mediated Cohesion 70
2. Inappropriate Expression of gp80 70
F. Identification of Contact Sites B 71
1. gp126 71
2. gp24 72
IV. Cohesion After Aggregation 73
A. gp95 73
B. gp150 74
V. Functional Effects of Cohesion 74
A. Cohesion Mediates Interspecies Sorting 74
B. Contact-Mediated Regulation of Gene Expression 75
C. Cohesion and Adhesion Regulate cAMP Relay 76
D. Sorting of Prespore and Prestalk Cells 79
VI. Acknowledgments 80
Addendum 80
References 80

I. INTRODUCTION TO *DICTYOSTELIUM discoideum*

Dictyostelium discoideum is a nonobligate metazoan that is found in moist, temperate environments where there is a supply of decaying matter.[1] *D. discoideum* amoebae phagocytize the bacteria associated with the decaying matter, digest them in food vacuoles, and utilize their cellular components as nutrients. In the laboratory, the amoebae are typically grown in association with Gram-negative bacteria or axenically in a broth composed of complex components. *D. discoideum* is a valuable model system for studying the role of cohesion in development because it is accessible to experimental manipulation, and during its life cycle it possesses many of the features associated with the development of more complex eucaryotes, i.e., cell-type-specific differentiation, morphogenesis, pattern formation, and selected cell death.

A. LIFE CYCLE

In the presence of an ample supply of nutrients, *D. discoideum* exists as a single amoeboid cell. Upon nutrient depletion, *D. discoideum* begins its developmental program (Figure 1). The first few hours of

0-8493-4559-6/95/$0.00+$.50

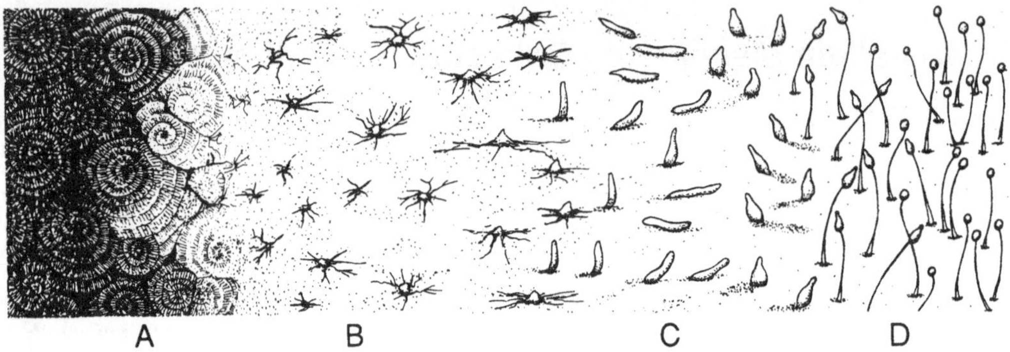

Figure 1 Life cycle of *Dictyostelium discoideum.*

development are a period of active protein synthesis. Some of the proteins synthesized are involved in cell–cell cohesion[2] or cell–substratum adhesion.[3] Other newly synthesized proteins participate in the cAMP communication system, which is instrumental in organizing the developmental program. Amoebae that have undergone this period of protein synthesis are called aggregation competent. If they are placed on a substratum, aggregation-competent amoebae aggregate in a highly organized fashion. The first sign of aggregation is the appearance of concentric circles and/or spirals (Figure 1A). Each spiral is composed of up to 100,000 amoebae and can be up to 1 cm in diameter. The center of the spiral ultimately will be the site occupied by the aggregate, and the aggregate will comprise those amoebae contained within the boundary of the spiral. Following the appearance of these patterns, the amoebae converge and form stream-like structures (Figure 1B). The amoebae in these streams move efficiently toward the aggregation center, where they form a single mound. The mound extends vertically so that it resembles a finger protruding from the surface. While extending, the mound becomes encased in a "slime" composed of carbohydrates, cellulose, and some proteinaceous material.[4] Subsequently, the elongated aggregate falls over onto the substratum and begins migrating (Figure 1C). This stage is called the pseudoplasmodium or slug stage. After a period of migration, the length of which is determined by genetic and environmental factors, the slug stops migrating and forms a fruiting body composed of a stalk, which contains vacuolated cells, and a sorus, which contains spores (Figure 1D). For a review of the life cycle see Reference 5.

B. cAMP RELAY REGULATES CELLULAR MOVEMENTS DURING DEVELOPMENT

In aggregation-competent amoebae, increasing the occupancy of the surface cAMP receptor elicits a transient activation of the adenylyl cyclase.[6] Some of the newly formed cAMP is secreted, with the rate of secretion directly proportional to the intracellular cAMP concentration.[6] The processes of cAMP binding, adenylyl cyclase activation, and cAMP secretion are collectively called cAMP relay. As a consequence of the cAMP relay response, a periodic pulse of cAMP secreted from an amoeba or small group of amoebae at the center of the aggregation territory is relayed; the result is waves of cAMP being propagated from the center to the edge of the territory.[7] As the cAMP wave reaches the amoebae, in addition to initiating a cAMP relay response, the amoebae respond to the cAMP gradient by moving chemotactically up the cAMP gradient toward the aggregation center.[7] As the peak of the wave passes, the amoebae again move in an apparently random manner. When observed under dark-field illumination, the bands of amoebae undergoing chemotactic movement appear light in contrast to the dark bands composed of randomly moving amoebae. Therefore, the light bands visible in Figure 1A are the result of amoebae undergoing directed motion. During slug and fruiting body formation, the amoebae again move in a pulsatile manner.[8-10] Evidence has been presented that suggests that cAMP relay and chemotaxis to cAMP regulate these morphogenetic steps.[9,11-14] This chapter contains a summary of the mechanisms through which *D. discoideum* adheres to particles and the substratum. The cohesion molecules that function during *D. discoideum* growth and development will be described. The participation of cohesion in the regulation of gene expression, cAMP relay, and cell sorting will be discussed.

II. ADHERENCE TO BACTERIA AND THE SUBSTRATUM

A. FUNCTION AND MECHANISMS OF ADHERENCE DURING GROWTH

Deposition of amoebae on a substratum results in attachment, spreading, and often the formation of ultrathin cytoplasmic lamellae that are 100 to 200 nm thick.[15] This adhesion, while necessary for motility, is most likely not essential for growth and cell division because it is possible to grow *D. discoideum* in liquid suspension. However in the presence of shear forces, *D. discoideum* amoebae must adhere to particles prior to their ingestion via phagocytosis. Ingested particles may include Gram-negative and Gram-positive bacteria, erythrocytes, yeast, and latex beads.

It has been suggested that adhesion to a particle or to the substratum involves the same or similar mechanisms, with the primary difference being that the particle is small enough to ingest while the substratum is not.[16] With adhesion to a particle and adhesion to a substratum, the adhesive interaction is uniformly distributed throughout the contact region, as evidenced by the apparent uniform distance between the amoebal membrane and the particle or surface.[17,18] Depending upon the surface, *D. discoideum* adhesion may be mediated by ligand–receptor interactions, hydrophobic interactions, and/or electrostatic interactions.

Mutant analysis revealed that *D. discoideum* amoebae are able to adhere to particles via a lectin-type ligand–receptor interaction in which the terminal glucose residues on the surface of the particle act as the ligand.[16,19] In order to study adhesion via lectin-type ligand–receptor interactions under more defined conditions, mono- and disaccharides were covalently linked to polyacrylamide gels.[20,21] Consistent with the mutant analysis, these studies revealed that amoebae will adhere to a gel utilizing a cell-surface receptor specific for glucose.[20] These studies further suggested that amoebae contain receptors for *N*-acetylglucosamine and mannose, and these receptors can mediate adhesion.[20] The glucose and mannose receptors are present in vegetative and aggregation-competent amoebae, while the ability to bind *N*-acetylglucosamine disappears during the acquisition of aggregation competence.[20]

Hellio and Ryter have presented evidence that suggests that a ligand–receptor type interaction may participate in yeast binding prior to ingestion.[22] Receptors for the lectin, wheat germ agglutinin, must be present on the amoebal surface for the amoebae to bind and ingest yeast. Clearing these receptors from the amoebal surface blocks the adhesion and uptake of yeast but not that of latex particles and bacteria. These results suggest that a relatively specific interaction occurs between the *D. discoideum* and the yeast. *D. discoideum* plasma membranes contain over 25 different proteins that can bind wheat germ agglutinin,[23] and the receptor(s) important in adhesion to yeast has not been identified.

Mutation analysis revealed that *D. discoideum* amoebae possess an adhesion system that discriminates on the basis of hydrophobicity.[16,19] Mutants lacking this adhesion system are deficient in adhesion to glass and plastic surfaces,[16,24] suggesting that this adhesion system may be instrumental in the cell–substratum adhesion typically observed in the laboratory. Hellio and Ryter compared the relative strength of hydrophobic and electrostatic interactions by quantitating the number of beads that remain bound to the amoebae after successive washings.[17] The relative strengths of adhesion are such that adhesion to hydrophobic particles is stronger than adhesion to positively charged particles, which is stronger than adhesion to negatively charged particles.[17] Consistent with this observation, Owens et al. observed that adhesion to a hydrophobic surface is stronger than adhesion to a surface modified to make it more hydrophilic.[25] These results suggest that hydrophobic interactions between amoebae and particles or the substratum are potentially stronger than electrostatic interactions, but either type of interaction is sufficient for adhesion.

Yabuno measured adhesion to a hydrophobic surface as a function of development. He observed that feeding amoebae are only weakly adhesive and that the strength of the adhesive interaction increases as *D. discoideum* progresses through development.[26] Based on enzymatic studies, it was suggested that the substance responsible for the increase in adhesiveness during development is a lipoprotein and that the amount of this cell surface lipoprotein increases during development. Enzymatic studies suggest that the same substance that is responsible for the increased adhesion also mediates the developmentally-regulated decrease in the negative charge associated with the *D. discoideum* surface.[26,27] Thus, amoebae appear capable of modulating their adhesiveness to hydrophobic surfaces by altering the charge associated with their surface.

B. ROLE OF ADHESION IN DEVELOPMENT

Cell–substratum interactions are important in the developmental cycle that follows growth because the complex cellular movements that accompany *D. discoideum* morphogenesis will not occur in suspension.

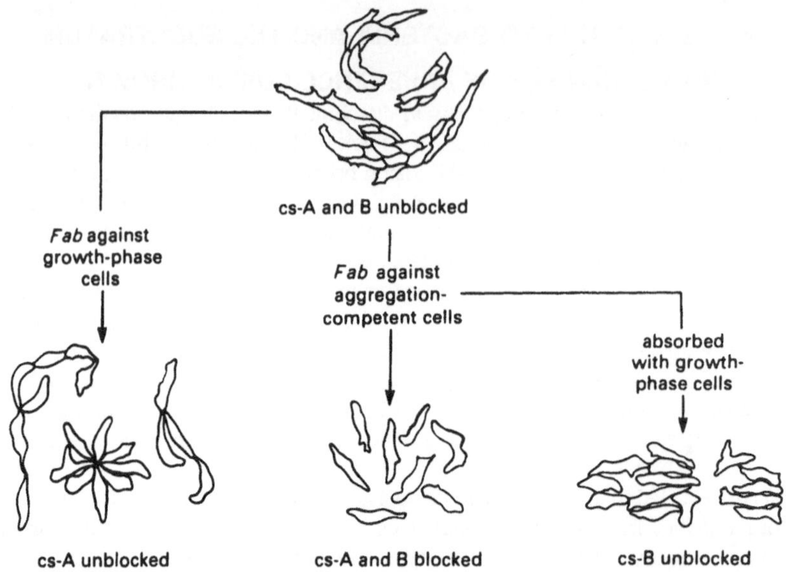

Figure 2 Two cohesion systems operate in *D. discoideum* aggregation streams. The amoebal tracings represent actual observations. This diagram is taken from Müller and Gerisch[39] and is based on the results of Beug, Katz, and Gerisch.[38] (From Müller, K. and Gerisch, G., *Nature,* 274, 446 1978, with permission.)

There is also some evidence that cell–substratum interactions may participate in the regulation of development. During late exponential growth in axenic culture or when their bacterial food source is depleted, amoebae begin to produce[28] and secrete[29,30] a lectin called discoidin I. Initially, it was proposed that this lectin might be involved in amoeba–amoeba cohesion. (For a discussion of this, see Reference 31.) Subsequent experimentation has led to the conclusion that discoidin I does not participate in cohesion, but secreted discoidin I may modify the substratum and assist in cell–substratum attachment and migration.[32] Discoidin I contains the amino acid sequence Gly-Arg-Gly-Asp (G-R-G-D),[33] and this sequence has been implicated in cell–substratum adhesion.[32] Springer et al. observed that the addition of discoidin I or short peptides that contain the G-R-G-D sequence inhibits the organized streaming that occurs during aggregation.[32] Blocking the carbohydrate-binding site of discoidin I does not alter its effectiveness in this assay. Higher concentrations of peptides inhibit the attachment and spreading of aggregation-competent amoebae.[32] Mutants lacking discoidin I fail to exhibit organized streaming when aggregating.[32,34] Thus, discoidin I may regulate cellular movements during *D. discoideum* development in a manner reminiscent of that displayed by fibronectin during the development of more advanced eucaryotes.[35] A 67 kD protein has been identified as the cellular receptor that interacts with discoidin I via its Gly-Arg-Gly-Asp binding site.[36] Univalent antibodies (fab fragments), prepared from antibodies that specifically stain this protein on immunoblots, also block organized streaming into aggregation centers.[36]

III. COHESION PRIOR TO AGGREGATION

A. DEFINITION OF CONTACT SITES A AND B

Using univalent antibody fragments (fab), Gerisch and colleagues provided evidence for the existence of discrete molecules on the surface of *D. discoideum* amoebae, molecules which are responsible for the cohesion observed in the aggregation streams.[37,38] Polyclonal antibodies were raised against lyophilized amoebae or amoebal homogenates. If vegetative amoebae were used to prepare the immunogen, the resulting fab would inhibit the side-to-side cohesion observed in aggregation streams (Figure 2; bottom, left).[37] If aggregation-competent amoebae were used to prepare the immunogen, the resulting fab would block completely the cohesion observed in the aggregation streams (Figure 2; bottom, middle). To isolate fab directed against components associated with aggregation-competent amoebae but not present on the surface of growing amoebae, the fab generated with aggregation-competent amoebae was preadsorbed with amoebae harvested during late exponential growth. This fab, which is more specific for components associated with developing amoebae, inhibits end-to-end cohesion (Figure 2; bottom, right).[38] In contrast, a fab primarily directed against cell-surface carbohydrates binds to the amoebal surface but does not block cohesion.[37,38] These results demonstrate that aggregating *D. discoideum* amoebae contain two independent cohesion systems. The cohesion molecules that appear during the late exponential phase of growth and mediate the side-to-side cohesion observed in aggregation streams are referred to as contact sites B (cs B).[38] Cs B-mediated cohesion is inhibited by EDTA and EGTA.[38] The cohesion molecules that appear as the amoebae begin to aggregate and that mediate the end-to-end cohesion observed in aggregation streams are referred to as contact sites A (cs A).[38] Cs A-mediated cohesion is insensitive to EDTA, affirming that the two cohesion systems are distinct.

B. IDENTIFICATION OF CONTACT SITES A

The protein responsible for cs A-mediated cohesion was isolated using, as an assay, the protein's ability to adsorb the fab that blocks the EDTA-insensitive cohesion.[39,40] The protein that was isolated is highly glycosylated (33% by weight carbohydrate) and runs as a wide band of approximately 80 to 90 kD when subjected to SDS-polyacrylamide gel electrophoresis (SDS-PAGE).[39,40] This protein is referred to as gp80.

To further implicate gp80 as cs A, several investigators attempted to generate specific, cell surface-binding antibodies against gp80. These attempts were frequently unsuccessful because the carbohydrate residues present on gp80 are highly antigenic, and antibodies against them cross-react with a number of other membrane proteins.[41-45] A monospecific monoclonal antibody (80L5C4) that recognizes a protein epitope on gp80 and binds to the amoebal surface was finally generated.[44] When aggregation-competent amoebae are dissociated in the presence of 10 m*M* EDTA, 80L5C4 blocks reassociation.[44] The antibody has no effect on EDTA-sensitive cohesion or the cohesion of amoebae that have already formed mounds.[44] If 80L5C4 is mixed with purified gp80 prior to its addition, no inhibition of reassociation is observed.[44] Thus, the monospecific anti-gp80 antibody blocks EDTA-insensitive reassociation when the dissociated amoebae are in a portion of development program where cs A is known to mediate the EDTA-resistant cohesion.

If gp80 is cs A, then it should be present on the surface of aggregation-competent amoebae at regions of cell–cell contact. Pulse-labeling studies suggest that the gp80 protein is synthesized during aggregation,[41,46] appearing when the amoebae begin to demonstrate cs A-mediated cohesion. Quantitative analysis of the mRNA for gp80 suggests that gp80 synthesis is regulated at the level of transcription and stimulated by oscillating levels of cAMP.[47-49] At its peak, there are approximately 2×10^5 molecules of gp80 per cell,[39,49] with greater than 90% of them on the cell surface.[49] Using the monoclonal antibody 80L5C4 and immunoelectron microscopy, it was established that gp80 is preferentially located at points of cell–cell contact and on filopodia.[50] When the membranes from regions of cell–cell contact were isolated from aggregation-competent amoebae, they were found to be rich in gp80.[51] Therefore, the cellular location of gp80 and the timing of its synthesis are consistent with gp80 being cs A.

The most convincing evidence for the identification of gp80 as cs A comes from the analysis of mutants. Mutants lacking gp80 were generated with a chemical mutagen[52] or by gene disruption via homologous recombination.[53] All the mutants lacking the protein component of gp80 demonstrate dramatically reduced cohesion when cs B-mediated cohesion is blocked with 10 m*M* EDTA.

C. gp80-MEDIATED COHESION IS HOMOPHILIC

The observation that only one protein is capable of adsorbing out the active adhesion-blocking fab from a polyspecific fab preparation[40] suggests that cs A-mediated cohesion is homophilic. Several approaches were used to confirm this conclusion. The first utilized gp80 conjugated to beads. The gp80-conjugated beads bind to each other and to the polar ends of aggregation-competent amoebae.[54] Soluble gp80 prevents the agglutination of these beads. Precoating amoebae with the monoclonal antibody 80L5C4 blocks the binding of gp80-coated beads to the amoebae.[54] Other approaches which suggested that gp80-mediated cohesion is homophilic involved purified, soluble gp80. Binding of soluble gp80 to amoebae is blocked by pretreating the amoebae with 80L5C4.[54] When purified gp80 is held on nitrocellulose, soluble gp80 binds to the nitrocellulose in a specific, saturable manner.[54] Amoebae transformed with a plasmid containing the gene for gp80 express greater amounts of gp80 on their surface and are able to bind more soluble gp80.[55] These results lead to the inescapable conclusion that gp80-mediated cohesion is homophilic.

D. STRUCTURE–FUNCTION ANALYSIS OF gp80

1. Predicted Structure of gp80

The cDNA for gp80 has been cloned and sequenced.[47,56] The deduced mol wt of the protein portion of gp80, minus its leader sequence of 19 amino acids, is 51.5 kD.[47,56] The carboxy terminus is composed of

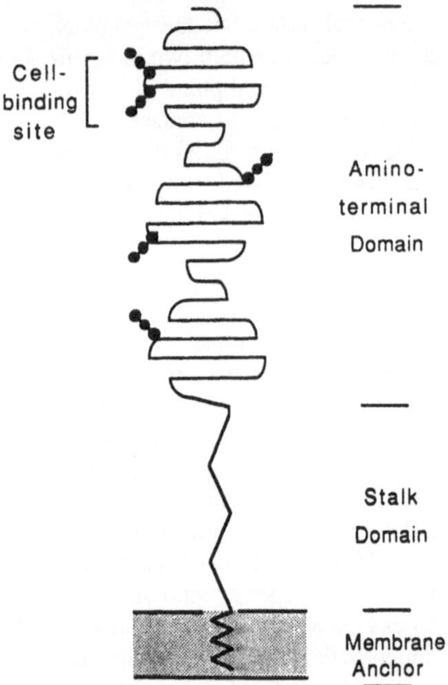

Figure 3 Model of gp80 showing the three homologous segments in the amino-terminal domain, supported by a stalk region and anchored in the plasma membrane (stippled bar) by the carboxy-terminal domain. The closed circles designate putative carbohydrate chains. (From Siu, C.-H. and Kamboj, R. K., *Dev. Genet.*, 11, 381, 1990, with permission.)

18 hydrophobic amino acids and may span the plasma membrane. This region has 30% sequence identity when it is compared to the carboxy terminal region of the *ssd* form of the chick neural cell adhesion molecule (NCAM).[57] Like the *ssd* form of NCAM,[58] at least some of the gp80 is anchored to the membrane via a covalently linked lipid moiety.[59,60] Whether this putative membrane-spanning region of gp80 is cleaved following the addition of the lipid anchor has not been determined. Adjacent to this putative membrane-spanning region is a 65 amino acid segment rich in proline and hydroxyamino acids; there are 9 Pro-Thr/Ser dipeptides in this region.[47,56,57] The analysis conducted by Siu and Kamboj,[57] using the method of Chou and Fasman,[61] predicts that this region has a "kinky" structure and might act like an extended stalk. This Pro-rich region shares substantial homology with a cytoplasmic region of NCAM and the hinge region of the human IgA-1 chain.[57] Because of the homology with the cytoplasmic region of NCAM, Siu and Kamboj used antibody and protease studies to confirm that this putative stalk region is extracellular.[57] The final domain contains 434 amino acids and five potential *N*-glycosylation sites, and consists of three homologous segments.[57] Each segment contains numerous short stretches of hydrophobic amino acids alternating with hydrophilic residues.[57] The composite hydropathy values for these segments are comparable to those of a globular protein. The analysis predicts that these segments are rich in β-sheets, with the sheets separated by regions with a high probability for turns.[57,61] It has been suggested that this outer domain consists of antiparallel β-sheets[57] similar to those characteristic of immunoglobulin domains.[62] When the sequence of this outer domain is compared to that of NCAM, several short regions of homology are observed. A model depicting the suggested structure of gp80 is shown in Figure 3. The circular dichroism spectrum of gp80 reveals a protein rich in β-sheets,[57] consistent with the predicted structure. The homology with NCAM and immunoglobulins suggests that gp80 may be a relative of the members of the immunoglobulin superfamily.

2. Localization of the Cell-Binding Site

In order to begin to determine which section of gp80 is involved in the homophilic interaction, Kamboj and Siu mapped the location of the 80L5C4 epitope.[63] 80L5C4 is the monoclonal antibody that specifically binds gp80 and blocks cohesion. Using the ability of 80L5C4 to bind to proteolyzed fragments as an assay, it was determined that the epitope is within 172 amino acids of the amino terminus. Fusion proteins containing fragments of gp80 were generated, and the 80L5C4 epitope was mapped to a 51 amino acid segment between Val-123 and Leu-173.[63] These amino acids are contained in the domain most distal to the amoebal surface (Figure 3). This segment, Val-123 to Leu-173, is composed of a hydrophilic region of 21 amino acids followed by a relatively hydrophobic stretch of 30 amino acids. Asn-128 and Asn-135 are potential sites for *N*-glycosylation.

To further implicate the region between Val-123 and Leu-173 in gp80-mediated cohesion, fusion proteins that contain this region were generated.[56] These fusion proteins are able to bind to aggregation-competent amoebae in a dose-dependent, saturable manner,[56] and the binding of the fusion proteins blocks EDTA-insensitive amoeba–amoeba cohesion.[56] The binding of the fusion proteins is inhibited by pretreating the amoebae with 80L5C4. When the ability to bind to gp80 immobilized on nitrocellulose is examined, the fusion proteins compete with native gp80. A fusion protein that does not contain the section of gp80 between Val-123 and Leu-173 is not active in these assays. To determine whether the Val-123 to Leu-173 region contains the "gp80 receptor" as well as the binding site, it was necessary to examine whether fusion proteins that contain this region are able to interact with each other. Such proteins, immobilized on nitrocellulose, are able to bind soluble fusion protein in a dose-dependent, saturable manner.[56] Fusion protein that do not contain this section of the gp80 are not able to bind to each other in this assay. Thus, the homophilic gp80–gp80 interaction is a protein–protein interaction and the cell-binding and "receptor" region of gp80 is contained within the Val-123 to Leu-173 segment.

In order to more precisely define the cell-binding region of gp80, Kamboj et al. used synthetic peptides.[64] They had shown previously that the ability of gp80 to undergo a homophilic interaction is sensitive to trypsin.[63] There are no Arg residues between Val-123 and Leu-173, but there is one Lys at position 133. This Lys is within the more hydrophilic section of the segment, consistent with it being on the surface of the protein and participating in cell binding. Therefore Kamboj et al. constructed a peptide that corresponds to the region between Asn-128 and Ile-140 (NISGYKLNVNDSI). Using immunoblot analysis, they demonstrated that this peptide contains the epitope for 80L5C4.[64] The 13 amino acid peptide is able to inhibit binding between purified gp80 proteins. The peptide also is able to inhibit EDTA-insensitive cohesion between aggregation-competent amoebae. When the peptide is conjugated to bovine serum albumin (BSA) and attached to beads, these beads bind to aggregation-competent amoebae. In the presence of excess BSA, the peptide–BSA conjugate is able to bind specifically to peptide–BSA conjugate immobilized on nitrocellulose. The demonstration that this 13 amino acid peptide is able to bind specifically to itself suggests that the peptide contains the "receptor" region of gp80 as well as the binding region. Examination of the sequence of the 13 amino acid peptide led to the suggestion that the positively charged Lys-133 on one molecule of gp80 might interact with the negatively charged Asp-138 on another.[64] In order to test this, Kamboj et al. tested the ability of shorter peptides and peptides with amino acid substitutions to inhibit the binding of gp80 to aggregation-competent amoebae and to inhibit the EDTA-insensitive cohesion between aggregation-competent amoebae.[64] The octapeptide YKLNVNDS is able to significantly inhibit the gp80–gp80 interactions in both assays. Removing the Tyr-Lys (Y-K) or the Asp-Ser (D-S) from the octapeptide abolishes the inhibitory activity. The decapeptide, SGYKLNVNDS, inhibits the binding of gp80 to aggregation-competent amoebae and inhibits the EDTA-resistant cohesion that occurs between aggregation-competent amoebae. If the Lys (K) and the Asp (D) are interchanged or if the Tyr (Y) is replaced by an Ala (A), the resulting decapeptides do not inhibit the binding of gp80 to aggregation-competent amoebae and do not inhibit EDTA-resistant cohesion. These studies suggest that a Lys-Asp interaction is important in gp80-mediated cohesion. If the gp80–gp80 interaction is ionic, then the two peptides should not bind each other in the presence of high salt. Kamboj et al. found that 1*M* NaCl completely abolished the binding between the 13 amino acid peptides. 1*M* NaCl inhibits the binding of gp80 to immobilized gp80 by 50%. This interaction is further inhibited by the nonionic detergent, Triton X-100. However, once the peptides are bound to each other or two molecules of gp80 are bound to each other, 1*M* NaCl appears to stabilize the interaction, and disruption of the complexes requires the nonionic detergent. Thus, Kamboj et al. concluded that there is an ionic interaction between Lys-133 on one molecule and Asp-138 on another, and this interaction is stabilized by hydrophobic interactions. These interactions are primarily responsible for gp80–gp80 binding and EDTA-insensitive cohesion between aggregation-competent amoebae.

3. Posttranslational Modifications of gp80

gp80 is heavily glycosylated.[40] In immunoblot analysis, the unglycosylated form of the protein appears to have a mol wt of 53 kD.[65-67] This is consistent with the calculated mol wt of 51.5 kD. The oligosaccharides are added in two discrete units,[65,67-69] designated type 1 and type 2.[69] Type 1 carbohydrate is *N*-linked, and it is added cotranslationally.[65] Type 2 is added posttranslationally,[67] presumably in the Golgi apparatus, where the type 1 carbohydrate is sulfated.[65,67,70] Neither type 1 nor type 2 carbohydrate appears to play a primary role in gp80-mediated cohesion. However, the carbohydrates present on gp80 are important. Glycosylation, either type 1 or type 2, is necessary for transport of the protein to the cell surface,[66,67] and type 2 glycosylation is necessary to protect the protein from proteolytic cleavage while it is on the cell surface.[66]

gp80 is phosphorylated[71,72] and anchored to the membrane via a lipid moiety.[59,60] The phosphorylation is cotranslational, and the phosphate is attached to serine residues.[72] No function has been attributed to the phosphorylation. The observation that gp80 could be radiolabeled by the addition of radiolabeled *myo*-inositol, ethanolamine, myristate, or palmitate to developing amoebae[59,60] suggested that gp80 may be anchored to the membrane via a phosphatidylinositol glycan moiety. Attempts to release the radiolabeled palmitate from gp80 with phosphatidylinositol phospholipase C have been unsuccessful.[59,60] Treatment with nitrous acid and alkali indicate that the radiolabeled palmitate might be attached to gp80 via an amide linkage.[60] Sphingomyelinase from human placenta and from *Staphylococcus aureus* are able to partially release the labeled palmitate, an observation that has led to the suggestion that gp80 is anchored to the membrane via an inositol ceramide.[60] Enriched plasma membrane preparations of *D. discoideum* contain an activity that, under physiological conditions, is able to cleave the anchor and release some gp80 from the membrane.[59,73] The function of the lipid anchor has not been established. However, such an anchor would increase the lateral mobility of gp80 within the plane of the membrane, enabling it to move quickly to a point of cell–cell contact. The observation that gp80 can be released under physiological conditions is tantalizing because it suggests a mechanism for the rapid down-regulation of gp80-mediated cohesion.

E. DEVELOPMENTAL RELEVANCE OF gp80-MEDIATED COHESION

1. Reduction or Elimination of gp80-Mediated Cohesion

Several approaches have been used to block gp80-mediated cohesion in order to determine its relevance to the development program of *D. discoideum*. Siu and Kamboj have shown that the addition of a fusion protein that contains the gp80 cell binding site to aggregation-competent amoebae on a substratum blocks gp80-mediated cohesion and the formation of organized aggregation streams.[57] The small aggregates that are formed in the presence of the protein are unable to progress further through the developmental program. Adding fab directed against the 13 amino acid peptide that contains the gp80 binding site elicits similar behavior.[57] In the presence of the 13 amino acid peptide that contains the gp80 cell binding site, rudimentary, noncontiguous aggregation streams are formed.[57,64] The few aggregates that are formed do not complete development but eventually dissociate to form single cells. Thus, these studies suggest that gp80-mediated cohesion is important for organized aggregation and subsequent development.

Noegel and her colleagues have disrupted the gene which encodes gp80.[53] They report that the mutants do not produce gp80 and yet form normal aggregation streams and develop into regularly proportioned fruiting bodies. These results are not consistent with the observations of Siu and his colleagues. Dr. Angelika Noegel has kindly provided us with one of the mutants and its parent. If the amoebae are placed on the substratum immediately after harvest, prior to any development, the aggregation streams formed by the mutant and the parent are small, and no mutant phenotype is apparent (Figure 4A). This is the protocol used by Noegel and colleagues. Siu and colleagues allowed their amoebae to develop in suspension for 6 h prior to plating. If the amoebae are allowed to develop in suspension for 6 h prior to plating on a substratum, a distinct mutant phenotype is observed (Figure 4C,D). Like its parent (Figure 4B), the mutant forms large aggregation streams. However, in the absence of gp80, the large streams are unstable and segment. The amoebae within each segment aggregate, independent of the other segments, and the aggregates complete development. The resulting slugs and fruiting bodies are smaller than those of the parental strain but otherwise appear normal. Therefore, this data demonstrates that gp80 is important in maintaining the stability of the aggregation stream. However once the mutant has aggregated, the absence of gp80 does not appear to influence development. Thus, there is still a discrepancy between our results and those of Siu and his colleagues because they report that when gp80-mediated cohesion is blocked by antibodies, fusion proteins, or peptides, the small aggregates that form do not complete development.[57,64] These differences may suggest that the amoebae are able to compensate for a lack of gp80 synthesis and thereby complete development, but they cannot adjust to a rapid abrogation of gp80-mediated cohesion.

2. Inappropriate Expression of gp80

D. discoideum amoebae have been transformed with an expression vector for gp80.[55,74,75] When expressed in growing cells, the gp80 is synthesized, properly glycosylated and anchored in the plasma membrane via the lipid moiety.[74,75] The transformed growing amoebae exhibit EDTA-insensitive cohesion.[55,74,75] Soluble gp80 binds to the transformed amoebae, and the antibody 80L5C4 blocks the binding.[55] Kamboj et al. observed that aggregates of their control strain tended to break down and form smaller aggregates.[55]

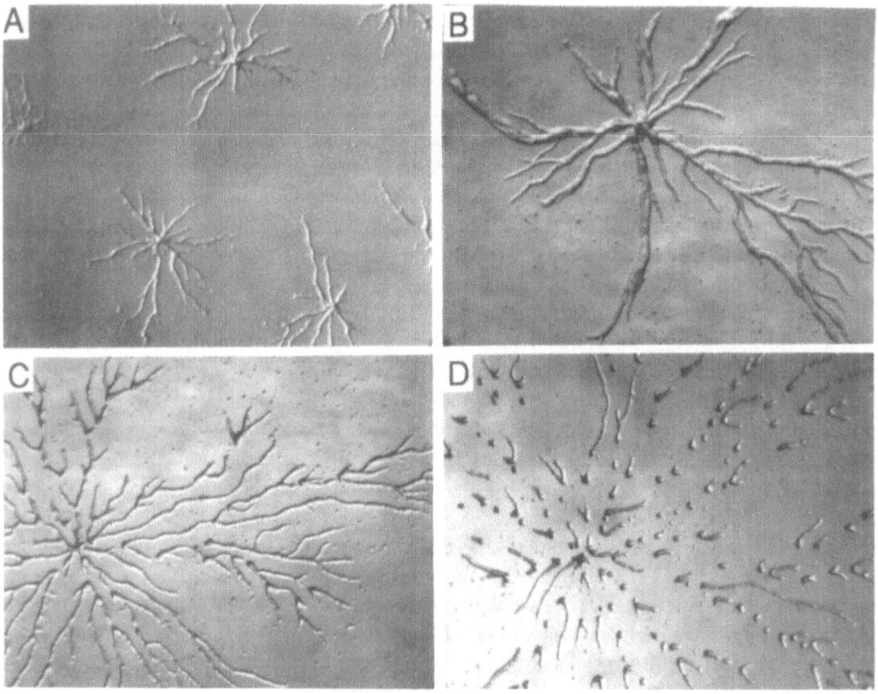

Figure 4 gp80 stabilizes large aggregation streams. Amoebae were harvested and plated immediately on nonnutrient agar (A) or allowed to develop for 6 h in suspension prior to plating (B,C,D). The amoebae were plated at a density of 1.3×10^5 cells per cm^2 and photographed at 200X magnification after the formation of aggregation streams. When amoebae were not allowed to develop prior to plating, the mutant T10 (shown in A) is indistinguishable from the parental strain, Ax-2 (not shown). When the amoebae were allowed to develop prior to plating, the parental strain formed large, stable aggregation streams (B). The mutant amoebae formed large aggregation streams, but these streams eventually segmented (C,D). The photograph shown in D was taken approximately 1 h after the one shown in C. The degree of segmentation seen in D is typical of that observed with the mutant.

The amoebae that had been transformed with the gp80 cDNA form aggregates that rarely break down. These larger aggregates develop into larger slugs. If the control strain is pulsed with cAMP during early development leading to an overproduction of gp80,[63,76] the control strain produces larger slugs. When comparing the length of the slugs formed by various laboratory strains of *D. discoideum,* Kamboj et al. found a strong positive correlation between the amount of gp80 they contained and their mean slug length. These results suggest that gp80-mediated cohesion participates in the stabilization of the aggregate as well as the aggregation stream. This conclusion is consistent with the observation of Loomis et al. that mutants with reduced levels of gp80 on their cell surface[66] form unstable aggregates that often divide before completing development.[77]

F. IDENTIFICATION OF CONTACT SITES B

Amoebae in the late exponential phase of axenic growth are able to cohere via an EDTA-sensitive cohesion system.[38] Amoebae that have been harvested during early exponential growth in axenic culture or amoebae grown in association with bacteria are not able to cohere immediately after harvesting.[2,78] If these amoebae are allowed to develop for 2 to 4 h, they demonstrate EDTA-sensitive cohesion.[2,78] Current evidence suggests that they use the same EDTA-sensitive cohesion system employed by amoebae harvested during late exponential growth,[2] and the cohesion molecules that mediate this cohesion have been designated contact sites B.[38]

1. gp126

Two proteins have been advanced as participants in contact sites B-mediated cohesion. Both have been identified using immunological methods. Chadwick and Garrod have proposed that a glycoprotein of 126 kD participates in cs B-mediated cohesion.[79,80] To identify this protein, antisera were generated against whole amoebae and amoebal extracts. Fab prepared from each type of antiserum blocks vegetative amoeba–amoeba cohesion and partially blocks cohesion between aggregation-competent amoebae. Intact vegetative amoebae were radioiodinated and extracted with detergent, and the extracts were immunoprecipitated with representatives of each type of antiserum. Those iodinated proteins that are present in all the immunoprecipitates were identified. Gel fragments, each containing one of these proteins, were tested for their ability to adsorb out the cohesion-blocking fab. A fragment that contains a 126 kD protein is able to adsorb the cohesion-blocking fab. The 126 kD protein is present in axenically grown amoebae, but its level is dramatically reduced in amoebae growing in association with bacteria,[80] observations consistent with gp126 participating in cs B-mediated cohesion.

In order to generate fab more specific to gp126, amoebae were extracted with detergent, and the extract was subjected to immunoprecipitation by an antiserum raised against lyophilized amoebae.[79] The immunoprecipitate was subjected to SDS-PAGE, and the 126 kD region was eluted and used as an immunogen. When examining surface proteins that can be iodinated when the amoebae are intact, the more specific fab recognizes only gp126.[80] The more specific fab is able to block cohesion,[79] the ingestion of *Escherichia coli*,[80] and adhesion to a glass substratum.[80] Control fab exhibits no blocking in these assays. Cells are able to complete development in the presence of the inhibitory fab, with the only obvious distinction being that the slugs and fruiting bodies are smaller than those of the controls.[80] The 126 kD protein and the ability of whole cells to adsorb the more specific cohesion-blocking fab persist in development through the period of slug migration.[81] When the prestalk and prespore cells are separated, the prestalk cells and not the prespore cells are able to adsorb the cohesion-blocking fab. These results suggest that gp126 is on the cell surface when cs B-mediated cohesion is observed, and it may participate in phagocytosis, cell-substratum adhesion, cs B-mediated cohesion, and the sorting of prestalk and prespore cells. The major difficulty with these studies is that the authors do not establish that their more specific serum is truly specific for the 126 kD glycoprotein. It may be that the antiserum cross-reacts with a cell surface protein that has no exposed tyrosine residues and therefore cannot be iodinated. Therefore, while the data suggest that gp126 participates in cs B-mediated cohesion, more needs to be done to confirm this tentative conclusion.

2. gp24

Loomis and colleagues have implicated a 24 kD glycoprotein in cs B-mediated cohesion. Initially, they immunized a rabbit with a membrane fraction isolated from a mutant that possesses reduced levels of gp80.[2,66] The resulting serum blocks cs B-mediated cohesion. A subset of membrane proteins was extracted with butanol and fractionated using DEAE-cellulose and Sephacryl S-200 chromatography. The fractions that are able to adsorb the inhibitory antibodies were identified.[2] SDS-PAGE revealed that all fractions that could adsorb the inhibitory activity contain a 24 kD protein. This protein was excised from the gel, and this gel-purified protein was able to adsorb the antibodies that inhibit cs B-mediated cohesion. Because the protein has a mol wt of 24 kD and binds specifically to Concanavalin A-Sepharose, it has been designated gp24.[2] Gel-purified gp24 was used as an immunogen, and as assessed by immunoblot analysis, highly specific anti-gp24 was generated. IgG from the specific serum is able to block cs B-mediated cohesion.

The appearance of the 24 kD glycoprotein is regulated in a manner consistent with it being involved in cs B-mediated cohesion.[2] It is present in amoebae growing axenically but not present in amoebae growing in association with bacteria. Within hours after the amoebae are washed free of their bacterial food source, gp24 appears. Thus, the appearance of gp24 coincides with the expression of cs B-mediated cohesion. When amoebae are grown on a bacterial food source, synthesis of gp24 is maximal during the first 4 h of development but continues at a lower level throughout the entire developmental program.

If amoebae are allowed to develop on filters, the addition of the IgG that eliminates cs B-mediated cohesion and is highly specific for gp24 blocks aggregation.[82] If this effect is specific for this IgG, it suggests that cell surface gp24 may be essential for aggregation. gp24 can be detected in both prestalk and prespore cells. Therefore if gp24 is sufficient for cs B-mediated cohesion, cs B-mediated cohesion may participate in the regulation of *D. discoideum* development after aggregation.

In summary, there are two independent cohesion systems responsible for amoeba–amoeba cohesion prior to *D. discoideum* aggregation. During the life cycle, the first apparent cohesion is mediated by cs B. This cohesion is observed in the late exponential phase if the amoebae are growing axenically and appears a few hours after the initiation of development if amoebae are grown in association with bacteria. Two glycoproteins have been advanced as participants in cs B-mediated cohesion, gp24 and gp126.

Shortly before aggregation, cs A-mediated cohesion becomes apparent. Unlike cs B-mediated cohesion, cs A-mediated cohesion is not sensitive to EDTA. A glycoprotein with an approximate mol wt of 80 kD is cs A. Cs A-mediated cohesion is homophilic and the result of a protein–protein interaction. Deleting gp80 results in the breakdown of large aggregation streams and large aggregates.

IV. COHESION AFTER AGGREGATION

After aggregation, cs A-mediated EDTA-resistant cohesion is supplanted by a new, seriologically distinct cohesion system that appears to participate in maintaining the integrity of the multicellular forms of *D. discoideum*.[83-85] Two proteins have been implicated as mediators of postaggregate cohesion, gp95 and gp150.

A. gp95

When *D. discoideum* slugs are disaggregated and the cells placed on a substratum, the cells reaggregate and complete development. If the slugs are disaggregated and the cells shaken in suspension in the presence of 10 m*M* EDTA, large aggregates are observed. If fab directed against components on the slug's plasma membrane is added to a suspension of disaggregated slug cells, reaggregation is inhibited.[83] A butanol extract of slug plasma membranes is able to adsorb the fab, which blocks EDTA-resistant slug cell cohesion.[83] Fab directed against the plasma membranes of vegetative or aggregation-competent amoebae has no effect in this assay, and vegetative and aggregation-competent amoebae are not able to adsorb the cohesion blocking fab.[83] Thus, the antiplasma membrane fab is blocking an EDTA-resistant slug cohesion system that is distinct from cs A.

Using antibodies directed against slug plasma membrane components, Parish et al. examined the major antigens in order to establish whether their synthesis varies as a function of stage in development. A 95 kD glycoprotein is synthesized at times consistent with its being involved in slug cell cohesion.[86] Amoebae begin to synthesize gp95 during late aggregation. Its rate of synthesis increases when the mound forms a tip and continues high through the remainder of development. gp95 is also the major antigen in the butanol extract that is able to adsorb the cohesion-blocking fab.[83] Slices of an SDS-polyacrylamide gel that contain gp95 are able to adsorb the cohesion-blocking fab. The remaining fab, greater than 98% of the total, has no effect on cohesion. Gel slices containing the remaining proteins do not adsorb cohesion-blocking fab. These results suggest that gp95 may participate in cohesion within the *D. discoideum* slug.

Mutant studies have further implicated gp95 in postaggregate cohesion. Wilcox and Sussman isolated a temperature-sensitive mutant, JC-5, which following aggregation at its nonpermissive temperature disperses into a smooth lawn of cells.[87] Cohesion assays with cells and ghosts suggest that JC-5 contains an unstable membrane-bound moiety that is required for the maintenance of EDTA-resistant cohesion. Immunological studies suggest that this moiety appears on the membrane surface after the amoebae have aggregated.[84] A comparison of the plasma membrane proteins of JC-5 and its parental strain using SDS-PAGE suggests that the unstable component is a minor component.[87] If these gels are stained with a [^{125}I]-labeled lectin, wheat germ agglutinin or Concanavalin A, it becomes apparent that the intensity of three protein bands is regulated by temperature in a manner consistent with any or all of them being involved in the temperature-labile cohesion. One of these bands, the one stained with wheat germ agglutinin (WGA), is a glycoprotein with a mol wt of 95 kD. In order to determine whether this 95 kD glycoprotein is the labile moiety in JC-5, plasma membrane proteins were fractionated on a WGA column. The addition of the WGA-reactive plasma membrane proteins harvested from JC-5 aggregates that were developing at the permissive temperature to JC-5 developing at the nonpermissive temperature rescues cohesion.[88] To identify the active component, the WGA-reactive preparation was fractionated by size using SDS-PAGE. Proteins from two gel slices possess the ability to rescue cohesion. One of the slices appears to contain a single component, a 95 kD glycoprotein, while the other slice contains several proteins with mol wts between 40 and 50 kD. Only the 95 kD protein is present at a reduced level in WGA-reactive plasma membrane protein preparations made from JC-5 harvested at the nonpermissive temperature when compared to preparations made from JC-5 harvested at the permissive temperature. The amount of gp95 recovered from the parental strain is temperature-independent, an observation consistent with gp95 being the labile component in JC-5. These observations suggest that gp95 is a cohesion molecule essential in maintaining the integrity of the multicellular forms of *D. discoideum*. However because the mutation in JC-5 is not in the structural gene for gp95,[89] it is possible that gp95 plays an indirect role in cohesion. More work needs to be done to establish conclusively that gp95 is a cohesion molecule.

B. gp150

gp150 was first noticed in a study that examined cell surface, Concanavalin A-binding proteins as a function of development.[90] Little gp150 is detected in vegetative amoebae, but its level increases dramatically at aggregation and remains high throughout the rest of the developmental program. Specific anti-gp150 was prepared by cutting a band containing gp150 from an SDS-polyacrylamide gel and injecting the gel-purified gp150 into rabbits.[91] When *D. discoideum* is dissociated just prior to fruiting body formation, the addition of anti-gp150 fab in the presence of EDTA blocks reassociation.[91] When the ability of anti-gp80 and anti-gp150 to block EDTA-resistant cohesion was compared as a function of development, it was observed that anti-gp150 is a weak inhibitor of EDTA-resistant cohesion during late aggregation when anti-gp80 is a potent inhibitor.[85] As development proceeds, the ability of anti-gp80 to inhibit EDTA-resistant cohesion drops rapidly. During this period, anti-gp150 becomes a potent inhibitor of EDTA-resistant cohesion. Fab generated from a polyspecific serum directed against the surface of vegetative amoebae and fab directed against a 130 kD glycoprotein present throughout development do not inhibit cohesion significantly in this assay. By performing indirect immunofluorescence microscopy and transmission electron microscopy, Geltosky et al. demonstrated that gp150 is associated with the plasma membrane and on the cell surface.[91,92] When aggregates that were about to form fruiting bodies were examined using scanning electron microscopy, it was observed that gp150 is preferentially localized at points of cell–cell contact.[92] Thus, the cellular location of gp150 and the ability of anti-gp150 to block EDTA-resistant cohesion suggest that gp150-mediated cohesion replaces gp80-mediated cohesion after *D. discoideum* has aggregated. More recent data suggest that gp150 is principally responsible for mediating cohesion between prespore cells and may participate in the sorting of prestalk and prespore cells.[93,94] This will be discussed in greater detail in Section V.D.

It is not known whether there is any relationship between gp95 and gp150. Even though the developmental time course for the appearance of gp150 suggests that it should be on the cell surface during the slug stage,[90,91] Steinemann and Parish found no evidence for gp150 or gp150-mediated cohesion while performing their studies on gp95.[83] They attempted to adsorb the cohesion-blocking anti-gp95 fab with a gel slice that contained proteins with mol wts of approximately 150 kD, but were unsuccessful.[83] Perhaps gp95 alone mediates EDTA-resistant cohesion during slug migration, and the developmental conditions under which gp150-mediated cohesion was studied did not result in prolonged slug migration. Also, nothing in these studies precludes the possibility that gp95 and gp150 interact with each other in order to mediate postaggregation EDTA-resistant cohesion in *D. discoideum*.

V. FUNCTIONAL EFFECTS OF COHESION

A. COHESION MEDIATES INTERSPECIES SORTING

In 1902, Olive reported that when two species of slime molds are mixed during growth, they are able to sort themselves out and form separate fruiting bodies.[95] Two mechanisms are known to mediate this segregation, differential cohesion and chemotaxis to different chemoattractants. Dictyostelium and Polysphondylium are two genera in the Dictyosteliaceae family.[96] When vegetative *D. discoideum* are mixed with vegetative *Polysphondylium violaceum* and placed on a substratum, they aggregate independently.[97] Streams of aggregating *D. discoideum* can cross streams of aggregating *P. violaceum* without disrupting either stream. The independent aggregation has been attributed to the distinct nature of their chemoattractants; *P. violaceum* uses a modified dipeptide, not cAMP, as a chemoattractant.[98] When cohesion is examined directly using amoebal suspensions, species of Dictyostelium are able to cohere to species of Polysphondylium.[99-101] If the mixed aggregates are allowed to remain in a shaken suspension, they sort themselves and form genus-specific regions within the aggregate. These sorted aggregates are unstable, and if left in suspension, they break down, forming aggregates composed of a single genus.

Gerisch and colleagues examined the segregation of *D. discoideum* and *P. pallidum* in greater detail.[101] *P. pallidum* appears also to have two cohesion systems that function prior to aggregation, with one present on vegetative amoebae and the other appearing as the amoebae begin to aggregate.[102] If vegetative *D. discoideum* and *P. pallidum* amoebae are mixed in a suspension lacking nutrients, they form a common aggregate.[101] Sorting is only observed after the amoebae have been in suspension for 4 h, with 4 h the length of time that this strain of *D. discoideum* takes to becomes aggregation-competent. If aggregation-competent *D. discoideum* and *P. pallidum* are mixed, sorting occurs within 1 h. Chemotaxis can be inhibited by the addition of 2,4-dinitrophenol, and sorting still occurs. If the fab that blocks cs A-mediated cohesion in *D. discoideum* is mixed with a comparable fab generated with *P. pallidum* and this mixture is added to the suspension of *D. discoideum* and *P. pallidum* amoebae, a mixed aggregate forms, but sorting does not occur. The addition of an anticarbohydrate fab does not block aggregate formation or sorting. These results suggest the following: (1) cohesion between vegetative *D. discoideum* and *P. pallidum* is not specific; (2) cohesion mediates the sorting observed with suspended mixtures of aggregation-competent *D. discoideum* and *P. pallidum;* (3) contact site A-mediated cohesion and its counterpart in *P. pallidum* are necessary for the sorting.

Members of the genus Dictyostelium sort when mixed. When vegetative *D. discoideum* and *D. mucoroides* are mixed and allowed to develop on a substratum, they form common aggregates but separate into species-specific entities prior to fruiting body formation.[97] The common aggregate is formed because both species of Dictyostelium use cAMP as a chemoattractant during aggregation.[103,104] The sorting that occurs after aggregation has been attributed to differential cohesion. When aggregation-competent *D. discoideum, D. mucoroides,* and *D. purpureum* are used to form binary mixtures in suspension, they form common aggregates and subsequently sort so that the aggregates contain regions that are populated predominantly by one species.[99,100] All these species use cAMP as a chemoattractant during aggregation,[103,104] and therefore, differences in cohesion are indirectly implicated as the source of the sorting behavior. When cohesion between aggregation-competent *D. discoideum* and *D. purpureum* is examined directly using amoebae immobilized on a dish, it is observed that *D. discoideum* preferentially coheres to *D. discoideum* and *D. purpureum* preferentially coheres to *D. purpureum.*[105] The degree to which the interaction is species-specific depends upon the magnitude of the applied shear force, with the species-specific cohesion being stronger than the nonspecific cohesion. Thus far, the molecular details of this interspecies sorting have not been determined.

B. CONTACT-MEDIATED REGULATION OF GENE EXPRESSION

Following aggregation, *D. discoideum* form compact mounds that contain cells in close contact with each other. Few intercellular spaces are observed within these mounds. The close contact that occurs in the compact mounds has been implicated as a regulator of developmental gene expression. An examination of changes in the synthesis of cellular proteins during development revealed that the formation of compact mounds results in a major shift in the pattern of protein synthesis.[106,107] If the cells contained in the mounds are dispersed, the accumulation of some of the contact-induced proteins stops.[108,109] Upon reaggregation, a new round of induction occurs.[108,110] When developmentally regulated mRNA species were examined, a similar picture emerged.[109,111,112] When compact mounds are dispersed, there is a rapid drop in the level of many of the mRNAs species that appear concomitant with the formation of the mounds. Upon reaggregation, the levels of these mRNAs rise. Changes in transcription and mRNA stability have been implicated in the loss of the mRNA.[111,113,114] With some mRNAs, the decrease upon dispersal can be reversed by the addition of cAMP.[111,112,114-117] In contrast to these proteins and mRNA species, the level discoidin I mRNA decreases when the amoebae form compact mounds.[119] If the compact mounds are dispersed, the level of the mRNA rises. Disaggregation in the presence of cAMP results in continually depressed levels of discoidin I mRNA.[119]

Several mutants that aggregate but do not form compact mounds have been examined. Genes whose inductions are correlated with compact mound formation are not expressed in mutants that form only loose mounds.[118] Mutants that form compact mounds but do not proceed through development express contact-dependent mRNA species. Discoidin I mRNA is overexpressed in a mutant that forms loose mounds but not compact mounds.[120] It is possible that the formation of the compact mound increases the level of a diffusible metabolite and this metabolite, not cell–cell cohesion, controls mRNA levels. However, because mutants that form loose mounds do not exhibit contact-mediated modulation of mRNA levels, it appears that cohesion and not just proximity is necessary to regulate gene expression. It has been proposed that, with regard to gene regulation, cAMP is the second messenger for tight cell–cell contact.[119-121]

The proposed role of close cell–cell contact in the regulation mRNA synthesis and stability has been controversial. The differentiation of amoebae plated in a dispersed monolayer on a substratum into vacuolated cells that resemble stalk cells can be achieved in the absence of cell–cell contact if a high concentration of cAMP and a morphogen, differentiation-inducing factor, is supplied.[122-126] If a permeant cAMP analogue is supplied, dispersed amoebae can be induced to form spores in the absence of cell–cell contact.[126] Thus, stalk cells and spores can be generated in the absence of cell–cell contact. However, in these monolayer experiments, the investigators are artificially controlling the intracellular and extracellular cAMP concentrations because cell-type-specific gene expression requires intracellular and extracellular cAMP at specific times during the developmental program.[127] In the aggregate, cell–cell contact may regulate intracellular and extracellular cAMP concentrations and thereby ensure appropriate gene expression and mRNA stability (see next section).

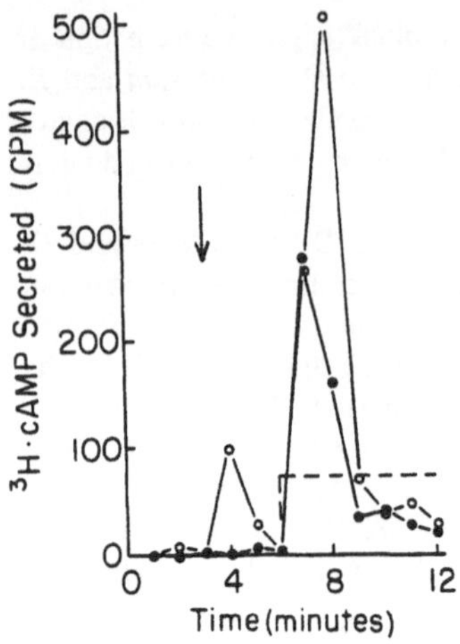

Figure 5 Contact induces transient cAMP secretion and alters the magnitude of a subsequent cAMP relay response. *D. discoideum*, strain NC-4, was used. 1×10^6 aggregation-competent amoebae, labeled with [^{3}H]adenosine, were placed on filters and perfused with buffer. At $t = 3$ min, either buffer (closed symbols) or 1×10^8 methanol-fixed *Enterobacter aerogenes* (open symbols) was added. At $t = 6$ min, the amoebae were stimulated with 1 μM cAMP (represented by the dashed lines). The amount of secreted [^{3}H]cAMP was determined[131] and plotted as a function of time.

C. COHESION AND ADHESION REGULATE cAMP RELAY

After observing that cross-linking the surface of aggregation-competent amoebae with lectins, various antibodies, or chemical cross-linking agents inhibits cAMP-induced adenylyl cyclase activation,[128] Fontana and Devreotes postulated that cell–cell contact might cross-link membrane components and modulate cAMP relay. If cell–cell contact could modulate the cAMP relay response, then cell–cell contact might regulate *D. discoideum* gene expression and mRNA stability by altering intracellular and extracellular cAMP concentrations. It was also suggested that bacterial–amoebal contact might inhibit cAMP relay, and this inhibition would facilitate the reversal of *D. discoideum* development following refeeding.

In the initial studies, the effect of cell–cell contact on cAMP relay was assayed directly using a perfusion device.[129,130] Aggregation-competent amoebae, labeled with [^{3}H]adenosine, were placed on filters such that they occupied only 0.1 to 0.2% of the available space within the filters. The amoebae were perfused with buffer and contact was initiated by the addition of methanol-fixed *E. aerogenes*, methanol-fixed *D. discoideum* amoebae, or latex beads to the perfusion filter. The cAMP relay response was stimulated by replacing the buffer with buffer containing 1 μM cAMP. The perfusate was collected and analyzed for [^{3}H]cAMP.[131] With this assay, the stimulus for the contact-mediated events could be the physical perturbation that results when two cells or a cell and a particle collide, or the stimulus could be cohesion or adhesion.

Cell–cell contact elicits two responses from aggregation-competent amoebae (Figure 5). These are (1) the immediate, but transient, secretion of cAMP and (2) an alteration in the magnitude of a subsequent cAMP relay response.[129,130,132] The initial characterization of these responses was conducted using methanol-fixed *E. aerogenes* to initiate contact.[129,130] With this stimulus, the subsequent cAMP relay response is enhanced when the buffer is rapidly replenished, and the cAMP relay response is diminished when the buffer is exchanged slowly.[130] These results demonstrate that cell–cell contact is able to regulate extracellular cAMP concentrations. Furthermore, the results suggest that a soluble molecule participates in the contact-induced alteration of cAMP relay.[130] In nature, *D. discoideum* amoebae do not experience a rapid removal of secreted products. Thus, under these conditions, bacteria–amoeba contact would inhibit cAMP relay. This observation is consistent with the hypothesis that the reversal of development by refeeding could be facilitated by a contact-mediated inhibition of cAMP relay.

Contact-induced cAMP secretion and contact-induced alterations in cAMP relay are independent responses, in that each can occur in the absence of the other.[130] Contact-induced cAMP secretion is not the result of a specific interaction because latex beads, charged and neutral, are able to elicit amoebal secretion of cAMP in an amount comparable to that observed following *E. aerogenes*–amoeba and amoeba–amoeba contact.[130] The addition of latex beads does not alter the magnitude of a cAMP relay response, suggesting more specificity in this contact-mediated response. Experimental conditions that affect the magnitude of a cAMP relay response or the number of beads ingested during a phagocytosis assay do not alter the magnitude of either of the contact-induced responses.[130] This suggests that the two contact-induced responses are separable from phagocytosis and cAMP relay.

More recently, cAMP relay has been studied using aggregation-competent amoebae in a buffered suspension.[133,134] In the suspension assay, 'contact' is initiated by the addition of fixed *E. aerogenes* or fixed *D. discoideum* to a suspension of aggregation-competent amoebae. The addition of *E. aerogenes*

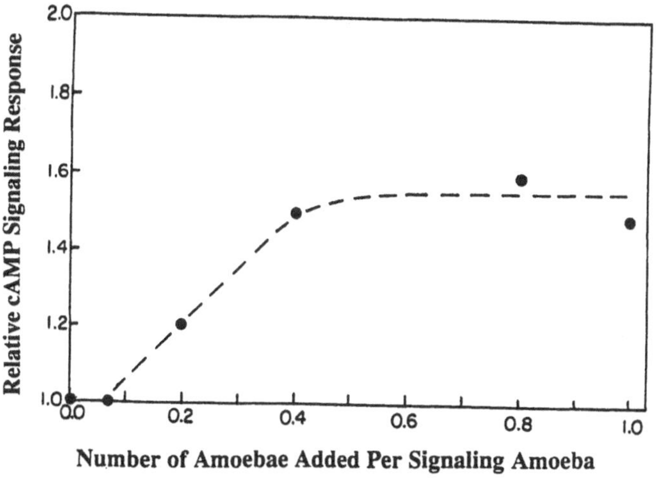

Figure 6 Amoeba–amoeba cohesion enhances cAMP relay. Aggregation-competent amoebae, strain NC-4, were suspended in buffer so that the final concentration was 5×10^7 cells per ml. Recently harvested amoebae, capable of cs B-mediated cohesion, were fixed in 100% methanol, washed, and added to the flasks in the amounts indicated. The amoebae were stimulated with a cAMP analogue, and the amount of cAMP produced was quantitated with an isotope dilution assay.[133]

or *D. discoideum* to the suspension does not elicit the secretion of cAMP, but the addition of either enhances the subsequent cAMP relay response (Figure 6). Because the methanol-fixed amoebae are smaller than living amoebae, it can be observed microscopically that these amoebae are cohering to their nonfixed counterparts. Because contact-induced cAMP secretion does not depend on the nature of the particle added to initiate contact and because contact-induced cAMP secretion is not observed when *E. aerogenes* or *D. discoideum* is added to an amoebal suspension, contact-induced cAMP secretion may be a response to the mechanical stimulation caused by contact, i.e., deformation of the plasma membrane or membrane skeleton may result in cAMP secretion. Because the addition of *E. aerogenes* or *D. discoideum* to a suspension of aggregation-competent amoebae elicits the other contact response, the enhancement of a subsequent cAMP relay response, this enhancement of cAMP relay must be a response to *E. aerogenes*–amoeba adhesion and amoeba–amoeba cohesion.

The correlation between the degree of cohesion within a suspension of aggregation-competent *D. discoideum* and the magnitude of their cAMP relay response was strengthened when various modulators of cAMP relay were found to influence cohesion.[134] Caffeine inhibits cAMP-induced activation of the adenylyl cyclase without altering cell viability, cAMP binding, cellular ATP and GTP levels, phosphodiesterase activity, the cAMP-induced rise in cGMP, or chemotaxis to cAMP.[135] Caffeine inhibits amoeba–amoeba cohesion in the same concentration range in which it inhibits cAMP relay.[134] Total inhibition of cAMP relay is accompanied by the total inhibition of cohesion. The addition of membranes harvested from aggregation-competent amoebae to a suspension of aggregation-competent amoebae inhibits a subsequent cAMP relay response.[132,134] When the membranes are fractionated, the inhibitory activity resides in the lipid fraction. Concentrations of membrane lipid that inhibit cAMP relay inhibit cohesion.[134] At membrane lipid concentrations that completely inhibit cAMP relay, no cohesion is observed. At this lipid concentration, there is no inhibition of cAMP binding or basal adenylyl cyclase activity or any alteration in basal cGMP levels.[134] The lipid-treated amoebae still are capable of cAMP-induced guanylyl cyclase activation, but lipid addition totally eliminates cAMP-induced activation of the adenylyl cyclase. With further fractionation, it was observed that the membrane phospholipids are weak inhibitors of cAMP relay and weak inhibitors of cohesion.[134] The membrane glycolipids enhance cohesion and cAMP relay. The neutral lipids are potent inhibitors of cohesion and cAMP relay. Thus, treatments that decrease cohesion decrease the magnitude of the cAMP relay response. The treatment that increases cohesion enhances cAMP relay. Caffeine and membrane lipid addition both block cAMP-induced activation of the adenylyl cyclase without inhibiting cAMP binding, suggesting that cohesion may modulate a step in the transduction pathway between the occupied cAMP receptor and the adenylyl cyclase.

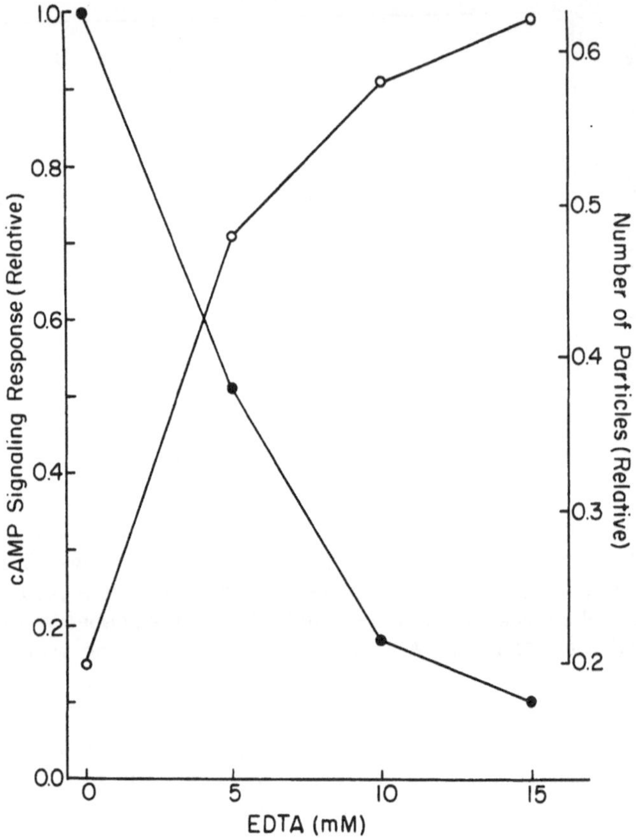

Figure 7 EDTA inhibits cAMP relay and cohesion with comparable dose–response curves. Aggregation-competent amoebae, strain NC-4, were suspended in a phosphate buffer so that the final concentration was 5×10^7 cells per ml. EDTA was added, and the amoebae were shaken gently for 15 min. At this time, the amoebae were either stimulated with a cAMP analogue and the amount of cAMP produced was determined (closed symbols), or the amoebae were diluted to 5×10^4 amoebae per ml and the number of particles per ml assayed with a particle counter (open signals). The relative cAMP relay response was calculated by dividing the amount of cAMP produced by the EDTA-treated amoebae by the amount of cAMP produced by amoebae suspended in buffer not containing EDTA. The particle counter determined the total number of particles present and did not discriminate between single cells and multicellular aggregates. The relative number of particles per ml was calculated by dividing the number of particles by the total number of amoebae present.

Because the aggregation-competent amoebae are cohering in suspension, adding more amoebae does not greatly increase the amount of cohesion experienced by a single aggregation-competent amoeba. Therefore, adding fixed amoebae to aggregation-competent amoebae in suspension will not greatly enhance the cAMP relay response, even if cohesion is an important modulator of the response. The effect of amoeba–amoeba cohesion on the magnitude of the cAMP relay response is illustrated best by observing the effect of disrupting the cohesion on the magnitude of the cAMP relay response. Some of these experiments have been performed using EDTA to inhibit cs B-mediated cohesion. EDTA inhibits cAMP relay (Figure 7). To allay the concern that the inhibition of cAMP relay by EDTA is the result of a direct effect of EDTA on cAMP relay and not the result of the EDTA-mediated inhibition of cs B, the dose–response curves for the EDTA-mediated inhibitions of cohesion and cAMP relay were examined (Figure 7). The concentration of EDTA that inhibits 50% of cs B-mediated cohesion is approximately 4 m*M*, and the concentration of EDTA that inhibits 50% of the cAMP relay response is approximately 5 m*M*. The fact that the concentrations of EDTA that inhibit cohesion are the same as those that inhibit cAMP relay suggests that EDTA may inhibit cAMP relay through its inhibition of cs B-mediated cohesion. This conclusion was reinforced when the developmental time course of cohesion and cAMP

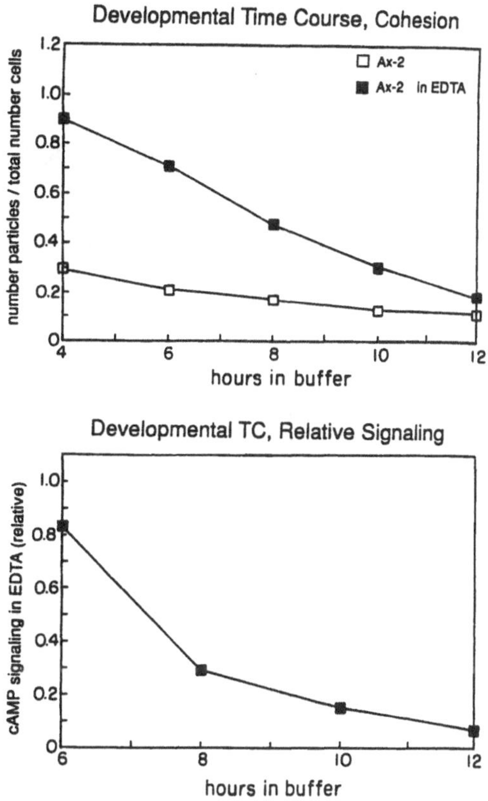

Figure 8 During *D. discoideum* development, the cAMP relay response progresses from a state where it is cohesion-independent to a state where it requires cs B-mediated cohesion. *D. discoideum* amoebae, strain Ax-2, were allowed to develop in a buffered suspension. At the various time points, amoebae were removed, washed, and suspended in buffer so that the final concentration was 5×10^7 cells per ml. EDTA was added to half of the samples. Following 15 min of gentle shaking, some of the amoebae were diluted and the number of particles per ml determined with a particle counter (top). Other amoebae were stimulated with a cAMP analogue, and the amount of cAMP produced was determined (bottom). The relative cAMP relay response was calculated by dividing the amount of cAMP produced by the EDTA-treated amoebae by the amount of cAMP produced by amoebae suspended in buffer not containing EDTA. cAMP relay was not detectable after only 4 h of development.

relay was examined (Figure 8). After 6 h of development in suspension, 10 m*M* EDTA is only a weak inhibitor of cAMP relay. As development proceeds, the cAMP relay response rapidly becomes sensitive to EDTA. By 12 h of development, cAMP relay is inhibited totally by EDTA. These results suggest that early in development, cAMP relay is relatively independent of cohesion. As development proceeds, the cAMP relay response becomes dependent upon cs B-mediated cohesion. This dependence is independent of the appearance of other EDTA-insensitive cohesion systems. These conclusions are being tested currently using anti-gp24 to block cs B-mediated cohesion. The observation that cAMP relay may become dependent upon cohesion as the amoebae enter aggregates supports the notion that cohesion may participate in the regulation of gene expression and mRNA stability by altering intracellular and extracellular concentrations of cAMP.

D. SORTING OF PRESPORE AND PRESTALK CELLS

In the *D. discoideum* slug, there is a segregation of prestalk and prespore cells, such that the anterior of the slug is composed of prestalk cells and the posterior contains the prespore cells. The various cell types become apparent in the aggregate, prior to tip formation, with cell sorting and positional information implicated in the formation of the pattern observed in the slug.[136,137] To study cell sorting under more defined conditions, *D. discoideum* slugs were disrupted and prestalk and prespore cells separated. When prestalk and prespore cells are mixed in suspension, they form a common aggregate and then sort into cell-type-specific zones within the aggregate.[138-140] Whether the prespore or the prestalk cells appear initially to be more cohesive seems to depend upon the method used to disaggregate the slug. When allowed to remain in suspension, the mixed aggregates form a sphere composed of a hemisphere of prestalk cells and a hemisphere of prespore cells. This sorting is blocked by the addition of fab directed against the putative late cohesion molecule, gp150,[140] suggesting that gp150 may participate in cell-type-specific sorting.

A role for cohesion in the maintenance of the pattern of prestalk cells and prespore cells within the slug has not been established. Examination of intact slugs reveals that the prestalk cells are in close contact.[141] In contrast, in the prespore region, the cells are loosely packed, leaving large intercellular spaces. This suggests that cohesion between prestalk cells is greater than the cohesion between prespore cells. This conclusion is consistent with the observation that the anterior of the slug is more difficult to dissociate than the posterior.[142] This difference in cohesion may assist in maintaining the patterns exhibited by the various cell types during *D. discoideum* morphogenesis.

VI. ACKNOWLEDGMENTS

I would like to thank Diane Schmidt for performing the dose–response curves that demonstrate the EDTA sensitivity of cAMP relay and cohesion. I would also like to thank Dr. Peter Plagemann for the use of his particle counter and Dr. Martin Dworkin for helpful discussion.

ADDENDUM

The research for this chapter was completed in July of 1991. Since that time a number of important publications have appeared. Barth et al. have reported that replacing the phospholipid anchor on gp80 with a transmembrane region does not alter gp80-mediated adhesion but reduces the amount of time a particular gp80 molecule resides on the cell surface (*J. Cell Biol.*, 124, 205, 1994). Faix et al. have observed that if gp80 is overexpressed, without altering the temporal regulation of its expression, aggregation is delayed and there is fragmentation of the aggregation streams (*J. Cell Sci.*, 102, 203, 1992). If the cells overexpressing gp80 are plated at high cell density, a substantial number of cells do not enter slugs or fruiting bodies. Regarding the nature of Cs B, Brar and Siu have reported that purified gp24 binds to cells in a dose-dependent, saturable manner (*J. Biol. Chem.*, 268, 24902, 1993). This binding can be blocked by the preaddition of anti-gp24 IgG and fab fragments suggesting that gp24-mediated adhesion is homophilic. These authors also demonstrated that gp24 binds Ca^{2+} and proposed that gp24-mediated adhesion is Ca^{2+}-based. I have discovered that two distinct systems are responsible for the EDTA-sensitive cohesion observed prior to *D. discoideum* aggregation (Fontana, D. R., *Differentiation*, 53, 139, 1993). Unlike Cs B-mediated adhesion, the second adhesion system is EGTA-resistant suggesting that the required cation is Mg^{2+}. The addition of Mg^{2+} rescues Cs C-mediated adhesion in the presence of EDTA. The molecules responsible for this second system have been designated Contact Sites C. It appears that cohesion via Contact Sites C, and not Contact Sites B, increases the amount of cAMP made during a cAMP relay response. In addition, the EDTA-insensitive relay reported in Figure 8 is observed only when amoebae that have recently germinated are studied. If the amoebae are allowed to develop after growing approximately 8 to 10 generations, all cAMP relay is EDTA sensitive. One of the surface glycoproteins implicated in the EDTA-insensitive adhesion observed following aggregation, gp150, has been purified (Gao, E. N., Shier, P., and Siu, C.-H., *J. Biol. Chem.*, 267, 9409, 1992). The purified protein binds to cells in a dose-dependent, saturable manner. Addition of the purified protein blocked the reassociation of postaggregation stage cells. The inhibition was protease sensitive, suggesting that the protein moiety is crucial for gp150 function.

REFERENCES

1. **Raper, K. B.,** *The Dictyostelids,* Princeton University Press, Princeton, NJ, 1984, chap. 2.
2. **Knecht, D. A., Fuller, D. L., and Loomis, W. F.,** Surface glycoprotein, gp24, involved in early adhesion of *Dictyostelium discoideum, Dev. Biol.,* 121, 277, 1987.
3. **Siu, C.-H., Lerner, R. A., Ma, G., Firtel, R. A., and Loomis, W. F.,** Developmentally regulated proteins of the plasma membrane of *Dictyostelium discoideum.* The carbohydrate binding protein, *J. Mol. Biol.,* 100, 157, 1976.
4. **Loomis, W. F.,** *Dictyostelium discoideum, A Developmental System,* Academic Press, New York, 1975, 59.
5. **Bonner, J. T.,** Comparative biology of cellular molds, in *The Development of Dictyostelium discoideum,* Loomis, W. F., Ed., Academic Press, New York, 1982, 1.
6. **Dinauer, M. C., MacKay, S. A., and Devreotes, P. N.,** Cyclic 3′,5′AMP relay in *Dictyostelium discoideum.* Part 3. The relationship of cAMP synthesis and secretion during the cAMP signaling response, *J. Cell Biol.,* 86, 537, 1980.
7. **Tomchik, K. J. and Devreotes, P. N.,** Adenosine 3′,5′-monophosphate waves in *Dictyostelium discoideum:* a demonstration by isotope dilution-fluorography, *Science,* 212, 443, 1981.
8. **Durston, A. J., Cohen, M. H., Drage, D. J., Potel, M. J., Robertson, A., and Wonio, D.,** Periodic movements of *Dictyostelium discoideum* sorocarps, *Dev. Biol.,* 52, 173, 1976.

9. **Schaap, P. and Wang, M.,** The possible involvement of oscillatory cAMP signaling in multicellular morphogenesis of the cellular slime molds, *Dev. Biol.,* 105, 470, 1984.
10. **Schaap, P.,** Regulation of size and pattern in the cellular slime molds, *Differentiation,* 33, 1, 1986.
11. **Schaap, P. and Spek, W.,** Cyclic-AMP binding to the cell surface during development of *Dictyostelium discoideum, Differentiation,* 27, 83, 1984.
12. **Schaap, P., Konijn, T. M., and van Haastert, P. J. M.,** cAMP pulses coordinate morphogenetic movement during fruiting body formation of *Dictyostelium minutum, Proc. Natl. Acad. Sci. U.S.A.,* 81, 2122, 1984.
13. **Schaap, P.,** cAMP relay during early culmination of *Dictyostelium minutum, Differentiation,* 28, 205, 1985.
14. **Schaap, P. and Wang, M.,** cAMP induces a transient elevation of cGMP levels during early culmination of *Dictyostelium minutum, Cell Diff.,* 16, 29, 1985.
15. **Owens, N. F., Gingell, D., and Bailey, J.,** Contact-mediated triggering of lamella formation by Dictyostelium amoebae on solid surfaces, *J. Cell Sci.,* 91, 367, 1988.
16. **Vogel, G., Thilo, L., Schwarz, H., and Steinhart, R.,** Mechanism of phagocytosis in *Dictyostelium discoideum:* phagocytosis is mediated by different recognition sites as disclosed by mutants with altered phagocytotic properties, *J. Cell Biol.,* 86, 456, 1980.
17. **Hellio, R. and Ryter, A.,** Role of particle electrostatic charge in adhesion and ingestion in *Dictyostelium discoideum* amoeboid cells, *J. Cell Sci.,* 79, 327, 1985.
18. **Todd, I., Mellor, J. S., and Gingell, D.,** Mapping cell-glass contacts of Dictyostelium amoebae by total internal reflection aqueous fluorescence overcomes a basic ambiguity of interference reflection microscopy, *J. Cell Sci.,* 89, 107, 1988.
19. **Vogel, G.,** Recognition mechanisms in phagocytosis in *Dictyostelium discoideum,* in *Endocytosis and Exocytosis in Host Defense,* Edebo, L. B., Enerbäck, L., and Stendahl, O. I., Eds., S. Karger, Basil, 1981, 1.
20. **Bozzaro, S. and Roseman, S.,** Adhesion of *Dictyostelium discoideum* cells to carbohydrates immobilized in polyacrylamide gels. I. Evidence for three sugar-specific cell surface receptors, *J. Biol. Chem.,* 258, 13882, 1983.
21. **Bozzaro, S. and Roseman, S.,** Adhesion of *Dictyostelium discoideum* cells to carbohydrates immobilized in polyacrylamide gels. II. Effect of D-glucose derivatives on development, *J. Biol. Chem.,* 258, 13890, 1983.
22. **Hellio, R. and Ryter, A.,** Relationships between anionic sites and lectin receptors in the plasma membrane of *Dictyostelium discoideum* and their role in phagocytosis, *J. Cell Sci.,* 41, 89, 1980.
23. **West, C. M., McMahon, D., and Molday, R. S.,** Identification of glycoproteins, using lectins as probes, in plasma membranes from *Dictyostelium discoideum* and human erythrocytes, *J. Biol. Chem.,* 253, 1716, 1978.
24. **Waddell, D. R., Duffy, K., and Vogel, G.,** Cytokinesis is defective in Dictyostelium mutants with altered phagocytic recognition, adhesion, and vegetative cell cohesion properties, *J. Cell Biol.,* 105, 2293, 1987.
25. **Owens, N. F., Gingell, D., and Rutter, P. R.,** Inhibition of cell adhesion by a synthetic polymer adsorbed to glass shown under defined hydrodynamic stress, *J. Cell Sci.,* 87, 667, 1987.
26. **Yabuno, K.,** Changes in cellular adhesiveness during the development of the slime mold *Dictyostelium discoideum, Dev. Growth Differ.,* 13, 181, 1971.
27. **Yabuno, K.,** Changes in electronegativity of the cell surface during the development of the cellular slime mold, *Dictyostelium discoideum, Dev. Growth Differ.,* 12, 229, 1970.
28. **Clarke, M., Kayman, S. C., and Riley, K.,** Density-dependent induction of discoidin-I synthesis in exponentially growing cells of *Dictyostelium discoideum, Differentiation,* 34, 79, 1987.
29. **Bartles, J. A., Santoro, B. C., and Frazier, W. A.,** Discoidin I-membrane interactions. III. Interaction of discoidin I with living *Dictyostelium discoideum* cells, *Biochim. Biophys. Acta,* 687, 137, 1982.
30. **Barondes, S. H., Cooper, D. N., and Haywood-Reid, P. L.,** Discoidin I and discoidin II are localized differently in developing *Dictyostelium discoideum, J. Cell Biol.,* 96, 291, 1983.
31. **Barondes, S. H., Springer, W. R., and Cooper, D. N.,** Cell adhesion, in *The Development of Dictyostelium discoideum,* Loomis, W. F., Ed., Academic Press, New York, 1982, 195.
32. **Springer, W. R., Cooper, D. N. W., and Barondes, S. H.,** Discoidin I is implicated in cell-substratum attachment and ordered cell migration of *Dictyostelium discoideum* and resembles fibronectin, *Cell,* 39, 557, 1984.
33. **Poole, S., Firtel, R. A., Lamar, E., and Rowekamp, W.,** Sequence and expression of the discoidin I gene family in *Dictyostelium discoideum, J. Mol. Biol.,* 153, 273, 1981.

34. **Crowley, T. E., Nellen, W., Gomer, R. H., and Firtel, R. A.,** Phenocopy of discoidin I-minus mutants by antisense transformation in Dictyostelium, *Cell,* 43, 633, 1985.
35. **Rovasio, R. A., Delouvee, A., Yamada, K. M., Timpl, R., and Thiery, J. P.,** Neural crest cell migration: requirements for exogenous fibronectin and high cell density, *J. Cell Biol.,* 96, 462, 1983.
36. **Gabius, H., Springer, W. R., and Barondes, S. H.,** Receptor for the cell binding site of discoidin I, *Cell,* 42, 449, 1985.
37. **Beug, H., Gerisch, G., Kempff, S., Riedel, V., and Cremer, G.,** Specific inhibition of cell contact formation in Dictyostelium by univalent antibodies, *Exp. Cell Res.,* 63, 147, 1970.
38. **Beug, H., Katz, F. E., and Gerisch, G.,** Dynamics of antigenic membrane sites relating to cell aggregation in *Dictyostelium discoideum, J. Cell Biol.,* 56, 647, 1973.
39. **Müller, K. and Gerisch, G.,** A specific glycoprotein as the target site of adhesion blocking FAB in aggregating Dictyostelium cells, *Nature (London),* 274, 445, 1978.
40. **Müller, K., Gerisch, G., Fromme, I., Mayer, H., and Tsugita, A.,** A specific glycoprotein as the target site of adhesion blocking FAB in aggregating Dictyostelium cells, *Eur. J. Biochem.,* 99, 419, 1979.
41. **Murray, B. A., Yee, L. D., and Loomis, W. F.,** Immunological analysis of a glycoprotein (contact sites A) involved in intercellular adhesion of *Dictyostelium discoideum, J. Supramol. Struct. Cell. Biochem.,* 17, 197, 1981.
42. **Murray, B. A., Niman, H. L., and Loomis, W. F.,** Monoclonal antibody recognizing gp80, a membrane glycoprotein implicated in intercellular adhesion of *Dictyostelium discoideum, Mol. Cell. Biol.,* 3, 863, 1983.
43. **Gerisch, G., Weinhart, U., Bertholdt, G., Claviez, M., and Stadler, J.,** Incomplete contact site A glycoprotein in HL220, a modB mutant of *Dictyostelium discoideum, J. Cell Sci.,* 73, 49, 1985.
44. **Siu, C.-H., Lam, T. Y., and Choi, A. H. C.,** Inhibition of cell–cell binding at the aggregation stage of *Dictyostelium discoideum* development by monoclonal antibodies directed against an 80,000-dalton surface glycoprotein, *J. Biol. Chem.,* 260, 16030, 1985.
45. **Springer, W. R. and Barondes, S. H.,** Protein-linked oligosaccharide implicated in cell–cell adhesion in two Dictyostelium species, *Dev. Biol.,* 109, 102, 1985.
46. **Parish, R. W. and Schmidlin, S.,** Synthesis of plasma membrane proteins during development of *Dictyostelium discoideum, FEBS Lett.,* 98, 251, 1979.
47. **Noegel, A., Gerisch, G., Stadler, J., and Westphal, M.,** Complete sequence and transcript regulation of a cell adhesion protein from aggregating Dictyostelium cells, *EMBO J.,* 5, 1473, 1986.
48. **Wong, L. M. and Siu, C.-H.,** Cloning of cDNA for the contact site A glycoprotein of *Dictyostelium discoideum, Proc. Natl. Acad. Sci. U.S.A.,* 83, 4248, 1986.
49. **Siu, C.-H., Lam, T. Y., and Wong, L. M.,** Expression of the contact site A glycoprotein in *Dictyostelium discoideum:* quantitation and developmental regulation, *Biochim. Biophys. Acta,* 968, 283, 1988.
50. **Choi, A. H. C. and Siu, C.-H.,** Filopodia are enriched in a cell cohesion molecule of M_r 80,000 and participate in cell–cell contact formation in *Dictyostelium discoideum, J. Cell Biol.,* 104, 1375, 1987.
51. **Ingalls, H. M., Goodloe-Holland, C. M., and Luna, E. J.,** Junctional plasma membrane domains isolated from aggregating *Dictyostelium discoideum* amoebae, *Proc. Natl. Acad. Sci. U.S.A.,* 83, 4779, 1986.
52. **Noegel, A., Harloff, C., Hirth, P., Merkl, R., Modersitzki, M., Stadler, J., Weinhart, U., Westphal, M., and Gerisch, G.,** Probing an adhesion mutant of *Dictyostelium discoideum* with cDNA clones and monoclonal antibodies indicates a specific defect in the contact site A glycoprotein, *EMBO J.,* 4, 3805, 1985.
53. **Harloff, C., Gerisch, G., and Noegel, A. A.,** Selective elimination of the contact site A protein of *Dictyostelium discoideum* by gene disruption, *Genes Dev.,* 3, 2011, 1989.
54. **Siu, C.-H., Cho, A., and Choi, A. H. C.,** The contact site A glycoprotein mediates cell–cell adhesion by homophilic binding in *Dictyostelium discoideum, J. Cell Biol.,* 105, 2523, 1987.
55. **Kamboj, R. K., Lam, T. Y., and Siu, C.-H.,** Regulation of slug size by the cell adhesion molecule gp80 in *Dictyostelium discoideum, Cell Regul.,* 1, 715, 1990.
56. **Kamboj, R. K., Wong, L. M., Lam, T. Y., and Siu, C.-H.,** Mapping of a cell-binding domain in the cell adhesion molecule gp80 of *Dictyostelium discoideum, J. Cell Biol.,* 107, 1835, 1988.
57. **Siu, C.-H. and Kamboj, R. K.,** Cell–cell adhesion and morphogenesis in *Dictyostelium discoideum, Dev. Genet.,* 11, 377, 1990.

58. **Hemperly, J. J., Edelman, G. M., and Cunningham, B. A.,** cDNA clones of the neural cell adhesion molecule (N-CAM) lacking a membrane-spanning region consistent with evidence for membrane attachment via a phosphatidylinositol intermediate, *Proc. Natl. Acad. Sci. U.S.A.*, 83, 9822, 1986.
59. **Sadeghi, H., da Silva, A. M., and Klein, C.,** Evidence that a glycolipid tail anchors antigen 117 to the plasma membrane of *Dictyostelium discoideum* cells, *Proc. Natl. Acad. Sci. U.S.A.*, 85, 5512, 1988.
60. **Stadler, J., Keenan, T. W., Bauer, G., and Gerisch, G.,** The contact site A glycoprotein of *Dictyostelium discoideum* carries a phospholipid anchor of a novel type, *EMBO J.*, 8, 371, 1989.
61. **Chou, P. Y. and Fasman, G. D.,** Prediction of β-turns, *Biophys. J.*, 26, 367, 1979.
62. **Amzel, L. M. and Poljak, R. J.,** Three-dimensional structure of immunoglobulins, *Annu. Rev. Biochem.*, 48, 961, 1979.
63. **Kamboj, R. K. and Siu, C.-H.,** Mapping of the monoclonal antibody 80L5C4 epitope on the cell adhesion molecule gp80 of *Dictyostelium discoideum, Biochim. Biophys. Acta,* 951, 78, 1988.
64. **Kamboj, R. K., Gariepy, J., and Siu, C.-H.,** Identification of an octapeptide involved in homophilic interaction of the cell adhesion molecule gp80 of *Dictyostelium discoideum, Cell,* 59, 615, 1989.
65. **Hohmann, H. P., Gerisch, G., Lee, R. W. H., and Huttner, W. B.,** Cell-free sulfation of the contact site A glycoprotein of *Dictyostelium discoideum* and of a partially glycosylated precursor, *J. Biol. Chem.*, 260, 13869, 1985.
66. **Hohmann, H. P., Bozzaro, S., Merkl, R., Wallraff, E., Yoshida, M., Weinhart, U., and Gerisch, G.,** Posttranslational glycosylation of the contact site A protein of *Dictyostelium discoideum* is important for stability but not for its function in cell adhesion, *EMBO J.*, 6, 3663, 1987.
67. **Hohmann, H. P., Bozzaro, S., Yoshida, M., Merkl, R., and Gerisch, G.,** Two-step glycosylation of the contact site A protein of *Dictyostelium discoideum* and transport of an incompletely glycosylated form to the cell surface, *J. Biol. Chem.*, 262, 16618, 1987.
68. **Ochiai, H., Stadler, J., Westphal, M., Wagle, G., Merkl, R., and Gerisch, G.,** Monoclonal antibodies against contact sites A of *Dictyostelium discoideum:* detection of modifications of the glycoprotein in tunicamycin-treated cells, *EMBO J.*, 1, 1011, 1982.
69. **Yoshida, M., Stadler, J., Bertholdt, G., and Gerisch, G.,** Wheat germ agglutinin binds to the contact site A glycoprotein of *Dictyostelium discoideum* and inhibits EDTA-stable adhesion, *EMBO J.*, 3, 2663, 1984.
70. **Stadler, J., Gerisch, G., Suchanek, C., and Huttner, W. B.,** *In vivo* sulfation of the contact site A glycoprotein of *Dictyostelium discoideum, EMBO J.*, 2, 1137, 1983.
71. **Coffman, D. S., Leichtling, B. H., and Richenberg, H. V.,** Phosphoproteins in *Dictyostelium discoideum, J. Supramol. Struct. Cell. Biochem.*, 15, 369, 1981.
72. **Schmidt, J. and Loomis, W. F.,** Phosphorylation of the contact site A glycoprotein (gp80) of *Dictyostelium discoideum, Dev. Biol.*, 91, 296, 1982.
73. **da Silva, A. M. and Klein, C.,** Characterization of a glycosyl-phosphatidylinositol degrading activity in *Dictyostelium discoideum* membranes, *Exp. Cell. Res.*, 185, 464, 1989.
74. **Faix, J., Gerisch, G., and Noegel, A. A.,** Constitutive overexpression of the contact site A glyco–protein enables growth-phase cells of *Dictyostelium discoideum* to aggregate, *EMBO J.*, 9, 2709, 1990.
75. **da Silva, A. M. and Klein, C.,** Cell adhesion in transformed *D. discoideum* cells: expression of gp80 and its biochemical characterization, *Dev. Biol.*, 140, 139, 1990.
76. **Gerisch, G., Fromm, H., Huesgen, A., and Wick, U.,** Control of cell-contact sites by cyclic AMP pulses in differentiating Dictyostelium cells, *Nature,* 255, 547, 1975.
77. **Loomis, W. F., Knecht, D. A., and Fuller, D. L.,** Adhesion mechanisms and multicellular control of cell-type divergence of *Dictyostelium discoideum,* in *Molecular Approaches to Developmental Biology,* Firtel, R. A. and Davidson, E. H., Eds., Alan R. Liss, New York, 1987, 339.
78. **Garrod, D. R.,** Acquisition of cohesiveness by slime mould cells prior to morphogenesis, *Exp. Cell Res.*, 72, 305, 1972.
79. **Chadwick, C. M. and Garrod, D. R.,** Identification of the cohesion molecule, contact sites B, of *Dictyostelium discoideum, J. Cell Sci.*, 60, 251, 1983.
80. **Chadwick, C. M., Ellison, J. E., and Garrod, D. R.,** Dual role for Dictyostelium contact site B in phagocytosis and developmental size regulation, *Nature,* 307, 646, 1984.
81. **Chadwick, C. M.,** Changes in the cell surface level of gp126 during development of *Dictyostelium discoideum, Dev. Growth Differ.*, 28, 203, 1986.
82. **Loomis, W. F.,** Cell–cell adhesion in *Dictyostelium discoideum, Dev. Genet.*, 9, 549, 1988.
83. **Steinemann, C. and Parish, R. W.,** Evidence that a developmentally regulated glycoprotein is target of adhesion-blocking Fab in reaggregating Dictyostelium, *Nature,* 286, 621, 1980.

84. **Wilcox, D. K. and Sussman, M.,** Serologically distinguishable alterations in the molecular specificity of cell cohesion during morphogenesis in *Dictyostelium discoideum, Proc. Natl. Acad. Sci. U.S.A.,* 78, 358, 1981.
85. **Siu, C.-H., Wong, L. M., Lam, T. Y., Kamboj, R. K., Choi, A., and Cho, A.,** Molecular mechanisms of cell–cell interaction in *Dictyostelium discoideum, Biochem. Cell Biol.,* 66, 1089, 1988.
86. **Parish, R. W., Schmidlin, S., and Parish, C. R.,** Detection of developmentally controlled plasma membrane antigens of *Dictyostelium discoideum* cells in SDS-polyacrylamide gels, *FEBS Lett.,* 95, 366, 1978.
87. **Wilcox, D. K. and Sussman, M.,** Defective cell cohesivity expressed late in the development of a *Dictyostelium discoideum* mutant, *Dev. Biol.,* 82, 102, 1981.
88. **Saxe, C. L., III, and Sussman, M.,** Induction of stage-specific cell cohesion in *D. discoideum* by a plasma-membrane-associated moiety reactive with wheat germ agglutinin, *Cell,* 29, 755, 1982.
89. **Saxe, C. L., III, and Firtel, R. A.,** Regulation of late gene expression in a temperature-sensitive cohesion-defective mutant of *Dictyostelium discoideum, Dev. Genet.,* 7, 99, 1986.
90. **Geltosky, J. E., Siu, C.-H., and Lerner, R. A.,** Glycoproteins of the plasma membrane of *Dictyostelium discoideum* during development, *Cell,* 8, 391, 1976.
91. **Geltosky, J. E., Weseman, J., Bakke, A., and Lerner, R. A.,** Identification of a cell surface glycoprotein involved in cell aggregation in *D. discoideum, Cell,* 18, 391, 1979.
92. **Geltosky, J. E., Birdwell, C. R., Weseman, J., and Lerner, R. A.,** A glycoprotein involved in aggregation of *D. discoideum* is distributed on the cell surface in a nonrandom fashion favoring cell junctions, *Cell,* 21, 339, 1980.
93. **Lam, T. Y., Pickering, G., Geltosky, J., and Siu, C.-H.,** Differential cell cohesiveness expressed by prespore and prestalk cells of *Dictyostelium discoideum, Differentiation,* 20, 22, 1981.
94. **Siu, C.-H., Des Roches, B., and Lam, T. Y.,** Involvement of a cell-surface glycoprotein in the cell-sorting process of *Dictyostelium discoideum, Proc. Natl. Acad. Sci. U.S.A.,* 80, 6595, 1983.
95. **Olive, E. W.,** Monograph of the Acrasieae., *Proc. Boston Soc. Nat. Hist.,* 30, 451, 1902.
96. **Raper, K. B.,** *The Dictyostelids,* Princeton University Press, Princeton, NJ, 1984, part 2.
97. **Raper, K. B. and Thom, C.,** Interspecific mixtures in the Dictyosteliaceae, *Am. J. Bot.,* 28, 69, 1941.
98. **Shimamura, O., Suthers, H. L. B., and Bonner, J. T.,** Chemical identity of the acrasin of the cellular slime mold *Polysphondylium violaceum, Proc. Natl. Acad. Sci. U.S.A.,* 79, 7376, 1982.
99. **Nichols, A. and Garrod, D. R.,** Mutual cohesion and cell sorting-out among four species of cellular slime moulds, *J. Cell Sci.,* 32, 377, 1978.
100. **Sternfeld, J.,** Evidence for differential cellular adhesion as the mechanism of sorting-out of various cellular slime mold species, *J. Embryol. Exp. Morphol.,* 53, 163, 1979.
101. **Gerisch, G., Krelle, H., Bozzaro, S., Eitle, E., and Guggenheim, R.,** Analysis of cell adhesion in Dictyostelium and Polysphondylium by the use of Fab, in *Cell Adhesion and Motility,* Curtis, A. S. G. and Pitts, J. D., Eds., Cambridge University Press, 1980, 293.
102. **Bozzaro, S. and Gerisch, G.,** Contact sites in aggregating cells of *Polysphondylium pallidum, J. Mol. Biol.,* 120, 265, 1978.
103. **Konijn, T. M., Van De Meene, J. G. C., Bonner, J. T., and Barkley, D. S.,** The acrasin activity of adenosine-3′,5′ cyclic phosphate, *Proc. Natl. Acad. Sci. U.S.A.,* 58, 1152, 1967.
104. **Konijn, T. M., Barkley, D. S., Chang, Y. Y., and Bonner, J. T.,** Cyclic AMP: a naturally occurring acrasin in the cellular slime molds, *Am. Nat.,* 102, 225, 1968.
105. **Springer, W. R. and Barondes, S. H.,** Direct measurement of species-specific cohesion in cellular slime molds, *J. Cell Biol.,* 78, 937, 1978.
106. **Alton, T. H. and Lodish, H. F.,** Developmental changes in messenger RNAs and protein synthesis in *Dictyostelium discoideum, Dev. Biol.,* 60, 180, 1977.
107. **Cardelli, J. A., Knecht, D. A., Wunderlich, R., and Dimond, R. L.,** Major changes in gene expression occur during at least four stages of development of *Dictyostelium discoideum, Dev. Biol.,* 110, 147, 1985.
108. **Newell, P. C., Longlands, M., and Sussman, M.,** Control of enzyme synthesis by cellular interaction during development of the cellular slime mold *Dictyostelium discoideum, J. Mol. Biol.,* 58, 541, 1971.
109. **Alton, T. H. and Lodish, H. F.,** Synthesis of developmentally regulated proteins in *Dictyostelium discoideum* which are dependent on continued cell–cell interaction, *Dev. Biol.,* 60, 207, 1977.
110. **Newell, P. C., Franke, J., and Sussman, M.,** Regulation of four functionally related enzymes during shifts in the developmental program of *Dictyostelium discoideum, J. Mol. Biol.,* 63, 373, 1972.

111. **Chung, S., Landfear, S. M., Blumberg, D. D., Cohen, N. S., and Lodish, H. F.,** Synthesis and stability of developmentally regulated Dictyostelium mRNAs are affected by cell–cell contact and cAMP, *Cell,* 24, 785, 1981.
112. **Mangiarotti, G., Ceccarelli, A., and Lodish, H. F.,** Cyclic AMP stabilizes a class of developmentally regulated *Dictyostelium discoideum* mRNAs, *Nature,* 301, 616, 1983.
113. **Mangiarotti, G., Lefebvre, P., and Lodish, H. F.,** Differences of mRNA stability in aggregated and disaggregated *Dictyostelium discoideum* cells, *Dev. Biol.,* 89, 82, 1981.
114. **Landfear, S. M., Lefebvre, P., Chung, S., and Lodish, H. F.,** Transcriptional control of gene expression during development of *Dictyostelium discoideum, Mol. Cell Biol.,* 2, 1417, 1982.
115. **Barklis, E. and Lodish, H. F.,** Regulation of *Dictyostelium discoideum* mRNAs specific for prespore or prestalk cells, *Cell,* 32, 1139, 1983.
116. **Mehdy, M. C., Ratner, D., and Firtel, R. A.,** Induction and modulation of cell-type-specific gene expression in Dictyostelium, *Cell,* 32, 763, 1983.
117. **Haribabu, B., Rajkovic, A., and Dottin, R. P.,** Cell–cell contact and cAMP regulate the expression of a UDP glucose pyrophosphorylase gene of *Dictyostelium discoideum, Dev. Biol.,* 113, 436, 1986.
118. **Blumberg, D. D., Margolskee, J. P., Barklis, E., Chung, S. N., Cohen, N. S., and Lodish, H. F.,** Specific cell–cell contacts are essential for induction of gene expression during differentiation of *Dictyostelium discoideum, Proc. Natl. Acad. Sci. U.S.A.,* 79, 127, 1982.
119. **Berger, E. A. and Clark, J. M.,** Specific cell–cell contact serves as the developmental signal to deactivate discoidin I gene expression in *Dictyostelium discoideum, Proc. Natl. Acad. Sci. U.S.A.,* 80, 4983, 1983.
120. **Berger, E. A., Bozzone, D. M., Berman, M. B., Morgenthaler, J. A., and Clark, J. M.,** Regulation of discoidin I gene expression in *Dictyostelium discoideum* by cell–cell contact and cAMP, *J. Cell. Biochem.,* 27, 391, 1985.
121. **Kaleko, M. and Rothman, F. G.,** Membrane sites regulating developmental gene expression in *Dictyostelium discoideum, Cell,* 28, 801, 1982.
122. **Town, C. D., Gross, J., and Kay, R. R.,** Cell differentiation without morphogenesis in *Dictyostelium discoideum, Nature,* 271, 717, 1976.
123. **Kay, R. R., Garrod, D., and Tilly, R.,** Requirement for cell differentiation in *Dictyostelium discoideum, Nature,* 271, 58, 1978.
124. **Town, C. D. and Stanford, E.,** An oligosaccharide-containing factor that induces cell differentiation in *Dictyostelium discoideum, Proc. Natl. Acad. Sci. U.S.A.,* 76, 308, 1979.
125. **Morris, H. R., Taylor, G. W., Masento, M. S., Jermyn, K. A., and Kay, R. R.,** Chemical structure of the morphogen differentiation inducing factor from *Dictyostelium discoideum, Nature,* 328, 811, 1987.
126. **Kay, R. R., Berks, M., Traynor, D., Taylor, G. W., Masento, M. S., and Morris, H. R.,** Signals controlling cell differentiation and pattern formation in Dictyostelium, *Dev. Genet.,* 9, 579, 1988.
127. **Williams, J. G.,** The role of diffusible molecules in regulating the cellular differentiation of *Dictyostelium discoideum, Development,* 103, 1, 1988.
128. **Fontana, D. R. and Devreotes, P. N.,** cAMP-stimulated adenylate cyclase activation in *Dictyostelium discoideum* is inhibited by agents acting at the cell surface, *Dev. Biol.,* 106, 76, 1984.
129. **Fontana, D. R. and Price, P. L.,** Contact alters cAMP metabolism in aggregation-competent Dictyostelium amoebae, *Dev. Genet.,* 9, 279, 1988.
130. **Fontana, D. R. and Price, P. L.,** Cell–cell contact elicits cAMP secretion and alters cAMP signaling in *Dictyostelium discoideum, Differentiation,* 41, 184, 1989.
131. **Salomon, Y.,** Adenylate cyclase assay, in *Advances in Cyclic Nucleotide Research,* Vol. 10, Brooker, G., Greengard, P., and Robison, G. A., Eds., Raven Press, New York, 1979, 35.
132. **Fontana, D. R., Price, P. L., and Phillips, J. C.,** Cell–cell contact mediates cAMP secretion in *Dictyostelium discoideum, Dev. Genet.,* 12, 54, 1991.
133. **van Haastert, P. J. M.,** A method for studying cAMP-relay in *Dictyostelium discoideum:* the effect of temperature on cAMP relay, *J. Gen. Micro.,* 130, 2559, 1984.
134. **Fontana, D. R., Luo, C., and Phillips, J. C.,** *Dictyostelium discoideum* lipids modulate cell–cell cohesion and cAMP signaling, *Mol. Cell Biol.,* 11, 468, 1990.
135. **Brenner, M. and Thoms, S.,** Caffeine blocks activation of cyclic AMP synthesis in *Dictyostelium discoideum, Dev. Biol.,* 101, 136, 1984.
136. **Williams, J. G., Duffy, K. T., Lane, D. P., McRobbie, S. J., Harwood, A. J., Traynor, D., Kay, R. R., and Jermyn, K. A.,** Origins of the prestalk-prespore pattern in Dictyostelium development, *Cell,* 59, 1157, 1989.

137. **Esch, R. K. and Firtel, R. A.,** cAMP and cell sorting control the spatial expression of a developmentally essential cell-type-specific ras gene in Dictyostelium, *Genes Dev.*, 5, 9, 1991.
138. **Sternfeld, J. and Bonner, J. T.,** Cell differentiation in Dictyostelium under submerged conditions, *Proc. Natl. Acad. Sci. U.S.A.*, 74, 268, 1977.
139. **Tasaka, M. and Takeuchi, I.,** Sorting out behaviour of disaggregated cells in the absence of morphogenesis in *Dictyostelium discoideum, J. Embryol. Exp. Morphol.*, 49, 89, 1979.
140. **Siu, C.-H., Des Roches, B., and Lam, T. Y.,** Involvement of a cell-surface glycoprotein in the cell-sorting process of *Dictyostelium discoideum, Proc. Natl. Acad. Sci. U.S.A.*, 80, 6596, 1983.
141. **Maeda, Y. and Takeuchi, I.,** Cell differentiation and fine structures in the development of the cellular slime molds, *Dev. Growth Differ.*, 11, 232, 1969.
142. **Takeuchi, I. and Yabuno, K.,** Disaggregation of slime mold pseudoplasmodia using EDTA and various proteolytic enzymes, *Exp. Cell Res.*, 61, 183, 1970.

Chapter 5

Adhesive Properties of Bacteria

Josef Beuth and Gerhard Uhlenbruck

CONTENTS

I. Introduction ... 87
II. Bacterial Lectins, Cell–Cell Recognition, and Infectious Diseases ... 88
A. Biology of Lectins ... 88
B. Bacterial Lectins — Mediators of Adhesion ... 88
1. Bacterial Fimbriae and Related Molecules ... 88
2. Carbohydrate Receptors and Recognition ... 90
3. Phase Variation, a Population Shift in the Fimbrial Phase of Bacteria ... 91
C. Experimental Evidence for Specific, Lectin-Mediated Bacterial Adhesion ... 94
1. Adhesion to Epithelial Cells and/or Tissues ... 94
2. Adhesion to Phagocytic Cells ... 95
3. Adhesion to Polymers ... 97
III. Nonspecific Factors Influencing Bacterial Adherence ... 98
A. Mechanical Cleansing of Mucosal Surfaces and Physicochemical Forces ... 98
B. Lipoteichoic Acids, Lipophilic Mediators of Microbial Adherence ... 99
C. Interactions Mediated by Bacterial Cell Surface Macromolecules ... 99
D. Extracellular Matrix and Bacterial Attachment ... 99
E. Bacteriopheresis: Surface Structure and Electrophoretic Mobility ... 100
IV. Concluding Remarks ... 100
References ... 101

I. INTRODUCTION

The concept of microbial attachment has recently stimulated extensive research concerning the pathogenesis of infectious diseases. Bacterial attachment to mucosal cells, obviously, is an important step in the infectious process and may be considered to be a prerequisite for the colonization of host tissues.[1-3] Recently, much evidence has been accumulated showing that the specificity of attachment appears to be the major factor that accounts for the organophilic tropism of a variety of microorganisms.[4,5]

Furthermore, microbial attachment protects bacteria from being swept away by the normal cleansing mechanisms operating on mucosal surfaces (e.g., urinary flow, peristalsis); it apparently permits subsequent penetration of the mucosal barrier and, thus, increases the efficiency of toxic microbial products towards target cells.[6,7] However, despite reasonable evidence that bacterial attachment to mucosal linings plays a major role in the infectious process, little information is available concerning the molecular mechanisms that confer an advantage to adherent and/or nonadherent bacteria in their survival, proliferation, and ability to cause tissue damage. Generally, nonspecific adherence mechanisms (e.g., net surface charge both of bacteria and cells; hydrophobic or lipophilic forces; fibronectin, vitronectin, laminin, collagen and lipoteichoic acid-mediated interactions) can be distinguished from specific attachment, mainly due to adhesins (lectins) or specific antibodies.[5,8]

We have suggested that lectin-mediated bacterial adhesion and the unspecific attachment to cells or other surfaces be designated "adherence".[5,9] Since the evaluation of microbial adhesive properties is a complex topic that demands a precise description, several terms need to be defined. Adhesins (lectins) are adhesive structures on bacterial surfaces, and receptors (in this context) are complementary carbohydrate moieties on the surfaces of host cells. Since microbial adhesive structures generally are composed of large polymeric molecules (e.g., filamentous appendages) the term "adhesins" (as established by Duguid[10]) should be preferred to the term "ligand", which usually refers to a small molecule.[11] Adhesive appendages should be called "fimbriae" rather than "pili" since the latter term should be reserved for those appendages involved in the conjugative transfer of genetic material. Since some filamentous appendages (e.g., of staphylococci and streptococci) lack regular size and shape, they will be referred to as "fibrillae".[11]

0-8493-4559-6/95/$0.00+$.50

While at present a detailed understanding of the importance of various molecules in host–parasite interaction is lacking, the material covered in this review emphasizes the way in which specific and/or nonspecific recognition has been shown to be involved and may provide the basis for further understanding.

II. BACTERIAL LECTINS, CELL–CELL RECOGNITION, AND INFECTIOUS DISEASES

A. BIOLOGY OF LECTINS

All living cells express surface carbohydrates that participate in intercellular interactions and in reactions of the cells with a variety of molecules. In the latter case, antibodies, agglutinins, toxins, or transmitter substances interact with cell-bound carbohydrates, triggering cell surface alterations, signal transmission, or metabolic activities. Because of their relative abundance at the cell surface, carbohydrates are preferentially selected as microbial receptors rather than peptides. This form of adhesion depends upon lectins (adhesins) on the surface of microorganisms and on the tip of filamentous appendages (fimbriae, fibrillae), respectively. It is considered to be highly specific and responsible for the tissue tropism of pathogenic and symbiotic microorganisms.[4,12-14]

Since the turn of the century, it has been known that cell agglutinating proteins (hemagglutinins) are widely distributed in nature. Such proteins were first found in plants and were therefore known as phytagglutinins, phytohemagglutinins, or, more recently, lectins. Lectins are ubiquitous proteins/glycoproteins that exhibit a specific and reversible carbohydrate-binding activity. They react with glycosylated macromolecules and cells, may coaggregate them, or lead to their lysis or other alterations.[12,15] The term "lectin" (Latin: legere, to select) was based on the observation that some seed extracts could discriminate among human blood groups. As an example, extracts from *Phaseolus limensis* were found to agglutinate type A erythrocytes selectively. The first description of hemagglutinating activity by the protein ricin of *Ricinus communis* was presented by Stillmark in 1888.[16] This hemagglutination was shown to be reversible and inhibitable by D-galactose. During the last decade, many lectins, specifically inhibitable by various glycoconjugates, were detected in plants and in micro- and macroorganisms.[12,17] Their specific binding was compared to that of enzymes and antibodies, which were both excluded from lectins by definition.[4,12,18] In the course of investigations, lectins have been shown to be useful tools for various scientific approaches, including detection and identification of blood groups and bacteria, mitogenic stimulation of mononuclear immune cells, detection of carbohydrates in solutions on macromolecules and cells, protein purification, and cell fractionation.[19,21] Furthermore, they could be shown to be involved in specific adhesion of symbiotic and pathogenic microorganisms to host tissue, in specific adhesion of tumor cells to organ cells in metastatic spread, in certain interaction with the cellular immune system, such as lysis and phagocytosis of microorganisms, and in the removal of senescent macromolecules and cells, respectively.[22-24]

B. BACTERIAL LECTINS — MEDIATORS OF ADHESION

1. Bacterial Fimbriae and Related Molecules

Consideration of the host–parasite interaction encompasses a wide range of phenomena, from adhesion to mucosal and/or solid surfaces to interactions with cells of the immune system. Bacterial attachment is a means of colonizing the appropriate ecological niche to avoid being swept away by mucosal secretions. It may be considered to be the first step in infectious diseases.[25-27]

Mucosal surfaces are characterized by an extensive carbohydrate coat contributed by membrane glycoproteins and glycolipids, as well as by more loosely associated mucin glycoproteins. Accordingly, it is not surprising that most bacterial adhesins have evolved to act as lectins, using carbohydrates as receptor sites. Some of these receptors represent glycoproteins containing D-mannose (Man), which is recognized by mannose-sensitive (MS) type 1 fimbriae. Other carbohydrate receptors occur on glycolipids, like the galactose (Gal) α1-4Gal moiety (a constituent of the P-blood group), which is present on the globoseries of glycolipids and recognized by the majority of pyelonephritic *E. coli* strains.[28-31] Accordingly, *E. coli* (among other enterobacteriaceae) can at least be divided into two classes since their hemagglutination and/or adhesion may be inhibited by Man or not. First mentioned microbial strains generally are called MS, the latter mannose-resistant (MR). Almost all strains of *E. coli,* as well as other enteric bacteria, express (or have the genetic potential to express) type 1 fimbriae.[4,8,32] Those fimbriae (as shown in Figure 1) generally are integral constituents of the microbial surface and vary in length (0.2 to 1 μm), width (2 to 7 nm), and number (20 to 400 per cell). They can be easily detached from the bacteria and characterized biochemically.[8,12] Because of their strong hydrophobicity, they have a tendency to

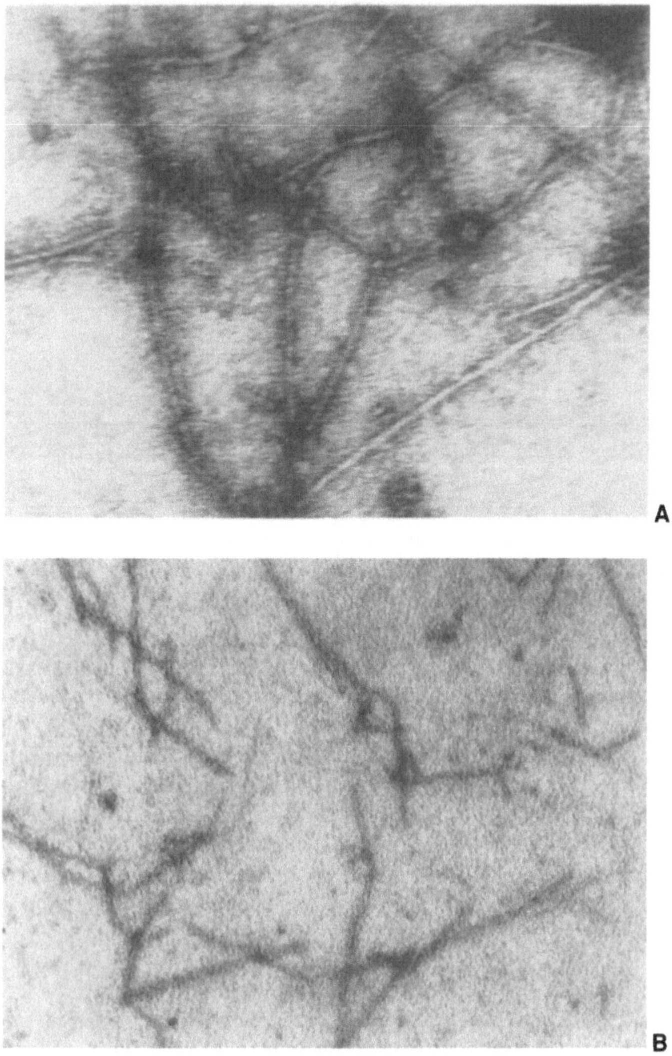

Figure 1 Electron micrographs of purified type 1 fimbriae isolated from *Klebsiella pneumoniae* (A) and type 3 fimbriae isolated from *Klebsiella oxytoca* (B) as recently presented. The |—| bar represents 100 nm. (From Przondo-Mordarska, A., et al., *Zbl. Bakt.*, 275, 521, 1991, with permission.)

aggregate, which may account for the observation that purified fimbriae agglutinate yeasts or erythrocytes in a mannose-specific manner. The fimbrillin subunit of the type 1 fimbriae of *E. coli* is composed of 158 amino acid residues. Evidence, mainly based on genetic experiments, strongly suggests the presence of an additional subunit in type 1 fimbriae, not easily detected in SDS-polyacrylamide gel electrophoresis (SDS-PAGE), which may carry the carbohydrate-binding site of the fimbriae.[33,34] This subunit could be shown to be located on the distal tip of the fimbriae in a position suitable for mediating the attachment of bacteria to target cells.[35] Whereas fimbrillin subunits of different *E. coli* strains have the same molecular size and amino acid composition, those of other enterobacteriaceae apparently are different both in size and composition.[4,12]

With respect to their structure and sugar specificity, several types of MR-fimbriae were identified besides MS type 1 fimbriae (Table 1). Almost all *E. coli* isolates associated with acute pyelonephritis in uncompromised hosts express MR-fimbriae that bind to the P-blood group antigens.[36-38] The most important portion of those glycolipid antigens necessary for binding was found to be a Galα1-4Gal moiety. Such digalactoside-specific binding is mediated by P-fimbriae.[37-39] The chromosomal pap gene cluster specifying P-fimbriae has recently been cloned from an uropathogenic *E. coli* strain and could be shown to confer MR hemagglutination and the expression of P-fimbriae onto other *E. coli* isolates.[28] The

Table 1 **Sugar specificity of bacterial surface agglutinins**

Saccharide	Bacteria	Fimbriae
Mannose	*Escherichia coli, Klebsiella pneumoniae, Pseudomonas aeruginosa, Salmonella spp., Serratia marcescens, Shigella flexneri*	type 1
Galactose	*Escherichia coli*	
L-fucose	*Vibrio cholerae*	
N-acetylglucosamine	*Escherichia coli*	type G
N-acetylgalactosamine	*Escherichia coli*	
Galα4Galβ-	*Escherichia coli*	type P
Galβ4Glc	*Actinomyces naeslundii, Actinomyces viscosus*	type 2
Galβ4GlcNAc	*Staphylococcus saprophyticus*	
GlcNAcβ3Gal	*Streptococcus pneumoniae*	
NeuAcα2-3Galβ3GalNAc	*Escherichia coli, Streptococcus mitis, S. sanguis, Mycoplasma gallisepticum, M. pneumoniae*	type S
NeuAcα2-3Galβ4Glc	*Escherichia coli*	K99

Note: For additional references and further details, see Reference 12.

From Sharon, N., *FEBS Lett.*, 217, 145, 1987, with permission.

genetic organization of the pap gene cluster, as well as the gene clusters encoding type 1, S, K88, and K 99 fimbriae, have been examined using molecular techniques.[28,40,41] These systems were found to resemble each other in several respects. Each cluster contains two genes involved in regulation that map at the promoter proximal end.

With the exception of the K88 cluster, the genes encoding the major fimbrillin subunit are located immediately distal to the regulatory region. Each cluster also contains several accessory genes that specify proteins involved in the transport and assembly of the fimbriae. The adhesin (lectin) is the product of a gene distinct from that which encodes fimbrillin, at least for the P, S, and type 1 fimbrial system.[12,40] Previous investigations have shown that formation of the fimbriae and expression of the carbohydrate-binding activity can be genetically separated: mutations outside the fimbrillin gene abolish the ability of the bacteria to bind to cells, but not the expression of fimbriae.[4,12]

2. Carbohydrate Receptors and Recognition

Certain pathogens display a predilection for specific tissues, which was previously shown in cases of microbial colonization, e.g., of the gastrointestinal tract by *Vibrio cholerae,* of lungs and meninges by *Streptococcus pneumoniae,* of the urinary tract by *Staphylococcus saprophyticus* and others.[42,43] However, it was a long time until the concept of receptor sites became recognized in microbiology. Specific receptors for microorganisms were first demonstrated in studies on the virulence of influenza virus.[44] Early investigations on the so-called receptor destroying enzyme (RDE) showed that it abolished the binding of influenza virus to the erythrocyte surface and, thus, prevented hemagglutination. The enzyme was subsequently recognized as neuraminidase, and its activity was shown to remove terminal sialic acid residues of sialoglycoproteins on the cell surface. Three of such sialoglycoproteins were identified on human erythrocyte membranes, and at least one protein involved in binding influenza virus, glycophorin A, was present in transmembrane configuration.[22] Many other studies on hemagglutinating viruses led to the definition of a number of virus receptors on erythrocytes.

Subsequently, it was proposed that similar glycoconjugates are also exposed on mucosal (epithelial) cell surfaces and, thus, recognized and infected by viruses. The presence of certain sugar sequences of glycoconjugates at the external cell surface places them in a key position to function in receptor phenomena.[45,46] Accordingly, carbohydrates are candidates for participation in recognition as receptors. The first substantial evidence for the key role of bacterial lectin–glycoconjugate receptor interactions in the pathogenesis of microbial infections was provided by Gibbons et al., who observed that specific glycoproteins in colostrum and saliva inhibited the hemagglutination of sheep erythrocytes by *E. coli.*[47] These studies suggested specific β-galactosyl residues of glycoproteins and glycolipids as possible natural receptors in the gastrointestinal tract. There is now considerable evidence for involvement of glycoconjugates in the receptors for bacterial adhesion molecules displaying lectin character. Detailed characterization of the combining site of bacterial surface lectins is important, not only for gaining a better insight into the nature of the interaction between bacteria and cell surfaces, but also for the design of more effective inhibitors of adhesion to mucosal surfaces and, thus, for the prevention of infection. Accordingly, the inhibitory effect of a large number of glycoconjugates and (oligo)saccharides on the agglutination of yeasts and/or erythrocytes by many bacterial species and strains was tested and previously published.[4,12]

Recent work has demonstrated that carbohydrate-specific bacterial adhesins (lectins, hemagglutinins) may recognize the carbohydrate receptor in both an internal and terminal position.[12,45] However, chemical groups distal as well as proximal to the oligosaccharide receptor site in the glycoprotein/glycolipid may in some cases enhance binding and in others sterically interfere with receptor binding.[25] Membranes from different tissues apparently differ in their composition (e.g., distribution of glycolipids). Accordingly, two lectins recognizing different epitopes on the same oligosaccharide may have a considerably different tissue-binding distribution due to different responses of neighboring groups. Previously, considerable evidence was accumulated showing that bacteria interact with components of the intercellular matrix, as well as with cell membrane receptors.[48-50] Apart from lectin-mediated specific adhesion, nonspecific adherence mechanisms, such as electrostatic forces, lipophilic or hydrophobic interactions, as well as more or less specific binding to fibronectin, vitronectin, laminin, and collagen, are common properties of both Gram-positive and Gram-negative bacteria. The various forms of specific and nonspecific attachment are outlined in Table 2 and do not need to be interpreted further, except for the phenomenon of a super or secondary coinfection, which may alter the receptor structure. As shown in Figure 2, various interactions are possible and may have clinical relevance. The receptor of one bacterium may be (totally or partly) identical with the receptor of another bacterium. Thus, *E. coli,* certain *Salmonella* species, and *Klebsiella pneumoniae* lectins (adhesins) are all Man-specific, although they slightly differ in their specificity.[4,12] However, there may be a competition on the *N*-linked mannosyl-oligosaccharides of membrane glycoproteins. Furthermore, the infection with certain microorganisms may destroy or uncover receptors for others. As shown in Figure 2, influenza virus neuraminidase destroys myxovirus receptors, MN blood group receptors, and receptors for certain mycoplasm strains (*Mycoplasma gallisepticum*), whereas blood group I/i receptors are uncovered and, thus, represent secondary uncovered receptors for *M. pneumoniae*.[51] In addition, myxovirus neuraminidase uncovers cryptic β-D-galactosido(1-3)-*N*-D-acetyl-D-galactosamine residues, which serve as receptors for *Actinomyces* strains or for certain strains of *Pseudomonas aeruginosa*.[52,53] Myxovirus sialidase and other bacterial neuraminidases (*V. cholerae*) could be shown to destroy human blood group M antigen,[52] which specifically binds certain pyelonephritic *E. coli* strains, a process which can be inhibited by the NH_2-terminal glycopeptide from blood group M glycoprotein.[54] Finally, microbial lectin receptors of glycolipid/glycoprotein nature can be destroyed or uncovered by bacterial proteases, as could be shown for MR pyelonephritic *E. coli* strains.[55]

3. Phase Variation, a Population Shift in the Fimbrial Phase of Bacteria

In the course of infectious diseases, the surface expression of the MS adhesins (lectins) by the infecting bacteria is likely to undergo phenotypic changes, as is the case in bacteria growing *in vitro*. The synthesis of MS-fimbrial adhesins (lectins) apparently is controlled by environmental and genetic factors. The genetic factor involves an off and on switch, which is responsible for the phenomenon of phase variation first described by Brinton and Duguid and is controlled at the transcriptional level.[56-58] The phenomenon occurs due to the periodic inversion of a specific DNA segment.[1] The phase variation rate from the phenotype expressing MS adhesins to that devoid of adhesins and vice versa was determined in *E. coli* to be about 1 per 1000 bacteria per generation.[59] Thus, a population of growing bacteria genetically capable of producing the MS-adhesin (lectin) will always contain a mixture of MS and MR phenotypes. The relative proportion of each phenotype in the growing population of bacteria depends on environmental factors such as composition of growth medium, growth phase, and number of subcultures.[56,57] These factors may exert their effect either on the rate of phase variation or, more likely, by causing overgrowth of one phenotype over the other, as was shown recently for the effect of glucose (Glc) on the growth of fimbriated bacteria.[59] Previous studies have shown that population shift from MS to MR phenotype during the early stages of infection suggests that the presence of MS adhesins (lectins) on the bacterial surface confers an advantage in initiating the infection, as well as in the survival of the microorganisms, probably by mediating adhesion to mucosal surfaces. No such advantage, however, was observed at subsequent stages of the infection, emphasizing the importance of phase variation phenomena in the survival at various sites of the host.[60]

Table 2 **Different forms of bacterial attachment**

Mode of attachment	Mediated by	Inhibited by	Unspecific inhibition by
		Specific Attachment	
Adhesion			
To cell surface carbohydrate groups 1) Genuine, primary sugar structures 2) Secondary uncovered sugar structures 3) Secondary acquired glycoconjugates of the membrane	A) Bacterial lectins (= adhesins, pili, fimbriae) B) Host organ lectins: membrane-integrated vertebrate lectins of the infected organ or organism C) Host macrophage lectins, lectins of neutrophils	a) Specific sugars or glycoconjugates reactive with the lectin or b) By soluble lectins (blocking lectins) with the same specificity from various sources c) Monoclonal antibodies against receptors for lectins	1) Receptor-destroying glycosidases or proteases 2) Covering of receptors (cryptic localization) 3) Topochemical misarrangement of lectins: a) Membrane-integrated b) Pili or fimbriae-associated c) Flagellae-associated
Fixation			
To a cell membrane of the host	Anti-bacterial antibody fixed to a cell membrane of the host via Fc fragment	Bacterial antigen (cell wall) or analogue glycoconjugates	Covering of the bacterial antigen (slime, exoglycoconjugates)

Agglutination Of suspended cells (or Co-agglutination)	Adhesins or heterologous lectins of plant, virus, vertebrate, and invertebrate origin	Lectin-specific sugars or glycoconjugates	1) Glycosidases, proteases 2) Zeta-potential of the particles 3) EDTA
	Specific antibodies reacting with bacterial membrane antigens	Bacterial antigens; fab fragments (receptor blocking)	1) Covering of the bacterial antigen 2) Shedding of the bacterial antigen
		Unspecific Attachment	
Adherence To cell membrane or other surface	1) Electrostatic forces, different charges: repulsion 2) Lipophilic or hydrophobic adherence: similia similibus aggregantur	1) Low mol-wt charged substances 2) Lipophilic or hydrophilic substances	1) Removal of the charged groups by enzymes 2) Neutralization of these groups 3) Covering of the groups
Aggregation Of suspended cells, also coaggregation	1) High-mol wt charged polymers (polylysine) 2) Lipophilic, hydrophobic, but also hydrophilic macromolecules	Low-mol wt analogous substances (see above); inhibition of bridging by ionic strength	1) Enzymatic removal of the charged groups 2) Enzymatic destruction of the aggregating macromolecule

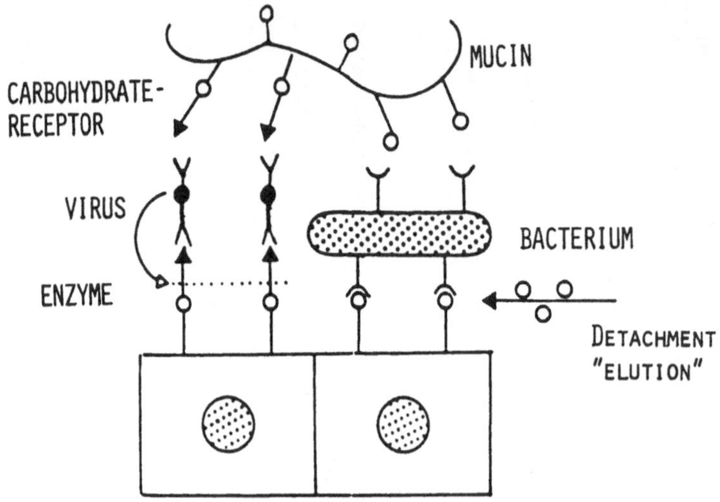

Figure 2 Enhanced and *de novo* creation of bacterial lectin receptors by a preinfection or coinfection of another, enzymatically active microorganism (virus, etc.) and the inhibition of adhesion by receptor-carrying mucins.

C. EXPERIMENTAL EVIDENCE FOR SPECIFIC, LECTIN-MEDIATED BACTERIAL ADHESION

1. Adhesion to Epithelial Cells and/or Tissues

The characteristic tissue tropism of many infectious agents can, at least in part, be explained by their recognition of specific epithelial receptor sites. Thus, attachment of bacteria to mucosal surfaces is considered to be a prerequisite for the colonization and infection of numerous tissues.[1-3] Although several nonspecific adherence mechanisms exist, specific adhesion seems to be most important. Generally, adhesion via lectins (adhesins) is considered as a specific interaction between bacteria and host tissue.[5] Accordingly, microbial lectins could be shown to be important virulence factors that mediate the interaction with complementary receptors.[4,12]

The hemagglutination assay has classically been used to demonstrate bacterial lectins (adhesins, hemagglutinins). In this assay, the simultaneous adhesion of bacteria to the surface receptors of two or more erythrocytes causes agglutination via a specific binding mechanism. This aggregation can easily be detected macroscopically on a slide, in a tube, or in a microtiter plate. Demonstration of lectin participation in this adhesion requires that the hemagglutination be sensitive to inhibition by carbohydrates. The ability of a sugar to interfere with hemagglutination is generally taken to indicate the involvement of this sugar in the receptor structure on the surface of the erythrocyte. During the last decade, hemagglutination tests were performed to specify surface lectins of a myriad of bacterial species and strains, respectively.[4,12] From the authors' laboratories, definite experimental evidence was presented for the presence of surface lectins on several microbial species, such as coagulase-negative *staphylococci* (CNS), *Klebsiella* bacteria, *S. pneumoniae*, and *P. aeruginosa*, which were chosen as examples to discuss the involvement of lectins in microbial adhesion to host tissue. Many other authors, however, have dealt with this topic recently, but it seems reasonable to limit the discussion of this chapter to the presentation of our own data, which generally confirm the majority of studies on bacterial adhesion.[3,4,12]

More than 30 strains of *S. saprophyticus* could be shown to agglutinate sheep erythrocytes, and this hemagglutination was inhibited by *N*-acetylglucosamine (GlcNAc) and additionally either in the presence of *N*-acetylgalactosamine (GalNAc) or *N*-acetylneuraminic acid (NeuAc). Interestingly, the GlcNAc-specificity was strictly confined to *S. saprophyticus* and could not be verified with other species of CNS, which might be of importance for a rapid test to characterize *S. saprophyticus* in future.[61] *S. pneumoniae* (GlcNAc-specificity) and *P. aeruginosa* (NeuAc-specificity) as well exhibited surface lectins, which could be shown to be involved in specific adhesion.[62-64]

To demonstrate the lectin-dependent mechanism of bacterial adhesion, cryotome sections were incubated with microorganisms suspended in phosphate-buffered saline (PBS) and in PBS-carbohydrate solutions, respectively. Sections of human lung and kidney both supported adherence of *S. saprophyticus* and *P. aeruginosa*. However, adherence of *S. pneumoniae* was only demonstrable to lung tissue and not to kidney tissue. This experimental observation is compatible with clinical experience of *S. saprophyticus* and *P. aeruginosa* infections, which both involve the respiratory and urinary tract, and pneumococcal infections, which primarily involve the respiratory tract. Receptor-mediated organotropism of infectious agents may be assumed by these findings and could be further confirmed in studies with tissue sections of liver, spleen, brain, meninges, and *S. pneumoniae*. PBS solutions of lectin-specific carbohydrates (NeuAc for *P. aeruginosa,* GlcNAc for *S. pneumoniae,* and GalNAc or NeuAc for *S. saprophyticus*) almost completely inhibited microbial adherence to tissue sections; nonspecific glycoconjugates did not show any inhibitory effect.[43,62-64]

To quantify the specific microbial adhesion and its inhibition by lectin-blocking carbohydrates, human uroepithelial cells (UECs) were incubated with *S. saprophyticus*. The total number of adherent bacteria to 100 UECs was determined, and the mean value was calculated for *S. saprophyticus* strain S1 (23.2 per UEC) and strain S35 (23.4 per UEC), respectively. Preincubation of the bacterial suspension with lectin-specific carbohydrates significantly decreased staphylococcal adhesion to UECs (81% decrease for strain S1 and 87% decrease for strain S35). Nonspecific sugars did not inhibit the adhesion process.[65] Analogous experiments with clinical isolates of *Klebsiella* bacteria and a variety of human epithelial cells yielded comparable results, since the type of fimbriation could be shown to determine the attachment. Thus, adhesion of *Klebsiella* strains carrying both MS and MR fimbriae was evidently stronger than adhesion of MS (type 1) or MR (type 3; MR/K) fimbriated strains and could be significantly inhibited in the presence of Man.[66]

In vivo, intratracheal administration of *S. pneumoniae* to experimental mice resulted in a diffuse settlement of lungs. However, intratracheal lavage with microbial lectin-specific GlcNAc solution inhibited streptococcal adhesion to pulmonary cells.[62]

Quantitative *in vivo* analysis with *P. aeruginosa* supported initial findings. Lung, kidney, and liver of experimental mice were heavily colonized with bacteria after intravenous inoculation. However, NeuAc treatment of the mice significantly reduced the *P. aeruginosa* burden of examined organs, whereas nonspecific carbohydrates did not interfere with bacterial adhesion to host tissue.[63] Since adhesion of *S. saprophyticus, S. pneumoniae, P. aeruginosa,* and *Klebsiella* bacteria to organ cells apparently can be inhibited when bacterial lectins (adhesins) are blocked with complementary (specific) carbohydrates, these substances may be useful for the prevention or treatment of (at least) certain infectious diseases.

Recently it was postulated that lectin-mediated bacterial adhesion requires intact (glyco)proteins (with respect to microbial lectins of glycoprotein character) for adequate interaction with carbohydrate moieties in the specific recognition process.[67,68] To evaluate the role of bacterial surface lectins of (glyco)protein nature in infectious diseases, subinhibitory concentrations of the antibiotic tunicamycin were utilized to modify them in order to assess their importance in the attachment process. Data obtained from these investigations confirm other researchers' and present experimental evidence that antibiotic (tunicamycin)-induced modification of surface glycoproteins (lectins) of *S. saprophyticus* and *S. pneumoniae* significantly reduced (more than 70%) microbial adherence to organ cells.[68-71] Accordingly, antibiotic treatment of microorganisms resulting in glycoprotein (lectin) dysfunction, as well as bacterial lectin blockade with complementary carbohydrates, significantly inhibited microbial adhesion to epithelial cells.

2. Adhesion to Phagocytic Cells

The recognition of physiologically relevant particles apparently involves specific mechanisms, since so far no direct evidence has been obtained for the postulation that the net surface charge or hydrophobicity of particles/bacteria determines whether or not they can be bound to phagocytes.[72] Serum factors (opsonins) enhance the uptake of microorganisms by phagocytic cells (opsonophagocytosis).[73,74] However, certain sites of macroorganisms (e.g., lung, kidney) have low serum opsonin concentrations, which may not be sufficient to effectively opsonize microorganisms.[75] Furthermore, not all microorganisms activate the alternative complement pathway, and complement-deficient states have been reported (e.g., during the neonatal period).[73,76,77] Recently, considerable evidence has been accumulated showing that phagocytosis of bacteria may be accomplished without opsonins by the interaction of carbohydrate-binding proteins (lectins) on the surface of one type of cell combining with complementary carbohydrate receptors of another (lectinophagocytosis), and it was speculated that lectinophagocytosis may be of particular importance in the defense against bacteria that evade opsonophagocytosis.[78] Accordingly, opsonophagocytosis and lectinophagocytosis seem to act as complementary processes in providing protection against invading bacteria.[79-81]

The methodology used to demonstrate the involvement of lectin–carbohydrate interactions in the recognition step of phagocytosis includes procedures commonly used in studies of ligand–cell receptor

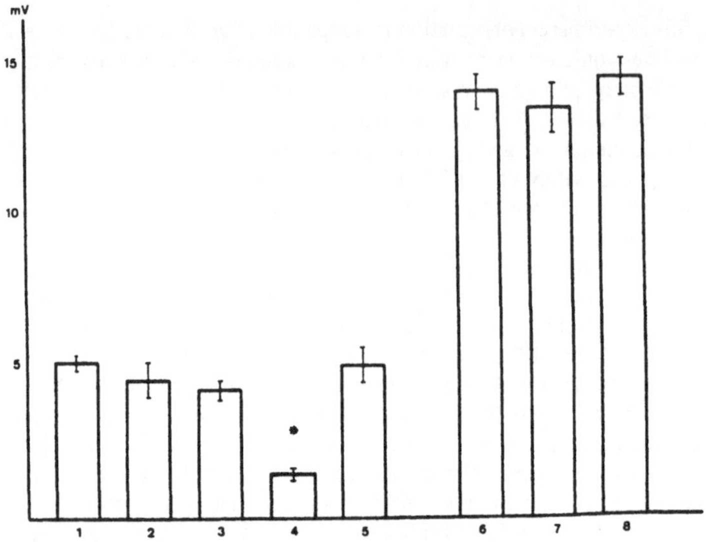

Figure 3 Human monocyte stimulation (chemiluminescence response) by nonopsonized and opsonized *S. saprophyticus* strain S 1. Microorganisms, respectively, phagocytes, were preincubated for 1 h in cell surface/membrane-specific (GalNAc for *S. saprophyticus* S1; Man for monocytes) and nonspecific glycoconjugate (0.02 *M*). All experiments were performed at least twice in duplicate and yielded reproducible and comparable results for phagocytes from five healthy donors and two staphylococcal strains. An individual experiment is presented. * $p < 0.05$ = statistically significant from control. Column 1: monocytes (Mo) + S1 (control); 2: Mo + S 1/Man; 3: Mo + S 1/GalNAc; 4: Mo/Man + S 1; 5: Mo/GalNAc + S 1; 6: Mo + S 1 opsonized; 7: Mo/Man + S 1 opsonized; 8: Mo + S 1 opsonized/Man. Man = D-mannose, GalNAc = N-acetylgalactosamine.

interactions and of lectin-mediated adherence of bacteria to organ cells, respectively.[1,3,12] As a first step, it requires testing the effect of a panel of carbohydrates on bacterial binding to (or agglutination of) the phagocytes. This step is based on the inhibition technique used in the study of specificity of antibodies and lectins. Since the recognition phase of bacterial phagocytosis is followed by a series of events (e.g., oxidative burst and degranulation) collectively termed "phagocytosis", specific agents that prevent recognition are expected to inhibit these events. Thus, control experiments should include bacteria coated with specific antibody to examine the effect of the inhibitory sugars on opsonophagocytosis, as well as mutants of bacteria lacking surface lectins, which show no activity in phagocytic assays.

The uptake of particles by phagocytes can be separated experimentally into two steps, attachment of the particle to the cell membrane and ingestion of the particle.[82] Previously, many authors have shown a correlation between luminol-enhanced chemiluminescence response and phagocytosis.[83,84] Chemiluminescence response is, therefore, used indirectly to quantify phagocytosis. Recent investigations on the chemiluminescence response of human monocytes and polymorphonuclear leukocytes (PMNLs) suggest that opsonophagocytosis is a considerably more effective process than lectinophagocytosis.[78] Pretreatment of opsonized microorganisms and phagocytes with various physiological glycoconjugates did not interfere with the phagocytic activity of the cells (opsonophagocytosis). To obtain evidence on lectinophagocytosis of *S. saprophyticus* and other bacteria by human monocytes and PMNLs, chemiluminescence assays were performed in the presence of lectin-specific and nonspecific carbohydrates.[85] Contrary to the PMNL stimulation, which was shown to be mediated by bacterial lectins, this activating mechanism proved to be negligible for monocytes.[85,86] Pretreatment of monocytes, however, with cell membrane lectin-specific Man significantly inhibited the phagocytic activity of the cells. Preincubation of the phagocytes in nonspecific carbohydrate solutions did not inhibit chemiluminescence activity (Figure 3). These data support the hypothesis that lectinophagocytosis of human monocytes is predominantly the result of specific interactions between phagocyte membrane lectins and bacterial carbohydrate receptors. Altogether, there appear to be two principal receptor-mediated mechanisms of recognition of microorganisms and/or particles, (1) lectin–carbohydrate interactions (lectinophagocytosis) and (2) bridging via opsonins (opsonophagocytosis). Both modes of phagocytosis should be distinguished strictly since competitive inhibition studies have demonstrated their independent character. Further investigations with

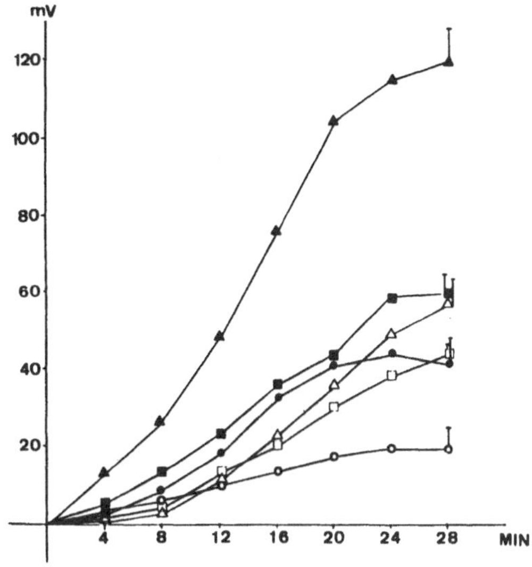

Figure 4 Chemiluminescence response of human PMNLs after stimulation with purified, latex-coated, nonopsonized, and opsonized *Klebsiella* fimbriae. ○—○, PMNLs + latex-coated type 1 fimbriae (L-T1F); ●—●, PMNLs + latex-coated type 3 fimbriae (L-T3F); ▲—▲, PMNLs + L-T1F serum opsonized; ■—■, PMNLs + L-T3F serum opsonized; Δ—Δ, PMNLs + L-T1F opsonized with heated serum; □—□, PMNLs + L-T3F opsonized with heated serum.

special emphasis on lectinophagocytosis should lead to a better understanding of the mechanisms through which microorganisms encounter/evade lectinophagocytosis and may lead to a more sophisticated approach in preventing or treating microbial infections.

Recently, extensive studies were initiated to evaluate the stimulatory effect of purified microbial fimbriae (e.g., of *Klebsiella* sp., *E. coli, P. aeruginosa*) on human PMNLs in opsonized and nonopsonized models.[87-90] Subsequent to isolation, bacterial fimbriae generally were purified by using deoxycholate and concentrated urea.[41] Since purified fimbriae per se did not effectively stimulate PMNL activity, phagocytes were challenged with latex particles coated with purified fimbriae.[91] In order to yield reliable and comparable results, specificity and amount of fimbriae coated to latex beads were checked by hemagglutination assays and determination of protein concentration, respectively.[91] In addition, electron microscopic studies were performed to confirm purity and morphological differences between various kinds of fimbriae.[41,91] With respect to the stimulatory activity on human PMNLs, purified (latex-coated) fimbriae could be shown to mimic faithfully the ability of fimbriated bacteria to stimulate PMNL oxidative metabolism.[91,92] Recent results from our laboratories generally support these findings, since purified, latex-coated *Klebsiella* fimbriae induced a potent stimulation of human PMNLs in chemiluminescence assays. Interestingly, in the absence of opsonin, activity was significantly more pronounced in response to latex-coated type 3 (MR/K) fimbriae as compared to type 1 fimbriae of *Klebsiella* origin.[91] This observation confirmed recent reports of Björksten and Wadström who found that MR-fimbriae of *E. coli* apparently facilitate adhesion to PMNLs, however, without stimulating phagocytic killing mechanisms.[92] Further investigations on the stimulatory activity of purified fimbriae confirmed recent studies that suggested that opsonophagocytosis is more effective than lectinophagocytosis. In contrast to nonopsonized *Klebsiella* fimbriae, opsonized type 1 fimbriae induced a significantly stronger PMNL chemiluminescence response than opsonized type 3 (MR/K) fimbriae. Furthermore, opsonization of both types of fimbriae with heated (complement-depleted) serum did not enhance PMNL activity in comparison to nonopsonized fimbriae (Figure 4). The observation that latex-coated type 1 fimbriae most effectively stimulated PMNL activity in the presence (but only moderately in the absence) of serum opsonins advanced the postulation that mainly type 1 fimbriae (at least of *Klebsiella* origin) mediate the interaction with PMNLs *in vivo.* Since opsonization of fimbriae with heat-inactivated serum induced less PMNL response than opsonization with untreated serum, the complement fraction apparently is more important than serum immunoglobulins in this respect. The data on lectino/opsonophagocytosis of purified *Klebsiella* fimbriae confirmed recent studies on the opsonin-independent/dependent interaction of *Klebsiella* bacteria and human PMNLs.[87,91,93]

3. Adhesion to Polymers

Bacterial adhesion to polymer surfaces is recognized as the initial step in the pathogenesis of foreign-body infections.[94,95] Whereas nonspecific interactions are well documented, knowledge about specific, lectin–receptor interactions between bacterial cell surfaces and polymer surfaces is poor.[96] It has recently been demonstrated that in the case of attachment of *S. aureus* and CNS to polymethylmethacrylate, fibronectin

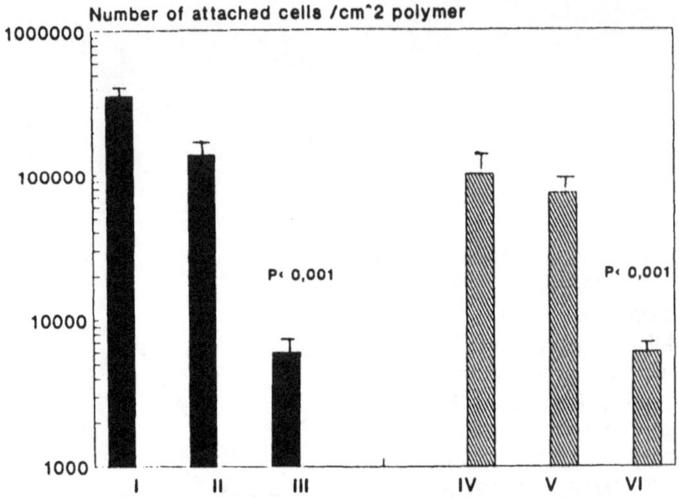

Figure 5 *S. saprophyticus* strain S 1 suspended in PBS (control; I, IV) or in PBS solutions containing 0.02 *M* unspecific Man (II, V) or specific GalNAc (III, VI), respectively. Man = D-mannose; GalNAc = *N*-acetylgalactosamine.

may play a role as mediator of adherence.[97] As well, tissue proteins like laminin or collagen may participate in the pathogenesis of foreign-body infection via attachment mechanisms.[95,97] A possible contribution of bacterial surface lectins in specific adhesion has recently been reported in investigations on the pathogenesis of foreign-body infections.[98] The results of these experiments offer some insight into bacterial adhesion to plastic devices, since lectin-mediated adhesion apparently plays a considerable role in the attachment to artificial surfaces that are in contact with body fluids. Suppression of adhesion by bacterial lectin blocking with specific carbohydrates was statistically significant for *S. saprophyticus* (Figure 5) and *P. aeruginosa*. Thus, lectin interactions might also be involved in the specific adhesion to and growth of microorganisms on surfaces of polymer devices such as intravascular, urethral, and other types of catheters. However, further studies on microbial lectin–receptor interactions are needed to support these initial findings and to evaluate the significance in the complex action of bacterial attachment to artificial surfaces. Accordingly, it must be taken into consideration, that adsorbed glycoconjugates from the serum or other body fluids may serve as a matrix for microbial attachment.

III. NONSPECIFIC FACTORS INFLUENCING BACTERIAL ADHERENCE

A. MECHANICAL CLEANSING OF MUCOSAL SURFACES AND PHYSICOCHEMICAL FORCES

Mucosal surfaces of healthy hosts are constantly bathed by secretions enriched with antibacterial enzymes and antibodies, which generally inhibit the colonization by pathogens.[1,4,12,99] Accordingly, nonattached microorganisms simply are washed away by mechanical forces (e.g., coughing, ciliary action, peristalsis). Pathogenic microorganisms that have taken advantage of impaired host defense mechanisms may furthermore be eliminated by desquamation of the colonized epithelial cells. Thus, successful pathogens are those capable not only of penetrating the local defense and attaching to mucosal cells but also of replenishing the new surfaces as colonized cells are desquamated and, eventually, of penetrating the epithelial cell barrier by invasion, either by the microorganisms themselves or by an excreted toxin.[4,12] The irreversibility of binding has indicated that microorganisms are held in place by many independent bonds between bacteria and host cell surfaces.[2,100] Several nonspecific factors are important in the formation of bonds between bacteria and host cell. The net surface charge of both bacteria and host cells generally is negative, generating repulsive forces between the cells. These repulsive forces, however, may be overcome by attractive forces between hydrophobic molecules on the two cell surfaces. It should be considered that two surfaces, although negatively charged, may nevertheless attract each other by forces due to atomic or molecular vibrations that produce fluctuating dipoles of similar frequencies on each surface.[5,101,102] Once the nonspecific attractive forces overcome the repulsive forces, the microorganism can attach to the host cell surface in a rather loose, reversible fashion. However, permanent anchoring to surfaces generally requires specific bonds between complementary molecules on each cell surface, which obviously are of major importance for the pathogenesis of infectious diseases.[2,5,103]

B. LIPOTEICHOIC ACIDS, LIPOPHILIC MEDIATORS OF MICROBIAL ADHERENCE

During recent decades, lipoteichoic acids (LTAs) were realized to be important and presumably essential components of most Gram-positive bacteria. Much of the impetus of studying them, however, has come from their immunological properties. Historically, it was the group D streptococcus that provided the first example for such an antigen, although the group F lactobacillus antigen was the first to be shown to contain the glycolipid component and to be located on the plasma membrane.[104] The presence of this glycolipid component also accounts for observations, extending over 20 years, of erythrocyte-sensitizing antigens present in, on the surface of, or in the culture fluid from a variety of Gram-positive bacteria. Apparently, it is the glycolipid component that accounts for the ability of the LTA to sensitize erythrocytes.[105,106] Recent studies suggest, however, that the carbohydrate part is responsible for the antigenicity, whereas the lipid fraction mediates the biological activity.[107]

LTAs are a group of phosphate-containing polymers associated with the cell walls and plasma membranes of Gram-positive bacteria.[104,108] However, they do not occur in all species of Gram-positive microorganisms, and they apparently are absent from Gram-negative bacteria.[104] LTAs are composed of a lipid portion covalently linked to a polyglycerophosphate chain and are believed to be adhesins of certain streptococci and staphylococci. Thus, the possible role of LTA in mediating the attachment of *S. saprophyticus, S. aureus,* and group A streptococci to host cell tissue was recently demonstrated.[65,109,110] To check and quantitate LTA-mediated staphylococcal attachment, human UECs were incubated with *S. saprophyticus*. The total number of adherent bacteria was determined by microscopic examination of 100 UECs, and the mean value was calculated. Preincubation of UECs with autologous LTA evidently decreased staphylococcal adherence (about 80% decrease). Heterologous LTA as well inhibited the attachment of *S. saprophyticus* to UECs, however, to a lesser extent.[65] Thus, nonspecific inhibition of microbial adherence to UECs by LTA from different staphylococcal strains suggests that these molecules obviously are quite similar structurally and functionally.

C. INTERACTIONS MEDIATED BY BACTERIAL CELL SURFACE MACROMOLECULES

The presence of capsule or slime substances in CNS (and other bacterial species) has been extensively demonstrated and could be shown to influence microbial interaction with host tissue, mainly due to physicochemical properties.[111,112] In our laboratories, a high mol wt cell surface complex (CSC) was obtained after elution of a crude phenol extract from *S. saprophyticus* from a Sepharose CL-6B column with ammonium acetate. Biochemical studies on this staphylococcal CSC revealed that it consisted of phosphate, protein, fatty acids, and carbohydrates; however, the composition was different from LTA and related antigens.[113]

In further studies, it was of special interest to define the interaction of CSC from *S. saprophyticus* with cellular elements of the nonspecific immune system. Contrary to LTA, the CSC of staphylococcal origin substantially enhanced human PMNL chemiluminescence response, whereas human monocytes could not be activated. Since heating as well as protease treatment of the CSC abolished its biological activity, it was suggested that the protein part of the molecule was responsible for its PMNL-activating potency.[113]

Since it was observed that formylmethionyl peptides (fMLP) can be generated by invading bacteria and apparently serve as ligands for appropriate receptors on PMNLs, our attention was focused on the fMLP receptor of human PMNLs as eventual target for staphylococcal CSC.[114] However, measurement of the PMNL chemiluminescence response towards fMLP, CSC, and nonopsonized *S. saprophyticus* cells prior to and after pretreatment of phagocytes with fMLP and CSC, respectively, suggested that the fMLP receptor of PMNLs is not involved in the interaction with CSC. Evaluation of the bactericidal capacity of PMNLs to opsonized *S. saprophyticus* cells positively correlated to chemiluminescence measurements. Pretreatment of the phagocytes with CSC evidently inhibited their bactericidal activity by interfering with attachment and phagocytosis. Apparently, CSC-mediated PMNL activation is dependent on immunoglobulin and/or complement receptors but not on fMLP receptors. These data suggest that nonspecific PMNL activation by *S. saprophyticus* is not only mediated by ligands like opsonins/immunoglobulins but may also be related to certain staphylococcal surface macromolecules.[113]

D. EXTRACELLULAR MATRIX AND BACTERIAL ATTACHMENT

Binding to various components of the extracellular matrix (ECM), such as fibronectin, collagen, laminin, and vitronectin, apparently allows certain pathogens to colonize host tissue.[95,97] Especially the role of fibronectin in bacterial adherence has been extensively studied recently; however, the complex interactions of fibronectin (and other components of the ECM) with antibodies, leukocytes, and microorganisms have not been completely elucidated. Fibronectin, which is a high mol wt glycoprotein and widely distributed in most body fluids and tissues, appears to be involved in the interaction between mammalian cells and certain microorganisms.[48,49] Two polypeptide chains are organized into structural domains having binding properties for different molecules and cells.[48] Staphylococci and streptococci bind to the amino-terminal region of fibronectin.[115] Thus, fibronectin promotes the adherence of staphylococcal cells, e.g., to polymethylmethacrylate (a material widely used in prosthetic devices) *in vitro* and *in vivo*.[95,97,116] However, there are conflicting data regarding the chemical structure and specificity of bacterial receptors for fibronectin. One fibronectin-binding protein was isolated from *S. aureus,* and the gene has been cloned and expressed in *E. coli*.[48,49] The attachment of *S. aureus* to immobilized fibronectin coated on glass, plastic, and PS surfaces has also been reported; however, the "specificity" of this process still remains to be determined.

Furthermore, the binding of *S. aureus,* group A streptococci, and tissue invasive pyelonephritic *E. coli* to laminin and the binding of type II collagen to *S. aureus,* CNS, and group A, C, and G streptococci have been reported.[95,97,117] However, criss-cross inhibition studies with bacteria, epithelial cells, and/or solid-phase fibronectin and ECM components did not prove the specificity of this process so far.[117]

E. BACTERIOPHORESIS: SURFACE STRUCTURE AND ELECTROPHORETIC MOBILITY

Polymers composing bacterial surfaces contain dissociable groups, such as carboxyl, phosphate, sulphate or amino groups. Since most cell surfaces contain both basic and acidic groups, the surface is amphoteric, with a negative net charge at high pH and a positive net charge at low pH.[118] Several experimental techniques have been used to demonstrate the net charge of individual bacteria and of bacterial populations. In the case of a limited number of microorganisms, electron microscopy proved to be an adequate method to localize charged molecules with a high topographical resolution.[119] The examination of large microbial populations, however, requires other methods, such as electrophoresis, in which the electrophoretic mobility of charged particles can be measured in an externally applied electric field.[120-122] The direction and rate of this movement depend on, among other factors, the polarity and density of the particles' net surface charge, the strength of the electric field, the ionic strength, and the temperature and pH of the medium.[121,122] Accordingly, the method of bacteriophoresis (this term is derived from bacterial cell electrophoresis) may be considered to be an approach for the classification of bacteria and other microorganisms. This is possible not only because of the bacterial glycoconjugate or protein membrane coat but also because of the possibility to determine microbial lectin or antibody-mediated coating (e.g., via Fc-receptors or bacteriophages). In addition, nonspecific acute phase reactants, such as C-reactive protein, could also be detected using specific antibodies or signals. Furthermore, secondary membrane alterations (e.g., by glycosidases, carbohydrate transferases, proteases) have revealed new aspects of the topochemical arrangement of different immunorelevant membrane components, similar to other membrane models.[123] Altogether, bacteriophoresis has been shown to be useful for the study of surface charges and the presumed topochemical arrangement of capsular polysaccharides in group B streptococci and of interactions between macromolecules and cells or artificial surfaces.[120-122]

IV. CONCLUDING REMARKS

Bacterial attachment (e.g., to mucosal and solid surfaces) is considered to be a prerequisite for the colonization and infection of numerous tissues and plastic devices. In general, attachment of microorganisms to surfaces of epithelial cells is the first of a series of events that may include the invasion of the host.

Bacterial agglutinins and/or hemagglutinins were first discovered in 1902, one year after the discovery of human blood groups by Landsteiner. In the mean time, considerable evidence accumulated showing the fundamental role that lectin–carbohydrate interactions play in cell–cell interactions, as well as in microbial pathogenicity. On the one hand, these lectins increase the virulence of bacteria by enhancing microbial adhesion to epithelial cells or solid surfaces; on the other, they decrease virulence by enhancing phagocytosis. It may be supposed that, in the near future, much more will be learned about bacterial lectins, and many others will probably be isolated, purified, and characterized. A detailed knowledge of the binding sites of these lectins and their receptors should lead to the design of potent inhibitors of adhesion, which will hopefully be suitable for testing in humans. We shall also gain a better understanding of lectinophagocytosis and/or opsonophagocytosis and its role in natural defense. This outlook, indeed, is promising. By understanding the molecular biology of adhesion and the different forms of nonspecific adherence, we can alter Paul Ehrlich's postulates originally meant for "chemical corpora" into "bacteria non agunt, nisi fixata." Accordingly, it is the task of the future to find out which events, signals, and messengers follow the fixation and what happens after the recognition of the specific and/or nonspecific receptor.

REFERENCES

1. **Beachey, E. H.,** *Bacterial Adherence,* Chapman and Hall, London, 1980.
2. **Beachey, E. H.,** Bacterial adhesion: adhesin–receptor interactions mediating the attachment of bacteria to mucosal surfaces, *J. Infect. Dis.,* 143, 325, 1981.
3. **Sharon, N.,** Lectin-like bacterial adherence to animal cells, in *Attachment of Organisms to the Gut Mucosa,* Boedecker, E. C., Ed., CRC Press, Boca Raton, FL, 1984, 129.
4. **Liener, E., Sharon, N., and Goldstein, I. J.,** *The Lectins,* Academic Press, London, 1986.
5. **Uhlenbruck, G.,** Bacterial lectins: mediators of adhesion, *Zbl. Bakt. Hyg. A,* 263, 497, 1987.
6. **Zafriri, D., Oron, Y., Eisenstein, B. I., and Ofek, I.,** Growth advantage and enhanced toxicity of *Escherichia coli* adherent to tissue culture cells due to restricted diffusion of products secreted by the cells, *J. Clin. Invest.,* 79, 1210, 1987.
7. **Gilboa-Garber, N. and Gilboa, N.,** Microbial lectin cofunction with lytic activities as a model for a general basic lectin role, *FEMS Microbiol. Rev.,* 63, 211, 1989.
8. **Sharon, N.,** Bacterial lectins, cell–cell recognition and infectious disease, *FEBS Lett.,* 217, 145, 1987.
9. **Uhlenbruck, G., Hartmann, C., and Koch, O. M.,** Adherence properties of *Pseudomonas* pili to epithelial cells of the human cornea: additional "lectinological" aspects, *Curr. Eye Res.,* 4, 641, 1983.
10. **Duguid, J. P.,** Fimbriae and adhesive properties in *Klebsiella* strains, *J. Gen. Microbiol.,* 21, 271, 1959.
11. **Schmidt, H., Naumann, G., and Putzke, G. B.,** Detection of different fimbriae-like structures on the surface of *Staphylococcus saprophyticus, Zbl. Bakt. Hyg. A,* 268, 228, 1988.
12. **Mirelman, D.,** *Microbial Lectins and Agglutinins,* John Wiley & Sons, New York, 1986.
13. **Wadström, T., Trust, T. J., and Brooks, D. E.,** Bacterial surface lectins, *Lectins,* 3, 479, 1983.
14. **Eshdat, Y. and Sharon, N.,** Recognitory bacterial surface lectins which mediate its mannose-specific adherence to eukaryotic cells, *Biol. Cell.,* 51, 259, 1984.
15. **Nowell, P. C.,** Phytohemagglutinin: an initiator of mitosis in cultures of normal human leukocytes, *Cancer Res.,* 20, 462, 1960.
16. **Stillmark, H.,** Ueber Ricin, ein giftiges Ferment aus dem Samen von Ricinus Comm. und einigen anderen Euphorbiaceen, *Inaug. Diss.,* Dorpat, 1888.
17. **Gold, E. R. and Balding, P.,** Receptor specific proteins. Plant and animal lectins, in *Excerpta Medica,* Amsterdam, 1975.
18. **Uhlenbruck, G.,** Die Biologie der Lektine: Eine biologische Lektion. Funkt, *Biol. Med.,* 2, 40, 1983.
19. **Prokop, O. and Uhlenbruck, G.,** A new source of antibody-like substances having anti-blood group specificity, *Vox Sang.,* 14, 321, 1968.
20. **Gabius, H. J., Rüdiger, H., and Uhlenbruck, G.,** Lektine, *Spektrum Wiss.,* 11, 50, 1988.
21. **Gabius, H. J.,** Endogenous lectins in tumors and the immune system, *Cancer Invest.,* 5, 39, 1987.
22. **Barondes, S. H.,** Lectins: their multiple endogenous cellular functions, *Annu. Rev. Biochem.,* 50, 207, 1981.
23. **Beuth, J., Ko, H. L., Schirrmacher, V., Uhlenbruck, G., and Pulverer, G.,** Inhibition of liver tumor cell colonization in two animal models by lectin blocking with D-galactose and arabinogalactan, *Clin. Exp. Metastasis,* 6, 115, 1988.
24. **Beuth, J., Ko, H. L., Gabius, H. J., and Pulverer, G.,** Influence of treatment with the immunomodulatory effective dose of the β-galactoside-specific lectin from mistletoe on tumor colonization in BALB/c-mice for two experimental model systems, *In vivo,* in press, 1991.
25. **Normark, S., Hultgren, S., Marklund, B. J., Nyberg, G., Olsen, A., and Strömberg, N.,** Bacterial adherence in pathogenicity, in *Current Topics in Infectious Diseases and Clinical Microbiology,* Jackson, G., Schlumberger, H. D., and Zeiler, H. J., Eds., Vieweg Verlag, Wiesbaden, 1989, 89.
26. **Uhlenbruck, G., Gross, R., Koch, O. M., and Chun-Kyung, L.,** Die Bedeutung von Lektinen für den Adhäsionsmechanismus von Bakterien, *Dtsch. Ärzteblatt,* 80, 27, 1983.

27. **Harber, M. J.,** Bacterial adherence, *Eur. J. Clin. Microbiol.*, 4, 257, 1985.
28. **Hull, S., Clegg, S., Svanborg-Eden, C., and Hull, R.,** Multiple forms of genes in pyelonephritic *Escherichia coli* encoding adhesins binding globoseries glycolipid receptors, *Infect. Immun.*, 47, 80, 1985.
29. **Orino, K. and Naiki, M.,** Two kinds of P-fimbrial variants of uropathogenic *Escherichia coli* recognizing Forssman glycosphingolipid, *Microbiol. Immunol.*, 34, 607, 1990.
30. **Kisielius, P. V., Schwan, W. R., Amundsen, S. K., Duncan, J. L., and Schaeffer, A. J.,** In vivo expression and variation of *Escherichia coli* type 1 and P pili in the urine of adults with acute urinary tract infections, *Infect. Immun.*, 57, 1656, 1989.
31. **Wold, A. E., Thorrsen, M., Hull, S., and Svanborg-Eden, C.,** Attachment of *Escherichia coli* via mannose or Gal 1→4 Gal-containing receptors to human colonic epithelial cells, *Infect. Immun.*, 56, 2531, 1988.
32. **Wadström, T. and Trust, T. J.,** Bacterial surface lectins, *Med. Microbiol.*, 4, 287, 1984.
33. **Maurer, L. and Orndorff, P. E.,** A new locus, pil E, required for the binding of type 1 piliated *Escherichia coli* to erythrocytes, *FEMS Microbiol. Lett.*, 30, 59, 1985.
34. **Minion, F. C., Abraham, S. N., Beachey, E. H., and Goguen, J. D.,** The genetic determinant of adhesive function in type 1 fimbriated *Escherichia coli* is distinct from the gene encoding the fimbrial subunit, *J. Bacteriol.*, 165, 1033, 1986.
35. **Lindberg, F., Lund, B., Johansson, L., and Normark, S.,** Localization of the receptor-binding protein adhesin at the tip of the bacterial pilus, *Nature*, 328, 84, 1987.
36. **Hultgren, S. J., Porter, T. N., Schaeffer, A. J., and Duncan, J. L.,** Role of type 1 pili and effects of phase variation on lower urinary tract infections produced by *Escherichia coli*, *Infect. Immun.*, 50, 370, 1985.
37. **Källenius, G., Svenson, S. B., Hultberg, H., Möllby, R., Helin, I., Cedergren, B., and Winberg, J.,** Occurrence of P-fimbriated *Escherichia coli* in urinary tract infections, *Lancet*, 12, 1369, 1981.
38. **Svanborg-Eden, C., Freter, R., Hagberg, L., Hull, R., Hull, S., Loeffler, H., and Schoolnik, G.,** Inhibition of experimental ascending urinary tract infection by an epithelial cell surface receptor analogue, *Nature*, 298, 560, 1982.
39. **Lomberg, H., Cedergren, B., Loeffler, H., Nilsson, B., Carlström, A. S., and Svanborg-Eden, C.,** Influence of blood group on the availability of receptors for attachment of uropathogenic *Escherichia coli*, *Infect. Immun.*, 51, 919, 1986.
40. **Normark, S., Lark, D., Hull, R., Norgren, M., Baga, M., O'Hanley, P., and Falkow, S.,** Genetics of digalactoside-binding adhesin from a uropathogenic *Escherichia coli* strain, *Infect. Immun.*, 41, 942, 1983.
41. **Gerlach, G. F. and Clegg, S.,** Cloning characterization of the gene cluster encoding type 3 (MR/K) fimbriae of *Klebsiella pneumoniae*, *FEMS Microbiol. Lett.*, 49, 377, 1988.
42. **Ofek, I., Beachey, E. H., and Sharon, N.,** Surface sugars of animal cells as determinants of recognition in bacterial adherence, *Trends Biochem. Sci.*, 3, 159, 1978.
43. **Beuth, J., Ko, H. L., Uhlenbruck, G., and Pulverer, G.,** Lectin mediated bacterial adhesion to human tissue, *Eur. J. Clin. Microbiol.*, 6, 591, 1987.
44. **Burnet, F. M.,** Mucoproteins in relation to virus action, *Physiol. Rev.*, 31, 131, 1951.
45. **Weir, D. M.,** Carbohydrates as recognition molecules in infection and immunity, *FEMS Microbiol. Immunol.*, 47, 331, 1989.
46. **Korhonen, T. K., Virkola, R., Westerlund, B., Tarkkanen, A. M., Lähteenmäki, K., Sareneva, T., Parkkinen, J., Kuusela, P., and Holthöfer, H.,** Tissue interactions of *Escherichia coli* adhesins, *Antonie van Leeuwenhoek*, 54, 411, 1988.
47. **Gibbons, T. R. A., Jones, G. W., and Sellwood, R.,** An attempt to identify the intestinal receptor for the K88 adhesin by means of haemagglutination. Inhibition test using glycoproteins and fractions of sow colostrum, *J. Gen. Microbiol.*, 186, 228, 1975.
48. **Ryden, C., Rubin, K., Speziale, P., Höök, M., Lindberg, M., and Wadström, T.,** Fibronectin receptors from *Staphylococcus aureus*, *J. Biol. Chem.*, 258, 3396, 1983.
49. **Hasty, D. L. and Simpson, W. A.,** Effects of fibronectin and other salivary macromolecules on the adherence of *Escherichia coli* to buccal epithelial cells, *Infect. Immun.*, 55, 2103, 1987.
50. **Miedzobrodzki, J., Naidu, A. S., Watts, J. L., Oborowski, P., Palm, K., and Wadström, T.,** Effect of milk on fibronectin and collagen type 1 binding to *Staphylococcus aureus* and coagulase-negative staphylococci isolated from bovine mastitis, *J. Clin. Microbiol.*, 27, 540, 1989.
51. **Basemann, B., Banai, M., and Kahane, J.,** Sialic acid residues mediated Mycoplasma pneumoniae attachment to human and sheep erythrocytes, *Infect. Immun.*, 38, 389, 1982.

52. **Uhlenbruck, G.,** Zur Definition der Panhämagglutination unter besonderer Berücksichtigung des Thomsen-Friedenreich'schen Phänomens, *Zbl. Bakt. Orig.*, 181, 285, 1961.
53. **Gilboa-Garber, N.,** *Pseudomonas aeruginosa* lectins, *Meth. Enzymol.*, 83, 378, 1982.
54. **Lis, H. and Sharon, N.,** Lectins as molecules and as tools, *Annu. Rev. Biochem.*, 55, 35, 1986.
55. **Parkinnen, J., Finne, J., Achtmann, M., Väisänen, V., and Korhonen, T. K.,** *Escherichia coli* strains binding neuraminyl α2-3 galactosides, *Biochem. Biophys. Res. Commun.*, 111, 456, 1983.
56. **Brinton, C. C.,** The piliation phase syndrome and the uses of purified pili in disease control, in *XIIIth Joint U.S.-Japan Conference on Cholera,* Miller, D., Ed., National Institutes of Health, Bethesda, MD, 1978, 34.
57. **Duguid, J. P., Clegg, S., and Wilson, M. J.,** The fimbrial and nonfimbrial haemagglutinins of *Escherichia coli, J. Med. Microbiol.*, 12, 213, 1979.
58. **Eisenstein, B.,** Phase variation by type 1 fimbriae in *Escherichia coli* is under transcriptional control, *Science,* 214, 337, 1981.
59. **Eisenstein, B. J. and Dodd, D. C.,** Pseudocatabolic repression of type 1 fimbriae of *Escherichia coli, J. Bacteriol.*, 151, 1560, 1982.
60. **Maayan, M. C., Ofek, I., Medalia, O., and Aronson, M.,** Population shift in mannose-specific fimbriated phase of *Klebsiella pneumoniae* during experimental urinary tract infection in mice, *Infect. Immun.*, 49, 785, 1985.
61. **Beuth, J., Ko, H. L., Schumacher-Perdreau, F., Peters, G., Heczko, P., and Pulverer, G.,** Hemagglutination by *Staphylococcus saprophyticus* and other coagulase-negative staphylococci, *Microb. Pathogen.*, 4, 379, 1988.
62. **Beuth, J., Ko, H. L., Schroten, H., Sölter, J., Uhlenbruck, G., and Pulverer, G.,** Lectin mediated adhesion of *Streptococcus pneumoniae* and its specific inhibition in vitro and in vivo, *Zbl. Bakt. Hyg. A,* 265, 160, 1987.
63. **Ko, H. L., Beuth, J., Sölter, J., Schroten, H., Uhlenbruck, G., and Pulverer, G.,** In vitro and in vivo inhibition of lectin mediated adhesion of *Pseudomonas aeruginosa* by receptor blocking carbohydrates, *Infection,* 15, 237, 1987.
64. **Beuth, J., Ko, H. L., Ohshima, Y., Schumacher-Perdreau, F., Peters, G., and Pulverer, G.,** Lectin-mediated cell attachment and phagocytosis of *Staphylococcus saprophyticus* strain S1, *Zbl. Bakt. Hyg. A,* 270, 22, 1988.
65. **Beuth, J., Ko, H. L., Ohshima, Y., Yassin, A., Uhlenbruck, G., and Pulverer, G.,** The role of lectins and lipoteichoic acid in adherence of *Staphylococcus saprophyticus, Zbl. Bakt.*, 268, 357, 1988.
66. **Würker, M., Beuth, J., Ko, H. L., Przondo-Mordarska, A., and Pulverer, G.,** Type of fimbriation determines adherence of *Klebsiella* bacteria to human epithelial cells, *Zbl. Bakt.*, 274, 239, 1990.
67. **Pulverer, G., Beuth, J., Ko, H. L., Sölter, J., and Uhlenbruck, G.,** Modification of glycosylation by tunicamycin treatment inhibits lectin-mediated adhesion of *Streptococcus pneumoniae* to various tissues, *Zbl. Bakt. Hyg. A,* 266, 137, 1987.
68. **Beuth, J., Ko, H. L., Pfeiffer, R., Yassin, A., Ohshima, Y., and Pulverer, G.,** Interference of tunicamycin-induced staphylococcal lectin dysfunction with specific adherence mechanisms and immune responses, in *The Influence of Antibiotics on the Host–Parasite Relationship,* Gillissen, G., Opferkuch, W., Peters, G., and Pulverer, G., Eds., Springer-Verlag, Berlin, 1989, 38.
69. **Vosbeck, K. and Mett, H.,** Bacterial adhesion: influence of drugs, *Med. Microbiol.*, 3, 21, 1983.
70. **Chugh, T. D., Babaa, E., Burns, G., and Shuhaiber, H.,** Effect of sublethal concentration of antibiotics on the adherence of *Staphylococcus epidermidis* to eukaryotic cells, *Chemotherapy,* 35, 113, 1989.
71. **Eisenstein, B. I., Beachey, E. H., and Ofek, I.,** Influence of sublethal concentrations of antibiotics on the expression of mannose-specific ligands of *Escherichia coli, Infect. Immun.*, 28, 154, 1980.
72. **Van Oss, C. J. and Gillman, C. F.,** Phagocytosis as a surface phenomenon. I. Contact angles and phagocytosis of non-opsonized bacteria, *Res. J. Reticuloendothel. Soc.*, 12, 283, 1972.
73. **Fearon, D. T. and Austen, K. F.,** The alternative pathway of complement: a system for host resistance to microbial infection, *N. Engl. J. Med.*, 303, 259, 1980.
74. **Griffin, F. M.,** Mononuclear cell phagocytic mechanisms and host defense, in *Host Defense Mechanisms,* Vol. 1, Gallin, J. I. and Fauci, A. J., Eds., Raven Press, New York, 1982, 31.
75. **Lipscomb, M. F., Ontario, J. M., Nash, E. J., Pierce, K. A., and Toews, G. B.,** Morphological study of the role of phagocytes in the clearance of *Staphylococcus aureus* from the lung, *Res. J. Reticuloendothel. Soc.*, 33, 429, 1983.
76. **Petersen, P. K. and Quie, P.,** Bacterial surface components and the pathogenesis of infectious diseases, *Annu. Rev. Med.*, 32, 29, 1981.

77. **Church, D. B. and Schlegel, R. J.,** Immune deficiency disorders, in *Immunology III,* Bellanti, J. A., Ed., W. B. Saunders, Philadelphia, 1985, 471.
78. **Ofek, T. and Sharon, N.,** Lectinophagocytosis: a molecular mechanism of recognition between cell surface sugars and lectins in the phagocytosis of bacteria, *Infect. Immun.,* 56, 539, 1988.
79. **Goetz, B. and Silverblatt, F. J.,** Stimulation of human polymorphonuclear leukocyte oxidative metabolism by type 1 pili from *Escherichia coli, Infect. Immun.,* 55, 534, 1987.
80. **Rainard, P.,** Phygocytosis by bovine polymorphs of glass adherent *Escherichia coli* in the presence and absence of serum, *FEMS Microbiol. Lett.,* 41, 217, 1987.
81. **Beuth, J., Ko, H. L., and Pulverer, G.,** The role of staphylococcal lectins in human granulocyte stimulation, *Infection,* 16, 46, 1988.
82. **Griffin, F. M., Griffin, J. A., Leider, J. E., and Silverstein, S. C.,** Studies on the mechanisms of phagocytosis. I. Requirement for circumferential attachment of particle-bound ligands to specific receptors on the macrophage plasma membrane, *J. Exp. Med.,* 142, 1263, 1975.
83. **Simoons-Smit, A. M., Verweigh-van Vught, A. M. J. J., Kanis, J. Y. R., and McLaren, D. M.,** Chemiluminescence of human leukocytes stimulated by clinical isolates of *Klebsiella, J. Med. Microbiol.,* 19, 333, 1985.
84. **Williams, A. J., Hastings, M. J. G., Easmon, C. S. F., and Cole, P. J.,** Factors affecting in vivo opsonization: a study of the kinetics of opsonization using the technique of phagocytic chemiluminescence, *Immunology,* 41, 903, 1980.
85. **Beuth, J., Ko, H. L., Quie, P., and Pulverer, G.,** Chemiluminescence response of human polymorphonuclear and mononuclear phagocytic cells induced by *Staphylococcus saprophyticus:* lectinophagocytosis vs. opsonophagocytosis, *Infection,* 18, 36, 1990.
86. **Beuth, J., Ko, H. L., Pfeiffer, R., Yassin A., and Pulverer, G.,** Lectinophagocytosis by *Staphylococcus saprophyticus* by human monocytes, *Zbl. Bakt.,* Suppl. 21, 267, 1990.
87. **Przondo-Mordarska, A., Ko, H. L., Matej-Melczynska, M., Beuth, J., Roszkowski, W., and Pulverer, G.,** Investigations on the opsonin-dependent interaction of *Klebsiella* bacteria and human polymorphonuclear leukocytes, *Zbl. Bakt. Hyg. A,* 270, 252, 1988.
88. **Spitznagel, J. K.,** Microbial interactions with neutrophils, *Rev. Infect. Dis.,* 5, 806, 1983.
89. **Silverblatt, F. J., Dreyer, J. S., and Schauer, S.,** Effect of pili on susceptibility of *Escherichia coli* to phagocytosis, *Infect. Immun.,* 24, 218, 1979.
90. **Rodriguez-Ortega, M., Ofek, I., and Sharon, N.,** Membrane glycoproteins of human polymorphonuclear leukocytes that act as receptors for mannose-specific *Escherichia coli, Infect. Immun.,* 55, 968, 1987.
91. **Przondo-Mordarska, A., Ko, H. L., Beuth, J., Gamian, A., and Pulverer, G.,** Chemiluminescence response of human polymorphonuclear leukocytes by purified, latex attached *Klebsiella* fimbriae, *Zbl. Bakt.,* 275, 521, 1991.
92. **Björksten, B. and Wadström, T.,** Interaction of *Escherichia coli* with different fimbriae and polymorphonuclear leukocytes, *Infect. Immun.,* 38, 298, 1982.
93. **Przondo-Hessek, A., Ko, H. L., Ciborowski, P., Roszkowski, W., and Pulverer, G.,** Opsonin independent interaction of *Klebsiella* strains with human polymorphonuclear leukocytes, *Zbl. Bakt. Hyg. A,* 262, 522, 1986.
94. **Sugarman, B. and Young, E. J.,** *Infections Associated with Prosthetic Devices,* CRC Press, Boca Raton, FL, 1984.
95. **Wadström, T.,** Molecular aspects of pathogenesis of wound and foreign body infection due to staphylococci, *Zbl. Bakt. Hyg. A,* 266, 191, 1987.
96. **Jansen, B., Peters, G., and Pulverer, G.,** Mechanisms and clinical relevance of bacterial adhesion to polymers, *J. Biomater. Appl.,* 2, 520, 1988.
97. **Hermann, M., Vandaux, P. E., Pittet, D., Auckenthaler, R., Lew, D. P., Schumacher-Perdreau, F., Peters, G., and Waldvogel, F.,** Fibronectin, fibrinogen, and laminin act as mediators of adherence of clinical staphylococcal isolates to foreign material, *J. Infect. Dis.,* 158, 693, 1988.
98. **Jansen, B., Beuth, J., and Ko, H. L.,** Evidence for lectin-mediated adherence of *S. saprophyticus* and *P. aeruginosa* to polymers, *Zbl. Bakt.,* 272, 437, 1990.
99. **Sanford, B. A., Thomas, V. L., and Ramsay, M. A.,** Binding of staphylococci to mucus in vivo and in vitro, *Infect. Immun.,* 57, 3735, 1989.
100. **Marrie, T. J., Lam, J., and Costerton, J. W.,** Bacterial adhesion to uroepithelial cells: a morphologic study, *J. Infect. Dis.,* 142, 239, 1980.

101. **Staedman, R., Knowlden, J., Lichodziejewska, M., and Williams, J.,** The influence of net surface charge on the interaction of uropathogenic *Escherichia coli* with human neutrophils, *Biochim. Biophys. Acta,* 1053, 37, 1990.
102. **Przondo-Hessek, A., Schumacher-Perdreau, F., and Pulverer, G.,** Cell surface hydrophobicity of *Klebsiella* strains, *Arch. Immunol. Ther. Exp.,* 35, 283, 1987.
103. **Beuth, J., Ko, H. L., Roszkowski, W., Roszkowski, K., and Oshima, Y.,** Lectins: mediators of adhesion for bacteria in infectious diseases and for tumor cells in metastasis, *Zbl. Bakt.,* 274, 350, 1990.
104. **Wicken, A. J. and Knox, K. W.,** Lipoteichoic acids: a new class of bacterial antigen, *Science,* 187, 1161, 1975.
105. **Beachey, E. H., Dale, J. B., Simpson, W. A., Evans, J. E., Knox, K. W., Ofek, I., and Wicken, A. J.,** Erythrocyte binding properties of streptococcal lipoteichoic acid, *Infect. Immun.,* 23, 618, 1979.
106. **Moskowitz, M.,** Separation and properties of a red cell sensitizing substance from streptococci, *J. Bacteriol.,* 81, 2200, 1966.
107. **Ohshima, Y., Beuth, J., Yassin, A., Ko, H. L., and Pulverer, G.,** Stimulation of human monocyte chemiluminescence by staphylococcal lipoteichoic acid, *Med. Microbiol. Immunol.,* 177, 115, 1988.
108. **Lambert, P. A., Hancock, I. C., and Baddiley, J.,** Occurrence and function of membrane teichoic acids, *Biochim. Biophys. Acta,* 472, 1, 1977.
109. **Beachey, E. H., Dale, J. B., Ahmed, A., and Ofek, I.,** Lymphocyte binding and T cell mitogenic properties of group A streptococcal lipoteichoic acid, *J. Immunol.,* 122, 189, 1979.
110. **Ofek, T., Beachey, E. H., Jefferson, W., and Campbell, G. L.,** Cell membrane binding properties of group A streptococcal lipoteichoic acid, *J. Exp. Med.,* 147, 990, 1975.
111. **Peters, G., Locci, R., and Pulverer, G.,** Adherence and growth of coagulase-negative staphylococci on surfaces of intravenous catheters, *J. Infect. Dis.,* 146, 479, 1982.
112. **Kristinsson, K. G.,** Adherence of staphylococci to intravascular catheters, *J. Med. Microbiol.,* 28, 249, 1989.
113. **Ko, H. L., Ohshima, Y., Beuth, J., Quie, P., and Pulverer, G.,** Granulocyte activation by a cell surface complex of *Staphylococcus saprophyticus:* a receptor-mediated phenomenon, *Zbl. Bakt. Hyg. A,* 271, 104, 1989.
114. **Marasco, W. A., Phan, S. A., and Krutzsch, H.,** Purification and identification of formylmethionyl-leucyl-phenylalanine as the major peptide neutrophil chemotactic factor produced by *Escherichia coli, J. Biol. Chem.,* 40, 5430, 1984.
115. **Switalski, L. M., Ryden, C., Rubin, K., Ljungh, A., Höök, M., and Wadström, T.,** Binding of fibronectin to *Staphylococcus* strains, *Infect. Immun.,* 42, 628, 1983.
116. **Velazco, M. J. and Waldvogel, F. A.,** Monosaccharide inhibition of *Staphylococcus aureus* adherence to human solid-phase fibronectin, *J. Infect. Dis.,* 155, 1069, 1987.
117. **Naidu, A. S., Paulsson, M., and Wadström, T.,** Particle agglutination assays for rapid detection of fibronectin, fibrinogen, and collagen receptors on *Staphylococcus aureus, J. Clin. Microbiol.,* 26, 1549, 1988.
118. **Lyklema, J.,** Interfacial electrochemistry of surfaces with biomedical relevance, in *Surface and Interfacial Aspects of Biomedical Polymers,* Andrade, J. D., Ed., Plenum Press, New York, 1985, 293.
119. **Magnusson, E. E. and Bayer, M. E.,** Anionic sites on the envelope of *Salmonella typhimurium* mapped with cationized ferritin, *Cell Biophys.,* 4, 163, 1982.
120. **Brinton, C. and Lauffer, M.,** The electrophoresis of viruses, bacteria, and cells and the microscope method of electrophoresis, in *Electrophoresis,* Bier, M., Ed., Academic Press, New York, 1959, 427.
121. **Uhlenbruck, G., Fröml, A., Lütticken, R., and Hannig, K.,** Cell electrophoresis of group B streptococci: separation of types Ia, Ib/c, II, III and IV before and after neuraminidase treatment, *Zbl. Bakt. Hyg. A,* 270, 28, 1988.
122. **Bayer, M. E. and Sloyer, J. L.,** The electrophoretic mobility of Gram-negative and Gram-positive bacteria: an electrokinetic analysis, *J. Gen. Microbiol.,* 136, 867, 1990.
123. **Van de Rijn, J. and Kessler, R. E.,** Growth characteristics of group A streptococci in new chemically defined medium, *Infect. Immun.,* 27, 444, 1980.

Chapter 6

Bioadhesives: DOPA and Phenolic Proteins as Components of Organic Composite Materials

Leszek M. Rzepecki, Karolyn M. Hansen, and J. Herbert Waite

CONTENTS

I. Overview 107
II. Introduction 108
A. Sclerotized Adhesives and Matrices 108
B. Biology of Sclerotized Matrices 109
1. Structural Adhesives 110
2. Composites 111
3. Varnishes 112
III. Essential Properties of Bioadhesives 112
A. Adherend Surface Properties 112
B. Composition and Chemical Properties of Phenolic Bioadhesives 113
IV. Variations on a Theme: Phenolic Protein Structure and Function 113
A. Motif Structure of Phenolic Adhesive Proteins 113
B. Mussel Adhesive DOPA Protein Precursors 119
C. Schistosome Eggcase Phenolic Protein Precursors 122
D. *Fasciola* Eggcase DOPA Protein Precursors 123
E. Diversity and Polymorphism in DOPA Proteins 123
1. Intraspecific Diversity 123
2. Intraspecific Polymorphism 123
3. Intraphyletic Diversity 125
F. Evolutionary Implications of DOPA Protein Diversity 125
G. Secondary Structure 127
1. Current Structural Models 127
2. Alternative Structural Proposals 129
H. Biosynthesis of Sclerotized Structures 130
I. Polypeptide Resins as Matrices for Biomineralization 131
J. Curing: the Maintenance of Structural Integrity 133
K. Permeability of Sclerotized Matrices 133
V. Summary 134
VI. Acknowledgments 134
References 135

I. OVERVIEW

Biological structural adhesives may be defined as thermosetting resins that bond two rigid adherends to form a permanent load-bearing joint. Such polymeric resins must not only form strong bonds with the adherends but also must become an integral component of the structure and may influence the macroscopic properties of the resulting material (e.g., laminates). It is apparent that the properties of biopolymers conducive to good adhesion also improve the qualities of composite materials and varnishes. Although structural and chemical properties of adhesive biopolymers often differ in detail, there appears to be a family resemblance that, though not necessarily a result of evolutionary homology, may reflect common design pathways or optimized solutions to the problems faced. Protein bioadhesives often incorporate 3,4-dihydroxyphenylalanine and tyrosine residues into the sequence as cross-linking or surface bonding agents, and are (thus far) invariably arranged as extensive linear arrays of tandem motifs. Since the design of these protein polymer resins is undoubtedly constrained both by the tasks for which

0-8493-4559-6/95/$0.00+$.50

they evolved and by the genetic material available, it is striking that such a wide range of biological applications has resulted in what may be termed a "structural thematic consensus".

The present review will focus primarily on the phenolic protein resins, especially the DOPA proteins and related peptides of parasitic invertebrates *(Fasciola, Schistosoma)* and marine invertebrates (mollusks and annelids) and will seek to draw inferences about the structure–function relationship from their comparative biochemistry and molecular biology. Recent results on other repetitive polypeptides from plants and animals, such as the extensins and involucrins, will also be compared. The questions to be addressed in this review include: What are the surface properties of hard structures in biology? What properties are required of phenolic protein resins to serve adhesive purposes? What is known about the evolution, composition, and structure of such resins, and are these factors related to specific function? How might the construction of an adhesive interface or composite be accomplished?

II. INTRODUCTION

Many of mankind's technological triumphs and commonplaces have been foreshadowed by the complex and contingent processes of biological evolution. Such is the case with the exploitation of adhesives and varnishes, where nature has long since handily solved the problems that beset attempts to join dissimilar structures in environments inimical to the often stringent requirements for long-term successful bonding. The superiority of adhesive bonding over merely mechanical fastening is manifold, especially for living organisms, which require topological continuity often coupled with joint flexibility.[1,2] The very success of the evolutionary enterprise is founded on the adhesive properties of biological polymers. Diverse uses of bioadhesives, ranging from the anchoring byssal threads of bivalved mollusks, to the eggshells of parasitic trematodes, to spider webs, are currently being explored at the molecular level, but our understanding still falls far short of that required to exploit the biopolymers involved in ways compatible with their evolutionary design. Indeed, the natural polymers mankind has pressed into use, for example, denatured collagens or celluloses as adhesives and plant phenolics as varnishes and tanning agents, often fulfill roles far removed from their natural ones.[3] Thus, although many biopolymers may have adhesive properties in appropriate contexts, relatively few of them have bona fide structural adhesive functions. Why, then, are some polymers adhesives and others merely sticky? The answers are often subtle: polymer composition, spatial configuration, and temporal processing (synthesis, secretion, and curing/setting) all play essential roles.

A. SCLEROTIZED ADHESIVES AND MATRICES

As exemplars of adhesion, we shall focus on the function of phenolic (or polyphenolic) proteins as structural adhesives, varnishes, and composite bonding agents. The historical terminology is somewhat confusing since the proteins discussed contain multiple tyrosine (a *para*-substituted phenol) and/or 3,4-dihydroxyphenylalanine (DOPA; a substituted *o*-diphenol or catechol) residues in their primary sequence, and "polyphenolic" is the adjective that has defined those proteins incorporating, or cross-linked to, *o*-diphenols and their quinones. We will use the term "DOPA protein" wherever the occurrence of DOPA residues is known or strongly implied, and "phenolic protein" as a generic term encompassing both DOPA proteins and polypeptides where the conversion of tyrosine to DOPA is unnecessary or uncertain. Synthetic phenolic resins have been used in bonding aluminum alloys in aircraft industries and in laminating and bonding wood but have been limited in application as a result of the high temperature and pressure conditions required for curing.[3,4] In contrast, natural phenolic proteins are thermosetting and do not require extraordinary curing treatments to function effectively.[5] In plants, where the phenolic proteins involved in the construction of cell walls include the proline- and hydroxyproline-enriched extensins and related proteins and the glycine-rich proteins (GRPs), curing is achieved following the peroxidase catalyzed one-electron oxidation of tyrosine by H_2O_2. The resultant free radical formation cross-links tyrosines to form isodityrosine *in vivo* and dityrosine *in vitro*. Further cross-links between celluloses such as the pectins or other phenolic polymers such as lignin (formed from low mol wt precursors) form complex composite materials of great strength and durability (i.e., wood).[6-9]

Among animals, especially the invertebrates, the most common form of curing is sclerotization, achieved usually by the O_2-dependent two-electron oxidation of *o*-diphenols to the resulting *o*-quinones, although oxidation of tyrosine by peroxidases may be an additional mechanism in marine arthropod cuticle. There are two classes of quinone-tanned sclerotin. The first to be characterized, in insect cuticles, is a flexible tanning strategy involving the cross-linking of miscellaneous protein and chitin components by highly diffusible low mol wt quinones, derived from oxidation of precursors such as *N*-acyldopamine

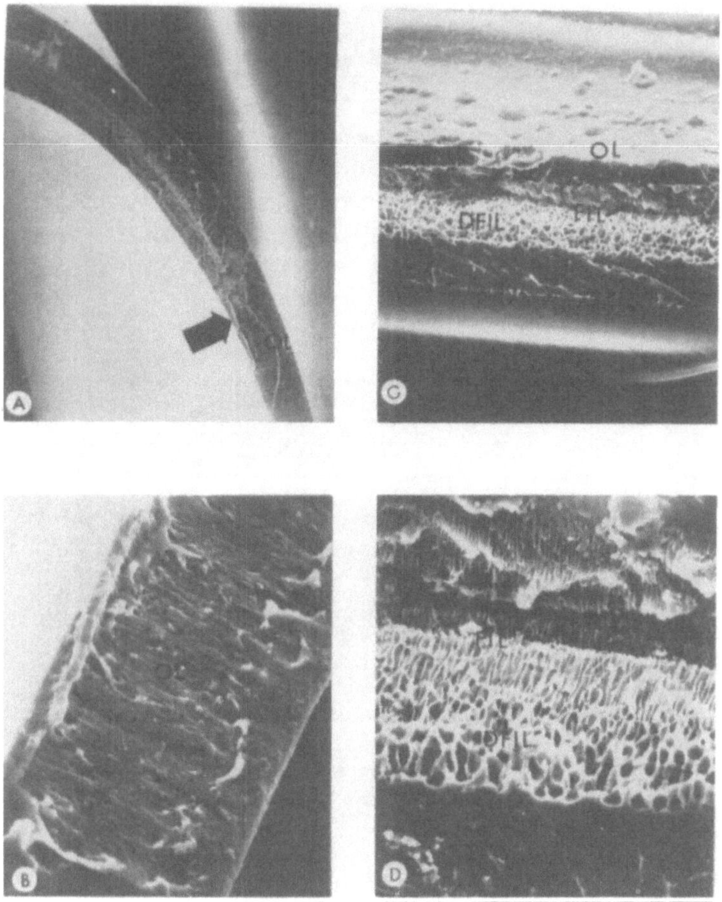

Figure 1A–D

or *N*-acyldehydrodopamine by polyphenoloxidases (also known as tyrosinases, phenolases, phenoloxidases, catechol oxidases, etc.).[10-12] The second class appears thus far to be restricted to exclusively aqueous environments where diffusion of a highly soluble tanning agent away from the desired site of action is a significant complication. Nature's solution has been to incorporate the tanning agent directly into the sclerotin precursors, in the form of integral DOPA residues formed by the post- or cotranslational hydroxylation of tyrosines, thus sacrificing the flexibility of low mol wt tanning agents for the assurance of reliable sclerotization. The activation of the tanning agent by the enzymic polyphenoloxidases implies that these catalysts must also form part of the resin, and their structure and properties must be dovetailed with those of the phenolic resin itself.[13]

B. BIOLOGY OF SCLEROTIZED MATRICES

Invertebrates are consummate DOPA protein exploiters.[14] The larvae of bivalves (phylum Mollusca) settle on an appropriate substratum before metamorphosis to adult form and anchor themselves by means of an extracorporeal tendon called the byssus, which consists of a thread terminating in a distal adhesive plaque.[15] DOPA proteins form the adhesive interface between the plaque and the substratum; they participate in the composite material of the plaque itself; and they form a protective seal over the entire structure, bonding to other plaque and thread components and protecting them from abrasion and biodegradation.[1,2,5] So useful is the byssal apparatus that many sessile bivalves, such as the mussels and fanshells, have retained it neotenously as a holdfast against the flux of their environment, although they may jettison it and relocate if necessary. In addition, nascent shell in mollusks is coated with a DOPA protein-tanned composite layer, the periostracum, which has a complex laminar structure (Figures 1A–D).[16,17] Another example of DOPA protein use is the eggcase of parasitic trematodes (phylum Platyhelminthes), which protects the ovum and associated cells during passage out of the primary host and into fresh water, where they hatch and continue

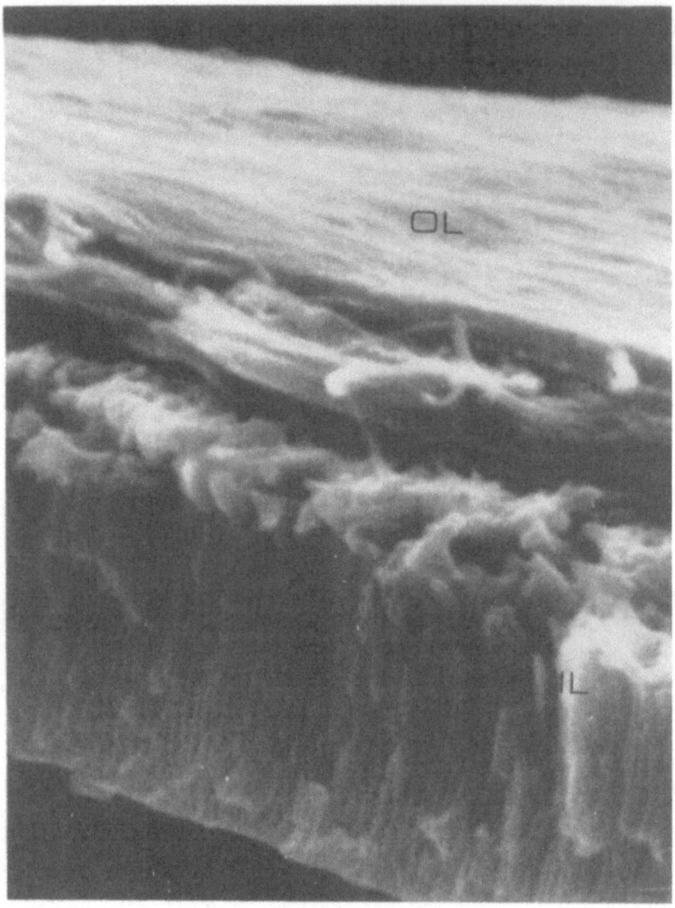

Figure 1E

their life cycle in the secondary host.[18] Upon passage of the egg from the host, the eggcase must protect the fragile contents from the instant osmotic stresses caused by the encounter with fresh water. Similar osmotic protection is required of the eggcases of *Bdelloura candida,* a turbellarian commensal on the gills of the horseshoe crab that inhabits the intertidal estuaries of the east coast and is often exposed to considerable variation in salinity. The laminar structure of the eggcase (Figures 1E and F) is ideally suited to withstand stretching and shear forces,[19] while sclerotization renders the eggcase impervious to the assault of proteases (and, frustratingly, to many other disruptive solvents or extraction treatments). DOPA proteins are used as cements in the construction of sandy dwelling tubes by reef building worms, such as *Phragmatopoma californica* (Annelida);[20] in tunicates (Chordata), they may participate in tunic repair and metal chelation;[21,22] they are probably components of the selachian eggcase (Chordata), their presence inferred from the detection of polyphenoloxidases and of DOPA, upon acid hydrolysis, in immature eggcases;[23,24] and they occur in the Cnidaria[25] and the Porifera,[26] indicating a respectably ancient evolutionary appreciation of the utility of DOPA proteins.[14]

The functions of these DOPA proteins are not all identical. In the Mollusca, they are used as adhesives and varnishes; in the Platyhelminthes, they are components of laminates; in the Annelids, they are used to cement inorganic mineral fragments to form tubular shelters. Thus, at least three classes of use may be distinguished.

1. Structural Adhesives

Various definitions of structural adhesives have been proposed. For convenience, we will adopt the ASTM definition, "a bonding agent used for transferring required loads between adherends exposed to service environments typical for the structure involved."[27] Implicit in this definition is the recognition that the adherends are hard, such as bone, shell, rock, wood, metal, or ceramic. Bonds between mussel byssus

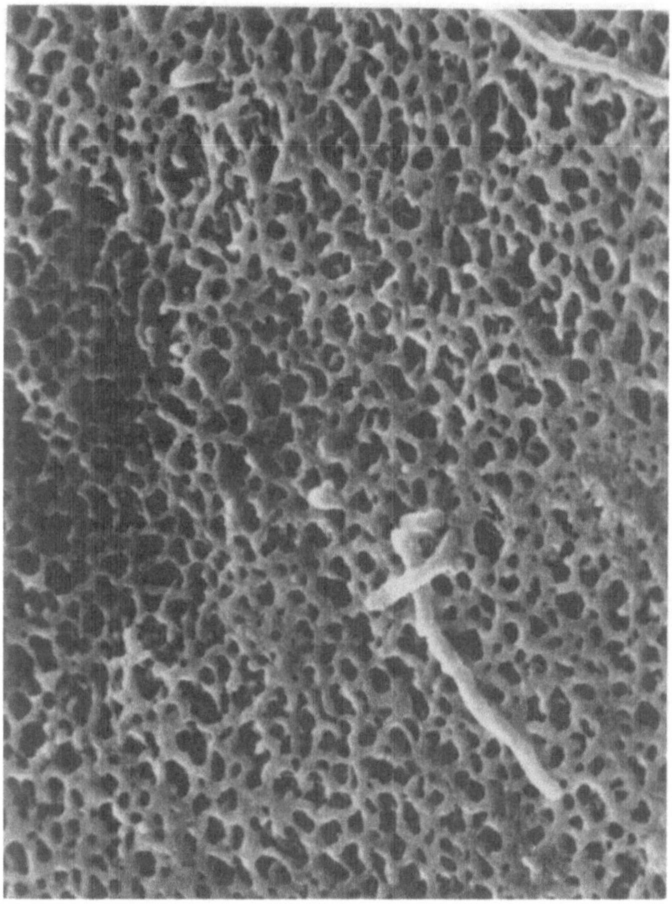

Figure 1F

Figure 1 (A–D) Scanning electron micrographs of *Geukensia demissa* marginal periostracum (growth region): (A) marginal periostracum showing the transition (arrow) from the single initial outer layer of pellicle (OL) to the laminated condition where the pellicle lies on top of a fibrous inner layer (IL) (Magnification ×182); (B) section through pellicle (OL) (Magnification ×1709); (C) section through marginal periostracum preceding the zone of shell mineralization — pellicle (OL), fibrous inner layer (FIL), distorted fibrous inner layer (DFIL) (Magnification ×524); (D) section through the fibrous inner layer illustrating transition in fiber conformation from straight to distorted (Magnification ×1709). Photos by J. H. Waite, unpublished. (E–F) Scanning electron micrographs of *Bdelloura candida* eggcase: (E) section of eggcase showing outer (OL) and inner (IL) layers (Magnification ×31,389);

and the substratum bear loads imposed by wave action or predatory seagulls, and worm tube cements must resist various concussive and shear forces. To accomplish their purpose, structural adhesives must spread over and bond tightly to the adherends, and to ensure the longevity of the joint they must be cured or hardened.[28] A common solution to these ends is the use of two-part adhesives consisting of soluble resin and hardener. Such adhesives are exemplified by man's use of epoxy or phenolic resins, which are cross-linked by free radical generating catalysts,[29] while in nature, the same purpose is met by applying a biopolymer (protein or carbohydrate), a cross-linking agent (*o*-diphenol), and a catalyst (polyphenoloxidase) together in solution.[13,14]

2. Composites

These are materials that consist of a stiff filler, such as chitin in arthropod cuticle, dispersed in a compliant, bonding, and fibrous matrix.[3] Composites offer greater degrees of stiffness than the bonding matrix itself, and many, such as molluscan shell or trematode eggshell, are often highly mineralized or sclerotized to increase their rigidity.[19] The strength of composites is conferred by strain amplification, which is the process of stressing the interactions between the filler and the matrix, leading to energy-absorbing breakdown and reorganization of local filler–matrix interactions. The greatest filler effect is seen when the filler bonds well with the matrix.[19] In dentistry, composite resins comprising borosilicate glass in an aromatic *bis*phenol glycidyl methacrylate polymer matrix exhibit greater compressive resistance than other cements.[30] Many structural bioadhesives are themselves composites, as are virtually all biological stiff structures.

3. Varnishes

These may be described as protective coatings that seal a surface and convey resistance to mechanical abrasion and chemical corrosion. As such, the components of the varnish film should bond well both to the surface and to each other, thus resisting sloughing and remaining impervious to environmental incursion. These properties are often incorporated directly into structural adhesives and composites by use of chemically appropriate compounds. For example, polyurethane and phenolic resins are used both as adhesives and as varnishes. Often, however, protection is conveniently afforded by a varnish, such as the periostracum of mollusk shells, the surface coatings of byssal threads and silk fibers, or various cavity liners used to shield sensitive dentine from corrosive dental pastes and cements.

III. ESSENTIAL PROPERTIES OF BIOADHESIVES

A. ADHEREND SURFACE PROPERTIES

Three basic types of hard structure feature as common adherends in biology. Crystalline or amorphous rocks and minerals that provide settlement substrates or building material for a host of invertebrates are frequently very rough, with many fractures, pits, and indentations. The ubiquitous silicon oxides offer sites on their surface for bonding to oxygen, by covalent or hydrogen bonding. Metal-bearing clays and minerals, such as the iron oxides hematite and goethite, provide opportunities for metal complexation in addition to the usual H-bonding potential.[31-33] In the biosphere, such substrata invariably bind one or more layers of water tightly and are often masked by an adsorbed layer of humic acids and other organic polymers and bacterial films. Consequently, considerable surface preparation is required of organisms wishing to anchor themselves securely. A second class includes biominerals, such as bone, dental enamel, invertebrate skeletal components, eggshells, and the molluscan shell matrix, and generally consists of various carbonate (calcite, aragonite) or phosphate (hydroxyapatite) salts of calcium, although the porifera also utilize magnesium salts and silica in spicule formation.[19] Other elements (sodium, aluminum, fluorine) may also be incorporated. Although the crystalline form of the biominerals is no different chemically from those produced geologically, the detailed structure, such as shape and crystallite shape and size, is usually very carefully controlled by the organism and of critical importance to its function.[34] In general, though, surface properties and problems of organic layers approximate those of the larger-scale geological cousins. The final class contains organic structures composed of various biopolymers, such as proteins and carbohydrates. These compounds provide opportunities for H-bonding, electrostatic interactions, and hydrophobic interactions, by virtue of their complex and variable composition. In most cases, this class of material provides at least one or both of the adherends to be bonded and is invariably part of biological composites. In common with the other classes, organic matrices present a complex, wet, and uneven surface to bioadhesives.

The properties of successful bioadhesives have been reviewed often[1,2] and so will be briefly summarized. Adhesives, varnishes, and composite bonding agents must

1. displace the bound layers of surface water that would undermine interfacial interactions,
2. have sufficiently low viscosity and surface tension to spread over the available substratum, penetrating the cracks and crazes offered by the surface,
3. maximize the number of bonding interactions, taking advantage of any H-bonding, van der Waal's, electrostatic, and hydrophobic interactions that may be presented,
4. repel encroachments by water, which will tend to plasticize the adhesive matrix and subvert interfacial interactions,
5. resist chemical, hydrolytic and abrasive degradation by various and unpredictable environmental forces.[35]

Conflicts between fluidity and final rigidity are resolved by using soluble polymers and curing agents in combination. Thus, any adhesive polymer must have adequate cross-linking potential. Moreover, in real life, as opposed to the factory environment, organisms may not always be able to choose an ideal surface to bond to. Although many larvae of sessile invertebrates tend to be recruited by specific substratum cues, resulting in the colonial habits of the reef building worm and the clustering of the ribbed mussel *Geukensia demissa* around the roots of the marsh grass *Spartina*, such recruitment is not always mandatory or possible.[36,37] *Geukensia* and other mussels will colonize varied other substrata, such as rocks, wood pilings, or metal. Indeed, the recent explosive infestation of the Great Lakes by the so-called zebra mussel, *Dreissena polymorpha*, is in part a tribute to its adhesive opportunism, smothering benthic organisms and clogging water intake pipes in the process.[38] In such cases where the organism cannot predict the nature of the substratum, it must produce adhesive polymers with interactive properties as diverse as remain compatible with their biosynthesis. Answers to the bioadhesive puzzle lie in a reductionist consideration of the composition, sequence, and higher-order structure of the proteins, together with an understanding of the levels of processing culminating in their physical application to or formation of a surface.

B. COMPOSITION AND CHEMICAL PROPERTIES OF PHENOLIC BIOADHESIVES

The striking compositional feature of purified sclerotin precursors is the presence of DOPA (Table 1). The chemistry of DOPA, discussed below, permits the formation of extensive cross-links with suitable partners, results in water displacement and H-bonding to surfaces offered by mineral or biopolymer substrates, and, by virtue of extremely stable complex formation with a wide variety of metals, allows very tight binding between the DOPA proteins and surfaces with exposed metal valencies.[5,13] The adhesive proteins are predominantly basic, with pI values ranging from 7 to >10 (Table 2). This basicity, usually conferred by lysine and hydroxylysine, although histidine and arginine often contribute, permits electrostatic interactions with the negatively charged organic layers on substrata. The high nucleophile content also offers candidates for cross-linking with DOPA residues during sclerotization.[12] The amount of lysine + hydroxylysine is generally proportional to the tyrosine + DOPA content of proteins isolated from a variety of organisms (Figure 2). DOPA proteins from other species have lower lysine/DOPA ratios, but other nucleophiles (histidine, cysteine) may replace lysine, especially among the Trematodes (Figure 2). Histidine forms cross-links with tanning agents in some insect cuticles.[60] In contrast, ferreascidin from the stolidobranch tunicate *Pyura stolonifera* contains over 50 mol% Tyr + DOPA but has relatively few nucleophilic or basic residues,[21] and other DOPA proteins (or protein fragments) isolated from the related *Halocynthia roretzi* also have iconoclastic compositions.[22] Periostracin from *Mytilus edulis* periostracum is also impoverished in eligible nucleophiles,[46] though not as drastically as ferreascidin. Such atypical compositions suggest either that other sclerotization reactions predominate in the latter systems or that they serve some other or additional function.

The composition of the proteins is otherwise dominated by relatively small and often interchangeable amino acids (serine, threonine, alanine, asparagine, aspartate, and glycine), which may be replaced by proline, especially in molluscan DOPA proteins. Proline is often hydroxylated in a sequence-specific manner to 4-hydroxyproline and, in *Mytilus* genera, 3-hydroxyproline.[54,61] The frequent hydroxylation of DOPA proteins affords extensive H-bonding opportunities, and proline hydroxylation might contribute to increased thermal stability.[62,63] The polypeptide length also maximizes van der Waals interactions with substrata and other polymers. DOPA proteins are notoriously insoluble at physiological pH, and will readily adsorb to glass or polypropylene unless borate is included in the buffer.[64] Oxidation by tyrosinases increases the rate of adsorption by an order of magnitude (Figure 3), perhaps by cross-linking proteins to form higher mol wt oligomers.[55,65]

* There has been some recent discussion about the nomenclature of repetitive subunits comprising larger protein structures, and expressions such as "motifs", "structural units", "domains", and "modules" have frequently been used.[66,67] The trend appears to be to associate domains with regions of a polypeptide chain that have all the characteristics of a globular protein, i.e., secondary and tertiary structure, and modules with the kind of domain that is clearly associated with exon shuffling and duplication. Since the tandem repeats here are smaller and of a lesser order of complexity, usage of the word "motif" appears more appropriate. Degeneracy of motifs often found at the N- and C-termini of proteins leads to some ambiguity about where to begin and end a functional motif. This may be referred to as a problem of punctuation, and is easily solved in cases such as the *M. edulis* protein, where the punctuation marker, lysine, is unambiguous and coincides with the tryptic peptides. In other cases, such as the *Schistosoma* proteins where motif degeneracy is greater, selection of an amino acid as the punctuation marker is less clear (e.g., is the repeating motif GGGY or YGGG?). In this review, we may differ from the original authors in order to maintain a (possibly spurious) consistency for the purposes of discussion.

Table 1 **Amino acid compositions of the DOPA proteins**

Amino acid	Be	Mg	Mms	Sb	Th	Gd	Me1	Mc	Aa	Cc	Mch	As
3-Hyp	0	0	0	0	0	0	27–42	30	0	0	43	0
4-Hyp	14–23	49	68–87	67–132	65–75	21	120–161	137	0	148–206	79	0
Asx	58	47	18	29	12	55	12	25	21	4	30	121
Thr	81	122	99	116	31	60	113	105	8	88	98	30
Ser	35	44	72	89	96	52	93	107	37	52	88	60
Glx	157	52	46	19	8	104	5	10	14	4	14	95
Pro	51–61	133	25–45	26–91	13–23	26	41–82	48	3	97–135	85	40
Gly	233	128	95	149	183	208	10	37	427	55	39	159
Ala	28	83	9	105	43	43	81	37	128	48	72	61
Cys/2	0	0	0	0	0	0	0	0	0	0	0	12
Val	29	15	27	5	21	42	5	11	92	61	21	28
Met	3	5	2	5	1	2	2	0	2	0	3	27
Ile	6	9	4	5	2	4	8	37	34	1	14	20
Leu	?	?	?	?	?	9	?	9	23	?	16	45
DOPA	83–119	89	107–152	83–120	97–146	86–115	110–181	126	79	161–176	131	50–62
Tyr	23–71	19	58–88	20–64	81–138	9–24	31–73	74	22	11–19	56	20–32
Phe	3	8	2	2	0	2	0	6	6	2	5	30
His	3	42	6	16	6	3	3	8	3	4	2	33
Hyl	7–34	14	7–17	5–13	12–16	17–55	trace	?	45	4	0	0–6
Lys	110–136	130	250–272	144–152	250–256	98–150	214	180	49	186	188	99–104
Arg	1	13	19	7	9	3	3	13	7	1	16	52

Amino acid	Me2	Me3	MePn	Pc1	Pc2	Fh B	Fh C	SmF4	Smf61-46	Sh2-1	Bc	PsFer
3-Hyp	0	0	0	0	0	0	0	—	—	—	0	0
4-Hyp	0	0	0	0	0	0	0	—	—	—	0	0
Asx	127	181	21	16	25	140	30	187	106	81	106	17
Thr	42	3	13	11	17	18	1	6	33	22	33	20
Ser	74	11	55	25	38	52	82	20	62	31	175	24
Glx	49	8	21	25	9	83	3	60	10	40	30	10
Pro	112	80	18	17	44	16	20	—	—	27	22	0
Gly	142	291	549	416	302	165	425	127	440	504	444	44
Ala	39	30	31	82	190	69	3	—	28	31	48	5
Cys/2	69	0	11	9	2	0	0	—	56	27	4	0
Val	43	0	32	53	38	9	1	—	11	9	6	4
Met	2	1	2	0	0	23	2	—	5	4	3	7
Ile	9	3	9	11	5	5	0	—	5	9	3	36
Leu	13	0	18	29	32	38	0	—	16	13	5	63
DOPA	29	231	22	98	73	106	198	*	*	*	32	169
Tyr	56	20	100	51	30	21	2	248	107	113	4	419
Phe	10	0	22	4	12	38	0	—	22	13	11	85
His	8	1	9	19	73	45	210	154	16	18	11	74
Hyl	0	0	0	0	0	0	0	—	—	—	4	0
Lys	135	45	15	124	63	120	0	174	56	40	35	28
Trp	n.d.	34	9	n.d.	n.d.	0	12	—	—	—	n.d.	0
Arg	44	62	35	11	18	60	12	20	—	9	25	0

Note: n.d. = Not determined. All amino acid compositions (residues/1000) were derived from isolated proteins except the schistosome protein compositions, which are calculated from DNA sequences. In the latter case, (—) indicates that the amino acid was not found in the DNA sequence and (*) indicates that the conversion of tyrosine to DOPA is unknown. For Pro/Hyp, Tyr/DOPA, and Lys/Hyl in mussel proteins, the composition ranges were derived from consecutive fractions obtained during HPLC purification, and presumably result from partial hydroxylation. Other amino acids occur in constant proportions. Letter codes: *Atrina serrata* (As),[45] *Aulacomya ater* (Aa),[42] *Bdelloura candida* (Bc),[53] *Brachidontes exustus* (Be),[39] *Choromytilus chorus* (Cc),[43] *Fasciola hepatica* (Fh B, C),[48,49] *Geukensia demissa* (Gd),[40] *Modiolus modiolus squamosus* (Mms),[39] *Mytella guyanensis* (Mg),[39] *Mytilus chilensis* (Mch),[44] *M. californianus* (Mc),[41] *M. edulis* (Me1, Me2, Me3 are, respectively, the 130,[39] 42 to 47,[74] and 11[75] kD proteins), *M. edulis* periostracin (MePn),[46] *Phragmatopoma californica* (Pc1, 2),[47] *Pyura stolonifera* ferreascidin (PsFer),[21] *Schistosoma mansoni* (SmF4,[50] Smf61-46[51]), *S. haematobium* (Sh2-1),[52] *Septifer bifurcatus* (Sb),[39] and *Trichomya hirsuta* (Th).[39]

Table 2 **Consensus motif sequences for phenolic proteins**

Phyla	Consensus
Phylum Mollusca	
	K-P-S---Y---p-g-Y-K
	t t t
Mytilus edulis (Linne, 1758)[54-57]	A-K-P-S---Y-P-P-T-Y-K**
Blue mussel	P K T
Mefp-1: 130 kD, pI = 10	A-K-P-----------T-Y-K**
Mytilus californianus (Conrad, 1837)[41, 56]	P-K-I-T---Y-P-P-T-Y-K*
Pacific mussel	R-K-P-S---Y-P-P-T-Y-K**
85 kD, pI = 10	
Geukensia demissa (Dillwyn, 1817)[40,56]	(G/A)-K-P-S-S-Y-D-P-G-Y-K** (α)
Ribbed mussel	P V
130 kD, pI = 8.1	G-Q-Q-K-Q-T-G-Y-D-P-G-Y-K** (β)
	V
	G-G-V-Q-K-T-G-Y-S-A-D-Y-K** (γ)
	G
Brachidontes exustus (Linne, 1758)[39]	Q-T-G-Y-D-P-G-Y-K-G-Q-Q-K*
Scorched mussel	V T
105 kD	(S/T)-G-Y-S-A-G-Y-K*
	G-K-P-S-S-Y-D-P-G-Y-K*
	P V
Septifer bifurcatus	Y-P-A---------K-P-T-S-Y-G-T-G-Y-K*
(Conrad, 1837)[39]	T P
Bifurcate mussel	S
99 kD	A-P-A-K-Y-T-T-K-P-S-S-Y-G-T-G-Y-K*
	P P S
	S-S-Y-G-T-G-Y-K*
	Y-P-A-K* P
Mytella guyanensis (Lamarck, 1819)[39]	(S/A/E)-H-K-P-----Y---T-G-Y-K-P-X*
Mangrove mussel	T-G-Y-K-P-?*
88 kD (X = [ASX, GLX, 2G, P, 2A, 2T]-K)	

Modiolus m. squamosus (Beauperthuy, 1967)[39]	S-**S**-Y-**Y**-P-K*
False tulip mussel	(**T**/K)-Y-**Y**-P-X-**G**-**Y**-G-------K*
120 kD (X = Q/K + $[G/Y/K/S/T/D]_{0\text{-}3}$)	
Trichomya hirsuta (Lamarck, 1819)[39]	(**S**/**T**/K)-Y-**Y**-P-K*
Hairy mussel	**G**-**Y**-G-----A-K*
70 kD	
Aulocomya ater (Molina, 1782)[43]	A-**G**-**Y**-G-G-G-L-K*
100 kD	V
Phylum Annelida	
Phragmatopoma californica[47]	
Reef building worm	
Pc 1: 20 kD, pI = 8.2	$(A/A\text{-}L/F)\text{-}(G\text{-}G\text{-}Y)_n\text{-}(G\text{-}A)_m\text{-}(W\text{-}G)_p\text{-}(H\text{-}P\text{-}A\text{-}V)_q\text{-}H\text{-}K^*$
n,m = 1–2; p,q = 0-1	
Pc 2: 20 kD, pI = 8.2	(W/L/K-V)-G-G-Y--------------------G-Y-G-----A-K*
Phylum Platyhelminthes	
Fasciola hepatica[48,49]	
Liver fluke	
vpB: 31 kD, pI = 7–8	Y-d-x-Y-G-(K/H/R)**
	FeF
vpC: 17 kD, pI = 6.9	$[G\text{-}(Y/H/S)]_?$*
Schistosoma mansoni[50,51]	
Blood fluke	
SmF4: 48 kD, pI > 7	Y-G-Y-D-K
SmF61-46:14 kD, pI > 7	Y-G-G-G
	C
Schistosoma haematobium[52]	
Blood fluke	
Sh2-1: 18–20 kD, pI > 7	$Y\text{-}(G)_{1\text{-}15}$
	C
	$Y\text{-}(G)_6\text{-}Y\text{-}E\text{-}(G)_3$

Table 2 **Consensus motif sequences for phenolic proteins (continued)**

Phyla	Consensus
	Miscellaneous
Loricrin (mouse keratinocytes)[58]	(G/V)-K-(Y/T)-S
37.8 kD, pI = 9.5	$(G)_{2-7}$-$(S/Y)_{1-2}$-(Y/C/S)
Hydroxyproline-rich proteins (Plants)[9]	P - P -V-Y-K
(selected sequences only)	E
	H
Extensins (Plants)[9]	S- P - P - P - P ----------------T-P-V-Y-K-(Y-K)
	S- P - P - P - P -K-K-P-Y-Y-P-P-H-T-P-V-Y-K
(selected sequences only)	
RNA Polymerase II (various phyla)[59]	
C-terminal domains, 17–52 repeats	Y-S-P-T-S-P-S
	K

Note: Standard amino acid symbols are used. Motif sequences are aligned, as far as possible, on Tyr/DOPA residues; dashes indicate gaps introduced to increase putative alignment. Underlining of Y and P indicates frequent hydroxylation: double underlining indicates invariable hydroxylation. Amino acid variants in any position are either given under the main sequence or are indicated thus: S/T. It is probable (but unknown) that tyrosine is converted to DOPA in the schistosomes. Molluscan sequences are those of the high mol wt DOPA proteins. Motifs marked (*) are derived from protein sequences obtained from tryptic peptides; those marked (**) are derived from protein and DNA sequences; all others derived from DNA sequences only. All consensus motifs are assigned on the basis of the most complete information available. However, it should not be assumed that given motifs necessarily represent the entire repertoire of the protein.

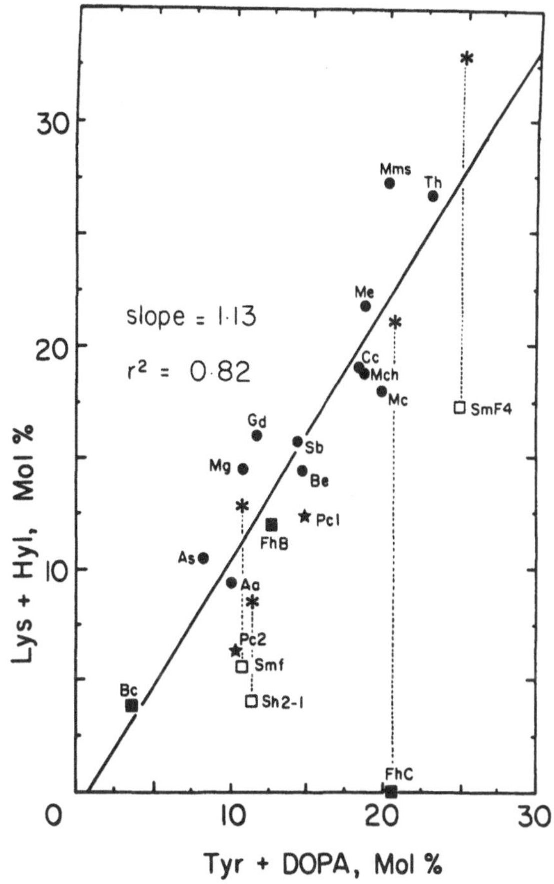

Figure 2 Correlation of the tyrosine + DOPA content with the lysine + hydroxylysine content of DOPA proteins (solid symbols) and other cloned proteins (open symbols) from various organisms. Compositions and letter codes from Table 1. The regression line is fitted to the solid symbols only, excepting the point Fh C. Asterisks (*) connected to certain points by fine dotted lines represent the total content of amino acids with reactive nucleophilic side chains, i.e., Lys + His + Cys, for trematode proteins with low lysine contents.

IV. VARIATIONS ON A THEME: PHENOLIC PROTEIN STRUCTURE AND FUNCTION

A. MOTIF STRUCTURE OF PHENOLIC ADHESIVE PROTEINS

In common with many structural polypeptides,[66-70] the salient feature of the DOPA proteins (mol wts = 10 to 130 kD), is their organization as linear arrays of tandem oligopeptide *motifs*,* frequently more than one type per protein, which are often punctuated by C-terminal nucleophilic residues. Consensus sequences of a variety of adhesive DOPA proteins and some structurally related peptides are given in Table 2, and although variations occur, functionality is conserved. There are probably two fundamental reasons for this distinctive arrangement. The multiplication of DNA sequences readily generates tandem arrays of motifs and results in a periodic organization of functional groups. This periodicity may be important to the correct interaction between polymers (e.g., by cross-linking) and ensures an even dispersal of required physicochemical properties. Second, since the posttranslational modification of these polymers is enzymatic, the target sequences must be conserved to assure maximal efficiency of modification. The frequency with which tyrosine in these motifs is flanked by small residues, such as glycine, may be related to steric requirements of tyrosyl hydroxylases. Notably, 2,4,5-trihydroxyphenylalanine (TOPA), an integral redox cofactor in certain copper amine oxidases (which comprise the only other class of protein known to incorporate *o*-diphenol derivatives) is also flanked by small amino acids in the sequence Leu-Asn-TOPA-Asp-Tyr.[71,72] The striking analogy between *M. edulis* (Pro-Pro-Thr-Dop-Lys) and plant cell wall phenolic protein (Pro-Pro-Val-Tyr-Lys) sequences (Table 2) curiously echoes Aristotle's presumption of the vegetable origin of mussel byssus.[73]

B. MUSSEL ADHESIVE DOPA PROTEIN PRECURSORS

A homocorrelation plot for a genomic clone[55] encoding a 100 kD fragment of the high mol wt *M. edulis* DOPA protein displays the nonrandom mosaic arrangement of motifs, and the expected periodicity of ten residues resulting from tandem repetition of the consensus motif AKPSYPPTYK is readily apparent (Figure 4). The N- and C-termini consist of degenerate motifs, and two major segments of hexa- and

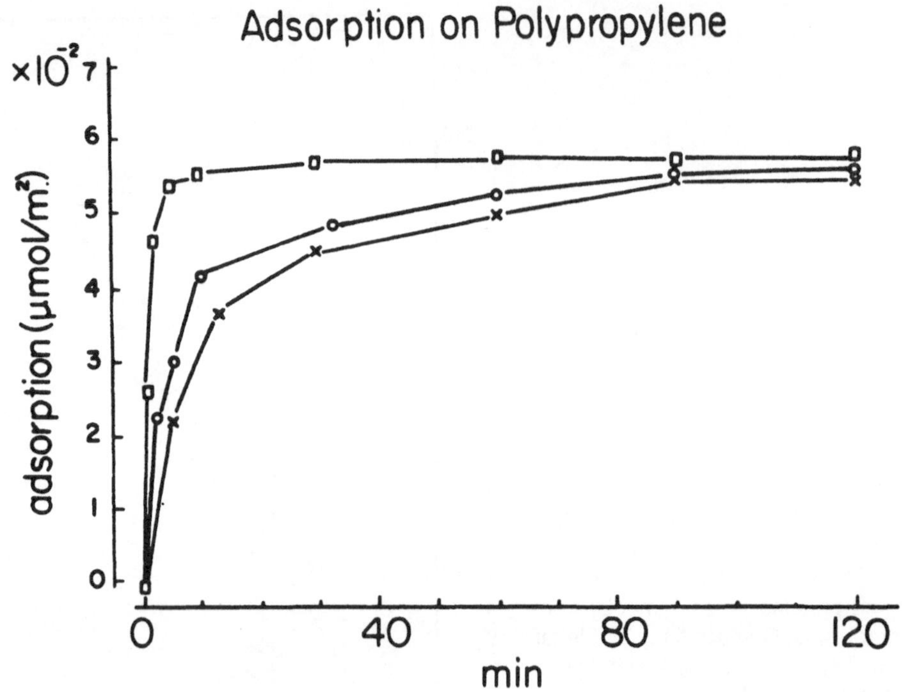

Figure 3 Adsorption of *M. edulis* high mol wt DOPA protein to polyp4ropylene in 40 m*M* sodium phosphate, pH 8.0.[65] Curves are DOPA protein, 0.12 mg/ml, $k_{adsorption}$ = 0.06 min^{-1} (○); DOPA protein + mushroom tyrosinase, 2.4 μg/ml, $k_{adsorption}$ = 0.54 min^{-1} (□); and DOPA protein + boiled mushroom tyrosinase (x); $k_{adsorption}$ is the apparent first order adsorption rate constant. Amount adsorbed was estimated by amino acid analysis of the buffer medium at different times.

decapeptide motifs are bracketed by shorter arrays of decapeptides, which have internally conserved protein sequences initiated by proline rather than alanine. Of the 14 hexapeptide motifs (AKPTYK), seven are consistently followed by a marginally deviant decapeptide (AKPTYPSTYK). Such smaller arrays may be termed *oligomotifs* (= conserved block of motifs). The two major segments are subdivided into oligomotifs of one to six consensus decapeptides punctuated by hexapeptide motifs. The DNA sequences of motifs and oligomotifs are also highly conserved. The putative C-terminal region of the *G. demissa* DOPA protein is also very highly structured and is characterized by tandem oligomotifs of form $\alpha\beta_{1\text{-}3}\gamma$, where α,β, and γ are motifs defined in Table 2.[56] [*Caveat:* Although the *M. edulis* genomic and cDNA clones are fairly homologous at their N- and C-termini, their relative motif distributions in the intervening zone are inconsistent.[55,56] In addition, the genomic clone lacks oligomotifs that occur in the cDNA clone, and vice versa. Since there seems to be only one exon for the single-copy gene of the *M. edulis* protein,[55] the source of these discrepancies is uncertain and may include recombination artifacts. The potential in repetitive proteins for alternative mRNA splicing (see Section IV.E) may further magnify apparent ambiguities. To compound these complications, we have recently uncovered heterogeneity in purified 130 kD protein, in the N-terminal sequence(s):

H_2N–N–I–Y*–K–A–K–V–	(major sequence)
y N h p	(variants)

where lower case letters represent minor amino acid alternatives, and Y* is DOPA. The N-terminus of the cloned gene is NIYNAHV,[56,57] and this clone may represent only one of several variants.]

The second family of *M. edulis* DOPA proteins (apparent M_r = 42,000 to 47,000) is depleted in DOPA but highly enriched in cystine[74] (Table 1). Electrophoretic analysis of both native and S-carboxymethylated protein indicated that the 42 to 47 kD protein occurred as a family of similar polypeptides with at least ten members. This family appears to be highly cross-linked by 25 to 30 internal disulfide bonds and forms oligomeric aggregates in neutral to alkaline solution that are resistant to sodium dodecyl sulphate treatment in the absence of a reducing agent (*intermolecular* disulfide cross-links were not evident). Only

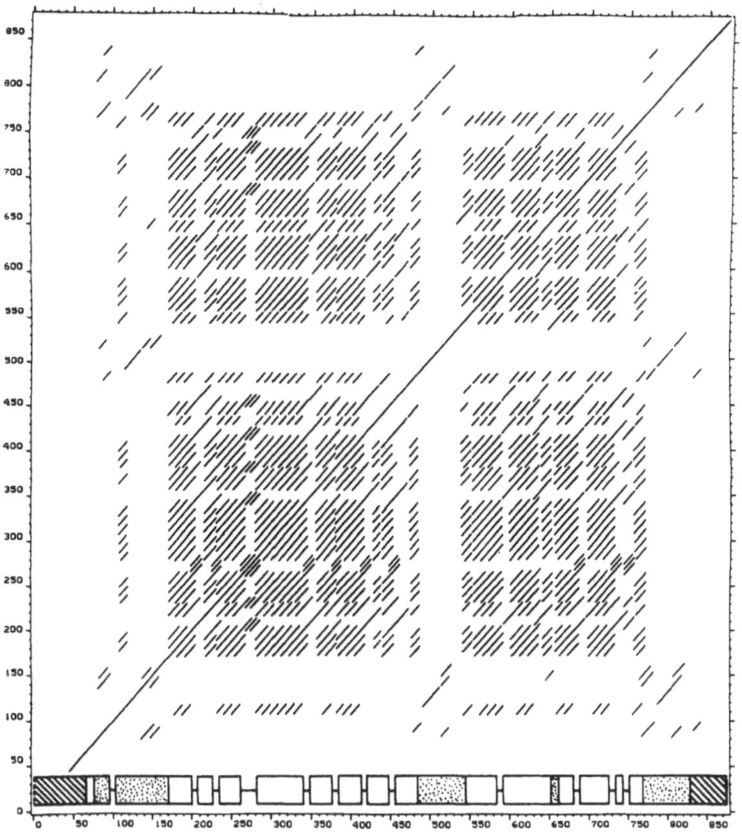

Figure 4 Homocorrelation diagram for the genomic clone of *M. edulis* 130 kD DOPA protein. The axes represent the amino acid sequence of the protein. Every segment of ten amino acids is compared with every other along the entire length of the protein. If a match is found, a mark is placed on the matrix. The stringency of the match here is ten identical amino acids out of ten in the correct sequence. In the linear representation of the gene at the bottom, open blocks represent consensus decapeptide oligomotifs; stippled blocks are proline-initiated decapeptide oligomotifs (see Table 2); and hatched blocks are degenerate regions. Hexapeptide motifs are represented by lines connecting the wide blocks.

one N-terminus was detected, with sequence H_2N-TSPxzY*DDDE-, i.e., homogeneous in positions 1 to 3 and 6 to 10, but heterogeneous in positions 4 and 5, where arginine and proline were the dominant variants. The apparent C-terminus was similar in sequence: KSPPSY*NDDDE(Y*), where the terminal DOPA residue was inferred from the amino acid composition of the pure peptide. Perhaps half of the central core of this family evidently consisted of (tandem?) repetitions of the sequence:

[CY[CV[GGYSG[PT[CQENACKPNPC[X]K]]]]]
T L GV V A

where X = 0 to 2 small polar residues, and the square brackets delineate optional extensions of the motif. A third family of proteins (apparent M_r = 11,000) has the highest DOPA content of any molluscan DOPA protein characterized to date and, unusually for molluscan DOPA proteins, contains considerable quantities of tryptophan (Table 1).[75] It is also electrophoretically heterogeneous. Similar but distinct N-termini of family members have been sequenced, but current overall sequence data do not seem to support periodic tandem repetition similar to the 130 kD DOPA protein. We have also tentatively identified at least one more family of independent DOPA proteins in *M. edulis,* but these have not yet been characterized to any degree.

This newly discovered plethora of DOPA proteins has greatly complicated a picture of the mussel byssus that had seemed relatively simple even 2 years ago. The 42 to 47 kD and 11 kD families have been extracted specifically from the byssal plaque,[74,75] where they evidently exist in significant quantities, but their function is uncertain. The glandular distribution of the 130 kD,[74] 42 to 47 kD,[74] and 11 kD[75] proteins and the byssal distribution of the latter two are consistent with the view that the 130 kD protein serves as the byssal varnish and that the other proteins are exclusively plaque components. Although the 130 kD protein has been localized immunochemically at the plaque–substratum interface,[98] recent findings of cross-reactivity of antibodies to the 42 to 47 kD family with the 130 kD proteins[74] suggest that this evidence must be re-examined urgently. Occurrence of 3-hydroxyproline in the plaque,[99] however, is consistent with some role for the 130 kD protein in that structure. The precise function of plaque DOPA proteins remains unclear, but candidate roles include surface primers, bonding agents and structural adhesives, and fillers. The presence also of non-DOPA proteins in the byssus, such as collagen,[98-100] emphasizes the composite nature of this sclerotized material and underscores the improbability of a simple solution to this structural puzzle.

C. SCHISTOSOME EGGCASE PHENOLIC PROTEIN PRECURSORS

It has not been proven rigorously that schistosome eggshell proteins do indeed contain DOPA, although DOPA is found in acid hydrolysates of schistosome eggcases.[75] However, by analogy with DOPA proteins from the related trematode *Fasciola hepatica*[48,49] and the turbellarian *Bdelloura candida*[53] and by the strong ability of certain proteins extracted from *Schistosoma mansoni* to reduce silver ions after polyacrylamide gel electrophoresis (a capacity mimicked by authentic DOPA proteins and *o*-diphenols, but not by tyrosine or sugars),[76] it is highly probable that at least some of the schistosome eggcase phenolic proteins are actually DOPA proteins. (This uncertainty underlines the fruitfulness of combined chemical and genetic studies of adhesive polypeptides, and the development of chromatographic techniques for the separation of DOPA from other amino acids in routine amino acid analysis or sequencing protocols, together with other old and new assays for DOPA in proteins,[76-81] should encourage a unification of the different strengths of both approaches.)

Schistosomes appear to synthesize at least two eggcase precursors. Cordingley and colleagues have cloned a tyrosine-rich polypeptide (cDNA clone F4) with a mol wt of 48 kD and a highly repetitive motif structure.[50] Although this protein has not been immunologically localized to eggcases, several lines of evidence imply that F4 is a true precursor. Clone F4 hybridized strongly with mature female-specific mRNA but not with mRNA from males, unisexual (i.e., nonegglaying) females, or eggs. Two open reading frames (ORFs) were present in F4, but only one, the tyrosine-rich ORF, could account for the observed incorporation of [^{3}H]-tyrosine in *in vitro* translation products of the appropriate mol wt. Neither male nor unisexual female mRNA translation products included a labeled 48 kD polypeptide. Tyrosine has been shown to be preferentially localized in schistosome vitelline cells,[82] which store eggcase precursors, and although both tyrosine and glycine are incorporated into eggcases of *S. mansoni* at comparable rates, the finished eggcase has a considerable excess of the latter.[83] Since dityrosine formation could not account for the missing tyrosine, it is probable that most of the tyrosine in the precursor proteins is converted to DOPA. Polyphenoloxidase has been implicated in eggcase maturation.[84]

The F4 protein, however, cannot alone account for the composition of the eggcase, and a glycine-rich eggcase precursor has been identified in *S. mansoni*.[51,85-88] LoVerde and colleagues have shown the gene (SMf61-46) is present in five copies, reminiscent of silkmoth chorion genes, although the detailed spatial arrangement differs.[87] Again, two ORFs are present in this gene, but antibodies raised against the glycine-rich *in vitro* translation product cross-reacted with the eggcase precursor globules of the vitelline cells.[88] An analogous glycine-rich protein, SH2-1, with two genomic copies, has been cloned from *S. haematobium*.[52]

Homocorrelation analysis of the *Schistosoma* phenolic proteins illustrates their considerable regularity of sequence, with motif periodicities of 5 for F4, 4 for SMf61-46, and 12 for SH2-1 (Figures 5A, B, and C). Clone F4 exhibits little homology with the glycine-rich proteins; both lysine and tyrosine are periodically dispersed throughout the N-terminus in the motif YGYDK, which gradually degenerates toward the C-terminus and is replaced by several interrupted polyhistidine sequences of 2 to 5 residues in length. Modest homology exists between the F4 N-terminal motif sequence and a segment of silkmoth chorion proteins.[89] Homology between the gly-rich proteins of *S. mansoni* and *S. haematobium* is considerable, especially at the N-terminus (Figure 5D). The central regions of these genes are dominated by polyglycine motifs punctuated by tyrosine or cysteine residues. The positively charged residues, mostly lysine, are concentrated at the C-termini of both, and (unlike F4) are not dispersed throughout the repetitive motif region. In the SMf61-46 motif, three glycines regularly follow the punctuator residue, while SH2-1 has a variable polyglycine chain length ranging from 1 to 15 residues. Other homologies have been identified between the deduced protein sequences and flanking DNA sequences of these clones and silkmoth and *Drosophila* chorion protein genes.[51,87] Clones SMf61-46 and SH2-1 are analogous to the glycine-rich protein loricrin derived from mouse keratinocytes,[58] where polyglycine sequences (2 to 7 residues) are punctuated by combinations of 1 to 3 serine, tyrosine, or cysteine residues (Figures 5E and F). Unsurprisingly, the degree of match is high and suggests that polyglycine, known as Nylon 2 to polymer chemists,[90] has some unique function in the formation of a variety of biological composites.

Vicinal glycine and DOPA residues are found in the motifs of proteins isolated from *F. hepatica, P. californica,* and certain mussels (Table 2). Periostracin and mussel periostracum in general also have high glycine contents (55% glycine and 2% DOPA for *M. edulis* periostracin;[46] 60 to 70% Gly and 1 to 2% DOPA for zebra mussel periostracum[91]), although the cautionary tale of the *F. hepatica* eggshell precursor Fh C (41% glycine), which appears to consist of $(GX)_n$ sequences where X = DOPA or histidine,[49] precludes us from rashly predicting universal polyglycine among glycine-rich DOPA proteins (except in Zebra mussel periostracum?). Some plant glycine-rich polypeptides (GRPs) also incorporate polyglycine sequences,[92] as does mouse epidermal keratin,[93] which has several runs of two to four glycines punctuated by tyrosine, phenylalanine, and serine residues at the N- and C-termini. Other animal glycine-rich proteins include the collagens, elastins, and the silk fibroins and sericins, but few of these contain unusual levels of tyrosine or polyglycine.

D. *FASCIOLA* EGGCASE DOPA PROTEIN PRECURSORS

Two DOPA proteins have been purified from *F. hepatica,* Fh B (31 kD)[48] and Fh C (17 kD),[49] and the cDNAs of two distinct genes coding for a 31 kD protein, VpB1 and VpB2, have been characterized, although at least six gene copies may be clustered in pairs in the genome,[94] reminiscent of *Schistosoma.* Antibodies to Fh B proteins selectively stained the eggcase precursor globules of the vitelline cells, and VpB transcripts have been localized to the vitelline cells. Mansour and colleagues isolated a similar gene, FGC, specifically expressed in the female genital complex, which shares extensive N-terminal homology with VpB1.[95] Although the FGC clone codes for a 20 kD peptide, antibodies raised against the *in vitro* translation product reacted with a 31 kD protein, which suggests that the FGC clone may have been truncated at the C-terminus (Table 3).[96] Comparison of the cDNA sequence of VpB1 to tryptic peptides derived from purified Fh B demonstrated that most of the tyrosines were converted to DOPA.[97] Homocorrelation analysis reveals that the apparent degeneracy of the sequence is limited (Table 3 and Figure 5G). The frequency of any interval length in residues between successive oligopeptide matches (of 3 identical residues out of 4) declines monotonically as the length increases but exhibits a very strong signal with a periodicity of 6 residues, extending to interval lengths as high as 168 residues between matches (Figure 5H). A hexapeptide listing of VpB and FGC reveals certain prominent features (Table 3). DOPA is predominantly found in position 1 or 4; Lys, His, and Arg favor positions 4 and 6; Asp and Glu favor positions 2 and 3; Gly dominates position 5; and hydrophobic amino acids prefer positions 1 and 3. Although several amino acid changes may be found in corresponding hexapeptides from different clones, the changes nevertheless tend to conform with the overall positional bias. The *F. hepatica* DOPA proteins exhibit little homology with the schistosome eggcase precursors, save for some vicinal DOPA and glycine residues, and the positively charged residues are distributed rather periodically through the protein, except for the first four hexapeptides (not counting the putative leader sequence), which have a net charge of +10, including the N-terminal amino group.

E. DIVERSITY AND POLYMORPHISM IN DOPA PROTEINS

1. Intraspecific Diversity

In *F. hepatica,* at least three types of DOPA protein have been identified, Fh A, B, and C, though only two have been characterized by protein chemistry, and only one has been cloned.[48,49,94] Among the schistosomes, two DOPA proteins, and perhaps other components, are required for eggcase synthesis.[50,51,85,86] Two DOPA proteins have been extracted from *Phragmatopoma californica* and may have different functions since a tanned sheath lines the inside of the worm's sandy tube.[47] Until recently, we believed that mussels were less prodigal, but the partial characterization of lower mol wt DOPA proteins from *M. edulis* has established intraspecific diversity as a common feature.[74] Compositional and immunological studies, using polyclonal antibodies to the 130 kD *M. edulis* DOPA protein, suggested that this protein was localized mostly at the interface between the plaque and the substrate and as a varnish around the thread, though some was evident in the plaque structure itself.[98-100] Since the lower mol wt DOPA proteins are also localized in the plaque,[74,75] assignment of function within the plaque matrix remains uncertain.

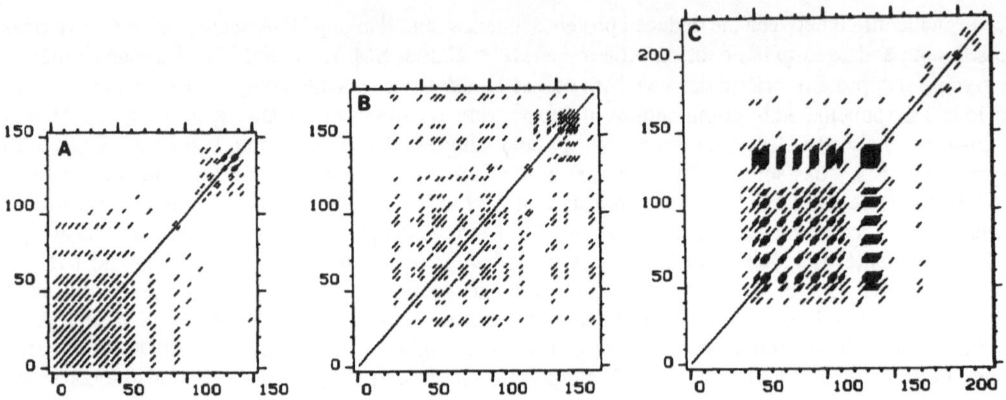

Figure 5A–C

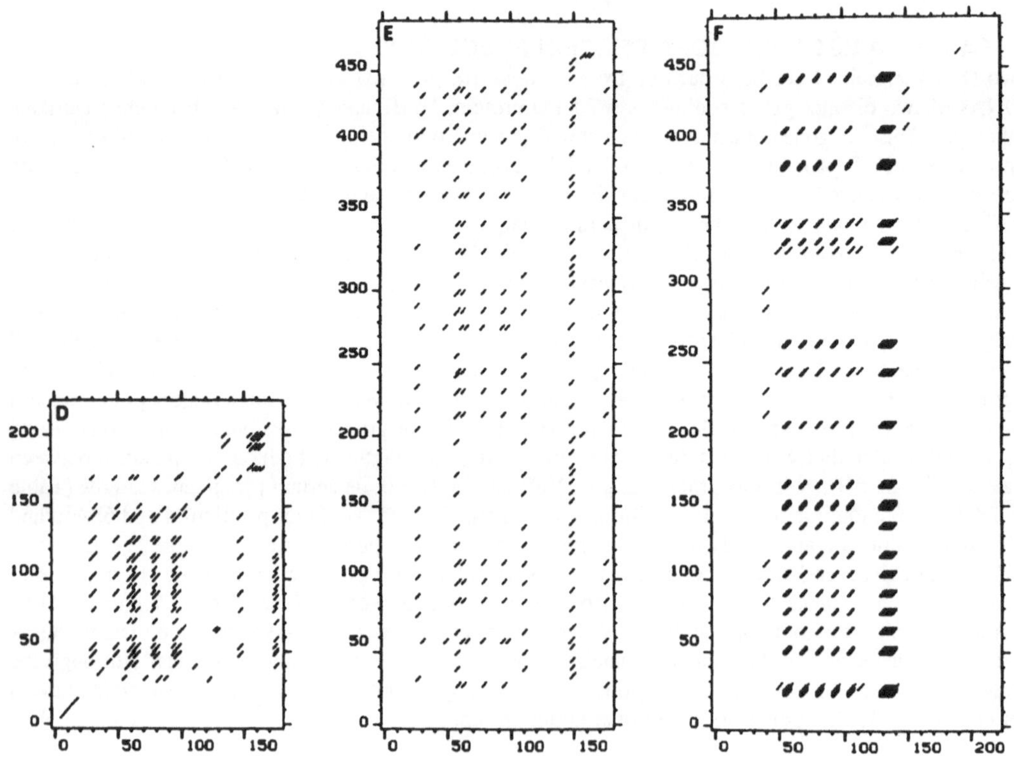

Figure 5D–F

2. Intraspecific Polymorphism

Interpretation of protein function is further complicated by the observed microheterogeneity, or polymorphism, of isolated proteins. The purified eggcase precursor Fh B was resolved into about 20 DOPA-containing peptides by isoelectric focusing (pI = 7 to 8), all in the 30 to 33 kD mol wt range, although only a single N-terminus was detected.[97] Similar polymorphism was found for the Fh C protein.[49] The 42 to 45 kD *M. edulis* DOPA proteins are also polymorphic upon isoelectric focusing, and protein extraction of *G. demissa* plaques has resulted in many electrophoretically distinct DOPA polypeptides, which may be proteolytically derived from higher mol wt precursors.[40] The 130 kD mussel DOPA protein may be polymorphic, but since its pI exceeds 10, resolution of variants is beyond the scope of common ampholytes. At least four possibilities might explain polymorphism.

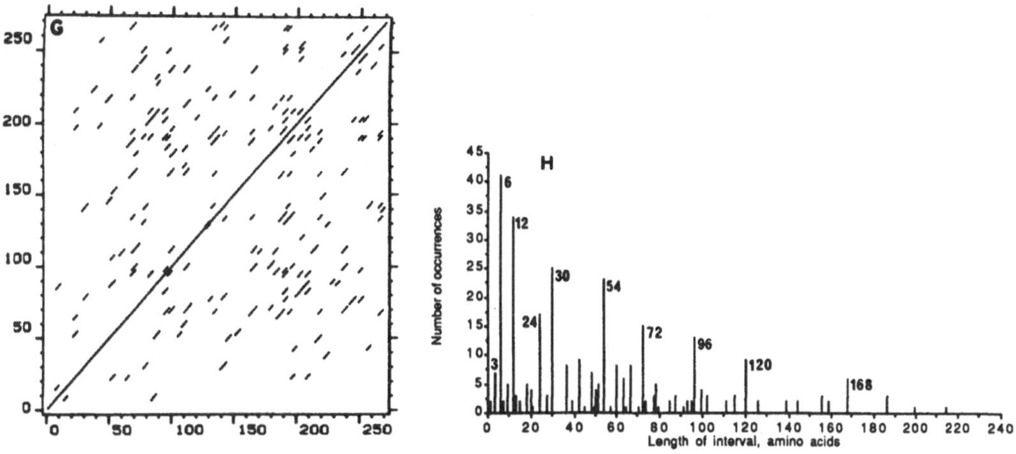

Figure 5G–H

Figure 5 Homocorrelation diagrams (see Figure 4 for explanation), where match stringency, *[x/y]* = *x* identical residues out of *y* residues in target peptide, in the correct sequence, of schistosome phenolic proteins: (A) SmF4 [3/3]; (B) Smf61-46 [3/3]; (C) Sh2-1 [4/4]. Heterocorrelation diagrams: (D) Smf61-46 (horizontal axis) and Sh2-1 (vertical axis), [4/4]; (E) Smf61-46 (horizontal axis) and loricrin (vertical axis) [4/4]; (F) Sh2-1 (horizontal axis) and loricrin (vertical axis), [5/5]. Homocorrelation diagram [3/4] for *Fasciola* DOPA protein clone VpB1 (G), and the frequency distribution of interval lengths in residues between adjacent matches (H).

1. Differing efficiencies are observed in the conversion of tyrosine to DOPA, although it is unknown whether modification is controlled in response to substratum chemical properties, is a function of maturation, or occurs in a random "shotgun" fashion.
2. There are at least 6 genomic copies of the VpB gene, and peptides specific to VpB1 and 2 and FGC have been isolated from purified Fh B, consistent with expression of all three genes.[97]
3. Alternative splicing of mRNAs, a common event in repetitive proteins,[101-103] might result in subtle differences in the hexapeptide population, and hence the size and properties of the mature protein.
4. Allelic polymorphism within the population may also contribute.

3. Intraphyletic Diversity

A third element of diversity arises from the adoption by species within a phylum of radically different protein designs to achieve the same purpose (Table 2). With the exception of *Brachidontes exustus* and *G. demissa,* which have virtually identical motifs, no two other mussels studied share a whole motif,[39] though they may share motif fractions *(submotifs),* such as PPTYK, common to *M. edulis* and *M. californianus.* Only submotifs tend to be polyspecific; motifs are generally monospecific. Mussels are classified into two subfamilies: the Mytilinae and the Modiolinae.[104] All the mussels listed in Table 2 belong to the Mytilinae except *Modiolus* and *Geukensia,* which are classified as Modiolinae. The degree of sequence homology observed between the supposedly unrelated species *B. exustus* and *G. demissa,* and between *Modiolus modiolus squamosus* and *Trichomya hirsuta,* seems taxonomically problematic. However, common submotifs also arise in species from different phyla, e.g., GYGAK in *T. hirsuta* and *P. californica,* and may reflect analogous solutions to conformational problems. The eggcase precursor proteins of the two trematode genera examined similarly have little in common.

F. EVOLUTIONARY IMPLICATIONS OF DOPA PROTEIN DIVERSITY

DOPA protein diversity (Table 2) raises questions concerning the evolutionary relationship of DOPA proteins. How does the adhesive precursor sequence change during speciation, and what is the significance of the change? Although extensive gene sequences are available solely for *M. edulis,*[55,56] the mussel motifs listed in Table 2 are limited in number and account for the bulk of the protein from each species. Ignorance of the precise order of the motifs need not forestall some general conclusions. Stochastic

Table 3 **VpB protein sequences arranged as hexads**

	vpB1	vpB2	FGC		vpB1	vpB2	FGC
–3	MKFTL	......	MKFHL	20	$^{-}$**¥DS¥GK**$^{+}$	$^{\circ}$**YGGYGK**$^{+}$	......
–2	VLLLAI	......	......		KKS^{+}	...	...
–1	VPLTLA	......	......		$^{-}$¥DD	...	...
					$^{-}$**YDTKGH**$^{+}$	......	......
					LKK^{+}	MKR^{+}	...
					$^{\circ}$**FANKGR**$^{+}$	$^{-}$**FADKGM**$^{+}$	......
1	$^{+}$**RHPHGK**$^{+}$	......	......		QSK^{+}	KSK^{+}	...
	$^{+}$**FNRHAS**$^{+}$	......	......	25	$^{-}$**FDM¥GN**$^{\circ}$	$^{-}$**FDLYGN**$^{\circ}$	......
	$^{-}$**YDDREK**$^{+}$	......	......		$^{+}$**VKADGQ**$^{-}$	$^{-}$**VEAKGK**$^{+}$	......
	$^{+}$**HRGYRK**$^{+}$	......	......		$^{\circ}$**AISNGN**$^{\circ}$	$^{-}$**¥DA¥GK**$^{+}$	......
5	$^{-}$**END¥LN**$^{\circ}$	......	......		$^{\circ}$**MNA¥GM**$^{\circ}$	$^{\circ}$**MGALGK**$^{+}$	MNA (TV
	$^{-}$**¥DLKGK**$^{+}$	......	......		$^{-}$**FDS¥GK**$^{+}$	......	VCLTPM
	$^{\circ}$**FAGRGK**$^{+}$	$^{\circ}$**FAGHGK**$^{+}$	$^{+}$**FSRTRQ**$^{+}$	30	$^{-}$**¥DQ¥GK**$^{+}$	......	ANTISM
	$^{\circ}$**A¥LHGS**$^{+}$	......	......		$^{-}$**MNDQGK**$^{+}$	......	AK).
	$^{\circ}$**FDKYGN**$^{\circ}$	......	......		$^{-}$**¥EEAGK**$^{+}$	......	
10	$^{-}$**ENERGR**$^{+}$	......	......		$^{\circ}$**YNAHGN**$^{+}$	......	
	$^{-}$**YDDQGK**$^{+}$	$^{\circ}$**YDHRGH**$^{+}$	......		$^{-}$**LDLYGH**$^{+}$	......	
	$^{\circ}$**YLLAGK**$^{+}$	$^{+}$**HSLAVK**$^{+}$	$^{\circ}$**YSLAGK**$^{+}$	35	$^{+}$**LRG¥GG**$^{\circ}$	......	
	$^{+}$**SAHDGK**$^{\circ}$	......	......		$^{\circ}$**SSAASK**$^{+}$	......	
	$^{\circ}$**YGMYGN**$^{\circ}$	$^{-}$**¥DM¥GR**$^{+}$	......		$^{-}$**SENYGN**$^{\circ}$	......	
15	$^{\circ}$**MYAKGD**$^{\circ}$	$^{\circ}$**MYAKAN**$^{+}$	......		$^{\circ}$**ARESGR**$^{+}$	......	
	$^{+}$**FKAYGN**$^{\circ}$	$^{-}$**FDAHGH**$^{+}$	......		$^{-}$**¥EP¥GR**$^{+}$	......	
	$^{-}$**EDEGAK**$^{+}$	$^{-}$**EKEGTK**$^{+}$	......	40	$^{\circ}$**YEK¥ED**–D^{-}	......	
	$^{-}$**FEEVTT**$^{\circ}$	$^{-}$**FEE¥TK**$^{+}$	......		$^{+}$**YARETP**$^{-}$	......	
	$^{+}$**FRRGGG**$^{\circ}$	......	......		$^{\circ}$**YDK¥S¥**$^{-}$	......	

Amino acid composition analysis of vpB1 (excluding signal peptide)

Amino acid type	**Hexad position — Number of incidences (% Total): 1**	**2**	**3**	**4**	**5**	**6**
+ ve	**4/2** (9.3)	**9/7** (21.4)	**7/7** (16.7)	**12/11** (28.6)	**1/1** (2.4)	**21/21** (50.0)
– ve	**4/3** (9.3)	**16/17** (38.1)	**10/9** (23.8)	**2/3** (4.8)	**4/3** (9.5)	**3/2** (7.1)
Tyr/DOPA	**13/14** (30.2)	**4/2** (9.5)	**0/0** (0)	**16/15** (38.1)	**1/0** (2.4)	**0/1** (0)
Phe	**8/9** (18.6)	**0/0** (0)	**0/0** (0)	**1/0** (2.4)	**0/0** (0)	**0/0** (0)
Other nonpolar	**10/9** (23.3)	**4/6** (9.5)	**13/12** (31.0)	**4/4** (9.5)	**4/3** (9.5)	**1/1** (2.4)
Gly	**0/0** (0)	**2/1** (4.8)	**3/3** (7.1)	**2/2** (4.8)	**29/30** (69.1)	**2/2** (4.8)
Pro, Polar	**4/3** (9.3)	**7/7** (16.7)	**9/9** (21.4)	**5/4** (11.9)	**3/4** (7.1)	**15/15** (35.7)
"Consensus"	Aromatic nonpolar	Charged (– ve > + ve)	Variable	Aromatic charged (+ ve)	Glycine	Charged (+ ve) polar

Note: Arrangement of *Fasciola hepatica* vpB and FGC protein sequences in a hexadic array. Hexads are numbered as if contiguous; negative numbers indicate the putative signal sequence of the vpB proteins. Symbols are in one-letter amino acid code, except (¥), which represents known hydroxylation of tyrosine to DOPA. The listings for vpB2 and FGC proteins give only specific hexads different from those of vpB1. The putative FGC C-terminus is given in brackets. Underlined sequences are those confirmed by peptide sequence analysis of purified protein from the vpB family. The symbols ($^{-}$), ($^{+}$), and ($^{\circ}$) give the sign of the net charge on the adjacent tripeptide sequence. For the amino acid composition analysis (excluding the putative signal sequence), two values are given thus, **a/b**: the first value in bold is calculated for an unadjusted (i.e., contiguous) hexadic vpB1$_{18\text{-}270}$ sequence; the second value is calculated for the hexad listing as given here but including only hexads emphasized in bold, i.e., excluding Asp-258 and motifs truncated to tripeptides. Values in brackets report the percent of residues in each hexad position contributed by each type of amino acid (assuming contiguous hexads); those contributing more than 20% of the total are underlined for emphasis. Positively charged amino acids (+ ve) are lysine, arginine, and histidine; negatively charged amino acids (– ve) are glutamate and aspartate; nonpolar amino acids include valine, alanine, leucine, isoleucine, and methionine; polar amino acids are serine, threonine, glutamine, and asparagine.

conversion of every motif in an array of 80 to 90 repeats into another single type of motif is implausible: mechanisms to cause this conversion during speciation must exist. Novel submotifs may arise quite unremarkably by mutation, deletion, or recombination, and may then be spliced or shuffled to generate new hybrid motifs. In related mussel species, however, we must also account for the complete rout, as in a palace revolution, of one motif family by another, although the usurpers may themselves be hybrids of conserved submotifs.

One possible mechanism would involve catastrophic deletions within a DOPA protein allele. The truncated, nonfunctional allele, bearing new hybrid motifs, might then be the seed for allelic regeneration by multiplication of the genetic remnant. A less drastic model is supplied by the theory of "molecular drive" developed to explain the stochastically unexpected conservation of DNA sequence and codon use between multiple copies of genes coding for single proteins.[105] Several individual mechanisms have been invoked to explicate this process:

1. unequal chromatid exchange within a tandem array, which will alter the copy number and may result in the fixation of one or other member of the array;
2. gene conversion, which results in the preferential replacement of one of a pair of mismatched alleles or gene sequences;
3. duplicative transposition, which involves the progressive duplication and movement of a sequence from one position to another.

Although these mechanisms have been invoked to explain the conservation of codon use in DNA motifs of involucrin (a motif-structured component of keratinocyte cell envelopes) in different primate species,[106,107] they may also be engines of creative protein design. The observational signature of minor deletions and insertions of one or a few motifs in a large protein would be variation, within experimental error, about a mean mol wt. This hypothesis further predicts hybrid proteins with domains enriched in one or another motif type, and it may well be that observation of various interspersed motifs and oligomotifs in a specific genetic clone (Figure 4) is a glimpse of a gene in the process of transformation. The fact that submotif themes such as PKITY are rare in *M. edulis*[54] but normative in *M. californianus*[41] may simply reflect continuous, stochastic substitution of one motif set by another. If this is correct, then considerable genetic variability of these proteins within a species should be expected, and our methods of distinguishing variants may simply be insufficiently sensitive or comprehensive. (It is instructive to remember that every clone represents a single DNA sequence from one cell of one organism; it is not necessarily identical in every other individual.) A suitable metaphor might be the Hollywood movie: examination of genes from species subpopulations would correspond to looking at neighboring frames in the middle of a film, while comparisons between species would be akin to comparing the opening and closing sequences — some of the actors (submotifs) may survive, but their relationships will differ. The marked motif differences between the otherwise homologous genes encoding the glycine-rich eggcase precursors in *S. mansoni* and *S. haematobium* suggest that motif substitution may be a general phenomenon. Multiple gene copies, though not demonstrated in *M. edulis*,[55] would facilitate motif variation and takeover.

Whatever the precise mechanism of motif conversion and shuffling, the implication for protein function is that within compositional parameters, considerable latitude may often exist as far as specific primary sequence is concerned. For example, there appears to be little obvious correlation between motif sequence and mussel habitat, since mussels with similar habitats may have divergent motifs (e.g., *M. edulis* and *Septifer bifurcatus*, which inhabit the intertidal rocky zone), while those with divergent habitats may have virtually identical motifs (e.g., *B. exustus* and *G. demissa*, which colonize oyster beds and tidal mudflats, respectively).[108]

G. SECONDARY STRUCTURE

Little hard data enlighten us about the secondary structure, if any, of DOPA proteins, but much ink and computer time has been devoted to speculation. The inequality of effort in these fields is in part due to the intractability of these proteins in physiologically relevant milieux, as well as the necessity of using highly denaturing methods in their purification. However, since hypothesis is often the mother of experimentation, it is worth examining existing knowledge and theoretical developments.

1. Current Structural Models

The suggestion that the *Mytilus* high mol wt DOPA proteins have a degree of secondary structure was prompted by the observation that a bias exists in the hydroxylation frequency of the two tyrosines in

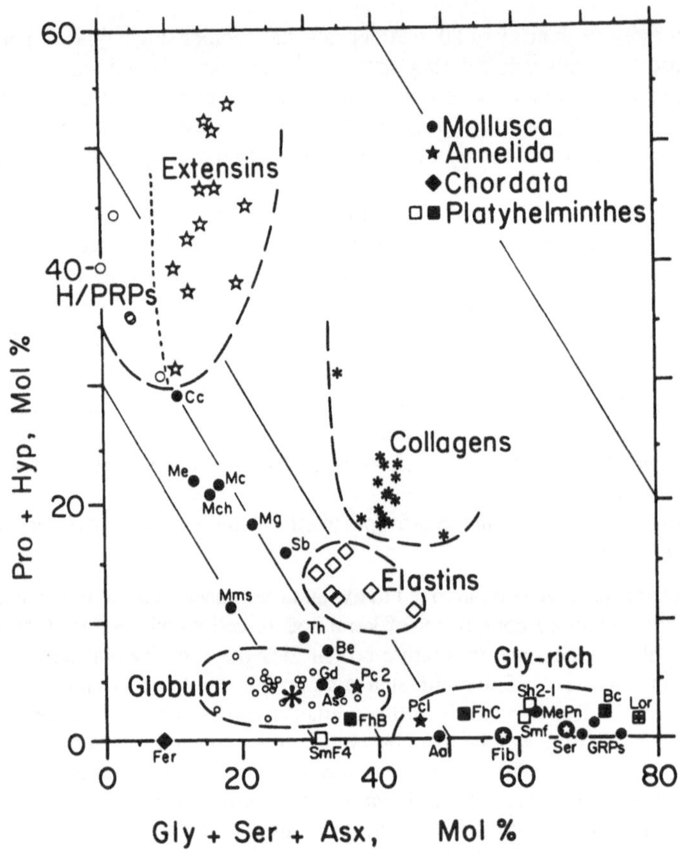

Figure 6 Correlation of the content of β-turn enforcing amino acids, proline and hydroxyproline (Pro + Hyp), with the content of small, β-turn facilitating amino acids, glycine, serine aspartate, and asparagine (Gly + Ser + Asx), for various proteins. DOPA proteins are given as solid symbols. The diagonal lines are isometric boundaries for the sum fraction of β-turn associated amino acids: from left to right, 30, 40, 50, and 100% respectively. Protein compositions are given for bivalves and trematodes, Table 1; plant extensins and H/PRPs;[9] vertebrate and invertebrate collagens;[115-117] vertebrate elastins;[118] various glycine-rich proteins; and miscellaneous globular proteins. [Asterisk (*) at the center of the Globular cluster is the Chou-Fasman average of 29 proteins — other points (small open circles) in the zone represent various individual globular proteins to assess the approximate extent of the zone.[114,119-121]] Letter codes are as in Table 1, except H/PRPs (plant hydroxyproline/proline-rich glycoproteins); Fib (silk fibroin);[122] Ser (silk sericin);[122] Lor (loricrin);[58] Fer (ferreascidin);[21] and GRPs (plant glycine-rich proteins).[92]

positions 5 and 9 of the decapeptide motif, with Tyr-9 almost invariably hydroxylated to DOPA and Tyr-5 rarely so.[41,54] *In vitro* hydroxylation by mushroom tyrosinase mimicked this bias (although *in vivo* hydroxylation is almost certainly affected by tyrosyl hydroxylase), which might be explained by differential accessibility of the two tyrosines.[109,110] The hydroxylation of proline is also sequence-specific, with Pro-3 and Pro-7 becoming 4-hydroxyproline, while Pro-6 is converted to the 3-hydroxy isomer.[41,54,61] Substrate secondary structure is believed to influence prolyl hydroxylase activity.[111,112] Circular dichroism (CD) spectra of purified 130 kD protein and a nonhydroxylated analogue of 20 tandem decapeptide motifs were similar, with a minimum at about 200 nm and a slight positive ellipticity between 220 to 230 nm.[110] These spectral features were essentially unaffected by salt and temperature (NaCl, KCl, KF, 4 to 37°C), though guanidine HCl at 60°C abolished the positive ellipticity at 220 to 230 nm. These results were interpreted to indicate a moderate (15 to 20%) content of β-turn, with the bulk of the protein assuming a random coil conformation, and were consistent with conformational energy minimization modeling methods (applied to an isolated decapeptide in an anhydrous state), which favored a 3_{10} helix over the polyproline II helix or α-helix.[113]

The method of Chou-Fasman was adduced to corroborate this interpretation and predicted a higher probability of β-turn than of α-helix or β-pleated sheet.[114] The basis for this prediction is apparent if one correlates, for a variety of proteins, the content of proline + hydroxyproline, which enforce a polypeptide turn, with the content of the four other amino acids (Gly, Ser, Asp, and Asn), which have a high probability of being found in β-turns by virtue of their small size, and which may be termed β-turn facilitators (Figure 6).[66] Various types of protein appear to map in distinct regions, and in general, the fibrous or extended proteins exceed the globular proteins in average content of β-turn enforcers + facilitators. Obviously, not all fibrous proteins consist of β-turns, but their various functional configurations favor the incorporation of β-turn-associated (i.e., small) amino acids. Most of the Platyhelminth phenolic eggcase precursors map in the glycine-rich zone and, in common with their neighbors, rarely contain more than 2 to 3 mol% proline. High glycine contents, in the absence of proline, often indicate β-sheet configurations, such as those of the prototypical silks,[122] and thus β-structures have been proposed for plant GRPs[92] and for glycine-rich periostracal proteins of marine bivalves (but not gastropods).[123] For the same reason, Simpson and co-workers,[124] analyzed a gene from *S. mansoni* very similar to Smf61-46, in which the N-terminal motifs might be organized into an array of five degenerate oligomotifs with consensus sequence YxxxYxxxCXxxxCxxx ("x" is usually Gly but sometimes Ser, Asp, or Asn). These workers suggested that each oligomotif might form a β-meander of three or four strands, with glycine or small residues at the centers (lower case letters) and tyrosine and cysteine residues at the turns (upper case letters). They pointed out that if these meanders were compactly stacked, an orderly alteration of cysteine and tyrosine residues would result on the two opposing faces formed by the turns between strands, though such details appear problematical in the light of the *S. haematobium* sequence, which has rather different oligomotif structure (Table 2). The mussel proteins, in contrast, generally contain 35 to 40 mol% of β-turn enforcers + facilitators (somewhat higher than the Chou-Fasman average of 31% for globular proteins), and display a Pro + Hyp range exceeding that of silks or globular proteins. Thus, they span a continuum between imino acid-rich plant proteins and glycine-rich proteins, which may here indicate functional interchangeability of Pro/Hyp residues with Gly/Ser/Asx residues. Such compositions, with the exception of *Aulacomya ater,* would seem to preclude a β-pleated sheet structure and encourage an extended conformation consistent with a series of β-turns, though the probability that those turns would form extensive regions of 3_{10} helix seems low, since such structures are rare and are believed to be unstable.[66] In *G. demissa* and *B. exustus,* the juxtaposition of Asx-Xaa may promote the formation of stable Asx turns, which may increase the flexibility of the peptide chain by competition with β-turns if Xaa = Pro, though Asx turns frequently accompany β-turns if Xaa is some other small residue.[125] Extended polypeptide chains seem advantageous in laminate formation where the resulting fibrous network, analogously to collagen, will resist stretching and compression in all directions. In mussel adhesive proteins, some elastin-like helical content might confer desirable shock-resistance and elasticity on the thread and plaque matrix.

2. Alternative Structural Proposals

Other feasible mussel DOPA protein configurations arise from the qualitative similarity between the CD spectra of polypeptides in random coil and those in polyproline II conformations. Polyhydroxyproline and repetitive hydroxyproline- or proline-rich polypeptides from plants (extensins) and animals (salivary proline-rich glycoproteins) also exhibit negative ellipticity around 200 to 205 nm and positive ellipticity in the 225 to 230 nm region.[126-128] The salivary proline-rich glycoproteins contain polyproline sequences of 5 or more residues, and these regions adopt the left-handed polyproline II helical conformation (pPII).[126] The extensins, in contrast, contain repetitive motifs of SP*P*P*P* (P* = hydroxyproline) punctuated by other motifs.[9] These hydroxyprolines are extensively arabinosylated, and the proteins themselves, visualized by electron microscopy following rotary shadowing, form kinked, extended rods about 80 nm long.[129] CD studies have confirmed that they adopt the pPII conformation. If, however, the extensins are deglycosylated, both CD and electron microscopy suggest that, although they probably retain a degree of pPII character in the proline-rich regions, the rest of the molecule adopts a more random conformation.[126,130] Thus, the pPII conformation of extensin is stabilized by internal H-bonding of the arabinosyl moieties to the pPII helix. The lesson of the extensins and salivary proteins suggests that a fairly long run of prolines (>4) is required to enforce a pPII conformation on other interspersed amino acids; however, the requirements may be eased if the pPII-type helix can be stabilized by H-bonding either to sugar residues on the helix itself or to other pPII-type helices, as occurs in collagen. Similar conformations have been postulated for the often phosphorylated, repetitive C-terminal regions of RNA polymerase II (Table 2).[59] Since high mol wt mussel DOPA proteins contain up to 20 mol% proline +

hydroxyproline (Figure 6), it is conceivable that segments might adopt at least a metastable pPII conformation, either for very short chain lengths or for longer lengths if stabilized by inter- or intrachain interactions. That interchain interactions occur is confirmed by the rapid precipitation of DOPA proteins in neutral or alkaline borate-free buffers.[13]

In exploring structural expectations for these protein families, the inability of unmodified Chou-Fasman parameters to predict the collagen triple helix is salutory, and the cautious speculator recognizes the danger of hiking across the wrong meadow. It is unnecessarily restrictive to shoehorn a motif-structured protein into conformations predicated by a relatively small range of globular proteins, since computer models predict only the expected and can be completely negated by unforeseen interactions. Synthetic repetitive polypeptides notoriously adopt alternative conformations in solution, depending on their experimental history.[131] Polyglycine is often a β-sheet (pGI) but may form a helix (pGII, structurally identical to pPII, except that pGII may be left- or right-handed), with 3 residues per turn and stabilized by interhelical H-bonds.[128] Proteins with exceptionally high glycine contents (>>50%) or, as in the case of some trematode eggcase precursors, with glycine-rich or polyglycine regions belied by their overall composition, may face similar structural choices. It seems improbable that the *Fasciola* Fh C protein would choose a β-structure, despite its high content of glycine. Silks containing pGII helix have been claimed,[122] and it is undecided whether the glycine-rich N- and C-terminal regions of cytoskeletal keratins adopt pGI or pGII structures, or both.[93] Moreover, silks may adopt various ordered and random coil conformations, either within a molecule, depending on the local composition/sequence and resulting in dispersed crystalline phases within an amorphous matrix, or at different stages of storage and deposition.[132,133] Though it seems intuitively reasonable that functional families of DOPA proteins should share structural parameters, their compositional range causes ambiguity. Must the *A. ater* DOPA protein (43 mol% glycine, see Tables 1, 2, and Figure 6), for example, conform to structural expectations for trematode eggcase precursors (β-meander, β-sheet, or otherwise), or does it lie in the same functional and structural continuum as other molluscan adhesive proteins?

If there is uncertainty about assigning DOPA proteins to traditional structure categories, are novel structures plausible? The orderly features of the VpB1 gene suggest that they may be. Since VpB1 lacks strong α-helix or β-sheet character, let us suppose a pGII (=pPII) helix similar to collagen, with 3 residues per turn (although a 3_{10} helical model could also be elaborated). Glycine then occurs every other turn and comprises almost 40% of one side of the resulting cylinder (Table 3, helix position 2). Glycine's position in the intervening turns is frequently occupied by Asp, Glu, or other small amino acid. DOPA and the hydrophobic amino acids predominate in position 1. The positional bias of positively and negatively charged residues in the hexapeptide results in an alternation of positively-charged, glycine-bearing turns with negatively-charged or neutral turns (Table 3). Two such fanciful helices might stabilize each other by aligning the positively charged glycine turns of one with the negatively-charged turns of the other. The binding energy of salt bridge formation exceeds that of protein H-bonds (-5 and -3 kcal mol^{-1}, respectively, assuming a dielectric constant (ε) of 4 for polyamides).[66] Collagen-like interhelix H-bond formation would be prevented by the bulky C_β atom of the nonglycine residues at the intervening turns. However, if we dare venture on quicksand, such steric hindrance might be mitigated if a *left-handed* and a *right-handed* helix together formed either a head-to-head or head-to-tail dimer stabilized by both H-bonds and electrostatic interactions. A right-handed pGII helix is energetically allowed for polyglycine and, to an extent, other (nonprolyl) poly amino acids.[66,128] The ambidexterity of such a putative Fh B pGII helical structure would be consistent with its avoidance of proline residues, which would enforce left-handedness, and the resultant regular distribution of functional groups would depend on the most favorable conformation of the dimer. VpB1 has two unbroken runs of hexapeptides with an invariant glycine in position 5 (motifs 6 to 16 and 25 to 35, Table 3), interrupted by a more degenerate (less structured?) region. A staggered alignment, where each run of glycine-conserved hexapeptides is bonded to a partner from a different molecule, would afford considerable fiber-forming potential. The heterogeneity of the Fh B protein population might then be related to differential heterodimer stability.

Although the above is a fanciful exercise in the exploration of structural novelty, such alternative models, including a stabilized 3_{10} helix[113] or left-handed α-helix,[134] still seem as plausible as the β-structures currently favored for some eggcase precursors and plant GRPs and should have distinguishable experimental consequences. Unfortunately, sophisticated *in situ* and *in vitro* structural analyses are lacking, and new methodologies are essential. (The precipitation of DOPA proteins in alkaline solution suggests that protein fibers might be induced to form slowly at an interface between acid and alkaline solutions in a manner analogous to eggcase biosynthesis.) However, any one model will probably fail to circumscribe the range of DOPA protein structures, especially considering the wide range of motif lengths, and diverse environments may impose different conformations on superficially similar polypeptides.

H. BIOSYNTHESIS OF SCLEROTIZED STRUCTURES

The participation of DOPA proteins and polyphenoloxidases in mussel byssal thread synthesis has been reviewed,[2,5,13] and so we will concentrate instead on an elegant model elaborated for schistosome eggcase formation, which may be generally relevant to the synthesis, storage, and secretion of other DOPA proteins.[135] Eggcase formation is an elegant process, which can be outlined as follows.[136,137] Ova are released periodically into the oviduct and subsequently encounter the vitelline (= shell) cells that have been simultaneously released from the vitellaria. The vitellaria are diffuse and extensive glands that stain intensively for *o*-diphenols and contain the DOPA protein Fh B in *F. hepatica*.[48,96] The eggcase precursors are stored in vitelline droplets in the cytoplasm of the vitelline cells. The cell mass containing an ovum and 30 to 40 companion vitelline cells then travels to the Mehlis gland, where it encounters Mehlis gland secretions, and the vitelline cells release their droplets, which fuse and harden into a sclerotized shell around the entire mass. The residue of the vitelline cells provides the yolk for the ovum.

Although it was long known that sclerotization involving polyphenoloxidases and phenolic precursors (now known to be DOPA proteins) was responsible for shell hardening,[138] molecular events had remained obscure. Wells and Cordingley,[135] experimenting with *S. mansoni,* have made the recent suggestion that the vitelline droplets are membrane-bound secretory vesicles enclosing what appears to be a protein emulsion, rather than a continuous included phase, a conclusion inferred from the mottled appearance of the material within the vesicles when tissue is prepared for electron microscopy.[135,139,140] Secretory bodies with analogous appearance have been observed in the cells of glands concerned with byssal thread synthesis[141] and have been found also, in secreted form, in byssal plaques under conditions of incomplete sclerotization.[99] The inversion of such emulsions was postulated to account for the foam-like appearance in electron micrographs of plaque cross-sections.[13] Wells and Cordingley additionally demonstrated that if adult egglaying females were treated with agents expected to raise the pH of cellular compartments (NH_4Cl, monensin, methylamine, and chloroquine), the vitellaria developed extensive autofluorescence analogous to that exhibited by mature eggcases. Yellow/green fluorescence has been observed during isolation and storage of *F. hepatica* DOPA proteins, although the phenomenon is poorly reproducible,[75] and yellow-fluorescent products derived from oxidized peptidyl DOPA analogues have been isolated but not yet fully characterized.[142] Wells and Cordingley suggested that the exocytotic vesicles containing eggcase precursor are probably acidic, since low pH would stabilize DOPA proteins against premature oxidation and cross-linking, and such stabilization of secreted low mol wt *o*-diphenols is well known.[143] Electron microscopy showed that the exocytotic vesicles had fused within the cytoplasm upon increasing the pH with NH_4Cl and had developed electron-dense microspines similar to those found on the mature eggcase.[144] Treatment with the Ca^{++} ionophore A23187, which induces secretion in other systems,[145] had similar effects, except that the fusion event appeared to occur outside the cells, presumably following exocytosis. The amino acid composition of isolated, fused vitelline droplets approximated that of mature eggcases.

The following hypothesis was proposed by Wells and Cordingley to account for these phenomena. The DOPA-containing eggcase precursor proteins are packaged at low pH (circa 5.5) in the vitelline vesicles in an emulsion, stabilized in a pH-dependent manner by the C-terminal histidine-rich regions of the protein F4, which will have a net positive charge at pH values below the histidine side group pKa of 6.5. Also within these vesicles is packaged an inactive propolyphenoloxidase, thus preventing premature oxidation of the DOPA proteins. Upon secretion, the polyphenoloxidase is activated by an unknown agent, and the process of sclerotization begins. Copackaging of inactive polyphenoloxidases and DOPA proteins is common,[141,146] and activation can often be mimicked by limited proteolysis or detergent treatment.[147] As the emulsion is expelled into the surrounding medium of the Mehlis gland, which has normal physiological pH of 7.4, the pH around the globules will slowly rise, the histidine residues will deprotonate, and the individual globules of the emulsion will fuse. This fusion event may occur around an apparent "interface" observed between the vitelline cells and the Mehlis gland secretions,[148] which may maintain a pH gradient to promote the orderly fusion of the globules into a shell around the cell mass. In the case of *F. hepatica,* the DOPA proteins are rather different in sequence, but protein Fh C is also histidine-rich and might serve a similar purpose. The mussel proteins, though uniformly basic, nevertheless contain little histidine. Although much of this model is speculative, it is consistent with known secretory phenomena, explains a variety of observations, and offers novel experimental ideas.

I. POLYPEPTIDE RESINS AS MATRICES FOR BIOMINERALIZATION

Proteinaceous organic matrices are common constituents of a wide variety of invertebrate mineralized composite structures, for example mollusk shell,[149,150] corals,[151-153] and echinoderm tests and spines.[154,155] While presence of an organic matrix, either soluble or insoluble, is readily determined, the actual role and/or function of these matrices in biomineral deposition and orientation remains an enigmatic problem. State-of-the-art reviews of biomineralization with specific attention devoted to the role of organic matrix are available,[156-158] and so we shall focus on the potential contribution of DOPA proteins to the three types of organic matrices that occur in molluscan shells: soluble and insoluble matrix and periostracum. Soluble and insoluble matrix components form a scaffold for mineralization of the inner shell layers (prismatic, nacreous, foliated calcite, myostracal), while the outer organic layer, the periostracum, is the template upon which nascent shell layers grow but is not itself extensively mineralized.[17,159] Organic matrix serves to delineate individual crystals and may also occur as inclusions within forming crystals.[34] The tissue source of matrix components is believed to be the molluscan mantle, a thin epithelial layer that envelops the body mass and is in close proximity to the inner shell surface. Periostracum is extruded from the outer fold of the mantle edge, whereas soluble and insoluble components are presumably secreted from the epithelial surface; epithelial secretion of matrix occurs in other phyla as well (Brachiopoda, Coelenterata).[158] Two components involved in sclerotization (*o*-diphenol and phenoloxidase activity) have been detected in mantle cellular inclusions and molluscan mantle edge and are presumably involved in periostracum formation.[160,161] Prevailing theory holds that insoluble sclerotized matrix is the hydrophobic framework of biomineral formation, while the soluble, charged components coat the framework and are involved in epitactic or ionotropic crystal nucleation and growth.[162-164] The regulation of crystal size, orientation, and morphology is attributed to the amino acid composition of the organic matrix.[165] Indeed, soluble matrix proteins from *Mytilus edulis* have been shown to influence orientation of crystal nucleation *in vitro*.[166] Intercalation of organic material within and around forming crystals provides a degree of flexibility to mineralized composites.[167] Therefore, inclusion of an organic matrix component not only provides a substratum for mineral orientation and deposition but also imparts structural integrity to the mineralized structure.[168,169]

Amino acid composition is strikingly different in soluble vs. insoluble matrix; for example in the oyster, *Crassostrea virginica,* the former is rich in acidic residues (30 mol% Asp), and the latter has a greater proportion of nonpolar residues (30 mol% Gly and 25 mol% Ala).[170] Although collagen has often been proposed as a matrix component, proline and hydroxyproline levels are extremely low in soluble and insoluble phases. Insoluble matrix also contains the sclerotization precursor DOPA (1 mol%). Degens et al.[149] provided early evidence for DOPA in molluscan shell matrix and periostracum, but unfortunately, DOPA content is not usually determined in studies of shell matrix amino acid composition.[150,171,172] Many studies refer to "tanned" components or conchiolin-like material in mineralized structures yet provide no direct evidence for the presence of catechols or phenoloxidase activity. Several soluble matrix components have been isolated and characterized, and DNA sequence information exists for a soluble component of sea urchin spicules.[155,173-175] Conversely, insoluble matrix components are not readily extracted or characterized. Periostracin, a highly unstable Gly/Tyr-rich (50 mol% and 10 mol%, respectively) protein, is a formic acid-soluble component of *Mytilus edulis* periostracum: at present it is the only reported sclerotin precursor isolated from molluscan periostracum.[46] Periostracin has not been localized to any particular layer of periostracum (Figure 1), although composition of whole *M. edulis* periostracum (50 mol% Gly, 10 mol% Tyr, 2 mol% DOPA) is similar to periostracin. The silk-like insoluble matrices of *Mytilus edulis* prismatic and nacreous shell layers are rich in Gly (30 mol%) as well as Ala (22 to 27 mol%) with low levels of DOPA and Tyr.[176] The intractability of the insoluble matrix presumably results from sclerotization, and no peptides or precursors involved in formation of insoluble matrix have yet been isolated or characterized, either from the mantle epithelial cells or the matrix itself. The presence of high levels of tyrosyl and lysyl residues are a possible source of covalent cross-links in shell matrix of *Mercenaria mercenaria,* and phenoloxidase activity is also present in the newly deposited organic phase. The rate of sclerotization in periostracum and insoluble matrix is not known, although the absence of sclerotization-associated growth rings for newly formed *M. mercenaria* periostracum suggests a slow rate at the mantle–periostracum interface.[177]

The polypeptide resins involved in shell formation may also function in the attachment mechanism of sessile species to substrata. Metamorphosis of larval oysters into the sedentary adult stage involves selection and irreversible attachment to a suitable hard substratum.[178] Following reversible attachment to the substratum with a byssal thread, the larva uses the foot to orient the left valve against the substratum and begins the process of permanent attachment. The presence of tyrosine, *o*-diphenol, and phenoloxidase activity in the larval cement and tissue has been documented[179-181] and occur as well in the juvenile shell and mantle responsible for production of the left (attached) valve. Tomaszewski[181] proposed that oyster periostracum secreted from the left mantle lobe is the primary mechanism of attachment by oysters to the substratum. Since periostracum is also involved in the initiation of shell formation, it might serve a dual purpose in oysters as an attachment cement or adhesive. Serpulid polychaetes also utilize an organic matrix associated with calcium granules as an adhesive, and matrix deposition and hardening in the calcareous tube was referred to as "setting".[158,182] Whether such setting involves quinone-mediated sclerotization remains to be determined.

The occurrence of DOPA in molluscan insoluble matrix and periostracum has been established. It remains uncertain, however, whether DOPA residues are involved primarily in sclerotization and adhesion or whether they also contribute to the processes of crystal nucleation and deposition.

J. CURING: THE MAINTENANCE OF STRUCTURAL INTEGRITY

Despite much effort, no chemical cross-link between DOPA and any other moiety has yet been demonstrated. Nevertheless, the ready oxidation of DOPA in physiological media and the chemical reactivity of the resultant quinones necessitate such cross-link formation, provided it is sterically permitted in the solid matrix of a composite structure.[12] The chemistry of cross-links involving model compounds has been extensively reviewed, with reference both to insect cuticle sclerotization[11] and to DOPA proteins.[13] Many amino acids have side groups that will yield nucleophilic Michael addition products with *o*-quinones, with the general reactivity order –SH > –NHR (pyrrolidine) > $-NH_2$ > –OH > –COOH.[185,186] Comproportionation of an *o*-diphenol and *o*-quinone will result in free radical formation,[187] resulting in yet another suite of reactions, including DOPA dimerization.[188] Aromatic free radicals, potentially derived from DOPA quinones, have been detected in periostracum.[189] Diverse other cross-linking reactions are possible.[12,13] Peptidyl DOPA quinone can also tautomerize to the enamide α,β-dehydroDOPA.[190] Such tautomerization may occur during biosynthesis of several unusual low molecular DOPA oligopeptides,[191] including the tunichromes and halocyamines, and celenamides from tunicates and sponges, respectively.[192-194] These oligopeptides contain DOPA, dehydroDOPA, and their 5-hydroxylated derivatives and may be involved in metal chelation or sclerotization. Since DOPA proteins isolated from two species of tunicate have very atypical compositions, their oxidation chemistries may be expected to be unique: indeed, they may serve as precursors for the DOPA oligopeptides.[191] DehydroDOPA or its quinone would be expected to increase the rotational freedom of peptide bond through resonance with the ring, resulting in a destabilization of some secondary structures[195] (although dehydrophenylalanine stabilizes β-turns[196]), and would also offer a new range of cross-linking opportunities, including participation of the DOPA α and β carbons and thus the polypeptide backbone itself. DehydroDOPA is also potentially susceptible to free radical-initiated chain propagation reactions, such as those responsible for the synthesis of acrylic polymers.[4] Clearly, the fate of DOPA in any biological matrix will be dependent on the size, accessibility, mobility, and reactivity of neighboring groups and cannot be predicted without such knowledge.

The lack of chemical identification of such cross-links may reflect both the variety of possible reaction products and their probable lability under the harsh extraction conditions required to dismember a sclerotized matrix. However, other methods of sclerotization cannot be excluded. The chelation of metals by DOPA residues will result in strong bonds between neighboring protein chains,[5,13] and indeed, the mussel byssus often contains high concentrations of heavy metals, although it appears that this is primarily a result of detoxification mechanisms.[197] Intriguingly, cross-linking of polypeptide chains by Fe(II) chelation can promote α-helix formation.[198] Dehydration induced by introduction of hydrophobic catechols has also been postulated as a factor in insect cuticle sclerotization,[199] but the application of this hypothesis to aquatic sclerotization is problematic, since hydroxylation of peptidyl tyrosines to DOPA would be expected to decrease the hydrophobicity of the matrix, rather than the reverse.

K. PERMEABILITY OF SCLEROTIZED MATRICES

Although sclerotization results in a structure resistant to chemical, abrasive, and microbial degradation, sclerotized matrices such as the eggcases of *F. hepatica* and the dogfish *Scyliorhinus canicula* must remain permeable to water, salts, nutrients, and metabolic wastes. Both eggcases appear fibrous when examined by electron microscopy. The *Fasciola* eggcase is relatively homogeneous and appears to contain randomly oriented fibrils (2.4 × 20 nm) leading to a reticulated appearance.[198] A reticulated structure is also evident in the immature Bdelloura eggcase, although on maturity, this network becomes coated with another protein layer (Figure 1, E and F).[53] The dogfish eggcase is composed of orthogonally stacked laminae of closely packed collagen fibrils,[199] which appear histochemically to be tanned by a protein precursor (DOPA protein?).[23,24,200]

The *Fasciola* eggcase is permeable to ions and low mol wt solutes,[198] and its permeability to alcohols (methanol to butanol) is inversely proportional to their mol wt.[201] The observed self-diffusion coefficient of water through the tanned dogfish eggcase was found to be consistent with the existence of pores 1.36 nm in radius,[202] assuming a channel model (although in the light of the eggcase ultrastructure, the validity of this simple model is not obvious). Preliminary experiments in our laboratory on the permeability and chemical reactivity of mussel byssus show that the undamaged thread reacts positively for DOPA (i.e., stains red) with the low mol wt reagents of the *o*-diphenol specific Arnow assay,[79] but no reaction occurs when the thread is tested with the redox cycling assay for quinones,[81] unless it has first been mechanically abraded or cut. In cut thread, however, the redox cycling assay results in the deposition of the purple formazan precipitate as a collar around the circumference of the cut end: the core of the thread, as well as regions of the cortex further away from the cut, remain clear. Thus the DOPA residues are not exposed on the thread surface, and it is conceivable that some barrier to diffusion, though immaterial to the inorganic Arnow reagents, must be breached to allow penetration by the organic redox cycling reagents.

The specific influence of DOPA protein sclerotization on matrix permeability is uncertain. A commercial preparation of *M. edulis* DOPA protein(s) (Cell-Tak™) has been shown to form thick layers (4 to 100 nm) on germanium surfaces in proportion to the protein concentration of the applied solution,[203] but paradoxically, no adverse effects of Cell-Tak on the permeability of coated 50 kD cut-off dialysis membrane or of coated corneal epithelium have been observed.[204] It is possible that cross-linking of proteins by sclerotization or specific interactions with other components of the natural composite are essential to develop the restrictive permeability properties found in natural matrices. Here, too, may lie a reason for the evident periodicity in the distribution of DOPA/tyrosine residues in most characterized sclerotins, since, if DOPA is truly involved in cross-linking, the periodicity may well influence both size and distribution of pores in the sclerotized mesh.

V. SUMMARY

Natural adhesives, varnishes, and composites often consist of protein polymers that incorporate the cross-linking agent DOPA in their primary sequence and enzymatic curing or hardening agents such as polyphenoloxidases. They are secreted as premixed multicomponent dispersions or emulsions. Upon secretion, in the case of sclerotized composite materials, the polyphenoloxidase is activated, and the process of hardening begins with the oxidation of the DOPA residues. Despite the diverse functions of the DOPA proteins, the details of secretion and cross-linking present common problems and have often been solved in similar ways in different animal phyla. However, various functions often require specialized interactions between the DOPA proteins and other system components, and these are provided by the incorporation of appropriate amino acids or by their posttranslational hydroxylation. It is not yet certain whether the structures of DOPA proteins will resemble each other, whether they will fall into any of the established structure categories, or even whether they have any long range and nonrandom structure at all. However, the study of the extraordinarily successful use, in the hostile aqueous environment, by various organisms' DOPA proteins has yielded clues about the minimal chemical and physical properties of such resins and offers the opportunity to apply that knowledge in biotechnology. The diversity of their biological applications points the way to exploitation in the fields of dentistry, surgery, ceramics, fabrics, and anticorrosives, to name but a few. The heterogeneity of their molecular structures indicates that they may serve better as models for designer polymers or polymer mixtures, tailored to a specific need, rather than merely as templates for the faithful reproduction of the original proteins. Not all the pieces of the puzzle are yet available. Vast areas of ignorance relevant to their biological success include the physicochemical interactions of DOPA proteins with their copolymer catecholoxidases and other scleropolymers (both in secretory vesicles and in maturing sclerotin), the nature and stability of secondary structure, the influence of quinone chemistry, and the mechanisms of biodegradation. Nevertheless, the last decade has greatly expanded our base of knowledge and rewarded us with pleasurable and often surprising insights.

VI. ACKNOWLEDGMENTS

We are indebted to our colleagues in this laboratory and our collaborators elsewhere for fruitful discussions and cross-fertilization of ideas, and to grant #N00014-84K-0290 from the Office of Naval Research.

REFERENCES

1. **Waite, J. H.,** Adhesion in byssally attached bivalves, *Biol. Rev. Cambridge Phil. Soc.*, 58, 209, 1983.
2. **Waite, J. H.,** Nature's underwater adhesive specialist, *Int. J. Adhesion Adhesives,* 7, 9, 1987.
3. **Grayson, M.,** *Encyclopaedia of Composite Materials and Components,* Wiley-Interscience, New York, 1983.
4. **Kinloch, A. J.,** *Adhesion and Adhesives,* Chapman and Hall, London, 1987.
5. **Waite, J. H.,** Marine adhesive proteins: natural composite thermosets, *Int. J. Biol. Macromol.,* 12, 139, 1990.
6. **Harkin, J. M.,** Lignin — a natural polymeric product of phenol oxidation, in *Oxidative Coupling of Phenols,* Taylor, W. I., and Battersby, A. R., Eds., Marcel Dekker, New York, 1967, 243.
7. **Fry, S. C.,** Cross-linking of matrix polymers in the growing cell walls of angiosperms, *Annu. Rev. Plant Physiol.,* 37, 165, 1986.
8. **Varner, J. E. and Lin, L.-S.,** Plant cell wall architecture, *Cell,* 56, 231, 1989.
9. **Showalter, A. M. and Rumeau, D.,** Molecular biology of plant cell wall hydroxyproline-rich glycoproteins, in *Organization and Assembly of Plant and Animal Extracellular Matrix,* Adair, W. S. and Mecham, R. P., Eds., Academic Press, 1990, 247.
10. **Andersen, S. O.,** Sclerotization and tanning of the cuticle, in *Comprehensive Insect Physiology, Biochemistry and Pharmacology,* Vol. 3, Kerkut, G. A. and Gilbert, L. I., Eds., Pergamon Press, Oxford, 1985, 59.
11. **Sugumaran, M.,** Molecular mechanisms for cuticular sclerotization, *Adv. Insect Physiol.,* 21, 179, 1988.
12. **Peter, M. G.,** Chemical modification of biopolymers by quinones and quinone methides, *Angew. Chem. Int. Ed. Engl.,* 28, 555, 1989.
13. **Rzepecki, L. M. and Waite, J. H.,** DOPA proteins: versatile varnishes and adhesives from marine fauna, in *Bioorganic Marine Chemistry,* Vol. 4, Scheuer, P. J., Ed., Springer-Verlag, Berlin, 1991, 119.
14. **Waite, J. H.,** The phylogeny and chemical diversity of the quinone-tanned glues and varnishes, *Comp. Biochem. Physiol. B,* 97, 19, 1990.
15. **Yonge, C. M. and Thompson, T. E.,** *Living Marine Molluscs,* Collins, London, 1976.
16. **Waite, J. H.,** Quinone-tanned scleroproteins, in *The Mollusca,* Vol. 1, Hochachka, P. W., Ed., Academic Press, New York, 1983, 467.
17. **Saleuddin, A. S. M. and Petit, H. P.,** The mode of formation and the structure of the periostracum, in *The Molluscs,* Vol. 4, Saleuddin, A. S. M. and Wilbur, K. M., Eds., Academic Press, New York, 1983, 199.
18. **Cordingley, J. S.,** Trematode eggshells: novel protein biopolymers, *Parasitol. Today,* 3, 341, 1987.
19. **Vincent, J. F. V.,** *Structural Biomaterials,* Halstead Press, New York, 1982.
20. **Jensen, R. A. and Morse, D. E.** The bioadhesive of *Phragmatopoma californica* tubes: a silk-like cement containing L-DOPA, *J. Comp. Physiol. B,* 158, 317, 1988.
21. **Dorsett, L. C., Hawkins, C. J., Grice, J. A., Lavin, M. F., Merefield, P. M., Parry, D. L., and Ross, I. L.,** Ferreascidin: a highly aromatic protein containing 3,4-dihydroxyphenylalanine from the blood cells of a stolidobranch ascidian, *Biochemistry,* 26, 8078, 1987.
22. **Azumi, K., Yokosawa, H., and Ishii, S.,** Presence of 3,4-dihydroxyphenylalanine-containing peptides in hemocytes of the ascidian, *Halocynthia roretzi, Experientia,* 46, 1020, 1990.
23. **Hunt, S.,** The selachian egg case collagen, in *Biology of Invertebrate and Lower Vertebrate Collagens,* Bairati, A. and Garrone, R., Eds., Plenum Press, New York, 1984.
24. **Cox, D. L., Mecham, R. P., and Koob, T. J.,** Site-specific variation in amino acid composition of skate egg capsule (*Raja erinacea* Mitchell, 1825), *J. Exp. Mar. Biol. Ecol.,* 107, 71, 1987.
25. **Waite, J. H. and Morse, D. E.,** unpublished data.
26. **Roche, J., Fontaine, M., and Leloup, J.,** Halides, *Chem. Zool.,* 5, 493, 1963.
27. **Hartshorn, S. R.,** Introduction, in *Structural Adhesives,* Hartshorn, S. R., Ed., Plenum Press, New York, 1986, 1.
28. **Pocius, A. V.,** Fundamentals of structural adhesive bonding, in *Structural Adhesives,* Hartshorn, S. R., Ed., Plenum Press, New York, 1986, 23.
29. **Robins, J.,** Phenolic resins, in *Structural Adhesives,* Hartshorn, S. R., Ed., Plenum Press, New York, 1986, 69.
30. **Smith, D. C.,** Dental cements, in *An Outline of Dental Materials and their Selection,* O'Brien, W. J. and Ryge, G., Eds., W. B. Saunders, Philadelphia, 1978, 152.

31. **McBride, M. B. and Wesselink, L. G.,** Chemisorption of catechol on gibbsite, boehmite, and noncrystalline aluminum surfaces, *Environ. Sci. Technol.*, 22, 703, 1988.
32. **Kung, K. H. and McBride, M. B.,** Electron transfer processes between hydroquinone and iron oxides, *Clays Clay Miner.*, 36, 303, 1988.
33. **LaKind, J. S. and Stone, A. T.,** Reductive dissolution of goethite by phenolic reductants, *Geochim. Cosmochim. Acta*, 53, 961, 1989.
34. **Berman, A., Addadi, L., Kvick, Å., Leiserowitz, L., Nelson, M., and Weiner, S.,** Intercalation of sea urchin proteins in calcite: study of a crystalline composite material, *Science*, 250, 664, 1990.
35. **Comyn, J.,** The relationship between joint durability and water diffusion, in *Developments in Adhesives*, Vol. 2, Kinloch, A. J., Ed., Applied Science Publishers, Barking, U.K., 1981, 279.
36. **Burke, R. D.,** The induction of marine invertebrate larvae: stimulus and response, *Can. J. Zool.*, 61, 1701, 1983.
37. **Morse, D. E.,** Recent progress in larval settlement and metamorphosis: closing the gaps between molecular biology and ecology, *Bull. Mar. Sci.*, 46, 465, 1990.
38. **Roberts, L.,** Zebra mussel invasion threatens U.S. waters, *Science*, 249, 1370, 1990.
39. **Rzepecki, L. M., Chin, S.-S., Waite, J. H., and Lavin, M. F.,** Molecular diversity of marine glues: polyphenolic proteins from five mussel species, *Mol. Mar. Biol. Biotech.*, 1, 78, 1991.
40. **Waite, J. H., Hansen, D. C., and Little, K. T.,** The glue protein of ribbed mussels *(Geukensia demissa):* a natural adhesive with some features of collagen, *J. Comp. Physiol. B*, 159, 517, 1989.
41. **Waite, J. H.,** Mussel glue from *Mytilus californianus:* a comparative study, *J. Comp. Physiol. B*, 156, 491, 1986.
42. **Saez, C., Pardo, J., Gutierrez, E., Brito, M., and Burzio, L. O.,** Immunological studies of the polyphenolic proteins of mussels, *Comp. Biochem. Physiol. B*, 98, 569, 1991.
43. **Waite, J. H. and Burzio, L. O.,** unpublished data.
44. **Pardo, J., Gutierrez, E., Saez, C., Brito, M., and Burzio, L. O.,** Purification of adhesive proteins from mussels, *Prot. Express. Purif.*, 1, 1, 47, 1990.
45. **Waite, J. H.,** unpublished data.
46. **Waite, J. H., Saleuddin, A. S. M., and Andersen, S. O.,** Periostracin — a soluble precursor of sclerotized periostracum in *Mytilus edulis* L., *J. Comp. Physiol. B*, 130, 301, 1979.
47. **Waite, J. H., Jensen, R. A., and Morse, D. E.,** Cement precursor proteins of the reef-building polychaete *Phragmatopoma californica* (Fewkes), *Biochemistry*, 31, 5733, 1992.
48. **Waite, J. H. and Rice-Ficht, A. C.,** Presclerotized eggshell protein from the liver fluke *Fasciola hepatica*, *Biochemistry*, 26, 7819, 1987.
49. **Waite, J. H. and Rice-Ficht, A. C.,** A histidine-rich protein from the liver fluke *Fasciola hepatica*, *Biochemistry*, 28, 6104, 1989.
50. **Johnson, K. S., Taylor, D. W., and Cordingley, J. S.,** Possible eggshell protein gene from *Schistosoma mansoni*, *Mol. Biochem. Parasitol.*, 22, 89, 1987.
51. **Bobek, L. A., Rekosh, D. M., van Keulen, H., and LoVerde, P. T.,** Characterization of a female-specific cDNA derived from developmentally regulated mRNA in the human blood fluke *Schistosoma mansoni*, *Proc. Natl. Acad. Sci. U.S.A.*, 83, 5544, 1986.
52. **Bobek, L. A., LoVerde, P. T., and Rekosh, D. M.,** *Schistosoma haematobium:* analysis of eggshell protein genes and their expression, *Exp. Parasitol.*, 68, 17, 1989.
53. **Huggins, L. and Waite, J. H.,** Eggshell formation in Bdelloura candida, an ectoparasite turbellarian of the horseshoe crab *Limulus polyphemus*, *J. Exp. Zool.*, in press, 1993.
54. **Waite, J. H., Housley, T. J., and Tanzer, M. L.,** Peptide repeats in mussel glue protein: theme and variations, *Biochemistry*, 24, 5010, 1985.
55. **Filpula, D. R., Lee, S.-M., Link, R. P., Strausberg, S. L., and Strausberg, R. L.,** Structural and functional repetition in a marine mussel adhesive protein, *Biotechnol. Prog.*, 6, 171, 1990.
56. **Laursen, R. A., Ou, J.-J., Shen, X.-T., and Connors, M. J.,** Characterization and structure of mussel adhesive proteins, in *Mat. Res. Soc. Symp. Proc.*, Vol. 174, Materials Research Society, 1990, 237.
57. **Laursen, R.,** Characterization of marine adhesive proteins, Report # N00014-86K-0217, Office of Naval Research Final, 1989.
58. **Mehrel, T., Hohl, D., Rothnagel, J. A., Longley, M. A., Bundman, D., Cheng, C., Lichti, U., Bisher, M. E., Steven, A. C., Steinert, P. M., Yuspa, S. H., and Roop, D. R.,** Identification of a major keratinocyte cell envelope protein, Loricrin, *Cell*, 61, 1103, 1990.

59. **Corden, J. L.,** Tails of RNA polymerase II, *Trends Biochem. Sci.,* 15, 383, 1990.
60. **Schaefer, J., Kramer, K. J., Garbow, J. R., Jacob, G. S., Stejskal, E. O., Hopkins, T. L., and Spier, R. D.,** Aromatic cross-links in insect cuticle: detection by solid state ^{13}C and ^{15}N NMR, *Science,* 235, 1200, 1987.
61. **Kassel, D. B. and Biemann, K.,** Differentiation of hydroxyproline isomers and isobars in peptides by tandem mass spectrometry, *Anal. Chem.,* 62, 1691, 1990.
62. **Ananthanarayanan, V. S.,** Structural aspects of hydroxyproline-containing proteins, *J. Biomol. Struct. Dynam.,* 1, 843, 1983.
63. **Piez, K. A. and Gross, J.,** The amino acid composition of some fish collagens: the relation between composition and structure, *J. Biol. Chem.,* 235, 995, 1960.
64. **Benedict, C. V. and Waite, J. H.,** Assay of dihydroxyphenylalanine (DOPA) in invertebrate structural proteins, *Meth. Enzymol.,* 107, 397, 1984.
65. **Nagafuchi, T. and Waite, J. H.,** unpublished data.
66. **Schulz, G. E. and Schirmer, R. H.,** *Principles of Protein Structure,* Springer-Verlag, New York, 1979.
67. **Barron, B., Norman, D. G., and Campbell, I. D.,** Protein modules, *Trends Biochem. Sci.,* 16, 13, 1991.
68. **Sandell, L. J. and Boyd, C. D.,** Conserved and divergent sequence and functional elements within collagen genes, in *Extracellular Matrix Genes,* Sandell, L. J. and Boyd, C. D., Eds., Academic Press, New York, 1990, 1.
69. **Schwarzbauer, J.,** The fibronectin gene, in *Extracellular Matrix Genes,* Sandell, L. J. and Boyd, C. D., Eds., Academic Press, New York, 1990, 195.
70. **Indik, Z., Yeh, H., Ornstein-Goldstein, N., and Rosenbloom, J.,** Structure of the elastin gene and alternative splicing of elastin mRNA, in *Extracellular Matrix Genes,* Sandell, L. J. and Boyd, C. D., Eds., Academic Press, New York, 1990, 221.
71. **Janes, S. M., Mu, D., Wemmer, D., Smith, A. J., Kaur, S., Maltby, D., Burlingame A. L., and Klinman, J. P.,** A new redox cofactor in eukaryotic enzymes: 6-hydroxydopa at the active site of bovine serum amine oxidase, *Science,* 248, 981, 1990.
72. **Brown, D. E., McGuirl, M. A., Dooley, D. M., Janes, S. M., Mu, D., and Klinman, J. P.,** The organic functional group in copper-containing amine oxidases, *J. Biol. Chem.,* 266, 4049, 1991.
73. **Aristotle,** *Historia Animalium* (transl.), Thompson, d'A. W., Interpreter, in *Works of Aristotle,* Vol. 4, Smith, J. A. and Ross, W. D., Eds., Oxford University Press, Oxford, 1910.
74. **Rzepecki, L. M., Hansen, K. M., and Waite, J. H.,** Characterization of a cysteine-rich polyphenolic protein family from the blue mussel *Mytilus edulis* L., *Biol. Bull.* 183, 123, 1992.
75. **Diamond, T. and Waite, J. H.,** unpublished data.
76. **Wells, K. and Cordingley, J. S.,** Detecting proteins containing 3,4-dihydroxyphenylalanine by silver staining of polyacrylamide gels, *Anal. Biochem.,* 194, 273, 1991.
77. **Waite, J. H.,** Detection of peptidyl-3,4-dihydroxyphenylalanine by amino acid analysis and microsequencing techniques, *Anal. Biochem.,* 192, 429, 1991.
78. **Waite, J. H., and Tanzer, M. L.,** Specific colorimetric detection of *o*-diphenols and 3,4-dihydroxyphenylalanine-containing peptides, *Anal. Biochem.,* 111, 131, 1981.
79. **Arnow, L.,** Colorimetric determination of components of 3,4-dihydroxyphenyl-L-alanine/tyrosine mixtures, *J. Biol. Chem.,* 118, 531, 1937.
80. **Waite, J. H.,** Determination of (catecholato)borate complexes using difference spectrophotometry, *Anal. Chem.,* 56, 1935, 1984.
81. **Paz, M. A., Flückiger, R., Boak, A., Kagan, H. M., and Gallop, P. M.,** Specific detection of quinoproteins by redox-cycling staining, *J. Biol. Chem.,* 266, 689, 1991.
82. **Erasmus, D. A.,** The subcellular localization of labelled tyrosine in the vitelline cells of *Schistosoma mansoni, Z. Parasitenk.,* 46, 75, 1975.
83. **Byram, J. E. and Senft, A. W.,** Structure of the schistosome eggshell: amino acid analysis and incorporation of labelled amino acids, *Am. J. Trop. Med. Hyg.,* 28, 539, 1979.
84. **Seed, J. L. and Bennett, J. L.,** *Schistosoma mansoni:* phenol oxidase's role in eggshell formation, *Exp. Parasitol.,* 49, 430, 1980.
85. **Simpson, A. G. J., Chaudri, M., Knight, M., Kelly, C., Rumjanek, F., Martin, S., and Smithers, S. R.,** Characterisation of the structure and expression of the gene encoding a major female specific polypeptide of *Schistosoma mansoni, Mol. Biochem. Parasitol.,* 22, 169, 1987.

86. **Kunz, W., Opatz, M., Finken, M., and Symmons, P.,** Sequences of two genomic fragments containing identical coding region for a putative eggshell precursor protein of *Schistosoma mansoni, Nucl. Acids Res.*, 15, 5894, 1987.
87. **Bobek, L. A., Rekosh, D. M., and LoVerde, P. T.,** Small gene family encoding an eggshell (chorion) protein of the human parasite *Schistosoma mansoni, Mol. Cell Biol.*, 8, 3008, 1988.
88. **Köster, B., Dargatz, H., Schröder, J., Hirzmann, J., Haarman, C., Symmons, P., and Kunz, W.,** Identification and localisation of a putative eggshell precursor gene in the vitellarium of *Schistosoma mansoni, Mol. Exp. Parasitol.*, 31, 183, 1988.
89. **Hamodrakas, S. J., Jones, C. W., and Kafatos, F. C.,** Secondary structure predictions for silkmoth chorion proteins, *Biochim. Biophys. Acta*, 700, 42, 1982.
90. **Moncrieff, R. W.,** *Man-made Fibres*, Newnes-Butterworths, London, 1975.
91. **Hansen, K. M. and Waite, J. H.,** unpublished data.
92. **Condit, C. M. and Keller, B.,** The glycine-rich cell wall proteins of higher plants, in *Organization and Assembly of Plant and Animal Extracellular Matrix*, Adair, W. S. and Mecham, R. P., Eds., Academic Press, New York, 1990, 119.
93. **Steinert, P. M., Rice, R. H., Roop, D. R., Trus, B. L., and Steven, A. C.,** Complete amino acid sequence of a mouse epidermal keratin subunit and implications for the structure of intermediate filaments, *Nature*, 302, 794, 1983.
94. **Rice-Ficht, A. C., Dusek, K. A., Kochevar, G. J., and Waite, J. H.,** Eggshell precursor proteins of *Fasciola hepatica*. I. Structure and expression of vitelline protein B, *Mol. Biochem. Parasitol.*, 54, 129, 1992.
95. **Zurita, M., Bieber, D., Ringold, G., and Mansour, T. E.,** Cloning and characterization of a female genital complex cDNA from the liver fluke *Fasciola hepatica, Proc. Natl. Acad. Sci. U.S.A.*, 84, 2340, 1987.
96. **Zurita, M., Bieber, D., and Mansour, T. E.,** Identification, expression and in situ hybridization of an eggshell protein gene from *Fasciola hepatica, Mol. Biochem. Parasitol.*, 37, 11, 1989.
97. **Waite, J. H. and Rice-Ficht, A. C.,** Eggshell precursor proteins of *Fasciola hepatica*. II. Microheterogeneity in vitelline protein B, *Mol. Biochem. Parasitol.*, 54, 129, 1992.
98. **Benedict, C. V. and Waite, J. H.,** Location and analysis of byssal structural proteins of *Mytilus edulis, J. Morphol.*, 189, 171, 1986.
99. **Benedict, C. V. and Waite, J. H.,** Composition and ultrastructure of the byssus of *Mytilus edulis, J. Morphol.*, 189, 261, 1986.
100. **Mascolo, J. M. and Waite, J. H.,** Protein gradients in byssal threads of some marine bivalve molluscs, *J. Exp. Zool.*, 240, 1, 1986.
101. **Pihlajaniemi, T. and Tamminen, M.,** The a1 chain of type XIII collagen consists of three collagenous and four noncollagenous domains, and its primary transcript undergoes complex alternative splicing, *J. Biol. Chem.*, 265, 16922, 1990.
102. **Michaille, J.-J., Couble, P., Prudhomme, J.-C., and Garel, A.,** A single gene produces multiple sericin messenger RNAs in the silk gland of *Bombyx mori, Biochimie*, 68, 1165, 1986.
103. **Waring, G. L., Hawley, R. J., and Schoenfeld, T.,** Multiple proteins are produced from the *dec-1* eggshell gene in *Drosophila* by alternative RNA splicing and proteolytic cleavage events, *Dev. Biol.*, 142, 1, 1990.
104. **Moore, R. C., Ed.,** *Treatise on Invertebrate Paleontology*, Part N, Vol. 3, University of Kansas and Geol. Society of America, Lawrence, KS, 1969, N270.
105. **Dover, G.,** Molecular drive: a cohesive mode of species evolution, *Nature*, 299, 111, 1982.
106. **Eckert, R. L. and Green, H.,** Structure and evolution of the human involucrin gene, *Cell*, 46, 583, 1986.
107. **Phillips, M., Djian, P., and Green, H.,** The involucrin gene of the Galago: existence of a correction process acting on its segment of repeats, *J. Biol. Chem.*, 265, 7804, 1990.
108. **Morton, B.,** Mangrove bivalves, in *The Mollusca*, Vol. 6, Russel-Hunter, W. D., Ed., Academic Press, New York, 1983.
109. **Marumo, K. and Waite, J. H.,** Optimization of hydroxylation of tyrosine and tyrosine-containing peptides by mushroom tyrosinase, *Biochim. Biophys. Acta*, 872, 98, 1986.
110. **Williams, T., Marumo, K., Waite, J. H., and Henkens, R. W.,** Mussel glue protein has an open conformation, *Arch. Biochem. Biophys.*, 269, 415, 1989.
111. **Tanaka, M., Shibata, H., and Uchida, T.,** A new prolyl hydroxylase acting on poly-L-proline from suspension cultured cells of *Vinca rosea, Biochim. Biophys. Acta*, 616, 188, 1980.

112. **Atreya, P. L. and Ananthanarayanan, V. S.,** Interaction of prolyl 4-hydroxylase with synthetic peptide substrata, *J. Biol. Chem.*, 266, 2852, 1991.
113. **Trumbore, M.,** Investigation of the Three Dimensional Conformation of the Adhesive Polyphenolic Protein of *Mytilus edulis* i using Small Angle Scattering and Molecular Modelling, Ph.D. thesis dissertation, University of Connecticut Health Center, Farmington, 1991.
114. **Chou, P. Y. and Fasman, G. D.,** Prediction of the secondary structure of proteins from their amino acid sequence, *Adv. Enzymol.*, 47, 45, 1978.
115. **Miller, E. J. and Gay, S.,** Collagen: an overview, *Meth. Enzymol.*, 82, 3, 1982.
116. **Murray, L. W., Waite, J. H., Tanzer, M. L., and Hauschka, P. V.,** Preparation and characterization of invertebrate collagens, *Meth. Enzymol.*, 82, 65, 1982.
117. **Sage, H. and Bornstein, P.,** Preparation and characterization of procollagens and procollagen–collagen intermediates, *Meth. Enzymol.*, 82, 96, 1982.
118. **Rosenbloom, J.,** Elastin: an overview, *Meth. Enzymol.*, 144, 172, 1987.
119. **Kivirikko, K. I. and Myllylä, R.,** Posttranslational enzymes in the biosynthesis of collagen: intracellular enzymes, *Meth. Enzymol.*, 82, 245, 1982.
120. **Kagan, H. M. and Sullivan, K. A.,** Lysyl oxidase: preparation and role in elastin biosynthesis, *Meth. Enzymol.*, 82, 637, 1982.
121. **Schroeder, W. A.,** *The Primary Structure of Proteins*, Harper and Row, New York, 1968.
122. **Lucas, F. and Rudall, K. M.,** Extracellular fibrous proteins: the silks, in *Comprehensive Biochemistry*, Vol. 26B, Florkin, M. and Stotz, E. H., Eds., Elsevier, New York, 1968, 475.
123. **Hunt, S. and Oates, K.,** Fine structure and molecular organization of the periostracum in a gastropod mollusc *Buccinum undatum* L. and its relation to similar structural protein systems in other invertebrates, *Phil. Trans. R. Soc. Lond.*, 283B, 417, 1978.
124. **Rodrigues, V., Chaudri, M., Knight, M., Meadows, H., Chambers, A. E., Taylor, W. R., Kelly, C., and Simpson, A. J. G.,** Predicted structure of a major *Schistosoma mansoni* eggshell protein, *Mol. Biochem. Parasitol.*, 32, 7, 1989.
125. **Abbadi, A., Mcharfi, M., Aubry, A., Prémilat, S., Boussard, G., and Marraud, M.,** Involvement of side functions in peptide structures: the Asx turn. Occurrence and conformational aspects, *J. Am. Chem. Soc.*, 113, 2729, 1991.
126. **van Holst, G.-V. and Varner, J. E.,** Reinforced polyproline II conformation in a hydroxyproline-rich cell wall glycoprotein from carrot root, *Plant Physiol.*, 72, 247, 1984.
127. **Isemura, T., Asakura, J., Shibata, S., Isemura, S., Saitoh, E., and Sanada, K.,** Conformational study of the salivary proline-rich polypeptides, *Int. J. Peptide Protein Res.*, 21, 281, 1983.
128. **Walton, A. G.,** *Polypeptides and Protein Structure*, Elsevier, New York, 1981.
129. **Stafstrom, J. P. and Staehelin, L. A.,** Cross-linking patterns in salt-extractable extensin from carrot cell walls, *Plant Physiol.*, 81, 234, 1986.
130. **Stafstrom, J. P. and Staehelin, L. A.,** The role of carbohydrate in maintaining extensin in an extended conformation, *Plant Physiol.*, 81, 242, 1986.
131. **Dolcet, C. and Tarazona, M. P.,** Influence of secondary structure of polyglycine on some conformational properties, *Int. J. Peptide Protein Res.*, 30, 548, 1987.
132. **Komatsu, K.,** Secretion and structure of liquid silk, *Leather Chem.* (Jpn., Engl. Abstr.), 27, 193, 1982.
133. **Komatsu, K.,** Silk. III. Sericin physical structure, *Serologia*, 22, 14, 1982.
134. **Cordingley, J. S.,** personal communication.
135. **Wells, K. E. and Cordingley, J. S.,** *Schistosoma mansoni:* eggshell formation is regulated by pH and calcium, *Exp. Parasitol.*, 295, 1991.
136. **Erasmus, D. A.,** The adult Schistosome: structure and reproductive biology, in *The Biology of Schistosomes from Genes to Latrines*, Rollinson, D. and Simpson, A. J. G., Eds., Academic Press, New York, 1987, 51.
137. **Smyth, J. D. and Halton, D. W.,** *The Physiology of Trematodes*, Cambridge University Press, Cambridge, 1983.
138. **Smyth, J. D. and Clegg, J. A.,** Egg-shell formation in Trematodes and Cestodes, *Exp. Parasitol.*, 8, 286, 1959.
139. **Erasmus, D. A., Popiel, I., and Shaw, J. R.,** A comparative study of the vitelline cell in *Schistosoma mansoni, S. haematobium, S. Japonicum,* and *S. mattheei, Parasitology*, 84, 283, 1982.
140. **Björkman, N. and Thorsell, W.,** On the fine morphology of the formation of egg-shell globules in the vitelline glands of the liver fluke (*Fasciola hepatica*, L.), *Exp. Cell Res.*, 32, 153, 1963.

141. **Vitellaro-Zuccarello, L.,** Ultrastructural and cytochemical study on the enzyme gland off the foot of a mollusc, *Tiss. Cell,* 13, 701, 1981.
142. **Rzepecki, L. M., Nagafuchi, T., and Waite, J. H.,** α,β-Dehydro-3,4-dihydroxyphenylalanine derivatives: potential sclerotization intermediates in natural composite materials, *Arch. Biochem. Biophys.,* 285, 17, 1991.
143. **Stewart, L. C. and Klinman, J. P.,** Dopamine beta-hydroxylase of adrenal chromaffin granules: structure and function, *Annu. Rev. Biochem.,* 57, 551, 1988.
144. **Neill, P. J. G., Smith, J. H., Doughty, B. L., and Kemp, M.,** The ultrastructure of the *Schistosoma mansoni* egg, *Am. J. Trop. Med. Hyg.,* 39, 52, 1988.
145. **Berridge, M. J. and Irvine, R. F.,** Inositol phosphates and cell signalling, *Nature,* 341, 197, 1989.
146. **Tamarin, A. and Keller, P.,** An ultrastructural study of the byssal thread forming system in *Mytilus, J. Ultrastruct. Res.,* 40, 401, 1972.
147. **Cox, D. L. and Koob, T. J.,** Latent egg capsule catechol oxidase in the little skate *(Raja erinacea), Comp. Biochem. Physiol. B,* 95, 767, 1990.
148. **Irwin, S. W. B. and Threadgold, L. T.,** Electron microscope studies of *Fasciola hepatica.* X. Egg formation, *Exp. Parasitol.,* 31, 321, 1972.
149. **Degens, E. T., Spencer, D. W., and Parker, R. H.,** Paleobiochemistry of molluscan shell proteins, *Comp. Biochem. Physiol.,* 20, 553, 1967.
150. **Travis, D. F., Francois, C. J., Bonar, L. C., and Glimcher, M. J.,** Comparative studies of the organic matrices of invertebrate mineralized tissues, *J. Ultrastruct. Res.,* 18, 519, 1967.
151. **Tidball, J. G.,** An ultrastructural and cytochemical analysis of the cellular basis for tyrosine-derived collagen crosslinks in *Leptogorgia virgulata* (Cnidaria: Gorgonacea), *Cell Tissue Res.,* 222, 635, 1982.
152. **Kingsley, R. J. and Watabe, N.,** Analysis of proteinaceous components of the organic matrices of spicules from the gorgonian *Leptogorgia virgulata, Comp. Biochem. Physiol.,* 76B, 443, 1983.
153. **Kingsley, R. J. and Watabe, N.,** The dynamics of spicule calcification in whole colonies of the gorgonian *Leptogorgia virgulata* (Lamarck) (Coelenterata: Gorgonacea), *J. Exp. Mar. Biol. Ecol.,* 133, 57, 1989.
154. **Benson, S., Jones, E. M. E., Crise-Benson, N., and Wilt, F.,** Morphology of the organic matrix of the spicule of the sea urchin larva, *Exp. Cell Res.,* 148, 249, 1983.
155. **Benson, S. C., Benson, N. C., and Wilt, F.,** The organic matrix of the skeletal spicule of sea urchin embryos, *J. Cell Biol.,* 102, 1878, 1986.
156. **Lowenstam, H. A. and Weiner, S.,** *On Biomineralization,* Oxford University Press, New York, 1989.
157. **Mann, S., Webb, J., and Williams, R. J. P.,** *Biomineralization: Chemical and Biochemical Perspectives,* VCH Publishers, New York, 1989.
158. **Simkiss, K. and Wilbur, K. M.,** *Biomineralization: Cell Biology and Mineral Deposition,* Academic Press, New York, 1989.
159. **Wilbur, K. M. and Simkiss, K.,** Calcified shells, *Comprehensive Biochemistry,* 26A, 229, 1968.
160. **Bubel, A.,** An electron-microscope investigation into the distribution of polyphenols in the periostracum and cells of the inner face of the outer fold of *Mytilus edulis, Mar. Biol.,* 23, 3, 1973.
161. **Waite, J. H. and Wilbur, K. M.,** Phenoloxidase in the periostracum of the marine bivalve *Modiolus demissus* Dillwyn, *J. Exp. Zool.,* 195, 359, 1976.
162. **Wheeler, A. P., Rusenko, K. W., and Sikes, C. S.,** Organic matrix from carbonate biomineral as a regulator of biomineralization, in *Chemical Aspects of Regulation of Mineralization,* Sikes, C. S. and Wheeler, A. P., Eds., University of South Carolina Publication Services, Mobile, AL, 1988, 3.
163. **Greenfield, E. M., Wilson, D. C., and Crenshaw, M. A.,** Ionotropic nucleation of calcium carbonate by molluscan matrix, *Am. Zool.,* 24, 925, 1984.
164. **Addadi, L., Moradian, J., Shay, E., Maroudas, N. G., and Weiner, S.,** A chemical model for the cooperation of sulfates and carboxylates in calcite crystal nucleation: relevance to biomineralization, *Proc. Natl. Acad. Sci. U.S.A.,* 84, 2732, 1987.
165. **Samata, T.,** Ca-binding glycoproteins in molluscan shells with different types of ultrastructure, *The Veliger,* 33, 190, 1990.
166. **Addadi, L. and Weiner, S.,** Interactions between acidic proteins and crystals: stereochemical requirements in mineralization, *Proc. Natl. Acad. Sci. U.S.A.,* 82, 4110, 1985.
167. **Emlet, R. B.,** Echinoderm calcite: a mechanical analysis from larval spicules, *Biol. Bull.,* 163, 264, 1982.
168. **Wainwright, S. A.,** Stress and design in bivalved mollusc shell, *Nature,* 224, 777, 1969.

169. **Wainwright, S. A., Biggs, W. D., Currey, J. D., and Gosline, J. M.,** *Mechanical Design in Organisms,* Edward Arnold Press, London, 1976.
170. **Wheeler, A. P., Rusenko, K. W., Swift, D. M., and Sikes, C. S.,** Regulation of *in vitro* and *in vivo* $CaCO_3$ crystallization by fractions of oyster shell organic matrix, *Mar. Biol.,* 98, 71, 1988.
171. **Meenakshi, V. R., Hare, P. E., and Wilbur, K. M.,** Amino acids of the organic matrix of neogastropod shell, *Comp. Biochem. Physiol.,* 40B, 1037, 1971.
172. **Samata, T., Sanguansri, P., Cazaux, C., Hamm, M., Engels, J., and Krampitz, G.,** Biochemical studies on components of molluscan shells, in *The Mechanisms of Biomineralization in Animals and Plants,* Omori, M. and Watabe, N., Eds., Tokai University Press, Tokyo, 1980, 37.
173. **Weiner, S.,** Mollusk shell formation: isolation of two organic matrix proteins associated with calcite deposition in the bivalve *Mytilus californianus, Biochemistry,* 57, 251, 1980.
174. **Benson, S., Sucov, H., Stephens, L., Davidson, E., and Wilt, F.,** A lineage-specific gene encoding a major matrix protein of the sea urchin embryo spicule. I. Authentication of the cloned gene and its developmental expression, *Dev. Biol.,* 120, 499, 1987.
175. **Sucov, H. M., Benson, S., Robinson, J. J., Britten, R. J., Wilt, F. and Davidson, E. H.,** A lineage-specific gene encoding a major matrix protein of the sea urchin embryo spicule. II. Structure of the gene and derived sequence of the protein, *Dev. Biol.,* 120, 507, 1987.
176. **Hansen, K. M. and Waite, J. H.,** unpublished data.
177. **Gordon, J. and Carriker, M. R.,** Sclerotized protein in the shell matrix of a bivalve mollusc, *Mar. Biol.,* 57, 251, 1980.
178. **Galtsoff, P. S.,** The American oyster, *Crassostrea virginica* Gmelin, *Fish. Bull.,* 64, 1964.
179. **Cranfield, H. J.,** A study of the morphology, ultrastructure, and histochemistry of the foot of the pediveliger of *Ostrea edulis, Mar. Biol.,* 22, 187, 1973.
180. **Cranfield, H. J.,** The ultrastructure and histochemistry of the larval cement of *Ostrea edulis* L., *J. Mar. Biol. Assoc. U.K.,* 55, 497, 1975.
181. **Tomaszewski, C.,** Cementation in the Early Dissoconch Stage of *Crassostrea virginica* (Gmelin), M.S. thesis, University of Delaware, Newark, 1982.
182. **Neff, J. M.,** Calcium Carbonate Tube Formation by Serpulid Polychaete Worms: Physiology and Ultrastructure, Ph.D. dissertation, Duke University, Durham, NC, 1967.
183. **Dryhurst, G., Kadish, K. M., Scheller, F., and Renneburg, R.,** *Biological Electrochemistry,* Academic Press, New York, 1982, 116.
184. **Kalyanaraman, B., Premovic, P. I., and Sealy, R. C.,** Semiquinone anion radicals from addition of amino acids, peptides and proteins to quinones derived from oxidation of catechols and catecholamines, *J. Biol. Chem.,* 262, 11080, 1987.
185. **Mason, H. S., Spencer, E., and Yamazaki, I.,** Identification by electron spin resonance spectroscopy of the primary product of tyrosinase-catalyzed catechol oxidation, *Biochem. Biophys. Res. Commun.,* 4, 236, 1961.
186. **Kandaswami, C. and Vaidyanathan, C. S.,** Oxidation of catechol in plants, *J. Biol. Chem.,* 248, 4035, 1973.
187. **Blanchard, S. C. and Chasteen, N. D.,** Electron paramagnetic resonance spectrum of a sea shell, *Mytilus edulis, J. Phys. Chem.,* 80, 1362, 1976.
188. **Rzepecki, L. M. and Waite, J. H.,** α,β-Dehydro-3,4-dihydroxyphenylalanine derivatives: rate and mechanism of formation, *Arch. Biochem. Biophys.,* 285, 27, 1991.
189. **Taylor, S. W., Molinski, T. F., Rzepecki, L. M., and Waite, J. H.,** Oxidation of peptidyl 3,4-dihydroxyphenylalanine analogs: implications for the biosynthesis of tunichromes and related oligopeptides, *J. Nat. Products,* 54, 918, 1991.
190. **Oltz, E. M., Bruening, R. C., Smith, M. J., Kustin, K., and Nakanishi, K.,** The tunichromes. A class of reducing blood pigments from Sea Squirts: isolation, structures and vanadium chemistry, *J. Am. Chem. Soc.,* 110, 6162, 1988.
191. **Azumi, K., Yokosawa, H., and Ishii, S.,** Halocyamines: novel antimicrobial tetrapeptide-like substances isolated from the hemocytes of the solitary ascidian *Halocynthia roretzi, Biochemistry,* 29, 159, 1990.
192. **Stonard, R. J. and Andersen, R. J.,** Celenamides A and B, linear peptide alkaloids from the sponge *Cliona celata, J. Org. Chem.,* 45, 3687, 1980.
193. **Ross, I. L.,** Studies of a DOPA-Protein and Associated Oxidase from the Ascidian *Pyura stolonifera,* Ph.D. thesis dissertation, University of Queensland, St. Lucia, Australia.

194. **Imazu, S., Shimohigashi, Y., Kodama, H., Sakaguchi, K., Waki, M., Kato, T., and Izumiya, N.,** Conformationally stabilized gramicidin S analog containing dehydrophenylalanine in place of D-phenylalanine[4,4'], *Int. J. Peptide Protein Res.*, 32, 298, 1988.
195. **George, S. G., Pirie, B. J. S., and Coombs, T. L.,** Kinetics of accumulation and excretion of ferric hydroxide in *Mytilus edulis* (L.) and its distribution in the tissues, *J. Exp. Mar. Biol. Ecol.*, 23, 71, 1976.
196. **Lieberman, M. and Sasaki, T.,** Iron (II) organizes a synthetic peptide into three-helix bundles, *J. Am. Chem. Soc.*, 113, 1470, 1991.
197. **Vincent, J. F. V. and Ablett, S.,** Hydration and tanning in insect cuticle, *J. Insect Physiol.*, 333, 973, 1987.
198. **Wilson, R. A.,** The structure and permeability of the shell and vitelline membrane of the egg of *Fasciola hepatica, Parasitology*, 57, 47, 1967.
199. **Knight, D. O. and Hunt, S.,** Fine structure of the dogfish egg case: a unique collagenous material, *Tissue Cell*, 8, 183, 1967.
200. **Brown, C. H.,** Egg-capsule proteins of Selachians and trout, *Q. J. Microsc. Sci.*, 96, 483, 1955.
201. **Rowan, W. B.,** The mode of hatching of the egg of *Fasciola hepatica, Exp. Parasitol.*, 5, 118, 1956.
202. **Hornsey, D. J.,** Permeability coefficients of the eggcase membrane of *Scyliorhinus canicula* L., *Experientia*, 34, 1596, 1978.
203. **Olivieri, M. P., Loomis, R. E., Meyer, A. E., and Baier, R. E.,** Surface characterization of mussel adhesive protein films, *J. Adhesion Sci. Technol.*, 4, 197, 1990.
204. **Green, K., Berdecia, R., and Cheeks, L.,** Mussel adhesive protein: permeability characteristics when used as a basement membrane, *Curr. Eye Res.*, 6, 835, 1987.

Chapter 7

Flow Effects on Leukocyte Adhesion to Vascular Endothelium

David A. Jones, C. Wayne Smith, and Larry V. McIntire

CONTENTS

I. Introduction ... 143
II. Properties of Receptors Involved in Leukocyte–Endothelial Cell Adhesion ... 144
A. Integrins ... 144
B. Immunoglobulin Superfamily ... 145
C. Selectins ... 146
D. Other Adhesion Molecules ... 147
III. Flow Effects on Adhesion Receptor Function ... 147
A. The Multistep Process of Leukocyte Extravasation ... 147
B. Two Mechanistically Distinct Adhesive Interactions in Leukocyte Extravasation ... 148
C. Properties of Selectin–Carbohydrate-Mediated Adhesion ... 148
D. Properties of Integrin–Peptide-Mediated Adhesion ... 149
E. Specificity in Leukocyte Recruitment ... 149
IV. Model Systems for Studying Leukocyte–Endothelial Cell Adhesion ... 150
A. Static Assays ... 150
B. *In vitro* Flow Assays ... 150
C. *In vivo* Flow Assays ... 151
V. Physiological Applications of the Principles of Adhesion Under Flow ... 151
A. Neutrophil Adherence in Inflammation ... 151
B. Lymphocyte Homing ... 152
C. Monocyte Adhesion in Atherosclerosis ... 153
VI. Acknowledgments ... 154
References ... 154

I. INTRODUCTION

The diffuse nature of the immune system makes specific cell–cell adhesion a critically important task, and the wide variety of cell types involved makes the process very complicated. Targeting of leukocytes to regions of the body where they are needed is one aspect of immune system function that is currently of great interest due to its potential relevance in the development of new medications. The targeting process consists of adhesion of a specific subpopulation of circulating leukocytes to a specific area of vascular endothelium via cell surface adhesion receptors. Attempts to understand this process are complicated by the large number of receptors involved and the complex regulation of their expression on particular cell subpopulations. A further very important complication arises from the fact that these interactions occur within the flowing bloodstream. Research on the effect of flow on leukocyte adhesion to endothelium has revealed that different types of receptors are capable of mediating distinct types of adhesive events, and this chapter will review current understanding of this subject.

We begin with an overview of the properties of three families of adhesion receptors involved in leukocyte targeting: the integrin family, the selectin family, and the immunoglobulin superfamily. This section describes the basic structure, distribution, and regulation of expression and binding for relevant members of each family. Next, we discuss the effects of flow on cell adhesion mediated by these receptors, showing how experimental evidence has led to the conclusion that selectins are important in an initial rolling interaction, while integrins are important in later development of firm adhesion and migration of neutrophils. We then review various experimental approaches to understanding aspects of cell adhesion under conditions of flow and finally examine how flow effects on cell adhesion are likely to be important in three important physiological situations: inflammation, leukocyte recirculation, and atherogenesis.

0-8493-4559-6/95/$0.00+$.50

Table 1 **Integrins found on leukocytes and implicated in specific leukocyte–endothelial cell interactions**

Molecule[a]	Ligands[b]	Distribution[c]
LFA-1, $\alpha_L\beta_2$, CD11a/CD18	ICAM-1,2,3	B, T, M, N, E
Mac-1, $\alpha_M\beta_2$, CD11b/CD18	ICAM-1, FX, Fb, iC3b	M, N, NK, E
p150,95, $\alpha_X\beta_2$, CD11c/CD18	Fb	M, N, NK, B′
VLA-4, $\alpha_4\beta_1$, CD49d/CD29	VCAM-1, FN	B, T, M, E
$\alpha_4\beta_7$	MAdCAM-1	B, T

[a] CD, cluster of differentiation designation; LFA, lymphocyte function-associated antigen; Mac, macrophage; VLA, very late activation antigen; [b] FX, factor X; Fb, fibrinogen; iC3b, complement component iC3b; FN, fibronectin; ICAM, intercellular adhesion molecule; VCAM, vascular cell adhesion molecule; MAdCAM, mucosal addressin cell adhesion molecule; [c] B, B lymphocytes; T, T lymphocytes; B′, activated B lymphocytes; M, monocytes; N, neutrophils; NK natural killer cells; E, eosinophils.

II. PROPERTIES OF RECEPTORS INVOLVED IN LEUKOCYTE–ENDOTHELIAL CELL ADHESION

A. INTEGRINS

The name of this receptor family refers to the fact that these molecules generally integrate the extracellular environment with intracellular cytoskeletal components by binding to extracellular matrix molecules or the surfaces of other cells. Integrins are heterodimers consisting of one of several α subunits and one of several β subunits bound noncovalently. Combinations of the 13 known α subunits and the seven known β subunits produce at least 19 receptors that have been extensively reviewed.[1-4] Three important groups of integrins are those that possess a β_1 (CD29), β_2 (CD18), or β_3 (CD61) subunit. The β_2 integrins are also known as the leukocyte integrins, due to their distribution, which is limited to white blood cells. The β_1 integrins are commonly referred to as the very late activation (VLA) antigens, since the first two of them discovered, VLA-1 and VLA-2, appear 2 to 4 weeks after stimulation of lymphocytes with antigen. β_3 integrins are commonly referred to as cytoadhesins and, like β_1 integrins, are generally involved in adhesion to extracellular matrix components and guide morphogenesis and wound healing. This review will focus on the β_2 integrins and those integrins containing the α_4 subunit, since these have been most strongly implicated in specific leukocyte–endothelial cell interactions (see Table 1).

The importance of integrins is dramatically illustrated by type 1 leukocyte adhesion deficiency (LAD-1), a rare immunodeficiency characterized by recurrent life-threatening bacterial and fungal infections.[5] The molecular defects underlying this disease are known to be defects in the β_2 subunit of the leukocyte integrins,[6] and pathology arises because neutrophils in these patients are unable to localize to sites of infection.

A schematic diagram of the structure of Mac-1 and VLA-4 is given in Figure 1. Other integrins are very similar in structure, with α subunits averaging approximately 1100 amino acids and β subunits averaging approximately 750 amino acids. The divalent cation binding site repeats give rise to the dependence on these cations for receptor function, and the I-domain (for "inserted" or "interacting" domain) seems to impart binding specificity to those integrins that have it. The I domain is also associated with conformational changes involved in affinity modulation.[7] In general, integrins without the I domain bind to Arg-Gly-Asp (RGD)-containing peptide sequences in extracellular matrix proteins, whereas presence of the I-domain seems to allow non-RGD binding. This is important in the binding of leukocyte integrins to ICAM-1 (intercellular adhesion molecule-1), which contains no RGD sequences. A noteworthy exception is VLA-4, which lacks an I-domain and yet can bind both VCAM-1 (vascular cell adhesion molecule-1) and fibronectin at non-RGD sites.

The broad distribution of integrins gives an indication of their importance in a wide variety of specific immune system interactions as well as other cell–cell and cell–matrix interactions. β_1 and β_3 integrins have an especially widespread distribution, including endothelial cells, epithelial cells, and fibroblasts. This is of particular importance to the proposed primary function of these integrins, which is to guide morphogenesis and wound healing. VLA-4 ($\alpha_4\beta_1$) is an unusual β_1 integrin, in that it is expressed on resting B-cells, T-cells, monocytes, and eosinophils and can bind VCAM-1 on endothelial cells. There is also evidence that $\alpha_4\beta_7$ is involved in lymphocyte homing,[8] leading to the possibility that α_4-containing integrins form a second class of integrins that function in leukocyte adhesion. Other β_1 integrins are found

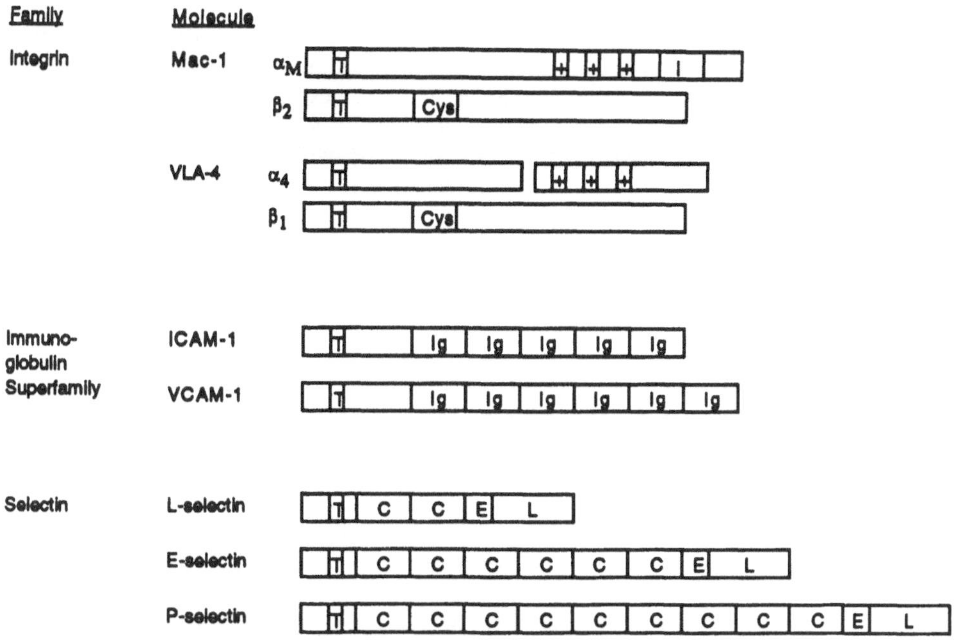

Nomenclature: T, hydrophobic transmembrane segment; +, divalent cation binding site; I, integrin I domain; Cys, cysteine-rich region; Ig, immunoglobulin domain; C, complement binding protein-like domain; E, epidermal growth factor-like domain; L, lectin-like carbohydrate binding domain.

Figure 1 Schematic diagrams of representative adhesion molecules.

variably on leukocytes,[4] but these bind only extracellular matrix proteins, so it is unlikely that they contribute to specific leukocyte–endothelial cell interactions. β_2 integrins are expressed only on leukocytes, as indicated in Table 1. Note that while all circulating leukocytes express LFA-1, expression of Mac-1 is limited to monocytes and granulocytes.

The regulation of integrin-mediated binding occurs at three levels: modulation of receptor synthesis, mobilization of an intracellular pool of receptors, and modulation of the binding affinity of individual receptors. Modulation of receptor synthesis is generally associated with cell differentiation rather than with the rapid response of terminally differentiated cells to particular challenges. For example, LFA-1, VLA-4, VLA-5, and VLA-6 increase two- to four-fold upon conversion of naive to memory T-lymphocytes,[9,10] which could influence localization and recirculation of these subpopulations. Also, differentiation of myeloid lineages is associated with early expression of LFA-1 and later expression of Mac-1, and the subsequent maturation of monocytes to tissue macrophages is associated with a decrease in Mac-1 and VLA-4 expression, along with an increase in p150,95 expression.[11,12]

Cell adhesion can be modulated within minutes by mobilizing intracellular pools of integrin receptors. For example, in neutrophils and monocytes, chemotactic factors result in up to a ten-fold increase in surface expression of Mac-1 and p150,95.[13] The mechanism of this form of up-regulation involves fusion of leukocyte granules containing the receptors with the cell surface.[14] An even more rapid mechanism of cell adhesion modulation is the ability of some integrins to increase their binding affinity through conformational changes upon stimulation of the cell. This effect has been shown for many integrins, and the details of this mechanism are an area of active research (reviewed in Reference 15). It is also important to note that in addition to serving as effector molecules of leukocyte activation, they can also serve as sensors, triggering a variety of leukocyte activation responses upon ligand binding.

B. IMMUNOGLOBULIN SUPERFAMILY

The immunoglobulin superfamily of receptors is defined by the presence of the immunoglobulin domain, which is composed of 70 to 110 amino acids arranged in a well-characterized structure (reviewed in References 1 and 16). This group of receptors is much more diverse than the integrins, ranging from soluble and membrane-bound immunoglobulins to the multireceptor T cell antigen receptor complex to single-chain cellular adhesion molecules. This review will focus on the cellular adhesion molecules ICAM-1 (CD54), VCAM-1, and MAdCAM-1, as these have been most strongly implicated in leukocyte–endothelial cell interactions.[8,17-20] Schematic diagrams of ICAM-1 and VCAM-1 are shown in Figure 1. Two alternatively spliced forms of the VCAM-1 molecule exist, with six and seven immunoglobulin domains.

The distribution of these members of the immunoglobulin superfamily is tightly regulated to enhance their function in specific cell–cell adhesion. ICAM-1 is constitutively expressed on only a few cell types but at the site of an inflammatory response is induced on a wide variety of cells. Endothelial cells up-regulate ICAM-1 within several hours of cytokine stimulation, and peak expression occurs after 24 to 48 h.[21] Since it is a counter-receptor for the leukocyte integrins LFA-1 and Mac-1, ICAM-1 induction provides an important means of recruiting leukocytes to inflammatory sites. LFA-1 binds primarily to the first immunoglobulin domain[22] and Mac-1 binds to the third.[23] ICAM-1 induction by cytokines generally requires new transcription and translation, but recent evidence suggests that ICAM-1 may be mobilized much more rapidly from intracellular stores with thrombin stimulation.[24] ICAM-2, a closely related molecule with two immunoglobulin domains, is another ligand for LFA-1, and is constitutively expressed on endothelial cells and a variety of blood cells. The closely related receptors ICAM-3 and ICAM-R have recently been cloned,[25,26] but these are not expressed on endothelial cells. VCAM-1 expression is induced by many of the same stimuli as ICAM-1, with a time course similar to ICAM-1,[19,27-29] although expression can be selectively induced *in vitro* by the cytokine IL-4.[30] VCAM-1 serves as a counter-receptor for leukocyte VLA-4[31] in the recruitment of leukocytes to inflammatory sites. MAdCAM-1 serves as the endothelial ligand for $\alpha_4\beta_7$ and is important in the preferential localization of lymphocyte subsets to mucosal lymph nodes.

A final immunoglobulin superfamily molecule of special interest in leukocyte–endothelial cell interactions is CD31 (platelet endothelial cell adhesion molecule-1, PECAM-1), which has six immunoglobulin homology domains and contains putative proteoglycan binding sequences.[32-34] Recent *in vivo* studies have demonstrated that PECAM-1 is essential for proper recruitment of neutrophils in a peritonitis model, although its exact role is not yet understood.[35] Additionally, it was recently reported that this molecule is expressed on unique T cell subsets and that CD31 binding has the capacity to induce integrin-mediated adhesion, indicating that CD31 signaling may provide another means of selective T cell recruitment.[36]

C. SELECTINS

The selectin family of adhesion receptors has three known members: L-selectin (LECCAM-1, LAM-1, murine gp90^{Mel14}, peripheral lymph node homing receptor), E-selectin (ELAM-1), and P-Selectin (GMP-140, PADGEM, CD62), which have been extensively reviewed.[1,37-39] They are grouped as selectins based on structural similarities diagrammed in Figure 1. The processed proteins have an NH_2-terminal domain homologous to the type C (calcium-dependent) lectins. Following this, there is a region homologous to conserved epidermal growth factor (EGF) motifs. Next is a chain of repeated domains similar to repeats found in certain complement regulatory proteins. Two of these repeats are found in L-selectin, six in E-selectin, and nine in P-selectin. Each molecule ends with a hydrophobic transmembrane region, followed by a short cytoplasmic tail. Splicing variants of L-selectin and P-selectin have been found,[40,41] including a soluble form of P-selectin. The lectin-like domain is the ligand binding site, and as with other lectins, ligands are carbohydrate structures (discussed in the next subsection). Binding is completely dependent on the conformational changes associated with the filling of two Ca^{2+} binding sites and is also sensitive to pH, with decreased binding below physiological pH. Interestingly, L-selectin binding also appears to be modulated by the cytoplasmic tail.[42]

Selectin expression is regulated by a variety of mechanisms. L-selectin is constitutively expressed on unactivated neutrophils, eosinophils, and monocytes, as well as naive T cells and one subset of memory T cells. On neutrophils, expression of L-selectin has been shown to be localized to the tips of microvillus-like projections on the cell surface.[43] This striking localization should facilitate the proposed role of neutrophil L-selectin as an adhesion receptor for endothelial cell carbohydrate structures. Neutrophils and lymphocytes also have the remarkable property of proteolytically cleaving L-selectin from the cell surface upon activation.[44-46] This shedding is quite rapid, with 50% of the protein shed within 5 min of stimulation with chemotactic factors such as formyl-Met-Leu-Phe (fMLP).[47]

E-selectin has been detected *in vivo* only in inflamed tissues (reviewed in Reference 48). It is induced on endothelial cells upon stimulation by the cytokines IL-1, TNFα, or LPS, and this induction appears to be predominantly localized to postcapillary venules,[49] the principal site of leukocyte emigration in inflammation. Expression requires new RNA and protein synthesis, which is under the control of an NFκB-like factor.[50,51] Peak levels of E-selectin are reached after approximately 4 h of cytokine stimulation and return to basal levels after 16 to 24 h.[52] However, much longer-lasting expression has been observed in the setting of chronic inflammation in the skin.[53]

P-selectin is found constitutively within the α granules of platelets and the Weibel-Palade bodies of endothelial cells.[54,55] In endothelial cells, histamine, thrombin, bradykinin, and substance P cause a rapid mobilization of the Weibel-Palade bodies to the cell periphery, where they fuse with the cell membrane, resulting in expression of the receptors on the cell surface.[56,57] Under these conditions, expression begins within seconds of stimulation, followed by peak expression at approximately 8 to 10 min and endocytosis of most receptors by 45 to 60 min. Peroxide stimulation of endothelial cells also results in mobilization of P-selectin, but in this case, expression lasts for more than 2 h, possibly the result of inhibited endocytosis.[58]

The physiological importance of selectin-mediated interactions is emphasized in the recently described syndrome called leukocyte adhesion deficiency type 2 (LAD-2).[59] This syndrome is characterized by impaired immunity and developmental anomalies, which can be traced to defective fucose metabolism. In particular, the fucose-containing oligosaccharide moiety sialyl-Lewis^x, which serves as a ligand for vascular selectins (see below), is missing in these patients.

D. OTHER ADHESION MOLECULES

Integrins, selectins, and members of the Ig superfamily are not the only structures that are utilized in specific adhesion of leukocytes to endothelial cells. Oligosaccharide ligands for selectins are the focus of considerable current research (reviewed in References 38, 39, 60, and 61). A large body of evidence suggests that the sialylated derivative of the α(1-3)fucosylated lactosaminoglycan, known as Lewis^x (CD15), serves as a primary ligand for endothelial E- and P-selectins. Although sialyl-Lewis^x (sLe^x) is a widespread glycocalyx component, it can contribute to specificity in leukocyte–endothelial cell interactions. In particular, neutrophils but not lymphocytes express sLe^x and are bound by endothelial E- and P-selectins.[43] One subset of lymphocytes, on the other hand, expresses the closely related cutaneous lymphocyte antigen (CLA), which is a sialylated oligosaccharide that can serve as an alternate ligand for E-selectin.[62] Three peptide core molecules that carry oligosaccharide binding sites for L-selectin have recently been characterized. These are GlyCAM-1 (Sgp50, the 50kD component of peripheral node addressin),[63] CD34 (Sgp90, the 90kD component of peripheral node addressin),[64] and MAdCAM-1.[8] MAdCAM-1 is apparently capable of serving as a ligand for both L-selectin and $\alpha_4\beta_7$ integrin at sites such as mucosal lymph nodes, where the necessary glycosylation takes place.[8]

CD44 (HCAM, the Hermes antigen) is another leukocyte adhesion molecule that can bind endothelial cell carbohydrate, but in this case binds hyaluronic acid (reviewed in Reference 65). It appears to function in the adhesion of lymphocytes to vascular endothelium and has also been implicated as a signaling molecule. A recently characterized endothelial cell adhesion molecule designated VAP-1[66] has also been shown to mediate adhesion of lymphocytes to high endothelial venules. The complete primary structure of this molecule is not yet known, although it does appear to be a novel receptor.

III. FLOW EFFECTS ON ADHESION RECEPTOR FUNCTION

A. THE MULTISTEP PROCESS OF LEUKOCYTE EXTRAVASATION

The adhesive events in leukocyte extravasation can be separated into four steps: initial contact, rolling, firm adhesion, and migration (see Figure 2). Initial contact with endothelium is aided by the size of postcapillary and high endothelial venules (which are the principle sites of selective leukocyte extravasation[67]) and by the increased vascular permeability found in inflammation. The size of postcapillary and high endothelial venules (approximately 20 to 60 μm) is small enough that contact between the leukocytes and endothelium is frequent but large enough that initial contact is brief, necessitating specific binding for further interactions.[68] In an inflammatory setting, increased vascular permeability leads to plasma leakage and an increase in local hematocrit, which changes the characteristics of the flow profile through the vessel[69,70] and allows more frequent contact between leukocytes and the vessel wall. Subsequent rolling of neutrophils along the endothelium[71,72] was first observed 100 years ago and is now believed to keep the cells in close enough contact with the endothelium that they can be stimulated by substances in the local environment and engage additional binding mechanisms that require close apposition. Rolling requires a definite adhesive interaction, which was first demonstrated by the observation that neutrophils roll with a much slower velocity than

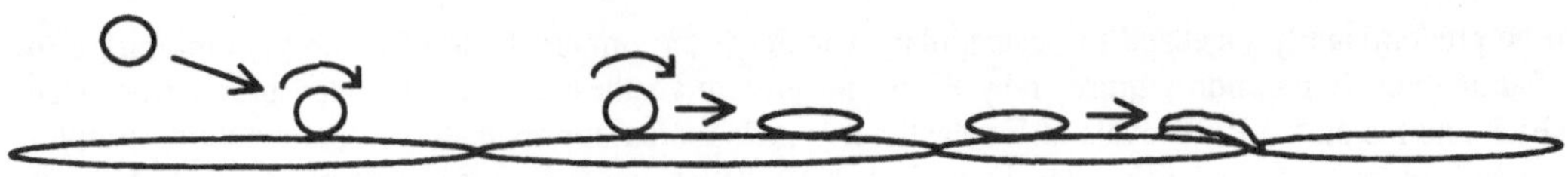

Figure 2 This diagram of the process of neutrophil extravasation can be viewed two ways. Mechanistically, four adhesive interactions are observed: initial contact, rolling, firm adhesion, and transmigration. Selectin–carbohydrate bonds appear more important in initial contact and rolling while integrin–peptide bonds appear more important in firm adhesion and transmigration. Physiologically, the process can be divided into three steps: primary adhesion, which is independent of leukocyte activation, activation of the leukocyte, and activation-dependent secondary adhesion. Specificity in leukocyte localization can be gained at any of these steps.

that predicted for cells tumbling in the fluid stream without any adhesion.[73,74] Firm adhesion follows the rolling interaction, generally utilizing activation-dependent mechanisms. Finally, the leukocytes migrate to interendothelial junctions and diapedese in response to chemokinetic substances, gradients of chemotactic substances, gradients of adhesion molecules, or some combination of these.[75]

B. TWO MECHANISTICALLY DISTINCT ADHESIVE INTERACTIONS IN LEUKOCYTE EXTRAVASATION

The four interactions involved in leukocyte extravasation (initial contact, rolling, firm adhesion, and migration) can be grouped mechanistically into two fundamentally distinct types of binding interactions: initial contact/rolling and firm adhesion/migration. It is becoming clear that in many systems studied to date, the binding interactions underlying initial contact/rolling and firm adhesion/migration rely on different families of receptors. This distinction remained entirely unknown until 1987, when Lawrence et al.[76] examined neutrophil adhesion to cytokine-stimulated endothelial cells under well-defined postcapillary venular flow conditions *in vitro.* These studies demonstrated that, under flow conditions, neutrophil adhesion to cytokine-stimulated endothelial cells is mediated almost exclusively by CD18-independent mechanisms but that subsequent neutrophil migration is CD18-dependent.[77] The initial flow studies were followed by many further studies both *in vitro*[47,78-83] and *in vivo*[84-88] that clearly distinguish separate mechanisms for initial adhesion/rolling and firm adhesion/leukocyte migration. This research has further shown that in the neutrophil system, selectin–carbohydrate interactions are primarily responsible for initial adhesion and rolling, while firm adhesion and leukocyte migration are mediated primarily by integrin–peptide interactions.

C. PROPERTIES OF SELECTIN–CARBOHYDRATE-MEDIATED ADHESION

There are several possible reasons why selectin–carbohydrate bonds are more effective in the initial arrest of leukocytes from the bloodstream. One of the most striking considerations is the location of L-selectin on the tips of microvilli of resting neutrophils.[43] This is especially important in light of the relatively small number of receptors on a neutrophil (approximately 2 to 4 × 10^4 sites per cell[47,82]) and the length of L-selectin (approximately 15 nm[89]) relative to E- and P-selectins (approximately 30 and 40 nm, respectively). Even though L-selectin is not a long molecule relative to other selectins, direct coupling to endothelial E- or P-selectins is predicted to form quite long tethering structures (on the order of 40 and 50 nm, respectively[89]). The intermembrane separation resulting from these tethers is large enough that energetically unfavorable glycocalyx interdigitation would not be necessary to form the selectin bonds.[90] In addition to avoiding glycocalyx steric hinderance, selectin–carbohydrate interactions avoid the steric hinderance that glycoprotein surface decorations can produce for integrin–peptide core interactions. The flexibility of the long single-chain selectins relative to the two-chain integrins may also be an advantage in mediating initial interactions. This has been suggested[82,91] based on molecular dynamic simulations and electron microscopy of similar proteins, but microscopic analysis of P-selectin shows only straight rod structures,[92] indicating that these molecules are at least reasonably stiff. Alternatively, some flexibility could arise in the carbohydrate chain containing the site bound by P-selectin.[93] In addition to buckling of receptors themselves, flexibility could result from deformation of the microvilli on which the receptors are located.[91] Any such deformations or buckling of receptors would have the effect of increasing the contact area between a spherical neutrophil and the planar endothelial cell surface, allowing more bonds to form. Flexibility might also have the important consequence of enhancing the mobility of binding sites so that enhanced diffusivity could increase the rate of formation of bonds.[82] An extremely rapid rate of bond formation is necessary to arrest leukocytes moving quickly in the bloodstream.

The actual bonds formed between selectins and their carbohydrate ligands may also have properties that enhance their function in mediating initial adhesion and rolling. These bond properties have been addressed in detail in recent mathematical models of leukocyte rolling, one developed by Tozeren and Ley[91] and one by Hammer and Apte.[94] Both model the bonds as springs, such that the force the bonds exert is directly proportional to the length the bond is stretched from its equilibrium configuration. In the model of Tozeren and Ley, unstrained bonds form and break as a simple bimolecular reaction. Strained bonds, on the other hand, break at a uniform rate, which is determined by fitting model predictions to experimental data. Their computational experiments point to an adhesion mechanism in which the rate of bond formation is high and the detachment rate low, except at the rear of the contact area where the stretched bonds detach at a high uniform rate. The strain resistance of the selectin bonds is set at a high value (10^{-5} dyn), which allows cell adhesion despite a relatively small number of L-selectin molecules available for bonding. Hammer and Apte reach similar conclusions using a more realistic model for the bonds based on the work of Dembo et al.[95] In this case, bonds are also modeled as springs, but the expressions for bond formation and breakage account for receptor–ligand separation distance. Fitting this model to experimental data leads to the conclusion that selectin bonds break very nearly as "ideal" bonds, which means that the breakage rate is nearly independent of the amount of strain on the bond. As a result, the bonds can withstand high strain (as in the model of Tozeren and Ley), which allows the cell to be arrested from the free stream and roll.

Another important property of selectin bonds that suits them to initial adhesion and rolling is that these bonds are functional without activation of the leukocyte. Activation-independent adhesion has been demonstrated for neutrophils, lymphocytes, and monocytes, and selectins contribute to this adhesion in all cases. This makes sense physiologically, since circulating leukocytes are normally unactivated but can bind to specific endothelial sites when necessary. It is nonetheless possible that later activation can affect selectin adhesion. It has been shown, for instance, that fMLP stimulation of neutrophils induces a brief increase in L-selectin on the cell surface, followed by rapid proteolytic shedding,[47] and that leukocyte activation can increase the affinity of L-selectin for a yeast-derived polysaccharide.[96]

D. PROPERTIES OF INTEGRIN–PEPTIDE-MEDIATED ADHESION

An important property of integrins that makes them well-suited to mediate firm adhesion and leukocyte migration is the fact that they are inducible upon cell activation. Activation produces morphological changes in the leukocyte, such that integrin adhesion is no longer disadvantaged. In particular, when the leukocyte is activated, it flattens out on the endothelial surface, greatly increasing the contact area. Increased contact area allows a large number of integrin bonds to form, taking advantage of the large number of integrin receptors (50×10^4 Mac-1 receptors on a stimulated neutrophil[47]) relative to selectin receptors (2 to 4×10^4 on an unstimulated neutrophil[47,82]). This effect, in combination with the greatly decreased fluid drag and torque on the flattened leukocyte,[97] allows activation-dependent integrin binding to produce adhesion that is up to 200-fold more shear-resistant than selectin-mediated binding.[82]

Additionally, one can speculate that the nature of the integrin bonds might make them well suited to leukocyte migration in a way analogous to the way selectin bonds are well-suited to leukocyte rolling. Whereas leukocyte rolling may be dependent on the ability of selectin bonds to withstand considerable strain (stretching) before failure, leukocyte migration might be dependent on an inability of integrin bonds to withstand strain. Thus, the many unstrained bonds in the large, uniformly spaced contact area of the flattened leukocyte would provide good adhesion, but bonds at the trailing edge could be easily broken by strain, allowing the cell to migrate. The selectin bonds, which are not easily broken by strain, are largely eliminated through proteolytic cleavage of L-selectin. Experimental evidence dealing with the strain responses of adhesion receptor bonds is scarce. In one study,[98] the force required to break integrin bonds was found to be 2.1×10^{-7} dyn for activated gpIIb-IIIa ($\alpha_{IIb}\beta_3$ integrin) and 5.7×10^{-8} dyn for nonactivated gpIIb-IIIa binding to fibrinogen. Corresponding data for selectins is not yet available, but in the model of Tozeren and Ley, a maximum selectin bond force of 10^{-5} dyn was found to be consistent with rolling adhesion.

E. SPECIFICITY IN LEUKOCYTE RECRUITMENT

We conclude this section on the mechanics of cell adhesion under flow conditions with a discussion of the implications of having distinctly different types of adhesive events. We have divided the process of leukocyte recruitment mechanistically into two types of adhesion: initial contact/rolling and firm adhesion/migration. Looking at this process from a slightly different point of view is more useful in examining how these events contribute specificity in leukocyte–endothelial cell interactions. If the process is divided into the three steps of activation-independent primary adhesion, activation, and activation-dependent

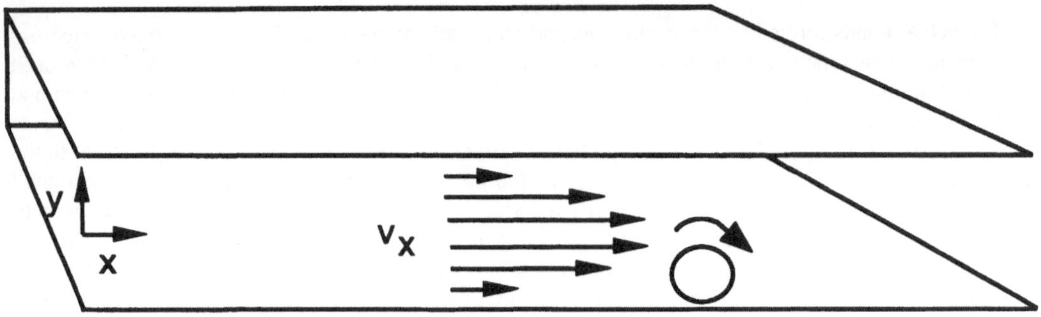

Figure 3 Flow through the parallel-plate flow chamber. Laminar flow through this geometry is well defined, and can be correlated to physiological flows by calculating parameters such as the shear rate or shear stress at the wall.

secondary adhesion, it is evident that specificity can be generated at any of the steps[99] (see Figure 2). For instance, a leukocyte that contains primary adhesion machinery to establish initial contact with a particular endothelial location might not emigrate there due to a lack of responsiveness to local activating substances or a lack of the proper activation-dependent adhesion machinery.

IV. MODEL SYSTEMS FOR STUDYING LEUKOCYTE–ENDOTHELIAL CELL ADHESION

A. STATIC ASSAYS

Leukocyte–endothelial cell adhesion is generally studied using three types of assays: static assays, *in vitro* flow assays, and *in vivo* flow assays. Static assays are generally the most straightforward and inexpensive systems available for basic adhesion studies. The Stamper-Woodruff assay is a good example.[100] In this system, a frozen section of the desired tissue is overlaid with a leukocyte suspension for some period of time. Nonadherent leukocytes are then rinsed or centrifuged away, and adherent leukocytes are quantified. Other popular static assay systems consist of a cultured monolayer of endothelial cells or transfectants on a solid or porous substrate exposed to leukocytes. These systems allow controlled stimulation of both the leukocytes and the endothelial cells, allow quantitation of leukocytes that transmigrate through the monolayer onto (or through) the substrate, and with centrifugation can give a measure of the strength of leukocyte–endothelial cell adhesion.[101] Variants utilize a multilayered artificial blood vessel wall construct, into which leukocytes can migrate following diapedesis[102,103] to examine transmigration and chemotactic/chemokinetic stimuli. The common disadvantage of all static assay systems is the impossibility of distinguishing initial adhesion events from later ones.

B. *IN VITRO* FLOW ASSAYS

In vitro flow assays provide a substantially more realistic rendering of leukocyte adhesion than static assays by allowing quantitation of initial binding events such as rolling, as well as subsequent binding events such as firm adhesion and transmigration. However, these systems require substantially more sophistication in design, construction, and use than static assays. The parallel-plate flow chamber is a commonly used configuration.[76,104,105] The parallel-plate geometry produces a well-defined laminar flow over the substrate, which can be ligand-coated plastic, an artificial lipid bilayer, a transfected cell line monolayer, or cultured endothelial cells. The flow field (see Figure 3) is typically characterized by either its shear rate (defined as dv_x/dy, the change of downstream velocity with change in distance from the wall with units of reciprocal time) or shear stress (defined as viscosity × shear rate with units of force per unit area) at the wall, and these parameters can be set equal to the wall shear rate or shear stress characteristic of flow through a blood vessel. Flow patterns corresponding to capillaries or large arteries are easily attained by adjusting either flow rate or chamber geometry according to the relation

$$\tau_w = \frac{6Q\mu}{wh^2}$$

where τ_w is the wall shear stress, Q is volumetric flow rate, μ is viscosity, w is channel width, and h is channel height. The devices are normally made out of a transparent material so that light microscopy can determine time courses of rolling, firmly adherent, and diapedesed leukocytes. Use of videomicroscopy with suitable digital image processing equipment greatly increases the flexibility and precision of measurements that can be made.[106] It is also relatively easy to add various cell-stimulating and receptor-blocking reagents to this system to investigate molecular mechanisms of leukocyte adhesion.[47,76-79,82,106]

A second flow configuration that can be used is radially outward flow between parallel discs. This configuration has been used in more general studies of receptor-mediated cell adhesion,[107] and could easily be adapted to study leukocyte–endothelial cell interactions. The flow pattern in this configuration has the characteristic of providing high shear near the center of the discs and progressively lower shear toward the outer edges, which could in some cases be an advantage over the parallel-plate geometry that provides only a single shear rate at a time. A serious potential problem in the study of leukocyte adhesion is that many adhesion events are dependent not only on the shear but also on leukocyte activation. Thus, increased cell adhesion near the periphery could be due either to the lower shear there or to longer exposure of the leukocytes to endothelial-derived activating substances. In addition, determination of leukocyte rolling velocities is difficult, since the wall shear rate is continuously changing. Finally, it is important to remember that in any assay that relies on cultured endothelial cells, important adhesive characteristics of the endothelial cells may be altered or lost *ex vivo*.

C. *IN VIVO* FLOW ASSAYS

A variety of systems have been used to study leukocyte adhesion *in vivo*. Intact translucent tissues, such as rodent ears, the hamster cheek pouch, and nude mouse skin,[72,108] are popular, due to the minimal stimulation of tissues involved and relative simplicity of the preparations. Exteriorized mesenteric venules are also a very popular model.[109-111] This system has considerable flexibility. For example, the mesentery can be superfused with a variety of stimulating cytokines, and neutrophils can be pretreated, labeled, and reintroduced to the vessel under study at known concentrations for quantitative measurements of adhesion.[85,112] The most serious potential drawback in this system is that exteriorization of the tissue can cause undesired stimulation of the endothelium. This often results in "spontaneous rolling" of leukocytes along the venules within minutes,[67] likely due to induction of endothelial P-selectin.[113]

V. PHYSIOLOGICAL APPLICATIONS OF THE PRINCIPLES OF ADHESION UNDER FLOW

A. NEUTROPHIL ADHERENCE IN INFLAMMATION

As discussed previously, the inflammatory response is a multistep process involving changes in vascular permeability, changes in expression of endothelial cell adhesion receptors, and the triggering of adherent leukocytes to emigrate into the tissues at the site of inflammation. This process is coordinated by the release of a wide variety of inflammatory mediators derived from pathogens, leukocytes, endothelial cells, and cells in the extravascular tissues. These mediators alter the local hemodynamics through vascular permeability changes, alter endothelial cell adhesion receptor expression, and induce production of leukocyte activating/chemotactic factors by endothelial cells. These changes combine to selectively recruit leukocytes to the area.

The molecular details of this recruitment have been investigated both *in vitro* and *in vivo*. *In vitro* experimental models commonly consist of purified peripheral blood neutrophils and cultured endothelial cell monolayers. The monolayers can be given controlled stimulation with cytokines such as IL-1, TNFα, LPS, and lymphotoxin, or with rapid acting inflammatory agonists such as histamine, thrombin, bradykinin, and substance P. As discussed earlier, it is well established that cytokine stimulation up-regulates E-selectin expression after several hours (as well as ICAM-1 and possibly other molecules), while rapid-acting agonist stimulation up-regulates P-selectin expression within minutes (along with ICAM-1 in the case of thrombin). These two selectins can then mediate the initial contact and rolling in conjunction with L-selectin on the neutrophil. The carbohydrate-containing ligands in these interactions are not yet completely defined, although there is evidence that L-selectin on neutrophils is decorated with sLe^x that is recognized by both E-selectin and P-selectin[43] and accounts for 50 to 70% of rolling adhesion.[47,106,114] Rolling velocities vary with shear stress, but for average postcapillary venular wall shear stresses, reported rolling velocities are approximately 10 to 40 μm/s, both *in vitro*[82,106,114] and *in vivo*.[68,73,85,97,108,115,116]

Rolling neutrophils in these systems can be activated by endothelial-derived factors to undergo firm adhesion, shape change, and migration. There is evidence that soluble IL-8 produced by endothelial cells may provide an activating and chemotactic stimulus for neutrophils,[102,117] and it has also been suggested that platelet-activating factor (PAF) on the endothelial cell surface can activate neutrophils in a "juxtacrine" mechanism.[118] Other soluble and cell-associated factors have been implicated, as has the transduction of activating signals by adhesion receptors themselves. To date, however, all evidence concerning activating factors has been acquired using static assay systems. It is essential that these experiments be checked in an assay that incorporates flow. Flow will greatly alter the mass transfer processes involved in activation by soluble factors, and rolling of neutrophils on the tips of microvilli could be very significant in determining the physiological importance of activation by cell-associated factors such as PAF and activation by adhesion receptor signal transduction.

Firm adhesion of neutrophils to cytokine-stimulated endothelium has been shown by monoclonal antibody blocking experiments to be highly dependent on β_2 integrin–ICAM-1 interactions. The relative contribution of the individual β_2 integrins to firm adhesion is not clear. Both Mac-1 and LFA-1 are known to adhere neutrophils to cytokine-stimulated endothelial cells under static conditions, but it is not known which of these integrins is more important in arresting rolling cells under conditions of flow. Both contribute to transendothelial migration.[119,120] Mechanistically, an important role for Mac-1 is appealing, since it is up-regulated by an increase in surface receptor expression from intracellular stores. This fits in well with the possibility proposed earlier that integrins work well as firm adhesion/migration receptors because they can utilize large numbers of bonds that break easily when strained. LFA-1, on the other hand, up-regulates its avidity through a conformational change. This type of regulation would be more efficient in cases such as the rapid binding and dissociation found in the delivery of a lethal hit by cytotoxic T-cells.[121]

The basic idea that selectin–carbohydrate bonds mediate initial adhesion and leukocyte rolling while integrin–peptide bonds mediate firm adhesion and migration has recently received dramatic confirmation in the case of the syndrome mentioned earlier, leukocyte adhesion deficiency type 2 (LAD-2). Experiments performed[122] using isolated neutrophils from both LAD-1 and LAD-2 patients in a mesenteric venule experimental model of inflammation in rabbits gave the following results: integrin-deficient LAD-1 neutrophils rolled along inflamed venules normally but were unable to bind firmly and transmigrate even under conditions of reduced flow. SLe^x-deficient LAD-2 neutrophils, on the other hand, were markedly deficient in their ability to bind to the endothelium and roll under flow conditions, but bound firmly and transmigrated when flow was reduced. Further characterization of these adhesion deficiencies should yield further insights into the physiological roles of these molecules.

B. LYMPHOCYTE HOMING

The second situation that we will discuss is the homing of lymphocytes to particular sites. The lymphoid components of the immune system can be functionally compartmentalized into primary, secondary, and tertiary lymphoid tissues, and the trafficking of lymphocytes among these compartments via the circulation requires specific cell adhesion events similar to the recruitment of neutrophils to inflammatory sites.[123] Primary lymphoid tissues are responsible for the production of differentiated, antigen-specific lymphocytes that have not yet been activated by encounter with antigen ("virgin" lymphocytes). Secondary lymphoid tissues include the peripheral lymph nodes (PLN), mucosal lymphoid tissues, and spleen. These tissues are specialized to accumulate and process antigen for efficient presentation to both virgin and memory lymphocytes, which recirculate regularly through these tissues. Tertiary lymphoid tissues include all other tissues in the body. These tissues normally contain few lymphoid elements but can recruit large numbers of predominantly memory lymphocytes when inflamed.

We begin with an example of lymphocyte homing to secondary lymphoid tissues: peripheral lymph nodes and mucosal lymphoid tissues. Homing to PLN involves the preferential binding of a subpopulation of lymphocytes to morphologically distinct venules, known as high endothelial venules (HEV). Specificity in emigration at these sites is evident in the observations that T cells predominate in PLN, while B cells predominate in Peyer's patches and the spleen, and that a subpopulation of memory T cells homes preferentially to PLN HEV, while another subpopulation homes preferentially to mucosal HEV (reviewed in Reference 123). The molecular mechanisms underlying this specific homing are likely to be quite complicated, although rapid progress is being made in the field. Initial adhesive events may be mediated by lymphocyte L-selectin binding to GlyCAM-1 or CD34 on PLN HEV,[124] and by L-selectin binding to MAdCAM-1 on mucosal HEV.[125] The mechanisms underlying firm adhesion in these cases are also not yet fully understood, but the integrins LFA-1, VLA-4, and $\alpha_4\beta_7$ binding to endothelial cell ICAM-1, VCAM-1, and MAdCAM-1 are all likely contributors. Recent work in our lab using purified peripheral blood T cells and cultured human umbilical vein endothelial cells (HUVEC) in an *in vitro* flow assay has demonstrated firm adhesion of T cells mediated by the LFA-1/ICAM-1 and VLA-4/VCAM-1 pathways, but we find that the initial adhesion and rolling seen in this system are mediated by a pathway that has not yet been characterized.[126] This work also indicates that patterns of adhesion may be more complicated with lymphocytes than with neutrophils. In particular, neutrophil adhesion to cytokine-stimulated HUVEC under flow generally follows the multistep adhesion cascade described earlier, and most neutrophils roll briefly and then adhere firmly and transmigrate. Lymphocytes, on the other hand, show more varied adhesion patterns, which include immediate arrest with no obvious rolling and extended rolling without eventual arrest. In one *in vivo* study,[127] no rolling was observed in HEV, while in another,[128] rolling was observed, and it was found that a G-protein-mediated activation event is associated with the establishment of firm adhesion.

Another example of lymphocyte homing involves homing to a tertiary lymphoid tissue, in this case chronically inflamed skin. Several adhesion molecules have been implicated in the recruitment of lymphocytes in this situation. E-selectin has been shown immunohistologically to be present in high levels in chronically inflamed cutaneous venules,[53] much higher than E-selectin expression seen in other sites of chronic inflammation. A carbohydrate ligand for E-selectin, designated cutaneous lymphocyte antigen (CLA), has been identified on a small subpopulation of memory T cells, and it has further been demonstrated that these lymphocytes preferentially localize to cutaneous inflammatory sites.[62] CLA is closely related to sLe^x, a known carbohydrate ligand for E-selectin, and studies in our laboratories have confirmed that E-selectin–CLA interactions mediate lymphocyte rolling under flow conditions.[126] VCAM-1 expression is also induced in chronically inflamed skin under certain conditions,[28] and its counter-receptor, VLA-4, is present in high levels on the CLA^+ subpopulation of memory lymphocytes. This receptor pair constitutes a likely activation-dependent mechanism for firm adhesion and migration analogous to β_2 integrin–ICAM-1 binding with neutrophils (see also Chapter 13 on homing of cells).

C. MONOCYTE ADHESION IN ATHEROSCLEROSIS

The final situation that we will discuss is the adhesion of monocytes to endothelium, which is particularly important in the pathogenesis of atherosclerosis.[129,130] This section begins with a brief overview of atherogenesis to demonstrate the importance of monocyte recruitment and then examines the possible mechanisms of specific monocyte binding in more detail. Atherosclerotic plaque formation begins with an accumulation of low-density lipoprotein (LDL) in the intima, which is enhanced at lesion-prone arterial sites. Monocytes are preferentially recruited to lesion-prone sites and take up intimal oxidized LDL (ox-LDL), forming foam cells. Excess ox-LDL further enhances both LDL uptake from the blood and macrophage recruitment, leading to the formation of a fatty streak containing many foam cells. Greatly increased migration of smooth muscle cells (SMC) across the intimal endothelial layer into the subendothelial space and their subsequent proliferation enhance monocyte recruitment through the production of monocyte chemotactic protein-1 (MCP-1). Foam cells eventually succumb to the cytotoxic effects of interstitial free radicals and ox-LDL, and the consequent release of lipid stores from foam cells leads to formation of a necrotic lipid core within the lesion. Accompanying these changes is fibrosis, which is mediated by the hyperproliferating SMC, and accumulation of lymphocytes at plaque borders, possibly representing an autoimmune response. The whole process is, of course, much more complex, but in keeping with the subject of this review, we will concentrate on the preferential recruitment of monocytes to lesion-prone arterial sites.

Several factors are likely to influence specificity of the monocyte–endothelial cell interaction, including receptor expression at the lesion site, alterations to the endothelial cells at these sites, activating substances released from the lesion, and the low fluid shear and longer blood cell residence times noted at lesion-prone sites. Receptors involved in the various adhesive events that lead to monocyte extravasation have not yet been thoroughly determined. It is appealing to think that, as with neutrophil binding, there is an initial selectin–carbohydrate interaction followed by integrin-mediated binding, but this may not be the case. L-selectin is known to be present on the unstimulated blood monocyte surface, but there is currently no evidence that E- or P-selectins are present on the endothelium overlying the lesion. Firm adherence could potentially involve β_2 integrins, which are present on monocytes, but there is no evidence that a suitable ligand such as ICAM-1 is present on the plaque. On the other hand, there is evidence that VCAM-1 is present on the surface of atherosclerotic plaques,[29] which mediates adhesion of monocytes and lymphocytes via the integrin VLA-4.[20,27,31,131] Presence of this integrin-binding mechanism without endothelial E- or P-selectin-binding mechanisms for initial attachment indicates the need for adhesion studies performed under appropriate flow conditions to assess functionality. While E-selectin-mediated adhesion of mononuclear cells to endothelial cells stimulated with cytokines for 4 h has been shown,[83,132-134] this mechanism is more likely to apply to leukocyte recruitment in inflammation rather than atherosclerosis.

The flow field over lesion-prone arterial sites is much different than the steady flow through postcapillary venules or high endothelial venules, and this may in part explain monocyte adhesion to atherosclerotic plaques in the absence of vascular selectins. Lesion-prone sites are known to be at low-shear locations within arteries, often just downstream from bifurcations. Flow can reverse at these locations, allowing for periodically alternating flow. Oscillatory wall shear stresses in this case have been shown in one study to range from –13 to 9 dyn/cm^2 with a time averaged mean of only –0.5 dyn/cm^2.[135] This is in marked contrast to nonlesion-prone arterial sites, where wall shear stresses range from approximately 10 to 50 dyn/cm^2. The effect of such flow conditions on leukocyte adhesion has not yet been studied.

In addition to affecting binding through direct forces on leukocytes, blood flow has dramatic effects on endothelial cell function, which could influence leukocyte adhesion. The most readily apparent effect is the elongation and orientation in the direction of flow characteristic of endothelial cells exposed to arterial shear stresses. This has been observed *in vivo* at high shear arterial locations[136] and with cultured endothelial cells exposed to steady arterial wall shear stresses.[137,138] Endothelial cells exposed to low shear stresses, on the other hand, take on a cobblestone appearance, which correlates with the more cobblestone endothelial morphology found at lesion-prone sites. Many other differences between endothelial cells subjected to high and low shear stress have been observed,[129,139-142] which could contribute to the increased susceptibility of low shear arterial surfaces to atherosclerotic lesions. A direct effect of shear stress on adhesion receptor expression is currently being investigated.[143] Finally, arterial endothelial cells are exposed to higher mean hydrostatic pressures and cyclic stretching, which can affect function as well.[144]

As in other leukocyte adhesion situations, specificity in the recruitment of monocytes to atherosclerotic sites could be enhanced at the point of monocyte activation. Both ox-LDL itself[145] and MCP-1[146] are demonstrated monocyte chemoattractants. MCP-1 is constitutively secreted by SMC and endothelial cells, and its synthesis and secretion are augmented by ox-LDL. Other mediators might come from the monocytes/macrophages themselves. Sporn et al.[147] have produced evidence that upon adhesion, monocytes increase transcription of a number of genes encoding members of a family of early host defense cytokines involved in inflammation and cell growth, including IL-8. The signaling of leukocyte transmigration by chemokines is an area of extensive research.[148]

VI. ACKNOWLEDGMENTS

This work was supported by NIH grants HL-18672, NS-23327, AI-233521, and HL-42550 and the Robert A. Welch Foundation grant C-938.

REFERENCES

1. **Bevilacqua, M. P.,** Endothelial-leukocyte adhesion molecules, *Annu. Rev. Immunol.*, 11, 67, 1993.
2. **Sanchez-Madrid, F. and Corbi, A. L.,** Leukocyte integrins: structure, function and regulation of their activity, *Semin. Cell Biol.*, 3(3), 199, 1992.
3. **Ruoslahti, E.,** Integrins, *J. Clin. Invest.*, 87(1), 1, 1991.
4. **Hemler, M. E.,** VLA proteins in the integrin family: structures, functions, and their role on leukocytes, *Annu. Rev. Immunol.*, 8, 65, 1990.
5. **Anderson, D. C. and Springer, T. A.,** Leukocyte Adhesion Deficiency: an inherited defect in the Mac-1, LFA-1, and p150,95 glycoproteins, *Annu. Rev. Med.*, 38, 75, 1987.
6. **Kishimoto, T. K., Hollander, N., Roberts, T. M., Anderson, D. C., and Springer, T. A.,** Heterogeneous mutations in the beta subunit common to the LFA-1, Mac-1, and p150,95 glycoproteins cause leukocyte adhesion deficiency, *Cell*, 50(2), 193, 1987.
7. **Landis, R. C., Bennett, R. I., and Hogg, N.,** A novel LFA-1 activation epitope maps to the I domain, *J. Cell Biol.*, 120(6), 1519, 1993.
8. **Berlin, C., Berg, E. L., Briskin, M. J., Andrew, D. P., Kilshaw, P. J., Holzmann, B., Weissman, I. L., Hamann, A., and Butcher, E. C.,** Alpha 4 beta 7 integrin mediates lymphocyte binding to the mucosal vascular addressin MAdCAM-1, *Cell*, 74(1), 185, 1993.
9. **Shimizu, Y., van Seventer, G. A., Horgan, K. J., and Shaw, S.,** Regulated expression and binding of three VLA (beta 1) integrin receptors on T cells, *Nature*, 345(6272), 250, 1990.
10. **Shimizu, Y., Newman, W., Gopal, T. V., Horgan, K. J., Graber, N., Beall, L. D., van Seventer, G. A., and Shaw, S.,** Four molecular pathways of T cell adhesion to endothelial cells: roles of LFA-1, VCAM-1, and ELAM-1 and changes in pathway hierarchy under different activation conditions, *J. Cell Biol.*, 113(5), 1203, 1991.

11. **Miller, L. J., Schwarting, R., and Springer, T. A.,** Regulated expression of the Mac-1, LFA-1, p150,95 glycoprotein family during leukocyte differentiation, *J. Immunol.*, 137, 2891, 1986.
12. **Kansas, G. S., Muirhead, M. J., and Dailey, M. O.,** Expression of the CD11/CD18, leukocyte adhesion molecule 1, and CD44 adhesion molecules during normal myeloid and erythroid differentiation in humans, *Blood,* 76(12), 2483, 1990.
13. **Miller, L. J., Bainton, D. F., Borregaard, N., and Springer, T. A.,** Stimulated mobilization of monocyte Mac-1 and p150,95 adhesion proteins from an intracellular vesicular compartment to the cell surface, *J. Clin. Invest.*, 80(2), 535, 1987.
14. **Singer, I. I., Scott, S., Kawka, D. W., and Kazazis, D. M.,** Adhesomes: specific granules containing receptors for laminin, C3bi/fibrinogen, fibronectin, and vitronectin in human polymorphonuclear leukocytes and monocytes, *J. Cell Biol.*, 109(6 Pt 1), 3169, 1989.
15. **Hogg, N. and Landis, R. C.,** Adhesion molecules in cell interactions, *Curr. Opin. Immunol.*, 5(3), 383, 1993.
16. **Buck, C. A.,** Immunoglobulin superfamily: structure, function and relationship to other receptor molecules. *Semin. Cell Biol.*, 3(3), 179, 1992.
17. **Rothlein, R., Dustin, M. L., Marlin, S. D., and Springer, T. A.,** A human intercellular adhesion molecule (ICAM-1) distinct from LFA-1, *J. Immunol.*, 137, 1270, 1986.
18. **Smith, C. W., Rothlein, R., Hughes, B. J., Mariscalco, M. M., Rudloff, H. E., Schmalstieg, F. C., and Anderson, D. C.,** Recognition of an endothelial determinant for CD18-dependent human neutrophil adherence and transendothelial migration, *J. Clin. Invest.*, 82(5), 1746, 1988.
19. **Osborn, L., Hession, C., Tizard, R., Vassallo, C., Luhowskyj, S., Chi-Rosso, G., and Lobb, R.,** Direct expression cloning of vascular cell adhesion molecule 1, a cytokine-induced endothelial protein that binds to lymphocytes, *Cell,* 59(6), 1203, 1989.
20. **Rice, G. E., Munro, J. M., and Bevilacqua, M. P.,** Inducible cell adhesion molecule 110 (INCAM-110) is an endothelial receptor for lymphocytes. A CD11/CD18-independent adhesion mechanism, *J. Exp. Med.*, 171(4), 1369, 1990.
21. **Luscinskas, F. W., Cybulsky, M. I., Kiely, J. M., Peckins, C. S., Davis, V. M., and Gimbrone, M. A.,** Cytokine-activated human endothelial monolayers support enhanced neutrophil transmigration via a mechanism involving both endothelial–leukocyte adhesion molecule-1 and intercellular adhesion molecule-1, *J. Immunol.*, 146(5), 1617, 1991.
22. **Staunton, D. E., Dustin, M. L., Erickson, H. P., and Springer, T. A.,** The arrangement of the immunoglobulin-like domains of ICAM-1 and the binding sites for LFA-1 and rhinovirus, *Cell,* 61(2), 243, 1990.
23. **Diamond, M. S., Staunton, D. E., Marlin, S. D., and Springer, T. A.,** Binding of the integrin Mac-1 (CD11b/CD18) to the third immunoglobulin-like domain of ICAM-1 (CD54) and its regulation by glycosylation, *Cell,* 65(6), 961, 1991.
24. **Sugama, Y., Chinnaswamy, T., Janakidevi, K., Andersen, T. T., Fenton, J. W., II, and Malik, A. B.,** Thrombin-induced expression of endothelial P-selectin and intercellular adhesion molecule-1: a mechanism for stabilizing neutrophil adhesion, *J. Cell Biol.*, 119(4), 935, 1992.
25. **Vazeux, R., Hoffman, P. A., Tomita, J. K., Dickinson, E. S., Jasman, R. L., St. John, T., and Gallatin, W. M.,** Cloning and characterization of a new intercellular adhesion molecule ICAM-R. *Nature,* 360(6403), 485, 1992.
26. **Fawcett, J., Holness, C. L., Needham, L. A., Turley, H., Gatter, K. C., Mason, D. Y., Simmons, D. L., and Gallatin, W. M.,** Molecular cloning of ICAM-3, a third ligand for LFA-1, constitutively expressed on resting leukocytes, *Nature,* 360(6403), 481, 1992.
27. **Rice, G. E. and Bevilacqua, M. P.,** An inducible endothelial cell surface glycoprotein mediates melanoma adhesion, *Science,* 246(4935), 1303, 1989.
28. **Rice, G. E., Munro, J. M., Corless, C., and Bevilacqua, M. P.,** Vascular and nonvascular expression of INCAM-110. A target for mononuclear leukocyte adhesion in normal and inflamed human tissues, *Am. J. Pathol.*, 138(2), 385, 1991.
29. **Cybulsky, M. I. and Gimbrone, M. A.,** Endothelial expression of a mononuclear leukocyte adhesion molecule during atherogenesis, *Science,* 251(4995), 788, 1991.
30. **Schleimer, R. P., Sterbinsky, S. A., Kaiser, J., Bickel, C. A., Klunk, D. A., Tomioka, K., Newman, W., Luscinskas, F. W., Gimbrone, M. A., and McIntyre, B. W.,** IL-4 induces adherence of human eosinophils and basophils but not neutrophils to endothelium. Association with expression of VCAM-1, *J. Immunol.*, 148(4), 1086, 1992.

31. **Elices, M. J., Osborn, L., Takada, Y., Crouse, C., Luhowskyj, S., Hemler, M. E., and Lobb, R. R.**, VCAM-1 on activated endothelium interacts with the leukocyte integrin VLA-4 at a site distinct from the VLA-4/fibronectin binding site, *Cell*, 60(4), 577, 1990.
32. **Newman, P. J., Berndt, M. C., Gorski, J., White, G. C., Lyman, S., Paddock, C., and Muller, W. A.**, PECAM-1 (CD31) cloning and relation to adhesion molecules of the immunoglobulin gene superfamily, *Science*, 247(4947), 1219, 1990.
33. **Muller, W. A., Ratti, C. M., McDonnell, S. L., and Cohn, Z. A.**, A human endothelial cell-restricted, externally disposed plasmalemmal protein enriched in intercellular junctions, *J. Exp. Med.*, 170(2), 399, 1989.
34. **Albelda, S. M., Muller, W. A., Buck, C. A., and Newman, P. J.**, Molecular and cellular properties of PECAM-1 (endoCAM/CD31): a novel vascular cell–cell adhesion molecule, *J. Cell Biol.*, 114(5), 1059, 1991.
35. **Vaporciyan, A. A., DeLisser, H. M., Yan, H.-C., Mendiguren, I. I., Thom, S. R., Jones, M. L., Ward, P. A., and Albelda, S. M.**, Involvement of Platelet-Endothelial Cell Adhesion Molecule-1 in neutrophil recruitment in vivo, *Science*, 262, 1580, 1993.
36. **Tanaka, Y., Albelda, S. M., Horgan, K. J., van Seventer, G. A., Shimizu, Y., Newman, W., Hallam, J., Newman, P. J., Buck, C. A., and Shaw, S.**, CD31 expressed on distinctive T cell subsets is a preferential amplifier of beta 1 integrin-mediated adhesion, *J. Exp. Med.*, 176(1), 245, 1992.
37. **McEver, R. P.**, Selectins: novel receptors that mediate leukocyte adhesion during inflammation, *Thromb. Haemost.*, 65(3), 223, 1991.
38. **Lasky, L. A.**, Selectins: interpreters of cell-specific carbohydrate information during inflammation, *Science*, 258, 964, 1992.
39. **Vestweber, D.**, Selectins: cell surface lectins which mediate the binding of leukocytes to endothelial cells, *Semin. Cell Biol.*, 3(3), 211, 1992.
40. **Camerini, D., James, S. P., Stamenkovic, I., and Seed, B.**, Leu-8/TQ1 is the human equivalent of the Mel-14 lymph node homing receptor, *Nature*, 342(6245), 78, 1989.
41. **Johnston, G. I., Cook, R. G., and McEver, R. P.**, Cloning of GMP-140, a granule membrane protein of platelets and endothelium: sequence similarity to proteins involved in cell adhesion and inflammation, *Cell*, 56(6), 1033, 1989.
42. **Kansas, G. S.**, Structure and function of L-selectin, *APMIS*, 100(4), 287, 1992.
43. **Picker, L. J., Warnock, R. A., Burns, A. R., Doerschuk, C. M., Berg, E. L., and Butcher, E. C.**, The neutrophil selectin LECAM-1 presents carbohydrate ligands to the vascular selectins ELAM-1 and GMP-140, *Cell*, 66(5), 921, 1991.
44. **Kishimoto, T. K., Jutila, M. A., Berg, E. L., and Butcher, E. C.**, Neutrophil Mac-1 and MEL-14 adhesion proteins inversely regulated by chemotactic factors, *Science*, 245(4923), 1238, 1989.
45. **Jutila, M. A., Kishimoto, T. K., and Butcher, E. C.**, Regulation and lectin activity of the human neutrophil peripheral lymph node homing receptor, *Blood*, 76(1), 178, 1990.
46. **Tedder, T. F., Penta, A. C., Levine, H. B., and Freedman, A. S.**, Expression of the human leukocyte adhesion molecule, LAM1. Identity with the TQ1 and Leu-8 differentiation antigens, *J. Immunol.*, 144(2), 532, 1990.
47. **Smith, C. W., Kishimoto, T. K., Abbassi, O., Hughes, B., Rothlein, R., McIntire, L. V., Butcher, E., and Anderson, D. C.** Chemotactic factors regulate lectin adhesion molecule 1 (LECAM-1)-dependent neutrophil adhesion to cytokine-stimulated endothelial cells in vitro, *J. Clin. Invest.*, 87(2), 609, 1991.
48. **Pober, J. S.**, Cytokine-mediated activation of vascular endothelium: physiology and pathology, *Am. J. Pathol.*, 133(3), 426, 1988.
49. **Rohde, D., Schluter-Wigger, W., Mielke, V., von den Driesch, P., von Gaudecker, B., and Sterry, W.**, Infiltration of both T cells and neutrophils in the skin is accompanied by the expression of endothelial leukocyte adhesion molecule-1 (ELAM-1): an immunohistochemical and ultrastructural study, *J. Invest. Dermatol.*, 98(5), 794, 1992.
50. **Montgomery, K. F., Osborn, L., Hession, C., Tizard, R., Goff, D., Vassallo, C., Tarr, P. I., Bomsztyk, K., Lobb, R., Harlan, J. M., and Pohlman, T. H.**, Activation of endothelial-leukocyte adhesion molecule 1 (ELAM-1) gene transcription, *Proc. Natl. Acad. Sci. U.S.A.*, 88(15), 6523, 1991.
51. **Whelan, J., Ghersa, P., van Houijsduijnen, R. H., Gray, J., Chandra, G., Talabot, F., and DeLamarter, J. F.**, An NF kappa B-like factor is essential but not sufficient for cytokine induction of endothelial leukocyte adhesion molecule 1 (ELAM-1) gene transcription, *Nucl. Acids Res.*, 19(10), 2645, 1991.

52. **Bevilacqua, M. P., Stengelin, S., Gimbrone, M. A., and Seed, B.,** Endothelial leukocyte adhesion molecule 1: an inducible receptor for neutrophils related to complement regulatory proteins and lectins, *Science,* 243(4895), 1160, 1989.
53. **Picker, L. J., Kishimoto, T. K., Smith, C. W., Warnock, R. A., and Butcher, E. C.,** ELAM-1 is an adhesion molecule for skin-homing T cells, *Nature,* 349(6312), 796, 1991.
54. **McEver, R. P., Beckstead, J. H., Moore, K. L., Marshall-Carlson, L., and Bainton, D. F.,** GMP-140, a platelet alpha-granule membrane protein, is also synthesized by vascular endothelial cells and is localized in Weibel-Palade bodies, *J. Clin. Invest.,* 84(1), 92, 1989.
55. **Larsen, E., Celi, A., Gilbert, G. E., Furie, B. C., Erban, J. K., Bonfanti, R., Wagner, D. D., and Furie, B.,** PADGEM protein: a receptor that mediates the interaction of activated platelets with neutrophils and monocytes, *Cell,* 59(2), 305, 1989.
56. **Hattori, R., Hamilton, K. K., Fugate, R. D., McEver, R. P., and Sims, P. J.,** Stimulated secretion of endothelial von Willebrand factor is accompanied by rapid redistribution to the cell surface of the intracellular granule membrane protein GMP-140, *J. Biol. Chem.,* 264(14), 7768, 1989.
57. **Geng, J. G., Bevilacqua, M. P., Moore, K. L., McIntyre, T. M., Prescott, S. M., Kim, J. M., Bliss, G. A., Zimmerman, G. A., and McEver, R. P.,** Rapid neutrophil adhesion to activated endothelium mediated by GMP-140, *Nature,* 343(6260), 757, 1990.
58. **Patel, K. D., Zimmerman, G. A., Prescott, S. M., McEver, R. P., and McIntyre, T. M.,** Oxygen radicals induce human endothelial cells to express GMP-140 and bind neutrophils, *J. Cell Biol.,* 112(4), 749, 1991.
59. **Etzioni, A., Frydman, M., Pollack, S., Avidor, I., Phillips, M. L., Paulson, J. C., and Gershoni-Baruch, R.,** Recurrent severe infections caused by a novel leukocyte adhesion deficiency, *N. Engl. J. Med.,* 327(25), 1789, 1992.
60. **Brandley, B. K., Swiedler, S. J., and Robbins, P. W.,** Carbohydrate ligands of the LEC cell adhesion molecules, *Cell,* 63(5), 861, 1990.
61. **Springer, T. A. and Lasky, L. A.,** Sticky sugars for selectins, *Nature,* 349(6306), 196, 1991.
62. **Berg, E. L., Yoshino, T., Rott, L. S., Robinson, M. K., Warnock, R. A., Kishimoto, T. K., Picker, L. J., and Butcher, E. C.,** The cutaneous lymphocyte antigen is a skin lymphocyte homing receptor for the vascular lectin endothelial cell–leukocyte adhesion molecule 1, *J. Exp. Med.,* 174(6), 1461, 1991.
63. **Dowbenko, D., Andalibi, A., Young, P. E., Lusis, A. J., and Lasky, L. A.,** Structure and chromosomal localization of the murine gene encoding GLYCAM 1. A mucin-like endothelial ligand for L selectin, *J. Biol. Chem.,* 268(6), 4525, 1993.
64. **Baumhueter, S., Singer, M. S., Henzel, W., Hemmerich, S., Renz, M., Rosen, S. D., and Lasky, L. A.,** Binding of L-selectin to the vascular sialomucin CD34, *Science,* 262, 436, 1993.
65. **Lesley, J., Hyman, R., and Kincade, P. W.,** CD44 and its interaction with extracellular matrix, *Adv. Immunol.,* 54, 271, 1993.
66. **Salmi, M. and Jalkanen, S.,** A 90-kilodalton endothelial cell molecule mediating lymphocyte binding in humans, *Science,* 257(5075), 1407, 1992.
67. **Fiebig, E., Ley, K., and Arfors, K. E.,** Rapid leukocyte accumulation by "spontaneous" rolling and adhesion in the exteriorized rabbit mesentery, *Int. J. Microcirc. Clin. Exp.,* 10(2), 127, 1991.
68. **Ley, K. and Gaehtgens, P.,** Endothelial, not hemodynamic, differences are responsible for preferential leukocyte rolling in rat mesenteric venules, *Circ. Res.,* 69(4), 1034, 1991.
69. **Chien, S.,** Rheology in the microcirculation in normal and low flow states, *Adv. Shock Res.,* 8, 71, 1982.
70. **Tangelder, G. J., Slaaf, D. W., Muijtjens, A. M., Arts, T., oude Egbrink, M. G., and Reneman, R. S.,** Velocity profiles of blood platelets and red blood cells flowing in arterioles of the rabbit mesentery, *Circ. Res.,* 59(5), 505, 1986.
71. **Cohnheim, J.,** *Lectures on General Pathology: A Handbook for Practitioners and Students,* The New Sydenham Society, London, 1889.
72. **Atherton, A. and Born, G. V. R.,** Quantitative investigation of the adhesiveness of circulating polymorphonuclear leukocytes to blood vessel walls, *J. Physiol.,* 222, 447, 1972.
73. **Atherton, A. and Born, G. V. R.,** Relationship between the velocity of rolling granulocytes and that of the blood flow in venules, *J. Physiol.,* 233, 157, 1973.
74. **Goldman, A. J., Cox, R. G., and Brenner, H.,** Slow viscous motion of a sphere parallel to a plane wall, *Chem. Eng. Sci.,* 22, 637, 1967.
75. **Smith, C. W.,** Transendothelial migration, in *Adhesion: Its Role in Inflammatory Disease,* Harlan, J. M. and Liu, D. Y., Eds., W. H. Freeman, New York, 1992, 85.

76. **Lawrence, M. B., McIntire, L. V., and Eskin, S. G.,** Effect of flow on polymorphonuclear leukocyte/endothelial cell adhesion, *Blood,* 70(5), 1284, 1987.
77. **Lawrence, M. B., Smith, C. W., Eskin, S. G., and McIntire, L. V.,** Effect of venous shear stress on CD18-mediated neutrophil adhesion to cultured endothelium, *Blood,* 75(1), 227, 1990.
78. **Abbassi, O., Lane, C. L., Krater, S., Kishimoto, T. K., Anderson, D. C., McIntire, L. V., and Smith, C. W.,** Canine neutrophil margination mediated by lectin adhesion molecule-1 in vitro, *J. Immunol.,* 147(7), 2107, 1991.
79. **Anderson, D. C., Abbassi, O., Kishimoto, T. K., Koenig, J. M., McIntire, L. V., and Smith, C. W.,** Diminished lectin-, epidermal growth factor-, complement binding domain–cell adhesion molecule-1 on neonatal neutrophils underlies their impaired CD18-independent adhesion to endothelial cells in vitro, *J. Immunol.,* 146(10), 3372, 1991.
80. **Kishimoto, T. K., Warnock, R. A., Jutila, M. A., Butcher, E. C., Lane, C., Anderson, D. C., and Smith, C. W.,** Antibodies against human neutrophil LECAM-1 (LAM-1/Leu-8/DREG-56 antigen) and endothelial cell ELAM-1 inhibit a common CD18-independent adhesion pathway in vitro, *Blood,* 78(3), 805, 1991.
81. **Spertini, O., Luscinskas, F. W., Kansas, G. S., Munro, J. M., Griffin, J. D., Gimbrone, M. A., and Tedder, T. F.,** Leukocyte adhesion molecule-1 (LAM-1, L-selectin) interacts with an inducible endothelial cell ligand to support leukocyte adhesion, *J. Immunol.,* 147(8), 2565, 1991.
82. **Lawrence, M. B. and Springer, T. A.,** Leukocytes roll on a selectin at physiologic flow rates: distinction from and prerequisite for adhesion through integrins, *Cell,* 65(5), 859, 1991.
83. **Hakkert, B. C., Kuijpers, T. W., Leeuwenberg, J. F. M., van Mourik, J. A., and Roos, D.,** Neutrophil and monocyte adherence to and migration across monolayers of cytokine-activated endothelial cells: the contribution of CD18, ELAM-1, and VLA-4, *Blood,* 78(10), 2721, 1991.
84. **Arfors, K. E., Lundberg, C., Lindbom, L., Lundberg, K., Beatty, P. G., and Harlan, J. M.,** A monoclonal antibody to the membrane glycoprotein complex CD18 inhibits polymorphonuclear leukocyte accumulation and plasma leakage in vivo, *Blood,* 69, 338, 1987.
85. **Perry, M. A. and Granger, D. N.,** Role of CD11/CD18 in shear rate-dependent leukocyte–endothelial cell interactions in cat mesenteric venules, *J. Clin. Invest.,* 87(5), 1798, 1991.
86. **von Andrian, U. H., Chambers, J. D., McEvoy, L. M., Bargatze, R. F., Arfors, K. E., and Butcher, E. C.,** Two-step model of leukocyte–endothelial cell interaction in inflammation: distinct roles for LECAM-1 and the leukocyte beta 2 integrins in vivo, *Proc. Natl. Acad. Sci. U.S.A.,* 88(17), 7538, 1991.
87. **Ley, K., Gaehtgens, P., Fennie, C., Singer, M. S., Lasky, L. A., and Rosen, S. D.,** Lectin-like cell adhesion molecule 1 mediates leukocyte rolling in mesenteric venules in vivo, *Blood,* 77(12), 2553, 1991.
88. **Watson, S. R., Fennie, C., and Lasky, L. A.,** Neutrophil influx into an inflammatory site inhibited by a soluble homing receptor–IgG chimaera, *Nature,* 349(6305), 164, 1991.
89. **Springer, T. A.,** Adhesion receptors of the immune system, *Nature,* 346(6283), 425, 1990.
90. **Bell, G. I., Dembo, M., and Bongrand, P.,** Cell adhesion: competition between nonspecific repulsion and specific bonding, *Biophys. J.,* 45, 1051, 1984.
91. **Tozeren, A. and Ley, K.,** How do selectins mediate leukocyte rolling in venules?, *Biophys. J.,* 63, 700, 1992.
92. **McEver, R. P.,** personal communication, 1992.
93. **Fukada, M., Spooncer, E. S., Oates, J. E., Dell, A., and Klock, J. C.,** Structure of sialylated fucosyl lactosaminoglycan isolated from human granulocytes, *J. Biol. Chem.,* 259, 10925, 1984.
94. **Hammer, D. A. and Apte, S. M.,** Simulation of cell rolling and adhesion on surfaces in shear flow: general results and analysis of selectin-mediated neutrophil adhesion, *Biophys. J.,* 63, 35, 1992.
95. **Dembo, M., Torney, D. C., Saxman, K., and Hammer, D. A.,** The reaction-limited kinetics of membrane-to-surface adhesion and detachment, *Proc. R. Soc. Lond. B. Biol. Sci.,* 234, 55, 1988.
96. **Spertini, O., Kansas, G. S., Munro, J. M., Griffin, J. D., and Tedder, T. F.,** Regulation of leukocyte migration by activation of the leukocyte adhesion molecule-1 (LAM-1) selectin, *Nature,* 349(6311), 691, 1991.
97. **Firrell, J. C. and Lipowsky, H. H.,** Leukocyte margination and deformation in mesenteric venules of rat, *Am. J. Physiol.,* 256(6 Pt 2), H1667, 1989.
98. **Sung, K.-L. P., Frojmovic, M. M., O'Toole, T. E., Zhu, C., Ginsberg, M. H., and Chien, S.,** Determination of adhesion force between single cell pairs generated by activated GpIIb-IIIa receptors, *Blood,* 81(2), 419, 1993.
99. **Butcher, E. C.,** Leukocyte–endothelial cell recognition: three (or more) steps to specificity and diversity, *Cell,* 67(6), 1033, 1991.

100. **Stamper, H. B. and Woodruff, J. J.,** Lymphocyte homing into lymph nodes: *in vitro* demonstration of the selective affinity of recirculating lymphocytes for high-endothelial venules, *J. Exp. Med.,* 144, 828, 1976.
101. **Charo, I. F., Yuen, C., and Goldstein, I. M.,** Adherence of human polymorphonuclear leukocytes to endothelial monolayers: effects of temperature, divalent cations, and chemotactic factors on the strength of adherence measured with a new centrifugation assay, *Blood,* 65, 473, 1985.
102. **Huber, A. R., Kunkel, S. L., Todd, R. F., and Weiss, S. J.,** Regulation of transendothelial neutrophil migration by endogenous interleukin-8, *Science,* 254(5028), 99, 1991.
103. **Hakkert, B. C., Rentenaar, J. M., van Aken, W. G., Roos, D., and van Mourik, J. A.,** A three-dimensional model system to study the interactions between human leukocytes and endothelial cells, *Eur. J. Immunol.,* 20, 2775, 1990.
104. **Hochmuth, R. M., Mohandas, N., and Blackshear, P. L.,** Measurement of the elastic modulus for red cell membrane using a fluid mechanical technique, *Biophys. J.,* 13, 747, 1973.
105. **Gallik, S., Usami, S., Jan, K.-M., and Chien, S.,** Shear stress-induced detachment of human polymorphonuclear leukocytes from endothelial cell monolayers, *Biorheology,* 26, 823, 1989.
106. **Jones, D. A., McIntire, L. V., McEver, R. P., and Smith, C. W.,** P-selectin supports neutrophil rolling on histamine-stimulated endothelial cells, *Biophys. J.,* 65, 1560, 1993.
107. **Cozens-Roberts, C., Quinn, J. A., and Lauffenburger, D. A.,** Receptor-mediated cell attachment and detachment kinetics. II. Experimental model studies with the radial-flow detachment assay, *Biophys. J.,* 58(4), 857, 1990.
108. **Mayrovitz, H. N.,** Leukocyte rolling: a prominent feature of venules in intact skin of anesthetized hairless mice, *Am. J. Physiol.,* 262(1 Pt 2), H157, 1992.
109. **Ley, K., Pries, A. R., and Gaehtgens, P.,** A versatile intravital microscope design, *Int. J. Microcirc. Clin. Exp.,* 6(2), 161, 1987.
110. **Zeintl, H., Sack, F. U., Intaglietta, M., and Messmer, K.,** Computer assisted leukocyte adhesion measurement in intravital microscopy, *Int. J. Microcirc. Clin. Exp.,* 8(3), 293, 1989.
111. **oude Egbrink, M. G., Tangelder, G. J., Slaaf, D. W., and Reneman, R. S.,** Influence of platelet-vessel wall interactions on leukocyte rolling in vivo, *Circ. Res.,* 70(2), 355, 1992.
112. **Yuan, Y. and Fleming, B. P.,** A method for isolation and fluorescent labeling of rat neutrophils for intravital microvascular studies, *Microvasc. Res.,* 40(2), 218, 1990.
113. **Dore, M., Korthuis, R. J., Granger, D. N., Entman, M. L., and Smith, C. W.,** P-selectin mediates spontaneous leukocyte rolling in vivo, *Blood,* 82(4), 1308, 1993.
114. **Abbassi, O., Kishimoto, T. K., McIntire, L. V., Anderson, D. C., and Smith, C. W.,** E-selectin supports neutrophil rolling in vitro under conditions of flow, *J. Clin. Invest.,* 92, 2719, 1993.
115. **Ley, K., Cerrito, M., and Arfors, K. E.,** Sulfated polysaccharides inhibit leukocyte rolling in rabbit mesentery venules, *Am. J. Physiol.,* 260(5 Pt 2), H1667, 1991.
116. **Barroso-Aanda, J. and Schmid-Schonbein, G. W.,** Pentoxifylline pretreatment decreases the pool of circulating activated neutrophils, in-vivo adhesion to endothelium, and improves survival from hemorrhagic shock, *Biorheology,* 27(3-4), 401, 1990.
117. **Smith, W. B., Gamble, J. R., Clark-Lewis, I., and Vadas, M. A.,** Interleukin-8 induces neutrophil transendothelial migration, *Immunology,* 72(1), 65, 1991.
118. **Lorant, D. E., Patel, K. D., McIntyre, T. M., McEver, R. P., Prescott, S. M., and Zimmerman, G. A.,** Coexpression of GMP-140 and PAF by endothelium stimulated by histamine or thrombin: a juxtacrine system for adhesion and activation of neutrophils, *J. Cell Biol.,* 115(1), 223, 1991.
119. **Smith, C. W., Marlin, S. D., Rothlein, R., Toman, C., and Anderson, D. C.,** Cooperative interactions of LFA-1 and Mac-1 with intercellular adhesion molecule-1 in facilitating adherence and transendothelial migration of human neutrophils in vitro, *J. Clin. Invest.,* 83(6), 2008, 1989.
120. **Anderson, D. C., Rothlein, R., Marlin, S. D., Krater, S. S., and Smith, C. W.,** Impaired transendothelial migration by neonatal neutrophils: abnormalities of Mac-1 (CD11b/CD18)-dependent adherence reactions, *Blood,* 76(12), 2613, 1990.
121. **Springer, T. A.,** The sensation and regulation of interactions with the extracellular environment: the cell biology of lymphocyte adhesion receptors, *Annu. Rev. Cell Biol.,* 6, 59, 1990.
122. **von Andrian, U. H., Berger, E. M., Ramezani, L., Chambers, J. D., Ochs, H. D., Harlan, J. M., Paulson, J. C., Etzioni, A., and Arfors, K. E.,** In vivo behavior of neutrophils from two patients with distinct inherited leukocyte adhesion deficiency syndromes, *J. Clin. Invest.,* 91(6), 2893, 1993.
123. **Picker, L. J. and Butcher, E. C.,** Physiological and molecular mechanisms of lymphocyte homing, *Annu. Rev. Immunol.,* 10, 61, 1992.
124. **Picker, L. J.,** Mechanisms of lymphocyte homing, *Curr. Opin. Immunol.,* 4(3), 277, 1992.

125. **Berg, E. L., McEvoy, L. M., Berlin, C., Bargatze, R. F., and Butcher, E. C.,** L-selectin-mediated lymphocyte rolling on MAdCAM-1, *Nature,* 366, 695, 1993.
126. **Jones, D. A., McIntire, L. V., Smith, C. W., and Picker, L. J.,** A two-step adhesion cascade for T cell/endothelial cell interactions under flow conditions, *J. Clin. Invest.,* submitted, 1994.
127. **Bjerknes, M., Cheng, H., and Ottaway, A.,** Dynamics of lymphocyte-endothelial interactions in vivo, *Nature,* 231, 402, 1986.
128. **Bargatze, R. F. and Butcher, E. C.,** Rapid G protein-regulated activation event involved in lymphocyte binding to high endothelial venules, *J. Exp. Med.,* 178(1), 367, 1993.
129. **Schwartz, C. J., Valente, A. J., Sprague, E. A., Kelley, J. L., and Nerem, R. M.,** The pathogenesis of atherosclerosis: an overview, *Clin. Cardiol.,* 14(2 Suppl. 1), 11, 1991.
130. **Ross, R., Masuda, J., and Raines, E. W.,** Cellular interactions, growth factors, and smooth muscle proliferation in atherogenesis, *Ann. N.Y. Acad. Sci.,* 598(6335), 2, 1990.
131. **Schwartz, B. R., Wayner, E. A., Carlos, T. M., Ochs, H. D., and Harlan, J. M.,** Identification of surface proteins mediating adherence of CD11/CD18-deficient lymphoblastoid cells to cultured human endothelium, *J. Clin. Invest.,* 85(6), 2019, 1990.
132. **Carlos, T. M., Dobrina, A., Ross, R., and Harlan, J. M.,** Multiple receptors on human monocytes are involved in adhesion to cultured human endothelial cells, *J. Leukoc. Biol.,* 48(5), 451, 1990.
133. **Carlos, T., Kovach, N., Schwartz, B., Rosa, M., Newman, B., Wayner, E., Benjamin, C., Osborn, L., Lobb, R., and Harlan, J.,** Human monocytes bind to two cytokine-induced adhesive ligands on cultured human endothelial cells: endothelial-leukocyte adhesion molecule-1 and vascular cell adhesion molecule-1, *Blood,* 77(10), 2266, 1991.
134. **Leeuwenberg, J. F., Jeunhomme, T. M., and Buurman, W. A.,** Role of ELAM-1 in adhesion of monocytes to activated human endothelial cells, *Scand. J. Immunol.,* 35(3), 335, 1992.
135. **Ku, D. N. and Giddens, D. P.,** Laser Doppler anemometer measurements of pulsatile flow in a model carotid bifurcation, *J. Biomech.,* 20(4), 407, 1987.
136. **Nerem, R. M., Levesque, M. J., and Cornhill, J. F.,** Vascular endothelial morphology as an indicator of patterns of blood flow, *J. Biomech. Eng.,* 103, 172, 1981.
137. **Ives, C. L., Eskin, S. G., and McIntire, L. V.,** Mechanical effects on endothelial cell morphology: in vitro assessment, *In Vitro Cell. Dev. Biol.,* 22(9), 500, 1986.
138. **Eskin, S. G., Ives, C. L., McIntire, L. V., and Navarro, L. T.,** Response of cultured endothelial cells to steady flow, *Microvasc. Res.,* 28, 87, 1984.
139. **Nollert, M. U., Diamond, S. L., and McIntire, L. V.,** Hydrodynamic shear stress and mass transport modulation of endothelial cell metabolism, *Biotechnol. Bioeng.,* 38, 588, 1991.
140. **Diamond, S. L., Sharefkin, J. B., Dieffenbach, C., Frasier-Scott, K., McIntire, L. V., and Eskin, S. G.,** Tissue plasminogen activator messenger RNA levels increase in cultured human endothelial cells exposed to laminar shear stress, *J. Cell. Physiol.,* 143(2), 364, 1990.
141. **Nollert, M. U., Panaro, N. J., and McIntire, L. V.,** Regulation of genetic expression in shear stress stimulated endothelial cells, *Ann. N.Y. Acad. Sci.,* 665, 94, 1992.
142. **Panaro, N. J. and McIntire, L. V.,** Flow and shear stress effects on endothelial cell function, in *Hemodynamic Forces in Modulating Vascular Cell Biology,* Bauer, E. S., Ed., R. G. Landes, Austin, TX, 1993, 47.
143. **Sampath, R., McIntire, L. V., and Eskin, S. G.,** Surface expression of leukocyte adhesion molecules on the endothelium under flow conditions, in Extended Abstracts, American Institute of Chemical Engineers 1992 Annual Meeting, 1992, page Abstract 169e.
144. **Carosi, J. A., Eskin, S. G., and McIntire, L. V.,** Cyclical strain effects on production of vasoactive materials in cultured endothelial cells, *J. Cell. Physiol.,* 151, 29, 1992.
145. **Quinn, M. T., Parthasarathy, S., Fong, L. G., and Steinberg, D.,** Oxidatively modified low density lipoproteins: a potential role in recruitment and retention of monocyte/macrophages during atherogenesis, *Proc. Natl. Acad. Sci. U.S.A.,* 84(9), 2995, 1987.
146. **Valente, A. J., Graves, D. T., Vialle-Valentin, C. E., Delgado, R., and Schwartz, C. J.,** Purification of a monocyte chemotactic factor secreted by nonhuman primate vascular cells in culture, *Biochemistry,* 27(11), 4162, 1988.
147. **Sporn, S. A., Eierman, D. F., Johnson, C. E., Morris, J., Martin, G., Ladner, M., and Haskill, S.,** Monocyte adherence results in selective induction of novel genes sharing homology with mediators of inflammation and tissue repair, *J. Immunol.,* 144(11), 4434, 1990.
148. **Miller, M. D. and Krangel, M. S.,** Biology and biochemistry of the chemokines: a family of chemotactic and inflammatory cytokines, *CRC Crit. Rev. Immunol.,* 12, 17, 1992.

Section II

The Molecular Biochemistry of Adhesion

Chapter 8

The Integrin Cell Adhesion Molecules

Jonathan M. Edelman, Paul A. DiMilla, and Steven M. Albelda

CONTENTS

I. Introduction: Integrin Cell Adhesion Receptors 163
  A. The Initial Descriptions of Integrins 163
  B. Integrin Nomenclature 164
II. Biochemistry: The Structure and Function of Integrins 165
  A. Integrin Structure 165
  B. Integrin Function 166
  C. Structure–Function Relationships of Integrins 169
III. Cell Biology: Normal Integrin Function 170
  A. Integrin Expression is Widespread 170
  B. Integrin Function is Regulated 171
  C. Integrins Mediate Cell–Cell Adhesion 172
  D. Integrins Mediate Cell Motility 172
  E. Integrin Expression is Developmentally Regulated 174
IV. Pathobiology: Integrins in Disease 174
  A. Congenital Integrin Dysfunction 174
  B. Integrins and their Ligands are Used by Infectious Agents 175
  C. Integrin Expression is Altered in Malignancy 175
  D. Integrins are Involved in Inflammatory and Healing Processes 176
V. Conclusions: Therapeutic Manipulation of Integrin Function 177
References 177

I. INTRODUCTION: INTEGRIN CELL ADHESION RECEPTORS

A. THE INITIAL DESCRIPTIONS OF INTEGRINS

The interaction of cells with each other and their underlying extracellular matrices has long been recognized as important for normal cell function. The means by which cells carry out these interactions has therefore been the subject of intense scientific investigation. The ability of cells to adhere to a limited number of specific adhesive substrates, such as the extracellular matrix proteins fibronectin, laminin, and vitronectin, led to the idea that there were specific cell surface proteins that functioned as "receptors" for these substrates.

Approximately 10 years ago, a number of investigators from widely different backgrounds began to describe such adhesion receptors. Among the first descriptions of such a receptor was the report by Neff et al.[1] that a monoclonal antibody named "CSAT" inhibited chick myoblast attachment to petri dishes and recognized a heterodimeric protein consisting of a 120-kD β and 160-kD α subunit. The CSAT antigen complex was later called "integrin" because of its integral role in mediating cell adhesion.[2] Using a fibronectin affinity column, Pytela et al.[3] identified a similar set of proteins from osteosarcoma cells, which they termed the human fibronectin receptor. Another heterodimeric glycoprotein receptor abundantly present on human platelets and involved in platelet aggregation, called gpIIb/IIIa, was shown to interact with fibrinogen by Bennett et al.[4] Springer[5] described a family of three sets of cell surface receptors found on human and mouse leukocytes, known as Leu-CAMs. These receptors were involved in normal T cell, macrophage, and neutrophil function and shared a common smaller subunit. Finally, five related antigens described by Hemler et al.,[6] the first two of which appeared on T lymphocytes 2 weeks after activation with antigen or mitogen (and hence named "very late activation" or VLA antigens), were also shown to be heterodimeric proteins with one common subunit.

The homology of protein sequences from several of these proteins prompted Hynes[7] to propose that these structurally related, noncovalently associated heterodimers belonged to a family of cell surface receptors, called the "integrins", linked by a common involvement in cell–cell and cell–substratum

0-8493-4559-6/95/$0.00+$.50

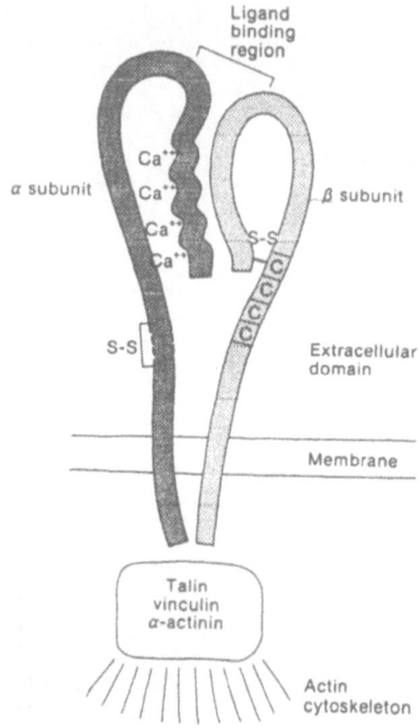

Figure 1 Schematic integrin structure. An integrin is comprised of a noncovalently associated larger α subunit and a smaller β subunit. Each subunit is anchored in the plasma membrane by a single short transmembrane region, with its C-terminus in the cytoplasm of the cell. All α subunits contain consensus sequences for divalent cation interaction in the extracellular portion (Ca^{++}). Some α subunits are cleaved to form two disulfide linked (S-S) fragments, while others contain an "I" domain between the N-terminus and the first divalent cation binding region. All β subunits contain five areas rich in cysteine (C) residues. Four of these regions are located in the C-terminal half of the extracellular region. All β subunits are probably folded and stabilized by disulfide linkages, as indicated in the figure. Both subunits contribute to an extracellular ligand binding region, while at least the β subunit is involved in cytoskeletal interactions.

adhesion. Integrins could be divided into three subfamilies that contained heterodimers comprised of a single β subunit paired with a unique set of α subunits. The largest subfamily consisted of the "β_1" integrins and included most of the cell–substratum adhesion receptors. The second, or β_2, subfamily contained the leukocyte integrins that were involved in cell–cell interactions. The final subfamily was comprised of those receptors that contained the "β_3" subunit and included the platelet integrin gpIIb/IIIa and the "vitronectin" receptor.

Over the ensuing years, cDNAs of β and α subunits from a wide variety of species (including human, rodent, avian, amphibian, and insect) have been cloned and sequenced, providing confirmation of the interrelatedness of these heterodimeric receptors. As of this writing, there are currently 7 described human β subunits and 12 human α subunits that combine to form 16 different heterodimers. Many of these have known homologues in other species.[8] Although most α subunits pair with a single β subunit, the α_4,[9,10] α_6,[11,12] and α_v[13,14] subunits have all been shown to have multiple β partners, thus blurring this original subfamily grouping as proposed by Hynes. Figure 1 summarizes the known integrin heterodimers and their ligands. The existence of other as yet uncharacterized integrins is undoubted.

B. INTEGRIN NOMENCLATURE

One of the difficulties in integrating all of the available information about integrins is the confusion caused by different nomenclatures. For example, the integrins found on platelets have been described classically by the glycoprotein naming system used by hematologists, while those found on leukocytes have been referred to by their "CD" number assignments or antigen names, given by immunologists. Thus, the major platelet integrin is known as gpIIb/IIIa, while the leukocyte type III complement receptor integrin is known as CR3, MAC-1, or CD11b/CD18. Although first described on T lymphocytes as VLA antigens or CD49a,b,c.../CD29, the β_1 integrins have a far wider distribution, and many are functionally important on cells outside the immune system. Other names have been used to describe individual members of this subfamily, such as the fibronectin receptor,[3] gpIa/IIa,[15] galactoprotein β_3,[16] and TSP-180.[17] For the purposes of clarity, simplicity, and accuracy, the nomenclature proposed by Hynes and modeled after the leukocyte integrin family has been adopted widely and will be used throughout this chapter. Each integrin is named by its component subunits and can be followed by its ligand. For example, the $\alpha_4\beta_1$ fibronectin receptor is identical to the $\alpha_4\beta_1$ VCAM-1 receptor in its components but functions in a different context. The $\alpha_{IIb}\beta_3$ receptor is platelet glycoprotein gpIIb/IIIa. The CR3 receptor is $\alpha_M\beta_2$. Because integrins appear to fall into subfamilies in which individual integrin members are composed of heterodimers with a common β subunit and a unique subset of α subunits, integrins are often described only in terms of their α subunits. However, the recent description of alternate β partners for three different α subunits emphasizes the need to include both subunits in the name.

II. BIOCHEMISTRY: THE STRUCTURE AND FUNCTION OF INTEGRINS

A. INTEGRIN STRUCTURE

Our knowledge of integrin structure is based primarily upon careful biochemical studies of the platelet membrane glycoprotein gpIIb/IIIa complex[18-23] and the fact that the DNA sequences for most integrin subunits are now known (Table 1). A schematic representation of a "typical" integrin is shown in Figure 2. Electron microscopic images of purified integrins[24] suggest an association of the amino terminal globular domains of the α and β subunits. Chemical and photoaffinity cross-linking experiments[25,26] have suggested that both subunits contain extracellular ligand-binding regions (see below). The seven known integrin β subunits are remarkably similar; their amino acid sequences are 40 to 48% homologous.[27] This homology is even higher, approaching 90%, when specific β subunits are compared across species.[28] All β subunits possess a large extracellular domain, one membrane-spanning domain, and most have a short cytoplasmic domain found at the carboxyl terminus of the molecule. The positions of the 56 cysteine residues found in the extracellular domain are conserved in all but two of the β subunits so far sequenced. Most of these cysteines are organized into four repeating units located in the C-terminal half of the extracellular domain. The β_4 subunit, whose mol wt is significantly greater than that of other known β subunits, possesses a cytoplasmic domain that is 20 times longer than any other known β subunit.[11,29] The functional importance of this domain remains unknown.

Careful biochemical characterization of the β_3 subunit has shown that the amino terminal portion of the molecule is folded into a loop that is stabilized by disulfide bonds formed between cysteine 5* near the amino terminus and cysteine 435 within the first cysteine-rich repeat, lending credence to the electron microscopic evidence of a globular head. In addition, this globular head is, itself, disulfide-linked between cysteine 406 and cysteine 655.[20] Given the relatively high degree of amino acid homology and the preservation of all the cysteine residues among the β subunits, this structural motif is probably found in others as well. Other than β_4, the highest degree of amino acid homology among the β subunits lies within a small region in the globular head and in the transmembrane and cytoplasmic domains. This suggests a generic role for these domains in the function of integrins, most likely involving interaction with extracellular ligands and cytoskeletal components of the cell.

Although the α subunits exhibit more sequence heterogeneity than the β subunits, they too are similar in structure, containing a large globular extracellular domain, a single membrane-spanning region, and a short carboxyl-terminal cytoplasmic domain. The extracellular domain of each known α subunit contains three to four consensus regions for calcium binding. With one exception, α subunits can be divided into two groups based upon their posttranslational modifications (Table 2). All known α subunits are synthesized as single polypeptide chains; however, one group is completely cleaved at a region in the extracellular domain near the membrane, resulting in a heavy and light chain joined by a single disulfide bond.[30] Mutations in the α_{IIb} that prevent this modification suggest that cleavage occurs after β subunit pairing and is not required for surface expression of the heterodimer.[31] It has been shown that the secondary structure of these disulfide-linked α subunits is less complicated than that of the β subunits, being constrained by nearest neighbor cysteine pairing that probably occurs during synthesis.[21] The second group of α subunits are not cleaved, and each contains an additional sequence of amino acids inserted between the amino terminus and the first Ca^{++} binding site.[27] Although the function of this "I" (inserted) domain is not known, it contains "type A" repeats like those found in collagen and other adhesive proteins. The lone exception to this grouping is α_4, which does not contain an I domain, but which is incompletely cleaved in the middle of its extracellular domain into two fragments that are not covalently linked. Together these fragments associate with the β subunit.[32]

Variations in integrin structure derived from alternative splicing have been identified for several subunits. At least two forms of human α_{IIb} mRNA exist in platelets, one coding for a subunit missing 34 amino acids of the extracellular domain near the C-terminus.[33] Alternatively spliced mRNA for human β_1[34] and β_3[35] and drosophila α_{PS2}[36] have also been described, but the cellular distribution and functional significance of these have not yet been determined.

It appears that high mannose glycosylation occurs cotranslationally for both subunits,[37] but that final subunit processing, by the addition of complex carbohydrates, occurs only after the heterodimer has been

---

* Amino acid numbers are exclusive of the signal peptide.

Table 1 **Known integrin subunit cDNA sequences. (The GenBank/EMBL accession number can be used to retrieve cDNA sequence information for the indicated integrin subunit from these databases using Genetics Computer Group software.[177] Complete sequence information is also available in the indicated reference.)**

Integrin subunit	Species	GenBank/EMBL accession #	Reference
β_1	Human	X07979	Argraves[178]
	Mouse	Y00818	Tominaga[179]
		Y00769	Holers[180]
	Chicken	M14049	Tamkun[2]
	Frog	J03736	DeSimone[8]
β_2	Human	M15395	Kishimoto[181]
	Mouse	X14951	Wilson[182]
		M31039	Zeger[183]
β_3	Human	J02703	Fitzgerald[184]
β_4	Human	X53587	Tamura[185]
			Hogervorst[186]
			Suzuki[29]
β_5	Human	J05633	McClean[187]
		M35011	Suzuki[188]
			Ramaswamy[189]
β_6	Human	J05522	Sheppard[190]
β_7	Human	M62880	Erle[10]
α_L	Human	Y00796	Larson[191]
	Mouse	M60778	Kaufmann[192]
α_M	Human	J03925	Corbi[193]
			Hickstein[194]
			Arnaout[195]
	Mouse	X07640	Pytela[196]
α_X	Human	J03925	Corbi[197]
α_1	Rat	X52140	Ignatius[198]
α_2	Human		Takada[199]
α_3	Hamster	J05281	Tsuji[16]
α_4	Human		Takada[200]
	Mouse	X53176	Neuhaus[201]
α_5	Human	X06256	Argraves[178]
	Mouse	X15203	Holers[180]
α_6	Human	X53586	Tamura[185]
	Chicken	X56559	deCurtis[202]
α_v	Human	M14648	Suzuki[203]
	Chicken	M60517	Bossy[204]
α_{IIb}	Human	J02764	Poncz[205]

formed.[38] β_1 subunits that are defective in their ability to form heterodimers are not completely processed and remain smaller than the mature form.[39] Incompletely processed heterodimers can be found on the cell surface but appear to have altered ligand-binding ability.[37] The glycosylation pattern appears to depend on the specific heterodimer, since the β_3 subunit is differentially glycosylated depending on whether it is paired with α_{IIb} or α_v.[40] The effect of this difference on ligand specificity is not known; however, glycosylation is an obvious potential integrin regulatory mechanism.

B. INTEGRIN FUNCTION

Although the integrins are expressed in an astonishing variety of cells throughout the body and interact with a diverse group of ligands, several common functional features are evident. First, the interaction of

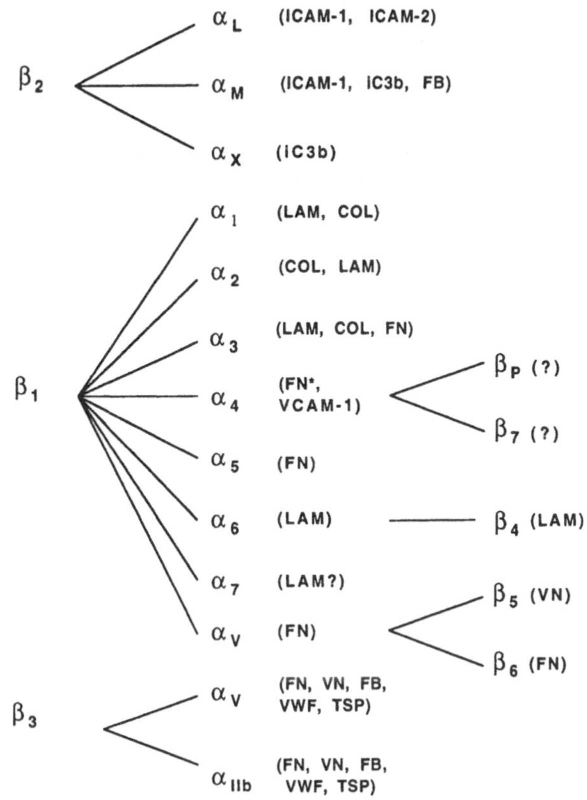

Figure 2 Integrin subunit associations and ligand specificities. Known integrin heterodimer pairs are indicated by connecting lines between α and β subunits. Described extracellular ligands for the specific heterodimer are found in parentheses. Abbreviations: ICAM-1, ICAM-2 is Intercellular adhesion molecule 1 and 2; iC3b is the cleavage fragment of complement protein C3; FB is fibrinogen; LAM is laminin; COL is collagen (all types); FN is fibronectin; FN* is fibronectin containing the alternatively splice CSIII region; VCAM-1 is vascular cell adhesion molecule 1; VN is vitronectin; VWF is von Willebrand factor; TSP is thrombospondin; ? indicates unknown ligand(s).

integrins with the cytoskeleton and extracellular matrix appears to require the presence of both subunits.[41] In fact, single integrin subunits have not been found on the cell surface.

Second, the association of integrins with cytoskeletal actin filaments has been shown to be via an indirect linkage, involving talin, vinculin, and perhaps other cytoskeletal-associated molecules.[42] Recent evidence suggests that α-actinin can bind directly to a β_1 cytoplasmic domain peptide and to β_1 and β_3 integrins, presumably by interacting with the β subunits alone.[43] The colocalization of integrins with these cytoskeletal proteins at the membrane–matrix junction of adherent cells in structures called "focal contacts" or "adhesion plaques"[44,45] is a hallmark of integrin-mediated adhesion (Figure 3).

Third, integrin extracellular ligand-binding specificity is determined by the αβ heterodimer. Although some integrins have a unique ligand, most are able to interact with more than one. The specificity for this binding appears to vary, depending on the cell type in which the receptor is expressed.[46] Among the integrins that mediate cell–matrix adhesion, there is considerable redundancy in ligand interaction. For example, there are no fewer than seven different integrins that can bind to fibronectin and six that bind to laminin (Figure 1).

The interaction of integrins with their extracellular ligands is a cation-dependent process, occurring with relatively low affinities (10^{-6} *M*). In general, *in vitro* measurements of integrin ligand affinity increase dramatically when Mn^{2+} or Mg^{2+} is substituted for Ca^{2+} (see References 47 and 48). Direct evidence that integrins are metaloproteins exists for the $\alpha_v\beta_3$ vitronectin receptor.[49] Isotopically labeled Mn^{2+} associates with the receptor in its ligand-binding pocket when vitronectin is present.

Integrins appear to recognize specific amino acid sequences in their ligands. The best studied is the arginine-glycine-aspartic acid (RGD) sequence found within several matrix proteins, including fibronectin,

Table 2 α **Subunit structure. (All α subunits are either posttranslationally cleaved or contain an "I" domain. Of those subunits that are cleaved, all except α_4 remain covalently linked by a disulfide bridge near the membrane. α_4 is cleaved near its midpoint and is not disulfide linked. In those α subunits that contain an "I" domain, it is inserted between the N-terminus of the molecule and the first divalent cation binding region.)**

α Subunit	Cleaved?	Disulfide linked?	"I" domain present?
α_1	No	—	Yes
α_2	No	—	Yes
α_L	No	—	Yes
α_M	No	—	Yes
α_X	No	—	Yes
α_4	Yes	No	No
α_3	Yes	Yes	No
α_5	Yes	Yes	No
α_6	Yes	Yes	No
α_7	Yes	Yes	No
α_v	Yes	Yes	No
α_{IIb}	Yes	Yes	No

fibrinogen, thrombospondin, vitronectin, laminin, and type I collagen.[50] However, not all integrins bind to ligands via RGD-containing domains. For example, the $\alpha_4\beta_1$ receptor binds to the first 25 amino acids (CS-1 peptide) and amino acids 90 to 109 (CS-5 peptide) of the variably spliced type III connecting segment (CSIII) region of fibronectin, which does not contain an RGD sequence.[51-53] Similarly, the $\alpha_2\beta_1$ receptor interacts with a tetrapeptide (DGEA) in a fragment of type I collagen.[54]

All of the integrins that mediate cell–cell adhesion for which the counter-receptor is known promote heterophilic association with members of the immunoglobulin superfamily. These ligands include ICAM-1 and ICAM-2 that bind to $\alpha_L\beta_2$ and $\alpha_M\beta_2$,[55,56] and VCAM-1 that binds to $\alpha_4\beta_1$.[57] The $\alpha_L\beta_2$ and $\alpha_M\beta_2$

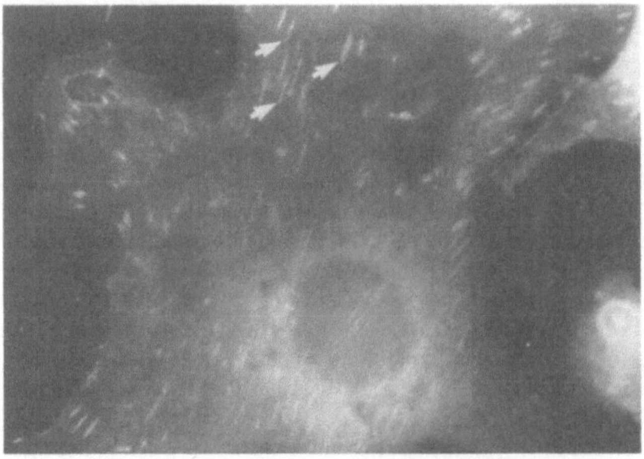

Figure 3 Focal contacts in an adherent fibroblast. A 3T3 mouse fibroblast was plated onto a fibronectin-coated coverslip and then visualized with an antibody directed against the β_1 integrin subunit by indirect immunofluorescence. The fluorescent streaks (arrows) are collections of β subunit at the ventral surface of the cell where it contacts the extracellular matrix and are called "focal contacts" or "adhesion plaques".

receptors bind to distinct domains on ICAM-1.[58] In addition, the leukocyte integrin $\alpha_M\beta_2$ binds to lipopolysaccharide and other material (pathogens, particles, etc.) that has been opsonized with the iC3b fragment of complement.[59]

C. STRUCTURE–FUNCTION RELATIONSHIPS OF INTEGRINS

From the above discussion, it is clear that integrins must possess functional domains within each subunit that regulate heterodimer formation and interaction with the cytoskeletal and extracellular ligands. Structure–function relationships have been most thoroughly studied for the β subunits. Photoaffinity and chemical cross-linking experiments using the β_3 subfamily integrins, $\alpha_{IIb}\beta_3$ and $\alpha_v\beta_3$, suggest that extracellular matrix binding occurs in a region at the amino terminus of the β subunit near amino acids 110 to 130. This region is more than 90% conserved in all known β subunits.[25,26] A naturally occurring point mutation at amino acid 119 of the β_3 subunit results in $\alpha_{IIb}\beta_3$ heterodimers that are unable to bind to fibrinogen, fibronectin, vitronectin, or von Willebrand factor.[60] The consequence of this mutation is a bleeding disorder known as Glanzman's thrombasthenia (see below). Recent studies have shown that a peptide fragment from a nearby region (aa 211 to 224) of the β_3 subunit blocks the binding of fibrinogen to gpIIb/IIIa and can prevent platelet aggregation.[61] Thus, it is likely that there are several extracellular ligand-binding domains of this β subunit located within the globular head.

A large body of evidence has accumulated to show that the cytoplasmic domain of the β subunit is required for normal integrin function. Although a β_1 subunit missing its cytoplasmic domain is processed normally, appears as a part of a heterodimer on the cell surface, and binds to its extracellular ligand, integrins containing such a subunit are unable to colocalize with cytoskeletal proteins such as vinculin or talin in adhesion plaques or to mediate adhesion.[62] Detailed deletional analysis of the 54 amino acid β_1 cytoplasmic domain suggests that proper folding of this domain is essential for focal contact formation. Several studies have shown that the C-terminal 5 amino acids, while not essential, greatly enhance cytoskeletal association.[63,64] Other studies have shown that the highly conserved transmembrane and cytoplasmic domains of the β_1 and β_3 subunits may be functionally interchanged, supporting the notion of a generic β cytoskeletal interaction domain.[65]

Less is known about the precise domains of the β subunits that are involved in heterodimer formation, although they probably reside in the extracellular domain. Several lines of evidence suggest that proper β conformation and membrane anchorage are necessary to permit heterodimer formation within the cell. Genetically defective β_2 subunits that result from mutations within the extracellular domain are incapable of pairing with α subunits.[66] Likewise, deletions within the globular head of the β_1 subunit result in a subunit that is unable to form a heterodimer and is incompletely processed.[39] Chimeric β subunits that contain the extracellular domain of β_1 but the membrane and cytoplasmic domains of β_3 pair exclusively with β_1 subfamily α subunits that normally associate with the β_1 subunit.[65] Mutated forms of β subunits that lack a transmembrane domain are secreted from the cell and are not capable of associating with membrane-bound α subunits.[31,65] However, when the extracellular domains of both subunits are present in a soluble form, they are secreted as heterodimers capable of binding to extracellular ligand. Thus, both subunits must be in the same compartment of the cell (membrane-bound or soluble) in order to pair, but the transmembrane and cytoplasmic domains are not required for this process.[67] The region of the β_1 subunit involved in α subunit selection has been further narrowed to include the N-terminal 500 amino acids by use of chimeric subunits containing portions of β_1 and β_3 protein.[39,68] The molecular defects in the β_2 subunit of several patients with the Leukocyte Adhesion Deficiency or LAD syndrome (see below) have provided some insight into potential heterodimer formation-controlling domains.[66] Point mutations at amino acid 149 (leucine to proline) or 169 (glycine to arginine), both conserved in all known β subunits, result in a β_2 protein of normal size that is incapable of pairing with its appropriate α subunit. These amino acids lie within the region of the β_1 subunit implicated in α subunit association by the chimeric studies and the β_3 subunit region involved in ligand binding by cross-linking studies. Indirect evidence suggests that the region of β_3 between amino acids 100 and 348 is not in contact with α_{IIb} since there was no protection of the subunit from proteolytic digestion in this region.[20] Thus, it appears that the globular head of the integrin β subunit contains distinct functional domains and that those involved with α subunit selection and α subunit association may be separate.

Similar structure–function correlations for α subunits are less well understood. Cross-linking studies with a fibrinogen peptide have implicated a region of α_{IIb} that contains the second calcium-binding domain as a ligand-binding domain.[69] A peptide corresponding to amino acids 296 to 306 of α_{IIb} blocks platelet aggregation and fibrinogen binding. This region is highly conserved among the cleaved α subunits and is therefore likely to play a role in other integrin–ligand interactions.[70] Thus, it seems that

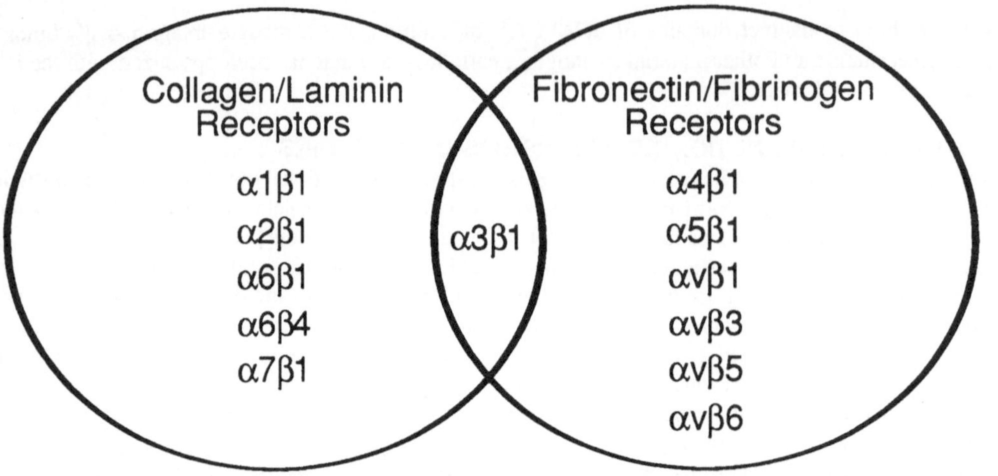

Figure 4 Extracellular matrix-binding integrin ligand specificity. The integrins of the β_1 subfamily, $\alpha_6\beta_4$ and those that contain α_v, are all involved in binding to components of the extracellular matrix. Of those integrins that bind to the collagens and laminin found in normal basement membranes or to the fibronectin and fibrinogen found elsewhere in matrices, only $\alpha_3\beta_1$ has overlapping specificity.

ligand specificity may reside outside of the ligand-binding domain in both α and β subunits. Detailed biochemical studies of $\alpha_{IIb}\beta_3$ have shown that a site between amino acids 558 to 747 of the α_{IIb} subunit is protected from proteolytic digestion in intact platelets but not in isolated subunits, suggesting that this region is protected by and therefore interacts with the β_3 subunit.[71] The role of the α subunit cytoplasmic domain remains unexplored.

Although their precise locations have not been identified, three distinct functional domains of α_4 have been identified by epitope mapping of bioactive monoclonal antibodies to this subunit.[72] $\alpha_4\beta_1$-mediated adhesion to VCAM-1 and fibronectin as well as homotypic aggregation could be independently inhibited by a panel of monoclonal antibodies that cluster in three distinct epitopes. It remains to be seen where on the extracellular domain these epitopes reside.

III. CELL BIOLOGY: NORMAL INTEGRIN FUNCTION

A. INTEGRIN EXPRESSION IS WIDESPREAD

The distribution of integrins has been studied on cells in culture and by immunohistochemical characterizations of various tissues. Every cell that has been studied to date has at least one integrin present on its surface; most have a repertoire of integrins. One fact that has become clear is the frequent but unpredictable alteration in this repertoire when cells are placed and passaged in culture.[73,74] For example, endothelial cells display a more restricted pattern of integrin expression *in situ* than they do in culture.[75] It is important, therefore, to draw conclusions cautiously about the functional implications of integrin expression from work done on cultured cells or cell lines. Nonetheless, there is extensive literature about the presence and function of integrins found in a variety of cells.

Surveys of human[76] and avian[77] tissue with monospecific antibodies for the β_1 subunit have shown members of this subfamily of extracellular matrix integrins to be present universally. More detailed studies using antibodies specific for the α subunits of this subfamily have been performed in a number of tissues. Because of the large numbers of integrin subunits and their multiple and overlapping ligand specificities, it is useful to have a scheme for classifying integrin cell–substratum receptors. Integrins from the β_1 subfamily as well as those that contain α_v and β_4 are all involved in interactions with proteins of the extracellular matrix. Since this includes the components of normal basement membranes (laminin and collagen IV) as well as proteins often associated with the provisional matrices found in areas of high cell movement (fibronectin, fibrinogen, vitronectin) typical of healing wounds, sites of inflammation, or developing tissues, it is helpful to group these integrin receptors according to their extracellular ligand specificities. One group includes those integrins that bind to the collagens and laminin. A second group of integrins binds to fibronectin, fibrinogen, and vitronectin (Figure 4). Only the promiscuous integrin $\alpha_3\beta_1$ overlaps using this grouping.

In the skin,[78,79] small bowel,[80] appendix,[81] colon,[82] breast,[83] lung,[84] lymph node, and tonsil,[81] epithelial cells show a predominance of collagen/laminin receptors. Several integrins, including $\alpha_6\beta_1$ and $\alpha_6\beta_4$ show a basilar distribution, colocalizing with the underlying basement membrane to which they presumably mediate adherence. Others, such as $\alpha_2\beta_1$ and $\alpha_3\beta_1$,[85,86] show a pericellular staining pattern more typical of cell–cell adhesion molecules. With several notable exceptions, normal epithelial cells show a restricted expression of those integrins that bind to fibronectin, vitronectin, and fibrinogen. Only fetal skin and ductal breast epithelium have been reported to show strong expression of the classic fibronectin receptor ($\alpha_5\beta_1$). If there is any epithelial cell interaction with fibronectin under normal conditions, it appears that $\alpha_3\beta_1$ or the α_v containing integrins must be the predominant receptors.

Capillary endothelial cells have a similar pattern of integrin expression throughout the body, with a predominance of collagen/laminin integrins and a lower amount of fibronectin/fibrinogen receptors.[75] Vessels of larger caliber, however, including those in kidney, lung, gut, skin, and breast, express abundant amounts of the $\alpha_5\beta_1$ fibronectin receptor in their endothelial cells. The reason for this difference between capillary and larger vessel endothelium in unclear, since fibronectin is not a major component of the extracellular matrix for any of these vessels.

Vascular and subepithelial smooth muscle cells express a heterogeneous pattern of collagen/laminin receptors. The α_1 and α_3 subunits are expressed strongly, while α_2 is less prominent, and α_6 is absent. These smooth muscle cells lack fibronectin/fibrinogen receptor subunit α_4 and α_5 expression but stain strongly for α_v. However, vascular smooth muscle cells express the β_3 subunit more intensely than subepithelial smooth muscle cells. Thus, it is likely that α_v is paired with one of the alternative β subunits (β_1, β_5, or β_6) in these cells.[39]

The expression of $\alpha_4\beta_1$ in the lymphoid organs deserves special mention because this integrin functions as a homing receptor for leukocyte–endothelial interactions, in addition to serving as a receptor for leukocyte fibronectin adhesion[87] and homotypic leukocyte aggregation.[88] This integrin is present on many white blood cell types, including T and B lymphocytes, but not neutrophils, as well as some cells derived from the neural crest. Thus, cellular staining with anti-α_4 antibody is evident in the bone marrow, spleen, lymph nodes, thymus, and various locations where these cells normally reside. Isolated staining in other tissue is usually attributable to resident or infiltrated leukocytes in the section, since there is no evidence that it is expressed by normal adult, nonlymphoid, nonneural crest tissue. Several malignant melanomas do express the α_4 subunit, however.[89]

The distribution of members of the β_2 subfamily is more limited than that of the β_1 and β_3 subfamilies. These three integrins ($\alpha_L\beta_2$, $\alpha_M\beta_2$, $\alpha_X\beta_2$) are found on leukocytes of all lineages, but not on erythroid or megakaryocyte derivatives.[90] $\alpha_M\beta_2$ is also found on several carcinoma cell lines, but again the significance of this is unclear. The major platelet integrin, $\alpha_{IIb}\beta_3$, accounts for up to 5% of its surface protein but is not found on any other normal cells.

B. INTEGRIN FUNCTION IS REGULATED

Despite the extensive presence of integrins on cells throughout the body, these receptors are not all equally or simultaneously active in their ability to bind to ligand. Intuitively, control mechanisms should exist that prevent unwanted platelet aggregation mediated by $\alpha_{IIb}\beta_3$ or leukocyte margination and migration mediated by $\alpha_L\beta_2$ yet permit these interactions at the specific sites where they are needed. An obvious control point for this regulation is the local availability of adhesive ligands (soluble fibrinogen or ICAM-1). For leukocyte–endothelial interactions, the up-regulation of the $\alpha_L\beta_2$ counter-receptor, ICAM-1, on endothelial cells in response to cytokine release has been shown.[91] Another control point for integrin function is the alteration of the ligand affinity state of the receptor. This type of regulation has been described for several integrins. The $\alpha_{IIb}\beta_3$ platelet receptor accounts for 5% of the surface protein of resting platelets but is incapable of binding to soluble fibrinogen. When platelets are activated by a number of different stimuli, this integrin undergoes a change in its conformation and is then able to bind to soluble fibrinogen and to participate in platelet aggregation. Several monoclonal antibodies have been described that are capable of changing the activation state of the integrin by binding or are activation-state dependent.[92] It appears that partial ligand occupancy may be a critical step in regulating the affinity state of this receptor, since purified receptor immobilized on plastic in the presence of RGD peptide bound to fibrinogen with the same avidity as activated platelets. When plated in the absence of RGD peptide, the purified receptor bound to fibrinogen with low affinity.[93] Furthermore, stimulation with platelet-activating agonists like thrombin or ADP of nucleated cells transfected with the cDNA from both subunits does not induce the high affinity state of the receptor,[94] but incubation of these cells with RGD peptide-ligand in the presence of fibrinogen induces them to aggregate.[95]

Similarly, the $\alpha_L\beta_2$ integrin exists in two states that differ in their avidity for ICAM-1. On T cells, the engagement of the T cell receptor causes a rapid and, in contrast to $\alpha_{IIb}\beta_3$, transient increase in the binding of $\alpha_L\beta_2$ to ICAM-1.[96] The signaling mechanism for this conversion is unknown but is presumed to involve the cytoskeletal domains, since this receptor will colocalize with talin in areas of T cell–antigen interaction.[97] Three other integrins present on T cells, $\alpha_4\beta_1$, $\alpha_5\beta_1$, and $\alpha_6\beta_1$, undergo an increase in ligand avidity after cross-linking of the T cell receptor.[98] Binding of fibroblasts with a β_1 specific monoclonal antibody, TASC, induces an increase in the binding kinetics of these cells to both laminin and vitronectin but decreases binding to collagen,[99] indicating that regulation by conformational change is ligand-specific.

A tyrosine phosphorylation site exists in the cytoplasmic sequences of several α and β subunits, suggesting a potential site for integrin regulation.[2] In virally transformed cells, the β_1 subunit is phosphorylated;[100] however, alteration of this site by targeted mutagenesis does not seem to affect the functioning of the β_1 subunit in transfected cells.[63] The binding of peritoneal macrophages to laminin requires stimulation with phorbol ester. This causes the $\alpha_6\beta_1$ receptor to associate with the insoluble cytoskeletal fraction of the cell and for the α_6 subunit to become phosphorylated.[101] Stimulation of Chinese hamster ovary cells with phorbol ester increases the avidity of $\alpha_5\beta_1$ fibronectin binding; however, no integrin phosphorylation has been detected.[102] Thus, both extracellular ligand binding and cytoskeletal organization of integrins may be regulated by a protein kinase-dependent pathway, either by direct receptor phosphorylation or phosphorylation of other associated cellular proteins.

Finally, the level of expression of both integrins and their ligands has been altered by exposure to a variety of cytokines.[103] For example, TGF-β is a potent inducer of fibronectin and $\alpha_5\beta_1$ in a variety of cells but causes a decrease in the $\alpha_3\beta_1$ and increase in the $\alpha_2\beta_1$ expression of osteosarcoma cells.[104] The resulting decrease in laminin adherence indicates that $\alpha_3\beta_1$ functions as the primary laminin receptor in this cell and that the regulation of integrin expression is an important control arm in the adhesive properties of cells.

C. INTEGRINS MEDIATE CELL–CELL ADHESION

Immunolocalization with α-specific monoclonal antibodies in cultured skin and endothelium has shown a pericellular distribution for several integrins. In epidermal cell culture, $\alpha_2\beta_1$ and $\alpha_3\beta_1$ colocalized with markers of cell–cell junctions. This pattern of localization was both receptor- and ligand-specific. On fibronectin and laminin but not on collagen IV keratinocyte $\alpha_2\beta_1$ appeared in a pericellular distribution.[105] In the presence of excess calcium, these cells formed multicell aggregates in which $\alpha_3\beta_1$ localized intercellularly.[85] While the relevance of these findings to the situation *in vivo* remains unclear, similar pericellular distribution patterns have been noted in tissue.[75]

Specific roles for integrin-mediated cell–cell adhesion in the process of leukocyte migration and aggregation, however, have been established. The $\alpha_4\beta_1$ integrin found on many white blood cells interacts in a heterophilic manner with VCAM-1, induced on endothelial cells by inflammatory cytokines.[106] In addition, $\alpha_4\beta_1$ mediates leukocyte aggregation through an unknown ligand that is independent of β_2 integrins.[88] Both $\alpha_L\beta_2$ and $\alpha_M\beta_2$ bind to ICAM-1, another immunoglobulin superfamily member, and participate in a cascade of adhesive events that enable circulating leukocytes to marginate and diapedese into sites of inflammation.[107] These two processes are precisely controlled by alterations in the number and affinity of β_2 integrins expressed by circulating white blood cells[96] and alterations in cell surface ligands, such as ICAM-1 and VCAM, on endothelial cells.[107]

D. INTEGRINS MEDIATE CELL MOTILITY

Cell movement is a complex phenomenon that depends on coordinated interactions among many underlying biophysical and biochemical processes, including integrin-mediated cell–matrix adhesion. Persistent migration over adhesive surfaces requires a cytoskeleton that is capable of both generating a contractile force and maintaining cell polarity and dynamic adhesion between the cell surface and underlying substratum.[108] For example, recent studies by Marks et al.[109] of neutrophils on highly adhesive substrata have demonstrated that cell movement speed is decreased when new lamellipodal attachment or uropodal detachment is prevented. Because integrins are transmembrane molecules that bind to both extracellular matrix ligands and cytoskeletal elements[42,110] with low affinity ($K_d \approx 10^{-6}$ M for both integrin–fibronectin[111] and integrin–talin interactions[110]), these molecules are attractive candidates for force transducers that transiently transmit force from the intracellular contractile apparatus to the extracellular environment and thus generate the traction necessary for movement.

Intuitively, the relationship between adhesive and motile behavior reflects the strength of both cell–substratum adhesion and contractile force generation: adhesive and contractile forces must balance to

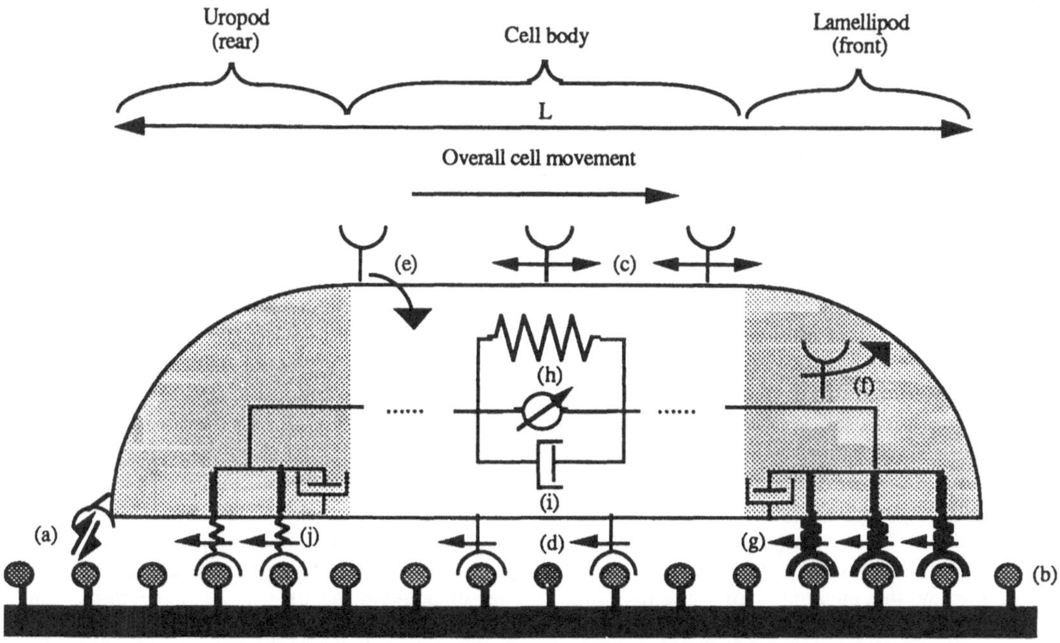

Figure 5 Schematic of integrin binding and trafficking events during cell migration over adhesive substrata. Adhesion receptors on the ventral cell surface bind reversibly with adhesive ligand (a), immobilized on the substratum at a uniform density (b). Free adhesion receptors can diffuse on the cell surface (c), while bound receptors drift backwards with respect to the cell at the cell's forward velocity (d). Two mechanisms for generating an asymmetric adhesion bound–number distribution are illustrated: free integrins can be endocytosed (e) and preferentially inserted at the front of the cell (f), or receptors at the front of the cell may bind more tightly to ligand than those at the rear (g). Contractile force, generated by cytoskeletal components in the cell body (h), deforms the cell (i) and is transmitted to the underlying substratum through integrin–matrix bonds (j).

allow movement. This hypothesis is supported by a number of recent *in vitro* observations that demonstrate an optimum cell–substratum adhesiveness exists for individual cell migration on rigid plastic substrata coated with different ligands for the integrins. Goodman et al.[112] found a biphasic relationship between movement speed and the concentration of adsorbed laminin or E8 (cell-binding) fragment of laminin for murine skeletal muscle myoblasts. Based on adhesion data for these same cells,[113] the maxima in speed occurred at approximately one third to one half of the maximum adhesive strength.[114] Similar results also have been reported for human adult smooth muscle cells on fibronectin and type IV collagen.[115] Duband et al.[116] observed that the extent of migration for individual neural crest cells decreased with increasing surface density of high-affinity antibodies for integrin β_1 but was enhanced by increasing concentrations of low-affinity antibodies. These results suggest that migration speed can be maintained by increase in ligand density compensated by decreases in adhesion receptor–ligand affinity.

Recently, a mathematical model has been developed (Figure 5) that can predict the effect of key cell-substratum adhesive properties on the rate of migration over matrix-coated surfaces.[114,117] This model is based on a chronological picture of a cell migration cycle consisting of lamellipodal extension, cytoskeletal contraction, and relaxation.[118] Net translocation requires a polarized asymmetry in the strength or number of low-affinity reversible adhesive interactions between lamellipod and uropod, for which two underlying mechanisms have been proposed previously: a spatial distribution of integrin number due to preferential endocytic recycling of adhesion receptors to the leading edge[119] and a spatial variation in integrin–matrix affinity between lamellipod and uropod.[96,120] Several groups have observed endocytosis of $\alpha_5\beta_1$ in Chinese hamster ovary (CHO) cells[36,121] and embryonic neural crest cells.[122] Replacement of Ca^{2+} with Mn^{2+} (see Reference 47) or addition of phorbol esters[96] can increase receptor–ligand affinity, while phosphorylation[123] has the opposite effect. Although either mechanism predicts the biphasic dependence of movement speed on ligand density and integrin–matrix affinity seen experimentally, the alternative asymmetry mechanisms may be discerned based on the effects of integrin number and adhesiveness on individual cell speed. Observations of decreased motility in CHO cells deficient in[124] or transfected with additional $\alpha_5\beta_1$[125] are consistent with these predictions but not sufficient to distinguish between mechanisms.

E. INTEGRIN EXPRESSION IS DEVELOPMENTALLY REGULATED

Integrin-mediated cell motility plays an important role in a number of physiological and pathological processes. During embryogenesis, spatial and temporal modulation of integrin expression[126] and extracellular matrix composition[127] regulate cell–matrix interactions, which may act to initiate and cease morphogenetic cell movements.[128]

The importance of the β_1 integrins in embryonic development has been examined primarily in avian systems.[129] In general, β_1 integrins tend to be codistributed with fibronectin and laminin; however, in the embryo as well as the adult, they are not found exclusively at sites of cell–matrix interactions. This again suggests that there are alternative ligands for these receptors on the surface of adjacent cells. Integrin expression is altered in the developing lung, for example, being widely distributed in all embryonic lung cell types, concentrating at areas of fibronectin deposition during airway development and diminishing as these structures mature.[130]

More direct evidence of the functional relevance of β_1 integrins during morphogenesis comes from experiments in which agents known to interfere with integrin–extracellular matrix interactions are administered during embryogenesis, resulting in the blockage of a particular developmental event. Two types of agents have been used for this purpose: (1) peptides containing the amino acid sequence RGD that prevent the interaction of integrins with fibronectin and vitronectin[50] and (2) the monoclonal antibodies CSAT[1] and JG22[131] that react with the β_1 subunit of avian integrins and prevent the seven β_1 subfamily integrins from binding to their ligands. When injected into amphibian[132] or insect embryos[133] RGD peptides interfered with gastrulation and, in some cases, neural crest migration, presumably by disturbing the interaction of integrins with their respective ligands. Similarly, when the antibodies JG22 and CSAT, or hybridomas producing these antibodies, were injected into chick embryos, the pattern of neural crest migration[134] as well as axon extension[135] was disrupted. These antibodies can also interfere with embryonic muscle formation.[136]

Thus, integrins are expressed in a wide variety of tissues and at very early times in embryonic development. In vertebrate embryos, they appear to be most important at times of morphogenetic movements, as it is only during these periods that antibodies or peptides reactive with the integrins seem to affect the developmental process. Their expression also seems be regulated during development, as most observers note a loss of integrin-specific antibody staining or a diminution of integrin function in cells once they have reached their final phase of differentiation.

IV. PATHOBIOLOGY: INTEGRINS IN DISEASE

A. CONGENITAL INTEGRIN DYSFUNCTION

The role of integrins in the inflammatory response and in coagulation is dramatically illustrated by several rare inherited deficiencies of the integrin β subunits. Patients with nonfunctional β_2 subunits have a clinical syndrome termed "leukocyte adhesion deficiency" (LAD), characterized by recurrent bacterial infections, delayed separation of the umbilical cord, leukocytosis, and abnormal leukocyte function (reviewed in Anderson and Springer[137]). These infections are nonsuppurative and appear to be the consequence of the inability of neutrophils to reach sites of extravascular inflammation. Neutrophils from LAD patients fail to adhere to endothelial cells and have severe impairments in cell aggregation, antibody-dependent cytotoxicity, and phagocytosis. Interestingly, lymphocyte migration is much less affected, probably because of the presence of $\alpha_4\beta_1$ on these cells, which can bind to VCAM-1 on activated endothelium.[138] Antibodies against the β_2 subunit can mimic the functional defect on neutrophils *in vitro*. Replacement of the defective β_2 subunit in LAD lymphocytes by retroviral gene transfer restores their activity to normal.[139]

Other patients with an inherited absence of the β_3 or α_{IIb} subunits have disordered thrombosis. These patients suffer from Glanzmann's thrombasthenia, a bleeding disorder characterized by the inability of platelets to aggregate by binding to fibrinogen once they are activated (reviewed in Newman[140]). At least one patient has also been described whose platelets lack the $\alpha_2\beta_1$ collagen/laminin receptor. This finding was associated with a clinically significant bleeding disorder characterized by the failure of the patient's platelets to aggregate in response to collagen in an *in vitro* assay.

The congenital absence of the β_1 subunit has not been described in humans, probably because it is a lethal defect. In drosophila, however, a defective β_1 homologue (position-specific antigen 1) has been

Table 3 **Pathogen adherence of eukaryotic cell integrins. (Certain bacterial and fungal surface proteins interact with cell surface integrins on target cells.)**

Pathogen	Surface protein	Target cell	Integrin	RGD involved[f]	Reference
B. pertussis	Pertactin	CHO[d]	Unknown	Yes	Leininger[144]
	FHA[a]	Macrophage	$\alpha_M\beta_2$	Yes	Relman[148]
E. coli (rough)	LPS[b] lipid IVa	PMN[e]	β_2 (unspecified)	No	Wright[59]
H. capsulatum	Unknown	M. phi[d]	All β_2	Unknown	Bullock[145]
L. mexicana	gp63	Macrophage	$\alpha_M\beta_2$	Yes	Russell[147]
	LPG[c]		$\alpha_M\beta_2$, $\alpha_X\beta_2$	No	Rohana-Talamas[146]
Y. pseudo-tuberculosis	Invasin	HEp-2[d]	$\alpha_5\beta_1$	Unknown	Leong[206]
		EJ[d]	$\alpha_3\beta_1$, $\alpha_5\beta_1$	Unknown	Isberg[142]
		K527[d]	$\alpha_5\beta_1$	Unknown	
		HPB MLT[d]	$\alpha_4\beta_1$	Unknown	
		Platelets	$\alpha_6\beta_1$	No	

[a] FHA is filamentous hemagglutinin; [b] LPS is lipopolysaccharide; [c] LPG is lipophosphoglycan; [d] CHO, M.phi, DEp-2, EJ, K527, and HPB MLT are all established cell lines; [e] PMN is polymorphonuclear leukocyte; [f] some of these interactions are inhibitable in the presence of RGD peptide.

identified.[141] This defect results in the myospheroid phenotype, a lethal developmental abnormality characterized by a failure of muscle fibers to insert into the exoskeleton.

B. INTEGRINS AND THEIR LIGANDS ARE USED BY INFECTIOUS AGENTS

In addition to their ability to bind ligands found in the extracellular matrix and on other eucaryotic cells, integrins have been shown to interact with microorganisms to facilitate adherence and internalization. A summary of these interactions is found in Table 3. Although binding of bacteria and fungi to cells via integrins has been documented, the relevance of this to the pathogenesis of infection remains to be established. Members of the β_1 subfamily of integrins have been shown to facilitate the adherence of *Yersinia pseudotuberculosis* to a variety of nonphagocytic eukaryotic cells.[142] The *Yersinia* ligand, "invasin", mediates this phenomenon by binding to β_1 integrins, including $\alpha_3\beta_1$, $\alpha_4\beta_1$, $\alpha_5\beta_1$, and $\alpha_6\beta_1$, but not to $\alpha_2\beta_1$ or the β_2 integrins. The active portion of the bacterial ligand is located in the carboxyl terminus of the protein and does not contain an RGD sequence.[143] Binding of invasin to $\alpha_6\beta_1$ is not inhibited by soluble RGD peptides. A 69 kD surface protein found on *Bordetella pertussis* is capable of attaching to the surface of CHO cells in a manner that is inhibitable by soluble RGD peptides, suggesting an integrin receptor as the target.[144] Although the specific cell receptor(s) involved in this event have not been confirmed, CHO cells express several members of the β_1 subfamily but no β_2 integrins on their surface.

Many bacteria are recognized by phagocytic cells once they have been opsonized with complement via the $\alpha_M\beta_2$ integrin. However, several organisms are capable of binding to phagocytic cells directly without first being opsonized. β_2 integrins found on neutrophils are able to bind to lipopolysaccharide lipid IVa, found on rough *E. coli.*[59] Attachment of *Histoplasma capsulatum* to the cell surface is apparently mediated by all three β_2 integrins.[145] The fungal ligand(s) for this process are unknown. The macrophage integrin, $\alpha_M\beta_2$, is also capable of binding directly to several bacterial surface proteins. For example, two proteins found on *Leishmania mexicana* promastigotes, gp63 and lipophosphoglycan (LPG), can serve as ligands for $\alpha_M\beta_2$.[146,147] In addition, LPG can bind to $\alpha_X\beta_2$ but not to $\alpha_L\beta_2$.[146] A second surface protein found on *B. pertussis,* filamentous hemagglutinin (FHA), serves as a ligand for macrophage $\alpha_M\beta_2$.[148] FHA can bind directly to $\alpha_M\beta_2$ in a manner that is inhibitable by RGD peptides.

C. INTEGRIN EXPRESSION IS ALTERED IN MALIGNANCY

Integrins are intimately involved in tumor invasion into healthy tissues, although in a complex manner. *In vivo* and *in vitro* studies have identified changes in the level of integrin expression[125,149] that correlated with tumorigenicity. CHO cells that overexpress the $\alpha_5\beta_1$ fibronectin receptor show a reduced ability to form tumors in nude mice,[125] while the metastatic frequency of rhabdomyosarcoma cells that overexpress $\alpha_2\beta_1$ is greatly increased.[150] Further, RGD-containing peptides that perturb cell–matrix interactions inhibit the movement of human melanoma cells through amniotic basement membranes[151] and also reduce the

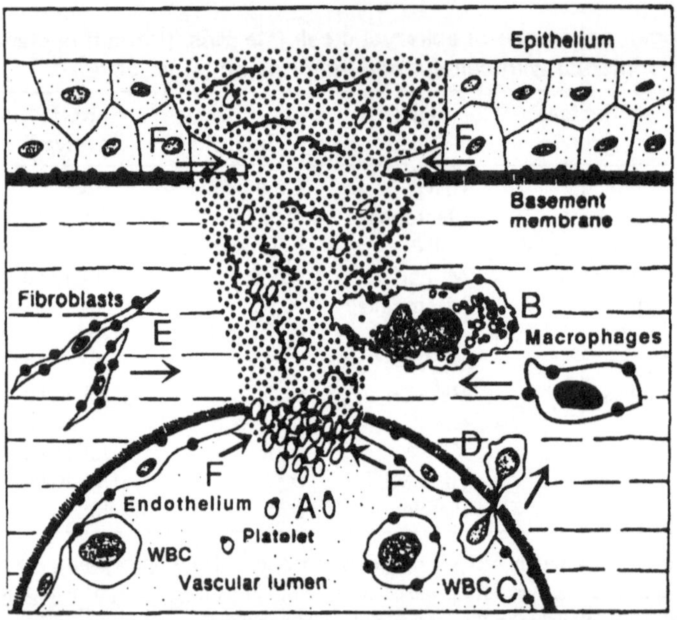

Figure 6 Integrin receptors involved in the response to an epithelial and endothelial wound. Disruption of tissue integrity by mechanical or other means initiates a host response in which integrins (•) are highly involved. Gaps in the endothelium are plugged by activated platelets expressing high affinity $\alpha_{IIb}\beta_3$ receptors that bind to soluble fibrinogen (A) and a provisional matrix comprised of fibronectin, fibrin, and fibrinogen (stippled area). Tissue resident macrophages up-regulate β_1 and β_2 integrins and release cytokines that drive the local inflammatory response (B). Circulating white blood cells (WBC) respond to the injury by marginating (C) and migrating (D) into the area using β_2 and β_1 integrins. As the provisional matrix matures, fibroblasts are recruited into the wound (E) and migrate over the fibronectin rich matrix using their $\alpha_5\beta_1$ receptors. At this point, collagen deposition may lead to scar formation, or the basement membrane may be reconstituted. In the latter case, the disrupted epithelium and endothelium migrate in (F) to close the wound using β_1 integrin receptors.

number of metastatic nodules found within the lungs of mice injected with $\beta_1$6F10 melanoma cells.[152] These data highlight the fact that the process of tumor progression is complex and requires malignant cells to display both decreased and increased adhesion properties at various times in the development of the tumor.[153]

At least three different patterns of integrin expression in tumors have been reported. In the development of malignant melanomas, a consistent and reproducible up-regulation of integrin expression accompanies the acquisition of metastatic potential.[154,155] In well-differentiated carcinomas, such as basal cell skin carcinomas,[78] colon adenocarcinomas,[156] and some breast carcinomas,[157] where the peripheral tumor cells are still in association with the basement membrane, integrin expression by these cells is similar to that seen in normal epithelium. In poorly differentiated carcinomas of skin,[78] colon,[158] breast,[157] and lung,[84] and in cells of more differentiated carcinomas separated from the basement membrane, integrin expression is highly variable but tends to be less strong than that of the normal epithelium. These results suggest that there may be a regulatory interplay between the cell and the extracellular matrix that is disrupted in tumor cells, either as a result of the destruction of the matrix or as a result of an alteration in the pattern of integrin expression.

D. INTEGRINS ARE INVOLVED IN INFLAMMATORY AND HEALING PROCESSES

Inflammation, wound healing, and thrombosis depend upon communication between resident cells of the injured tissue and circulating effector cells. One avenue of communication is through direct cell–cell contact, utilizing specific cell adhesion receptors. A highly simplified scenario for the role of integrins in a wound is illustrated in Figure 6.

The inflammatory response to disruption of normal tissue integrity by physical, chemical, or microbial means begins when platelets bind to exposed matrix (A) via integrins, including the $\alpha_2\beta_1$ collagen/laminin receptor, the $\alpha_5\beta_1$ fibronectin receptor, and the $\alpha_6\beta_1$ laminin receptor[159] and with the migration of tissue-based macrophages or mast cells into the area through a matrix–protein rich stroma (B). The activation of the blood coagulation cascade generates thrombin, which activates platelet $\alpha_{IIb}\beta_3$ and, in turn, promotes further platelet aggregation and granule release. A provisional wound matrix is thus formed, which contains platelets, fibrinogen, fibrin, and fibronectin. Resident tissue macrophages release biologically active mediators, such as interleukin-1, tumor necrosis factor, histamine, and TGF-β (B), which, along with bacterial endotoxins and other chemoattractants, stimulate the expression of integrin counter-ligands, such as ICAM-1 and VCAM-1, and other adhesion molecules on endothelial cells (C) and up-regulate the expression and/or avidity of β_2 integrins and ICAM-1 on white blood cells in the vascular lumen (C). White blood cells then adhere to the "activated" vessel wall, migrate through the endothelium, and enter into areas of infection or tissue damage (D), where they may phagocytose opsonized material. A similar series of events operates in the healing process. Cytokines released by the activated white blood cells and platelets that accumulate at the site of injury (E) stimulate the up-regulation of integrins on macrophages and fibroblasts that promote their migration into the wound site.[104] Epithelial cell migration over the fibronectin-containing wound matrix (F) is also dependent on induction of new integrins.[160] In a porcine model of healing skin, for example, migrating but not resting epidermis expressed the $\alpha_5\beta_1$ receptor. Tissue fibroblasts also up-regulated this integrin just prior to wound contracture.[161] Other examples of integrin up-regulation in a disease state include an increase in the expression of $\alpha_1\beta_1$ in the colonic epithelium of patients with active but not quiescent inflammatory bowel disease[162] and the presence of increased numbers of α_v-containing integrins in the microglia of patients with Alzheimers disease.[163]

V. CONCLUSIONS: THERAPEUTIC MANIPULATION OF INTEGRIN FUNCTION

The recognition that integrins are involved in so many biological processes has led to several novel therapeutic approaches in experimental models and in human trials (reviewed in Carlos and Harlan[164]). Monoclonal antibodies against specific β_2 integrins have been used to inhibit neutrophil recruitment into areas of skin inflammation,[165] peritonitis,[166] meningitis,[167] and pneumonitis,[168] to increase survival in animal models of septic shock,[169] and to ameliorate tissue damage caused by an ischemia-reperfusion injury in the intestine,[170] heart,[171] and lung.[172] A number of investigators have begun using antibodies against α_{IIb}/β_3 in patients with active heart disease[173] or RGD-containing peptides to prevent thrombus formation.[174] Others have used antibodies against α_L to ameliorate graft rejection in bone marrow transplant recipients.[175] The potential use of similar reagents as adjuvant to cancer surgery or chemotherapy to limit metastatic spread[176] to treat conditions of disordered wound repair that result in fibrosis are just on the horizon. As more complete and thorough information about the functional roles and structural domains of integrins becomes available, the number of such trials is likely to dramatically increase. The consequences of such information will undoubtedly lead to new and innovative ways to treat the many human conditions in which integrins are intimately involved.

REFERENCES

1. **Neff, N. T., Lowrey, C., Decker, C., Tovar, A., Damsky, C., Buck, C., and Horwitz, A. F.,** Monoclonal antibody detaches embryonic skeletal muscle from extracellular matrices, *J. Cell Biol.,* 95, 654, 1982.
2. **Tamkun, J. W., DeSimone, D. W., Fonda, D., Patel, R. S., Buck, C., Horwitz, A. F., and Hynes, R. O.,** Structure of integrin, a glycoprotein involved in the transmembrane linkage between fibronectin and actin, *Cell,* 46, 271, 1986.
3. **Pytela, R., Pierschbacher, M. D., and Ruoslahti, E.,** Identification and isolation of a 140 kd cell surface glycoprotein with properties expected of a fibronectin receptor, *Cell,* 40, 191, 1985.
4. **Bennett, J. S., Hoxie, J. A., Leitman, S. F., Vilaire, G., and Cines, D. B.,** Inhibition of fibrinogen binding to stimulated human platelets by a monoclonal antibody, *Proc. Natl. Acad. Sci. U.S.A.,* 80, 2417, 1983.
5. **Springer, T. A.,** The LFA-1, Mac-1 glycoprotein family and its deficiency in an inherited disease, *Fed. Proc.,* 44, 2660, 1985.
6. **Hemler, M., Ware, C., and Strominger, J.,** Characterization of a novel differentiation antigen complex recognized by a monoclonal antibody ($\alpha_1\alpha_5$): unique activation-specific molecular forms on stimulated T cells, *J. Immunol.,* 131, 334, 1983.
7. **Hynes, R. O.,** Integrins, a family of cell surface receptors, *Cell,* 48, 549, 1987.

8. **DeSimone, D. W. and Hynes, R. O.,** Xenopus laevis integrins: structural conservation and evolutionary divergence of integrin β subunits, *J. Biol. Chem.*, 263, 5333, 1988.
9. **Holzmann, B. and Weissman, I. L.,** Integrin molecules involved in lymphocyte homing to Peyer's patches, *Immunol. Rev.*, 108, 45, 1989.
10. **Erle, D. J., Rüegg, C., Sheppard, D. and Pytela, R.,** Complete amino acid sequence of an integrin β subunit (β_7) identified in leukocytes, *J. Biol. Chem.*, 266, 11009, 1991.
11. **Hemler, M. E., Crouse, C., and Sonnenberg, A.,** Association of the VLA α_6 subunit with a novel protein, *J. Biol. Chem.*, 264, 6529, 1989.
12. **Kajiji, S., Tamura, R. N., and Quaranta, V.,** A novel integrin ($\alpha_E\beta_4$) from human epithelial cells suggests a fourth family of integrin adhesion receptors, *EMBO J.*, 8, 673, 1989.
13. **Freed, E., Gailit, J., Van der Geer, P., Ruoslahti, E., and Hunter, T.,** A novel integrin β subunit is associated with the vitronectin receptor α subunit (α_v) in a human osteosarcoma cell line and is a substrate for protein kinase C, *EMBO J.*, 8, 2955, 1989.
14. **Cheresh, D., Smith, J., Cooper, H., and Quaranta, V.,** A novel vitronectin receptor integrin ($\alpha_v\beta_x$) is responsible for distinct adhesive properties of carcinoma cells, *Cell*, 57, 59, 1989.
15. **Santoro, S. A., Rajpara, S. M., Staatz, W. D., and Woods, V. L., Jr.,** Isolation and characterization of a platelet surface collagen binding complex related to VLA-2, *Biochem. Biophys. Res. Commun.*, 153, 217, 1988.
16. **Tsuji, T., Hakomori, S., and Osawa, T.,** Identification of human galactoprotein β_3, an oncogenic transformation-induced membrane glycoprotein, as VLA-3 α subunit: the primary structure of human integrin α_3, *J. Biochem. (Tokyo)*, 109, 659, 1991.
17. **Kennel, S. J., Foote, L. J., Falcioni, R., Sonnenberg, A., Stringer, C. D., Crouse, C., and Hemler, M. E.,** Analysis of the tumor-associated antigen TSP-180. Identity with α_6-β_4 in the integrin superfamily, *J. Biol. Chem.*, 264, 15515, 1989.
18. **Phillips, D. R., Charo, I. F., Parise, L. V., and Fitzgerald, L. A.,** The platelet membrane glycoprotein IIb-IIIa complex, *Blood*, 71, 831, 1988.
19. **Beer, J. and Coller, B. D.,** Evidence that platelet glycoprotein IIIa has a large disulfide-bonded loop that is susceptible to proteolytic cleavage, *J. Biol. Chem.*, 264, 17564, 1989.
20. **Calvete, J. J., Henschen, A., and González-Rodríguez, J.,** Assignment of disulphide bonds in human platelet GPIIIa. A disulphide pattern for the β-subunits of the integrin family, *Biochem. J.*, 274, 63, 1991.
21. **Calvete, J. J., Henschen, A., and González-Rodríguez, J.,** Complete localization of the intrachain disulphide bonds and the N-glycosylation points in the α-subunit of human platelet glycoprotein IIb, *Biochem. J.*, 261, 561, 1989.
22. **Calvete, J. J., Shafer, W., Henschen, A., and González-Rodríguez, J.,** Characterization of the β-chain N-terminus heterogeneity and the α-chain C-terminus of human platelet GPIIb, *FEBS Lett.*, 272, 37, 1990.
23. **Calvete, J. J., Schafer, W., Henschen, A., and González-Rodríguez, J.,** C-terminal amino acid determination of the transmembrane subunits of the human platelet fibrinogen receptor, the GPIIb/IIIa complex, *FEBS Lett.*, 263, 43, 1990.
24. **Nermut, M. V., Green, N. M., Eason, P., Yamada, S. S., and Yamada, K. M.,** Electron microscopy and structural model of human fibronectin receptor, *EMBO J.*, 7, 4093, 1988.
25. **D'Souza, S. E., Ginsberg, M. H., Lam, S. C.-T., and Plow, E.,** Chemical cross-linking of arginyl-glycyl-aspartic acid peptides on adhesion receptors on platelets, *Science*, 242, 91, 1988.
26. **Smith, J. W. and Cheresh, D. A.,** The Arg-Gly-Asp-binding domain of the vitronectin receptor, *J. Biol. Chem.*, 263, 18726, 1988.
27. **Hemler, M. E.,** VLA proteins in the integrin family: structures, functions and their role in leukocytes, *Annu. Rev. Immunol.*, 8, 365, 1990.
28. **Marcantonio, E. E. and Hynes, R. O.,** Antibodies to the conserved cytoplasmic domain of the integrin β_1 subunit react with proteins in vertebrates, invertebrates, and fungi, *J. Cell Biol.*, 106, 1765, 1988.
29. **Suzuki, S. and Naitoh, Y.,** Amino acid sequence of a novel integrin β_4 subunit and primary expression of the mRNA in epithelial cells, *EMBO J.*, 9, 757, 1990.
30. **Humphries, M.,** The molecular basis and specificity of integrin–ligand interactions, *J. Cell Sci.*, 97, 585, 1990.
31. **Bennett, J. S.,** Integrin structure and function in hemostasis and thrombosis, *Ann. N.Y. Acad. Sci.*, 614, 214, 1991.

32. **Hemler, M. E., Crouse, C., and Sonnenberg, A.**, Association of the VLA α_6 subunit with a novel protein. A possible alternative to the common VLA β_1 subunit on certain cell lines, *J. Biol. Chem.*, 264, 6529, 1989.
33. **Bray, P. F., Leung, C. S.-I., and Shuman, M. A.**, Human platelets and megakaryocytes contain alternately spliced glycoprotein IIb mRNAs, *J. Biol. Chem.*, 265, 9587, 1990.
34. **Altruda, F., Cervella, P., Tarone, G., Botta, C., Balzac, F., Stefanuto, G., and Silengo, L.**, A human integrin β_1 subunit with a unique cytoplasmic domain generated by alternative mRNA processing, *Gene*, 95, 261, 1990.
35. **van Kuppevelt, T. H., Languino, S. M., Gailit, L. R., Suzuki, S., and Ruoslahti, E.**, An alternative cytoplasmic domain of the integrin β_3 subunit, *Proc. Natl. Acad. Sci.*, 86, 5415, 1989.
36. **Brown, N. H., King, D. L., Wilcox, M., and Kafatos, F. C.**, Developmentally regulated alternative splicing of Drosophila integrin PS2 α transcripts, *Cell*, 59, 185, 1989.
37. **Akiyama, S. K., Yamada, S. S., and Yamada, K. M.**, Analysis of the role of glycosylation of the human fibronectin receptor, *J. Biol. Chem.*, 264, 18011, 1989.
38. **Hynes, R. O., Marcantonio, E. E., Stepp, M. A., Urry, L. A., and Yee, G. H.**, Integrin heterodimer and receptor complexity in avian and mammalian cells, *J. Cell Biol.*, 109, 409, 1989.
39. **Buck, C., Albelda, S., Damjanovich, L., Edelman, J., Shih, D.-T., and Solowska, J.**, Cell–matrix contacts and pericellular proteolysis. Immunohistochemical and molecular analysis of β_1 and β_3 integrins, *Cell Differ. Dev.*, 32, 189, 1990.
40. **Troesch, A., Duperray, A., Polack, B., and Marguerie, G.**, Comparative study of the glycosylation of platelet glycoprotein GPIIb/IIIa and the vitronectin receptor. Differential processing of their β-subunit, *Biochem. J.*, 268, 129, 1990.
41. **Buck, C., Shea, E., Duggin, K., and Horwitz, A.**, Integrin, the CSAT antigen: functionality requires oligomeric integrity, *J. Cell Biol.*, 103, 2421, 1986.
42. **Burridge, K., Fath, K., Kelly, T., Nuckolis, B., and Turner, C.**, Focal adhesions: transmembrane junctions between the extracellular matrix and the cytoskeleton, *Annu. Rev. Cell Biol.*, 4, 487, 1988.
43. **Otey, C., Pavalko, F., and Burridge, K.**, An interaction between α-actinin and the β_1 integrin subunit in vitro, *J. Cell Biol.*, 111, 721, 1990.
44. **Singer, I. I., Kazazis, D. M., and Scott, S.**, Scanning electron microscopy of focal contacts on the substratum attachment surface of fibroblasts adherent to fibronectin, *J. Cell Sci.*, 93, 147, 1989.
45. **Burridge, K., Nuckolls, G., Otey, C., Pavalko, F., Simon, K., and Turner, C.**, Actin–membrane interaction in focal adhesions, *Cell Differ. Dev.*, 32, 337, 1990.
46. **Elices, M. J. and Hemler, M. E.**, The human integrin VLA-2 is a collagen receptor on some cells and a collagen/laminin receptor on others, *Proc. Natl. Acad. Sci. U.S.A.*, 86, 9906, 1989.
47. **Gailit, J. and Ruoslahti, E.**, Regulation of the fibronectin receptor affinity by divalent cations, *J. Biol. Chem.*, 263, 12927, 1988.
48. **Elices, M. J., Urry, L. A., and Hemler, M. E.**, Receptor functions for the integrin VLA-3: fibronectin, collagen, and laminin binding are differentially influenced by ARG-GLY-ASP peptide and by divalent cations, *J. Cell Biol.*, 112, 169, 1991.
49. **Smith, J. W. and Cheresh, D. A.**, Labeling of integrin $\alpha_v\beta_3$ with ^{58}Co(III). Evidence of metal ion coordination sphere involvement in ligand binding, *J. Biol. Chem.*, 266, 11429, 1991.
50. **Ruoslahti, E. and Pierschbacher, M. D.**, New perspectives in cell adhesion: RGD and integrins, *Science*, 238, 491, 1987.
51. **Garcia-Pardo, A., Wayner, E. A., Carter, W. G., and Ferreira, O. C., Jr.**, Human B lymphocytes define an alternative mechanism of adhesion to fibronectin. The interaction of the $\alpha_4\beta_1$ integrin with the LHGPEILDVPST sequence of the type III connecting segment is sufficient to promote cell attachment, *J. Immunol.*, 144, 3361, 1990.
52. **Nojima, Y., Humphries, M. J., Mould, A. P., Komoriya, A., Yamada, K. M., Schlossman, S. F., and Morimoto, C.**, VLA-4 mediates CD3-dependent CD4+ T cell activation via the CS1 alternatively spliced domain of fibronectin, *J. Exp. Med.*, 172, 1185, 1990.
53. **Mould, A. P., Komoriya, A., Yamada, K. M., and Humphries, M. J.**, The CS5 peptide is a second site in the IIICS region of fibronectin recognized by the integrin alpha 4 beta 1. Inhibition of alpha 4 beta 1 function by RGD peptide homologues, *J. Biol. Chem.*, 266, 3579, 1991.
54. **Staatz, W. D., Fok, K. F., Zutter, M. M., Adams, S. P., Rodriguez, B. A., and Santoro, S. A.**, Identification of a tetrapeptide recognition sequence for the $\alpha_2\beta_1$ integrin in collagen, *J. Biol. Chem.*, 266, 7363, 1991.

55. **Kishimoto, T. K., Larson, R. S., Corbi, A. L., Dustin, M. L., Staunton, D. E., and Springer, T. A.,** The leukocyte integrins, *Adv. Immunol.,* 46, 149, 1989.
56. **de Fougerolles, A., Stacker, S., Schwarting, R., and Springer, T.,** Characterization of ICAM-2 and evidence for a third counter–receptor for LFA-1, *J. Exp. Med.,* 174, 253, 1991.
57. **Elices, M. J., Osborn, L., Takada, Y., Course, C., Luhowskyl, S., Hemler, M., and Lobb, R.,** VCAM-1 on activated endothelium interacts with the leukocyte integrin VLA-4 at a site distinct from the VLA-4/fibronectin binding site, *J. Immunol.,* 60, 577, 1990.
58. **Diamond, M. S., Staunton, D. E., Marlin, S. D., and Springer, T. A.,** Binding of the integrin Mac-1 (CD11b/CD18)) to the third immunoglobulin-like domain of ICAM-1 (CD54) and its regulation by glycosylation, *Cell,* 65, 961, 1991.
59. **Wright, S. D., Levin, S. M., Jong, M. C. T., Chad, Z., and Kabbash, L. G.,** CR3 (CD11b/CD18) expresses one binding site for Arg-Gly-Asp-containing peptides and a second site for bacterial lipopolysaccharide, *J. Exp. Med.,* 169, 175, 1989.
60. **Loftus, J. C., O'Toole, T. E., Plow, E. F., Glass, A., Frelinger, A. L., III, and Ginsberg, M. H.,** A β_3 integrin mutation abolishes ligand binding and alters divalent cation-dependent conformation, *Science,* 249, 915, 1990.
61. **Charo, I. F., Nannizzi, L., Phillips, D. R., Hsu, M. A., and Scarborough, R. M.,** Inhibition of fibrinogen binding to GP IIb-IIIa by a GP IIIa peptide, *J. Biol. Chem.,* 266, 1415, 1991.
62. **Solowska, J., Guan, J. G., Marcandonio, E., Buck, C. A., and Hynes, R. O.,** Expression of normal and mutant avian integrin subunits in rodent cells, *J. Cell Biol.,* 109, 853, 1989.
63. **Hayashi, Y., Haimovich, B., Reszka, A., Boettiger, D., and Horwitz, A.,** Expression and function of chicken integrin β_1 subunit and its cytoplasmic domain mutants in mouse NIH3T3 Cells, *J. Cell Biol.,* 110, 175, 1990.
64. **Marcantonio, E. E., Guan, J.-L., Trevithick, J. E., and Hynes, R. O.,** Mapping of the functional determinants of the integrin β_1 cytoplasmic domain by site-directed mutagenesis, *Cell Regul.,* 1, 597, 1990.
65. **Solowska, J., Edelman, J. M., Albelda, S. M., and Buck, C. A.,** Cytoplasmic and transmembrane domains of integrin β_1 and β_3 subunits are functionally interchangeable, *J. Cell Biol.,* 114, 1079, 1991.
66. **Wardlaw, A. J., Hibbs, M. L., Stacker, S. A., and Springer, T. A.,** Distinct mutations in two patients with leukocyte adhesion deficiency and their functional correlates, *J. Exp. Med.,* 172, 335, 1990.
67. **Dana, N., Fatallah, D. M., and Amaout, M. A.,** Expression of a soluble and functional form of the human β_2 integrin CD11b/CD18, *Proc. Natl. Acad. Sci. U.S.A.,* 88, 3106, 1991.
68. **Edelman, J. M., Shih, D.-T., Albelda, S. M., and Buck, C. A.,** Structure–function analysis of the extracellular domain of the β_1 integrin subunit, *Am. Rev. Respir. Dis.,* 143, 51, 1991.
69. **D'Souza, S. E., Ginsberg, M. H., Burke, T. A., and Plow, E. F.,** The ligand binding site of the platelet integrin receptor GPIIb-IIIa is proximal to the second calcium binding domain of its α subunit, *J. Biol. Chem.,* 265, 3440, 1990.
70. **D'Souza, S. E., Ginsberg, M. H., Matsueda, G. R., and Plow, E. F.,** A discrete sequence in a platelet integrin is involved in ligand recognition, *Nature,* 350, 66, 1991.
71. **Calvete, J. J., Arias, J., Alvarez, M. C., Lopez, M. M., Henschen, A., and Gonzalez-Rodriguez, J.,** Further studies on the topography of human platelet glycoprotein IIb, *Biochem. J.,* 273, 767, 1991.
72. **Pulido, R., Elices, M. J., Campanero, M. R., Osborn, L., Schiffer, S., Garcia Pardo, A., Lobb, R., Hemler, M. E., and Sanchez-Madrid, F.,** Functional evidence for three distinct and independently inhibitable adhesion activities mediated by the human integrin VLA-4. Correlation with distinct alpha 4 epitopes, *J. Biol. Chem.,* 266, 10241, 1991.
73. **Albelda, S. M., Daise, M., Levine, E. N., and Buck, C. A.,** Identification and characterization of cell–substratum adhesion receptors on cultured human adult large vessel endothelial cells, *J. Clin. Invest.,* 83, 1992, 1989.
74. **Horton, M.,** Vitronectin receptor: tissue specific expression or adaptation to culture?, *Int. J. Exp. Pathol.,* 71, 741, 1990.
75. **Albelda, S. M.,** Endothelial and epithelial cell adhesion molecules, *Am. J. Respir. Cell Mol. Biol.,* 4, 195, 1991.
76. **De Strooper, B., Van der Schueren, B., Jaspers, M., Saison, M., Spaepen, M., Van Leuven, F., Van den Berghe, H., and Cassiman, J.-J.,** Distribution of the β_1 subgroup of the integrins in human cells and tissues, *J. Histochem. Cytochem.,* 37, 299, 1989.
77. **Krotoski, D. M., Domingo, C., and Bronner-Fraser, M.,** Distribution of a putative cell surface receptor for fibronectin and laminin in the avian embryo, *J. Cell Biol.,* 103, 1061, 1986.

78. **Peltonen, J., Larjave, H., Jaakkola, S., Gralnick, H., Akiyama, S., Yamada, S., Yamada, K., and Uitto, J.,** Localization of integrin receptors for fibronectin, collagen and laminin in human skin, *J. Clin. Invest.,* 84, 196, 1989.
79. **Hertle, M. D., Adams, J. C., and Watt, F. M.,** Integrin expression during human epidermal development *in vivo* and *in vitro, Development,* 112, 193, 1991.
80. **Choy, M.-Y., Richman, P. I., Horton, M. A., and MacDonald, T. T.,** Expression of the VLA family of integrins in human intestine, *J. Pathol.,* 160, 35, 1990.
81. **Zutter, M. M.,** Immunolocalization of integrin receptors in normal lymphoid tissues, *Blood,* 77, 2231, 1991.
82. **Koretz, K., Schlag, P., Boumsell, L., and Moller, P.,** Expression of VLA-α_2, VLA-α_6, and VLA-β_1 chains in normal mucosa and adenomas of the colon, and in colon carcinomas and their liver metastases, *Am. J. Pathol.,* 138, 741, 1991.
83. **Gould, V. E., Koukoulis, G. K., and Virtanen, I.,** Extracellular matrix proteins and their receptors in the normal, hyperplastic and neoplastic breast, *Cell Differ. Dev.,* 32, 409, 1990.
84. **Damjanovich, L., Albelda, S. M., Mette, S. A., and Buck, C. A.,** The distribution of integrin cell adhesion receptors in normal and malignant lung tissues, *Am. J. Respir. Cell Mol. Biol.,* in press, 1991.
85. **Carter, W. G., Wayner, E. A., Bouchard, T. S., and Kaur, P.,** The role of integrins $\alpha_2\beta_1$ and $\alpha_3\beta_1$ in cell–cell and cell–substrate adhesion of human epidermal cells, *J. Cell Biol.,* 110, 1387, 1990.
86. **Larjave, H., Pelsonen, J., Akiyama, S., Yamada, S., Gralnick, J., Uitto, J., and Yamada, K.,** Novel function for beta 1 integrins in keratinocyte cell–cell interactions., *J. Cell Biol.,* 110, 803, 1990.
87. **Williams, D. A., Rios, M., Stephens, C., and Patel, V. P.,** Fibronectin and VLA-4 in haematopoietic stem cell–microenvironment interactions, *Nature,* 352, 438, 1991.
88. **Campanero, M. R., Pulido, R., Ursa, M. A., Rodriguez-Moya, M., deLandazuri, M. O., and Sanchez-Madrid, F.,** An alternative leukocyte homotypic adhesion mechanism, LFA-1/ICAM-1-independent, triggered through the human VLA-4 integrin, *J. Cell Biol.,* 110, 2157, 1990.
89. **Albelda, S. M., Mette, S. A., Elder, D. E., Stewart, R. M., Damjanovich, L., Herlyn, M., and Buck, C. A.,** Integrin distribution in malignant melanoma: association of the β_3 subunit with tumor progression, *Cancer Res.,* 50, 6757, 1991.
90. **Soligo, D., Schiro, R., Luksch, R., Manara, G., Quirici, N., Parravicini, C., and Deliliers, G. L.,** Expression of integrins in human bone marrow, *Br. J. Haematol.,* 76, 323, 1990.
91. **Dustin, M. L., Rothlein, R., Bhan, A. K., Dinarello, C. A., and Springer, T. A.,** Induction by IL1 and interferon-γ: tissue distribution, biochemistry, and function of a natural adherence molecule (ICAM-1), *J. Immunol.,* 137, 245, 1986.
92. **Frelinger, A. L., Cohen, I., Plow, E. F., Smith, M. A., Roberts, J., Lam, S. C.-T., and Ginsberg, M. H.,** Selective inhibition of integrin function by antibodies specific for ligand-occupied receptor conformers, *J. Biol. Chem.,* 265, 6346, 1990.
93. **Du, X., Plow, E. F., Frelinger, A. L., III, O'Toole, T. E., Loftus, J. C., and Ginsberg, M. H.,** Ligands "activate" integrin $\alpha_{IIb}\beta_3$ (platelet GPIIb-IIIa), *Cell,* 65, 409, 1991.
94. **O'Toole, T. E., Loftus, J. C., Du, X. P., Glass, A. A., Ruggeri, Z. M., Shattil, S. J., Plow, E. F., and Ginsberg, M. H.,** Affinity modulation of the alpha IIb beta 3 integrin (platelet GPIIb-IIIa) is an intrinsic property of the receptor, *Cell Regul.,* 1, 883, 1990.
95. **Frojmovic, M. M., O'Toole, T. E., Plow, E. F., Loftus, J. C., and Ginsberg, M. H.,** Platelet glycoprotein II_b-III_a ($\alpha_{IIb}\beta_3$ integrin) confers fibrinogen- and activation-dependent aggregation on heterologous cells, *Blood,* 78, 369, 1991.
96. **Dustin, M. L. and Springer, T. A.,** T cell receptor cross-linking transiently stimulates adhesiveness through LFA-1, *Nature,* 341, 619, 1989.
97. **Kupfer, A. and Singer, S. J.,** The specific interaction of helper T cells and antigen-presenting B cells. IV. Membrane and cytoskeletal reorganizations in the bound T cell as a function of antigen dose, *J. Exp. Med.,* 170, 1697, 1989.
98. **Shimizu, Y., Van Seventer, G., Horgan, K., and Shaw, S.,** Regulated expression and binding of three VLA (β_1) integrin receptors on T cells, *Nature,* 345, 250, 1990.
99. **Neugebauer, K. M. and Reichardt, L. F.,** Cell-surface regulation of β_1-integrin activity on developing retinal neurons, *Nature,* 350, 68, 1991.
100. **Hirst, R., Horwitz, A., Buck, C., and Rohrschneider, L.,** Phosphorylation of the fibronectin receptor complex in cells transformed by oncogenes that encode tyrosine kinases, *Proc. Natl. Acad. Sci. U.S.A.,* 83, 6470, 1986.

101. **Shaw, L. M., Messier, J. M., and Mercurio, A. M.,** The activation dependent adhesion of macrophages to laminin involves cytoskeletal anchoring and phosphorylation of the $\alpha_6\beta_1$ integrin, *J. Cell Biol.,* 110, 2167, 1990.
102. **Danilov, Y. N. and Juliano, R. L.,** Phorbol ester modulation of integrin-mediated cell adhesion: a postreceptor event, *J. Cell Biol.,* 108, 1925, 1989.
103. **Defilippi, P., Truffa, G., Stefanuto, G., Altruda, F., Silengo, L., and Tarone, G.,** Tumor necrosis factor α and interferon gamma modulate the expression of the vitronectin receptor (integrin β_3) in human endothelial cells, *J. Biol. Chem.,* 266, 7638, 1991.
104. **Heino, J. and Massague, J.,** Transforming growth factor-β switches the pattern of integrins expressed in MG-63 human osteosarcoma cells and causes a selective loss of cell adhesion to laminin, *J. Biol. Chem.,* 264, 21806, 1989.
105. **Marchisio, P. C., Bondanza, S., Cremona, O., Cancedda, R., and De Luca, M.,** Polarized expression of integrin receptors (alpha 6 beta 4, alpha 2 beta 1, alpha 3 beta 1, and alpha v beta 5) and their relationship with the cytoskeleton and basement membrane matrix in cultured human keratinocytes, *J. Cell Biol.,* 112, 761, 1991.
106. **Carlos, T., Kovach, N., Schwartz, B., Rosa, M., Newman, B., Wayner, E., Benjamin, C., Osborn, L., Lobb, R., and Harlan, J.,** Human monocytes bind to two cytokine-induced adhesive ligands on cultured human endothelial cells: endothelial–leukocyte adhesion molecule-1 and vascular cell adhesion molecule-1, *Blood,* 77, 2266, 1991.
107. **Osborn, L.,** Leukocyte adhesion to endothelium in inflammation, *Cell,* 62, 3, 1990.
108. **Singer, S. J. and Kupfer, A.,** The directed migration of eukaryotic cells, *Annu. Rev. Cell Biol.,* 2, 337, 1986.
109. **Marks, P. W., Hendey, B., and Maxfield, F. R.,** Attachment to fibronectin or vitronectin makes human neutrophil migration sensitive to alterations in cytosolic free calcium concentration, *J. Cell Biol.,* 112, 149, 1991.
110. **Horwitz, A., Duggan, K., Buck, C., Beckerle, M. C., and Burridge, K.,** Interaction of plasma membrane fibronectin receptor with talin — a transmembrane linkage, *Nature,* 320, 531, 1986.
111. **Akiyama, S. K. and Yamada, K. M.,** The interaction of plasma fibronectin with fibroblastic cells in suspension, *J. Biol. Chem.,* 260, 4492, 1985.
112. **Goodman, S. L., Risse, G., Von der Mark, K., and Grinnell, F.,** The E8 subfragment of laminin promotes locomotion of myoblasts over extracellular matrix, *J. Cell Biol.,* 109, 799, 1989.
113. **Goodman, S. L., Deutzmann, R., and Von der Mark, K.,** Two distinct cell-binding domains in laminin can independently promote nonneuronal cell adhesion and spreading, *J. Cell Biol.,* 105, 589, 1987.
114. **DiMilla, P. A., Barbee, K., and Lauffenburger, D. A.,** Mathematical model for the effects of adhesion and mechanics on cell migration speed, *Biophys. J.,* 60, 15, 1991.
115. **DiMilla, P. A., Albelda, S. M., Quinn, J. A., and Lauffenburger, D. A.,** Discerning underlying mechanisms for cell migration over adhesive surfaces using models of individual cell movement, *J. Cell Biol.,* 111, 288a, 1990.
116. **Duband, J.-L., Dufour, S., Yamada, S. S., Yamada, K. M., and Thiery, J. P.,** Neural crest cell locomotion induced by antibodies to β_1 integrins. A tool for studying the roles of substratum molecular avidity and density in migration, *J. Cell Sci.,* 98, 517, 1991.
117. **Lauffenburger, D. A.,** A simple model for the effects of receptor-mediated cell–substratum adhesion on cell migration, *Chem. Eng. Sci.,* 44, 1903, 1989.
118. **Trinkaus, J. P.,** *Cells into Organs: The Forces that Shape the Embryo,* Prentice-Hall, Englewood Cliffs, NJ, 1984.
119. **Bretscher, M. S.,** Endocytosis and recycling of the fibronectin receptor in CHO cells, *EMBO J.,* 8, 1341, 1989.
120. **Grinnell, F.,** Focal adhesion sites and the removal of substratum-bound fibronectin, *J. Cell Biol.,* 103, 2697, 1986.
121. **Sczekan, M. M. and Juliano, R. L.,** Internalization of the fibronectin receptor is a constitutive process, *J. Cell. Physiol.,* 142, 574, 1990.
122. **Duband, J.-L., Dufour, S., and Thiery, J. P.,** Extracellular matrix–cytoskeleton interactions in locomoting embryonic cells, *Protoplasma,* 145, 112, 1988.
123. **Tapley, P., Horwitz, A., Buck, C., Duggan, K., and Rohrschneider, L.,** Integrins isolated from Rous sarcoma virus-transformed chicken embryo fibroblasts, *Oncogene,* 4, 325, 1989.

124. **Schreiner, C. L., Bauer, J. S., Danilov, Y. N., Hussein, S., Sczekan, M. M., and Juliano, R. L.,** Isolation and characterization of Chinese hamster ovary cell variants deficient in the expression of fibronectin receptor, *J. Cell Biol.,* 109, 3157, 1989.
125. **Giancotti, F. G. and Ruoslahti, E.,** Elevated levels of the $\alpha_5\beta_1$ fibronectin receptor suppress the transformed phenotype of Chinese hamster ovary cells, *Cell,* 60, 1990.
126. **Albelda, S. M. and Buck, C. A.,** Integrins and other cell adhesion molecules, *FASEB J.,* 4, 2868, 1990.
127. **Sternberg, J. and Kimber, S. J.,** Distribution of fibronectin, laminin and entactin in the environment of migrating neural crest cells in early mouse embryos, *J. Embryol. Exp. Morphol.,* 91, 267, 1986.
128. **Tucker, G. C., Duband, J. L., Dufour, S., and Thiery, J.-P.,** Cell-adhesion and substrate-adhesion molecules: their instructive roles in neural crest cell migration, *Development,* 103 (Suppl.), 81, 1988.
129. **Duband, J.-L., Rocher, S., Chen, W.-T., Yamada, K., and Thiery, J.-P.,** Cell adhesion and migration in the early vertebrate embryo: location and possible role of the putative fibronectin receptor complex, *J. Cell Biol.,* 102, 160, 1986.
130. **Chen, W.-T., Chen, J.-M., and Mueller, S. C.,** Coupled expression and colocalization of 140K cell adhesion molecules, fibronectin, and laminin during morphogenesis and cytodifferentiation of chick lung cells, *J. Cell Biol.,* 103, 1073, 1986.
131. **Greve, J. M. and Gottlieb, D. I.,** Monoclonal antibodies which alter the morphology of chick myogenic cells, *J. Cell Biol.,* 18, 221, 1982.
132. **Boucaut, J. C., Darribere, T., Poole, T. J., Aoyama, H., Yamada, K. M., and Thiery, J. P.,** Biological active synthetic peptides as probes of embryonic development: a competitive peptide inhibitor of fibronectin function inhibits gastrulation in amphibian embryos and neural crest cell migration in the avian embryo, *J. Cell Biol.,* 99, 1822, 1984.
133. **Naidet, C., Semeriva, M., Yamada, K. M., and Thiery, J. P.,** Peptides containing the cell-attachment recognition signal Arg-Gly-Asp prevent gastrulation in Drosophila embryos, *Nature,* 325, 348, 1987.
134. **Bronner-Fraser, M.,** Alterations in neural crest migration by a monoclonal antibody that affects cell adhesion, *J. Cell Biol.,* 101, 610, 1985.
135. **Cohen, J., Burne, J. F., Winter, G., and Bartlett, P.,** Retinal ganglion cells lose response to laminin with maturation, *Nature,* 322, 465, 1986.
136. **Jaffredo, T., Horwitz, A. F., Buck, C. A., Rong, P. M., and Dieterlen-Lievre, F.,** Myoblast migration specifically inhibited in the chick embryo by grafted CSAT hybridoma cells secreting an anti-integrin antibody, *Development,* 103, 431, 1988.
137. **Anderson, D. C. and Springer, T. A.,** Leukocyte adhesion deficiency: an inherited defect in the Mac-1, LFA-1 and p150,95 glycoproteins, *Annu. Rev. Med.,* 38, 175, 1987.
138. **Springer, T. A., Dustin, M. L., Kishimoto, T. K., and Marlin, S.,** The lymphocyte function-associated LFA-1, CD2, and LFA-3 molecules: cell adhesion receptors of the immune system, *Annu. Rev. Immunol.,* 5, 223, 1987.
139. **Wilson, J. M., Ping, A. J., Krauss, J. C., Mayo-Bond, L., Rogers, C. E., Anderson, D. C., and Todd, R. F., III,** Correction of CD18-deficient lymphocytes by retrovirus-mediated gene transfer, *Science,* 248, 1413, 1990.
140. **Newman, P. J.,** Platelet GPIIb-IIIa: molecular variations and alloantigens, *Thromb. Haemostasis,* 66, 111, 1991.
141. **Wilcox, M., DiAntonio, A., and Leptin, M.,** The function of PS integrins in Drosophila wing morphogenesis, *Development,* 107, 891, 1989.
142. **Isberg, R. R. and Leong, J. M.,** Multiple β_1 chain integrins are receptors for invasin, a protein that promotes bacterial penetration into mammalian cells, *Cell,* 60, 861, 1990.
143. **Leong, J. M., Fournier, R. S., and Isberg, R. R.,** Identification of the integrin binding domain of the Yersinia pseudotuberculosis invasin protein, *EMBO J.,* 9, 1979, 1990.
144. **Leininger, E., Roberts, M., Kenimer, J. G., Charles, I. G., Fairweather, N., Novotny, P., and Brennan, M. J.,** Pertactin, an Arg-Gly-Asp-containing *Bordetella pertussis* surface protein that promotes adherence of mammalian cells, *Proc. Natl. Acad. Sci. U.S.A.,* 88, 345, 1991.
145. **Bullock, W. E. and Wright, S. D.,** Role of the adherence-promoting receptors, CR3, LFA-1, and p150,95, in binding of Histoplasma capsulatum by human macrophages, *J. Exp. Med.,* 165, 195, 1987.
146. **Rohana-Talamas, P., Wright, S. D., Lennartz, M. R., and Russell, D. G.,** Lipophosphoglycan from *Leishmania mexicana* promastigotes binds to members of the CR3, p150,95 and LFA-1 family of leukocyte integrins, *J. Immunol.,* 144, 4817, 1991.

147. **Russell, D. G. and Wright, S. D.,** Complement receptor type 3 (CR3) binds to an Arg-Gly-Asp-containing region of the major surface glycoprotein, gp63, of Leishmania promastigotes, *J. Exp. Med.,* 168, 279, 1988.
148. **Relman, D., Tuomanen, E., Falkow, S., Golenbock, D. T., Saukkonen, K., and Wright, S. D.,** Recognition of a bacterial adhesin by an integrin: macrophage CR3 ($\alpha_M\beta_2$, CD11b/CD18) binds filamentous hemagglutinin of Bordetella pertussis, *Cell,* 61, 1375, 1990.
149. **Plantefaber, L. and Hynes, R. O.,** Changes in integrin receptors on oncogenically transformed cells, *Cell,* 56, 281, 1989.
150. **Chan, B. M. C., Matsuura, N., Takada, Y., Zetter, B. R., and Hemler, M. E.,** In vitro and in vivo consequences of VLA-2 expression on rhabdomyosarcoma cells, *Science,* 251, 1600, 1991.
151. **Gehlsen, K. R., Argraves, W. S., Pierschbacher, M. D., and Ruoslahti, E.,** Inhibition of in vitro tumor cell invasion by Arg-Gly-Asp-containing synthetic peptides, *J. Cell Biol.,* 106, 925, 1988.
152. **Humphries, M. J., Olden, K., and Yamada, K.,** A synthetic peptide from fibronectin inhibits experimental metastasis of murine melanoma cells, *Science,* 233, 467, 1986.
153. **Ruoslahti, E. and Giancotti, F. G.,** Integrins and tumor cell dissemination, *Canc. Cells,* 1, 119, 1989.
154. **Albelda, S. M., Mette, S. A., Elder, D. E., Stewart, R., Herlyn, M., and Buck, C. A.,** Expression of an integrin cell–substratum adhesion receptor as a marker of tumorigenic malignant melanoma, *Cancer Res.,* 50, 6757, 1990.
155. **McGregor, B., McGregor, J. L., Weiss, L. M., Wood, G. S., Hu, C.-H., Boukerche, H., and Warnke, R. A.,** Presence of cytoadhesins (IIb/IIIa-like glycoproteins) on human metastatic melanomas but not on benign melanocytes, *Am. J. Clin. Pathol.,* 92, 495, 1989.
156. **Pignatelli, M. and Bodmer, W. F.,** Integrin cell adhesion molecules and colorectal cancer, *J. Pathol.,* 162, 95, 1990.
157. **Zutter, M. M., Mazoujian, G., and Santoro, S. A.,** Decreased expression of integrin adhesive protein receptors in adenocarcinoma of the breast, *Am. J. Pathol.,* 137, 863, 1990.
158. **Pignatelli, M., Smith, M. E. F., and Bodmer, W. F.,** Low expression of collagen receptors in moderate and poorly differentiated colorectal adenocarcinomas, *Br. J. Cancer,* 61, 636, 1990.
159. **Ginsberg, M. H., Loftus, J. C., and Plow, E. F.,** Cytoadhesins, integrins, and platelets, *Thromb. Haemostasis,* 59, 1, 1988.
160. **Grinnell, F., Toda, K.-I., and Takashima, A.,** Role of fibronectin in epithelialization and wound healing, *Prog. Clin. Biol. Res.,* 266, 259, 1988.
161. **Clark, R. A. F.,** Fibronectin matrix deposition and fibronectin receptor expression in healing and normal skin, *J. Invest. Dermatol.,* 94 (Suppl.), 128S, 1990.
162. **MacDonald, T. T., Horton, M. A., Choy, M. Y., and Richman, P. I.,** Increased expression of laminin/collagen receptor (VLA-1) on epithelium of inflamed human intestine, *J. Clin. Pathol.,* 43, 313, 1990.
163. **Akiyama, H., Kawamata, T., Dedhar, S., and McGeer, P. L.,** Immunohistochemical localization of vitronectin, its receptor and β_3 integrin in Alzheimer brain tissue, *J. Neuroimmunol.,* 32, 19, 1991.
164. **Carlos, T. M. and Harlan, J. M.,** Membrane proteins involved in phagocyte adherence to endothelium, *Immunol. Rev.,* 114, 5, 1990.
165. **Lewinsohn, D. M., Bargatze, R. F., and Butcher, E. C.,** Leukocyte–endothelial cell recognition: evidence of a common molecular mechanism shared by neutrophils, lymphocytes, and other leukocytes, *J. Immunol.,* 138, 4313, 1987.
166. **Mileski, W., Harlan, J., Rice, C., and Winn, R.,** Streptococcus pneumoniae-stimulated macrophages induce neutrophils to emigrate by a CD18-independent mechanism of adherence, *Circ. Shock,* 31, 259, 1990.
167. **Tuomanen, E. I., Saukkonen, K., Sande, S., Cioffe, C., and Wright, S. D.,** Reduction of inflammation, tissue damage, and mortality in bacterial meningitis in rabbits treated with monoclonal antibodies against adhesion-promoting receptors of leukocytes, *J. Exp. Med.,* 170, 959, 1989.
168. **Rosen, H. and Gordon, S.,** The role of the type 3 complement receptor in the induced recruitment of myelomonocytic cells to inflammatory sites in the mouse, *Am. J. Respir. Cell Mol. Biol.,* 3, 3, 1990.
169. **Vedder, N. B., Winn, R. K., Rice, C. L., Chi, E. Y., Arfos, K.-E., and Harlan, J. M.,** A monoclonal antibody to the adherence-promoting leukocyte glycoprotein, CD18, reduces organ injury and improves survival from hemorrhagic shock and resuscitation in rabbits, *J. Clin. Invest.,* 81, 939, 1988.
170. **Hernandez, L. A., Grisham, M. B., Twohig, B., Arfors, K. E., Harlan, J. M., and Granger, D. N.,** Role of neutrophils in ischemia-reperfusion-induced microvascular injury, *Am. J. Physiol.,* 253, H699, 1991.

171. **Simpson, P. J., Todd, R., III, Fantone, J. C., Mickelson, J. K., Griffin, J. D., and Lucchesi, B. R.,** Reduction of experimental canine myocardial reperfusion injury by a monoclonal antibody (anti-Mo1, anti-CD11b) that inhibits leukocyte adhesion, *J. Clin. Invest.*, 81, 624, 1988.

172. **Horgan, M. J., Wright, S. D., and Malik, A. B.,** Antibody against leukocyte integrin (CD18) prevents reperfusion-induced lung vascular injury, *Am. J. Physiol.*, 259, L315, 1990.

173. **Gold, H. K., Gimple, L. W., Yasuda, T., Leinbach, R. C., Werner, W., Holt, R., Jordan, R., Berger, H., Collen, D., and Coller, B. S.,** Pharmacodynamic study of $F(ab')_2$ fragments of murine monoclonal antibody 7E3 directed against human platelet glycoprotein IIb/IIIa in patients with unstable angina pectoris, *J. Clin. Invest.*, 86, 651, 1990.

174. **Cadroy, Y., Houghten, R. A., and Hanson, S. R.,** RGDV peptide selectively inhibits platelet-dependent thrombus formation *in vivo, J. Clin. Invest.*, 84, 939, 1989.

175. **Fischer, A., Friedrich, W., Fasth, A., Blanche, S., Le Deist, F., Girault, D., Veber, F., Vossen, J., Lopez, M., Griscelli, C., and Hirn, M.,** Reduction of graft failure by a monoclonal antibody (anti-LFA-1 CD11a) after HLA nonidentical bone marrow transplantation in children with immunodeficiencies, osteopetrosis, and Fanconi's anemia: a European group for immunodeficiency/European group for bone marrow transplantation report, *Blood,* 77, 249, 1991.

176. **Saiki, I., Murata, J., Lida, J., Nishi, N., Sugimura, K., and Azuma, I.,** The inhibition of murine lung metastasis by synthetic polypeptides [poly(arg-gly-asp) and poly(tyr-ile-gly-ser-arg)] with a core sequence of cell adhesion molecules, *Br. J. Cancer,* 59, 194, 1988.

177. **Devereux, J., Haeberli, P., and Smithies, O.,** A comprehensive set of sequence analysis programs for the VAX, *Nucl. Acids Res.*, 12, 387, 1984.

178. **Argraves, W. S., Suzuki, S., Arai, H., Thompson, K., Pierschbacher, M. D., and Ruoslahti, E.,** Amino acid sequence of the human fibronectin receptor, *J. Cell Biol.*, 105, 1183, 1987.

179. **Tominaga, S.-I.,** Murine mRNA for the β-subunit of integrin is increased in BALB/c-3T3 cells entering the G_1 phase from the G_0 state, *FEBS Lett.*, 238, 315, 1988.

180. **Holers, V. M., Ruff, T. G., Parks, D. L., McDonald, J. A., Ballard, L. L., and Brown, E. J.,** Molecular cloning of a murine fibronectin receptor and its expression during inflammation: expression of VLA-5 is increased in activated peritoneal macrophages in a manner discordant from major histocompatibility complex class II, *J. Exp. Med.*, 169, 1589, 1989.

181. **Kishimoto, T. K., O'Connor, K., Lee, A., Roberts, T. M., and Springer, T. A.,** Cloning of the beta subunit of the leukocyte adhesion proteins: homology to an extracellular matrix receptor defines a novel supergene family, *Cell,* 48, 681, 1987.

182. **Wilson, R. W., O'Brien, W. E., and Beaudet, A. L.,** Nucleotide sequence of the cDNA from mouse leukocyte adhesion protein CD18, *Nucl. Acids Res.*, 17, 5397, 1989.

183. **Zeger, D. L., Osman, N., Hennings, M., Mckenzie, I. F., Sears, D. W., and Hogarth, P. M.,** Mouse macrophage beta-subunit (CD11b) cDNA for the CR3 complement receptor/MAC-1 antigen, *Immunogenetics,* 31, 191, 1990.

184. **Fitzgerald, L. A., Steiner, B., Rall, S. C., Jr., Lo, S., and Phillips, D. R.,** Protein sequence of endothelial glycoprotein IIIa derive from a cDNA clone: identity with platelet glycoprotein IIIa and similarity to "integrin", *J. Biol. Chem.*, 262, 3936, 1987.

185. **Tamura, R. N., Rozzo, C., Starr, L., Chambers, J., Reichardt, L. F., Cooper, H. M., and Quaranta, V.,** Epithelial integrin α_6/β_4: complete primary structure of α_6 and variant forms of β_4, *J. Cell Biol.*, 111, None, 1990.

186. **Hogervorst, F., Kuikman, I., von dem Borne, A. E. G. R., and Sonnenberg, A.,** Cloning and sequence analysis of beta-4 cDNA: an integrin subunit that contains a unique 118 kD cytoplasmic domain, *EMBO J.*, 9, 765, 1990.

187. **McClean, J. W., Vestal, D. J., Cheresh, D. A., and Bodary, S. C.,** cDNA sequence of human integrin β_5 subunit, *J. Biol. Chem.*, 265, 17126, 1990.

188. **Suzuki, S., Huang, Z.-S., and Tanihara, H.,** Cloning of an integrin β subunit exhibiting high homology with integrin β_3 subunit, *Proc. Natl. Acad. Sci. U.S.A.*, 87, 5354, 1990.

189. **Ramaswamy, H. and Hemler, M.,** Cloning, primary structure and properties of a novel human integrin β subunit, *EMBO J.*, 9, 1561, 1991.

190. **Sheppard, D., Rozzo, C., Starr, L., Quaranta, V., Erle, D. J., and Pytela, R.,** Complete amino acid sequence of a novel integrin β subunit (β_6) identified in epithelial cells using the polymerase chain reaction, *J. Biol. Chem.*, 265, 11502, 1990.

191. **Larson, R. S., Corbi, A. L., Berman, L., and Springer, T.,** Primary structure of the leukocyte function-associated molecule-1 α subunit: an integrin with an embedded domain defining a protein superfamily, *J. Cell Biol.*, 108, 703, 1989.

192. **Kaufmann, Y., Tseng, E., and Springer, T. A.,** Cloning of the murine lymphocyte function-associated molecule-1 (LFA-1) alpha subunit and its expression in Cos cells, *J. Immunol.*, in press, 1991.
193. **Corbi, A. L., Kishimoto, T. K., Miller, L. J., and Springer, T. A.,** The human leukocyte adhesion glycoprotein Mac-1 (complement receptor type 3, CD11b) α subunit. Cloning, primary structure, and relation to the integrins, von Willebrand factor and factor B, *J. Biol. Chem.*, 263, 12403, 1988.
194. **Hickstein, D. D., Hickey, M. J., Ozols, J., Baker, D. M., Back, A. L., and Roth, G. J.,** cDNA sequence for the αM subunit of the human neutrophil adherence receptor indicates homology to integrin α subunits, *Proc. Natl. Acad. Sci. U.S.A.*, 86, 257, 1989.
195. **Arnaout, M. A., Gupta, S. K., Pierce, M. W., and Tenen, D. G.,** Amino acid sequence of the alpha subunit of human leukocyte adhesion receptor Mo1 (complement receptor type 3), *J. Cell Biol.*, 106, 2153, 1988.
196. **Pytela, R.,** Amino acid sequence of the murine Mac-1 α chain reveals homology with the integrin family and an additional domain related to von Willebrand factor, *EMBO J.*, 7, 1371, 1988.
197. **Corbi, A. L., Miller, L. J., O'Connor, K., Larson, R. S., and Springer, T. A.,** cDNA cloning and complete primary structure of the alpha subunit of a leukocyte adhesion glycoprotein, *EMBO J.*, 6, 4023, 1987.
198. **Ignatius, M. J., Large, T. M., Houde, M., Tawil, J. W., Barton, A., Esch, F., Carbonetto, S., and Reichardt, L. F.,** Molecular cloning of the rat integrin α_1-subunit: a receptor for laminin and collagen, *J. Cell Biol.*, 111, 709, 1990.
199. **Takada, Y. and Hemler, M. E.,** The primary structure of the VLA-2/collagen receptor α_2 subunit (platelet GPIa): homology to other integrins and the presence of a possible collagen-binding domain, *J. Cell Biol.*, 109, 397, 1989.
200. **Takada, Y., Elices, M. J., Crouse, C., and Hemler, M. E.,** The primary structure of the α_4 subunit of VLA-4: homology to other integrins and a possible cell–cell adhesion function, *EMBO J.*, 8, 1361, 1989.
201. **Neuhaus, H., Hu, M. C.-T., Hemler, M. E., Takada, Y., Holamann, B., and Weissman, I. L.,** Cloning and expression of cDNAs for the α subunit of the murine lymphocyte-peyer's patch homing receptor, *J. Cell Biol.*, in press, 1991.
202. **deCurtis, I., Quaranta, V., Tamura, R. N., and Reichardt, L. F.,** Laminin receptors in the retina: sequence analysis of the chick integrin α_6 subunit, *J. Cell Biol.*, 113, 405, 1991.
203. **Suzuki, S., Argraves, W. S., Arai, H., Languino, L. R., Pierschbacher, M. D., and Ruoslahti, E.,** Amino acid sequence of the vitronectin receptor alpha subunit and comparative expression of adhesion receptor mRNAs, *Proc. Natl. Acad. Sci. U.S.A.*, 83, 8614, 1986.
204. **Bossy, B. and Reichardt, L. F.,** Chick integrin α_v subunit molecular analysis reveals high conservation of structural domains and association with multiple β subunits in embryo fibroblasts, *Biochemistry*, 29, 10191, 1990.
205. **Poncz, M., Eisman, R., Heidenteich, R., Silver, S. M., Vilaire, G., Surrey, S., Schwartz, E., and Bennett, J. S.,** Structure of the platelet membrane glycoprotein IIb, *J. Biol. Chem.*, 262, 8476, 1987.
206. **Leong, J. M., Fournier, R. S., and Isberg, R. R.,** Identification of the integrin binding domain of the *Yersinia pseudotuperculosis* invasin protein, *EMBO J.*, 9, 1979, 1990.

Chapter 9

The Structure and Biological Function of the Neural Cell Adhesion Molecule N-CAM

John J. Hemperly

CONTENTS

I. Introduction 187
II. Structural Studies of N-CAM 187
III. Molecular Biological Studies of N-CAM 190
IV. Biological Studies of N-CAM 194
V. N-CAM in Disease 194
VI. Summary 195
References 195

I. INTRODUCTION

The proper recognition, adhesion, and signal transduction between cells is of fundamental importance to the development and maintenance of tissue form and function. It had been observed for some time in a multitude of developing systems that individual cells have the capability of sorting out and adhering specifically with appropriate partners. Although a number of molecules that mediate cell-to-cell adhesion have been identified,[1] one of the first to be identified as fundamental during embryonic development is the neural cell adhesion molecule, N-CAM.

N-CAM was initially identified by a number of laboratories using at least two conceptually different approaches. One approach, pursued by Dr. Gerald Edelman and his colleagues, involved the identification of particular molecular components mediating the adhesion of chicken neural retina cells.[2-4] In tissue culture, these cells can be dissociated, and if incubated under the appropriate conditions, the cells can reaggregate (Figure 1). Antisera prepared against the whole cell surface (presumably containing antibodies against particular adhesion components) can inhibit this reaggregation. This experimental system was used as an assay for cell adhesion molecules by examining which purified membrane components could neutralize the antibodies. Following this approach, a membrane component termed the "neural cell adhesion molecule (N-CAM)" was identified. At about the same time, other laboratories, in particular those of Bock and Goridis, were identifying and characterizing brain and neural cell surface components using monoclonal antibodies.[5-7] Independently, all of the laboratories identified new molecules, which were given names such as D2, BSP-2, etc. Eventually it became clear that these various laboratories were pursuing what was almost certainly the same molecule or the corresponding proteins in chicken, rodents, and man. The name "neural cell adhesion molecule" (abbreviated N-CAM or even more succinctly NCAM) has become generally accepted and is the one I will use in this review. As will become apparent, many if not all of the findings made in one species hold across other species, and I will not make an attempt to contrast and compare them. In all cases, N-CAM is a major neural antigen of profound functional significance. I will first discuss some structural studies on N-CAM and then attempt to relate its structure(s) to biological function. In the concluding section, I will describe some of the exciting developments suggesting N-CAM involvement in the pathogenesis and diagnosis of disease.

II. STRUCTURAL STUDIES OF N-CAM

N-CAM was first identified both as a neural antigen and a component that could inhibit the inhibition of cell aggregation by polyspecific antisera. As the molecule was analyzed using the techniques of protein chemistry, several characteristics became readily apparent. In particular, N-CAM comprised multiple, relatively large polypeptide chains, and N-CAM from embryonic tissue migrated on SDS polyacrylamide gels as a diffuse, possibly heterogeneous component of a mol wt greater than 200 kD. Treatment of this embryonic N-CAM "smear" with endoglycosidases or neuraminidase to remove sialic acid revealed three

0-8493-4559-6/95/$0.00+$.50

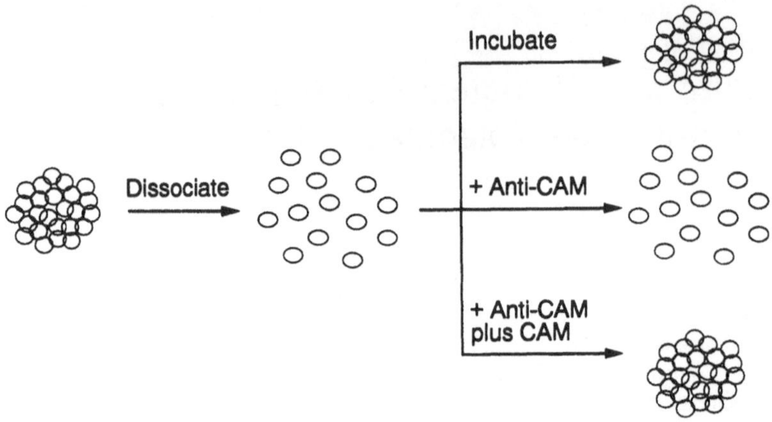

Figure 1 Cell-cell adhesion assay used initially to identify N-CAM. Dissociated cells are incubated in the absence (top) or presence (middle) of antibodies to molecules on the cell surface that inhibit aggregation. In the presence of exogenous cell adhesion molecules (bottom), the aggregation is inhibited.

polypeptides of approximately 180, 140, and 120 kD, with slight variations from species to species, from tissue to tissue, and with age of development. An important result that emerged from these studies and the molecular biological approaches described below is that N-CAM consists of multiple polypeptides subject to extensive and varied posttranslational modification.[6,8-10]

One of the more striking posttranslational modifications of N-CAM is glycosylation. Treatment with endoglycosidase F, which removes carbohydrates linked to the polypeptide chain through the amino acid asparagine (N-linked glycosylation), generates three distinct polypeptides.[10] Similarly, growth of tissue culture cells in the presence of the drug tunicamycin, which prevents the metabolic addition of N-linked sugars, gives rise to three similar bands, suggesting that most of the sugars contributing to the migration of N-CAM are N-linked. There has yet been no clear demonstration of sugars on N-CAM linked through the hydroxy groups of serine or threonine except for the muscle-specific variant described below. Other apparent posttranslational modifications of N-CAM include the attachment of sulfate and phosphate moieties.[11] Neither of these latter modifications have been clearly shown to have biological consequences, but it appears that at least some of the ability of N-CAM to incorporate radioactive sulfate is through sulfation of carbohydrate. For example, the monoclonal antibody L2/HNK-1, which recognizes N-CAM as well as a number of other cell surface molecules involved in cell-cell adhesion,[12] is specific for a sulfated glucuronic acid epitope.[13] Phosphorylation of N-CAM apparently occurs intracellularly on serine and threonine residues but not on tyrosine,[11] and two protein kinases capable of phosphorylating N-CAM have been identified.[14] There have also been reports that N-CAM can incorporate radioactive fatty acids, presumably near the cell membrane, but again, the significance of this acylation is not known.

As noted above, one of the first recognized features of N-CAM from embryonic animals was its unusual migration on SDS polyacrylamide gels.[10] In contrast, N-CAM from older animals does not show this migration, although some embryonic-like N-CAM clearly persists in adults in restricted parts of the brain and in disease. The unusual migration of N-CAM is due to the presence of alpha-2,8-linked polysialic acid.[15] This sugar is extremely rare in higher organisms,[16] enough so that its presence is largely limited to N-CAM. (There have been reports that polysialic acid is also found on sodium channels,[17] so the presence of the sugar is not totally diagnostic for N-CAM). In the embryonic form of N-CAM, the polysialic acid accounts for about 30% by weight of the molecule.[10] In chemical terms, this implies sugar chains consisting of over a hundred monomer units! Although the exact structure of the sugar is not known (for example, exactly how many such chains there are and where precisely they are attached to the protein), experiments with glycosidases suggest three sites of attachment in the extracellular portion of the chicken molecule.[18] In the more "adult" form of N-CAM, there is still extensive polysialic acid, accounting for about 10% of the mass of the molecule.[19] Studies using purified N-CAM from embryos and adults reconstituted into lipid vesicles have shown that the adult, less polysialic acid-rich form of N-CAM is more effective in mediating adhesion.[20] A probable explanation for this is that the polysialic acid forms a large sphere of hydration around the molecule, preventing appropriate polypeptide binding sites from approaching each other efficiently. In fact, the large volume of space that must be occupied by the

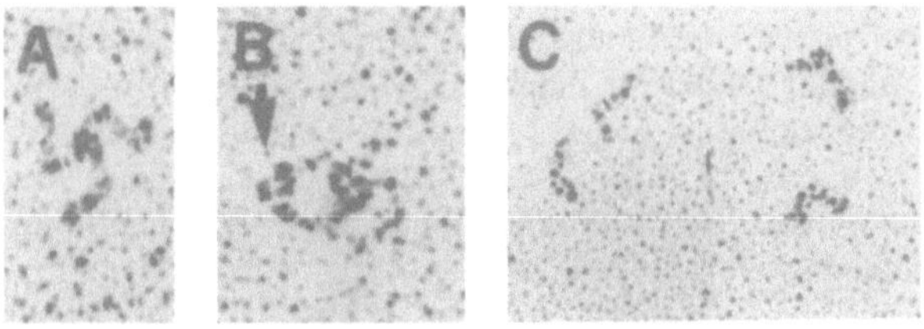

Figure 2 Electron micrographs of N-CAM. Images were selected from pictures of rotary-shadowed embryonic chicken N-CAM.[131] **A** N-CAM, 180 and 140 kD isoform mixture; **B** As A, with added monoclonal antibody to N-CAM (arrowhead); and **C** N-CAM, 120 kD isoforms lacking transmembrane segment. Note the triskelion in A and the lack of multimer formation in C.

polysialic acid of N-CAM can prevent the proper apposition of N-CAM-containing cell membranes[21] and perturb other cell-cell interactions, giving rise to the suggestion that N-CAM can have quite profound effects on cellular communication apart from its direct binding activities.[22,23] In further support for this notion, highly polysialylated N-CAM (which is less adhesive per se) better supports neurite outgrowth.[24] The conversion of the embryonic to the adult form of N-CAM appears to occur through the synthesis of new molecules rather than the removal of polysialic acid residues from embryonic forms and most likely reflects parallel expression of particular glycosyltransferases.[25]

Detergents are required to release the bulk of N-CAM from brain tissue, suggesting that it is an integral membrane protein.[3,26] This supposition is well supported by the molecular biological studies described below. What was not immediately clear, however, was that various N-CAM isoforms can differ in the way they interact with the membrane. Although the larger, 180 and 140 kD isoforms insert directly through the cell membrane to form extracellular and intracellular domains (For example, the phosphorylation of N-CAM occurs intracellularly, and the 180 and 140 kD forms differ intracellularly.), the 120 kD isoform is attached to the cell via a phosphatidylinositol (PI) linkage.[27-29] In particular, the 120 kD form can be released by a PI-specific phospholipase, where the 180 and 140 kD forms cannot. The released fragment is entirely water soluble and does not reassociate with the membrane. This is consistent with earlier observations that N-CAM can exist in multiple soluble forms, albeit in relatively small amounts.[30,31] Such forms could clearly arise through endogenous phospholipase and/or protease activities. Additionally, there is molecular biological evidence that some soluble N-CAM isoforms may arise as direct translation products.[32] Although there is no direct data on the biosynthetic processing of PI-linked N-CAM, other PI-linked cell surface proteins, such as the Thy-1 glycoprotein and alkaline phosphatase, are synthesized initially with a hydrophobic tail that presumably holds them to an intracellular membrane.[33] The polypeptide is then combined through a transamidation with a preformed PI-linked glycoconjugate.

The three major forms of N-CAM can therefore interact with the cell in quite different ways: the smaller, PI-linked isoform does not transverse the lipid bilayer, and consistent with this, shows a relatively rapid diffusion in the plane of the membrane, similar to that observed for cell surface lipids.[34,35] The middle-sized, 140 kD isoform shows a cell-surface mobility similar to other cell surface proteins, suggesting that it is largely free to diffuse in the plane of the membrane. In contrast, the largest, 180 kD form of N-CAM shows a slower lateral diffusion, suggesting that it is interacting with some intracellular component, most likely the cytoskeleton.[34] In fact, there is some evidence that the 180 kD but not the 140 kD form of N-CAM will bind to brain spectrin *in vitro*[36] and is localized to postsynaptic membranes.[37] It is interesting to speculate that different signals could be transmitted to the cell upon the same extracellular binding event through these different methods of anchorage. Even more intriguing are the soluble forms of N-CAM. Could these function as anti-adhesion molecules by nonproductive binding to and blocking of cell surface sites? There is also evidence for N-CAM in the extracellular matrix that is neither strictly soluble nor cell surface associated,[38,39] but the exact molecular nature of this material is as yet unclear. It has also recently been shown biochemically that N-CAM can interact with collagen *in vitro.*[40]

N-CAM has been examined in the electron microscope to investigate how it might actually "look" on the cell surface.[41] Studies performed in the absence of detergents reveal the presence of N-CAM multimers, the most commonly observed of which is the "triskelion", in which three bent arms protrude from a central hub (Figure 2). Using monoclonal antibodies that bind to specific regions of the molecule, it appears that the amino-terminal region involved in cell-cell binding is located at the ends of the bent arms, and the carboxy-terminal region, including the hydrophobic transmembrane segments, comprises the central hubs.[41] The reason for the bend in the middle of the N-CAM is not known, but it could arise from primary structure features combined with the presence of the polysialic acid hydration sphere. Other studies, performed in the *presence* of nonionic detergents, suggest an alternate structure in which the C-terminal regions were at the distal ends of N-CAM dimers and the dimers were interacting via their amino-terminal binding domains.[42]

As mentioned above, binding mediated by purified N-CAM was studied by the aggregation of reconstituted lipid vesicles and the binding of N-CAM immobilized on polystyrene beads. Such studies indicated that one molecule of N-CAM can bind to another N-CAM molecule in what is termed a "homophilic" or like-to-like interaction. The rate of vesicle aggregation is extremely sensitive to the amount of the N-CAM present; for example, a 2-fold increase in N-CAM leads to a 30-fold increase in aggregation.[20] Moreover, the binding is sensitive to the amount of polysialic acid. However, separate studies using soluble, amino-terminal fragments of N-CAM showed that the binding was mediated by residues in this region, and because these fragments are totally devoid of polysialic acid, the interaction was via polypeptide contacts.[9] This is somewhat in contrast to the observations that N-CAM can bind heparin-like molecules in a heterophilic interaction[43] but may be suggesting that, just as there are multiple isoforms of N-CAM, there are multiple binding mechanisms that may operate *in vivo.* To attempt to understand more fully how N-CAM functions at the molecular level, studies were initiated in a number of laboratories, for the most part using different species, to clone and sequence the N-CAM mRNA and gene.

III. MOLECULAR BIOLOGICAL STUDIES OF N-CAM

The antibody tools used to identify and characterize N-CAM at the protein level were used to isolate cDNA clones encoding N-CAM.[44-46] Other N-CAM clones were identified using oligonucleotide probes[47-49] based on known protein sequences.[50] From a multitude of such studies, the major conclusions are (1) N-CAM is a member of the immunoglobulin supergene family,[29] (2) various N-CAM isoforms arise by the alternate splicing of RNA transcribed from a single gene,[51] and (3) other N-CAM isoforms not detectable at the protein level abound and are conserved across species. It has been possible to identify homologues of N-CAM in lower vertebrates, such as frogs,[52,53] and in invertebrates, such as flies and grasshoppers.[54,55] Here I will review the major structural aspects of N-CAM revealed by such studies and the beginnings of attempts to understand the transcriptional control elements of the gene.

One of the first facets of N-CAM revealed by cDNA cloning and sequencing was that it was similar in some ways to immunoglobulins. For example, the sequence of human N-CAM, which is prototypic of all N-CAMs but by no means the first to be isolated, is shown in Figure 3, and the overall structure of the N-CAM polypeptides based on this and other sequences is shown schematically in Figure 4.[32,56] In particular, there are five extracellular repeating sequences, each of which contains a pair of cysteine residues spaced about 50 amino acid residues apart. Also conserved are a limited number of additional amino acid residues flanking the cysteines. This has allowed the placement of N-CAM into what is now referred to as the immunoglobulin gene superfamily,[57] and this type of repeat (called the C2-SET) has subsequently been found in a number of other CAMs. Many of these, like N-CAM, appear to mediate calcium-independent adhesion. Another structural feature of N-CAM is a series of fibronectin Type-III repeats, located between the immunoglobulin domain region and the presumptive hydrophobic membrane-spanning region. Recent NMR[58] and X-ray crystallographic[59] analyses of the fibronectin Type III repeats have suggested that they, too, form structural domains capable of biological activities. In the extracellular region are also found a number of consensus sites for the attachment of N-linked carbohydrates. By analogy with glycosylation in chicken N-CAM,[18] it is probable that the polysialic acid of human N-CAM would be bound to one or all of the sites at residues 410, 425, and 460 in the human.

In addition to the above conserved features of N-CAM, there are a plethora of forms that arise by alternate splicing of RNA. A particularly pronounced aspect of this splicing accounts for the multiple isoforms observed at the protein level (Figure 5). For example, the major difference between the 180 kD and 140 kD polypeptides is the inclusion of a 265 amino block of sequence in the intracellular region of the molecule.[60,61] This block arises by the inclusion of a single large exon (exon 18) in the mRNA encoding the 180 kD form; the 140 KD form lacks this exon. It should be noted as well that the cDNA sequence predicts mol wts much smaller than the approximately 170 and 130 kD observed for the deglycosylated isoforms. This anomalous migration in SDS polyacrylamide gels appears quite common for membrane proteins of this type. In contrast to the transmembrane forms of N-CAM, the PI-linked, 120

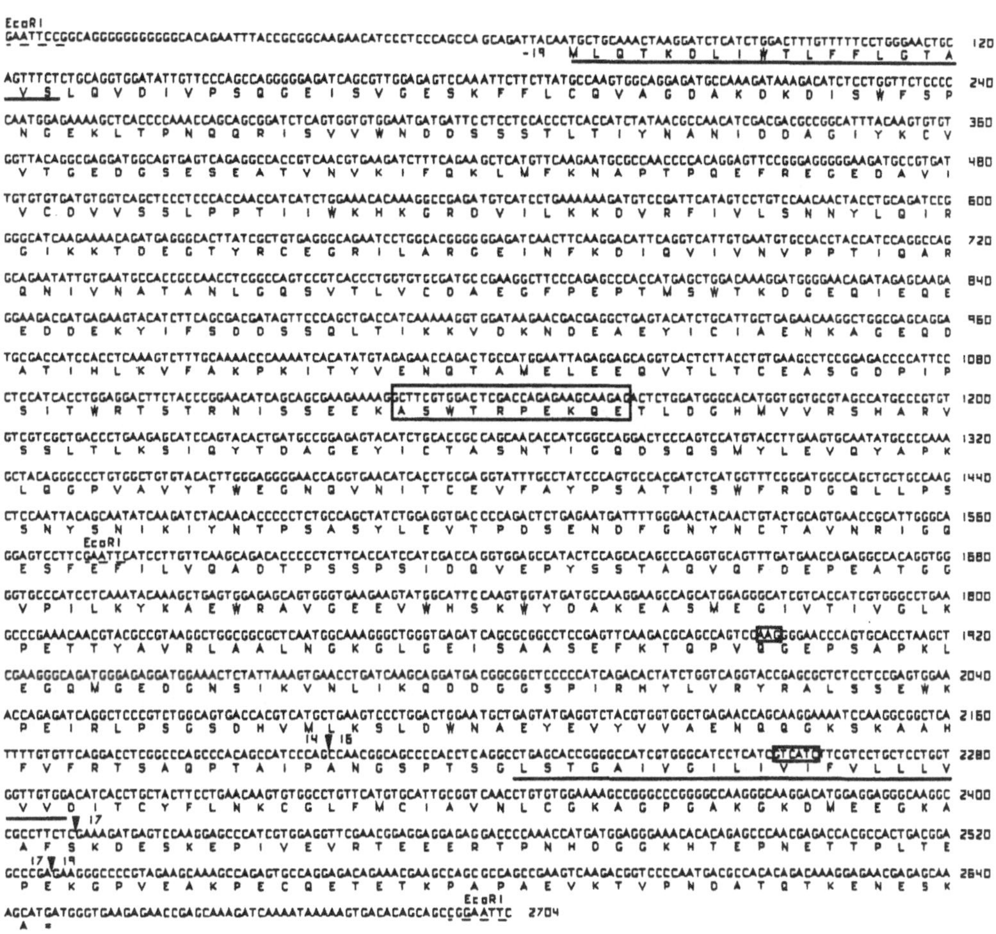

Figure 3 Nucleotide and predicted amino acid sequence of the 140 kD form of human N-CAM. Nucleotides are numbered on the right and amino acids are numbered on the left; the amino terminal leucine of the mature protein is labeled as residue 1. Presumptive signal and transmembrane segments are underlined, and three regions of sequence variance are enclosed in boxes; nucleotides 1130 to 1159 correspond to the VASE exon. Several intron junctions are indicated by triangles and are numbered according to Reference 72. (From Hemperly, J. J., DeGuglielmo, J. K., and Reid, R. A., *J. Mol. Neurosci.*, 2, 71, 1990, with permission.)

kD form arises by the inclusion in its mRNA of an alternative hydrophobic domain, i.e., a sequence of hydrophobic amino acids different from those observed in the 180 and 140 kD isoforms, followed immediately by a stop codon.[29] This would give rise to a membrane-associated molecule with no intracellular domain. Other N-CAM cDNA clones appear to include another block of amino acids that are lacking any hydrophobic amino acid segment and could hence give rise to a soluble form of N-CAM.[32] The factors and sequences controlling the alternate splicing are just now being addressed; for example, the inclusion or exclusion of the exon giving rise to the difference between the 180 and 140 Kd isoforms depends on particular intronic sequences.[62]

In contrast to the relatively large alternative RNA splices that change membrane associations and intracellular sequences, there are also a number of smaller splice sites. One of these, called alternatively the VASE[63] or pi[64] region, leads to the inclusion of a 30 base pair (10 amino acid) sequence between exons 7 and 8 (see, for example, Figure 3). This is in the fourth immunoglobulin-like domain. A number of elegant studies have been undertaken to probe the occurrence and/or physiological significance of this splice. For example, in embryonic rats, the splice is largely absent, but as development proceeds, N-CAM containing the 10 amino acid segment increases to comprise about 40% of brain N-CAM.[65] Similar segments have now been found in chickens, rats, mice, and humans, as well as in frogs. The function of

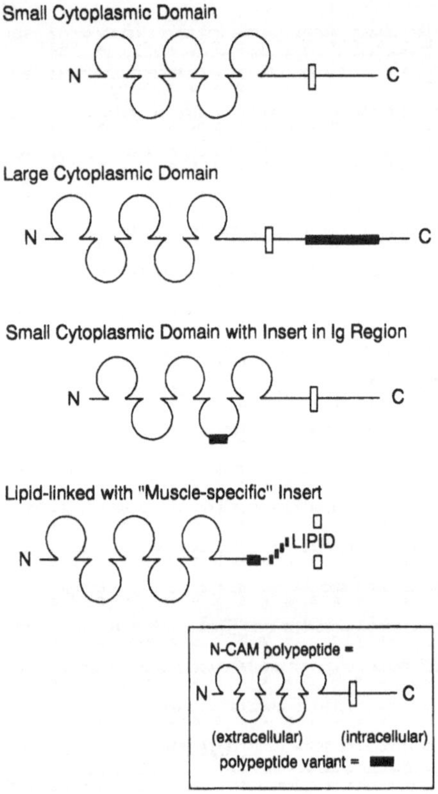

Figure 4 Schematic representations of N-CAM polypeptides. The extracellular immunoglobulin-like domains are indicated by loops, and the amino- and carboxyl-termini are indicated by N and C, respectively. The transmembrane region is represented by a vertical box. Areas of protein variation are indicated by solid boxes. The bottom image represents N-CAM, linked to the surface via a glycan-phosphatidylinositol lipid anchor.

the VASE segment is also unknown, although it has recently been reported that it can affect neurite outgrowth on a N-CAM substratum[66] and that similar mechanisms may modulate neurite outgrowth *in vivo.*[67] In particular, N-CAM containing the VASE exon is a poorer substrate for outgrowth.

An additional region of alternate RNA splicing is at the exon 12–13 junction. This region is in the extracellular domain between the two fibronectin type III repeats. At least four exons can be spliced into this region in various combinations.[68,69] The largest exon, so far only described in humans, would give rise to a soluble form of N-CAM.[32] More commonly observed is a set of four small exons that occur in all combinations. Although additional exons have been looked for using polymerase chain reaction techniques, no others have yet been found.[70] The various patterns of N-CAM splices are shown schematically in Figure 6, but the factors controlling the splicing are as yet unknown. Although the splices are quite

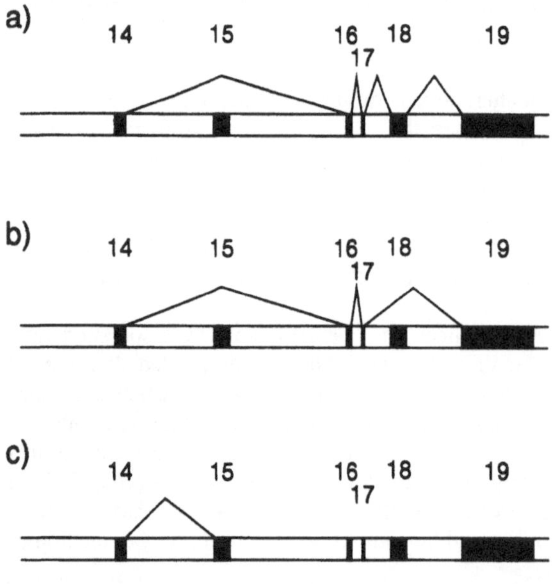

Figure 5 Alternate RNA splicing patterns give rise to the 180, 140, and 120 kD isoforms of N-CAM. (a) Exons 1 to 14 and 16 to 19 are spliced to generate the 180 kD form, (b) the same exons as in (a) except for exon 18 are used to generate the 140 kD form, and (c) alternate splicing of exon 15 gives rise to the 120 kD, lipid-anchored form of the molecule.

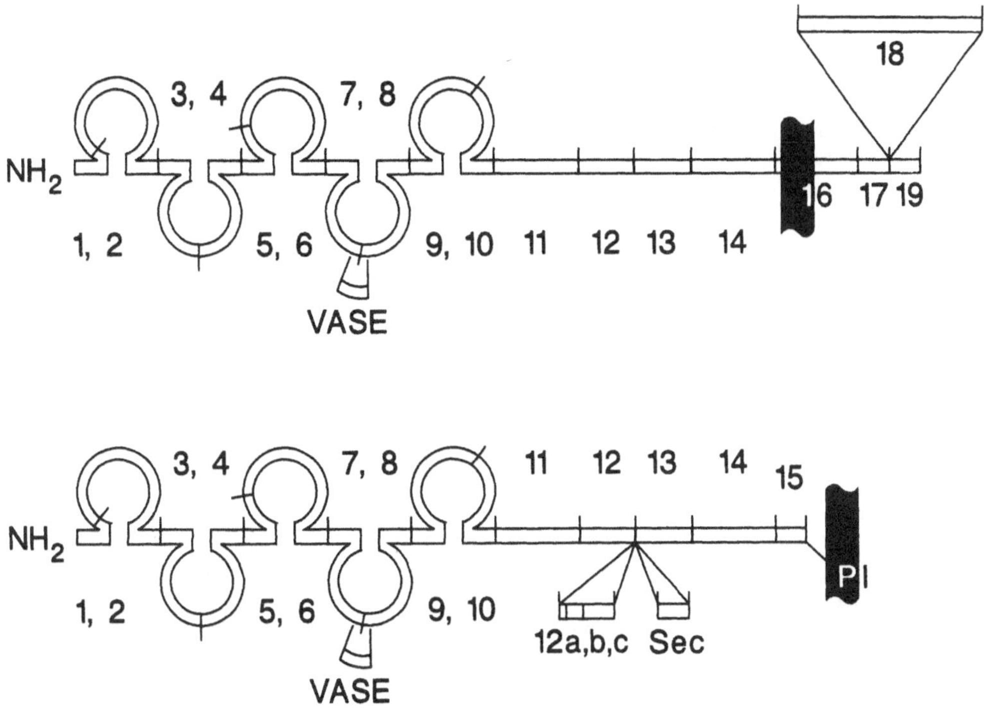

Figure 6 Intron-exon structure and RNA splicing give rise to additional N-CAM isoforms. Each of the five immunoglobulin-like domains are encoded by two exons (exons 1 to 10). The VASE or pi exon occurs between exons 7 and 8 and generates N-CAM with a 10 amino acid insert at this position. Splicing at the exon 12-13 junction can give rise to secreted forms of N-CAM by utilization of the "Sec" exon. Use of exons 12a, 12b, or 12c, in combination with a minor AAG exon in various combinations, generates additional isoforms. As in Figure 5, the use of exon 15 results in a lipid-anchored form of N-CAM, and the inclusion of exon 18 gives rise to the 180 kD large cytoplasmic domain polypeptide.

small, the proteins that arise from them may have special characteristics. For example, in human muscle N-CAM, the inclusion of exons 12a, 12b, and 12c gives rise to a form of N-CAM that has a new O-linked oligosaccharide.[71] This form of N-CAM can be distinguished from other isoforms by its binding to peanut agglutinin, a plant lectin. Another of the exons at the exon 12-13 junction is particularly rich in proline residues and may lead to quite different distances of projection from the cell membrane for the two forms.[41] Again, the biological significance of these splices is not at all clear, and in fact it is possible that some of them are irrelevant artifacts of RNA splicing. However, the conservation of the major splices across all species surveyed almost demands that they subserve important biological functions.

In concert with cDNA studies on chicken N-CAM cDNA,[44] an analysis of the intron/exon structure of the gene was performed.[72] These studies indicated that chicken N-CAM consisted of a least 19 exons arranged over >50 kb of genomic DNA. The current N-CAM exon numbering is based on this study. Subsequent to these studies, the additional splicing events above were discovered, and the current intron/exon structure of N-CAM is shown in Figure 6. N-CAM appears to be encoded by a single gene. In mouse the gene (called *ncam*) is located on chromosome 9,[73] and in humans it is on chromosome 11.[74,75] This same human chromosome includes a number of immunoglobulin superfamily genes. The *ncam* locus is in band 11q23,[76] a band frequently involved in genetic rearrangements. While an intriguing possibility, there is no evidence for *ncam* rearrangement in human disease.

The transcriptional control regions of the human, mouse, and rat N-CAM genes have now been partially characterized.[77-80] As expected, consensus sequences for a number of transcription factors have been found, but there is no classical TATA box. Further studies will be needed to determine the factors involved in the transcriptional control of the *ncam* gene and how they interact to regulate the spatial and temporal expression of the protein. A particularly exciting recent observation is that the N-CAM promoter contains a binding site for homeodomain proteins, in specific *Hox2.4* and *Hox2.5*.[79,81]

IV. BIOLOGICAL STUDIES OF N-CAM

Early studies on the function of the N-CAM molecule took two basic paths: studies on the *in vitro* association of cells and the ability of antibodies to N-CAM to interfere with those associations. As mentioned above, antibodies to N-CAM can prevent the aggregation of neural retina cells. Moreover, in a tissue culture environment, they can perturb histotypic retina formation[82] and *in vivo* can perturb developing projections from retina to tectum during embryogenesis.[83] It was observed early that N-CAM disappears from migrating neural crest cells,[84] and recent results have shown N-CAM to be expressed in the pathways used by luteinizing-hormone-releasing neurons during development.[85] A particularly pronounced effect of anti-N-CAM antibodies is their ability to inhibit fasciculation, that is, the side-to-side bundling of individual neurites. The overall conclusion is that N-CAM functions *in vivo* as part of a highly complex and somewhat redundant biological system involving a multitude of cell-to-cell and cell-to-extracellular matrix adhesion molecules. A particularly striking example of this is in the fly, where the N-CAM protein can be totally eliminated, but the flies are largely normal, except when coupled to another mutation in the Abelson proto-oncogene.[86] This is of course not to say that N-CAM is without function, only that it appears that organisms may be viable without it.

N-CAM, despite its initial localization to the nervous system, has been detected in a number of tissues. It was appreciated quite early that N-CAM, particularly the 140 kD form, is found on muscle and appears to aid in the interaction of nerves with muscle.[87,88] In adult muscle, however, the level of N-CAM is quite low and is confined to the region of the neuromuscular junction. More recently, N-CAM has also been detected in endocrine tissue and tumors,[89-91] including pancreatic non-beta cells.[92] N-CAM has also been detected on thymocytes,[93] and most intriguingly, on cells of the immune system. The antibody anti-Leu19, which is commercially available and is used as a marker for natural killer cells, recognizes N-CAM.[94] Structural analysis of the N-CAM on leukocytes has shown it to be identical to the 140 kD form[95] and that it contains polysialic acid.[96] What has been disappointing so far is the inability to clearly demonstrate the involvement of N-CAM in the immunobiology of leukocytes. This may reflect the overshadowing of N-CAM by other more robust adhesion molecules, such as integrins[97] or cadherins,[98] except in particular situations.[99]

Although it is clear that N-CAM is capable of mediating adhesion by itself, for example in reconstituted vesicles or bound to plastic beads, these is also good evidence that N-CAM in some circumstances may be capable of heterophilic binding to either other molecules, such as heparin-like structures,[43] proteoglycans,[100] collagen,[40] or others.[101] It is also possible that *cis* interactions of N-CAM on the cell surface with other CAMs such as the L1 antigen can augment binding.[102] There is also growing evidence that N-CAM may be acting through the activation of second messenger systems: the binding may be relatively weak, perhaps too weak to generate shear resistant adhesion in all but the most gentle experimental paradigms, yet may be sufficient to generate a transmembrane signal. If this were the case, N-CAM would be similar to the T cell receptor on lymphocytes, where molecular complexes may mediate the actual adhesion, but the specific T cell receptor interaction leads to signal generation.[103] At least two lines of evidence suggest N-CAM acts through the activation of second messenger systems. First it has been observed that the binding of antibodies to N-CAM or fragments of N-CAM itself can lead to changes in calcium fluxes.[104-106] In another paradigm, the outgrowth of neurites in culture on substrata of N-CAM-transfected fibroblasts, it has been shown that calcium channel blockers can interfere with the outgrowth.[107] Alternatively, the addition of drugs can mimic N-CAM effects.[66] If this indeed is the case, it may be necessary to further subdivide the cell adhesion molecules into those that mediate strong, structural adhesion such as the cadherins, those that mediate weaker, but signal-transducing interactions, and those that mediate cell adhesion via a substrate-enzyme interaction, for example, glycosyl transferases.

Another fascinating development is the observation that N-CAM-like molecules can be modulated during learning in the sea snail *Aplysia.*[108] This would be consistent with N-CAM being located in postsynaptic densities where it is subject to proteolytic modification by the protease calpain.[109] In higher vertebrates, there has also been the observation that N-CAM sialylation can change during a passive avoidance response[110] and that infusions of antibodies to N-CAM can modulate the response.[111]

V. N-CAM IN DISEASE

An especially exciting area of N-CAM research has been the recognition of N-CAM as a marker in disease. In addition to the recognition of N-CAM on natural killer leukocytes, it has also been found on multiple myeloma cells,[112,113] and it has been suggested that N-CAM levels may be prognostic of the course of the disease.[114] In particular, it has been suggested that the decrease in tumor cell N-CAM in late stages of the disease leads to increased metastasis and a poorer prognosis. Similarly, N-CAM is found in lymphomas with a propensity to metastasize toward neuronal sites.[115]

N-CAM has long been recognized as a marker in muscle disease.[116] As noted above, in adult muscle, the level of N-CAM is low and confined to regions of the neuromuscular junction. Upon injury to the muscle or to the nerve innervating the muscle, the level of N-CAM expression increases. Afterward, presumably when as much regeneration as possible has occurred, the levels of N-CAM return to their normal, low level.

N-CAM has also been recognized as a oncofetal antigen in a number of systems, exemplified by Wilms' tumor of kidney.[117] Although N-CAM is found in embryonic kidney, it is largely absent from the adult. However, the N-CAM reappears in Wilms' tumor, and in particular, the polysialic acid-rich form of the molecule predominates and may be useful diagnostically.[117]

N-CAM has also been long recognized, although as an uncharacterized epitope, in tumors of neuroectodermal origin, such as neuroblastoma, rhabdomyosarcoma, and Ewing's sarcoma.[118,119] These small round tumors of childhood are often difficult to distinguish, and there is hope that anti-N-CAM reagents may be useful in this regard. The presence of N-CAM in these tumors is also consistent with a prolonged expression of a fetal antigen. A particularly exciting recent observation is the identification of N-CAM as the "cluster 1 antigen" of small cell lung carcinoma,[120] and there are reports that here, too, N-CAM may be a useful marker for tumor typing and for the following of metastasis to bone marrow and other sites.[121,122] Reagents to the polysialic acid of N-CAM may be even more useful. In addition to these diagnostic opportunities in small cell lung carcinoma, there have also been preliminary reports of using antibodies to N-CAM for imaging tumors *in vivo*[123] and to detect[124] or therapeutically purge neuroblastoma cells from autologous bone marrow transplants. Antibodies have also been tried therapeutically *in vivo.*[125]

N-CAM has also been looked at in a number of neurological conditions, such as hydrocephalus,[126] amyotrophic lateral sclerosis,[127] and even in schizophrenia[128] and affective disorders.[129] Levels of N-CAM polysialylation are affected by chronic lead poisoning, and it has been suggested that this could contribute to the pathology.[130] Although it has not yet been possible to show an overwhelmingly clear involvement of N-CAM in neurological disease, it is obvious that this is an area of much promise and excitement.

VI. SUMMARY

N-CAM is one of the best-studied molecules involved in neural cell adhesion. It has become clear that the name, improperly interpreted, can be confusing. N-CAM is not found only in the nervous system but throughout very specific locations in both normal and diseased tissues. N-CAM is not a single molecule, but is rather a family of polypeptides, which differ radically in their membrane and intracellular segments. Presumably this reflects different functions as well.

It has become increasingly clear that N-CAM does not function in isolation. There are more and more cell adhesion molecules being identified, and there is increasing evidence that they interact not only in functionally but perhaps also at the molecular level. It is also becoming evident that adhesion per se is not the whole story behind N-CAM. Recent reports have indicated that subsequent to ligand binding, there are changes in intracellular second messenger systems that also impact on the final biological effect. In summary, we have learned much about the structure of N-CAM and have begun to see dimly how it may function. Much work remains, however, in relating N-CAM to other macromolecules, both inside and outside the cell.

REFERENCES

1. **Edelman, G. M. and Crossin, K. L.,** Cell adhesion molecules — Implications for a molecular histology, *Annu. Rev. Biochem.,* 60, 155, 1991.
2. **Brackenbury, R., Thiery, J. P., Rutishauser, U., and Edelman, G. M.,** Adhesion among neural cells of the chick embryo. I. An immunological assay for molecules involved in cell–cell binding, *J. Biol. Chem.,* 252, 6835, 1977.
3. **Thiery, J. P., Brackenbury, R., Rutishauser, U., and Edelman, G. M.,** Adhesion among neural cells of the chick embryo. II. Purification and characterization of a cell adhesion molecule from neural retina, *J. Biol. Chem.,* 252, 6841, 1977.

4. **Edelman, G. M.,** Cell adhesion molecules, *Science,* 219, 450, 1983.
5. **Rasmussen, S., Ramlau, J., Axelsen, N. H., and Bock, E.,** Purification of the synaptic membrane glycoprotein D2 from rat brain, *Scand. J. Immunol.,* 15, 179, 1982.
6. **Hirn, M., Deagostini-Bazin, H., Gennarini, M. J., Santoni, M. J., He, H. T., Hirsch, M. R., and Goridis, C.,** Structural and functional studies on N-CAM neural cell adhesion molecules, *J. Physiol.,* 80, 247, 1985.
7. **Ellis, L., Wallis, I., Abreu, E., and Pfenninger, K. H.,** Nerve growth cones isolated from fetal rat brain., *J. Cell Biol.,* 101, 1977, 1985.
8. **Hansen, O. C., Nybroe, O., and Bock, E.,** Cell-free synthesis of the D2-Cell adhesion molecule: evidence for three primary translation products, *J. Neurochem.,* 44, 712, 1985.
9. **Cunningham, B. A., Hoffman, S., Rutishauser, U., Hemperly, J. J., and Edelman, G. M.,** Molecular topography of the neural cell adhesion molecule N-CAM: surface orientation and location of sialic acid-rich and binding regions, *Proc. Natl. Acad. Sci. U.S.A.,* 80, 3116, 1983.
10. **Hoffman, S., Sorkin, B. C., White, P. C., Brackenbury, R., Mailhammer, R., Rutishauser, U., Cunningham, B. A., and Edelman, G. M.,** Chemical characterization of a neural cell adhesion molecule purified from embryonic brain membranes, *J. Biol. Chem.,* 257, 7720, 1982.
11. **Sorkin, B. C., Hoffman, S., Edelman, G. M., and Cunningham, B. A.,** Sulfation and phosphorylation of the neural cell adhesion molecule, N-CAM, *Science,* 225, 1476, 1984.
12. **Kruse, J., Mailhammer, R., Wernecke, H., Faissner, A., Sommer, I., Goridis, C., and Schachner, M.,** Neural cell adhesion molecules and myelin-associated glycoprotein share a common carbohydrate moiety recognized by monoclonal antibodies L2 and HNK-1, *Nature,* 311, 153, 1984.
13. **Chou, D. K., Ilyas, A. A., Evans, J. E., Costello, C., Quarles, R. H., and Jungalwala, F. B.,** Structure of sulfated glucuronyl glycolipids in the nervous sytem reacting with HNK-1 antibody and some IgM paraproteins in neuropathy, *J. Biol. Chem.,* 261, 11717, 1986.
14. **Mackie, K., Sorkin, B. C., Nairn, A. C., Greengard, P., Edelman, G. M., and Cunningham B. A.,** Identification of two protein kinases that phosphorylate the neural cell adhesion molecule, N-CAM, *J. Neurosci.,* 9, 1883, 1989.
15. **Finne, J., Finne, U., Deagostini-Bazin, H., and Goridis, C.,** Occurrence of alpha-2,8-linked polysialosyl units in a neural cell adhesion molecule, *Biochem. Biophys. Res. Commun.,* 112, 482, 1983.
16. **Troy, F. A.,** Polysialylation — from bacteria to brains, *Glycobiology,* 2, 5, 1992.
17. **Zuber, C., Lackie, P. M., Catterall, W. A., and Roth, J.,** Polysialic acid is associated with sodium channels and the neural cell adhesion molecule N-CAM in adult rat brain, *J. Biol. Chem.,* 267, 9965, 1992.
18. **Crossin, K. L., Edelman, G. M., and Cunningham, B. A.,** Mapping of three carbohydrate attachment sites in embryonic and adult forms of the neural cell adhesion molecule (N-CAM), *J. Cell Biol.,* 99, 1848, 1984.
19. **Rothbard, J. B., Brackenbury, R., Cunningham, B. A., and Edelman, G. M.,** Differences in the carbohydrate structures of neural cell-adhesion molecules from adult and embryonic chicken brains, *J. Biol. Chem.,* 257, 11064, 1982.
20. **Hoffman, S. and Edelman, G. M.,** Kinetics of homophilic binding by embryonic and adult forms of the neural cell adhesion molecule, *Proc. Natl. Acad. Sci. U.S.A.,* 80, 5762, 1983.
21. **Yang, P. F., Yin, X. H., and Rutishauser, U.,** Intercellular space is affected by the polysialic acid content of NCAM, *J. Cell Biol.,* 116, 1487, 1992.
22. **Edelman, G. M.,** Cell adhesion molecules in the regulation of animal form and tissue pattern, *Annu. Rev. Cell Biol.,* 2, 81, 1986.
23. **Rutishauser, U., Acheson, A., Hall, A. K., Mann, D. M., and Sunshine, J.,** The neural cell adhesion molecule (NCAM) as a regulator of cell-cell interactions, *Science,* 240, 53, 1988.
24. **Doherty, P., Cohen, J., and Walsh, F. S.,** Neurite outgrowth in response to transfected N-CAM changes during development and is modulated by polysialic acid, *Neuron,* 5, 209, 1990.
25. **Friedlander, D. R., Brackenbury, R., and Edelman, G. M.,** Conversion of embryonic form to adult forms of N-CAM in vitro results from de novo synthesis of adult forms, *J. Cell Biol.,* 101, 412, 1985.
26. **Gennarini, G., Rougon, G., Deagostini-Bazin, H., Hirn, M., and Goridis, C.,** Studies on the transmembrane disposition of the neural cell adhesion molecule N-CAM, *Eur. J. Biochem.,* 142, 57, 1984.
27. **He, H. T., Barbet, J., Chaix, J. C., and Goridis, C.,** Phosphatidylinositol is involved in the membrane attachment of NCAM-120, the smallest component of the neural cell adhesion molecule, *EMBO J.,* 5, 2489, 1986.

28. **Sadoul, K., Meyer, A., Low, M. G., and Schachner, M.,** Release of the 120 kDa component of the mouse neural cell adhesion molecule N-CAM from cell surfaces by phosphatidylinositol-specific phospholipase C, *Neurosci. Lett.,* 72, 341, 1986.
29. **Hemperly, J. J., Edelman, G. M., and Cunningham, B. A.,** cDNA clones of the neural cell adhesion molecule (N-CAM) lacking a membrane-spanning region consistent with evidence for membrane attachment via a phosphatidylinositol intermediate., *Proc. Natl. Acad. Sci. U.S.A.,* 83, 9822, 1986.
30. **Nybroe, O., Linnemann, D., and Bock, E.,** Heterogeneity of soluble neural cell adhesion molecule, *J. Neurochem.,* 53, 1372, 1989.
31. **Krog, L., Olsen, M., Dalseg, A. M., Roth, J., and Bock, E.,** Characterization of soluble neural cell adhesion molecule in rat brain, CSF, and plasma, *J. Neurochem.,* 59, 838, 1992.
32. **Gower, H. J., Barton, C. H., Elsom, V. L., Thompson, J., Moore, S. E., Dickson, G., and Walsh, F. S.,** Alternative splicing generates a secreted form of N-CAM in muscle and brain, *Cell,* 55, 955, 1988.
33. **Gerber, L. D., Kodukula, K., and Udenfriend, S.,** Phosphatidylinositol glycan (PI-G) anchored membrane proteins, *J. Biol. Chem.,* 267, 12168, 1992.
34. **Pollerberg, G. E., Schachner, M., and Davoust, J.,** Differentiation state-dependent surface mobilities of two forms of the neural cell adhesion molecule, *Nature,* 324, 462, 1986.
35. **Gall, W. E. and Edelman, G. M.,** *Science* 219, 903, 1981.
36. **Pollerberg, G. E., Burridge, K., Krebs, K. E., Goodman, S. R., and Schachner, M.,** The 180-kD component of the neural cell adhesion molecule N-CAM is involved in cell-cell contacts and cytoskeleton-membrane interactions., *Cell Tissue Res.,* 250, 227, 1987.
37. **Persohn, E., Pollerberg, G. E., and Schachner, M.,** Immuno-electron-microscopic localization of the 180 kD component of the neural cell adhesion molecule N-CAM in postsynaptic membranes, *J. Comp. Neurol.,* 288, 92, 1989.
38. **Rieger, F., Nicolet, M., Pincon-Raymond, M., Murawsky, M., Levi, G., and Edelman, G. M.,** Distribution and role in regeneration of N-CAM in the basel laminae of muscle and Schwann cells, *J. Cell Biol.,* 107, 707, 1988.
39. **Booth, C. M. and Brown, M. C.,** Localization of neural cell adhesion molecule in denervated muscle to both the plasma membrane and extracellular compartments by immuno-electron microscopy, *Neuroscience,* 27, 699, 1988.
40. **Probstmeier, R., Fahrig, T., Spiess, E., and Schachner, M.,** Interactions of the neural cell adhesion molecule and the myelin-associated glycoprotein with collagen type-I — involvement in fibrillogenesis, *J. Cell Biol.,* 116, 1063, 1992.
41. **Becker, J. W., Erickson, H. P., Hoffman, S., Cunningham, B. A., and Edelman, G. M.,** Topology of cell adhesion molecules, *Proc. Natl. Acad. Sci. U.S.A.,* 86, 1088, 1989.
42. **Hall, A. K. and Rutishauser, U.,** Visualization of neural cell adhesion molecule by electron microscopy, *J. Cell Biol.,* 104, 1579, 1987.
43. **Cole, G. J., Lowey, A., and Glaser, L.,** Neuronal cell-cell adhesion depends on interactions of N-CAM with heparin-like molecules, *Nature,* 320, 445, 1986.
44. **Cunningham, B. A., Hemperly, J. J., Murray, B. A., Prediger, E. A., Brackenbury, R., and Edelman, G. M.,** Neural cell adhesion molecule: structure, immunoglobulin-like domains, cell surface modulation, and alternative RNA splicing, *Science,* 236, 799, 1987.
45. **Murray, B. A., Hemperly, J. J., Gallin, W. J., MacGregor, J. S., Edelman, G. M., and Cunningham, B. A.,** Isolation of cDNA clones for the chicken neural cell adhesion molecule (N-CAM), *Proc. Natl. Acad. Sci. U.S.A.,* 81, 5584, 1984.
46. **Goridis, C., Hirn, M., Santoni, M. J., Gennarini, G., Deagostini-Bazin, H., Jordan, B. R., Kiefer, M., and Steinmetz, M.,** Isolation of mouse N-CAM-related cDNA: detection and cloning using monoclonal antibodies, *EMBO J.,* 4, 631, 1985.
47. **Barthels, D., Santoni, M. J., Wille, W., Ruppert, C., Chaix, J. C., Hirsch, M. R., Fontecilla-Camps, J. C., and Goridis, C.,** Isolation and nucleotide sequence of mouse NCAM cDNA that codes for a Mr 79,000 polypeptide without a membrane spanning region, *EMBO J.,* 6, 907, 1987.
48. **Santoni, M. J., Barthels, D., Barbas, J. A., Hirsch, M. R., and Steinmetz, M.,** Analysis of cDNA clones that code for the transmembrane forms of the mouse neural cell adhesion molecule (NCAM) and are generated by alternative RNA splicing., *Nucl. Acids Res.,* 15, 8621, 1987.
49. **Small, S. J., Shull, G. E., Santoni, M. J., and Akeson, R.,** Identification of a cDNA clone that contains the complete coding sequence for a 140-kD rat NCAM polypeptide, *J. Cell Biol.,* 105, 2335, 1987.

50. **Rougon, G. and Marshak, D. R.,** Structural and immunological characterization of the amino-terminal domain of mammalian neural cell adhesion molecules, *J. Biol. Chem.,* 261, 3396, 1986.
51. **Walsh, F. S. and Dickson, G.,** Generation of multiple n-cam polypeptides from a single gene, *Bioessays,* 11, 83, 1989.
52. **Levi, G., Crossin, K. L., and Edelman, G. M.,** Expression sequences and distribution of two primary cell adhesion molecules during embryonic development of Xenopus laevis, *J. Cell Biol.,* 105, 2359, 1987.
53. **Balak, K., Jacobson, M., Sunshine, J., and Rutishauser, U.,** Neural cell adhesion molecule expression in Xenopus embryos, *Dev. Biol.,* 119, 540, 1987.
54. **Harrelson, A. L. and Goodman, C. S.,** Growth core guidance in insects: fasciclin II is a member of the immunoglobulin superfamily, *Science,* 242, 700, 1988.
55. **Seeger, M. A., Haffley, L., and Kaufman, T. C.,** Characterization of amalgam: a member of the immunoglobulin superfamily from Drosophila, *Cell,* 55, 589, 1988.
56. **Hemperly, J. J., DeGuglielmo, J. K., and Reid, R. A.,** Characterization of cDNA clones defining variant forms of human neural cell adhesion molecule N-CAM, *J. Mol. Neurosci.,* 2, 71, 1990.
57. **Williams, A. F.,** A year in the life of the immunoglobulin superfamily, *Immunol. Today,* 8, 298, 1987.
58. **Baron, M., Main, A. L., Driscoll, P. C., Mardon, H. J., Boyd, J., and Campbell, I. D.,** H-1 NMR assignment and secondary structure of the cell adhesion type III module of fibronectin, *Biochemistry,* 31, 2068, 1992.
59. **Leahy, D. J., Hendrickson, W. A., Aukhil, I., and Erickson, H. P.,** Structure of a fibronectin type-III domain from tenascin phased by mad analysis of the selenomethionyl protein, *Science,* 258, 987, 1992.
60. **Murray, B. A., Hemperly, J. J., Prediger, E. A., Edelman, G. M., and Cunningham, B. A.,** Alternatively spliced mRNAs code for different polypeptide chains of the chicken neural cell adhesion molecule (N-CAM), *J. Cell Biol.,* 102, 189, 1986.
61. **Barthels, D., Vopper, G., and Wille, W.,** NCAM-180, the large isoform of the neural cell adhesion molecule of the mouse, is encoded by an alternatively spliced transcript, *Nucl. Acids Res.,* 16, 4217, 1988.
62. **Tacke, R. and Goridis, C.,** Alternative splicing in the neural cell adhesion molecule pre-messenger RNA: regulation of exon 18 skipping depends on the 5′-splice site, *Genes Dev.,* 5, 1416, 1991.
63. **Small, S. J. and Akeson, R.,** Expression of the unique ncam vase exon is independently regulated in distinct tissues during development, *J. Cell Biol.,* 111, 2089, 1990.
64. **Barthels, D., Vopper, G., Boned, A., Cremer, H., and Wille, W.,** High degree of NCAM diversity generated by alternative RNA splicing in brain and muscle, *Eur. J. Neurosci.,* 4, 327, 1992.
65. **Small, S. J., Haines, S. L., and Akeson, R. A.,** Polypeptide variation in an N-CAM extracellular immunoglobulin-like fold is developmentally regulated through alternative splicing, *Neuron,* 1, 1007, 1988.
66. **Doherty, P., Moolenaar, C. E. C. K., Ashton, S. V., Michalides, R. J. A. M., and Walsh, F. S.,** The VASE exon downregulates the neurite growth-promoting activity of NCAM 140, *Nature,* 356, 791, 1992.
67. **Walsh, F. S., Furness, J., Moore, S. E., Ashton, S., and Doherty, P.,** Use of the neural cell adhesion molecule vase exon by neurons is associated with a specific down-regulation of neural cell adhesion molecule-dependent neurite outgrowth in the developing cerebellum and hippocampus, *J. Neurochem.,* 59, 1959, 1992.
68. **Prediger, E. A., Hoffman, S., Edelman, G. M., and Cunningham, B. A.,** Four exons encode a 93-base-pair insert in three neural cell adhesion molecule mRNAs specific for chicken heart and skeletal muscle, *Proc. Natl. Acad. Sci. U.S.A.,* 85, 9616, 1988.
69. **Reyes, A. A., Small, S. J., and Akeson, R.,** At least 27 alternatively spliced forms of the neural cell adhesion molecule mRNA are expressed during rat heart development, *Mol. Cell. Biol.,* 11, 1654, 1991.
70. **Hamshere, M., Dickson, G., and Eperon, I.,** The muscle specific domain of mouse N-CAM — structure and alternative splicing patterns, *Nucl. Acids Res.,* 19, 4709, 1991.
71. **Walsh, F. S., Parekh, R. B., Moore, S. E., Dickson, G., Barton, C. H., Gower, H. J., Dwek, R. A., and Rademacher, T. W.,** Tissue specific O-linked glycosylation of the neural cell adhesion molecule (N-CAM), *Development,* 105, 803, 1989.
72. **Owens, G. C., Edelman, G. M., and Cunningham, B. A.,** Organization of the neural cell adhesion molecule (N-CAM) gene: alternative exon usage as the basis for different membrane-associated domains, *Proc. Natl. Acad. Sci. U.S.A.,* 84, 294, 1987.

73. **D'Eustachio, P., Owens, G. C., Edelman, G. M., and Cunningham, B. A.,** Chromosomal location of the gene encoding the neural cell adhesion molecule (N-CAM) in the mouse, *Proc. Natl. Acad. Sci. U.S.A.*, 82, 7631, 1985.
74. **Nguyen, C., Mattei, M. G., Mattei, J. F., Santoni, M. J., Goridis, C., and Jordan, B. R.,** Localization of the human NCAM gene to band q23 of chromosome 11: the third gene coding for a cell interaction molecule mapped to the distal portion of the long arm of chromosome 11, *J. Cell Biol.*, 102, 711, 1986.
75. **Walsh, F. S., Putt., W., Dickson, J. G., Quinn, C. A., Cox, R. D., Webb, M., Spurr, N., and Goodfellow, P. N.,** Human N-CAM gene: mapping to chromosome 11 by analysis of somatic cell hybrids with mouse and human cDNA probes, *Mol. Brain Res.*, 1, 197, 1986.
76. **Mietus-Snyder, M., Charmley, P., Korf, B., Ladias, J. A. A., Gatti, R. A., and Karathanasis, S. K.,** Genetic linkage of the human apolipoprotein AI-CIII-AIV gene cluster and the neural cell adhesion molecule NCAM gene, *Genomics*, 7, 633, 1990.
77. **Mann, D. A., Barton, C. H., and Walsh, F. S.,** Characterization of a regulatory region with the human neural cell adhesion molecule gene, *Biochem. Soc. Trans.*, 18, 410, 1990.
78. **Chen, A., Reyes, A., and Akeson, R.,** Transcription initiation sites and structural organization of the extreme 5′ region of the rat neural cell adhesion molecule gene, *Mol. Cell. Biol.*, 10, 3314, 1990.
79. **Hirsch, M. R., Gaugler, L., Deagostinibazin, H., Ballycuif, L., and Goridis, C.,** Identification of positive and negative regulatory elements governing cell-type-specific expression of the neural cell adhesion molecule gene, *Mol. Cell. Biol.*, 10, 1959, 1990.
80. **Barton, C. H., Mann, D. A., and Walsh, F. S.,** Characterization of the human N-CAM promoter, *Biochem. J.*, 268, 161, 1990.
81. **Jones, F. S., Prediger, E. A., Bittner, D. A., Derobertis, E. M., and Edelman, G. M.,** Cell adhesion molecules as targets for hox genes — neural cell adhesion molecule promoter activity is modulated by cotransfection with Hox-2.5 and Hox-2.4, *Proc. Natl. Acad. Sci. U.S.A.*, 89, 2086, 1992.
82. **Buskirk, D. R., Thiery, J. P., Rutishauser, U., and Edelman, G. M.,** Antibodies to a neural cell adhesion molecule disrupt histogenesis in cultured chick retinae, *Nature*, 285, 488, 1980.
83. **Fraser, S. F., Carhart, M. S., Murray, B. A., Chuong, C. M., and Edelman, G. M.,** Alterations in the Xenopus retinotectal projection by antibodies to Xenopus N-CAM, *Dev. Biol.*, 129, 217, 1988.
84. **Thiery, J. P., Dubard, J. L., Rutishauser, U., and Edelman, G. M.,** Cell adhesion molecules in early chicken embryogenesis, *Proc. Natl. Acad. Sci. U.S.A.*, 79, 6737, 1982.
85. **Schwanzel-Fukuda, M., Abraham, S., Crossin, K. L., Edelman, G. M., and Pfaff, D. W.,** Immunocytochemical demonstration of neural cell adhesion molecule (NCAM) along the migration route of luteinizing hormone-releasing hormone (LHRH) neurons in mice, *J. Comp. Neurol.*, 321, 1, 1992.
86. **Elkins, T., Zinn, K., McAllister, L., Hoffmann, F. M., and Goodman, C. S.,** Genetic analysis of a Drosophila neural cell adhesion molecule: interaction of fasciclin I and Abelson tyrosine kinase mutations, *Cell*, 60, 565, 1990.
87. **Reiger, F., Grumet, M., and Edelman, G. M.,** N-CAM at the vertebrate neuromuscular junction, *J. Cell Biol.*, 101, 285, 1985.
88. **Covault, J., Merlie, J. P., Goridis, C., and Sanes, J. R.,** Molecular forms of N-CAM and its RNA in developing and denervated skeletal muscle, *J. Cell Biol.*, 102, 731, 1986.
89. **Langley, O. K., Aletsee, M. C., and Gratzl, M.,** Endocrine cells share expression of N-CAM with neurones, *FEBS Lett.*, 220, 108, 1987.
90. **Jin, L., Hemperly, J. J., and Lloyd, R. V.,** Expression of neural cell adhesion molecule in normal and neoplastic human neuroendocrine tissues, *Am. J. Pathol.*, 138, 961, 1991.
91. **Aletsee-Ufrecht, M. C., Langley, K., Gratzl, O., and Gratzl, M.,** Differential expression of the neural cell adhesion molecule NCAM 140 in human pituitary tumors, *FEBS Lett.*, 272, 45, 1990.
92. **Rouiller, D., Cirulli, V., and Halban, P. A.,** Differences in aggregation properties and levels of the neural cell adhesion molecule NCAM between islet cell types, *Exp. Cell Res.*, 191, 305, 1990.
93. **Brunet, J. F., Hirsch, M. R., Naquet, P., Uberla, K., Diamantstein, T., Lipinski, M., and Goridis, C.,** Developmentally regulated expression of the neural cell adhesion molecule (NCAM) by mouse thymocytes, *Eur. J. Immunol.*, 19, 837, 1989.
94. **Lanier, L. L., Testi, R., Bindl, J., and Phillips, J. H.,** Identity of Leu-19 (CD56) leukocyte differentiation antigen and neural cell adhesion molecule, *J. Exp. Med.*, 169, 2233, 1989.
95. **Lanier, L. L., Chang, C. W., Azuma, M., Ruitenberg, J. J., Hemperly, J. J., and Phillips, J. H.,** Molecular and functional analysis of human natural killer cell-associated neural cell adhesion molecule (N-CAM/CD56), *J. Immunol.*, 146, 4421, 1991.

96. **Husmann, M., Pietsch, T., Fleischer, B., Weisgerber, C., and Bitter-Suermann, D.,** Embryonic neural cell adhesion molecules on human natural killer cells, *Eur. J. Immunol.,* 19, 1761, 1989.
97. **Albelda, S. M. and Buck, C. A.,** Integrins and other cell adhesion molecules, *FASEB J.,* 4, 2868, 1990.
98. **Takeichi, M.,** Cadherin cell adhesion receptors as a morphogenetic regulator, *Science,* 251, 1451, 1991.
99. **Poggi, A. and Zocchi, M. R.,** Cultured human thymocytes lacking CD2 and CD11a/CD18 antigens are functional and adhere to endothelial cells via CD56 or CDW49d molecules, *Cell. Immunol.,* 140, 319, 1992.
100. **Cole, G. J. and Burg, M.,** Characterization of a heparan sulfate proteoglycan that copurifies with the neural cell adhesion molecule, *Exp. Cell Res.,* 182, 44, 1989.
101. **Murray, B. A. and Jensen, J. J.,** Evidence for heterophilic adhesion of embryonic retinal cells and neuroblastoma cells to substratum-adsorbed NCAM, *J. Cell Biol.,* 117, 1311, 1992.
102. **Kadmon, G., Kowitz, A., Altevogt, P., and Schachner, M.,** The neural cell adhesion molecule N-CAM enhances L1-dependent cell-cell interactions, *J. Cell Biol.,* 110, 193, 1990.
103. **Williams, A. F. and Beyers, A. D.,** At grip with interactions, *Nature,* 356, 746, 1992.
104. **Schuch, U., Lohse, M. J., and Schachner, M.,** Neural cell adhesion molecule influences second messenger systems, *Neuron,* 3, 1, 1989.
105. **Sontheimer, H., Kettenmann, H., Schachner, M., and Trotter, J.,** The neural cell adhesion molecule (N-CAM) modulates K+ channels in cultured glial precursor cells, *Eur. J. Neurosci.,* 3, 230, 1991.
106. **Atashi, J. R., Klinz, S. G., Ingraham, C. A., Matten, W. T., Schachner, M., and Maness, P. F.,** Neural cell adhesion molecules modulate tyrosine phosphorylation of tubulin in nerve growth cone membranes, *Neuron,* 8, 831, 1992.
107. **Doherty, P., Ashton, S. V., Moore, S. E., and Walsh, F. S.,** Morphoregulatory activities of NCAM and N-cadherin can be accounted for by G protein-dependent activation of L-type and N-type neuronal Ca2+ channels, *Cell,* 67, 21, 1991.
108. **Mayford, M., Barzilai, A., Keller, F., Schacher, S., and Kandel, E. R.,** Modulation of an NCAM-related adhesion molecule with long-term synaptic plasticity in Aplysia, *Science,* 256, 638, 1992.
109. **Sheppard, A., Wu, J., Rutishauser, U., and Lynch, G.,** Proteolytic modification of neural cell adhesion molecule (NCAM) by the intracellular proteinase calpain, *Biochim. Biophys. Acta,* 1076, 156, 1991.
110. **Doyle, E., Nolan, P. M., Bell, R., and Regan, C. M.,** Hippocampal-NCAM 180 transiently increases sialylation during the acquisition and consolidation of a passive avoidance response in the adult rat, *J. Neurosci. Res.,* 31, 513, 1992.
111. **Doyle, E., Nolan, P. M., Bell, R., and Regan, C. M.,** Intraventricular infusions of anti-neural cell adhesion molecules in a discrete posttraining period impair consolidation of a passive avoidance response in the rat, *J. Neurochem.,* 59, 1570, 1992.
112. **Leo, R., Boeker, M., Peest, D., Hein, R., Bartl, R., Gessner, J. E., Selbach, J., and Wackerg-Deicher, H.,** Multiparameter analyses of normal and malignant human plasma cells: CD38-positive, CD56-positive, CD54-positive, CIG-positive is the common phenotype of myeloma cells, *Ann. Hematol.,* 64, 132, 1992.
113. **Van Riet, I., De Waele, M., Remels, L., Lacvob, P., Schots, R., and Van Camp, B.,** Expression of cytoadhesion molecules (CD56, CD54, CD18, and CD29) by multiple myeloma plasma cells, *Br. J. Haematol.,* 79, 421, 1991.
114. **Van Camp, B., Durie, B. G. M., Spier, C., De Waele, M., Van Riet, I., Vela, E., Frutiger, Y., Richter, L., and Grogan, T. M.,** Plasma cells in multiple myeloma express a natural killer cell-associated antigen: CD56 (NHK-1; Leu-19), *Blood,* 2, 377, 1990.
115. **Kern, W. F., Spier, C. M., Hanneman, E. H., Miller, T. P., Matzner, M., and Grogan, T. M.,** Neural cell adhesion molecule-positive peripheral T cell lymphoma — a rare variant with a propensity for unusual sites of involvement, *Blood,* 79, 2432, 1992.
116. **Cashman, N. R., Covault, J., Wollman, R. L., and Sanes, J. R.,** Neural cell adhesion molecule in normal, denervated, and myopathic human muscle, *Ann. Neurol.,* 21, 481, 1987.
117. **Roth, J., Zuber, C., Wagner, P., Blaha, I., Bitter-Suermann, D., and Heitz, P. U.,** Presence of the long chain form of polysialic acid of the neural cell adhesion molecule in Wilms' tumor, *Am. J. Pathol.,* 133, 227, 1988.

118. **Bourne, S. P., Patel, K., Walsh, F., Popham, C. J., Coakham, H. B., and Kemshead, J. T.,** A monoclonal antibody (ERIC-1) raised against retinoblastoma that recognizes the neural cell adhesion molecule (NCAM) expressed on brain and tumours arising from the neuroectoderm, *J. Neuro-Oncol.,* 10, 111, 1991.
119. **Garin-Chesa, P., Fellinger, E. J., Huvos, A. G., Beresford, H. R., Melamed, M. R., Triche, T. J., and Rettig, W. J.,** Immunohistochemical analysis of neural cell adhesion molecules — differential expression in small round cell tumors of childhood and adolescence, *Am. J. Pathol.,* 139, 275, 1991.
120. **Patel, K., Moore, S. E., Dickson, G., Rossell, R. J., Beverley, P. C., Kemshead, J. T., and Walsh, F. S.,** Neural cell adhesion molecule (NCAM) is the antigen recognized by monoclonal antibodies of similar specificity in small-cell lung carcinoma and neuroblastoma, *Int. J. Cancer,* 44, 573, 1989.
121. **Rogers, D. W., Treleaven, J. G., Kemshead, J. T., and Pritchard, J.,** Monoclonal antibodies for detecting bone marrow invasion by neuroblastoma, *J. Clin. Pathol.,* 42, 422, 1989.
122. **Molenaar, W. M., Deleij, L., and Trojanowski, J. Q.,** Neuroectodermal tumors of the peripheral and the central nervous system share neuroendocrine N-CAM-related antigens with small cell lung carcinomas, *Acta Neuropathol.,* 83, 46, 1991.
123. **Kemshead, J. T., Lashford, L. S., Jones, D. H., and Coakham, H. B.,** Diagnosis and therapy of neuroectodermally associated tumours using targeted radiation, *Dev. Neurosci.,* 9, 69, 1987.
124. **Combaret, V., Favrot, M. C., Kremens, B., Philip, I., Bailly, C., Fontaniere, B., Gentilhomme, O., Chauvin, F., Zucker, J. M., Bernard, J. L., and Philip, T.,** Immunological detection of neuroblastoma cells in bone marrow harvested for autologous transplantation, *Br. J. Cancer,* 59, 844, 1989.
125. **Lashford, L., Jones, D., Pritchard, J., Gordon, I., Breatnach, F., and Kemshead, J. T.,** Therapeutic application of radiolabeled monoclonal antibody UJ13A in children with disseminated neuroblastoma, *Natl. Cancer Inst. Monogr.,* 3, 53, 1987.
126. **Sorensen, P. S., Gjerris, F., Ibsen, S., and Bock, E.,** Low cerebrospinal fluid concentration of brain-specific protein D2 in patients with normal pressure hydrocephalus, *J. Neurol. Sci.,* 62, 59, 1983.
127. **Werdelin, L., Gjerris, A., Boysen, G., Fahrenkrug, J., Jorgensen, O. S., and Rehfeld, J. F.,** Neuropeptides and neural cell adhesion molecule (N-CAM) in CSF from patients with ALS, *Acta Neurol. Scand.,* 79, 188, 1989.
128. **Lyons, F., Martin, M. L., Maguire, C., Jackson, A., Regan, C. M., and Shelley, R. K.,** The expression of an N-CAM serum fragment is positively correlated with severity of negative features in type II schizophrenia, *Biol. Psychiatry,* 23, 769, 1988.
129. **Jorgensen, O. S.,** Neural cell adhesion molecule (NCAM) and prealbumin in cerebrospinal fluid from depressed patients, *Acta Psychiatr. Scand. Suppl.,* 345, 29, 1988.
130. **Regan, C. M. and Keegan, K.,** Neuroteratological consequences of chronic low-level lead exposure, *Dev. Pharmacol. Ther.,* 15, 189, 1990.
131. **Edelman, G. M. and Crossin, K. L.,** Cell adhesion molecules in neural morphogenesis, in *Volume Transmission in the Brain: Novel Mechanisms for Neural Transmission,* Fuxe, K. and Agnati, L. F., Eds., Raven Press, New York, 1991, 25.

Chapter 10

The Selectin Family of Adhesion Molecules

Eric Larsen

CONTENTS

I. Introduction 203
II. The Molecular Structure of the Selectins 203
III. Characteristics and Functions of the Selectins 205
A. L-Selectin 205
B. E-Selectin 206
C. P-Selectin 207
IV. Carbohydrate Ligands for Selectin-Mediated Binding 208
A. L-Selectin Ligand 208
B. E-Selectin Ligand 209
C. P-Selectin Ligand 210
V. Structure-Function Relationships 210
VI. Role of Selectins in Health and Disease 211
A. Inflammation/Immunity 211
B. Hemostasis 212
C. Metastasis 213
D. Atherosclerosis 213
VII. Summary 213
References 214

I. INTRODUCTION

Cell-cell and cell-matrix interactions are critical events in several basic physiological processes, including inflammation, hemostasis, and the immune response. These interactions are often mediated by specific membrane glycoprotein receptors on circulating or fixed vascular cells. These adhesion molecules are classified on the basis of structural homology. Well-characterized families of adhesion molecules include the integrins, the immunoglobulin superfamily, the cadherins, and the recently defined selectin family. This chapter will review our current understanding of the structure and function of the selectins. The other families of adhesion molecules are discussed in detail in other chapters of this volume.

The selectins were discovered independently by investigators in different areas of study; therefore, the terminology for these proteins has been confusing. The term "selectin" was proposed, since these proteins mediate *selective* cellular interactions, and there is accumulating evidence that these proteins interact with carbohydrate structures on target cells through their *lectin* domain. The selectins are also known as the LEC-CAMs by many investigators.[1,2] This acronym is based on the three extracellular domains of the molecules: Lectin, EGF, Complement — cell adhesion molecule. The term "selectin" will be used in this review. Recently, several investigators have agreed to a common designation, which is outlined in Table 1.[3] The selectins are prefaced by a capital letter, representing the cell in which each molecule was originally discovered; lymphocyte, endothelial cell, and platelet.

This overview will begin with a description of the molecular structure of the selectins, since it is the structure of these molecules that defines the family. Next, the characteristics and functions of the individual proteins will be discussed. The rapidly progressing field involving the ligands for these adhesion molecules will be presented. Finally, speculation on the role of these interesting proteins in health and disease will conclude this chapter.

II. THE MOLECULAR STRUCTURE OF THE SELECTINS

Adhesion molecules are traditionally categorized on the basis of structural homology. As discussed in detail elsewhere in this volume, the individual members of both the integrin family and the immunoglobulin

0-8493-4559-6/95/$0.00+$.50

Table 1 **Terminology of the selectin family of adhesion molecules**

Term	Alternative Terms
L-selectin	LECAM-1, LAM-1, MEL-14, Leu-8, TQ1, DREG.56, LHR
E-selectin	ELAM-1, LECAM 2
P-selectin	GMP-140, PADGEM, CD62, LECAM 3

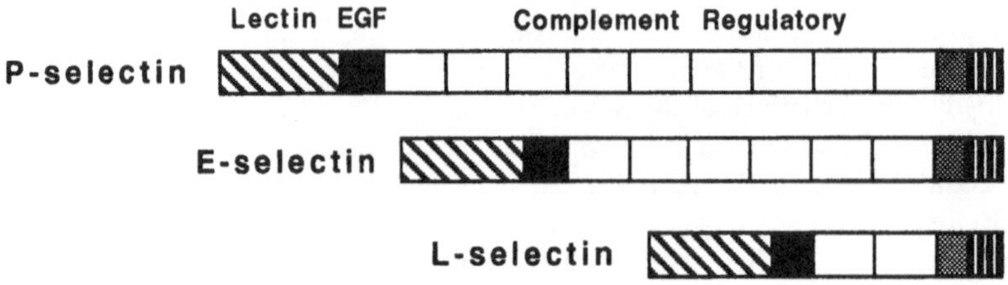

Figure 1 Domain structure of the selectins.

superfamily are characterized by similar molecular structures. In addition to structural homology, the members of these families also share a certain degree of functional homology.

The selectins are defined on the basis of similar structure. Currently, there are three members of the selectin family. The cDNA for these proteins were independently cloned within a short time period, and their sequence homology was immediately recognized. The structure of the selectins based on their predicted amino acid sequences are shown schematically in Figure 1. Following a characteristic signal peptide, there is an NH_2 terminal lectin domain, an epidermal growth factor (EGF) domain, a series of tandem complement regulatory (CR) domains, a putative transmembrane domain, and a short COOH terminal cytoplasmic tail.

The NH_2 terminal lectin domain is comprised of ~120 residues and demonstrates 23 to 30% homology to calcium-dependent animal lectins, including the human IgE receptor, chicken hepatic lectin, and rat mannose-binding protein C.[4-6] There are four invariant cysteine residues within the lectin domain that may be critical for the folding of the native polypeptide. Within the selectin family there is 60% homology between lectin domains.[7-9] The presence of a lectin domain at the NH_2 terminus of these proteins has stimulated the investigation of the potential carbohydrate-binding properties of the selectins.

The second major extracellular domain, the EGF domain contains ~30 residues and displays a high degree of sequence homology with the EGF domains of protein C and coagulation factors IX and X.[7-9] Like the lectin domains, the EGF domains of the selectin family are 60% homologous within the selectin family. The precise role of EGF domains within proteins is not entirely clear, however, there is data that implicates these domains in protein-protein interactions.

The next domain of the selectins consists of a series of short tandem CR repeats. The CR repeats are 40% identical to one another and consist of approximately 62 amino acids. This structure has been observed in a small group of proteins that function to bind complement proteins or to regulate complement activity.[10] As demonstrated in Figure 1, the selectins differ in the number of complement regulatory domains; two in L-selectin, six in E-selectin, and nine in P-selectin.

Following the CR domain is a putative transmembrane domain, including several hydrophobic residues followed by a cytoplasmic domain characterized by highly charged residues, potential phosphorylation sites, and possible targets for posttranslational processing of the proteins. Interestingly, during the cloning of the cDNA for P-selectin, Johnston et al. identified alternatively spliced cDNA clones characterized by deletions of the transmembrane and cytoplasmic domains.[7] Similarly, mRNA transcripts have been identified suggesting the presence of soluble forms of P-selectin. Although soluble forms of L-selectin and E-selectin have been demonstrated at a protein level, there has been no documentation of a similar deletion at the nucleic acid level. These forms may result from proteolysis of the transmembrane forms of the proteins. The function of soluble forms of the selectins will be discussed below.

The human genes for all three selectins are clustered in a small region at bands q21 to 24 on the long arm of chromosome 1.[11] Interestingly, these genes all map to an equivalent region of the mouse chromosome 1, indicating a close association throughout evolution.

III. CHARACTERISTICS AND FUNCTIONS OF THE SELECTINS

The selectins share functional homology as well as structural homology. The adhesive functions of L-selectin and E-selectin were well characterized prior to to the appreciation of their molecular structure. The analogous adhesive function of P-selectin was inferred from its structural homology with these known adhesion proteins. All three selectins are capable of mediating calcium-dependent binding between various normal human vascular cells. Each selectin has a unique cellular distribution and mechanism of expression. These common and distinct properties of the selectins will be discussed below.

A. L-SELECTIN

L-selectin is expressed on the surface of circulating B-lymphocytes, neutrophils, and monocytes. It functions as an adhesion molecule mediating the binding of these circulating blood cells with specialized structures on the endothelial surface.

This protein was initially discovered through the study of lymphocyte recirculation in murine models. Lymphocytes normally traffic from the circulation to various lymphoid organs. An integral component of this process is the adherence of the lymphocyte to specialized cuboidal endothelium, referred to as "high endothelial venules" (HEVs). Different lymphoid organs have unique HEVs, which allow for certain lymphocytes to be targeted to these areas.[12] This process of lymphocytes binding to HEVs of lymphoid organs can be studied *in vitro* by a method developed by Stamper and Woodruff.[13] In this technique, thinly cut frozen cross-sections of lymphoid organs are bathed with lymphocyte populations, and the extent and pattern of cell adherence is analyzed microscopically. To study the critical structure on the lymphocyte that is involved in this binding, a murine monoclonal antibody, Mel 14, was generated that blocked the binding of lymphocytes to peripheral lymph node HEVs.[14] The Mel 14 antibody is directed against a lymphocyte surface structure that has specificity for the peripheral lymph node. Mel 14 antibody does not inhibit the binding of lymphocytes to HEVs from Peyer's patches. The antigen for this antibody, the Mel 14 antigen, is termed gp $90^{\mathrm{Mel\ 14}}$, or the peripheral lymph node homing receptor. The corresponding human structure, originally termed "LAM-1", is identified by the antibodies Leu 8 and TQ/1.[15] The murine (MEL-14) and human (LAM-1) homologues of L-selectin are very similar in both structure and function. There is 77% homology at the amino acid level.[16] Within the lectin domains of these molecules there is 83% homology.

L-selectin, immunoprecipitated from human lymphocytes, has a M_r of 74 kD, whereas, the molecule precipitated from neutrophils has a M_r of 90 kD.[15,17,18] This mol wt difference raises the possibility that there may be fundamental differences in the protein produced by these two cells. However, cDNAs isolated from neutrophils and lymphocytes after PCR have identical sequences. Furthermore, Northern blot analysis reveals that neutrophils and lymphocytes express the same size mRNA.[18] Therefore, it is most likely that the differences between neutrophil L-selectin and lymphocyte L-selectin are due to differences in posttranslational modification.

The mature protein begins with a tryptophan, an unusual residue for the NH_2 terminus of a protein. There are ten potential asparagine-linked glycosylation sites. Glycosylation appears to be limited to N-linked with no evidence of O-linked glycosylation at the protein level, and there are no clusters of serine or threonine in the primary sequence, which is characteristic of O-linked glycosylation. Further, there are 22 cysteine residues, which are concentrated in the pretransmembrane region of the molecule.

The cDNA encoding L-selectin was initially cloned by screening a mouse spleen cDNA library with an oligonucleotide based on NH_2 terminal sequence data.[9] As outlined above, the amino acid sequence derived from the cDNA clone demonstrated the typical NH_2 terminal lectin domain, followed by a short EGF domain, and two complement regulatory type domains prior to the transmembrane domain (see Figure 1).

A series of investigations demonstrated that L-selectin-dependent binding of lymphocytes to peripheral lymph node HEVs is effectively inhibited by high concentrations of certain monomeric sugars, mannose-6-phosphate and fucose-1-phosphate, and by relatively lower concentrations of polymers of charged sugars, fucoidin and polyphosphomannan ester (PPME). In addition, the pretreatment of endothelium with neuraminidase completely abolishes lymphocyte binding, suggesting the involvement of sialic acid in this binding. Further experiments demonstrated that PPME-coated beads bind to lymphocytes and that the Mel 14 antibody effectively inhibits this binding. These observations of carbohydrate-protein interactions with L-selectin provided a foundation for the subsequent analysis of the ligand for this adhesion molecule.

Recent data suggest that changes in the affinity of the L-selectin molecule result from cellular activation. Stimulation of T lymphocytes through cross-linking of the T cell receptor complex or activation of neutrophils with specific cytokines (G-CSF, GM-CSF, TNF-α) leads to enhanced binding to PPME without a concomitant increase in L-selectin expression.[19] This cytokine-mediated enhanced affinity of the L-selectin molecule is transient and returns to baseline within 15 min at 37°C. It is postulated that cell activation induces a conformational change in the molecule, which results in increased affinity. This effect may play a central role in the control of leukocyte migration into areas of inflammation.

In addition to functioning as a homing receptor for circulating lymphocytes, L-selectin is also expressed by neutrophils and monocytes, cells which do not ordinarily recirculate through lymphoid organs.[20] L-selectin promotes the adhesion of neutrophils to inflamed endothelium. Neutrophil-mediated inflammation in certain animal models is inhibited by the Mel 14 antibody and by removing the L-selectin molecule from the leukocyte surface.[21,22] Intravenous administration of an L-selectin-IgG chimeric molecule results in significant inhibition of neutrophil influx in a murine peritonitis model.[23] Taken together, these *in vivo* experiments are consistent with an important role of L-selectin in the homing of circulating neutrophils to inflamed endothelial surfaces.

A soluble form of L-selectin has been detected *in vivo* and can be constructed recombinantly. The recombinant form is capable of inhibiting lymphocyte binding to HEV of peripheral lymph nodes and also binds directly to these structures in a specific manner. There is evidence that L-selectin is shed from the leukocyte surface, presumably by proteolysis.[15,20] This process may act to down-regulate further intercellular adhesion as a negative feedback mechanism.

B. E-SELECTIN

Over the past two decades, there has been a growing appreciation of the endothelium as more than a simple conduit for the flow of blood. There is an emerging perspective of endothelium as a dynamic organ, which actively participates in coagulation, inflammation, and immunity.[24] Vascular injury or the stimulation of the endothelium by certain mediators leads to modulation of constitutive endothelial functions and may lead to the induction of new molecules or functions. E-selectin, also known as ELAM-1, and LECAM 2 is expressed on the surface of cytokine-stimulated endothelium and mediate the binding of neutrophils.

Pretreatment of human umbilical vein endothelial cells *in vitro* with IL-1 or TNF markedly increases the adherence of neutrophils, monocytes, HL60, and U937 cells.[25,26] This enhanced adherence is due to a specific change in the endothelial cell and is blocked by the inhibition of RNA or protein synthesis. The time sequence of these changes reveals maximal effect 4 to 6 h from stimulation, with a return to basal levels at 24 h. To identify potential novel adhesion molecules on the surface of cytokine-treated endothelial cells, monoclonal antibodies have been generated. E-selectin was originally identified through the development of monoclonal antibodies in this manner. Standard hybridoma technology was applied using IL-1-treated endothelial cells as the immunogen. The resultant hybridomas were screened for selective binding to cytokine-treated endothelial cells with minimal binding to resting cells. Using this method, two monoclonal antibodies were produced, H4/18 and H18/7.[27,28] One of these, H18/7, effectively inhibits the adhesion of human neutrophils and HL60 cells to stimulated endothelial cells, thus leading to the original designation of endothelial leukocyte adhesion molecule-1 (ELAM-1).[27] These antibodies identify a protein unique to stimulated endothelium of mol wt 115 kD.

The cDNA for E-selectin was initially cloned by Bevilacqua et al. in 1989.[8] The strategy involved the construction of a cDNA library using RNA from IL-1-treated human endothelial cells, transfection of the library into COS cells, and then screening the cells expressing the E-selectin epitope with the antibodies H18/7 and H4/18. In this manner, a full length cDNA for E-selectin was isolated that supports the adhesion of neutrophils and HL60 cells. The rapid expression of the protein and high turnover rate can be correlated with the time sequence of mRNA transcripts. Biosynthesis studies reveal that the protein undergoes posttranslational modification, including N-glycosylation. Analysis of the predicted amino acid sequence based upon the nucleotide sequence reveals a mosaic structure characteristic of the selectin family (see Figure 1). The E-selectin has an N-terminal lectin domain, followed by an EGF domain, and six tandem complement regulatory domains.

Hession et al. have also isolated the cDNA for E-selectin using an alternative approach of direct expression cloning.[29] A subtraction cDNA library from IL-1-induced endothelial cells is transfected into COS-7 cells, and transfected cells are screened for adhesion to HL60 cells. This method avoids the requirement for antibody or ligand and enables investigators to embark on initial functional studies of the protein. Using this technique, Hession et al. demonstrated that E-selectin-expressing COS cells bind to

HL60 cells and human neutrophils in a calcium-dependent but temperature-independent manner. Since the N-terminal domain of this protein has homology to lectin domains of other molecules, attempts to inhibit binding with carbohydrate structures were attempted. Several soluble carbohydrates, including mannose 6-phosphate, fucoidan, galactose, galactose 6-phosphate, fucose, mannose, and N-acetylglucosamine, as well as carbohydrates coupled to albumin, including HSA-fucose, HSA-mannose, and HSA-galactose, were screened and had no inhibitory activity in this system.

The underlying molecular mechanism of E-selectin expression is an area of active investigation. Sequence analysis for the 5′ flanking region of the E-selectin gene demonstrates consensus DNA binding sequences for two known transcriptional factors, AP-1 and NF-κB.[30-32] Stimulation of human umbilical artery endothelial cells with cytokines known to induce E-selectin expression leads to the activation of NF-κB and thus increased gene transcription. This activation of NF-κB is independent of protein kinase C. There are currently no available data to suggest a direct role for the AP-1 transcription factor in the regulation of E-selectin expression.

Taken together, *in vitro* data demonstrate that E-selectin is a receptor on cytokine-stimulated endothelial cells, which mediates the binding of neutrophils. *In vivo,* E-selectin is not expressed in the microvascular endothelium of normal tissues but is expressed on the surface of endothelium at sites of inflammation.[33] This can be demonstrated by standard immunohistochemical techniques using anti-E-selectin antibodies to probe human tissue sections. This expression is most prominent in the postcapillary venules, a critical area of leukocyte extravasation during the inflammatory process.

C. P-SELECTIN

P-selectin is expressed on the surface of activated platelets and endothelium, where it mediates the binding of these cells to monocytes and neutrophils. P-selectin was originally discovered by the development of monoclonal antibodies that selectively recognized activated but not resting platelets.[34,35] In contrast to L-selectin and E-selectin, which were discovered through the generation of antibodies that blocked a specific intercellar interaction, P-selectin was initially determined to be a novel platelet activation antigen without a known function. Through subsequent cloning and structural analysis, the adhesive function of this protein was appreciated.

P-selectin is constituitively synthesized in resting, noninflamed tissues, including megakaryocytes, platelets, and endothelial cells. P-selectin is a transmembrane protein of the α-granule in resting platelets and upon cell activation undergoes translocation to the plasma membrane surface.[36] During this process, the membrane of the α-granule fuses with and becomes incorporated into the plasma membrane. Agonists that stimulate the rapid expression of this protein on the platelet surface include thrombin, ADP, histamine, collagen, and epinephrine. P-selectin in endothelial cells has a similar pattern of expression. In resting endothelium, the molecule is a membrane protein of the Weibel-Palade body, a storage granule for von Willebrand's factor. Upon activation with a variety of agonists, including thrombin, histamine, calcium ionophores, complement C5b-9 complex, or peroxides, P-selectin is redistributed to the endothelial plasma membrane.[37-40] The expression of P-selectin upon cell activation in the platelet and the endothelial cell occurs within seconds. This rapid redistribution of this molecule from internal stores in the platelet and the endothelium to the plasma membrane suggests that it may play an important role in the inflammatory or hemostatic response to vascular injury. In the endothelial cell, surface P-selectin levels return to baseline within 60 min of cell stimulation. The exact mechanism of this decreased expression is unclear, but there are limited data to implicate endocytosis.[40] There are currently no data to suggest that platelet P-selectin undergoes endocytosis.

P-selectin can be purified from human platelets through serial combination of membrane fractionation, glycoprotein purification, and affinity chromatography.[34,35,41] The purified protein has a M_r of 140kD. Amino acid analysis of purified P-selectin reveals the protein to be rich in cysteine, proline, and tryptophan residues.[42] The protein is highly glycosylated, containing 29% carbohydrate by weight consisting exclusively of N-linked oligosaccharides.[42]

The cloning of the cDNA for P-selectin by Johnston et al. provided valuable insight into the function of this protein.[7] This cloning was achieved by screening a λgt11 human endothelial cell cDNA library with oligonucleotides based on NH_2 terminal sequence data. The amino acid sequence predicted from the cDNA clone revealed the mosaic structure characteristic of the selectin family.

Interestingly, transcripts have been identified that lack the 40 residue transmembrane domain, raising the possibility of a soluble form of the molecule.[7] Although there is evidence at the RNA level for a soluble form of P-selectin, it is yet to be clearly identified and characterized at the protein level. Upon activation, platelets undergo microvesiculation, and Sims et al. have shown clearly that platelet microvesicles

Table 2 **Characteristics of the selectins**

	L-selectin	E-selectin	P-selectin
Tissue	Lymphocyte Neutrophil Monocyte	Endothelium	Platelet Endothelium
Expression	Constitutive	Stimulated 4–6 h	Stimulated Seconds
Target Cells	Endothelium	Neutrophil	Neutrophil Monocyte

are enriched for P-selectin.[43] It may be difficult to distinguish clearly the native form of this molecule inserted into a microvesicle from a truly soluble form of the protein. As outlined above, soluble forms of L-selectin and E-selectin have been described, setting a precedent for a soluble form of P-selectin.

P-selectin is a receptor for neutrophils and monocytes. Activated but not resting platelets bind to human neutrophils, monocytes, and human leukemia cell lines HL60 and U937.[44-46] This binding is calcium-dependent and is selectively inhibited by anti-P-selectin antibodies or by purified P-selectin.[44,45] P-selectin incorporated into phospholipid vesicles binds specifically and in a saturable manner to neutrophils.[44] In addition, radiolabeled P-selectin binds specifically and reversibly to neutrophils and monocytes.[47] Analogous studies demonstrate that endothelial P-selectin serves an identical function as a receptor for neutrophils and monocytes. Neutrophils bind to stimulated endothelial cells, and this binding can be inhibited effectively with anti-P-selectin antibodies.[48]

In summary, the three members of the selectin family share function as well as structural homology. These proteins all function as adhesion receptors mediating the interaction between various vascular cells. It is likely yet not proven at this time that the ligands for the selectins share certain structural components. The selectins are individually unique in their cellular distribution and in their pattern of expression. The important similarities and differences between these molecules are summarized in Table 2.

IV. CARBOHYDRATE LIGANDS FOR SELECTIN-MEDIATED BINDING

The determination that selectins function as adhesion molecules on various vascular cells has stimulated the search for the natural ligand on target cells. Several observations suggest that the ligands for the selectins involve a carbohydrate structure. The presence of a lectin domain at the NH_2 terminus of the protein places it in an optimal position for access to carbohydrate structures on leukocytes. Lectin-carbohydrate interactions are often calcium-dependent as are selectin-mediated interactions. Selectin-mediated binding, particularly L-selectin binding, can be inhibited by various soluble carbohydrates. Furthermore, sialidase treatment of target cells abolishes selectin-mediated binding. These observations generate the hypothesis that the selectin receptors interact with specific carbohydrate structures on target cells.

There are numerous examples of lectin-carbohydrate interactions in cell-cell adhesion throughout nature, across a broad phylogenetic spectrum.[49-51] The extensive diversity of carbohydrate structure confers a high level of specificity for cell-cell interactions. Relative to peptides and nucleotides, carbohydrates have an enormous capacity to encode biological information. Whereas 4 different amino acids can form 24 different peptides, 4 different monosaccharides can form 35,560 distinct tetrasaccharides.[50] This diversity presents the intriguing possibility that minor changes in molecular structure may effect major changes in binding and function. For these reasons, the lectin-carbohydrate interface is a particularly intriguing aspect of the selectins.

A. L-SELECTIN LIGAND

L-selectin-mediated binding of lymphocytes to HEVs can be effectively inhibited by certain soluble phosphorylated carbohydrates, as discussed above. Two monosaccharides, mannose-6-phosphate and fucose-1-phosphate, selectively inhibit the attachment of lymphocytes to HEV in both the mouse and the human.[52,53] In addition, the related polysaccharides, fucoidin and PPME are even more potent inhibitors of lymphocyte binding. In addition to blocking experiments, direct binding of L-selectin to a carbohydrate structure has been clearly demonstrated. Immunopurified L-selectin, immobilized on plastic, binds PPME in a saturable manner.[54] This interaction can be inhibited by mannose-6-phosphate, fucose-1-phosphate,

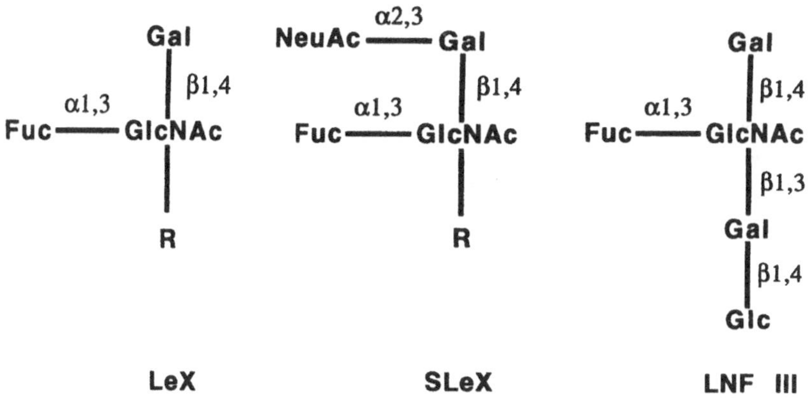

Figure 2 Carbohydrate ligands for the selectins.

sulfated glycolipid, sulfatide, and fucoidin. The importance of carbohydrates on target tissue is further supported by the observation that treatment of HEVs with neuraminidase abolishes lymphocyte adhesion.[55] These data taken together suggest that certain carbohydrate structures and a terminal sialic acid residue are important components of the ligand for L-selectin.

Investigators have focused on the lectin domain of the protein as the potential binding domain of L-selectin, since carbohydrate structures are implicated in binding. There are currently two lines of evidence that suggest that the lectin domain of L-selectin is critical in intercellular binding. First, antibodies directed against the lectin domain inhibit binding to L-selectin. Second, molecular deletion or substitution of the lectin domain results in loss of characteristic L-selectin binding. The original Mel 14 antibody, which blocks L-selectin binding, has been mapped to the lectin domain of the molecule.[56] Using chimeric molecules, Bowen et al. have shown that the Mel 14 antibody binds to the NH_2-terminal 53 amino acid residues.[56] Three additional monoclonal antibodies that inhibit the binding of lymphocytes to HEV have been developed. These antibodies, LAM1-1, LAM1-3, and LAM1-6, seem to recognize three independent epitopes, all within the lectin domain of L-selectin.[57] Within the limitation of this methodology, these data suggest that the lectin domain of the L-selectin molecule is involved in the binding of L-selectin to carbohydrate ligands. Kansas et al. have provided further evidence to implicate the lectin domain of L-selectin in carbohydrate binding. Using chimeric proteins with P-selectin domains substituted into the L-selectin molecule, the unique binding of L-selectin to PPME and fucoidan was shown to be dependent only on the lectin domain of the L-selectin protein.[57]

Streeter et al. employed traditional hybridoma technology to further identify the L-selectin ligand. Antibodies directed against HEV, which selectively blocked the adhesion of lymphocytes *in vitro,* were developed.[58] The MECA 79 antibody was identified and characterized by this method. The antigen recognized by the MECA 79 antibody has been termed the "peripheral lymph node vascular addressin (PNAd)." The PNAd is expressed only in peripheral lymph node HEV and is not observed in mucosal lymph node HEV. PNAd had been purified by immunological methods and is molecularly distinct from the mucosal lymph node vascular addressin (MAd).[59] Using a soluble recombinant form of L-selectin, Imai et al. have identified a ~50kD sulfated, fucosylated, and sialylated glycoprotein, Sgp^{50}, which is an endothelial ligand for L-selectin.[60] Several biochemical features of Sgp^{50}, such as calcium dependence, precipitation by MECA-79, and inhibition by specific carbohydrates, neuraminidase, and Mel 14, suggest that this protein is the ligand for L-selectin on the endothelial surface. A related higher mol wt glycoprotein, Sgp^{90} has also been identified, which may also be involved in L-selectin binding.[60]

B. E-SELECTIN LIGAND

In a rapid series of publications in the latter part of 1990, several laboratories reported the importance of a carbohydrate structure, sialylated Lewis X antigen (SLe^X), in E-selectin binding. SLe^X and the desialylated form, Le^X, are expressed on neutrophils, monocytes, and certain cell lines in the form of glycoproteins and glycolipids. The structures of these and related carbohydrates are illustrated in Figure 2. There is a direct correlation between the degree of E-selectin-dependent binding and the expression of SLe^X on cell lines and mutants.[61-63] In addition, antibodies directed against SLe^X inhibit the binding of neutrophils to stimulated endothelium or to purified E-selectin.[61,62] These experiments demonstrate that anti-SLe^X antibodies are potent inhibitors of E-selectin binding, while anti-Le^X antibodies have minimal inhibition and control antibodies have no effect. In addition to antibody inhibition, there have been successful attempts to use the antigen to block binding. Two laboratories have demonstrated that SLe^x-containing glycolipids, when incorporated into liposomes, effectively inhibit E-selectin-mediated binding.[62,64] Therefore, observations from several sources clearly document that anti-SLe^X and SLe^X are effective inhibitors of E-selectin binding. Additional evidence that this carbohydrate structure may be critical in this interaction is evident through molecular techniques.

The structure of SLe^X is characterized by an α2,3 sialic acid and an α1,3 fucose linked to a core N-acetyllactosamine sequence (see Figure 2). These critical sugar moieties are sequentially added to the carbohydrate core through specific enzymes, α2,3 sialyltansferase and α1,3 fucosyltransferases. Transfection of a specific human fucosyltransferase cDNA into nonbinding cell lines confers E-selectin-dependent adhesion.[63,65] The acquisition of binding in these cells correlates with the surface expression of SLe^X.[63]

C. P-SELECTIN LIGAND

There is growing evidence that SLe^X is also a critical component of the P-selectin ligand. Anti-CD15 antibodies and the CD15 antigen inhibit P-selectin-mediated binding.[66] The CD15 antigen, known as lacto-N-fucopentaose III (LNF III), is expressed on various glycolipids, glycoproteins, and proteoglycans. The core component of LNF III is Le^X, as shown in Figure 2. The observation that neuraminidase treatment of CD15 positive cells strongly inhibits P-selectin-mediated adhesion suggests that a terminal sialic acid may also be a critical component of the ligand.[47,67] Although anti-CD15 antibodies typically have higher affinity for the desialylated forms of the antigen, they still bind to sialylated CD15, which likely explains the inhibitory effect of these antibodies. The effect of the terminal sialic acid has been examined in detail by Polley et al.[64] Soluble SLe^X is a more potent inhibitor of the adhesion of activated platelets to neutrophils than is Le^X. In these experiments, 50% inhibition was achieved with SLe^X at 2 mg/ml compared to Le^X at 54 mg/ml. These findings have led to the concept of SLe^x as the high affinity ligand for P-selectin.

Although the ligands for P-selectin and E-selectin appear to be structurally similar, recent data have determined important differences in these structures.[68] Glycosidase treatment of HL60 cells abolishes both P-selectin- and E-selectin-dependent binding, whereas protease treatment only affects P-selectin binding. These data suggest that P-selectin binding may depend on an additional protein structure within the ligand, not necessary for E-selectin binding, to confer specificity. In addition, the *Sambucus nigra* lectin, specific for the sialyl-2-6 β-Gal linkage, inhibits P-selectin but not E-selectin binding. Taken together, these data suggest that a protein component and a sialyl-2,6 β-Gal structure of the P-selectin ligand may contribute to specificity to P-selectin.

V. STRUCTURE-FUNCTION RELATIONSHIPS

The molecular structure of the selectins incorporates domains from seemingly unrelated proteins. The serial combination of lectin, EGF, and CR domains is unique to this family of adhesion molecules. The function of these three extracellular domains and the potential for cooperativity between the domains is an area of active investigation.

The investigation of the function of the lectin domain is driven by accumulating evidence that selectins bind to carbohydrate-containing structures on target cells. Although certain components of the selectin ligands have been identified, the entire molecular species that is involved in binding is unclear at this time. It has been presumed that binding to carbohydrates occurs via the lectin domain of the molecules. The Mel 14 antibody, which inhibits L-selectin-mediated cell binding, can be mapped to the N-terminal 53 amino acids of the lectin domain.[56] Further evidence that the lectin domain is directly involved in adhesion was demonstrated by the development of a panel of monoclonal antibodies directed against E-selectin.[69] Seven of the eight antibodies generated map to the lectin domain, with one mapping to the CR domain. One of the antibodies, BBIB-E2, directed against an epitope in the lectin domain of E-selectin, inhibits the binding of U937 cells to L-selectin-expressing COS cells and to human umbilical vein endothelial cells.[69] In studying L-selectin, Kansas et al. have demonstrated that three monoclonal antibodies directed against the lectin domain are inhibitory of lymphocyte binding to HEV.[57]

In addition to mapping inhibitory antibodies to the lectin domain, mutagenesis experiments lend support to the hypothesis that the lectin domain is the selectin-binding domain. COS cells expressing E-selectin mutants lacking the lectin domain demonstrate no adhesion to U937 cells.[69] Further, as discussed above, chimeras of P-selectin domains substituted into L-selectin fail to bind the L-selectin ligands PPME and fucoidan.[57]

There are similar data to suggest the importance of the EGF domain in selectin-mediated adhesion. Although few antibodies directed against the EGF domains of the selectins exist, one has been shown to be inhibitory. A monoclonal antibody, Ly-22, which recognizes an epitope within the EGF domain of L-selectin, inhibits the binding of lymphocytes to HEVs.[70] In molecular constructs expressing only the lectin domain, there is no Mel 14 binding; however, inclusion of the EGF domain with the lectin domain confers binding.[56] In addition, the deletion of the EGF domain of E-selectin markedly decreases the binding of the monoclonal antibodies directed against the lectin domain as well as binding to U937 target cells.[69] These results provide a more direct demonstration of the potential role of the EGF domain in binding; however, alterations in the conformation of the molecule could also explain the decreased antibody binding. It is impossible to determine from the data available whether the EGF domain acts to stabilize the conformation of the molecule and thus the lectin binding domain, or if the EGF domain is more directly involved in binding. Since EGF domains of proteins are traditionally considered important in protein-protein interactions, it is conceivable that the EGF domain recognizes a critical protein structure in the selectin ligand in concert with the recognition of a carbohydrate structure by the lectin domain. As discussed above, initial studies have demonstrated that a protein component of the P-selectin ligand may be necessary for binding.

The potential role of the CR domains of the selectins has received relatively little attention. This may be due to the lower degree of homology in the CR domain compared to the homology in the EGF and lectin domains. Interestingly, the few experiments that have addressed the function of the CR domain have revealed its importance in selectin adhesion. Binding of the Mel 14 antibody to L-selectin constructs is significantly decreased when the complement regulatory domain is deleted.[23] In addition to alteration of the Mel 14 epitope with deletion of the CR domain, there was a significant loss of carbohydrate-binding capacity, as evidenced by reduced binding to PPME-coated microtiter plates. These data are consistent with a model of the CR domain contributing important structural support of a functional lectin conformation. In addition, the CR repeats may be involved in the regulation or binding of complement and thus play a critical role in the interactions between inflammatory and vascular cells.

VI. ROLE OF SELECTINS IN HEALTH AND DISEASE

There is an emerging understanding of the cellular function of selectins as mediators of intercellular binding between leukocytes and endothelial cells. However, the role that this family of adhesion molecules plays in health and disease is largely speculative and a fertile area of investigation. Selectins are conceivably involved in several physiological and pathological processes, including inflammation, immunity, hemostasis, metastasis, and atherosclerosis.

A. INFLAMMATION/IMMUNITY

Since selectins mediate the binding between leukocytes and endothelial cells, they likely play a central role in the inflammatory process. Prior to the discovery of the selectins and their recognition as adhesion molecules, attachment of leukocytes to endothelium was felt to be largely a result of the integrins. Antibodies directed against integrins LFA-1 and Mac-1 inhibit firm attachment of neutrophils to the endothelium and subsequent extravasation. There are compelling clinical data to support the importance of integrins in this process. The genetic absence of functional β2 integrins is associated with a severe inability to produce an adequate inflammatory response. This disorder, leukocyte adhesion deficiency (LAD), is characterized by frequent, severe, and often life-threatening infections in the young infant. Although integrins are important in leukocyte-endothelial interactions, investigators have also recognized "integrin-independent" binding. The emergence of the selectins has accounted for at least a portion of integrin-independent binding.

There are three basic events during the early phases of the acute inflammatory process. First, circulating leukocytes attach to the activated endothelium of the postcapillary venule and "roll" along the endothelial surface. Second, there is firm adhesion to the endothelial surface, causing the rolling neutrophils to arrest. Third, the leukocytes extravasate into the surrounding tissues. There is accumulating evidence that selectins play a major role in the first step of this process, the initial attachment of leukocytes and rolling on the endothelium. The subsequent step involving firm adhesion appears to involve the integrin family of adhesion molecules.

The contribution of adhesion molecules to rolling of leukocytes has been studied extensively by Lawrence and Springer.[71] Artificial phospholipid bilayers were made, containing either P-selectin or ICAM-1, the endothelial ligand for the leukocyte integrins LFA-1 and Mac-1. In a flow cell system under physiological conditions, the P-selectin-containing surface supports the rolling of neutrophils on endothelium, whereas the ICAM-1 surface does not. If neutrophils were allowed to interact with these surfaces under static conditions, the ICAM-1-containing bilayer supported firm adhesion that could not be displaced, whereas the P-selectin surface supported a weaker adhesion, as seen with the rolling phenomenon. This second step, mediated by ICAM-1, could also be reproduced by including fMLP, an activator that up-regulates the ICAM-1 receptors on the neutrophil, LFA-1 and Mac-1. The initial rolling phenomenon mediated by a selectin, followed by the subsequent adhesion strengthening through the action of integrins, leads to diapedesis of the white cell across the endothelial layer.

Since P-selectin is expressed within seconds of stimulation, it is considered central in the rolling phenomenon on endothelium. However, there is evidence that both E-selectin and L-selectin are also involved in integrin-independent cell adhesion in the early phases of inflammation. L-selectin and E-selectin both mediate the binding of neutrophils to activated endothelial cells *in vitro.*[72] Smith et al. have investigated the role of these two selectins in a flow system. The rolling of neutrophils on cytokine-activated endothelium is integrin-independent and blocked by antibodies directed against L-selectin or E-selectin.[73]

Interestingly, the binding of neutrophils to E-selectin results in neutrophil activation and expression of CD18 integrins, which act to further strengthen the attachment.[74] Further cooperativity is evidenced by the finding that activation of neutrophils leads to an increase in the affinity of L-selectin for its carbohydrate ligand.[74] Thus, selectin-mediated initial attachment and rolling of leukocytes results in increased affinity and increased expression of adhesion molecules, representing a positive feedback loop.

During inflammation, there is also cooperativity within the selectin family, in addition to cooperativity with the integrin family of adhesion molecules. As outlined in Table 2, the selectins differ in both tissue distribution and mechanism of expression. L-selectin is constituitively expressed on lymphocytes, monocytes, and neutrophils. P-selectin is newly expressed on the endothelial surface within seconds of stimulation with a variety of agonists. E-selectin is expressed on the endothelial surface 4 to 6 h following cytokine stimulation. Thus, there is constituitive, immediate, and delayed expression of selectins during inflammation. These fundamental differences in expression may be important in sustaining an inflammatory response.

In addition to the *in vitro* data above demonstrating a potential role for selectins in leukocyte-endothelial interactions, there are accumulating *in vivo* data that lend further support to the hypothesis that selectins are critical in the inflammatory response. Much of the *in vivo* work with the selectins has involved the study of E-selectin. Agents that stimulate the expression of endothelial E-selectin *in vitro,* such as TNFα, IL-1β, and lipopolysaccharide (LPS), are well-known mediators of inflammation. Thus, E-selectin may be involved in the processes mediated by these biologically active substances. LPS is well known to induce the clinical manifestations of septic shock, including multiorgan failure. The administration of LPS to baboons causes increased expression of E-selectin in the postcapillary venules of several organs, as well as increased levels of TNF and IL-1, cytokines known to up-regulate E-selectin. Anti-E-selectin antibodies as well as anti-TNF antibodies reduce neutrophil binding and vascular damage in rat models of lung and skin injury.[75] In addition to expression during acute processes, increased expression of E-selectin has been documented in certain chronic inflammatory conditions, such as rheumatoid arthritis, psoriasis, and allergic cutaneous inflammation. The persistent expression of E-selectin in certain chronic inflammatory conditions is an interesting finding, since *in vitro* experiments demonstrate transient expression of E-selectin following cytokine stimulation. This discrepancy may relate to fundamental differences between the *in vitro* and the *in vivo* expression and regulation of E-selectin.

Naturally, the focus of selectin-mediated inflammatory response is on leukocyte-endothelial interactions. However, platelet P-selectin may also contribute significantly to inflammation. P-selectin is expressed on activated platelets, where it mediates the binding of platelets to monocytes and neutrophils. The functional consequence of platelet-leukocyte interactions is not currently known. This interaction may simply represent a mechanism to clear activated platelets by circulating phagocytes. Alternatively, this binding may lead to a modulation in leukocyte function. Platelet-neutrophil interactions result in unique leukotriene products, which may be important mediators in the inflammatory process.[76] If soluble P-selectin in fact exists, then actual cell-cell binding may not be necessary to effect this modulation.

B. HEMOSTASIS

In addition to playing a critical role in the early phases of inflammation, the selectins may also participate in hemostasis and thrombosis. Effective hemostasis includes primary hemostasis, involving vasoconstriction and platelet adhesion/aggregation, and secondary hemostasis, involving the formation of coagulation factor complexes on the surfaces of cells and matrices. Selectins may be involved at any step of hemostasis.

The common stimuli involved in the promotion of primary hemostasis result in the rapid expression of P-selectin on the endothelial surface. This serves to recruit circulating leukocytes to the area of vascular injury. Simultaneously, platelets are recruited through selectin-dependent and independent mechanisms. The rapid recruitment of these cells provides important phospholipid surfaces for the development of prothrombinase complex and may have additional effects. Monocytes participate in coagulation by providing coagulation factors, membrane surface, and possibly activated surface-bound coagulation complexes.[77-79] Platelet-neutrophil interactions result in the synthesis of novel leukotriene products through transcellular metabolism that may have important modulatory effects on inflammation and hemostasis.[80] Parmentier et al. have reported data suggesting that P-selectin may be involved in platelet aggregation as well as platelet-leukocyte interactions. A monoclonal antibody directed against P-selectin, LYP20, effectively inhibits both thrombin- and collagen-induced aggregation of washed platelets and platelet-rich plasma.[81] This is the first report implicating a role of P-selectin in platelet aggregation and represents an area requiring further investigation.

After the initial events of hemostasis, the generation of inflammatory cytokines may have important effects on hemostasis. In addition to the synthesis and expression of E-selectin, these cytokines induce the expression of tissue factor on cell surfaces, which promotes thrombosis and simultaneously down-regulates thrombomodulin, a natural anticoagulant.[82]

Platelets become activated during normal and pathological events, thereby triggering many hemostatic functions. An efficient mechanism must exist that clears activated platelets from the circulation as part of the delicate balance between the promotion and the control of thrombosis. Monocytes and/or neutrophils may function to clear activated platelets by binding via P-selectin, followed by direct phagocytosis or signaling for clearance to occur elsewhere.

C. METASTASIS

Several human malignancies, both hematological and solid tumors, express CD15 or Le^x.[83,84] The binding of circulating tumor cells to endothelium through Selectin/Le^x interactions may be a critical step in the metastatic cascade. The human colon carcinoma cell line HT-29 binds to cytokine-treated endothelium in an E-selectin-dependent manner.[85] In addition, lymphoma cells are capable of expressing the L-selectin and may bind to vascular addressins on the endothelium.[86] This is clearly an area deserving further investigation.

In addition to tumor cell-endothelial interactions, tumor cell-platelet interactions may participate in metastasis. Observations from several sources have suggested that platelets play an important role in metastasis and tumor progression.[87,88] Since many tumors express the ligand for P-selectin, there is a molecular basis for the binding of platelets to certain tumor cells. CD15-expressing human tumor cell lines bind activated platelets, and this binding can be inhibited by P-selectin antibodies and purified P-selectin.[89] This interaction may be important in forming aggregates of tumor cells, which may be critical during certain steps of metastasis. In addition to the physical effect of creating a critical mass of tumor cells, platelet binding to tumor cells may have functional consequences and affect the metastatic potential of the tumor. In summary, there is potential for selectin-mediated binding at both the tumor cell-platelet and the tumor cell-endothelial cell interfaces to play a significant role in the metastatic process.

D. ATHEROSCLEROSIS

Atherogenesis involves a complex interaction between certain substrates and cells, including monocytes, platelets, endothelial cells, and smooth muscle cells.[90,91] The role of adhesion molecules in atherogenesis is not clear. Selectin-mediated cell-cell interactions represent a potential mechanism for recruitment of the critical cellular elements of atherosclerosis. The binding of selectins to their ligands may function to associate platelets, monocytes, and endothelial cells in areas of vascular injury, thus acting to promote the atherosclerotic lesion. Although selectin-mediated events are traditionally thought to occur within the confines of the vascular system, there has not been adequate investigation of possible selectin binding within the extracellular matrix. As outlined above, accumulating evidence suggests that selectins bind to sialylated, fucosylated lactoaminoglycans. The exact molecular ligands for the selectins on known target cells are being actively investigated and their DNA cloned. The potential role of selectins outside the vasculature should be investigated.

VII. SUMMARY

There has been rapid progress in the characterization of the selectin family of adhesion molecules. The study of the selectins is a fertile area of investigation. Although our understanding of these molecules has progressed, there are still large gaps in our knowledge. Although specific carbohydrates have been identified as key components of the ligands for the selectins, the precise molecular ligands have not been identified and isolated. There is an emerging model of leukocyte-endothelial interactions in which all three selectins are involved in the initial rolling phenomenon. We need to gain a greater understanding of the interplay between the receptor/ligand pairs involved in various stages of inflammation. We have an incomplete understanding of the role of the selectins in hemostasis. In addition to normal physiological processes, the potential role of the selectins in pathological processes like atherosclerosis and metastasis deserve further investigation. The current three members of the selectin family were all identified initially through the development of monoclonal antibodies. There may exist other yet undiscovered selectin molecules that are less immunogenic and less likely to be detected by immunological methods.

As our knowledge of these molecules and their ligands advances, it becomes increasingly important to apply this knowledge to the clinical setting. The effect of selectin chimeras or synthetic compounds mimicking selectin or ligand structure on physiological processes *in vivo* needs to be investigated. Information from *in vitro* experiments and animal models hopefully will lead to the generation of compounds that are capable of modulating the inflammatory process in a clinically significant way.

REFERENCES

1. **Lasky, L. A.,** Lectin cell adhesion molecules (LEC-CAMs): a new family of cell adhesion proteins involved with inflammation, *J. Cell Biochem.,* 45, 139, 1991.
2. **McEver, R. P.,** Selectins: novel receptors that mediate leukocyte adhesion during inflammation, *Thromb. Haemost.,* 65, 223, 1991.
3. **Bevilacqua, M., Butcher, E., Furie, B. C., Furie, B., Gallatin, M., Gimbrone, M., Harlan, J., Kishimoto, K., Lashy, L., McEver, R., Paulson, J., Rosen, S., Seed, B., Siegelman, M., Springer, T., Stoolman, L., Tedder, T., Varki, A., Wagner, D., Weissman, I., and Zimmerman, G.,** Selectins: a family of adhesion receptors, *Cell,* 67, 233, 1991.
4. **Drickamer, K., Dordal, M. S., and Reynolds, L.,** Mannose-binding proteins isolated from rat liver contain carbohydrate-recognition domains linked to collagenous tails, *J. Biol. Chem.,* 261, 6878, 1986.
5. **Drickamer, K.,** Two distinct classes of carbohydrate-recognition domains in animal lectins, *J. Biol. Chem.,* 263, 9557, 1988.
6. **Krusius, T., Gehlsen, K. R., and Ruoslahti, E.,** A fibroblast chondroitin sulfate proteoglycan core protein contains lectin-like and growth factor-like sequences, *J. Biol. Chem.,* 262, 13120, 1987.
7. **Johnston, G. I., Cook, R. G., and McEver, R. P.,** Cloning of GMP-140, a granule membrane protein of platelets and endothelium: sequence similarity to proteins involved in cell adhesion and inflammation, *Cell,* 56, 1033, 1989.
8. **Bevilacqua, M. P., Stengelin, S., Gimbrone, M. A., Jr., and Seed, B.,** Endothelial leukocyte adhesion molecule 1: an inducible receptor for neutrophils related to complement regulatory proteins and lectins, *Science,* 243, 1160, 1989.
9. **Siegelman, M. H., van de Rijn, M., and Weissman, I. L.,** Mouse lymph node homing receptor cDNA clone encodes a glycoprotein revealing tandem interaction domains, *Science,* 243, 1165, 1989.
10. **Hourcade, D., Holers, V. M., and Atkinson, J. P.,** The regulators of complement activation (RCA) gene cluster, *Adv. Immunol.,* 45, 381, 1989.
11. **Watson, M. L., Kingsmore, S. F., Johnston, G. I., Siegelman, M. H., Le Beau, M. M., Lemons, R. S., Bora, N. S., Howard, T. A., Weissman, I. L., and McEver, R. P.,** Genomic organization of the selectin family of leukocyte adhesion molecules on human and mouse chromosome 1, *J. Exp. Med.,* 172, 263, 1990.
12. **Yednock, T. and Rosen, S. D.,** Lymphocyte homing, *Adv. Immunol.,* 44, 313, 1989.
13. **Woodruff, J., Katz, H., Lucas, L., and Stamper, H.,** An in vitro model of lymphocyte homing, *J. Immunol.,* 119, 1603, 1977.
14. **Gallatin, W. M., Weissman, I. L., and Butcher, E. C.,** A cell-surface molecule involved in organ-specific homing of lymphocytes, *Nature,* 304, 30, 1983.
15. **Tedder, T. F., Penta, A. C., Levine, H. B., and Freedman, A. S.,** Expression of the human leukocyte adhesion molecule, LAM1. Identity with the TQ1 and Leu-8 differentiation antigens, *J. Immunol.,* 144, 532, 1990.
16. **Tedder, T. F., Isaacs, C. M., Ernst, T. J., Demetri, G. D., Adler, D. A., and Disteche, C. M.,** Isolation and chromosomal localization of cDNAs encoding a novel human lymphocyte cell surface molecule, LAM-1. Homology with the mouse lymphocyte homing receptor and other human adhesion proteins, *J. Exp. Med.,* 170, 123, 1989.

17. **Tedder, T. F., Matsuyama, T., Rothstein, D., Schlossman, S. F., and Morimoto, C.,** Human antigen-specific memory T cells express the homing receptor (LAM-1) necessary for lymphocyte recirculation, *Eur. J. Immunol.,* 20, 1351, 1990.
18. **Ord, D. C., Ernst, T. J., Zhou, L. J., Rambaldi, A., Spertini, O., Griffin, J., and Tedder, T. F.,** Structure of the gene encoding the human leukocyte adhesion molecule-1 (TQ1, Leu-8) of lymphocytes and neutrophils, *J. Biol. Chem.,* 265, 7760, 1990.
19. **Spertini, O., Kansas, G. S., Munro, J. M., Griffin, J. D., and Tedder, T. F.,** Regulation of leukocyte migration by activation of the leukocyte adhesion molecule-1 (LAM-1) selectin, *Nature,* 349, 691, 1991.
20. **Jutila, M., Rott, L., Berg, E., and Butcher, E.,** Function and regulation of the neutrophil Mel 14 antigen in vivo: comparison with LFA 1 and MAC 1, *J. Immunol.,* 143, 3318, 1989.
21. **Jutila, M.** Low dose chymotrypsin treatment inhibits neutrophil migration into sites of inflammation in vivo: effects on Mac-1 and Mel 14 adhesion protein expression and function, *Cell. Immunol.,* 132, 201, 1991.
22. **Lewinsohn, D. M., Bargatze, R. F., and Butcher, E. C.,** Leukocyte-endothelial cell recognition: evidence of a common molecular mechanism shared by neutrophils, lymphocytes, and other leukocytes, *J. Immunol.,* 138, 4313, 1987.
23. **Watson, S. R., Fennie, C., and Lasky, L. A.,** Neutrophil influx into an inflammatory site inhibited by a soluble homing receptor-IgG chimaera, *Nature,* 349, 164, 1991.
24. **Cotran, R. S.,** New roles for the endothelium in inflammation and immunity, *Am. J. Pathol.,* 129, 407, 1987.
25. **Bevilacqua, M. P., Pober, J. S., Wheeler, M. E., Cotran, R. S., and Gimbrone, M. A., Jr.,** Interleukin 1 acts on cultured human vascular endothelial cells to increase the adhesion of polymorphonuclear leukocytes, monocytes, and related leukocyte cell lines, *J. Clin. Invest.,* 76, 2003, 1985.
26. **Gamble, J. R., Harlan, J. M., Klebanoff, S. J., and Vadas, M. A.,** Stimulation of the adherence of neutrophils to umbilical vein endothelium by human recombinant tumor necrosis factor, *Proc. Natl. Acad. Sci. U.S.A.,* 82, 8667, 1985.
27. **Bevilacqua, M. P., Pober, J. S., Mendrick, D. L., Cotran, R. S., and Gimbrone, M. A., Jr.,** Identification of an inducible endothelial-leukocyte adhesion molecule, *Proc. Natl. Acad. Sci. U.S.A.,* 84, 9238, 1987.
28. **Pober, J. R., Bevilacqua, M. P., Mendrick, D. L., Lapierre, L. A., Fiers, W., and Gimbrone, M. A., Jr.,** Two distinct monokines, interleukin 1 and tumor necrosis factor, each independently induce biosynthesis and transient expression of the same antigen on the surface of cultured human vascular endothelial cells, *J. Immunol.,* 136, 1680, 1986.
29. **Hession, C., Osborn, L., Goff, D., Chi Rosso, G., Vassallo, C., Pasek, M., Pittack, C., Tizard, R., Goelz, S., and McCarthy, K.,** Endothelial leukocyte adhesion molecule 1: direct expression cloning and functional interactions, *Proc. Natl. Acad. Sci. U.S.A.,* 87, 1673, 1990.
30. **Collins, T., Williams, A., Johnston, G. I., Kim, J., Eddy, R., Shows, T., Gimbrone, M. A., Jr., and Bevilacqua, M. P.,** Structure and chromosomal location of the gene for endothelial-leukocyte adhesion molecule 1, *J. Biol. Chem.,* 266, 2466, 1991.
31. **Montgomery, K. F., Osborn, L., Hession, C., Tizard, R., Goff, D., Vassallo, C., Tarr, P. I., Bomsztyk, K., Lobb, R., and Harlan, J. M.,** Activation of endothelial-leukocyte adhesion molecule 1 (ELAM-1) gene transcription, *Proc. Natl. Acad. Sci. U.S.A.,* 88, 6523, 1991.
32. **Whelan, J., Ghersa, P., Hooft van Huijsduijnen, R., Gray, J., Chandra, G., Talabot, F., and DeLamarter, J. F.,** An NF kappa B-like factor is essential but not sufficient for cytokine induction of endothelial leukocyte adhesion molecule 1 (ELAM-1) gene transcription, *Nucl. Acids Res.,* 19, 2645, 1991.
33. **Cotran, R. S., Gimbrone, M. A., Jr., Bevilacqua, M. P., Mendrick, D. L., and Pober, J. S.,** Induction and detection of a human endothelial activation antigen in vivo, *J. Exp. Med.,* 164, 661, 1986.
34. **Hsu-Lin, S. C., Berman, C. L., Furie, B. C., August, D., and Furie, B.,** A platelet membrane protein expressed during platelet activation, *J. Biol. Chem.,* 259, 9121, 1984.
35. **McEver, R. P. and Martin, M. N.,** A monoclonal antibody to a membrane glycoprotein binds only to activated platelets, *J. Biol. Chem.,* 259, 9799, 1984.
36. **Berman, C. L., Yeo, E. L., Wencel Drake, J. D., Furie, B. C., Ginsberg, M. H., and Furie, B.,** A platelet alpha granule membrane protein that is associated with the plasma membrane after activation. Characterization and subcellular localization of platelet activation-dependent granule-external membrane protein, *J. Clin. Invest.,* 78, 130, 1986.

37. **Bonfanti, R., Furie, B. C., Furie, B., and Wagner, D. D.,** PADGEM (GMP140) is a component of Weibel-Palade bodies of human endothelial cells, *Blood,* 73, 1109, 1989.
38. **Patel, K. D., Zimmerman, G. A., Prescott, S. M., McEver, R. P., and McIntyre, T. M.,** Oxygen radicals induce human endothelial cells to express GMP-140 and bind neutrophils, *J. Cell Biol.,* 112, 749, 1991.
39. **Hattori, R., Hamilton, K. K., McEver, R. P., and Sims, P. J.,** Complement proteins C5b-9 induce secretion of high mol wt multimers of endothelial von Willebrand factor and translocation of granule membrane protein GMP-140 to the cell surface, *J. Biol. Chem.,* 264, 9053, 1989.
40. **Hattori, R., Hamilton, K. K., Fugate, R. D., McEver, R. P., and Sims, P. J.,** Stimulated secretion of endothelial von Willebrand factor is accompanied by rapid redistribution to the cell surface of the intracellular granule membrane protein GMP-140, *J. Biol. Chem.,* 264, 7768, 1989.
41. **Berman, C. L., Yeo, E. L., Furie, B. C., and Furie, B.,** PADGEM protein, *Meth. Enzymol.,* 169, 311, 1989.
42. **Johnston, G. I., Kurosky, A., and McEver, R. P.,** Structural and biosynthetic studies of the granule membrane protein, GMP-140, from human platelets and endothelial cells, *J. Biol. Chem.,* 264, 1816, 1989.
43. **Sims, P. J., Faioni, E. M., Wiedmer, T., and Shattil, S. J.,** Complement proteins C5b-9 cause release of membrane vesicles from the platelet surface that are enriched in the membrane receptor for coagulation factor Va and express prothrombinase activity, *J. Biol. Chem.,* 263, 18205, 1988.
44. **Larsen, E., Celi, A., Gilbert, G. E., Furie, B. C., Erban, J. K., Bonfanti, R., Wagner, D. D., and Furie, B.,** PADGEM protein: a receptor that mediates the interaction of activated platelets with neutrophils and monocytes, *Cell,* 59, 305, 1989.
45. **Hamburger, S. A. and McEver, R. P.,** GMP-140 mediates adhesion of stimulated platelets to neutrophils, *Blood,* 75, 550, 1990.
46. **Jungi, T. W., Spycher, M. O., Nydegger, U. E., and Barandun, S.,** Platelet-leukocyte interactions: selective binding of thrombin-stimulated platelets to human monocytes, polymorphonuclear leukocytes, and related cell lines, *Blood,* 67, 629, 1986.
47. **Moore, K. L., Varki, A., and McEver, R. P.,** GMP-140 binds to a glycoprotein receptor on human neutrophils: evidence for a lectin-like interaction, *J. Cell Biol.,* 112, 491, 1991.
48. **Geng, J. G., Bevilacqua, M. P., Moore, K. L., McIntyre, T. M., Prescott, S. M., Kim, J. M., Bliss, G. A., Zimmerman, G. A., and McEver, R. P.,** Rapid neutrophil adhesion to activated endothelium mediated by GMP-140, *Nature,* 343, 757, 1990.
49. **Brandley, B. K. and Schnaar, R. L.,** Cell-surface carbohydrates in cell recognition and response, *J. Leuk. Biol.,* 40, 97, 1986.
50. **Sharon, N. and Lis, H.,** Lectins as cell recognition molecules, *Science,* 246, 227, 1989.
51. **DiCorleto, P. E. and de le Motte, C. A.,** Role of cell surface carbohydrate moieties in monocytic cell adhesion to endothelium in vitro, *J. Immunol.,* 143, 3666, 1989.
52. **Stoolman, L. M., Yednock, T. A., and Rosen, S. D.,** Homing receptors on human and rodent lymphocytes — evidence for a conserved carbohydrate-binding specificity, *Blood,* 70, 1842, 1987.
53. **Yednock, T. A., Butcher, E. C., Stoolman, L. M., and Rosen, S. D.,** Receptors involved in lymphocyte homing: relationship between a carbohydrate-binding receptor and the MEL-14 antigen, *J. Cell Biol.,* 104, 725, 1987.
54. **Imai, Y., True, D. D., Singer, M. S., and Rosen, S. D.,** Direct demonstration of the lectin activity of gp90MEL, a lymphocyte homing receptor, *J. Cell Biol.,* 111, 1225, 1990.
55. **Rosen, S. D., Singer, M. S., Yednock, T. A., and Stoolman, L. M.,** Involvement of sialic acid on endothelial cells in organ-specific lymphocyte recirculation, *Science,* 228, 1005-1007. 1985.
56. **Bowen, B. R., Fennie, C., and Lasky, L. A.,** The Mel 14 antibody binds to the lectin domain of the murine peripheral lymph node homing receptor, *J. Cell Biol.,* 110, 147, 1990.
57. **Kansas, G. S., Spertini, O., Stoolman, L. M., and Tedder, T. F.,** Molecular mapping of functional domains of the leukocyte receptor for endothelium, LAM-1, *J. Cell Biol.,* 114, 351, 1991.
58. **Streeter, P. R., Rouse, B. T. N., and Butcher, E. C.,** Immunohistologic and functional characterization of a vascular addressin involved in lymphocyte homing into peripheral lymph nodes, *J. Cell Biol.,* 107, 1853, 1988.
59. **Berg, E. L., Robinson, M. K., Warnock, R. A., and Butcher, E. C.,** The human peripheral lymph node vascular addressin is a ligand for LECAM-1, the peripheral lymph node homing receptor, *J. Cell Biol.,* 114, 343, 1991.
60. **Imai, Y., Singer, M. S., Fennie, C., Lasky, L. A., and Rosen, S. D.,** Identification of a carbohydrate-based endothelial ligand for a lymphocyte homing receptor, *J. Cell Biol.,* 113, 1213, 1991.

61. **Walz, G., Aruffo, A., Kolanus, W., Bevilacqua, M., and Seed, B.,** Recognition by ELAM-1 of the sialyl-Lex determinant on myeloid and tumor cells, *Science,* 250, 1132, 1990.
62. **Phillips, M. L., Nudelman, E., Gaeta, F. C., Perez, M., Singhal, A. K., Hakomori, S., and Paulson, J. C.,** ELAM-1 mediates cell adhesion by recognition of a carbohydrate ligand, sialyl-Lex, *Science,* 250, 1130, 1990.
63. **Lowe, J. B., Stoolman, L. M., Nair, R. P., Larsen, R. D., Berhend, T. L., and Marks, R. M.,** ELAM-1-dependent cell adhesion to vascular endothelium determined by a transfected human fucosyltransferase cDNA, *Cell,* 63, 475, 1990.
64. **Polley, M. J., Phillips, M. L., Wayner, E., Nudelman, E., Singhal, A. K., Hakomori, S., and Paulson, J. C.,** CD62 and endothelial cell-leukocyte adhesion molecule 1 (ELAM-1) recognize the same carbohydrate ligand, sialyl-Lewis x, *Proc. Natl. Acad. Sci. U.S.A.,* 88, 6224, 1991.
65. **Goelz, S. E., Hession, C., Goff, D., Griffiths, B., Tizard, R., Newman, B., Chi Rosso, G., and Lobb, R.,** ELFT: a gene that directs the expression of an ELAM-1 ligand, *Cell,* 63, 1349, 1990.
66. **Larsen, E., Palabrica, T., Sajer, S., Gilbert, G. E., Wagner, D. D., Furie, B. C., and Furie, B.,** PADGEM-dependent adhesion of platelets to monocytes and neutrophils is mediated by a lineage-specific carbohydrate, LNF III (CD15), *Cell,* 63, 467, 1990.
67. **Corral, L., Singer, M. S., Macher, B. A., and Rosen, S. D.,** Requirement for sialic acid on neutrophils in a GMP-140 (PADGEM) mediated adhesive interaction with activated platelets, *Biochem. Biophys. Res. Commun.,* 172, 1349, 1990.
68. **Sajer, S. A., Erban, J. K., Gibson, R. M., Sako, D., Ahern, T., Wagner, D. D., Furie, B. C., Fure, B., and Larsen, G.,** PADGEM and ELAM-1: distinct but overlapping leukocyte ligand specificities, *Blood,* 78, 377a, 1991.
69. **Pigott, R., Needham, L. A., Edwards, R. M., Walker, C., and Power, C.,** Structural and functional studies of the endothelial activation antigen endothelial leukocyte adhesion molecule-1 using a panel of monoclonal antibodies, *J. Immunol.,* 147, 130, 1991.
70. **Siegelman, M. H., Cheng, I. C., Weissman, I. L., and Wakeland, E. K.,** The mouse lymph node homing receptor is identical with the lymphocyte cell surface marker Ly-22: role of the EGF domain in endothelial binding, *Cell,* 61, 611, 1991.
71. **Lawrence, M. B. and Springer, T. A.,** Leukocytes roll on a selectin at physiologic flow rates: distinction from and prerequisite for adhesion through integrins, *Cell,* 65, 859, 1991.
72. **Kishimoto, T. K., Warnock, R. A., Jutila, M. A., Butcher, E. C., Lane, C., Anderson, D. C., and Smith, C. W.,** Antibodies against human neutrophil LECAM-1 (LAM-1/Leu-8/DREG-56 antigen) and endothelial cell ELAM-1 inhibit a common CD18-independent adhesion pathway in vitro, *Blood,* 78, 805, 1991.
73. **Smith, C. W., Kishimoto, T. K., Abbass, O., Hughes, B., Rothlein, R., McIntire, L. V., Butcher, E. C., and Anderson, D. C.,** Chemotactic factors regulate lectin adhesion molecule 1 (LECAM-1)-dependent neutrophil adhesion to cytokine-stimulated endothelial cells in vitro, *J. Clin. Invest.,* 87, 609, 1991.
74. **Lo, S. K., Lee, S., Ramos, R. A., Lobb, R., Rosa, M., Chi Rosso, G., and Wright, S. D.,** Endothelial-leukocyte adhesion molecule 1 stimulates the adhesive activity of leukocyte integrin CR3 on human neutrophils, *J. Exp. Med.,* 173, 1493, 1991.
75. **Mulligan, M. S., Varani, J., Dame, M. K., Lane, C. L., Smith, C. W., Anderson, D. C., and Ward, P. A.,** Role of endothelial-leukocyte adhesion molecule 1 (ELAM-1) in neutrophil-mediated lung injury in rats, *J. Clin. Invest.,* 88, 1396, 1991.
76. **Gamble, J. R., Skinner, M. P., Berndt, M. C., and Vadas, M. A.,** Prevention of activated neutrophil adhesion to endothelium by soluble adhesion protein GMP140, *Science,* 249, 414, 1990.
77. **Altieri, D. C. and Edgington, T. S.,** The saturable high affinity association of factor X to ADP-stimulated monocytes defines a novel function of the Mac-1 receptor, *J. Biol. Chem.,* 263, 7007, 1988.
78. **Altieri, D. C. and Edgington, T. S.,** Sequential receptor cascade for coagulation proteins on monocytes, *J. Biol. Chem.,* 264, 2969, 1989.
79. **Gregory, S. A. and Edgington, T. S.,** Tissue factor induction in human monocytes, *J. Clin. Invest.,* 76, 2440, 1985.
80. **Henson, P. M.,** Interactions between neutrophils and platelets, *Lab. Invest.,* 62, 391, 1990.
81. **Parmentier, S., McGregor, L., Catimel, B., Leung, L. L., and McGregor, J. L.,** Inhibition of platelet functions by a monoclonal antibody (LYP20) directed against a granule membrane glycoprotein (GMP-140/PADGEM), *Blood,* 77, 1734, 1991.
82. **Esmon, C. T.,** The roles of protein C and thrombomodulin in the regulation of blood coagulation, *J. Biol. Chem.,* 264, 4743, 1989.

83. **Fukoshima, K., Hirota, M., Terasaki, P. I., Wakisaka, A., Togashi, H., Shia, D., Suyama, N., Fukushi, Y., Nudleman, E., and Hakamori, S.,** Characterization of sialosylated Lewis x as a new tumor-associated antigen, *Cancer Res.*, 44, 5279, 1984.
84. **Hakamori, S.,** Tumor-associated carbohydrate antigens, *Annu. Rev. Immunol.*, 2, 103, 1990.
85. **Rice, G. E. and Bevilacqua, M. P.,** An inducible endothelial cell surface glycoprotein mediates melanoma adhesion, *Science*, 246, 1303, 1989.
86. **Bargatze, R. F., Wu, N. W., Weissman, I. L., and Butcher, E. C.,** High endothelial venule binding as a predictor of the dissemination of passaged murine lymphomas, *J. Exp. Med.*, 166, 1125, 1987.
87. **Metha, P.,** The potential role of platelets in the pathogenesis of tumor metastasis, *Blood*, 63, 55, 1983.
88. **Al-Mondhiry, H., McGarvey, V., and Leitzel, K.,** Interaction of human tumor cells with human platelets and the coagulation system, *Thromb. Haemostas.*, 50, 726, 1983.
89. **Larsen, E.,** P-selectin dependent binding of platelets to tumor cells, in press, 1992.
90. **Hoak, J.,** Platelets and atherosclerosis, *Semin. Thromb. Hemostas.*, 14, 202, 1988.
91. **Ross, R.,** The pathogenesis of atherosclerosis, *N. Engl. J. Med.*, 314, 488, 1986.

Chapter 11

Structure and Assembly of Basement Membrane and Related Extracellular Matrix Proteins

Konrad Beck and Tanja Gruber

CONTENTS

I. Basement Membranes as a Natural Substrate for Cell Adhesion 219
II. Structure of Basement Membrane-Specific Proteins 221
A. Laminin 222
B. Nidogen/Entactin 225
C. Laminin/Nidogen-Complex 227
D. Collagen Type IV 227
E. Basement Membrane Proteoglycans 228
F. Other Basement Membrane Proteins 229
III. Structure of Other Selected Extracellular Matrix Proteins 230
A. Collagen VII 230
B. Fibronectin 230
C. SPARC/BM-40/Osteonectin 231
D. Tenascin/Cytotactin/Hexabrachion 233
IV. Interaction of Basement Membrane Proteins 234
A. Laminin Self-Interaction 235
B. Collagen IV Self-Interaction 235
C. Interactions of Basement Membrane Protein Mixtures 236
V. Cell Adhesion to Extracellular Matrix Proteins 237
VI. Outlook 239
VII. Acknowledgments 240
References 240

I. BASEMENT MEMBRANES AS A NATURAL SUBSTRATE FOR CELL ADHESION

The extracellular matrix is the natural environment in which the cells of multicellular organisms are embedded. Both *in vivo* as well as *in vitro* their constituent molecules modulate cell adhesion, spreading, growth, morphology, differentiation, and life span. In cell culture work, it is a common observation that cells adhere differently to glass, plastic, or other surfaces, and thus an understanding of how matrix factors influence cellular behavior is also of direct relevance in many clinical disciplines like surgery and prosthetics. Therefore, it is not surprising that the interest in the anatomy and function of the extracellular matrix is nearly as old as the cell concept itself. (For short reviews on the early history and the molecular oriented era, see References 1 and 2). Due to the development of many histological stains, the specific glycosylation of the "fundamental substance" was an early observation. Meanwhile, many molecules located specifically within the extracellular matrix have been characterized. Quantitatively, the majority of them belong to the classes of collagens, proteoglycans, and glycosaminoglycans, but also several other glycoproteins with different spatial and temporal distributions have been found. In order to perform their special tasks, they must be organized in a defined but presently only weakly understood meshwork.

The extracellular matrix not only serves as a mechanical scaffolding to which cells adhere or as a biological glue to connect and separate different tissues, but it forms highly specialized structures, such as the connective tissue, tendons, bones, and teeth. Excellent introductions to the field are presented in the textbooks of Alberts et al.[3] and Darnell et al.[4] For a rigorous, current overview, the book of Hay is highly recommended.[5] Besides some book series (e.g., Reference 6) and other journals, the annual progress on different aspects is reviewed in the October issues of *Current Opinion in Cell Biology*. In this chapter, we will focus on some molecular aspects of the structure and assembly of basement membranes

0-8493-4559-6/95/$0.00+$.50

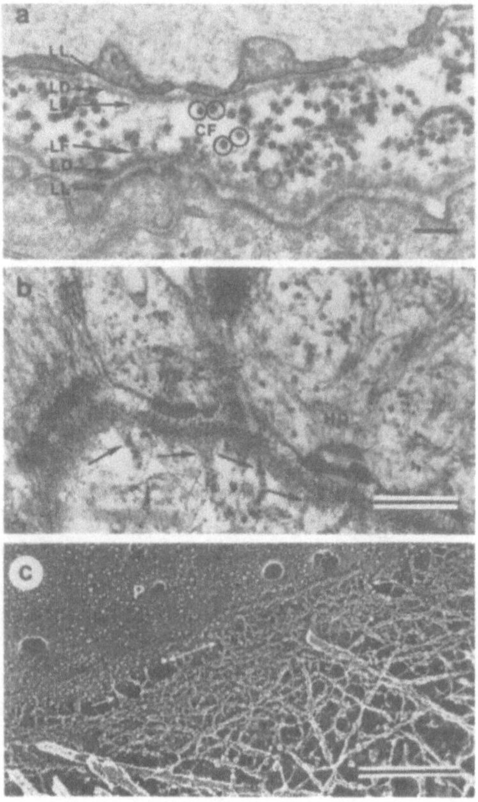

Figure 1 Electron micrographs of basement membranes. (a) Two basement membranes underlying endothelial (top) and epithelial cells (bottom) in guinea pig colonic mucosa separate the cell layers from the connective tissue with collagen fibrils (CF). Arrows point to the *lamina lucida* (LL), *lamina densa* (LD), and *lamina fibroreticularis* (LF), respectively. (b) Basement membranes (human skin epidermis) below hemidesmosomes (HD) exhibit a peripheral density, called *subbasal dense plaque.* The major component of anchoring fibrils (arrows) is collagen type VII. (c) After quick-freezing and deep-etching cell-matrix attachment sites become visible, presumably involving receptors for laminin. Arrows denote the interconnection between cell membrane and basement membrane; P: external face of the membrane bilayer; bars: 250 nm. (a, b from Reale, E., in *Ultrastructure of the Connective Tissue Matrix,* Martinus Nijhoff, Boston, 1984, chap. 11, p. 192, with permission; c, original micrograph by courtesy of J.E. Heuser, Washington University School of Medicine.)

and related proteins. Finally, we will give some hints on difficulties that can arise in mapping distinct functions to structural elements.

Basement membranes are abundant, proteinous sheets underlying epithelial and endothelial cells and surrounding muscle fibers, fat cells, and peripheral nerves. Therefore, these special kinds of extracellular matrices might serve as a prominent site to direct cell adhesion processes. In most cases, their thickness is in the range of 50 to 300 nm, and thus they represent only a small fraction of the total body mass. When analyzed by different thin section electron microscopic techniques, basement membranes appear divided into three layers, which are named with respect to their staining behavior and localization. The *lamina lucida,* about 10 to 50 nm thick, is tightly apposed to the plasma membrane (Figure 1a). It is followed by the 20 to 300 nm thick *lamina densa,* to which the *lamina fibroreticularis* (also called *reticular lamina*), containing the anchoring fibrils (Figure 1b), is attached. Sometimes the term "basal lamina" is used as a synonym for "basement membrane", but usually this refers to the *lamina densa* (for nomenclature see Reference 7). The distinct morphology is frequently thought to reflect a different composition and/or architecture of these layers. Recent advances in electron microscopy sample preparation avoiding dehydration by freeze substitution, however, show only a *Lamina densa,* and it is argued that the conventional layered appearance of basement membranes is an artifact.[8]

In contrast to other biological membranes, basement membranes seem to consist exclusively of proteins, and many of them are highly glycosylated. Aside from the low total mass, the characterization is hampered by their insolubility, due to many covalent and noncovalent interactions. Estimates based on two-dimensional gel-electrophoresis of basement membrane material suggest the presence of more than 100 different polypeptide chains.[10] Although some spots might reflect nongenuine proteins and various mature proteins consist of several protein chains, a total number in the range of about 50 different proteins is expected. The major mass, however, might be built up by only a few of them, including laminin, nidogen/entactin, collagen type IV, and different proteoglycans. Whereas these proteins are common to and specific for most basement membranes, the distribution of some others is restricted to defined places or stages in development, indicating their crucial functions.

Cellular activities influenced by basement membrane molecules include cell adhesion, migration, locomotion, spreading, differentiation, polarization, chemotactic responses, and the expression of distinct gene products. They also play a role in metastasis, in which tumor cells have to cross the basement

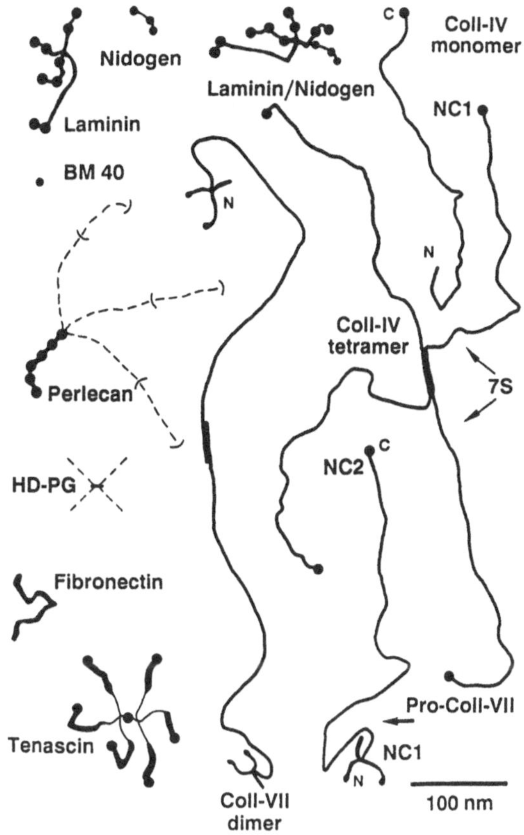

Figure 2 Redrawings of extracellular matrix proteins from electron micrographs covered in this chapter. All molecules are drawn to the scale indicated. Thickness dimensions are overestimated due to the metal decoration occurring in rotary shadowing. The dashed line in perlecan and the high density proteoglycan (HD-PG) indicate heparan sulfate chains whose lengths are variable (parentheses). N and C at the collagens (Coll) mark the amino- and carboxytermini, respectively. In contrast to collagen type IV, NC2 of pro-collagen VII is cleaved off in the mature protein. Note the extreme length, especially of collagen type IV in comparison to the thickness of basement membranes. (NC1, NC2: noncollagenous domains; 7S: disulfide-linked region of the Coll-IV tetramer, which can be prepared as a fragment sedimenting with 7S).

membrane "frontier" to penetrate into other tissues.[11,12] Some of these functions have been assigned to rather small epitopes on the different large proteins, but for others, complex mixtures seem necessary.

The close proximity of the *lamina lucida* to the plasma membrane of adjacent cell layers and the morphoregulatory function of basement membranes implicate an intimate contact between them (Figure 1c). Indeed, several cell surface receptors for basement membrane molecules, which exhibit different specificities and binding strengths, have been identified. Although some receptors belong to rather diverse classes, integrins (Chapter 8) are now thought of as the main group of integral membrane proteins that bind to extracellular matrix components.[13-17] Interestingly, at least some integrins physically interlink the extracellular matrix with the intracellular cytoskeleton.[18] This relates directly to the early observation that transformed cells grown on organized fibronectin filaments flatten and assemble actin filament bundles similar to untransformed cells.[19] In analogy to the influence of the cytoskeleton and membrane skeleton on the restricted lateral mobility of many integral membrane proteins, which cannot be understood on the basis of pure hydrodynamical theories,[20] extracellular matrix proteins might transiently interact with their ectodomains.[21]

In this chapter, we will first give an overview on the structure of the major basement membrane components and a few other extracellular matrix molecules, followed by some ideas on how they assemble into defined meshworks. In the final part, we will relate this structural basis to data on cell adhesion assays, with special reference to their limitations. For more detailed recent reviews covering similar aspects, see References 22 to 27. The relation of vertebrate extracellular matrix constituents to their counterparts in Drosophila is excellently reviewed in References 28 and 29.

II. STRUCTURE OF BASEMENT MEMBRANE-SPECIFIC PROTEINS

Some of the better characterized extracellular matrix proteins possess extremely high molecular masses, and in contrast to what is frequently thought of as the "normal" (i.e., globular) structure of proteins, they are often rather extended and ramified. Therefore, electron microscopy in conjunction with other biophysical and biochemical techniques as well as the determination of their amino acid sequence are valuable tools for a detailed structural analysis.[30-37] Figure 2 presents redrawings from electron micrographs of the shapes of molecules covered in this chapter, all at the same scale. The rather elongated shapes compared with the thickness of basement membranes suggest an ordered arrangement of its components. The branching and high flexibility of such molecules, expressed by the bended appearance on electron micrographs, hinder the crystallization for a high resolution structural analysis. When the primary structures of these molecules are analyzed and compared with the rapidly increasing number of known sequences, typically many regions show internal homologies and similarities to other proteins. This domain structure of proteins is thought to reflect their evolutionary relationship and suggests possible genetic mechanisms of reassortment of modular units.[38-42] In the figures representing the structure of some extracellular molecules, we have highlighted this aspect by relating the electron microscopic shape to the repeating sequence motifs. In some cases, a modeling of these motifs as individual independent domains arranged linearly like beads on a string leads to a surprisingly good agreement with the dimensions evaluated by electron microscopy.[34-37,47] Table 1 lists sequence motifs appearing in the proteins covered here, with some of their specificities. The increasing zoo of sequence motifs is reviewed in References 41 to 44, and powerful computer approaches for the detection of such similarities have been developed (see References 44 to 46). Fast progress in the determination of the high resolution structure of such modular proteins can be expected from the combination of the information revealed by n-dimensional NMR analysis of characteristic individual domains (see Figures 3, 4, and 8).[43,48]

A. LAMININ

Laminins now spread out to a rapidly growing family of mosaic proteins.[47,52]* The prototype of this most abundant noncollagenous glycoprotein of basement membranes was at first identified in a mouse teratocarcinoma endothelial cell line and purified from the transplantable mouse Engelbreth-Holm-Swarm (EHS) tumor.[69,70] The comparatively large amount of basement membrane molecules extractable from this tumor facilitate their characterization, and thus the most detailed data are available for proteins derived from it. A rather mild extraction protocol taking advantage of the Ca^{2+}-dependence of laminin aggregation has been established and general application has been found for laminin purification for a broad variety of tissues and species, ranging from man down to the level of medusae.[71-80] EHS laminin (M_r ~900,000) consists of three disulfide-linked genetically different polypeptide chains named A (M_r ~400,000), B1, and B2 (both M_r ~200,000). On electron micrographs, all laminins depict the unique shape of an asymmetric cross, with terminal globular domains at each arm and inner globules within the short arms (Figure 3a).[30] Two of the short arms measure about 34 nm; the third one displays two internal globules and measures about 45 to 50 nm.[73,74,81] In some laminin isoforms, especially those from rat Schwannoma, one of the short arms is truncated, and the polypeptide pattern on SDS-polyacrylamide gels upon reduction can vary considerably, indicating laminin variants of different chain composition.[72-75,79,82-89] The long arm of laminin differs in length for different species origin, from about 75 up to 110 nm, and sometimes shows two globular termini separated by a short rod.[30,72-74,77,87,88,90]

Meanwhile a complete set (A, B1, B2) of the primary structures of human, mouse, and Drosophila laminin polypeptide chains are known.[49-51,91-99] Furthermore, tissue-specific isoforms, namely s-laminin homologous to the B1 chain, and merosin, a laminin A chain isoform, have been sequenced.[79,80,84,86,100] Recently, an epithelium-specific protein termed kalinin, which is a component of the anchoring filaments and is localized adjacent to hemidesmosomes (cf. Section III.A),[87] has been characterized. Kalinin consists of three different disulfide-linked protein chains (M_r ~140,000, 155,000, and 165,000), which form a 107-nm long rod with one large and three small globular domains at opposite termini. cDNA sequences now reveal that kalinin is built up of laminin homologous protein chains, which contain only short portions of the N-terminal short arm regions.[88] The A-chain homologue of kalinin, which in its native form has an M_r of 200 kDa, can assemble with normal B1 and B2 chains to form a trimer of M_r ~600,000, called kalinin. In this complex, the kalinin A-chain, however, is shortened to an M_r of 190,000.[88] Sequences and immunolocalization studies showed that kalinin is identical to BM-600/Nicein, the GB3 antigen, which is a marker of junctional epidermolysis bullosa.[89,102]

Mouse EHS laminin sequences have been confirmed by a large set of protein-derived partial sequences.[103-105] In contrast to many other proteins whose hypothetical structures are mainly based on prediction algorithms applied on cDNA-derived sequences, protein sequencing, the characterization of proteolytic fragments and the mapping of epitopes by means of antibodies, has led to a rather accurate

* While this paper was in press, a new nomenclature for laminins has been proposed and is widely accepted: laminins are numbered with arabic numerals, and the polypeptide chains are called α, β, γ instead of A, B1, B2. Laminin-1 to laminin-6 refer to (in this order): EHS-laminin, merosin, S-laminin, a merosin/S-laminin complex, kalinin/nicein, and kalinin.[277]

Table 1 **Sequence motifs of extracellular matrix proteins covered in this chapter (see Figures 3 to 6, 8). More detailed lists of motifs, their consensus sequences, and occurrence in other proteins are given in References 39 to 46. If the sequence motif corresponds to an independent structural domain, a rough estimate of its size can be derived from $d = (6\ M_r v_2/\pi\ N_A)^{1/3}$, which assumes a compact spherical shape (d, diameter of sphere; M_r, molecular mass; v_2, partial specific volume, N_A, Avogadro's number; cf. References 34, 37).**

Symbol	Motif name	# Conserved cysteines	Mean motif length (# amino acids)	Domain size d [nm]*	Specific references and remarks
	Laminin N-terminus	6 or 8	250	4.0	49–51
	Laminin-like 8-Cys (domains III/V)	8	30–70	2.5	Cys-1 to Cys-6 region similar to EGF pattern, Refs. 37, 47, 52
	Laminin IV	8	250	2.5	As Laminin-like 8-Cys (V/III), but long insert between Cys-3 and Cys-4, Ref. 47
	Sex hormone binding globulin (SHBG)	2	150–180	3.5	37, 51, 53, 54
	Epidermal growth factor (EGF)	6	30–50	2.2	For 2D-NMR structure of EGF, see Ref. 55
	Thyroglobulin-like	6	65	2.5	Similar to EGF pattern, Ref. 56
	LDL-Receptor	6	40	2.2	57
	N-CAM, Immunoglobulin	2	95	4*2.5*2.5	Average translation 3.5 nm, Ref. 58
	Fibronectin type I	4	50	3.2*1.7*1.6	2D-NMR dimensions, Ref. 59
	Fibronectin type II	4	60	2.5	For 2D-NMR dimensions, see Ref. 60
	Fibronectin type III	0	90	2.9	Conserved Trp, Phe, Thr; Refs. 36, 61, 62 For 2D-NMR structures, see Refs. 43, 63, 64
	Fibrinogen β, γ	4	220	3.8	36, 65, 66
	Coiled-coil heptad	0	$[abcdefg]_n$	1 nm translation/ heptad	Residues a, d hydrophobic; e, g frequently charged, Refs. 47, 67, 68

* If not otherwise stated, approximate dimensions are calculated by the equation given above or result from length measurements of rod-like regions on electron micrographs.

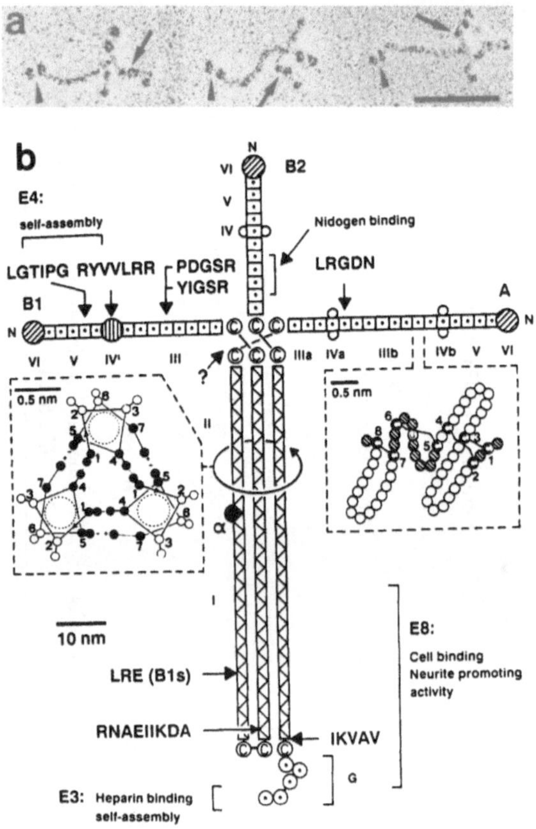

Figure 3 Structure of laminin. (a) On electron micrographs (rotary shadowing technique) laminin exhibits the shape of an asymmetric cross with terminal globular domains. In contrast to other laminins, sea urchin laminin shown here has a comparably long arm of 110 nm. Arrows point to one of the short arms with three globular domains (A-chain), arrowheads highlight the short rod-domain separating terminal globules at the tip of the long arm (bar: 100 nm; original electron micrographs of K.B. in collaboration with Robert A. McCarthy, Harvard Medical School, Boston). (b) Model of laminin based on sequences from mouse. Sequence motifs are modeled and depicted as outlined in Table 1. Different sequence regions are signed by roman numerals.[50] Synthetic peptides applied in functional assays are indicated by capital letters at their putative locus. The LRE peptide belongs to the B1-related s-laminin chain.[101] In other assays, proteolytic fragments have been used: elastase (E) fragment E8 corresponds to the lower half of the long arm, but lacks the terminal two SHBG homologous regions that comprise fragment E3. E4 consists of the N-terminal part of the B1 chain. Several other fragments produced especially by elastase and pepsin contain the short arms, which are truncated at the N-termini to different degrees (not marked).

structural model of laminin, which also allows a precise localization of functional domains (Figure 3b).[34,71,81,103-109]

The three laminin chains, A, B1, and B2, share several characteristics with each other and demonstrate a high degree of homology. The N-terminal, about 250-residue long sequence regions VI, are still unique to laminin and form the terminal globular domains of the short arms. They are followed by a different number of tandem repeating 8-cysteine motifs (regions V and III), in which the main part, covering cysteines numbers 1 to 6, is similar to the epidermal growth factor (EGF) motif, and thus an analogous disulfide arrangement is assumed. Cys-7 and Cys-8 might form a separate loop, which increases the translation per modular unit. The right inset in Figure 3b shows the putative folding of these domains, drawn in analogy to the 2D-NMR structure of EGF.[55] The shaded residues, which in EGF are rather immobile, might form the backbone of the rod-like structures of the short arms, and specific residues and distances between cysteines are conserved. Regions IV in the B2 and A chain contain a long insert of about 180 to 200 residues between the third and fourth cysteine without any further Cys, resulting in their globular appearance on electron micrographs. The sequence part of the B1 chain responsible for the inner short arm globular domain (region IV′), however, is absent in the B2 and A chain. The last 8-Cys motifs are directly followed by two further closely spaced cysteines, which most probably are involved in the disulfide linkage between all three chains similar to the disulfide knob established for fibrinogen.[110] The exact linkage pattern is not known presently. The covalent cross-linking between all three chains near this position is directly evident from the existence of three-armed fragments, which exhibit a correspondingly high molecular mass on SDS polyacrylamide gels running without reduction.[81,107] Sequence regions III to VI show high sequence identities between the chains of different species origin and tissue-specific isoforms, but a different shuffleing of modules is reported for the Drosophila A-chain.[51,100]

The adjacent regions II and I are rather different for the known laminins when compared by homology criteria. Most of their parts, however, can be arranged in repeating heptads with hydrophobic residues in their first and fourth position. Positions five and seven are frequently occupied by charged amino acids. This kind of sequence pattern (Figure 3b, left inset) is well known as a characteristic of coiled-coils of α-helices and has been specifically worked out for myosin.[67,68] Indeed, a high amount of α-helical structure has been observed for long arm fragments of laminin.[81,105,108] De- and renaturing experiments on these fragments have shown that the assembly of the different laminin chains is guided by the specific interaction of coiled-coil domains.[108,109] The selectivity for forming heterodimeric and -trimeric structures could be attributed to the distinct distribution of charged residues, especially in positions 5 and 7.[111] In the B1 chain, the heptad pattern is interrupted by an approximately 30-residue long domain α containing 6 cysteines. Near their carboxy termini, the B-chains are disulfide linked.[105] The mouse and human A-chains but not that of Drosophila contain a cysteine residue in an equivalent position as well. It might be speculated that this cysteine could form a disulfide bond with one B-chain if the other B-chain happens to lack that certain cysteine. A variable number of heptads (about 85 in mouse laminin) might be responsible for the variety observed for the long arm lengths of laminins of different species origin.[72-74]

In contrast to the B-chains, the A chain is extended by about 1000 more residues, constituting the long arm terminal globular domain G. The fivefold internal homology G1 to G5 shows similarities to the sex hormone binding globulin (SHBG), crumbs, slit, agrin, the human anticoagulation protein S, perlecan (see below), and neurexins.[37,51,53,112] Due to the homology with the Drosophila gene product crumbs, which also contain EGF-like repeats, it was suggested that domain G might be involved in epithelial polarization.[54] In mouse laminin, the outermost repeats G4 and G5 are separated by a special sequence segment from G1 to G3, which is highly susceptible to proteases.[103,107] This part varies in length in Drosophila laminin and might correspond to the rod-like segment separating the terminal globules in sea urchin (Figure 3a), leech, and Drosophila laminin.[51,72-74,90]

EHS mouse laminin is highly glycosylated (12 to 27%), nearly exclusively by Asn-linked oligosaccharides with an increased occurrence within the long arm.[69,113,114] As glycosylation is a posttranslational modification and for several tumors overglycosylations have been reported, these data cannot be transferred directly to other laminins for which data are not available. Laminin glycosylation was reported to influence tumor cell adhesion, cell spreading, neurite outgrowth, and also laminin-integrin interaction but does not seem to influence chain assembly, stability, and heparin binding.[115-120]

B. NIDOGEN/ENTACTIN

Nidogen and entactin were originally characterized as two different basement membrane proteins with a few common but several different properties. As entactin it was originally detected in matrices produced by endodermal cells in culture as a sulfated single chain glycoprotein of M_r ~158,000.[121] Later it was purified from the EHS tumor as an endogenous proteolytic fragment of M_r ~80,000 and termed nidogen.[122] The purification of its mature form (M_r ~150,000) needed strict control on protease activities, which results in a protein with a similar behavior as entactin.[71,123] The final proof that both proteins are identical is derived from the evaluation of the primary structures (about 1200 amino acids).[56,124] In the following, we refer to this molecule as "nidogen".

On a first view, the modular sequence structure seemed to support the dumbbell model derived from electron microscopy in conjunction with hydrodynamic analysis.[71,123,124] Recently, however, the structure of recombinant nidogen was investigated and, in contrast to EHS nidogen, this protein was much less susceptible to proteases, and in the electron microscope it displays the shape of three globules connected by rods (Figure 4a, b).[125] This structure could be converted to the dumbbell shape by denaturing with 2 *M* guanidine-HCl, which formerly was applied during purification.

At a first glance, a direct relation between the shape and the modular sequence arrangement seemed somehow arbitrary, but the analysis of a nidogen-related ascidian basement membrane protein reveals that the latter one does not contain a stretch of about 100 residues (domain X) formerly attributed to the aminoterminal domain (Figure 4c), suggesting that this part has evolved at a very late stage.[126] This view is supported by the sequence of human nidogen, which is 85% homologous to mouse nidogen and contains region X.[127] This region X might form the link-region between domains G1 and G2. The adjacent EGF-like domain contains a seventh cysteine residue not matching the EGF-motif consensus pattern. This cysteine is disulfide linked with one of the cysteines of the G2-region. Whereas domain X is highly susceptible for different proteases, the EGF-like/G2 region is rather stable, suggesting that they together form the inner globular domain of nidogen.[128] G2 is followed by five tandem repeats, each containing six cysteines. Whereas within the first four the cysteine pattern matches that of EGF, the last one is more similar to that found in thyroglobulin.[56] In the ascidian nidogen-like protein, the four EGF-like repeats are exchanged by two thyroglobulin-like repeats.[126] This suggests that the rod-like portion might function mainly as a spacer element. The RGD peptide found in the mouse sequence is not present in the ascidian molecule. The C-terminus is again homologous to EGF. After negative staining, on electron micrographs no indications for a separate domain could be found that might indicate its integration within domain G3. The preceding sequence shows some homology to the EGF-precursor and LDL receptor, which also in

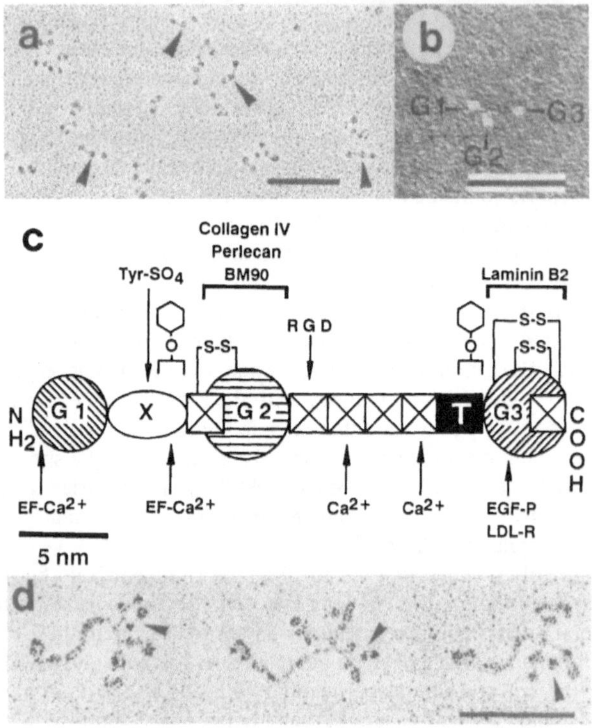

Figure 4 Structure of nidogen/entactin and the laminin/nidogen complex. (a) On electron micrographs, recombinant nidogen after rotary shadowing and (b) negative staining exhibits the shape of three globules G connected by 5 (G1/G2) and 11 nm (G2/G3) long rods. (c) The dimension of the sequence-based model (cf. Table 1) fit with those determined by electron microscopy when a sequence part X, which is absent in a nidogen-like protein of ascidian, is regarded as a separate domain of elongated shape. Binding sites for laminin, collagen IV, perlecan, and BM-90/fibulin are indicated, as well as the tyrosine sulfatation site (Tyr-SO_4) and the locations of sequence motifs that might be involved in Ca^{2+}-binding; G3 has some similarity to the EGF precursor and the LDL receptor. (d) Nidogen (arrowheads) binds to one of the short arms of laminin near the center of the cross (bars in a, d: 100 nm, b: 50 nm). (a, b reproduced from Fox, J. W. et al., *EMBO J.*, 10, 3137, 1991, with permission; d from Chiquet, M., Masuda-Nakagawa, L., and Beck, K., *J. Cell Biol.*, 107, 1189, 1988, with permission.)

these molecules is flanked by EGF-like repeats.[56] For G3 a disulfide linkage of 1-4 and 2-3 has been established.[128] Whereas G3 is responsible for the binding of nidogen to the laminin B2 chain (see below), domain G2 binds especially to collagen IV, perlecan, and BM-90/fibulin.[129,130] When region X is assumed to be of ellipsoid shape, a linear arrangement of the different sequence motifs, modeled as indicated in Table 1, results in the dimensions evaluated from electron micrographs (Figure 4c).[125] Denaturing might lead to irreversible cross-links between residues of domain G1 and G2, resulting in the increased diameter of the formerly observed aminoterminal globule of the dumbbell model.

In contrast to laminin, nidogen contains not only N- but also O-linked oligosaccharides, the latter mainly concentrated N-terminal to domains G2 and G3. Together they account for about 5% of the molecular mass.[123] A sulfated tyrosine residue is located within the G1/G2 link-region X.[123] Evidence is provided that in mouse nidogen the two sequence-predicted asparagine acceptor sites are indeed glycosylated. Seven serine and threonine residues within domain X and the thyroglobulin motif are O-linked glycosylated by di- and tetrasaccharides.[125,130] Very near the N-terminus (G1) and within region X, a 12-residue pattern of negatively charged residues matching the Ca^{2+}-binding EF-hand homology are found. Further specific Ca^{2+}-binding sites might be localized within the second and fourth EGF-homologous regions of the rod connecting G2 and G3, which contain aspartic acid residues in the critical positions preceding the first and third motif cysteines (Figure 4c).[131]

C. LAMININ/NIDOGEN COMPLEX

Soon after the discovery of nidogen, some observations pointed to its colocalization with laminin.[121,122] Both molecules appear together in guandine-HCl extracts and show interactions in a variety of recombination assays.[121,132] When laminin is extracted by EDTA as a calcium chelating agent, nidogen can be found as a 1:1 stoichiometric complex with laminin.[71,73,74] Whereas in the EHS tumor laminin and nidogen occur in approximately equimolar amounts, in culture media from various tissue explants and cell lines laminin is often found in a molar excess.[71,133] For complete dissociation of this noncovalent complex, 6 *M* guanidine-HCl is necessary. On electron micrographs, nidogen appears bound to the inner rod-like segment of one of the 35-nm long short arms of laminin via one of its terminal globules, indicating a binding site at a B-chain (Figure 4c).[71,73] Recently the nidogen binding site on EHS laminin could be mapped to the single eight-cysteine repeat four of domain III of the B2 chain, which interacts with the C-terminal domain G3 of nidogen (Figures 3b and 4b).[106] The laminin/nidogen complex is assembled shortly after translation, suggesting that complexation is necessary for transport to the extracellular compartment.[120] Due to their close association both in structure and biosynthesis, it has been suggested to regard nidogen as a laminin C chain.[22,134] In evolution the laminin/nidogen complex seems to have formed at a later stage since it could neither be found in Hydromedusae nor Drosophila, whereas it is present in sea urchin and leech.[28,73,74,90] Although the Drosophila basement membrane-related protein glutactin shares several similarities with nidogen, including a dumbbell shape agreeing in dimensions (unpublished), its primary structure is rather different.[135]

D. COLLAGEN TYPE IV

Type IV collagen[86,136,137] is the principal collagenous constituent of basement membranes. The majority of vertebrate collagen type IV monomers consists as $\alpha 1(IV)_2\alpha 2(IV)$ heterotrimers (M_r ~550 to 600,000). Each of the genetically different α chains is about 1700 amino acids long (M_r ~180,000). From the earlier known interstitial collagens I to III, they differ in that they contain several interruptions of the Gly-X-Y repeat and the terminal domains are not proteolytically processed upon secretion.[138] The chains form an approximately 400-nm long flexible rod with a noncollagenous domain NC1 at the C-terminus and a special 30-nm long collagenous triplehelical domain at the N-terminus, named "7S" due to the sedimentation coefficient of a correspondent fragment.[139,140] The complete amino acid sequences of both α chains have been determined for human, mouse, and caernor-habditis and additionally the $\alpha 1$ chains from chick and drosophila have been sequenced.[141-147] Whereas the separate α chains exhibit high sequence similarities between human and mouse ($\alpha 1$: 90%, $\alpha 2$: 84% identity), $\alpha 1$ and $\alpha 2$ are quite different, indicating an early divergence during evolution (about 40% identity; this is a rather low value as about one third of it results from the regular glycine pattern). Those sequence parts contributing to the major triple helical region contain about 20 sites deviating from the Gly-X-Y repeat in Gly→Ala substitutions and 1- to 24-residue long interruptions.[148] Although evenly distributed along the chains, interestingly most of them are coaligned along $\alpha 1$ and $\alpha 2$. These disturbances might provide the especially flexible sites analyzed by electron microscopy.[149] When mouse and Drosophila $\alpha 1$ chains are compared, about half of the interruptions are conserved with respect to their localization but not to the special interrupting sequence.[145]

On the protein level, the sequence-structure relationship is especially well characterized for collagen IV purified from the EHS mouse tumor, from which it can be purified in comparatively high amounts.[150] The analysis of several fragments was most useful in order to understand the interaction between collagen IV monomers. Interchain disulfide bonds are localized within domains 7S, NC1, and the triple helix, about one third distant from the aminoterminus.[148,151] *In vitro* studies suggest that dimerization occurs via rearrangement of disulfide bonds between adjacent NC1 domains during maturation of the assembled monomers.[152] A striking feature of the NC1 sequence is the presence of two equally-sized, homologous subdomains, each with a set of six cysteine residues in invariant positions.[151] Within the 7S region, four monomers are laterally aligned over a 30-nm distance in an alternating parallel/antiparallel fashion (Figure 2) and covalently cross-linked by disulfide and lysine/hydroxylysine bridges.[153]

Collagen IV is glycosylated by N-linked oligosaccharide and hydroxylysine-linked disaccharide units, both at the $\alpha 1$ and $\alpha 2$ chains.[154,155] At the 7S domain, Asn-linked biantennary N-acetyllactosamine type chains are found.[156] The carbohydrate units within the 7S region are thought to play an important role in the tetramerization of the collagen IV monomers by constituting a hydrophilic face and determination of the longitudinal overlap of promoters.[156]

In contrast to vertebrate collagen IV and that of *C. elegans*, for Drosophila only one α chain has been detected, which points to its arrangement as a homotrimer.[142,143,157] There are also indications for an

$\alpha 1(IV)_3$ homotrimeric molecular species in vertebrates.[158] Meanwhile, four genetically different α(IV) chains have been detected in man, which were found through investigation of the Goodpasture (α3, α4) and Alports (α5) syndrome.[159-162] The Goodpasture antigen could be mapped to the NC1 domain of α3(IV).[159] A more extended search for α3(IV) and α4(IV) epitopes showed that similar to the localization of laminin isoforms, these collagen isoforms are found enriched within several specific basement membranes.[163] Recent studies on the NC1 domain organization using HPLC, Western blotting, SDS-electrophoresis, and ELISA techniques suggest that α3 and α4 chains form $\alpha 3(IV)_2\alpha 4(IV)$ complexes, and all three possible head-to-head aggregates, $[\alpha 1(IV)_2\alpha 2(IV)]$-$[\alpha 1(IV)_2\alpha 2(IV)]$, $[\alpha 1(IV)_2\alpha 2(IV)]$-$[\alpha 3(IV)_2\alpha 4(IV)]$, and $[\alpha 3(IV)_2\alpha 4(IV)]$-$[\alpha 3(IV)_2\alpha 4(IV)]$, are possible.[164]

E. BASEMENT MEMBRANE PROTEOGLYCANS

Proteoglycans are complex aggregates consisting of glycosaminoglycan side chains covalently linked to a core protein. The polysaccharide chains consist mainly of repeating disaccharides. Most glycosaminoglycans are linked via a short specific carbohydrate sequence (three to four units) in an O-glycosidic linkage to a serine residue of the core protein. In the extracellular matrix, many different types are present, and some can form extremely large complexes several micrometers in length. In basement membranes, some specific heparan sulfate proteoglycans have been found, which show a considerable diversity with respect to their mass and the mass ratio of carbohydrate to protein.[165-171]

The mouse EHS tumor is also a valuable source for the isolation of proteoglycans, and the structure of two forms differing in their buoyant density, namely the low density and high density proteoglycans, has been characterized.[169-172] The low density proteoglycan consists of a single chain core protein ($M_r \sim 500{,}000$) with only three heparan sulfate chains attached, whose lengths can vary between tumor strains ($M_r \sim 30{,}000$ to 80,000 per chain).[172] On electron micrographs, the core protein appears as a nodular 80-nm long rod with six globular domains of different sizes. The heparan sulfate chains arise from one pole of the core (Figure 5a).[172,173] Due to this "beads on a string" appearance ("perle") and the variable posttranslational glycosaminoglycan modifications ("can"), this molecule was termed perlecan.[174]

Recently the core protein sequences for mouse and human perlecan have been established.[174,175] They both show a surprisingly clear pattern of already well-known sequence motifs (Figure 5b). Only the approximately 170 aminoterminal residues form a still unique domain I, which is rich in negatively charged amino acids and does not contain any cysteine. Within a short stretch, the tripeptide SGD occurs three times, which is common to syndecan, glypican, and versican, and therefore they are assumed as the putative attachment site for the glycosaminoglycan chains. The adjacent ~210 residues (domain II) contain four internal repeats with striking similarities to the LDL-receptor family with an about 40-residue long interruption after the first motif. After a single N-CAM motif, 11 8-cysteine repeats very similar to those of the laminin chains follow. Interestingly this domain III also contains three units with a long insert between the third and fourth cysteine, as discussed for domain IV of laminin. The following domain IV has a different number of tandem repeats of the immunoglobulin gene family with great similarity to the neural cell adhesion molecule, N-CAM.[176,177] The mouse sequence contains 14 repeats in contrast to 21 found in the human sequence. By comparison of similarities in addition to the consensus sequence, Kallunki and Tryggvason concluded that the seven human-specific motifs might be regarded as an insert into the fifth repeat of the mouse protein.[175] Whether this is due to either a very recent duplication or alternative splicing is unclear, but no alternative mRNAs have been observed in mouse and man.[175] The carboxyterminal domain V is similar to the C-terminus of the laminin A chain, but the three motifs homologous to SHBG are separated by two doublets of EGF-like cysteine motifs (cf. Table 1). The relation of domains III and V to laminin raises several speculations both on the evolutionary origin and common functions.[174,175] The sequence homology to the established heparin-binding site at domain G5 of the laminin A-chain and some preliminary N-terminal protein sequence data of proteolytic perlecan fragments raise the question of whether it is the C-terminal domain of perlecan to which the heparan sulfate chains are bound instead of the N-terminal domain.[178]

In contrast to laminin, nidogen, fibronectin, tenascin, and other proteins, a simple "beads on a string" arrangement of the different sequence motifs fails to predict the length correctly, as observed on electron micrographs. This calculation would result in about 110 nm instead of 80 nm length (Figure 5a).[172] In analogy to laminin domains IV, the perlecan domains III a to c might form three of the six observed nodules. Domains I and II together can be thought to correspond to one of the terminal globules. The opposite terminal domain could result from homophilic interactions between the SHBG homologous regions. In laminin this part looks like a homogeneous domain as well when analyzed by the rotary shadowing technique. The N-CAM-related motifs might also show some back-folding to form the fifth

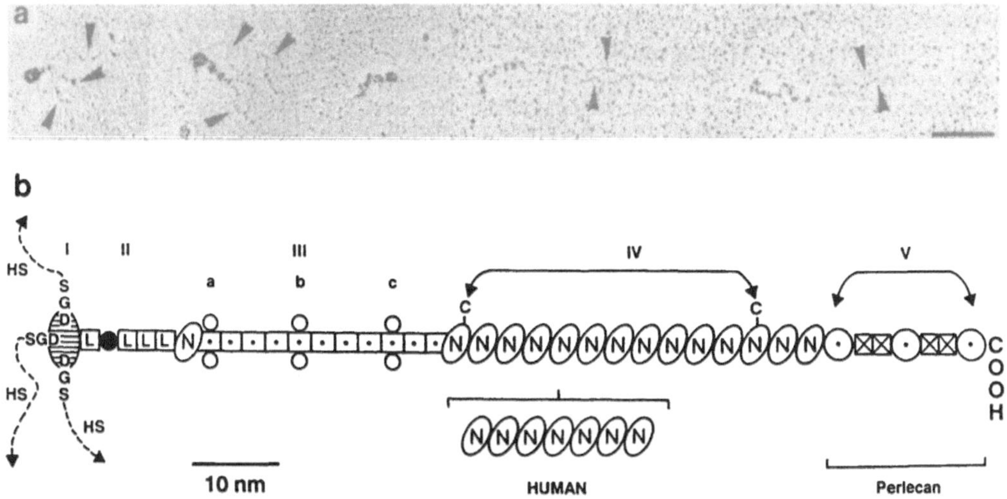

Figure 5 Structure of perlecan. (a) On electron micrographs, the low density heparan sulfate proteoglycan exhibits a nodular rod with up to six subdomains. To one terminus, three very thin strands of variable length are attached (arrowheads, bar: 100 nm). (Reproduced from Paulsson, M. et al., *J. Mol. Biol.*, 197, 297, 1987, with permission.) (b) Model of perlecan based on sequences from mouse and human (cf. Table 1). The amino terminal domain I is unique to perlecan. The peptides SGD appearing thrice might serve as the heparan sulfate attachment sites.[174] Some biochemical data, however, suggest that the heparan sulfate chains are bound to the C-terminal domain V. Domains II to V consist of repeating motifs already known from other proteins. The row of LDL receptor-like motifs L is interrupted by a short stretch unrelated to other motifs (black sphere). The human sequence contains seven N-CAM motifs, in addition to those of mouse. To fit the total length evaluated by electron microscopy, a backfolding within different sequence regions must be assumed, possibly within domains IV and V, as indicated above the model. Independent of whether the glycosaminoglycan chains are attached to the C- or N-terminus, the opposite domain is involved in the self-aggregation of perlecan.[173]

of the six observed globules. Interestingly, the first and third N-CAM motif counted from the N- and C-terminus of domain IV, respectively, of both sequences contain an extraneous uneven cysteine, which could stabilize such a putative loop structure by a disulfide bond.[174]

The structure of the high density proteoglycans is less well understood. For the EHS tumor, a proteoglycan has been described with a core protein of M_r ~5,000–12,000, to which four instead of three heparan sulfate chains are connected, each with a length of about 30 nm and a mean mass of M_r ~ 29,000 (Figure 2).[169] Several other high density proteoglycans have been found in the EHS tumor and also normal basement membranes.[166-168,170,171] Besides heparan sulfate, some additionally contain chondroitin sulfate chains, which have not been observed for low density proteoglycans.[22,179]

F. OTHER BASEMENT MEMBRANE PROTEINS

Whereas laminin, nidogen, collagen IV, and some proteoglycans can be regarded as the principal components of basement membranes, it has now become evident that specific isoforms of these molecules are on the one hand localized only in special basement membranes and on the other hand can occur with several diseases.[86] Other basement membrane proteins show a restricted distribution, pointing to their special functions.

Agrin is highly concentrated in synaptic basement membranes and probably secreted at the nerve terminal.[180-182] It has the ability to aggregate acetylcholine receptors and acetylcholinesterase on the surface of muscle cells.[183] Recently, the full length sequence of rat agrin (1940 residues) has been reported.[184] The amino terminal half is built up by eight ten-cysteine motifs, homologous to the Kazel-type protease inhibitor domain. Then four eight-cysteine motifs similar to those of laminin domain III and a further ten-cysteine motif follow. The carboxy terminal part, which is involved in the clustering of the acetylcholine receptor and in membrane binding, contains four six-cysteine EGF-like motifs and three laminin type G regions.[37,51,53,184] A novel noncollagenous protein (M_r ~58,000) has been found in the tubular basement membrane of kidneys in patients suffering from tubulointerstitial nephritis where it appears in considerable amounts, but structural details are not known.[185] This is also the case for *Bamin* (M_r ~72,000–80,000), which has been purified from the EHS tumor and can be found in normal glomerular basement membranes.[186]

Many other proteins are present in basement membranes that are not, however, specific components. Thus, basement membranes are thought of as a storage compartment for the basic fibroblast growth factor, which binds to heparan sulfate proteoglycans.[187,188] Complement factors C3d, C3g, von Willebrand factor (factor VIII), and also thrombospondin appear in different basement membranes.

III. STRUCTURE OF OTHER SELECTED EXTRACELLULAR MATRIX PROTEINS

A. COLLAGEN VII

Collagen type VII[189] is one of the largest known collagens. It has been recognized as the major component of anchoring fibrils that seems to secure the *lamina densa* of certain epithelia to their subjacent stroma (Figure 1b).[190,191] The protein exist as an $\alpha 1$(VII) homotrimer. The pro-α-chain (M_r ~350,000) contains a noncollagenous domain NC1 (M_r ~145,000), an approximately 425-nm long triple helical collagenous rod with a major discontinuity near the middle (M_r ~170,000), and a small globular domain NC2 (M_r ~32,000, cf. Figure 2).[192,193] NC2 serves to align two monomers, which assemble in an antiparallel manner within an approximately 60-nm long overlapping region. Here the dimer is stabilized by disulfide bonds, and NC2 is proteolytically removed.[193-195] Such 800-nm long dimers are thought to form the anchoring fibrils by unstaggered lateral associations.[193] Based on ultrastructural and biochemical studies, it was assumed that NC1 and NC2 correspond to the carboxy- and aminoterminus, respectively. Recent sequence data and mapping of monoclonal antibody epitopes, however, suggest that the opposite nomenclature is correct.[196,197] On electron micrographs, NC1 exhibits three fuzzy-like 36-nm long strands, each of which is most probably formed by one α chain.[195] The sequence of this region starts with a domain homologous to the cartilage matrix protein, followed by nine consecutive fibronectin type III domains and a region homologous to the Von Willebrand factor type A motif continued by [Gly-X-Y]$_n$ repeats with several interruptions.[61,196,197] The NC1 termini are not processed but seem to plunge into the *lamina densa* and the anchoring plaques (Figure 1b).[192] In basement membranes, NC1 was found in close proximity to collagen IV and laminin, and *in vitro* it binds especially to the NC1 domain of collagen IV, laminin, kalinin, and K-laminin, but not to the helical domains of collagens I, II, or VII.[189]

B. FIBRONECTIN

Fibronectin[198,199] is the most extensively studied noncollagenous protein of the extracellular matrix. Obviously, it is not a basement membrane protein in the sense that it is predominantly present in these structures. A critical comparison of the localization studies of matrix fibronectin showed that fibronectins are widely distributed in basement membranes and connective tissue matrices throughout the body.[198] Here we will only summarize the gross structure of fibronectin and mention the interaction sites for other molecules and cells (Figure 6). Fortunately, the huge numbers of studies on all aspects relating to fibronectin have been critically reviewed.[198,199] Fibronectin consists of two similar protein chains (M_r ~250,000), which are held together by a pair of interchain disulfides at the C-terminal end. On electron micrographs it appears as an approximately 120-nm long flexible strand with a pronounced kink near the middle (Figure 6a), but shape and dimensions depend critically on the preparation technique, ionic strength, and pH used.[30-32,34,35] At specific conditions some nodularity can be seen.[200] Meanwhile, the protein sequences have been determined for several species and tissue isoforms, including human, rat, bovine, and chicken (see Reference 198). Molecules that share several biochemical similarities with fibronectin are also described for Drosophila, Coelenterates, and even some Porifera, but sequence data for these proteins are not available.[78,201-203] Especially for Drosophila, an extensive search for a fibronectin cDNA using PCR techniques has thus far failed, and it remains an open question whether Drosophila does or does not have a fibronectin homologue (for discussion see Reference 29).

An intriguing finding arising from the sequence at that time was that it is composed of a series of three different homologous repeating units, named type I, II, and III (Table 1, Figure 6b).[204] Especially in the N-terminal half, these are separated by short spacer elements. In contrast to the type I and II sequence regions, type III repeats do not contain any cysteines, and the sequence homology of these motifs is much less pronounced than for others. When each segment is thought to be folded into small globular domains and they are linearly arranged, the length fits perfectly with the electron microscopical data.[34,35] This kind of modeling also holds for computing the hydrodynamical behavior of fibronectin.[30,31,34,205] Meanwhile,

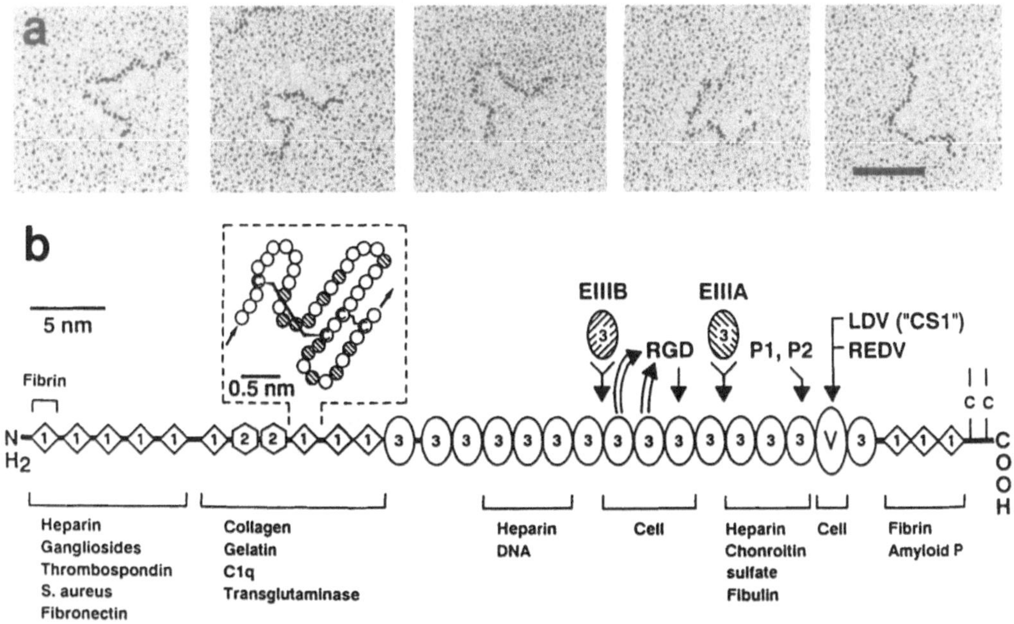

Figure 6 Structure of fibronectin. (a) On electron micrographs, fibronectin appears as a flexible rod-like structure with a length of about 120 nm (bar: 50 nm). (Reproduced from Engel, J., et al., *J. Mol. Biol.*, 150, 97, 1981, with permission.) (b) Model of fibronectin based on amino acid sequences. Only one of the two similar chains that are disulfide linked (C) near the C-terminus is shown. The sequence is made up of fibronectin type I, II, and III modules (cf. Table 1). Sequences are modified by alternative splicing, resulting in a different number of type III repeats (EIIIA, EIIIB). Further modifications occur in the variable domain V, which is not homologous to the other domains. The inset depicts the secondary structure of the seventh type I domain, as analyzed by 2D-NMR.[59] The loci of synthetic peptides proved to mediate cell adhesion are indicated above the model. Peptides P1 and P2 comprise 19 and 15 residues, respectively.[16] Open arrows pointing to the integrin-binding peptide RGD mark the synergistic effect arising from the adjacent type III motifs. Some of the binding sites for cells and other molecules evaluated by proteolytic fragments are shown below the model. A site for fibronectin self-interaction consists of the N-terminal five type I domains.

the atomic structure of examples of all three types of fibronectin modules has been evaluated by 2D-NMR and/or X-ray crystallography (Table 1).[59,60,63,64] Type I and II sequences are thought to be encoded by single exons, whereas type III sequences usually are encoded by two exons.[206-208] Whereas the number of type I (12) and II (2) domains is constant, a different number of type III repeats results due to alternative splicing. Two extra type III domains, EIIIA and EIIIB (also mentioned as ED-A, ED-B), that are encoded by a single exon can be either spliced out or included during RNA processing. Another domain V (or IIICS) is not homologous to other parts, and its size is rather variable. With these small variations (in relation to the entire gene product), the cell can produce a large set of different isoforms, probably to serve different functions, but these presently are not understood.

Fibronectin undergoes several posttranslational modifications. It contains both N- and O-linked carbohydrate chains, accounting for about 5 to 7% molar mass, and it is sulfated and phosphorylated.

C. SPARC/BM-40/OSTEONECTIN

A calcium binding extracellular matrix glycoprotein of M_r ~40,000[209] has been partially characterized originally by several sources, and thus it appears under different names in the literature. It was called "osteonectin" when first purified from bone, where it was thought to influence mineralization.[200] As a Secreted Protein Acidic and Rich in Cysteine (SPARC) it is produced from endodermal and endothelial cells in large quantities, which also typically synthesize other basement membrane components.[211,212] When purified from the EHS tumor, it was named "BM-40", suggesting it as a basement membrane-specific protein.[213] cDNA sequencing revealed that all these proteins made up of 285 to 287 residues are nearly identical.[212,214,215] Meanwhile, this protein has been localized in several connective tissues, but yet,

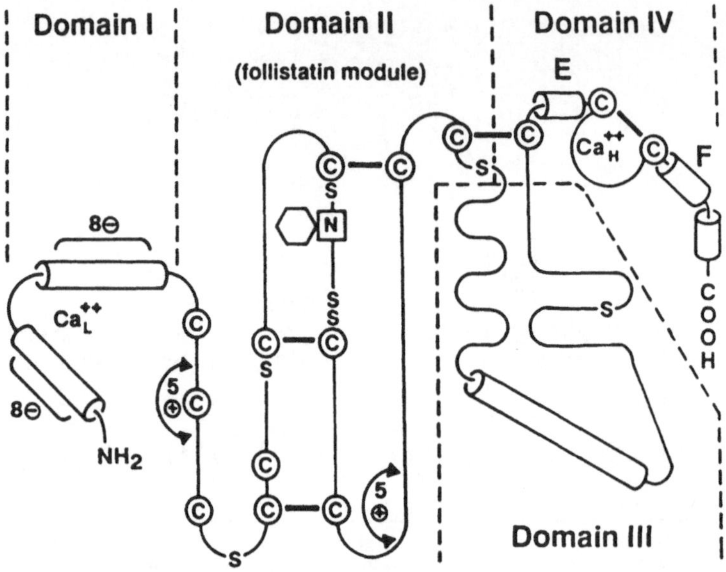

Figure 7 Model of SPARC/BM-40/Osteonectin with some of the established or proposed disulfide bonds. Dashed lines outline the domain boundaries, according to Engel et al.[220] α-helical parts are drawn as cylinders. Domain I consists of two α-helices, each with eight negatively charged residues, and has a low affinity for calcium ions. Domain II exhibits some similarity to follistatin, is rich in cysteines C and serines S as putative phosphorylation sites, contains the known glycosylation site (hexagon) and two clusters of positively charged residues. One part of domain III is predicted as α-helical. Domain IV has a high affinity calcium-binding site, probably due to the sequence similarity to the EF-hand motif.

besides calcium binding, the biological functions remain an open question. For simplicity in the following, we refer to this molecule as SPARC.

On electron micrographs, SPARC adopts a globular to worm-like shape with some nodularity, indicating the presence of two to three domains arranged in a row.[218] By analytical ultracentrifugation, drastic Ca^{2+}-induced shape changes could be detected: assuming a prolate ellipsoid of revolution, data fit to an axial ratio of 5 or 10 in the presence or absence of Ca^{2+}, respectively.[218] The 285 residues make up a molecular mass of about 33,000, and one diantennary carbohydrate complex linked to Asn_{98} accounts for a $M_r \sim 1{,}300$.[219] The domain structure of the protein is less evident than for the other proteins covered in this chapter, but a four domain model has been established from sequence analysis and is in agreement with the cleavage pattern resulting from trypsin digestion (Figure 7).[218,220,221]

The N-terminal domain I consists of two approximately 20-residue long α-helices, each with eight negatively charged amino acids. Domain II contains eleven cysteines, two clusters of five positively charged residues, the glycosylation site, and six of the seven serine residues, which might act as phosporylation sites. This sequence part has a clear homology to the follistatin module, which also occurs in ovomucoid and the pancreatic secretory trypsin inhibitor.[41,222] Domain III is free of cysteine and proline, and one part is predicted as α-helical. The carboxyterminal domain IV contains a helix-loop-helix sequence motif, known as "EF-hand structure", where the loop in contrast to other such motifs is stabilized by a disulfide bridge. The Ca^{2+}-binding function of such sequence regions is best established for parvalbumin and calmodulin.[223] Other extracellular proteins with putative EF-hand Ca^{2+}-binding sites are fibrinogen and nidogen (see above), and the ectodomains of integrin α-chains also contain this motif three or four times.[224] The somewhat similar sequence repeats in the C-terminal part of thrombospondin are thought to form ω loops.[225] Besides the EF-hand domain IV, representing the high affinity Ca^{2+}-binding site (K_d ~0.3 μ*M*), domain I is also involved in the Ca^{2+}-binding activity of SPARC ($K_d \geq 10$ m*M*) and undergoes drastic structural changes upon binding.[218,220,226]

Recently, recombinant human SPARC was analyzed and found very similar to authentic SPARC, especially with respect to its binding activity for calcium and collagen IV.[227] Site-directed mutagenesis showed that the cysteine linkage in domain IV is essential for Ca^{2+}-binding.[226] In other proteins with EF-hand Ca^{2+}-binding sites, this stabilization might not be necessary, as there the pairwise occurrence of such domains could help strengthen the conformation. Deletion mutants lacking either the C- or N-terminal half of the α-helical stretch of domain III do not bind to collagen IV, suggesting that this part contains the binding site.[221,226] For collagen type IV-binding the high-affinity Ca^{2+}-binding site must be saturated.[221]

A further calcium-binding glycoprotein, named BM-90, has been purified from the EHS tumor.[228] It is of M_r ~90,000, and additionally to basement membranes, it appears in substantial amounts in serum. N-terminal sequences of a set of BM-90 proteolytic fragments, some common biochemical characteristics, and finally the complete cDNA derived sequence coding for 676 residues indicated that it is identical to fibulin.[229,230] Initially due to its interaction capability with the cytoplasmic domain of integrin $\alpha_5\beta_1$, fibulin was thought to be an intracellular protein. Expression and immunofluorescence studies, however, showed that fibulin is a secreted glycoprotein which becomes incorporated into a fibrillar extracellular matrix.[229] Fibulin consists of a unique 539-amino acid long cysteine-rich polypeptide with three further isoforms which differ in their carboxytermini due to alternative splicing. BM-90 purified from the EHS tumor has the longest C-terminal domain (137 residues).[230] The N-terminal domain contains three repetitive six-cysteine motifs resembling those found in the complement component anaphylatoxins, C3a, C4a, and C5a. They are followed by nine EGF-like repeats which (except for the first one) start with a calcium binding consensus sequence.[131,229,230] On electron micrographs, BM-90 appears as a 30-nm long rod which agrees well with an expected length of 25 to 30 nm (see Table 1).

D. TENASCIN/CYTOTACTIN/HEXABRACHION

Tenascin was originally discovered by several groups, and thus different names exist[231-233] including (in historical order) CSFN oligomer/hemagglutinin, GP-250, Glioma Mesenchymal Extracellular Matrix antigen (GMEM), Myotendinous antigen, Hexabrachion/brachionectin, J1 glycoprotein, and cytotactin (for references see References 231 and 233). All biochemical data point to the fact that these groups have worked on the same class of protein. With the intention to overcome this babylonic speech disaster and to avoid some misleading implications, Chiquet-Ehrismann et al. tried to establish "tenascin" as a new name but with limited success.[234] This name refers to the frequent localization of the protein in Tendons and its increased expression in development and during wound healing (nascere). However, a broad tissue distribution of this molecule with different turnover rates in adult and embryonic tissues has been found, and it seems clear that tenascin is *not* a basement membrane constituent (for a comprehensive review see Reference 231). In early studies, some confusion on the originality of this molecule arose from functional and structural similarities to fibronectin.[233] A major breakthrough in its establishment as a unique molecule comes from the electron microscopic evaluation, where it appears as a star-like structure with six arms arising in two triplets from a central globular domain (cf. Figure 8a).[33] Further analysis has shown that, aside from the hexameric structure ("hexabrachion"), molecules with 3, 9, and even 12 arms also exist in different amounts, but recently from adult chicken gizzard single-armed tenascin molecules have also been characterized.[231,235] Six-armed "tenascin-like" proteins have also been described for leech and Porifera, and transcripts very similar to tenascin were found in Drosophila.[73,76,236-238]

Since complete or major parts of the tenascin protein sequence are available for chick, mouse, porcine, and human (M_r ~190,000 to 240,000), the related behavior of tenascin to fibronectin but also its differences appear obvious.[36,239-242] Except for a unique N-terminal sequence (cf. perlecan) all other parts share striking similarities with already known structures (Figure 8b). The approximately 150 aminoterminal residues are followed by 4 α-helical heptad-repeats, proposed to serve for the assembly to the trimeric basic unit ("T-junction") of the hexamer. This triple coiled-coil structure could be stabilized by the uneven number of cysteine residues adjacent to this domain.[36] When hexameric tenascin is reduced and acetylated, single-armed molecules result.[243] The size of the adjacent 13.5 or 14.5 EGF-like 6-cysteine motifs is limited to the minimum of residues observed so far for this sequence pattern. Nearly all of the repeats comprise 31 residues, matching in glycines and cysteines. The mouse and porcine sequences contain exactly one repeat more than the chick sequence. This part is followed by fibronectin type-III repeats with conserved tryptophan, leucine, and threonine residues. The number of repeats is variable with respect to splicing variants, as was already observed for the extra domains in fibronectin. The total number known so far varies between 8 repeats in the minimal variants up to 16 repeats in the maximal variant of human tenascin. Common to fibronectin, the borders of the extra domains fit directly to the formal sequence motifs, indicating the value of this kind of interpretation. Whereas in fibronectin alternative splicing occurs at different sites, the special repeats of tenascin appear consecutively as one insert between repeats common to all known isoforms. After an eight-residue long stretch, the C-terminal domain (about 220 residues) follows, which exhibits a significant homology to a fibrinogen β and γ domain. A linear alignment of the sequence motifs as globular units results in the correct length

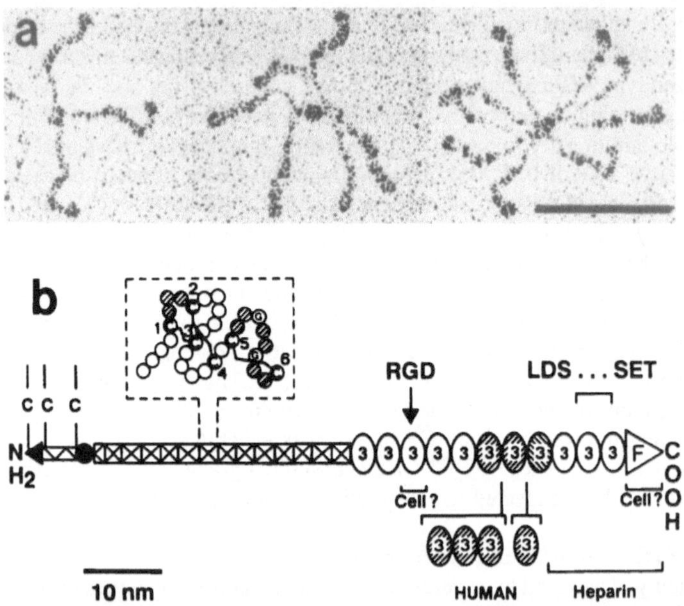

Figure 8 Structure of tenascin. (a) On electron micrographs, tenascin (here from chicken) appears as a star-like structure with 3, 6, 9, and even 12 arms, each consisting of one polypeptide chain. Three arms are connected in a T-junction-like manner near the central globular domain. The inner rod-like structures are thinner than the distal portions, which terminate in globules (bar: 100 nm). (b) Model of one polypeptide chain. The N-terminal tenascin-unique motif is followed by a short heptad repeating sequence, which bears cysteines C near the ends. These might form interchain disulfide bonds to adjacent tenascin chains. After an intermediate domain, 13.5 or 14.5 EGF-like motifs follow, which might represent the inner thin rod-like parts. The adjacent 8 to 16 sequence repeats similar to the fibronectin type III motif are thought to build up the thick distal rods. The localization of motifs present in alternatively spliced variants are shown by different shading. The C-terminal part of the sequence (F) has some homology to fibrinogen. Several sites are proposed for cell binding (see Section V). A C-terminal proteolytic fragment was shown to bind to heparin.[235]

dimension, as determined by electron microscopy.[36,37] When the extra fibronectin type III sequence repeats are compared, they show a higher similarity with each other but also more differences to the other common repeats. Furthermore, they contain more putative sites for N-linked glycosylation, and rough estimates on the carbohydrate content have substantiated that in chicken tenascin, the extra domains indeed are highly glycosylated.[36,243]

Recently, for the neural extracellular matrix, a tenascin-homologue protein, called restrictin has been characterized. This molecule is different in its aminoterminal portion but rather similar in its parts starting from the EGF-like repeats.[66] Furthermore, an approximately 80-kb segment, which codes for heptad repeats, EGF-like, fibronectin type III, and a fibrinogen-like domain, has been found in the centromeric portion of CYP21 within the human MHC class III region. Two additional tenascin-like genomic DNAs have been found in chicken. These data implicate that at least three different genes for tenascin-like proteins exist.[244]

IV. INTERACTION OF BASEMENT MEMBRANE PROTEINS

The supramolecular architecture and the assembly of the proteins to form the sheet-like basement membrane arrays have been studied both by *in vitro* as well as *in vivo* approaches (for a specialized review see Reference 26). In principle, in the first instance purified proteins are mixed at different stoichiometries and conditions in order to look for any interaction products. These are frequently characterized by electron microscopy, centrifugation, turbidity measurements, and several biochemical techniques. In these studies, it is presumed that the used components are sufficient to form at least the basic network. As a guideline, informations from biosynthetic and developmental studies are considered. Although the major constituents of basement membranes are now characterized in several aspects, it cannot be excluded that components present only in rather small amounts might play important roles in the self-assembly process. Furthermore, cell culture experiments suggest some influence of cellular cooperation on the formation of basement membranes. Most of the experiments described below were performed with proteins purified from the EHS tumor.

In most *in vivo* studies, the distribution of the different proteins along the thickness of the basement membrane is analyzed by immunoelectronmicroscopy, where antibodies are labeled with gold, ferritin, or latex beads (reviewed in Reference 245). Aside from a number of technical limitations, the large size of the proteins frequently hampers a high-resolution mapping. To overcome this problem, antibodies against defined epitopes have been used in a few studies, which allowed determination of the orientation of some of the proteins.

A. LAMININ SELF-INTERACTION

Since in early embryogenesis laminin is the only secreted basement membrane protein, studying its polymerization in the absence of other components might mimic a physiologically relevant situation. *In vitro* laminin aggregates to large complexes, depending on temperature, time, concentration, and the presence of divalent cations; upon addition of EDTA this process is reversible.[246] Polymerization in solution shows a critical laminin concentration of about 60 n*M*.[246] When aggregation, however, is studied in the presence of a planar bilayer of sulfatides as a laminin-binding matrix, this concentration is significantly smaller.[247] In both cases, divalent cations are essential for polymerization, and in solution half-maximal aggregation was found at about 10 μ*M* calcium or a tenfold higher amount of magnesium.[248] Indeed laminin has about 16 binding sites for calcium ions, but only 1 to 3 of them (at least one within the short arm regions) are of sufficiently high affinity to account for polymerization.[248] These results are in accordance with the good extractability of laminin with EDTA-containing buffers from several tissues.[71-80] In adult mature tissues, the dissociation of laminin aggregates is restricted due to nonreducible cross-links.[75] *In vitro* such a covalent stabilization has been observed due to the action of transglutaminase, which might induce isopeptide bond formation.[249]

Laminin self-aggregation proceeds via the terminal globular domains.[246,248,250] Whereas initially both long and short arms were thought to be involved, more recent studies point particularly to the importance of the short arms. Fragments E4 and E1′ (cf. Figure 3; E1′ consists of the three short arms but lacks the E4 portion) inhibit laminin aggregation. Whereas E4 binds to laminin as well as E1′ but not to itself, E1′ alone is able to self-interact. Therefore, the short arms seem sufficient for aggregation.[250] When laminin orientation is studied *in situ* by fragment-specific antibodies, short arm epitopes are mainly found pointing to the center of mouse cornea basement membranes, whereas the distal portion of the long arm (corresponding to fragment E3) is often directed to one of the surfaces.[251] Also in mouse kidney, long and short arms show highly variable orientations, suggesting a dynamic or heterogeneous structure of basement membranes.[251,252] The frequently observed proximity of the long arm to the adjacent cell membrane fits with the localization of a cell binding site to the long arm (compare the chapter on integrins; for reviews see also References 14 to 17). Recently the collagen-free basement membrane of cultured embryonal carcinoma cells and the laminin organization in the EHS tumor after collagenase treatment were investigated by freeze-etching and platinum/carbon replication electron microscopy as well as biochemical techniques.[253] This study confirmed the *in vitro* findings that laminin forms a network of its own independent of type IV collagen. In the EHS basement membrane, about 80% of laminin is anchored through noncovalent bonds between laminin monomers, whereas 20% is anchored through a combination of these bonds and laminin-collagen bridges probably mediated by nidogen (cf. Figure 11).[253]

B. COLLAGEN IV SELF-INTERACTION

Collagen IV monomers exhibit different modes of self-interaction. Linear disulfide-linked dimers are formed via binding between NC1 domains.[140,151,152] Spider-like tetramers assemble via lateral alignments within a 30-nm overlap region at the N-terminal portions, forming the 7S region (Figure 2).[153-155,254] The combination of these two interaction types has led to a first collagen IV meshwork model, which in its most extended conformation would result in a 800-nm mesh size.[140] In gelation experiments starting with NC1-linked dimers, further side-by-side interactions were found. The formation of the resulting rather complex three-dimensional polygonal network depends on temperature and concentration.[254-256] The lattice sides consist of two to three triple helical strands, and the involved dimers are irregularly staggered with a median of 170 nm. Several alternative models for this type of assembly have been worked out.[255,256] A major role for this kind of collagen polymerization was attributed to the NC1 domain, as antibodies against NC1 effectively block lateral associations and free NC1 reduces network formation in a concentration-dependent

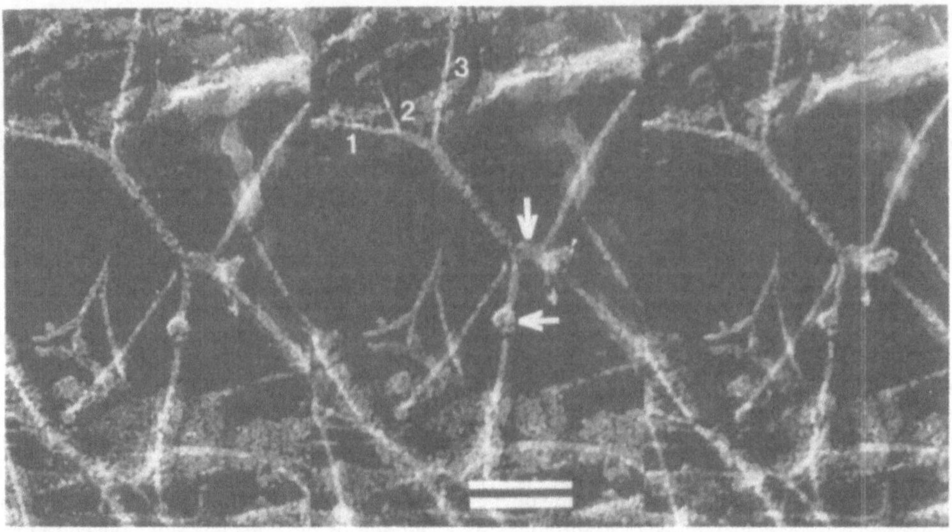

Figure 9 Three-frame stereo view of the collagen IV network in amniotic basement membranes. After extraction of all noncollagenous glycoproteins by guanidine-HCl, the remaining structure was freeze-dried and unidirectionally metal-shadowed at a high angle. Three filaments (1 to 3) twist around each other and form a laterally associated imperfect helical supercoil. Vertical arrow marks a branch point of several fibers. Horizontal arrow points to a globular domain, which might correspond to NC1 domains. (bar: 50 nm) (Reproduced from Yurchenco, P. D. and Ruben, G. C., *Am. J. Pathol.*, 132, 278, 1988, with permission.)

manner.[257] A synthetic peptide corresponding to residues 49 to 60 of the α1 chain was reported to inhibit network formation specifically.[258]

A genuine breakthrough on the *in situ* organization of collagen type IV has been achieved by Yurchenco and Ruben using freeze-drying unidirectional metal shadow casting electron microscopy in combination with antibody decoration and stereo reliefs.[259,260] After removal of nearly all other components by high salt and guanidine treatment, human amniotic and EHS tumor basement membranes show an irregular polygonal network of collagen IV with similarities to those formed *in vitro* (Figure 9). Laterally associated molecules form supertwisted helices with many branching points. This branching might be due to several noncollagenous interruptions in the triple helical domain of the monomer. Although this meshwork is qualitatively rather similar to that formed *in vitro* starting from collagen dimers, it shows a considerably tighter packing density, with a mesh size of about 50 nm. A working model on the collagen network structure integrating both the *in vitro* and the *in situ* studies is presented in Figure 11A (see also Reference 253).

C. INTERACTIONS OF BASEMENT MEMBRANE PROTEIN MIXTURES

Several studies have been performed with complex mixtures of basement membrane proteins. First experiments using mainly electron microscopy especially focused on the localization of any binding sites on collagen IV monomers for laminin resulted in rather contradictory observations.[261-263] Whereas these reports agree in that interaction occurs within the collagenous rod domain, they suggest preferential binding activities for the terminal globules of either laminin short or long arms, and furthermore the favored collagen epitopes are in direct disagreement. Experiments using the laminin/nidogen complex and nidogen alone showed binding to collagen for both components. When nidogen, however, was dissociated, the remaining laminin was unable to bind to collagen IV.[264] These findings suggest that nidogen mediates laminin-collagen IV interaction. By recent experiments carried out with recombinant nidogen and complementary deletion fragments thereof, this view has been substantiated.[125,129] Whereas intact nidogen binds to a short arm laminin fragment lacking any terminal globular domains (Figure 10a) as well as to the rod portion of collagen IV (Figure 10b), the aminoterminal nidogen fragment could only interact with collagen (Figure 10c). In contrast to this, the carboxyterminus of nidogen binds to laminin (Figures 3b, 4c).[106,125,264,265] In ternary mixtures, the laminin fragment is connected to collagen via intact nidogen (Figure 10d).[125]

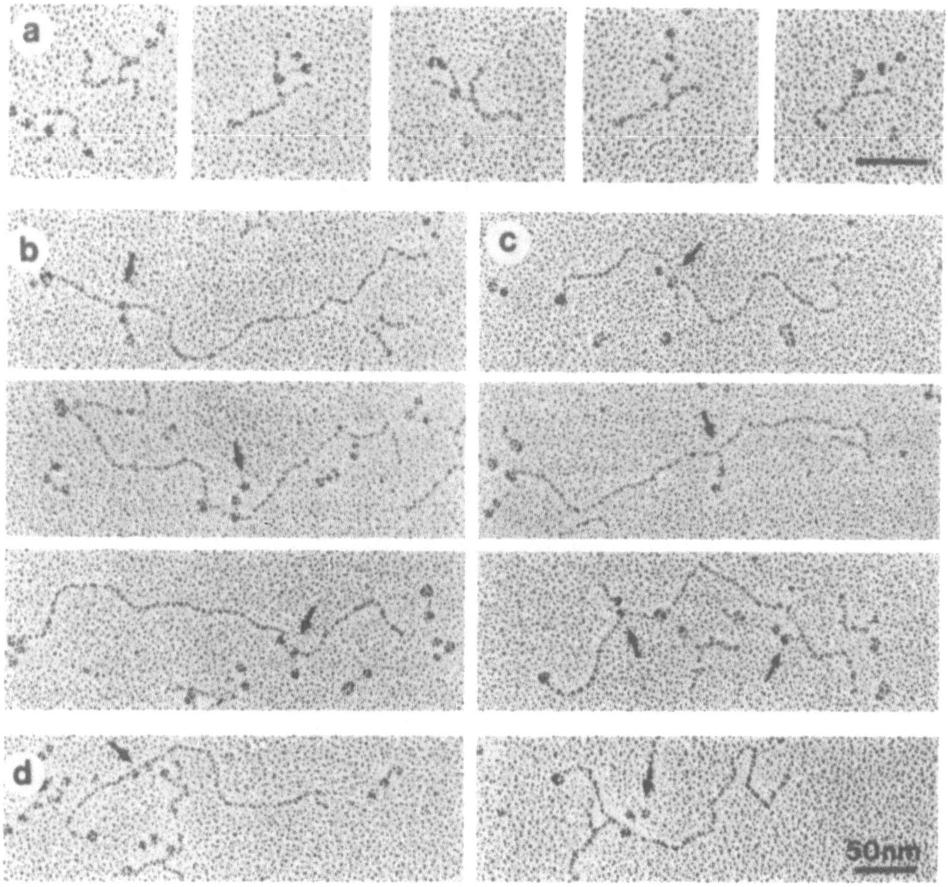

Figure 10 *In vitro* assembly of laminin, recombinant nidogen and collagen IV. (a) A laminin fragment consisting of the central part of the short arms was incubated with nidogen, which binds to a distinct site in a 1:1 stoichiometry. (b) Nidogen or (c) a fragment of it lacking domain G3 binds to collagen IV monomers at two distinct sites of the triple helical strand (arrows). (d) In a ternary mixture of all three components, interaction between collagen IV and the laminin fragment is mediated by nidogen. (bars: 50 nm) (Reproduced from Fox, J. W., et al., *EMBO J.,* 10, 3137, 1991, with permission.)

In Figure 11B, a model for an interdigitating heteropolymer of independent laminin and collagen meshworks with some cross-connections, which tries to integrate most of the known morphological and biochemical data is depicted. On this model, a further network formed of perlecan is superimposed (Figure 11C). Perlecan was shown to form dimers and to a lesser extent oligomers by head-to-head interaction of the sides opposite to the heparansulfate chains (see above, Figure 5b).[173] As laminin binds to heparin and perlecan weakly associates to collagen, such a proteoglycan lattice might also have some cross-links to these arrays.[169,266]

V. CELL ADHESION TO EXTRACELLULAR MATRIX PROTEINS

The close proximity of basement membranes to cell membranes suggests several interactions between them. Therefore, it is common practice that each of its constituent molecules is subjected to different test systems for providing any information on its cell binding capability already at a rather early stage of its characterization. Due to the different assay systems, it is not surprising that one comes across quite contradictory results in the literature. Unfortunately, it sometimes seems harder to disprove any cell binding activity than to establish a new binding site. For all of the molecules described in this chapter, except SPARC and collagen type VII, several cell binding sites have been outlined. It is beyond the scope of this article to summarize even the most relevant work in this field, so we will only highlight some

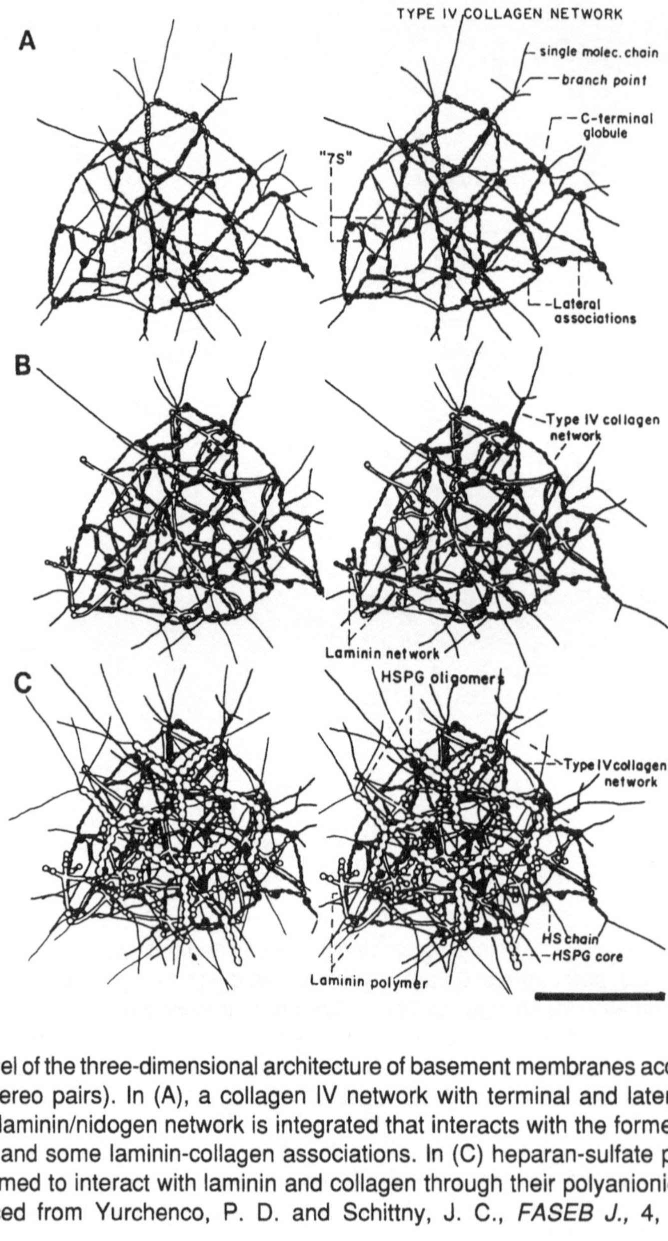

Figure 11 Model of the three-dimensional architecture of basement membranes according Yurchenco and Schittny (stereo pairs). In (A), a collagen IV network with terminal and lateral associations is shown. In (B) a laminin/nidogen network is integrated that interacts with the former one via nidogen (see Figure 10) and some laminin-collagen associations. In (C) heparan-sulfate proteoglycan complexes are assumed to interact with laminin and collagen through their polyanionic chains (bar: 200 nm). (Reproduced from Yurchenco, P. D. and Schittny, J. C., *FASEB J.*, 4, 1577, 1990, with permission.)

difficulties that can arise in correlating specific functions to defined structural epitopes. For details on matrix receptors see the recent reviews in References 14 to 17.

The most widely used techniques for studying cell-matrix interaction are *cell attachment assays*, in which the protein under investigation is unspecifically adsorbed to a glass or plastic surface to which cells should adhere. For both the method of surface coating as well as evaluation of adhesion, many protocols are in use, some accompanied with mysterious "tricks". To qualify as a specific test, aside from using strictly standardized conditions, related proteins must be included in these studies, and preferably inhibition by ligands or antibodies against the substrate molecule in solution should be investigated.

The majority of the rather large extracellular matrix molecules are responsible for several biological functions, suggesting their direct or indirect interactions with cells. In accordance with the rapidly growing knowledge on their primary structures, some functions have been attributed to rather short sequence regions by means of testing synthetic peptides of decreasing length. Together with the sequence relation to the electron microscopic shape, a rather defined localization (up to about 2 to 3 nm resolution) of these epitopes on some of the extended proteins is possible.[36,37] A few of such determined sites thought to be responsible for cell adhesion and binding to other proteins are marked in the figures (Figures 3 to 6, 8). In many cases, these peptides are only active at molar concentrations several orders of magnitude higher than that of the mature protein. Although not very much is known about the conformation of the peptides (For the structure of cyclic RGD-containing pentapeptides, see Reference 267), at least for the very short ones it seems justified to assume that it differs significantly from that embedded in the intact molecule. This problem is circumvented by using proteolytic or recombinant fragments, especially in combination with related antibodies. The critical dependence of function on native conformation has been demonstrated for the laminin long arm fragment E8. After denaturation and dissociation, its activity for cell binding, spreading, and neurite outgrowth stimulation was completely lost.[268] Whereas protease-derived fragments have the advantage of correct posttranslational modifications, their activity could be reduced by harsh conditions applied in purification. This was shown in particular for nidogen, which aside from its different appearance in the electron microscope, demonstrates a drastic reduction of laminin-binding activity when compared to recombinant nidogen.[125]

Originally identified as one important cellular recognition site in fibronectin, the tripeptide Arg-Gly-Asp has been found to act in a similar manner in some other extracellular matrix and platelet adhesion proteins.[269,270] Therefore, each new sequence is scanned for this motif. Due to the presence of glycine, the probability of finding it is considerably high in the repeating $[Gly\text{-}X\text{-}Y]_n$ unit of collagens. Probably only a few RGD sites are indeed involved in cell binding. As a minimal requirement, it should be apposed to the protein surface. As still no algorithm is established to model the tertiary structure on the basis of the sequence, this criterion is hard to prove. As a simple minimal but not satisfactory requirement for the activity of RGD, it was found that binding occurs favorably if the peptide is enclosed by two or more closely spaced residues that have a high probability of initiating a β-bend, as it is also realized in the original fibronectin type III module.[62] The 2D-NMR structure of this domain shows that here the RGD is separately looped out and rather flexible.[63] The laminin A-chains of mouse and human contain RGD at different positions, either within the short-arm domain IIIb or the long-arm domain G. For mouse laminin, it has been shown that the short arm RGD that fulfills the above criterion is only active in cell binding after proteolytic degradation.[271] Such a "cryptic" site might be used during tissue remodeling and tumor cell invasion.

It is still controversial whether the RGD site of chicken and human tenascin (Figure 8) mediates cell attachment activity. (For discussion see References 231 to 233.) The isolation of a human RGD-dependent tenascin-binding integrin has been described.[272] In experiments with chicken tenascin, however, there was no indication for a cell binding site in the RGD-containing fibronectin type III repeat. By mapping with an inhibitory antibody and with a functionally active recombinant protein containing this epitope, a putative cell-binding site has been localized on the two most C-terminal fibronectin type III repeats.[36] This is in agreement with studies on a C-terminal proteolytic fragment that does not contain RGD.[235] In mouse and porcine tenascin, this RGD site is not conserved but reads RVD and RAD, respectively.[241,242]

Many extracellular matrix proteins possess cysteine-rich domains with close or distant homology to EGF (cf. Table 1 with Figures 3 to 5, and 8). As the small diffusible growth factors are important in embryonic development, it is thought that such domains embedded in the large proteins might act as localized signals for growth and differentiation. Indeed, for laminin, tenascin, and thrombospondin, growth-promoting functions have been found.[27] Laminin has a mitogenic effect on a number of cells, which in dose-response and time-dependence is comparable to EGF.[273,274] This function has been attributed to the inner short arm regions III, and no relation to its cell attachment properties was determined.[274]

Whereas many extracellular matrix proteins have been characterized on the basis of their ability to promote cell adhesion and spreading, it becomes more and more clear that many of them can also induce the reverse effect, *antiadhesiveness*. The best studied examples are tenascin, SPARC, and thrombospondin.[216,217] Now it turns out that also nidogen, together with laminin, and laminin, in concert with fibronectin, can inhibit cellular adhesion for a variety of cell types.[275,276] Due to insufficient data, the modulation and function of this effect can only be subjects of speculation. Antiadhesiveness might be important in wound repair, tissue remodeling, or guiding of neuronal growth.

VI. OUTLOOK

The combination of biophysical, biochemical, cell biological, and genetic techniques has led to rather detailed *models* of the structure and localization of functional epitopes of some prototype extracellular matrix molecules. It should not be forgotten that these models only attempt to present a snapshot integrating many of the current available data, although some are uncertain and in contradiction to other studies. The many sites active in different processes are only partially characterized. Especially little is known about which cell binding site is occupied by which cellular receptor. At least six integrins have been identified that bind to laminin and six others that bind to fibronectin but also additional receptors, including lectins, glycolipids, and sulfatides, are known.[13-17] Recently, synergistic effects of two distant sequence regions necessary for full activity of the RGD of fibronectin have been established (Figure 6).[16] Not very much is known about the significance of the changes occurring in the growing number of isoforms, particularly for laminin, collagen type IV, and tenascin.[86,244] Recombinant proteins might help to overcome some of the problems encountered in biochemical purification.[36,125,128,226] Site-directed and deletion mutagenesis can be regarded as a powerful tool in future research to map functional sites in their native conformation. A useful guideline for a definite test of the biological relevance of any adhesive recognition sequences has recently been proposed by Yamada.[16]

Knowledge of the *in situ* architecture of the extracellular matrix is still rather fragmentary, and recent reports on cell adhesion/antiadhesion effects of protein *mixtures* have highlighted the importance of bridging the gap between the single molecule and the ensemble level.[216,217,275,276] The *in situ* architecture of the laminin and collagenous network (Figures 9 to 11) and the information on the orientation of different molecules in basement membranes should be taken into account in further efforts to reconstitute artificial substrates for cell adhesion.[251-253,259,260] The repolymerization of laminin at a membrane surface suggests that this system allows the formation of protein layers resembling basement membranes more closely than random aggregation in solution.[247] Many efforts, however, are still necessary to mimic the native structure more realistically.

VII. ACKNOWLEDGMENTS

We are grateful to many colleagues who have informed us about their most recent results prior to publication, namely Jürgen Engel (Basel, Switzerland), Robert Garrone (Villeurbanne, France), Ulrike Mayer (Munich, Germany), Mats Paulsson (Bern, Switzerland), Rupert Timpl (Munich, Germany), and Yoshihiko Yamada (Bethesda, MD). Special thanks go to Drs. John Heuser (St. Louis, MO), Enrico Reale (Hannover, Germany), and Peter D. Yurchenco (Piscataway, NJ) for providing originals of electron micrographs presented in the figures. This chapter was written when the authors shared a lab at the Institute for Biophysics, Johannes Kepler University, Linz, Austria.

REFERENCES

1. **Robert, L.,** Structural glycoproteins: historical remarks, *Front. Matrix Biol.,* 11, 1, 1986.
2. **Hay, E. D.,** Extracellular matrix, *J. Cell Biol.,* 91, 205s, 1981.
3. **Alberts, B., Bray, D., Lewis, J., Raff, M., Roberts, K., and Watson, J. D.,** *Molecular Biology of the Cell,* 2nd ed., Garland Publ. Inc., New York, 1989, chap. 14.
4. **Darnell, J., Lodish, H., and Baltimore, D.,** *Molecular Cell Biology,* 2nd ed., Scientific American Books, W. H. Freeman, New York, 1990, chap. 23.
5. **Hay, E. D., Ed.,** *Cell Biology of Extracellular Matrix,* 2nd ed., Plenum Press, New York, 1991.
6. **Mecham, R. P., Ed.,** *Biology of Extracellular Matrix: A Series,* Academic Press, San Diego, 1989.
7. **Laurie, G. W. and Leblond, C. P.,** Basement membrane nomenclature, *Nature,* 313, 272, 1985.
8. **Chan, F. L. and Inoue, S.,** Lamina lucida of basement membrane: an artifact, *Micros. Res. Technol.,* 28, 48, 1994.
9. **Reale, E.,** Electron microscopy of the basement membrane, in *Ultrastructure of the Connective Tissue Matrix,* Ruggeri, A. and Motta, P. M., Eds., Martinus Nijhoff, Boston, 1984, chap. 11, p. 192.
10. **Timpl, R. and Dziadek, M.,** Structure, development, and molecular pathology of basement membranes, *Int. Rev. Exp. Pathol.,* 29, 1, 1986.
11. **Tryggvason, K., Höyhtyä, M., and Salo, T.,** Proteolytic degradation of extracellular matrix in tumor invasion, *Biochim. Biophys. Acta,* 907, 191, 1987.
12. **Stetler-Stevenson, W. G., Aznavoorian, S., and Liotta, L. A.,** Tumor cell interactions with the extracellular matrix during invasion and metastasis, *Annu. Rev. Cell Biol.,* 9, 541, 1993.
13. **Ruoslahti, E. and Pierschbacher, M. D.,** New perspectives in cell adhesion. RGD and integrins, *Science,* 238, 491, 1987.
14. **Mecham, R. P.,** Receptors for laminin on mammalian cells, *FASEB J.,* 5, 2538, 1991.

15. **Hynes, R. O.,** Integrins — versatility, modulation, and signaling in cell adhesion, *Cell,* 69, 11, 1992.
16. **Yamada, K. M.,** Adhesive recognition sequences, *J. Biol. Chem.,* 266, 12809, 1991.
17. **McDonald, J. A. and Mecham, R. P., Eds.,** *Receptors for Extracellular Matrix,* Academic Press, San Diego, 1991.
18. **Sastry, S. K. and Horwitz, A. F.,** Integrin cytoplasmic domains: mediators of cytoskeletal linkages and extra- and intracellular initiated transmembrane signalling, *Curr. Opin. Cell Biol.,* 5, 819, 1993.
19. **Ali, I. U., Mautner, V. M., Lanza, R. P., and Hynes, R. O.,** Restoration of normal morphology, adhesion and cytoskeleton in transformed cells by addition of a transformation-sensitive surface protein, *Cell,* 11, 115, 1977.
20. **Beck, K.,** Mechanical concepts of membrane dynamics: diffusion and phase separation in two dimensions, in *Cytomechanics, The Mechanical Basis of Cell Form and Structure,* Bereiter-Hahn, J. and Anderson, O. R., Eds., Springer-Verlag, Berlin, 1987, 79.
21. **Zhang, F., Crise, B., Su, B., Hou, Y., Rose, J. K., Bothwell, A., and Jacobson, K.,** Lateral diffusion of membrane-spanning and glycosylphosphatidylinositol-linked proteins: towards establishing rules governing the lateral mobility of membrane proteins, *J. Cell Biol.,* 115, 75, 1991.
22. **Paulsson, M.,** Basement membrane proteins: structure, assembly and cellular interactions, *Crit. Rev. Biochem. Mol. Biol.,* 27, 93, 1992.
23. **Timpl, R.,** Structure and biological activity of basement membranes, *Eur. J. Biochem.,* 180, 487, 1989.
24. **Rohrbach, D. H. and Timpl, R., Eds.,** *Molecular and Cellular Aspects of Basement Membranes,* Academic Press, San Diego, 1993.
25. **Timpl, R. and Dziadek, M.,** Structure, development, and molecular pathology of basement membranes, *Int. Rev. Exp. Pathol.,* 29, 1, 1986.
26. **Yurchenco, P. D. and Schittny, J. C.,** Molecular architecture of basement membranes, *FASEB J.,* 4, 1577, 1990.
27. **End, P. and Engel, J.,** Multidomain proteins of the extracellular matrix and cellular growth, in *Receptors for Extracellular Matrix,* McDonald, J. A. and Mecham, R. P., Eds., Academic Press, San Diego, 1991, 79.
28. **Fessler, J. H. and Fessler, L. I.,** Drosophila extracellular matrix, *Annu. Rev. Cell Biol.,* 5, 309, 1989.
29. **Hortsch, M. and Goodman, C. S.,** Cell and substrate adhesion molecules in drosophila, *Annu. Rev. Cell Biol.,* 7, 505, 1991.
30. **Engel, J., Odermatt, E., Engel, A., Madri, J. A., Furthmayr, H., Rohde, H., and Timpl, R.,** Shapes, domain organization and flexibility of laminin and fibronectin, two multifunctional proteins of the extracellular matrix, *J. Mol. Biol.,* 150, 97, 1981.
31. **Odermatt, E. and Engel, J.,** Physical properties of fibronectin, in *Fibronectin,* Mosher, D. F., Ed., Academic Press, San Diego, 25.
32. **Erickson, H. P., Carrell, N., and McDonagh, J.,** Fibronectin molecule visualized in electron microscopy: a long, thin flexible strand, *J. Cell Biol.,* 91, 673, 1981.
33. **Erickson, H. P. and Inglesias, J.,** A six-armed oligomer isolated from cell surface fibronectin preparations, *Nature,* 311, 267, 1984.
34. **Engel, J. and Furthmayr, H.,** Electron microscopy and other physical methods for the characterization of extracellular matrix components: laminin, fibronectin, collagen IV, collagen VI and proteoglycans, *Meth. Enzymol.,* 145, 3, 1987.
35. **Engel, J.,** Domain organization of laminin and fibronectin, two multifunctional proteins of the extracellular matrix, in *Multidomain Proteins,* Patthy, L. and Friedrich, P., Eds., Akademiai Kiado, Budapest, 1986, 89.
36. **Spring, J., Beck, K., and Chiquet-Ehrismann, R.,** Two contrary functions of tenascin: dissection of the active sites by recombinant tenascin fragments, *Cell,* 59, 325, 1989.
37. **Beck, K., Spring, J., Chiquet-Ehrismann, R., Engel, J., and Chiquet, M.,** Structural motifs of the extracellular matrix proteins laminin and tenascin, in *Patterns in Protein Sequence and Structure,* Taylor, W. R., Ed., Springer-Verlag, Berlin, 1991, 229.
38. **Doolittle, R. F.,** The geneology of some recently evolved vertebrate proteins, *Trends Biochem. Sci.,* 10, 233, 1985.
39. **Doolittle, R. F., Peng, D. F., Johansson, M. S., and McClure, M. A.,** Relationship of human protein sequences to those of other organisms, *Cold Spring Harbor Symp. Quant. Biol.,* 51, 447, 1986.
40. **Dorit, R. I., Schoenbach, I., and Gilbert, W.,** How big is the universe of exons?, *Science,* 250, 1377, 1990.
41. **Patthy, L.,** Modular exchange principles in proteins, *Curr. Opin. Struct. Biol.,* 1, 351, 1991.

42. **Bork, P.**, Shuffled domains in extracellular proteins, *FEBS Lett.*, 286, 47, 1991.
43. **Campbell, I. D. and Spitzfaden, C.**, Building proteins with fibronectin type III modules, *Structure*, 2, 333, 1994.
44. **Taylor, W. R., Ed.**, *Patterns in Protein Sequences and Structure*, Springer-Verlag, Berlin, 1991.
45. **Bairoch, A.**, PC/Gene: A Protein and Nucleic Acid Sequence Analysis Microcomputer Package, PROSITE: A Dictionary of Sites and Patterns in Proteins, and SWISS-PROT: A Protein Sequence Data Bank, Ph.D. dissertation, University of Geneva, 1990.
46. **Taylor, W. R. and Jones, D. T.**, Templates, consensus pattern and motifs, *Curr. Opin. Struct. Biol.*, 1, 327, 1991.
47. **Beck, K., Hunter, I., and Engel, J.**, Structure and function of laminin: anatomy of a multidomain glycoprotein, *FASEB J.*, 4, 148, 1990.
48. **Engel, J.**, Common structural motifs in proteins of the extracellular matrix, *Curr. Opin. Cell Biol.*, 3, 779, 1991.
49. **Sasaki, M. and Yamada, Y.**, The laminin B2 chain has a multidomain structure homologous to the B1 chain, *J. Biol. Chem.*, 262, 17111, 1987.
50. **Sasaki, M., Kleinman, H. K., Huber, H., Deutzmann, R., and Yamada, Y.**, Laminin, a multidomain protein. The A chain has a unique globular domain and homology with the basement membrane proteoglycan and the laminin B chains, *J. Biol. Chem.*, 263, 16536, 1988.
51. **Kusche-Gullberg, M., Garrison, K., MacKrell, A. J., Fessler, L. I., and Fessler, J. H.**, Laminin A chain: expression during *Drosophila* development and genomic sequence, *EMBO J.*, 11, 4519, 1992.
52. **Engel, J.**, Structure and function of laminin, in *Molecular and Cellular Aspects of Basement Membranes*, Rohrbach, D. H. and Timpl, R., Eds., Academic Press, San Diego, 1993, 147.
53. **Ushkaryov, Y. A., Petrenko, A. G., Geppert, M., and Südhof, T. C.**, Neurexins: synaptic cell surface proteins related to the α-latrotoxin receptor and laminin, *Science*, 257, 50, 1992.
54. **Patthy, L.**, Laminin A-related domains in crb protein of Drosophila and their possible role in epithelial polarization, *FEBS Lett.*, 289, 99, 1991.
55. **Cooke, R., Wilkinson, A., Baron, M., Pastore, A., Tappin, M., Campbell, I. D., Gregory, H., and Sheard, B.**, The solution structure of human epidermal growth factor, *Nature*, 327, 339, 1987.
56. **Durkin, M. E., Chakravarti, S., Bartos, B. B., Liu, S.-H., Friedman, R. L., and Chung, A. E.**, Amino acid sequence and domain structure of entactin. Homology with epidermal growth factor precursor and low density lipoprotein receptor, *J. Cell Biol.*, 107, 2749, 1988.
57. **Yamamoto, T., Davis, C., Brown, M., Schneider, W., Casey, M., Goldstein, J., and Russell, D.**, The human LDL receptor: a cysteine-rich protein with multiple Alu sequences in its mRNA, *Cell*, 39, 27, 1984.
58. **Becker, J. W., Erickson, H. P., Hoffman, S., Cunningham, B. A., and Edelman, G. M.**, Topology of cell adhesion molecules, *Proc. Natl. Acad. Sci. U.S.A.*, 86, 1088, 1989.
59. **Baron, M., Norman, D., Willis, A., and Campbell, I. D.**, Structure of the fibronectin type 1 module, *Nature*, 345, 642, 1990.
60. **Constantine, K. L., Madrid, M., Banyi, L., Trexler, K., Patthy, L., and Llinas, M.**, Refined solution structure and ligand-binding properties of PDC-109 domain β. A collagen-binding type II domain, *J. Mol. Biol.*, 223, 281, 1992.
61. **Bork, P.**, The molecular architecture of vertebrate collagens, *FEBS Lett.*, 307, 49, 1992.
62. **Reed, J., Hull, W. E., Lieth, C.-W., Kübler, D., Suhai, S., and Kinzel, V.**, Secondary structure of the Arg-Gly-Asp recognition site in proteins involved in cell-surface adhesion. Evidence for the occurrence of nested β-bends in the model hexapeptide GRGDSP, *Eur. J. Biochem.*, 178, 141, 1988.
63. **Baron, M., Main, A. L., Driscoll, P. C., Mardon, H. J., Boyd, J., and Campbell, I. D.**, ^{1}H NMR assignment and secondary structure of the cell adhesion type III module of fibronectin, *Biochemistry*, 31, 2068, 1992.
64. **Leahy, D. J., Hendrickson, W. A., Aukhil, I., and Erickson, H. P.**, Structure of a fibronectin type III domain from tenascin phased by MAD analysis of the selenomethionyl protein, *Science*, 258, 987, 1992.
65. **Xu, X. and Doolittle, R.**, Presence of a vertebrate fibrinogen-like sequence in an echinoderm, *Proc. Natl. Acad. Sci. U.S.A.*, 87, 2097, 1990.
66. **Nörenberg, U., Wille, H., Wolff, J. M., Frank, R., and Rathjen, F. G.**, The chicken neural extracellular matrix molecule restrictin: similarity with EGF-, fibronectin type III-, and fibrinogen-like motifs, *Neuron*, 8, 849, 1992.

67. **McLachlan, A. D. and Stewart, M.,** Tropomyosin coiled-coil interactions: evidence for an unstaggered structure, *J. Mol. Biol.*, 98, 293, 1975.
68. **Cohen, C. and Parry, D. A. D.,** α-Helical coiled coils and bundles: how to design an α-helical protein, *Proteins*, 7, 1, 1990.
69. **Chung, A. E., Jaffe, R., Freeman, I. L., Vergnes, J. P., Braginski, J. E., and Carlin, B.,** Properties of a basement membrane-related glycoprotein synthesized in culture by a mouse embryonal carcinoma-derived cell line, *Cell*, 16, 277, 1979.
70. **Timpl, R., Rohde, H., Gehron Robey, P., Rennard, S. I., Foidart, J.-M., and Martin, G. R.,** Laminin — a glycoprotein from basement membranes, *J. Biol. Chem.*, 254, 9933, 1979.
71. **Paulsson, M., Aumailley, M., Deutzmann, R., Timpl, R., Beck, K., and Engel, J.,** Laminin-nidogen complex. Extraction with chelating agents and structural characterization, *Eur. J. Biochem.*, 166, 11, 1987.
72. **McCarthy, R. A., Beck, K., and Burger, M. M.,** Laminin is structurally conserved in the sea urchin basal lamina, *EMBO J.*, 6, 1587, 1987.
73. **Chiquet, M., Masuda-Nakagawa, L., and Beck, K.,** Attachment to an endogenous laminin-like protein initiates sprouting by leech neurons, *J. Cell Biol.*, 107, 1189, 1988.
74. **Beck, K., McCarthy, R. A., Chiquet, M., Masuda-Nakagawa, L., and Schlage, W. K.,** Structure of the basement membrane protein laminin: variation on a theme, in *Cytoskeletal and Extracellular Proteins: Structure, Interactions and Assembly*, Aebi, U. and Engel, J., Eds., Springer-Verlag, Berlin, 1989, 102.
75. **Paulsson, M. and Saladin, K.,** Mouse heart laminin. Purification of the native protein and structural comparison with Engelbreth-Holm-Swarm tumor laminin, *J. Biol. Chem.*, 264, 18726, 1989.
76. **Humbert-David, N.,** Evolution des glycoproteines de la matrice extracellulaire: Caracterisation de proteines de type laminine chez Helix asper a (Mollusque) et de tenascine chez Oscarella tuberculata (Spongiaire), Ph.D. thesis, Universite Claude-Bernard-Lyon 1, Villeurbonne, France, 1993.
77. **Kuffler, D. P. and Luethi, T.,** Identification of molecules in a muscle extracellular matrix extract that promotes process outgrowth from cultured adult frog motoneurons, *J. Neurobiol.*, 24, 515, 1993.
78. **Schmid, V., Bally, A., Beck, K., Haller, M., Schlage, W. K., and Weber, Ch.,** The ECM *(mesoglea)* of hydrozoan jellyfish and its ability to support cell adhesion and spreading, *Hydrobiologia*, 216/217, 3, 1991.
79. **Engvall, E., Earwicker, D., Haaparanta, T., Ruoslahti, E., and Sanes, J. R.,** Distribution and isolation of four laminin variants, tissue restricted distribution of heterotrimers assembled from five different subunits, *Cell Regul.*, 1, 731, 1990.
80. **Paulsson, M., Saladin, K., and Engvall, E.,** Structure of laminin variants. The 300-kDa chains of murine and bovine heart laminin are related to the human placenta merosin heavy chain and replace the A chain in some laminin variants, *J. Biol. Chem.*, 266, 17545, 1991.
81. **Bruch, M., Landwehr, R., and Engel, J.,** Dissection of laminin by cathepsin G into its long-arm and short-arm structures and localization of regions involved in calcium dependent stabilization and self-association, *Eur. J. Biochem.*, 185, 271, 1989.
82. **Davis, G. E., Manthorpe, M., Engvall, E., and Varon, S.,** Isolation and characterization of rat Schwannoma neurite-promoting factor: evidence that the factor contains laminin, *J. Neurosci.*, 5, 2662, 1985.
83. **Edgar, D., Timpl, R., and Thoenen, H.,** Structural requirements for the stimulation of neurite outgrowth by two variants of laminin and their inhibition by antibodies, *J. Cell Biol.*, 106, 1299, 1988.
84. **Vuolteenaho, R., Nissinen, M., Sainio, K., Byers, M., Eddy, R., Hirvonen, H., Shows, T. B., Sariola, H., Engvall, E., and Tryggvason, K.,** Human laminin M chain (merosin): complete primary structure, chromosomal assignment and expression of the M and A chains in human fetal tissues, *J. Cell Biol.*, 124, 381, 1994.
85. **Tokida, Y., Aratani, Y., Morita, A., and Kitagawa, Y.,** Production of two variant laminin forms by endothelial cells and shift of their relative levels by angiostatic steroids, *J. Biol. Chem.*, 265, 18123, 1990.
86. **Paulsson, M.,** Laminin and collagen IV variants and heterogeneity in basement membrane composition, in *Molecular and Cellular Aspects of Basement Membranes*, Rohrbach, D. H. and Timpl, R., Eds., Academic Press, San Diego, 1993, 117.
87. **Rousselle, P., Lunstrum, G. P., Keene, D. R., and Burgeson, R. E.,** Kalinin: an epithelium-specific basement membrane adhesion molecule that is a component of anchoring filaments, *J. Cell Biol.*, 567, 1991.

88. **Gerecke, D. R., Wagman, D. W., Champliaud, M.-F., and Burgeson, R. E.,** The complete primary structure for a novel laminin chain, the laminin B1k chain, *J. Biol. Chem.*, 269, 11073, 1994.
89. **Verrando, P., Blanchet-Bardon, C., Pisani, A., Thomas, L., Cambazard, F., Eady, R. A., Schofield, O., and Ortonne, J.-P.,** Monoclonal antibody GB3 defines a widespread defect of several basement membranes and a keratinocyte dysfunction in patients with lethal junctional epidermolysis bullosa, *Lab. Invest.*, 64, 85, 1991.
90. **Fessler, L. I., Campbell, G., Duncan, K. G., and Fessler, J. H.,** Drosophila laminin: characterization and localization, *J. Cell Biol.*, 105, 2383, 1987.
91. **Sasaki, M., Kato, S., Kohno, K., Martin, G. R., and Yamada, Y.,** Sequence of the cDNA encoding the laminin B1 chain reveals a multidomain protein containing cysteine-rich repeats, *Proc. Natl. Acad. Sci. U.S.A.*, 84, 935, 1987.
92. **Durkin, M. E., Bartos, B. B., Liu, S.-H., Phillips, S. L., and Chung, A. E.,** Primary structure of the mouse laminin B2 chain and comparison with laminin B1, *Biochemistry*, 27, 5198, 1988.
93. **Pikkarainen, T., Eddy, R., Fukushima, Y., Byers, M., Shows, T., Pihlajaniemi, T., Saraste, M., and Tryggvason, K.,** Human laminin B1 chain. A multidomain protein with gene (LAMB1) locus in the q22 region of chromosome 7, *J. Biol. Chem.*, 263, 6751, 1988.
94. **Pikkarainen, T., Kallunki, T., and Tryggvason, K.,** Human laminin B2 chain. Comparison of the complete amino acid sequence with the B1 chain reveals variability in sequence homology between different structural domains, *J. Biol. Chem.*, 263, 6751, 1988.
95. **Haaparanta, T., Uitto, J., Ruoslahti, E., and Engvall, E.,** Molecular cloning of the cDNA encoding human laminin A chain, *Matrix*, 11, 151, 1991.
96. **Nissinen, M., Vuolteenaho, R., Boot-Handford, R., Kallunki, T., and Tryggvason, K.,** Primary structure of the human laminin A chain, *Biochem. J.*, 276, 369, 1991.
97. **Montell, D. J. and Goodman, C. S.,** Drosophila substrate adhesion molecule: sequence of laminin B1 chain reveals domains of homology with mouse, *Cell*, 53, 463, 1988.
98. **Chi, H.-C. and Hui, C.-F.,** Primary structure of the drosophila laminin B2 chain and comparison with human, mouse, and Drosophila B1 and B2 chains, *J. Biol. Chem.*, 264, 1543, 1989.
99. **Montell, D. J. and Goodman, C. S.,** Drosophila laminin: sequence of B2 subunit and expression of all three subunits during embryogenesis, *J. Cell Biol.*, 109, 2441, 1989.
100. **Hunter, D. D., Shah, V., Merlie, J.-P., and Sanes, J. R.,** A laminin-like adhesive protein concentrated in the synaptic cleft of the neuromuscular junction, *Nature*, 338, 229, 1989.
101. **Hunter, D. D., Porter, B. E., Bulock, J. W., Adams, S. P., Merlie, J. P., and Sanes, J. R.,** Primary sequence of a motor neuron-selective adhesive site in the synaptic basal lamina protein s-laminin, *Cell*, 59, 905, 1989.
102. **Vailly, J., Verrando, P., Champliaud, M.-F., Gerecke, D., Wagman, D. W., Baudoin, C., Aberdam, D., Burgeson, R. E., Bauer, E., and Ortonne, J.-P.,** The 100-kDa chain of nicein/kalinin is a laminin B2 chain variant, *Eur. J. Biochem.*, 219, 209, 1994.
103. **Deutzmann, R., Huber, H., Schmetz, K. A., Schmetz, K. A., Oberbäumer, I., and Hartl, L.,** Structural study of long arm fragments of laminin. Evidence for repetitive C-terminal sequences in the A-chain, not present in the B-chains, *Eur. J. Biochem.*, 177, 35, 1988.
104. **Hartl, L., Oberbäumer, I., and Deutzmann, R.,** The N terminus of laminin A chain is homologous to the B chains, *Eur. J. Biochem.*, 173, 629, 1988.
105. **Paulsson, M., Deutzmann, R., Timpl, R., Dalzoppo, D., Odermatt, E., and Engel, J.,** Evidence for coiled-coil α-helical regions in the long arm of laminin, *EMBO J.*, 4, 309, 1985.
106. **Mayer, U., Nischt, R., Pöschl, E., Mann, K., Fukuda, K., Gerl, M., Yamada, Y., and Timpl, R.,** A single EGF-like motif of laminin is responsible for high affinity nidogen binding, *EMBO J.*, 12, 1879, 1993.
107. **Ott, U., Odermatt, E., Engel, J., Furthmayr, H., and Timpl, R.,** Protease resistance and conformation of laminin, *Eur. J. Biochem.*, 123, 63, 1982.
108. **Hunter, I., Schulthess, T., Bruch, M., Beck, K., and Engel, J.,** Evidence for a specific mechanism of laminin assembly, *Eur. J. Biochem.*, 188, 205, 1990.
109. **Hunter, I., Schulthess, T., and Engel, J.,** Laminin chain assembly by triple- and double-stranded coiled-coil structures, *J. Biol. Chem.*, 267, 6006, 1992.
110. **Doolittle, R. F.,** Fibrinogen and fibrin, *Annu. Rev. Biochem.*, 53, 195, 1984.
111. **Beck, K., Dixon, T. W., Engel, J., and Parry, D. A. D.,** Ionic interactions in the coiled-coil domain of laminin determine the specificity of chain assembly, *J. Mol. Biol.*, 231, 311, 1993.

112. **Morel, Y., Bristow, J., Gitelman, S., and Miller, W.,** Transcript encoded on the opposite strand of the human steroid 21-hydroxylase/complement component C4 gene locus, *Proc. Natl. Acad. Sci. U.S.A.*, 86, 6582, 1989.
113. **Arumugham, R. G., Hsieh, T. C.-Y., Tanzer, M. L., and Laine, R. A.,** Structure of the asparagine-linked sugar chains of laminin, *Biochim. Biophys. Acta,* 883, 112, 1986.
114. **Knibbs, R. N., Perini, F., and Goldstein, I. J.,** Structure of the major concanavalin A reactive oligosaccharides of the extracellular matrix component laminin, *Biochemistry,* 28, 6379, 1989.
115. **Chammas, R., Veiga, S. S., Line, S., Potocnjak, P., and Brentani, P. R.,** Asn-linked oligosaccharide-dependent interaction between laminin and gp120/140, an α_6/β_1 integrin, *J. Biol. Chem.,* 266, 3349, 1991.
116. **Tanzer, M. L., Chandrasekaran, S., Dean, J. W., III, and Giniger, M. S.,** Role of laminin carbohydrates on cellular interactions, *Kidney Intern.*, 43, 66, 1993.
117. **Dennis, J. W., Waller, C. A., and Schirrmacher, V.,** Identification of asparagine-linked oligosaccharides involved in tumor cell adhesion to laminin and type IV collagen, *J. Cell Biol.*, 99, 1416, 1984.
118. **Bouzon, M., Dussert, C., Lissitzky, J. C., and Martin, P. M.,** Spreading of B16 F1 cells on laminin and its proteolytic fragments P1 and E8: involvement of laminin carbohydrate chains, *Exp. Cell Res.*, 190, 47, 1990.
119. **Howe, C. C.,** Functional role of laminin carbohydrate, *Mol. Cell. Biol.*, 4, 1, 1984.
120. **Wu, C., Friedman, R., and Chung, A. E.,** Analysis of the assembly of laminin and the laminin entactin complex with laminin chain specific monoclonal and polyclonal antibodies, *Biochemistry,* 27, 8780, 1988.
121. **Carlin, B., Jaffe, R., Bender, B., and Chung, A. E.,** Entactin, a novel basal lamina-associated sulfated glycoprotein, *J. Biol. Chem.,* 5209, 1981.
122. **Timpl, R., Dziadek, M., Fujiwara, S., Nowack, H., and Wick, G.,** Nidogen: a new self-aggregating basement membrane protein, *Eur. J. Biochem.*, 137, 455, 1983.
123. **Paulsson, M., Deutzmann, R., Dziadek, M., Nowack, H., Timpl, R., Weber, S., and Engel, J.,** Purification and structural characterization of intact and fragmented nidogen obtained from a tumor basement membrane, *Eur. J. Biochem.*, 467, 1986.
124. **Mann, K., Deutzmann, R., Aumailley, M., Timpl, R., Raimondi, L., Yamada, Y., Pan, T.-C., Conway, D., and Chu, M.-L.,** Amino acid sequence of mouse nidogen, a multidomain basement membrane protein with binding activity for laminin, collagen IV and cells, *EMBO J.*, 8, 65, 1989.
125. **Fox, J. W., Mayer, U., Nischt, R., Aumailley, M., Reinhardt, D., Wiedemann, H., Mann, K., Timpl, R., Krieg, T., Engel, J., and Chu, M.-L.,** Recombinant nidogen consists of three globular domains and mediates binding of laminin to collagen type IV, *EMBO J.*, 10, 3137, 1991.
126. **Nakae, H., Sugano, M., Ishimori, Y., Endo, T., and Obinata, T.,** Ascidian entactin/nidogen: implication of evolution by shuffling two kinds of cysteine-rich motifs, *Eur. J. Biochem.*, 1993, 213, 11, 1993.
127. **Nagayoshi, T., Sanborn, D., Hickok, N. J., Olson, D. R., Fazio, M. J., Chu, M.-L., Knowlton, R., Mann, K., Deutzmann, R., Timpl, R., and Uitto, J.,** Human nidogen: complete amino acid sequence and structural domains deduced from cDNAs, and evidence for polymorphism of the gene, *DNA,* 8, 581, 1989.
128. **Mayer, U., Mann, K., Timpl, R., and Murphy, G.,** Sites of nidogen cleavage by proteases involved in tissue homeostasis and remodeling, *Eur. J. Biochem.*, 217, 877, 1993.
129. **Reinhardt, D., Mann, K., Nischt, R., Fox, J. W., Chu, M.-L., Krieg, T., and Timpl, R.,** Mapping of nidogen binding sites for collagen type IV, heparan sulfate proteoglycan, and zinc, *J. Biol. Chem.*, 268, 10881, 1993.
130. **Fujiwara, S., Shinkai, H., Mann, K., and Timpl, R.,** Structure and localization of O- and N-linked oligosaccharide chains on basement membrane protein nidogen, *Matrix,* 13, 215, 1993.
131. **Handford, P. A., Mayhew, M., Baron, M., Winship, P. R., Campbell, I. D., and Brownlee, G. G.,** Key residues in calcium-binding motifs in EGF-like domains, *Nature,* 351, 164, 1991.
132. **Dziadek, M., Paulsson, M., and Timpl, R.,** Identification and interaction repertoire of large forms of the basement membrane protein nidogen, *EMBO J.*, 4, 2513, 1985.
133. **Dziadek, M. and Timpl, R.,** Expression of nidogen and laminin in basement membranes during mouse embryogenesis and in teratocarcinoma cells, *Dev. Biol.*, 111, 372, 1985.
134. **Cooper, A. R., Kurkinen, M., Taylor, A., and Hogan, B. L. M.,** Studies on the biosynthesis of laminin by murine parietal endoderm cells, *Eur. J. Biochem.*, 119, 189, 1981.

135. **Olson, P. F., Fessler, L. I., Nelson, R. E., Sterne, R. E., Campbell, A. G., and Fessler, J. H.,** Glutactin, a novel *drosophila* basement membrane-related glycoprotein with sequence similarity to serineesterases, *EMBO J.*, 9, 1219, 1990.
136. **Hudson, B. G., Reeders, S. T., and Tryggvason, K.,** Type IV collagen: structure, gene organization, and role in human diseases. Molecular basis of Goodpasture and Alpert syndromes and diffuse leiomyomatosis, *J. Biol. Chem.*, 268, 26033, 1993.
137. **Kühn, K.,** Basement membrane (Type IV) collagen — its molecular and macromolecular structure, in *Cytoskeletal and Extracellular Proteins: Structure, Interactions and Assembly,* Aebi, U. and Engel, J., Eds., Springer-Verlag, Berlin, 1989, 69.
138. **Minor, R. R., Clark, C. C., Strause, E. L., Koszalka, T. R., Brent, R. L., and Kefalides, N. A.,** Basement membrane procollagen is not converted to collagen in organ cultures of parietal yolk sac endoderm, *J. Biol. Chem.*, 251, 1789, 1976.
139. **Timpl, R., Risteli, J., and Bächinger, H. P.,** Identification of a new basement membrane collagen by the aid of a large fragment resistant to bacterial collagenase, *FEBS Lett.*, 101, 265, 1979.
140. **Timpl, R., Wiedemann, H., van Delden, V., Furthmayr, H., and Kühn, K.,** A network model for the organization of type IV collagen molecules in basement membranes, *Eur. J. Biochem.*, 120, 203, 1981.
141. **Brazel, D., Oberbäumer, I., Dieringer, H., Babel, W., Glanville, R. W., Deutzmann, R., and Kühn, K.,** Completion of the α1 chain of human basement membrane collagen (type IV) reveals 21 non-triplet interruptions located within the collagenous domain, *Eur. J. Biochem.*, 168, 529, 1987.
142. **Guo, X., Johnson, J. J., and Kramer, J. M.,** Embryonic lethality caused by mutations in basement membrane collagen of *C. elegans, Nature*, 349, 707, 1991.
143. **Sibley, M. H., Johnson, J. J., Mello, C. C., and Kramer, J. M.,** Genetic identification, sequence, and alternative splicing of the *Caenorhabditis elegans* α2(IV) collagen gene, *J. Cell Biol.*, 123, 255, 1993.
144. **Muthukumaran, G., Blumberg, B., and Kurkinen, M.,** The complete primary structure for the α1-chain of mouse collagen IV, *J. Biol. Chem.*, 264, 6310, 1989.
145. **Blumberg, B., MacKrell, A. J., and Fessler, J. H.,** Drosophila basement membrane procollagen α1(IV). II. Complete cDNA sequence, genomic structure, and general implications for supramolecular assemblies, *J. Biol. Chem.*, 263, 18328, 1988.
146. **Hostikka, S. L. and Tryggvason, K.,** The complete primary structure of the α2 chain of human type IV collagen and comparison with the α1(IV) chain, *J. Biol. Chem.*, 263, 19488, 1988.
147. **Saus, J., Quinones, S., MacKrell, A., Blumberg, B., Muthukumaran, G., Pihlajaniemi, T., and Kurkinen, M.,** The complete primary structure of mouse α2(IV) collagen. Alignment with mouse α1(IV) collagen, *J. Biol. Chem.*, 264, 6318, 1989.
148. **Brazel, D., Pollner, R., Oberbäumer, I., and Kühn, K.,** Human basement membrane collagen (Type IV). The amino acid sequence of the α2(IV) chain and its comparison with the α1(IV) chain reveals deletions in the α1(IV) chain, *Eur. J. Biochem.*, 172, 35, 1988.
149. **Hoffmann, H., Voss, T., Kühn, K., and Engel, J.,** Localization of flexible sites in thread-like molecules from electronmicrographs — comparison of interstitial, basement membrane and intima collagens, *J. Mol. Biol.*, 172, 325, 1984.
150. **Timpl, R., Martin, G. R., Bruckner, P., Wick, G., and Wiedemann, H.,** Nature of the collagenous protein in a tumor basement membrane, *Eur. J. Biochem.*, 84, 43, 1978.
151. **Siebold, B., Deutzmann, R., and Kühn, K.,** The arrangement of intra- and intermolecular disulfide bonds in the carboxyterminal, non-collagenous aggregation and cross-linking domain of basement membrane type IV collagen, *Eur. J. Biochem.*, 176, 617, 1988.
152. **Weber, S., Dölz, R., Timpl, R., Fessler, J., and Engel, J.,** Reductive cleavage and reformation of the interchain and intrachain disulfide bonds in the globular hexameric domain NC1 involved in network assembly of basement membrane collagen (type IV), *Eur. J. Biochem.*, 175, 229, 1988.
153. **Risteli, J., Bächinger, H. P., Engel, J., Furtmayr, H., and Timpl, R.,** 7-S collagen: characterization of an unusual basement membrane structure, *Eur. J. Biochem.*, 108, 239, 1980.
154. **Glanville, R. W., Qian, R.-q., Siebold, B., Risteli, J., and Kühn, K.,** Amino acid sequence of the N-terminal aggregation and cross-linking region (7S domain) of the α1(IV) chain of human basement membrane collagen, *Eur. J. Biochem.* 152, 213, 1985.
155. **Siebold, B., Qian, R.-q., Glanville, R. W., Hoffmann, H., Deutzmann, R., and Kühn, K.,** Construction of a model for the aggregation and cross-linking region (7S domain) of type IV collagen based upon an evaluation of the primary structure of the α1 and α2 chains in this region, *Eur. J. Biochem.*, 168, 569, 1987.

156. **Langeveld, J. P. M., Noelken, M. E., Hård, K., Todd, P., Vliegenthart, J. F. G., Rouse, J., and Hudson, B. G.,** Bovine glomerular basement membrane. Location and structure of the asparagine-linked oligosaccharide units and their potential role in the assembly of the 7 S collagen IV tetramer, *J. Biol. Chem.*, 266, 2622, 1991.
157. **Lunstrum, G. P., Bächinger, H.-P., Fessler, L. I., Duncan, K. G., Nelson, R. E., and Fessler, J. H.,** Drosophila basement membrane procollagen IV. I. Protein characterization and distribution, *J. Biol. Chem.*, 263, 18318, 1988.
158. **Haralson, M. A., Federspiel, S. J., Martinez-Hernandez, A., Rhodes, R. K., and Miller, E. J.,** Synthesis of [pro $\alpha1(IV)]_3$ collagen molecules by cultured embryo-derived parietal yolk sac cells, *Biochemistry*, 24, 5792, 1985.
159. **Saus, J., Wieslander, J., Langeveld, J. P. M., Quinones, S., and Hudson, B. G.,** Identification of the Goodpasture antigen as the α3(IV) chain of collagen IV, *J. Biol. Chem.*, 263, 13374, 1988.
160. **Oohashi, T., Sugimoto, M., Mattei, M.-G., and Ninomiya, Y.,** Identification of a new collagen IV chain, α6(IV), by cDNA isolation and assignment of the gene to chromosome Xq22, which is the same locus for COL4A5, *J. Biol. Chem.*, 269, 7520, 1994.
161. **Hostikka, S. L., Eddy, R. L., Byers, M. G., Hoyhtya, M., Shows, T. B., and Tryggvason, K.,** Identification of a distinct type IV collagen α chain with restricted kidney distribution and assignment of its gene to the locus of X chromosome-linked Alport syndrome, *Proc. Natl. Acad. Sci. U.S.A.*, 87, 1606, 1990.
162. **Pihlajaniemi, T., Pohjolainen, E.-R., and Myers, J.,** Complete primary structure of the triple-helical region and the carboxy-terminal domain of a new type IV collagen chain, α5(IV), *J. Biol. Chem.*, 265, 13758, 1990.
163. **Sanes, J. R., Engvall, E., Butkowski, R., and Hunter, D. D.,** Molecular heterogeneity of basal laminae: isoforms of laminin and collagen IV at the neuromuscular junction and elsewhere, *J. Cell Biol.*, 111, 1685, 1990.
164. **Johansson, C., Butkowski, R., and Wieslander, J.,** The structural organization of type IV collagen: identification of 3 NC1 populations in the glomerular basement membrane, *J. Biol. Chem.*, 267, 24533, 1993.
165. **Timpl, R.,** Proteoglycans of basement membranes, *Experientia*, 49, 417, 1993.
166. **Dziadek, M., Fujiwara, S., Paulsson, M., and Timpl, R.,** Immunological characterization of basement membrane types of heparan sulphate proteoglycan, *EMBO J.*, 4, 905, 1985.
167. **Kanwar, Y. S., Veis, A., Kimura, J. H., and Jakubowski, M. L.,** Characterization of heparan sulfate proteoglycan of glomerular basement membranes, *Proc. Natl. Acad. Sci. U.S.A.*, 81, 762, 1984.
168. **Parthasarathy, N. and Spiro, R. G.,** Characterization of the glycosaminoglycan component of the renal glomerular basement membrane and its relationship to the peptide portion, *J. Biol. Chem.*, 256, 507, 1981.
169. **Fujiwara, S., Wiedemann, H., Timpl, R., Lustig, A., and Engel, J.,** Structure and interactions of heparan sulfate proteoglycans from a mouse tumor basement membrane, *Eur. J. Biochem.*, 143, 145, 1984.
170. **Hassell, J. R., Leyshon, W. C., Ledbetter, S. R., Tyree, B., Suzuki, S., Kato, M., Kimata, K., and Kleinman, H. K.,** Isolation of two forms of basement membrane proteoglycan, *J. Biol. Chem.*, 260, 8098, 1985.
171. **Tyree, B., Horigan, E. A., Klippenstein, D. L., and Hassell, J. R.,** Heterogeneity of heparan sulfate proteoglycans synthesized by PYS-2 cells, *Arch. Biochem. Biophys.*, 231, 328, 1984.
172. **Paulsson, M., Yurchenco, P. D., Ruben, G. C., Engel, J., and Timpl, R.,** Structure of the low density heparan sulfate proteoglycan isolated from a mouse tumor basement membrane, *J. Mol. Biol.*, 197, 297, 1987.
173. **Yurchenco, P. D., Cheng, Y.-S., and Ruben, G. C.,** Self-assembly of a high mol wt basement membrane proteoglycan into dimers and oligomers, *J. Biol. Chem.*, 262, 17668, 1987.
174. **Noonan, D. M., Fulle, A., Valente, P., Cai, S., Horigan, E., Sasaki, M., Yamada, Y., and Hassell, J. R.,** The complete sequence of perlecan, a basement membrane heparan sulfate proteoglycan, reveals extensive similarity with laminin A chain, LDL receptor and N-CAM, *J. Biol. Chem.*, 266, 22939, 1991.
175. **Kallunki, P. and Tryggvason, K.,** Human basement membrane heparan sulfate proteoglycan core protein: a 467-kD protein containing multiple domains resembling elements of the low density lipoprotein receptor, laminin, neural cell adhesion molecules and epidermal growth factor, *J. Cell Biol.*, 116, 559, 1992.

176. **Barthels, D., Santoni, M.-J., Wille, W., Ruppert, C., Chaix, J.-C., Hirsch, M.-R., Fontecilla-Camps, J. C., and Goridis, C.**, Isolation and nucleotide sequence of mouse NCAM cDNA that codes for a M_r 79000 polypeptide without a membrane-spanning region, *EMBO J.*, 6, 907, 1987.
177. **Edelman, G. M. and Crossin, K. L.**, Cell adhesion molecules: implications for a molecular histology, *Annu. Rev. Biochem.*, 60, 155, 1991.
178. **Timpl, R., Aumailley, M., Gerl, M., Mann, K., Nurcombe, V., Edgar, D., and Deutzmann, R.**, Structure and function of the laminin-nidogen complex, *Ann. N.Y. Acad. Sci.*, 580, 311, 1990.
179. **Kato, M., Koike, Y., Ito, Y., Suzuki, S., and Kimata, K.**, Multiple forms of heparan sulfate proteoglycan in the Engelbreth-Holm-Swarm mouse tumor. The occurrence of high density forms bearing both heparan sulfate and chondroitin sulfate side chains, *J. Biol. Chem.*, 262, 7180, 1987.
180. **Fallon, J. R., Nitkin, R. M., Reist, N. E., Wallace, B. G., and McMahan, U. J.**, Acetylcholine receptor-aggregating factor is similar to molecules concentrated at neuromuscular junctions, *Nature*, 315, 571, 1985.
181. **Reist, N. E., Magill, C., and McMahan, U. J.**, Agrin-like molecules at synaptic sites in normal, denervated, and damaged skeletal muscles, *J. Cell Biol.*, 105, 2557, 1987.
182. **Magill, S. C. and McMahan, U. J.**, Motor neurons contain agrin-like molecules, *J. Cell Biol.*, 107, 1825, 1988.
183. **Nitkin, R. M., Smith, M. A., Magill, C., Fallon, J. R., Yao, M. E., Wallace, B. G., and McMahan, U. J.**, Identification of agrin, a synaptic organizing protein from *Torpedo* electric organ, *J. Cell Biol.*, 105, 2471, 1987.
184. **Rupp, F., Payan, D. G., Magill-Solc, C., Cowan, D. M., and Scheller, R. H.**, Structure and expression of a rat agrin, *Neuron*, 6, 811, 1991.
185. **Butkowski, R. J., Langeveld, J. P. M., Wieslander, J., Brentjens, J. R., and Andres, G. A.**, Characterization of a tubular basement membrane component reactive with autoantibodies associated with tubulointerstitial nephritis, *J. Biol. Chem.*, 265, 21091, 1990.
186. **Robinson, L. A. K., Murrah, V. A., Moyer, M. P., and Rohrbach, D. H.**, Characterization of a novel glycoprotein isolated from the basement membrane matrix of the Engelbreth-Holm-Swarm tumor, *J. Biol. Chem.*, 264, 5141, 1989.
187. **Folkman, J., Klagsbrun, M., Sasse, J., Wadzinski, M., Ingber, D., and Vlodavsky, I.**, A heparin-binding angiogenic protein, basic fibroblast growth factor, is stored within basement membrane, *Am. J. Pathol.*, 130, 393, 1988.
188. **Vigny, M., Ollier-Hartmann, M. P., Lavigne, N., Fayein, N., Jeanny, J. C., Laurent, M., and Courtois, Y.**, Specific binding of basic fibroblast growth factor to basement membrane-like structures and purified heparan sulfate proteoglycan of the EHS tumor, *J. Cell Physiol.*, 137, 321, 1988.
189. **Burgeson, R. E.**, Type VII collagen, anchoring fibrills, and Epidermolysis Bullosa, *J. Invest. Dermatol.*, 101, 252, 1993.
190. **Sakai, L., Keene, D. R., Morris, N. P., and Burgeson, R. E.**, Type VII collagen is a major structural component of anchoring fibrils, *J. Cell Biol.*, 103, 1577, 1986.
191. **Keene, D. R., Sakai, L. Y., Lunstrum, G. P., Morris, N. P., and Burgeson, R. E.**, Type VII collagen forms an extended network of anchoring fibrils, *J. Cell Biol.*, 104, 611, 1987.
192. **Lunstrum, G. P., Sakai, L. Y., Keene, D. R., Morris, N. P., and Burgeson, R. E.**, Large complex domains of type VII procollagen contribute to the structure of anchoring fibrils, *J. Biol. Chem.*, 261, 9042, 1986.
193. **Lunstrum, G. P., Kuo, H.-J., Rosenbaum, L. M., Keene, D. R., Glanville, R. W., Sakai, L. Y., and Burgeson, R. E.**, Anchoring fibrils contain the carboxy-terminal globular domain of type VII procollagen but lack the amino-terminal globular domain, *J. Biol. Chem.*, 262, 13706, 1987.
194. **Morris, N. P., Keene, D. R., Glanville, R. W., Bentz, H., and Burgeson, R. E.**, The tissue form of type VII collagen is an antiparallel dimer, *J. Biol. Chem.*, 261, 5638, 1986.
195. **Bächinger, H. P., Morris, N. P., Lunstrum, G. P., Keene, D. R., Rosenbaum, L. M., Compton, L. A., and Burgeson, R. E.**, The relationship of the biophysical and biochemical characteristics of type VII collagen to the function of anchoring fibrils, *J. Biol. Chem.*, 265, 10095, 1990.
196. **Parente, M. G., Chung, L. C., Ryynänen, J., Woodley, D. T., Wynn, K. C., Bauer, E. A., Mattei, M.-G., Chu, M.-L., and Uitto, J.**, Human type VII collagen: cDNA cloning and chromosomal mapping of the gene, *Proc. Natl. Acad. Sci. U.S.A.*, 88, 6931, 1991.
197. **Christiano, A. M., Rosenbaum, L. M., Chung-Honet, L. C., Parente, M. G., Woodley, D. T., Pan, T.-C., Zhang, R. Z., Chu, M.-L., Burgeson, R. E., and Uitto, J.**, The large non-collagenous domain (NC-1) of type VII collagen is amino-terminal and chimeric. Homology to cartilage matrix protein, the type III domains of fibronectin and the A domains of von Willebrand factor, *Human Mol. Genet.*, 7, 475, 1992.

198. **Hynes, R. O.,** *Fibronectins,* Rich, A., Ed., Springer Series in Molecular Biology, Springer-Verlag, New York, 1990.
199. **Yamada, K. M.,** Fibronectin domains and receptors, in *Fibronectin,* D. F. Mosher, Ed., Biology of the Extracellular Matrix: A Series, Mecham, R. P., Series Ed., Academic Press, San Diego, 1989, 47.
200. **Price, T. M., Rudee, M. L., Pierschbacher, M., and Ruoslahti, E.,** Structure of fibronectin and its fragments in electron microscopy, *Eur. J. Biochem.,* 129, 359, 1982.
201. **Gratecos, D., Naidet, C., Astier, M., Thiery, J. P., and Sémériva, M.,** Drosophila fibronectin: a protein that shares properties similar to those of its mammalian homologue, *EMBO J.,* 7, 215, 1988.
202. **Schlage, W. K.,** Isolation and characterization of a fibronectin from marine coelenterates, *Eur. J. Cell Biol.,* 47, 395, 1988.
203. **Labat-Robert, J., Robert, L., Auger, C., Lethias, C., and Garrone, R.,** Fibronectin-like protein in porifera: its role in cell aggregation, *Proc. Natl. Acad. Sci. U.S.A.,* 78, 6261, 1981.
204. **Petersen, T. E., Thogersen, H. C., Skorstengaard, K., Vibe-Pedersen, K., Sahl, P., Sottrup-Jensen, L., and Magnusson, S.,** Partial primary structure of bovine plasma fibronectin: three types of internal homology, *Proc. Natl. Acad. Sci. U.S.A.,* 80, 137, 1983.
205. **Rocco, M., Infusini, E., Daga, M. G., Gogioso, L., and Cuniberti, C.,** Models of fibronectin, *EMBO J.,* 6, 2343, 1987.
206. **Owens, R. J. and Baralle, F. E.,** Exon structure of the collagen-binding domain of human fibronectin, *FEBS Lett.,* 204, 318, 1986.
207. **Odermatt, E., Tamkun, J. W., and Hynes, R. O.,** The repeating modular structure of the fibronectin gene: Relationship to protein structure and subunit variation, *Proc. Natl. Acad. Sci. U.S.A.,* 82, 6571, 1985.
208. **Oldberg, A. and Ruoslahti, E.,** Evolution of the fibronectin gene: exon structure of cell attachment domain, *J. Biol. Chem.,* 261, 2113, 1986.
209. **Hogan, B. L., Holland, P. W. H., and Engel, J.,** Structure and expression of SPARC (Osteonectin, BM-40): a secreted calcium-binding glycoprotein associated with extracellular matrix production, in *Morphoregulatory Molecules,* Edelman, G. M., Cunningham, B. A., and Thiery, J. P., Eds., John Wiley & Sons, New York, 1990, 245.
210. **Termine, J. D., Kleinman, H. K., Whitson, S. W., Conn, K. M., McGarvey, M. L., and Martin, G. R.,** Osteonectin, a bone-specific protein linking mineral to collagen, *Cell,* 26, 99, 1981.
211. **Sage, H., Johnson, C., and Bornstein, P.,** Characterization of a novel serum albumin-binding glycoprotein secreted by endothelial cells in culture, *J. Biol. Chem.,* 259, 3993, 1984.
212. **Mason, I. J., Taylor, A., Williams, J. G., Sage, H., and Hogan, B. L. M.,** Evidence from molecular cloning that SPARC, a major product of mouse embryo parietal endoderm, is related to an endothelial cell 'culture shock' glycoprotein of M_r 43,000, *EMBO J.,* 5, 1465, 1986.
213. **Dziadek, M., Paulsson, M., Aumailley, M., and Timpl, R.,** Purification and tissue distribution of a small protein (BM-40) extracted from a basement membrane tumor, *Eur. J. Biochem.,* 161, 455, 1986.
214. **Mann, K., Deutzmann, R., Paulsson, M., and Timpl, R.,** Solubilization of protein BM-40 from a basement membrane tumor with chelating agents and evidence for its identity with osteonectin and SPARC, *FEBS Lett.,* 218, 167, 1987.
215. **Bolander, M. E., Young, M. F., Fisher, L. W., Yamada, Y., and Termine, J. D.,** Osteonectin cDNA sequence reveals potential binding regions for calcium and hydroxyapatite and shows homologies with both a basement membrane protein (SPARC) and a serine protease inhibitor (ovomucoid), *Proc. Natl. Acad. Sci. U.S.A.,* 85, 2919, 1988.
216. **Chiquet-Ehrismann, R.,** Anti-adhesive molecules of the extracellular matrix, *Curr. Opin. Cell Biol.,* 3, 800, 1991.
217. **Sage, H. and Bornstein, P.,** Extracellular proteins that modulate cell-matrix interactions, *J. Biol. Chem.,* 266, 14831, 1991.
218. **Maurer, P., Mayer, U., Bruch, M., Jenö, P., Mann, K., Landwehr, R., Engel, J., and Timpl, R.,** High-affinity and low-affinity calcium binding and stability of the multidomain extracellular 40-kDa basement membrane glycoprotein (BM-40/SPARC/osteonectin), *Eur. J. Biochem.,* 205, 233, 1992.
219. **Hughes, R. C., Taylor, A., Sage, H., and Hogan, B. L. M.,** Distinct patterns of glycosylation of colligin, a collagen-binding glycoprotein, and SPARC (osteonectin) a secreted Ca^{++}-binding glycoprotein, *Eur. J. Biochem.,* 163, 57, 1987.
220. **Engel, J., Taylor, W., Paulsson, M., Sage, H., and Hogan, B.,** Calcium binding domains and calcium-induced conformational transition of SPARC/BM-40/osteonectin, an extracellular glycoprotein expressed in mineralized and non-mineralized tissues, *Biochemistry,* 26, 6958, 1987.

221. **Mayer, U., Aumailley, M., Mann, K., Timpl, R., and Engel, J.,** Calcium-dependent binding of BM-40 (osteonectin, SPARC) to basement membrane collagen type IV, *Eur. J. Biochem.*, 198, 141, 1991.
222. **Laskowski, M., Jr., Kato, I., Ardelt, W., Cook, J., Denton, A., Empie, M. W., Kohr, W. J., Park, S. J., Parks, K., Schatzley, B. L., Schoenberger, O. L., Tashiro, M., Vichot, G., Whatley, H. F., Wieczorek, A., and Wieczorek, M.,** Ovomucoid third domains from 100 avian species: isolation, sequences and hypervariability of enzyme-inhibitor contact residues, *Biochemistry*, 26, 202, 1987.
223. **Kretsinger, R. H.,** The informational role of calcium in the cytosol, *Adv. Cyclic Nucleotide Res.*, 11, 1, 1979.
224. **Dang, C. V., Erbert, R. F., and Bell, W. O.,** Localization of a fibrinogen calcium binding site between α-subunit positions 311 and 336 by terbium fluorescence, *J. Biol. Chem.*, 260, 9713, 1985.
225. **Leszczynski, J. F. and Rose, G. D.,** Loops in globular proteins: a novel category of secondary structure, *Science*, 234, 849, 1986.
226. **Pottgiesser, J., Maurer, P., Mayer, U., Nischt, R., Mann, K., Timpl, R., Krieg, T., and Engel, J.,** Changes in calcium and collagen IV binding caused by mutations in the EF hand and other domains of extracellular matrix protein BM-40 (SPARC, osteonectin), *J. Mol. Biol.*, 238, 563, 1994.
227. **Nischt, R., Pottgiesser, J., Krieg, T., Mayer, U., Aumailley, M., and Timpl, R.,** Recombinant expression and properties of the human calcium-binding extracellular matrix protein BM-40, *Eur. J. Biochem.*, 200, 529, 1991.
228. **Kluge, M., Mann, K., Dziadek, M., and Timpl, R.,** Characterization of a novel calcium-binding 90-kDa glycoprotein (BM-90) shared by basement membranes and serum, *Eur. J. Biochem.*, 193, 651, 1990.
229. **Argraves, W. S., Tran, H., Burgess, W. H., and Dickerson, K.,** Fibulin is an extracellular matrix and plasma glycoprotein with repeated domain structure, *J. Cell Biol.*, 111, 3155, 1990.
230. **Pan, T.-C., Kluge, M., Zhang, R.-Z., Mayer, U., Timpl, R., and Chu, M.-L.,** Sequence of extracellular mouse protein BM90/fibulin and its calcium-dependent binding of other basement-membrane ligands, *Eur. J. Biochem.*, 215, 733, 1993.
231. **Erickson, H. P. and Lightner, V. A.,** Hexabrachion protein (tenascin, cytotactin, brachionectin) in connective tissues, embryonic brain, and tumors, *Adv. Cell Biol.*, 2, 55, 1988.
232. **Erickson, H. P. and Bourdon, M. A.,** Tenascin: extracellular matrix protein prominent in specialized embryonic tissues and tumors, *Annu. Rev. Cell Biol.*, 5, 71, 1989.
233. **Chiquet-Ehrismann, R.,** What distinguishes tenascin from fibronectin?, *FASEB J.*, 4, 2598, 1990.
234. **Chiquet-Ehrismann, R., Mackie, E. J., Pearson, C. A., and Sakakura, T.,** Tenascin: an extracellular matrix protein involved in tissue interactions during fetal development and oncogenesis, *Cell*, 47, 131, 1986.
235. **Chiquet, M., Vrucinic-Filipi, N., Schenk, S., Beck, K., and Chiquet-Ehrismann, R.,** Isolation of chick tenascin variants and fragments. A C-terminal heparin-binding fragment produced by cleavage of the extra domain from the largest subunit splicing variant, *Eur. J. Biochem.*, 199, 379, 1991.
236. **Masuda-Nakagawa, L., Beck, K., and Chiquet, M.,** Identification of molecules in leech extracellular matrix that promote neurite outgrowth, *Proc. R. Soc. Lond. B*, 235, 247, 1988.
237. **Humbert-David, N. and Garrone, R.,** A six-armed, tenascin-like protein extracted from the porifera *Oscarella tuberculata* (Homosclerophorida), *Eur. J. Biochem.*, 216, 255, 1993.
238. **Baumgartner, S. and Chiquet-Ehrismann, R.,** *Ten^{a}*, a Drosophila gene related to tenascin, shows selective transcript localization, *Mech. Dev.*, 40, 165, 1993.
239. **Gulcher, J. R., Nies, D. E., Marton, L. S., and Stefansson, K.,** An alternatively spliced region of the human hexabrachion contains a repeat of potential N-glycosylation sites, *Proc. Natl. Acad. Sci. U.S.A.*, 86, 1588, 1989.
240. **Jones, F. S., Hoffman, S., Cunningham, B. A., and Edelman, G. M.,** A detailed structural model of cytotactin: protein homologies, alternative RNA splicing, and binding regions, *Proc. Natl. Acad. Sci. U.S.A.*, 86, 1905, 1989.
241. **Weller, A., Beck, S., and Ekblom, P.,** Amino acid sequence of mouse tenascin and differential expression of two tenascin isoforms during embryogenesis, *J. Cell Biol.*, 112, 355, 1991.
242. **Nishi, T., Weinstein, J., Gillespie, W. M., and Paulson, J. C.,** Complete primary structure of porcine tenascin. Detection of tenascin transcripts in adult submaxillary glands, *Eur. J. Biochem.*, 202, 643, 1991.
243. **Taylor, H. C., Lightner, V. A., Beyer, W. F., Jr., McCaslin, D., Briscoe, G., and Erickson, H. P.,** Biochemical and structural studies of tenascin/hexabrachion proteins, *J. Cell. Biochem.*, 41, 71, 1989.
244. **Erickson, H. P.,** Tenascin-C, tenascin-R and tenascin-X: a family of talented proteins in search of functions, *Curr. Opin. Cell Biol.*, 5, 869, 1993.

245. **Inoue, S.,** Ultrastructure of basement membranes, *Int. Rev. Cytol.,* 117, 57, 1989.
246. **Yurchenco, P. D., Tsilibary, E. C., Charonis, A. S., and Furthmayr, H.,** Laminin polymerization *in vitro.* Evidence for a two-step assembly with domain specificity, *J. Biol. Chem.,* 260, 7636, 1985.
247. **Kalb, E. and Engel, J.,** Binding and calcium-induced aggregation of laminin onto lipid bilayers, *J. Biol. Chem.,* 266, 19047, 1991.
248. **Paulsson, M.,** The role of Ca^{2+} binding in the self-aggregation of laminin-nidogen complexes, *J. Biol. Chem.,* 263, 5425, 1988.
249. **Aeschlimann, D. and Paulsson, M.,** Cross-linking of laminin-nidogen complexes by tissue transglutaminase. A novel mechanism for basement membrane stabilization, *J. Biol. Chem.,* 266, 15308, 1991.
250. **Schittny, J. C. and Yurchenco, P. D.,** Terminal short arm domains of basement membrane laminin are critical for its self-assembly, *J. Cell Biol.,* 110, 825, 1990.
251. **Schittny, J. C., Timpl, R., and Engel, J.,** High resolution immunoelectron microscopic localization of functional domains of laminin, nidogen, and heparan sulfate proteoglycan in epithelial basement membrane of mouse cornea reveals different topological orientations, *J. Cell Biol.,* 107, 1599, 1988.
252. **Abrahamson, D. R., Irwin, M. H., St. John, P. L., Perry, E. W., Accavitti, M. A., Heck, L. W., and Couchman, J. R.,** Selective immunoreactivities of kidney basement membranes to monoclonal antibodies against laminin: localization of the ends of the long arm and the short arms to discrete microdomains, *J. Cell Biol.,* 109, 3477, 1989.
253. **Yurchenco, P. D., Cheng, Y.-S., and Colognato, H.,** Laminin forms an independent network in basement membranes, *J. Cell Biol.,* 117, 1119, 1992.
254. **Yurchenco, P. D. and Furthmayr, H.,** Self-assembly of basement membrane collagen, *Biochemistry,* 23, 1839, 1984.
255. **Madri, J. A., Pratt, B. M., Yurchenco, P. D., and Furthmayr, H.,** The ultrastructural organization and architecture of basement membranes, *CIBA Found. Symp.,* 108, 6, 1984.
256. **Yurchenco, P. D., Tsilibary, E. C., Charonis, A. S., and Furthmayr, H.,** Models for the self-assembly of basement membrane, *J. Histochem. Cytochem.,* 34, 93, 1986.
257. **Tsilibary, E. C. and Charonis, A. S.,** The role of the main noncollagenous domain (NC1) in type IV collagen self-assembly, *J. Cell Biol.,* 103, 401, 1986.
258. **Tsilibary, E. C., Reger, L. A., Vogel, A. M., Koliakos, G. G., Anderson, S. S., Charonis, A. S., Alegre, J. N., and Furcht, L. T.,** Identification of a multifunctional, cell binding peptide sequence from the a1 (NC1) of type IV collagen, *J. Cell Biol.,* 111, 1583, 1990.
259. **Yurchenco, P. D. and Ruben, G. C.,** Basement membrane structure *in situ:* evidence for lateral associations in the type IV collagen network, *J. Cell Biol.,* 105, 2559, 1987.
260. **Yurchenco, P. D. and Ruben, G. C.,** Type IV collagen lateral associations in the EHS tumor matrix. Comparison with amniotic and in vitro networks, *Am. J. Pathol.,* 132, 278, 1988.
261. **Charonis, A. S., Tsilibary, E. C., Yurchenco, P. D., and Furthmayr, H.,** Binding of laminin to type IV collagen: a morphological study, *J. Cell Biol.,* 100, 1848, 1985.
262. **Laurie, G. W., Bing, J. T., Kleinman, H. K., Hassel, J. R., Aumailley, M., Martin, G., and Feldman, R. J.,** Localization of binding sites for laminin, heparan sulfate proteoglycan and fibronectin on basement membrane (type IV) collagen, *J. Mol. Biol.,* 189, 205, 1986.
263. **Rao, N. C., Margulies, I. M. K., and Liotta, L. A.,** Binding domain for laminin on type IV collagen, *Biochem. Biophys. Res. Commun.,* 128, 45, 1985.
264. **Aumailley, M., Wiedemann, H., Mann, K., and Timpl, R.,** Binding of nidogen and the laminin/nidogen complex to basement membrane collagen type IV, *Eur. J. Biochem.,* 184, 241, 1989.
265. **Mann, K., Deutzmann, R., and Timpl, R.,** Characterization of proteolytic fragments of the laminin-nidogen complex and their activity in ligand-binding assays, *Eur. J. Biochem.,* 178, 71, 1988.
266. **Sakashita, S., Engvall, E., and Ruoslahti, E.,** Basement membrane glycoprotein laminin binds to heparin, *FEBS Lett.,* 116, 243, 1980.
267. **Aumailley, M., Gurrath, M., Müller, G., Calvete, J., Timpl, R., and Kessler, H.,** Arg-Gly-Asp constrained with cyclic pentapeptides. Strong and selective inhibitors of cell adhesion to vitronectin and laminin fragment P1, *FEBS Lett.,* 291, 50, 1991.
268. **Deutzmann, R., Aumailley, M., Wiedemann, H., Pysny, W., Timpl, R., and Edgar, D.,** Cell adhesion, spreading and neurite stimulation by laminin fragment E8 depends on maintenance of secondary and tertiary structure in its rod and globular domain, *Eur. J. Biochem.,* 191, 513, 1990.
269. **Pierschbacher, M. D. and Ruoslahti, E.,** Cell attachment activity of fibronectin can be duplicated by small synthetic fragments of the molecule, *Nature,* 309, 30, 1984.

270. **Pierschbacher, M. D. and Ruoslahti, E.,** Variants of the cell recognition site of fibronectin that retain attachment-promoting activity, *Proc. Natl. Acad. Sci. U.S.A.*, 81, 5985, 1984.
271. **Aumailley, M., Gerl, M., Sonnenberg, A., Deutzmann, R., and Timpl, R.,** Identification of the Arg-Gly-Asp sequence in laminin A chain as a latent cell-binding site exposed in fragment P1, *FEBS Lett.*, 262, 82, 1990.
272. **Bourdon, M. A. and Ruoslahti, E.,** Tenascin mediates cell attachment through an RGD-dependent receptor, *J. Cell Biol.*, 108, 1149, 1989.
273. **Kleinman, H. K., Cannon, F. B., Laurie, G. W., Hassell, J. R., Aumailley, M., Terranova, U. P., Martin, G. R., and Dubois-Dalc, M.,** Biological activities of laminin, *J. Cell. Biochem.*, 27, 317, 1985.
274. **Panayotou, G., End, P., Aumailley, M., Timpl, R., and Engel, J.,** Domains of laminin with growth-factor activity, *Cell*, 56, 93, 1989.
275. **Dedhar, S.,** Entactin inhibits the attachment and spreading of human carcinoma cells to laminin, *J. Cell Biol.*, 115, 135a, 1991.
276. **Calof, A. L. and Lander, A. D.,** Relationship between neuronal migration and cell-substratum adhesion: laminin and merosin promote olfactory neuronal migration but are anti-adhesive, *J. Cell Biol.*, 115, 779, 1991.
277. **Burgeson, R. E., Chiquet, M., Deutzmann, R., Ekblom, P., Engel, J., Kleinman, H. K., Martin, G. R., Meneguzzi, G., Paulsson, M., Sanes, J., Timpl, R., Tryggvason, K., Yamada, Y., and Yurchenco, P. D.,** A new nomenclature for lamins, *Matrix Biol.*, 14, 209, 1994.

Chapter 12

The Interaction of the Cytoskeleton with Adhesive Receptors: Effects on Cell Adhesion

Joan E. B. Fox

CONTENTS

I. Introduction 253
II. Cadherins 254
A. Regulation of Adhesive Events by Cadherin-Cytoskeleton Interaction 254
B. Identification of Proteins Mediating Cadherin-Cytoskeleton Interactions 254
III. Integrins 257
A. Transmembrane Signaling Mediated by Integrin-Ligand Association 257
B. Regulation of Adhesive Interactions by Integrin-Cytoskeleton Association 257
C. Identification of Proteins Mediating Integrin-Cytoskeleton Interactions 259
IV. Glycoprotein Ib-IX Complex 261
A. Potential Role of the Cytoskeleton in Regulating Glycoprotein Ib-IX Function 261
B. Mechanism of Association of the Glycoprotein Ib-IX Complex with the Cytoskeleton 262
Summary 264
References 264

I. INTRODUCTION

It has long been suggested that the adhesion of cells to each other or to the substratum can be regulated by the cytoskeleton. Much of the evidence for this suggestion has been indirect. However, in recent years, with the identification and characterization of receptors mediating cell-cell and cell-substratum adhesion and of intracellular cytoskeletal proteins, it has become increasingly apparent that the cytoskeleton may indeed play an important role in regulating adhesive events. As discussed elsewhere in this book, several families of adhesion receptors have now been identified. These include

1. the cadherins, Ca^{2+}-dependent homophilic cell-cell adhesion receptors that have an important role in tissue development,
2. the Ca^{2+}-independent adhesive receptors of the immunoglobulin superfamily that are involved in cell-cell adhesion and are important in tissue development, inflammation, and hemostasis,
3. members of the integrin superfamily that mediate both cell-cell and cell-substratum interactions and are important in a wide variety of adhesive processes, including inflammation, hemostasis, thrombosis, and tumor metastasis,
4. members of the selectin superfamily that mediate interactions of blood cells with the endothelium and with each other,
5. miscellaneous proteins, such as the glycoprotein (gp) Ib-IX complex on platelets, a receptor that plays an important role in hemostasis and thrombosis.

The receptors that have been studied most extensively in terms of their potential cytoskeletal interactions are the cadherins, the integrins, and the platelet gpIb-IX complex. Less is known about the possibility that members of the immunoglobulin or selectin superfamilies are associated with the cytoskeleton.* This chapter will therefore focus on cadherins, integrins, and the gpIb-IX complex. It will review what is

* Publications that appeared after this review was written indicate that PECAM-1 and ICAM-1, members of the IgG family of adhesion receptors, also associate with the cytoskeleton (Newman, P. J., Hillery, C. A., Albrecht, R., Parise, L. V., Berndt, M. C., Mazurov, A. V., Dunlop, L. C., Zhang, J., and Rittenhouse, S. J., *Cell Biol.*, 119, 239, 1992; Carpéno, O., Pallai, P., Staunton, D. E., and Springer, T. A., *J. Cell Biol.*, 118, 1223, 1992.)

0-8493-4559-6/95/$0.00+$.50

known about the interaction of these receptors with cytoskeletal proteins both in intact cells and *in vitro* and will summarize the evidence that adhesive interactions mediated by these receptors may be regulated by their attachment to the cytoskeleton.

II. CADHERINS

Cadherins are members of a family of adhesion molecules involved in Ca^{2+}-dependent homotypic cell-cell interactions. Each cadherin consists of a single transmembrane glycoprotein with a high level of sequence identity to other members of the family. The highest level of sequence identity is found in the cytoplasmic domains (see Reference 1 for a review). The best-studied cadherin in terms of its potential cytoskeletal interaction is E-cadherin (otherwise known as uvomorulin), which mediates cell-cell adhesion in epithelial cells. This receptor was shown immunocytochemically to be concentrated in the adherens junction,[2,3] where it colocalized with actin filaments.[4-6] It was assumed that the colocalization resulted from an association of the cadherin with the cytoskeleton because upon lysis of adherent cells with detergent, the cadherin remained in the detergent-insoluble fraction, where it continued to colocalize with actin.[4] Recent experiments in which E-cadherin was transfected into cadherin-negative cells have provided support for the conclusion that the cytoplasmic domain of E-cadherin associates with the cytoskeleton. Thus, transfection of a full-length cDNA for E-cadherin resulted in expression of a molecule that reorganized the cytoskeleton and was recovered in the detergent-insoluble fraction upon cell lysis.[7] However, truncated forms lacking much of the cytoplasmic domain did not reorganize the cytoskeleton[7] and were no longer recovered in the detergent-insoluble fraction.[7-9]

A. REGULATION OF ADHESIVE EVENTS BY CADHERIN-CYTOSKELETON INTERACTION

Several lines of evidence suggest that the cadherin-cytoskeleton interaction allows communication between the extracellular domain of cadherins and the intracellular cytoskeleton. For example, disruption of the interaction of the extracellular domains of E-cadherin on adjacent epithelial cells by chelation of Ca^{2+} alters cytoskeletal organization and leads to a loss of cadherin at the sites of cell-cell contact.[5,10] When full-length cDNA for E-cadherin is transfected into heterologous cells, the protein that is expressed associates with the cytoskeleton and mediates Ca^{2+}-dependent aggregation of the cells,[8] forms of cadherin in which the cytoplasmic domain has been truncated do not associate with the cytoskeleton[7-9] and are unable to induce aggregation of cells.[8] Thus, it appears that adhesion can regulate the organization of the cytoskeleton and that the cytoskeleton-associated cytoplasmic domain of E-cadherin can in turn regulate the adhesive function of the extracellular domain of the molecule.

In one study,[11] cells were transfected with full-length E-cadherin, cytoplasmic-truncated E-cadherin, or a chimeric molecule containing the extracellular domain of E-cadherin but the transmembrane and cytoplasmic domain of the Ca^{2+}-independent immunoglobulin-like adhesion molecule N-CAM. As expected from previous work, cells containing the truncated form of E-cadherin did not aggregate. However, cells containing the chimeric molecule aggregated similarly to those containing the full-length E-cadherin. Interestingly, although incubation of cells containing full-length E-cadherin with cytochalasin D resulted in inhibition of aggregation, incubation of cells containing the chimeric E-cadherin-N-CAM molecule had no effect on the ability of these cells to aggregate. Furthermore, when cells containing full-length E-cadherin were mixed with those containing the chimeric molecule, the cells containing full-length E-cadherin sorted out from those containing the chimeric molecule. These results are intriguing in that they point to an important role of the cytoplasmic domain of the cadherins in regulating not only the ability of the extracellular domain to participate in adhesive interactions but also the selectivity of the interaction. This presumably has important implications for events such as cell sorting. The differential effects of cytochalasins on the aggregation of the different cells suggest that the importance of the cytoplasmic domain of E-cadherin may result from its interaction with the cytoskeleton.

B. IDENTIFICATION OF PROTEINS MEDIATING CADHERIN-CYTOSKELETON INTERACTIONS

The cumulative evidence discussed so far points to an important role of the cytoskeleton in regulating the functional activity of cadherins. Considerable effort has gone into identifying cytoskeletal proteins to

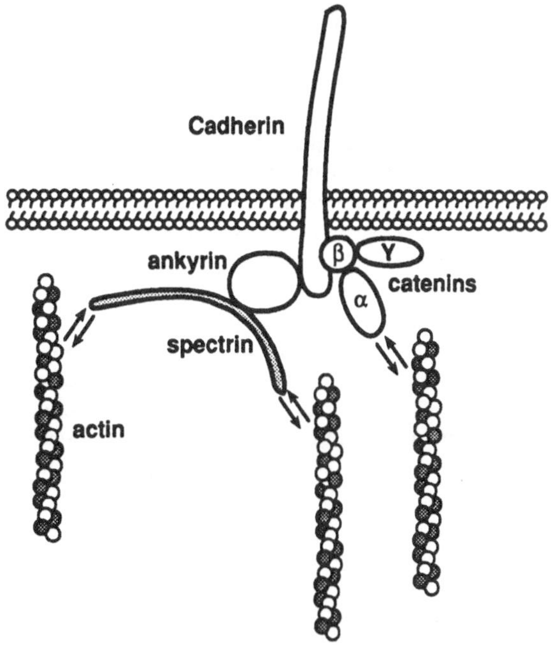

Figure 1 Schematic representation of E-cadherin and potential mechanisms (indicated by arrows) involved in its association with the cytoskeleton.

which cadherins bind. Because of the high degree of homology between the cytoplasmic domains of cadherins, it has been assumed that a common mechanism may exist. To identify linkage proteins, investigators have lysed cells with detergents and immunoprecipitated cadherins from the cell lysates. Using this approach, they have identified two major groups of proteins as candidates for linking cadherins to the cytoskeleton, catenins[9] and fodrin/ankyrin.[12]

Catenins were first identified as three proteins of M_r = 102,000, 88,000, and 80,000 (α, β, and γ catenins, respectively) that coimmunoprecipitate with a number of different cadherin.[13-16] When E-cadherin was expressed in a cell type that lacked endogenous cadherin, the distribution of the catenins changed from a cytoplasmic one to a submembranous one.[16] Catenins were coimmunoprecipitated with full-length cadherin[9] but did not co-immunoprecipitate with forms of cadherin in which most of the cytoplasmic domain had been truncated.[9] The observation that full-length E-cadherin associated with catenins and with the detergent-insoluble cellular fraction while the truncated form of the receptor was not able to associate with either catenins or the detergent-insoluble fraction suggested that the catenins mediate the interaction of cadherins with the cytoskeleton.[9]

The recent cloning of α-catenin demonstrated that this protein has significant homology to vinculin,[16] a protein known to bind to actin filaments in a variety of cell types. Further studies will be needed to determine directly whether α-catenin, like vinculin, can bind to the cytoskeleton.

It will also be necessary to elucidate the precise function of the other two catenins. It appears that β-catenin may be the component of the catenin complex that binds directly to the cytoplasmic domain of the cadherin.[17,18] Initial studies in which various truncated and chimeric forms of E-cadherin's cytoplasmic domain were expressed revealed that the binding site for catenins on E-cadherin is contained within a highly conserved 64-amino acid domain at the carboxy terminus of the molecule.[17,18] It is hoped that by identifying specific amino acid residues involved in the binding of catenins, it will be possible to express forms of E-cadherin that still contain most of the cytoplasmic domain but have specific mutations that prevent catenin binding. Experiments such as this would allow a rigorous test of the idea that catenins mediate the interaction of cadherins with the cytoskeleton in the intact cell.

As shown in Figure 1, the second mechanism that has been proposed for linking cadherins to the cytoskeleton is one involving fodrin (spectrin) and ankyrin, proteins that are involved in linking the membrane skeleton of other cells to plasma membrane glycoproteins.[12] It has been observed that these proteins have a diffuse distribution in Madin-Darby canine kidney epithelial cells.[19-22] However, as E-cadherin molecules initiate cellular interactions and accumulate at sites of cell-cell contact, fodrin and ankyrin also begin to accumulate at these sites.[19-22] When these cells are lysed with detergent, E-cadherin, ankyrin, and fodrin remain detergent-insoluble, making it difficult to analyze interactions. To circumvent this problem, Nelson and Hammerton [21] utilized Madin-Darby canine kidney epithelial cells grown in the absence of cell-cell contacts. In these cells, neither E-cadherin nor fodrin or ankyrin had been incorporated into the detergent-insoluble cytoskeletal fraction and could therefore be recovered primarily in the soluble fraction of lysed cells. Analysis of the detergent-soluble E-cadherin on sucrose gradients revealed that it coisolated in fractions that also contained fodrin, ankyrin, and Na^+-K^+-ATPase. Upon immunoprecipitation of cadherin from the sucrose gradient fractions, it was found that E-cadherin coprecipitated in a complex containing ankyrin and fodrin. The Na^+-K^+-ATPase appeared to associate with fodrin and ankyrin in independent complexes. These findings led to the suggestion that even in nonadherent cells, E-cadherin is associated directly or indirectly with either fodrin or ankyrin and that upon initiation of cell-cell contact, these cytoskeletal proteins become linked into a submembranous network that restrains the receptor and promotes cell-cell interactions.

Further support for this model came from experiments in which full-length E-cadherin was transfected into fibroblasts.[7] In control fibroblasts, fodrin had a diffuse distribution, but it became concentrated at sites of cell-cell contact in the E-cadherin-containing cells. In contrast, the distribution of fodrin was not altered significantly in cells transfected with E-cadherin containing a truncated cytoplasmic domain. Figure 1 shows one possible mechanism by which a spectrin-ankyrin complex could mediate the association of cadherins with actin filaments. Many other possible mechanisms exist; e.g., spectrin could bind directly to the cytoplasmic domain of cadherins, or the spectrin-ankyrin complex could bind via the catenin complex. Future studies will be needed to elucidate the molecular details of the interactions.

Because fodrin and ankyrin are known cytoskeletal proteins that have been shown to bind to membrane glycoproteins in a variety of other cell types,[12] a model in which a spectrin/ankyrin complex mediates the association of cadherins with actin filaments is an attractive one. However, the model has not been universally accepted. McCrea and Gumbiner[23] found that much of the E-cadherin from Madin-Darby canine epithelial cells was soluble in detergent even when these cells had undergone cell-cell contact. In contrast, most of the fodrin remained associated with the Triton-insoluble fraction. Under more stringent conditions, these workers were able to extract fodrin from the detergent-insoluble fraction but could find no evidence for coimmunoprecipitation of fodrin with E-cadherin. They found that the protein most tightly associated with immunoprecipitated E-cadherin was one of $M_r = 92{,}000$ that may represent one of the previously described catenins.

It is not clear why one group of workers recovers fodrin/ankyrin in association with cadherins and another does not. It appears likely that this may be a result of using different lysis conditions. There are always problems inherent in the study of cytoskeletal interactions in detergent lysates. Detergent insolubility of a protein is not an absolute indicator that the protein is associated with the cytoskeleton; problems with nonspecific trapping, incomplete solubilization of the lipid bilayer, or association with proteins other than cytoskeletal ones that are also inherently insoluble in detergents are always problems. Conversely, the absence of an association in a detergent lysate does not mean that it did not occur in the intact cell. The lysis conditions may be such that key linkage proteins are cleaved by proteases, dissociated by buffer components, or are of such low affinity that they are dissociated when local concentrations of the proteins are decreased upon cell lysis. These problems have been recognized by McCrea and Gumbiner.[23] They suggested that the cytoplasmic domain of E-cadherin is associated most strongly with a protein of $M_r = 92{,}000$ that probably represents one of the previously described catenins. Other proteins of $M_r = 103{,}000$, 85,000, and 78,000 are more loosely associated. Because they have not been able to detect association with known cytoskeletal proteins such as fodrin and ankyrin, they suggested that these associations, if they occur, may be of lower affinity or transient in nature. If interactions are of low affinity, they will prove harder to identify and characterize definitively. However, such interactions are in many ways of more interest than higher-affinity ones because they suggest the possibility of dynamic regulation that could conceivably be of major importance in allowing rapid reorganizations within the cell. During development, cells must continually change their interactions with other cells; thus, mechanisms of regulating cadherin function must exist. The possibility that altered interactions of the cytoplasmic domain of cadherins with the cytoskeleton regulate their function is an attractive one.

Given the cumulative evidence that E-cadherin colocalizes with actin and other cytoskeletal proteins in intact cells, that E-cadherin can be recovered with the detergent-insoluble fraction from lysed cells, that cytochalasins can affect adhesion, and that disruption of cell-cell contacts results in reorganization of cadherins and cytoskeletal proteins, it appears probable that cytoskeletal interactions do occur and that they are important in regulating adhesive interactions. Undoubtedly, studies in which native and mutated forms of E-cadherin are expressed in cultured cells will help to identify specific amino acid residues that are important in mediating the interaction of cadherins with the cytoskeleton. Together with *in vitro* binding assays of purified proteins, such studies should elucidate the details of the interactions of cadherins with cytoskeletal proteins and their importance in regulating adhesion between adjacent cells.

III. INTEGRINS

Integrins are members of a family of transmembrane glycoproteins that play important roles in mediating cell-matrix and cell-cell interactions.[24] Each integrin consists of one α- and one β-subunit. There are at least 16 α-subunits and 8 β-subunits, which exist in a variety of combinations in different cell lines. The β-subunits are highly homologous to each other and are highly conserved between species, whereas the α-subunits show more variation and play important roles in conferring ligand specificity.

A. TRANSMEMBRANE SIGNALING MEDIATED BY INTEGRIN-LIGAND ASSOCIATION

The first integrin to be identified and characterized was the gpIIb-IIIa complex on platelets.[25] This integrin has subsequently been shown to consist of a unique α-chain (gpIIb) associated with the β_3 integrin subunit gpIIIa. The gpIIb-IIIa complex cannot bind adhesive ligands on unstimulated platelets. However, when platelets are activated, an intracellular event, which appears to involve tyrosine kinases, serine-threonine kinases, and G proteins,[26] results in a conformational change in the extracellular domain of the complex such that it can now bind fibrinogen. Fibrinogen is a bivalent molecule that cross-links adjacent activated platelets, leading to the formation of a platelet aggregate. The cross-linking of platelets by fibrinogen results in a reorganization of the platelet cytoskeleton.[27] One of the manifestations of this reorganization is that whereas the gpIIb-IIIa is recovered in the soluble fraction when unstimulated platelets are lysed with Triton X-100, a portion of it is recovered in association with detergent-insoluble cytoplasmic actin filaments from aggregated platelets.[28] The observation of this aggregation-induced association of gpIIb-IIIa with the cytoskeleton provided the first demonstration that an integrin can, under certain conditions, associate with cytoskeletal elements.

As other members of the integrin superfamily have been identified over the years, it has become apparent that, like the gpIIb-IIIa complex, their cytoplasmic domains can associate with the cytoskeleton. As with gpIIb-IIIa, this interaction is regulated by the binding of adhesive ligands to the extracellular domain. Most of the evidence for the interaction of other integrins with the cytoskeleton has come from the study of focal contacts. Focal contacts form in many adherent cells and are sites at which the plasma membrane is in closest contact with the substrate. At these sites, bundles of actin filaments (known as stress fibers) terminate at the membrane, and integrins accumulate at the end of the filaments.[29] When the cells are plated on different substrates, the integrin that binds that particular substrate moves into the focal contact.[30] Antibodies that reverse the integrin-substrate interaction cause disruption of focal contacts. Thus, as in platelets, binding of ligand to an integrin causes a reorganization of the cytoskeleton. The integrin-cytoskeleton interaction presumably allows this transmission of information to occur. In platelets, antibodies that prevent ligand binding also prevent agonist-induced activation of calpain,[31] a protease that normally cleaves several cytoskeletal proteins in activated platelets,[32,33] and prevent the phosphorylation of specific proteins that normally become phosphorylated on tyrosine residues in aggregating platelets.[34,35] It appears possible that the integrin-cytoskeleton interaction may also be important in allowing this integrin-mediated transduction of signals from the outside of the cell to the inside.

B. REGULATION OF ADHESIVE INTERACTIONS BY INTEGRIN-CYTOSKELETON ASSOCIATION

The gpIIb-IIIa complex on platelets was the first integrin shown to exist in an inactivated form and to become competent to bind its adhesive ligand only after cellular activation. However, it is becoming increasingly obvious that the activity of integrins in other blood cells can be regulated by intracellular mechanisms. For example, the ability of LFA-1 ($\alpha_L\beta_2$) on lymphocytes to bind to endothelial cell ICAM is up-regulated by treatment of lymphocytes with phorbol esters.[36,37] Recent work expressing truncated forms of LFA-1 showed that the ability of LFA-1 to bind ICAM and the ability of phorbol esters to induce binding depend on the presence of the cytoplasmic domain of the integrin.[38] Similarly, there is evidence that the cytoplasmic domain of gpIIb plays a role in regulating the ability of expressed gpIIIa to bind ligand.[39] An attractive hypothesis is that the functional importance of the cytoplasmic domain of an integrin results from its association with cytoskeletal proteins. Evidence for this possibility comes from the study of focal contacts: if stress fibers in focal contacts are disrupted with cytochalasins, adhesion is decreased.[40]

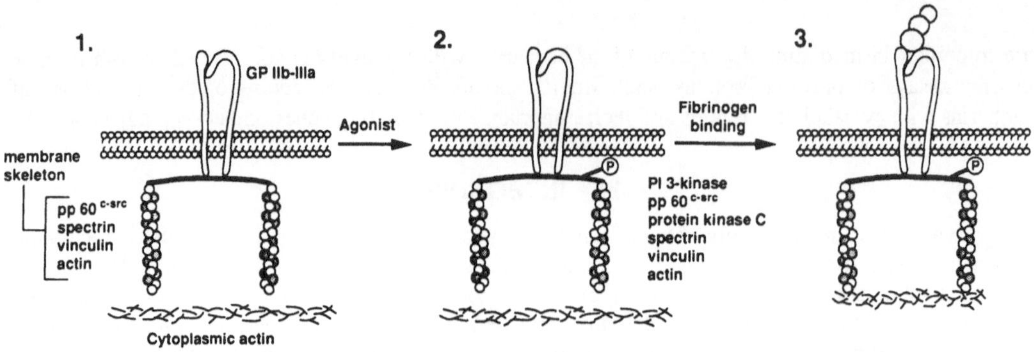

Figure 2 Schematic representation of associations of the glycoprotein (gp) IIb-IIIa complex with the cytoskeleton of unstimulated (1) and activated (2 and 3) platelets. P = phosphate groups.

Until recently, it seemed rather unlikely that association of the cytoplasmic domain of integrins with the cytoskeleton of cells such as platelets and leukocytes could regulate the ability of the integrin to bind adhesive ligand. Although there was evidence that integrins may associate with the cytoskeleton in activated cells,[28] there was no clear demonstration that integrins were associated with the cytoskeleton in the unstimulated cell. Thus, it has been assumed that the association of integrins with the cytoskeleton was the result rather than the cause of adhesive interactions in these cells.[28] However, recent work suggests that the failure to detect integrin-cytoskeleton interactions in the unstimulated cell may result from inappropriate methods. It is now apparent that platelets contain a membrane skeleton that does not sediment at the low *g* forces traditionally used to sediment cytoskeletons from detergent-lysed cells.[41,42] Recent studies showed that a subpopulation (10 to 40%) of the total gpIIb-IIIa in platelets isolates with the membrane skeleton from unstimulated platelets.[43] With platelet activation, the gpIIIb-IIIa-containing membrane skeletal complex shows increased sedimentation with the rest of the cytoskeleton,[43] indicating that the previously detected association of the gpIIb-IIIa complex with the detergent-insoluble cytoplasmic filaments from aggregating platelets may not result from ligand-induced association of gpIIb-IIIa with the cytoskeleton but rather from an increased association of membrane skeletal proteins with the underlying cytoskeletal core. Thus, it now appears quite feasible that, at least in the platelet, association of integrin with the cytoskeleton may be involved in agonist-induced integrin activation.

Further characterization of the membrane skeleton in platelets has provided some reason to believe that the association of the gpIIb-IIIa complex with the membrane skeleton may regulate the function of the integrin. Thus, the pool of gpIIb-IIIa that is detergent-insoluble in unstimulated platelets cosediments with a component of the platelet membrane skeleton that is enriched in spectrin, vinculin, and, most interestingly, pp $60^{c\text{-}src}$.[43] Upon activation of platelets, other signaling molecules (including protein kinase C and phosphoinositide 3-kinase) are present in this integrin-rich cytoskeletal fraction.[44] These findings led us to propose the model shown in Figure 2. In this model, we propose the following:

- Panel 1: A subpopulation of gpIIb-IIIa is associated with a component of the membrane skeleton that contains signalling molecules, such as pp $60^{c\text{-}src}$.
- Panel 2: Upon platelet activation, there is rapid phosphorylation of components of the membrane skeleton and recruitment of additional signaling molecules to the skeletal complex. One or more of these changes is required for the gpIIb-IIIa complex to become competent to bind fibrinogen.
- Panel 3: Fibrinogen binding results in an increased association of membrane skeletal proteins with the underlying cytoplasmic actin filaments. Contraction of the cytoplasmic filaments leads to selective movement of fractional gpIIb-IIIa toward the center of the platelet.

One can envisage several ways in which the association of integrin with the membrane skeleton could regulate integrin function. For example, by localizing signaling enzymes, the cytoskeleton may allow the generation of regulatory molecules at the appropriate site. One candidate molecule is phosphatidic acid, which is generated in activated platelets and has been shown recently to be able to render purified gpIIb-IIIa competent to bind fibrinogen.[45] Similarly, a lipid generated in neutrophils can activate purified LFA-1. Another possibility is that the cytoskeleton localizes kinases that phosphorylate the cytoplasmic domain of the gpIIb-IIIa complex and that this alters the activity of the complex. Like other integrin β-subunits, gpIIIa contains potential phosphorylation sites for both protein kinase C and tyrosine kinases.[46]

A recent study revealed that gpIIIa underwent increased phosphorylation of serine and threonine residues with platelet activation.[47] The significance of this increased phosphorylation has been questioned because the stoichiometry was only about 0.03 mol of phosphate per mol of gpIIb-IIIa.[48] However, because only the detergent-soluble fraction of gpIIIa was analyzed in these studies, it is possible that selective phosphorylation of the membrane skeleton-associated component could have gone undetected. Yet another mechanism by which the membrane skeleton could allow activation of the gpIIb-IIIa to which it is attached is if platelet activation causes modifications of components of the membrane skeleton that could in turn induce a conformational change in the cytoskeleton-associated integrin. The observation that several components of the membrane skeleton became rapidly phosphorylated on tyrosine residues upon platelet activation[43] indicates that alterations to the membrane skeleton do occur with a time course that could allow them to play a role in inducing the adhesive function of the gpIIb-IIIa complex. Clearly, future experiments will be required in order to test the hypothesis that the membrane skeleton is important in regulating the functional activity of the gpIIb-IIIa complex in platelets and to identify mechanisms involved. Furthermore, studies in which additional cell types are lysed under the conditions shown to maintain the platelet gpIIb-IIIa complex in association with the membrane skeleton will be needed to determine whether integrins are associated with skeletal elements in other cells in which integrin function can be regulated by intracellular events.

C. IDENTIFICATION OF PROTEINS MEDIATING INTEGRIN-CYTOSKELETON INTERACTIONS

Because of the potential importance of the cytoskeleton-integrin interaction in allowing communication in both directions across the cell membrane (i.e., in regulating adhesive interactions and in allowing signal transduction and ligand-induced cytoskeletal reorganization), there is considerable interest in elucidating the molecular details of the interaction between integrins and cytoskeletal structures. Insight into which cytoskeletal proteins are involved in the interactions has come from the study of focal contacts. Proteins present at these sites include α-actinin, talin, fimbrin, vinculin, tensin, and unidentified proteins of M_r = 200,000 and 82,000.[29] When cultured cells are lysed with detergents, most of the integrin is recovered with the detergent-soluble fraction; thus, approaches designed to determine whether any cytoskeletal proteins coisolate with integrins from detergent-lysed cells have not been fruitful. Focal contacts form when transmembrane adhesion receptors (in this case, integrins such as the fibronectin receptor and the vitronectin receptor) bind ligand and are thus induced to accumulate in integrin-rich domains where they associate with stress fibers.[29] It has been assumed that before its accumulation in focal contacts, an integrin is not associated with the cytoskeleton. However, based on the work on the platelet gpIIb-IIIa complex described in the previous section and also on the work described in the section on cadherins, it is possible that integrins may already be associated with submembranous skeletal proteins in nonadherent cells and that these integrin-skeletal protein(s) complexes may be induced to associate with the stress fibers as a result of adhesive ligand-induced association of the integrin-skeletal protein(s) with the underlying filamentous structures. Future studies in which cultured cells are lysed with buffers that maintain integrin-skeleton associations in platelets (e.g., Reference 43) may provide insight into this possibility and may result in identification of skeletal proteins that form the ultimate contact of the cytoskeleton with the cytoplasmic domain of integrins.

One approach used to gain insight into the cytoskeletal protein(s) that serves to link the cytoskeleton to the cytoplasmic domain of integrins has been to determine whether any of the cytoskeletal proteins that are present in focal contacts can bind to the cytoplasmic domain of integrins in *in vitro* binding assays. This approach has identified two cytoskeletal proteins that are able to associate with integrins. One is talin, a protein that was shown to bind to the fibronectin receptor ($\alpha_5\beta_1$) in an assay using equilibrium gel filtration.[49] Because talin is known to bind to vinculin, which in turn can bind to α-actinin, which can bind to actin, a model containing this chain of proteins was proposed.[29]

Recent studies suggest a second potential candidate for mediating the actin-integrin interaction.[50] When cell lysates were passed over an affinity column containing the cytoplasmic domain of the β_1-integrin subunit, several proteins bound. One of them was α-actinin.[50] An assay in which the synthetic peptide was coated onto microtiter wells confirmed that α-actinin was able to bind to the integrin's cytoplasmic domain. The affinity of this interaction was approximately 1.6×10^{-8} *M*, which was much higher than that of the binding of integrin to talin, and apparently involved the second of four homologous repeats in the rod-shaped portion of the α-actinin molecule.[51] Further studies using purified chicken gizzard integrin (a β_1-containing integrin) and purified platelet gpIIb-IIIa complex ($\alpha_{IIb}\beta_3$) showed that α-actinin was able to bind to the intact integrins. Interestingly, the affinity of the interaction with the intact

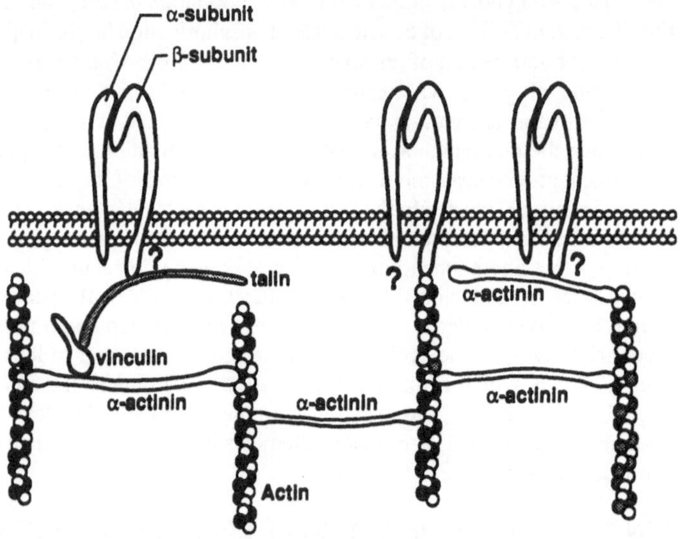

Figure 3 Model showing potential interactions of integrins with cytoskeletal elements in focal contacts. Talin and α-actinin are present in focal contacts and bind to integrin in *in vitro* assays.[49,50] A direct contact of integrins with actin has been suggested from work on platelet membranes.[52]

integrins was about 100-fold lower than that with the isolated β_1 cytoplasmic domain, suggesting that conformational restraints may exist in the intact complexes.[50]

Yet another mechanism by which integrins might associate with the cytoskeleton is through direct interaction with actin. Some evidence for this suggestion comes from work on the platelet. Painter and co-workers[52] isolated membranes from unstimulated platelets and showed that approximately 10% of the gpIIb-IIIa was associated with actin filaments, apparently in the absence of other detectable proteins. Further studies will be needed to determine which of the potential interactions occurs in intact cells. It is conceivable that more than one interaction might occur and that this could provide a means for regulating the functional activity of the integrins.

The current information available from focal contacts and the information on cytoskeletal proteins with which integrins can associate *in vitro* are summarized in Figure 3. There are clearly a number of potential interactions that can occur between integrins and cytoskeletal proteins. It is conceivable, as shown in Figure 3, that more than one cytoskeletal protein binds to the integrins in the integrin-rich domains of the membrane. It is also possible that the cytoskeletal protein to which the integrins bind is regulated by intracellular events, such as phosphorylation reactions. The presence of protein kinases in both focal contacts[29] and in the integrin-rich skeletal fractions from platelets[43,44] suggests this as a possibility. A dynamic switching of one cytoskeletal protein for another could conceivably provide a mechanism for regulating integrin function or transmembrane-signaling events.

An increased understanding of linkage proteins involved in an intact cell may be gained by identifying the sequences on the integrins that are involved in the interaction of integrin with each of the proteins with which it will interact *in vitro* and then determining whether these sequences are involved in the intact cell. This could be carried out either by introducing peptides or antibodies against the sequence into cells or by expressing integrin subunits lacking the identified sequence. Recently, several groups of investigators expressed the avian β_1 integrin subunit in mouse cell lines.[53-55] The avian β-subunit associates with mouse α-subunit and becomes associated with focal contacts. Truncated forms of the β-subunit still form complexes with the mouse α-subunit but no longer localize in focal contacts. These findings suggest that the cytoskeletal binding site is localized entirely in the β-subunit, a conclusion consistent with the observation that the cytoplasmic domains of β-subunits are involved in the interaction with α-actinin[50] and talin.[56]

Based on inhibition studies using peptides derived from the sequence of the β_1-subunit, it has been suggested that a sequence between residues 780 and 789 is involved in the binding to talin.[56] Studies using synthetic peptides encompassing the entire cytoplasmic domain of the β_1 subunit immobilized on microtiter wells have indicated that a region much closer to the membrane insertion site of the β_1-subunit is involved in the binding to α-actinin.[51] To date, studies in which mutated forms of the β_1 subunit have been expressed have been somewhat confusing. Thus, mutants in which only the four carboxy-terminal amino acid residues had been deleted localized to focal contacts.[54] Several other mutants with longer truncations did not localize to the focal contacts. However, a mutant in which most of the cytoplasmic domain was missing showed some focal contact distribution. One interpretation of these results is that the region closest to the membrane insertion site mediates cytoskeletal attachment but that conformational changes induced when other truncations are made can interfere with its normal function. It is hoped that future work in which specific mutations in domains shown in the *in vitro* experiments to be involved in associations of the purified integrins with talin and with α-actinin will minimize conformational changes induced by major deletions will provide an increased understanding of the exact molecular interactions mediating the interaction of integrins with the cytoskeleton.

IV. GLYCOPROTEIN IB-IX COMPLEX

The gplb-IX complex is one of the major glycoprotein complexes on the platelet surface.[57] The complex consists of three transmembrane components: the disulfide-linked α- and β-chains of gplb (M_r ~135,000 and ~24,000, respectively) and the noncovalently associated gpIX (M_r ~22,000). The extracellular domain of gplbα contains a binding site for von Willebrand factor. The gplb-IX complex does not normally bind to von Willebrand factor present in the plasma. However, at a site of injury, von Willebrand factor in the extracellular matrix is exposed. This surface-bound von Willebrand factor is in a conformation that can bind to the platelet gplb-IX complex. This interaction provides the major adhesive interaction of the platelet with the extracellular matrix.

A. POTENTIAL ROLE OF THE CYTOSKELETON IN REGULATING GLYCOPROTEIN IB-IX FUNCTION

As with other adhesive receptors, it appeared likely that if the gplb-IX complex were to effectively mediate the adhesion of platelets at a site of injury in the circulation, it would have to be anchored to an intracellular cytoskeleton. Without such an attachment, the platelets would probably be washed away, leaving fragments of adherent membranes. The first indication that the gplb-IX complex is indeed associated with the cytoskeleton in unstimulated platelets came from experiments in which platelets were lysed in detergent. The gplb-IX complex was found to be retained in a Triton X-100-insoluble fraction.[41] Several lines of evidence showed that the detergent insolubility of the gplb-IX complex results from an association of the complex with the cytoskeleton and is not just the result of trapping or an inherent insolubility of the complex. First, depolymerization of actin filaments in the Triton-insoluble residues with a combination of Ca^{2+} and DNAse I resulted in solubilization of the complex,[41] second, if platelets were lysed under conditions in which the Ca^{2+}-dependent protease was active, the gplb-IX complex was released into the soluble fraction.[41] Because the major substrates for the Ca^{2+}-dependent protease in platelets are all components of the cytoskeleton, this pointed to an attachment between gplb-IX and the cytoskeleton. Furthermore, it suggested that the association with the cytoskeleton was mediated by one of the substrates for the Ca^{2+}-dependent protease. The third piece of evidence that the gplb-IX complex was associated with the cytoskeleton came from the morphological demonstration that the complex was selectively retained with the detergent-insoluble structure, referred to as the membrane skeleton, that coats the inner plasma membrane of the platelet.[42] The complex was retained under conditions clearly shown to allow solubilization of the lipid bilayer and of many other glycoproteins.

When von Willebrand factor associates with the gplb-IX complex, intracellular changes that result in platelet activation are induced.[58,59] These changes include reorganization of the cytoskeleton, secretion, and activation of the fibrinogen receptor.[58,59] As with other adhesive receptors, it appears probable that the association of the cytoplasmic domain of the transmembrane receptor with the cytoskeleton may provide a mechanism for allowing transmission of the signals necessary to induce these changes across the membrane.

Attempts have been made to determine whether the association of the cytoplasmic domain of the gplb-IX complex with the cytoskeleton is important in regulating the ability of the glycoprotein to bind its adhesive ligand, as it is with other adhesive receptors. With cadherins and integrins, the primary evidence that the cytoskeleton regulates binding of adhesive ligand comes from observations that cytochalasins inhibit adhesion or from a correlation between truncation of the cytoplasmic domain and loss of adhesive functions. In the platelet, there is little information on the effects of cytochalasins on the adhesive function of the gpIb-IX complex, nor are there any reports on the effects of deletions of the cytoplasmic domain of the complex on the ability of expressed gpIb-IX to bind von Willebrand factor in cultured cells.

The one approach that has been taken to determine whether the gpIb-IX-cytoskeletal interaction is important in regulating binding of adhesive ligand has been to use a variety of methods of dissociating the complex from the cytoskeleton within intact platelets and to determine the effect that this has on the ability of the platelets to bind von Willebrand factor. Because the gpIb-IX complex is linked to the cytoskeleton by a protein that can be cleaved by the Ca^{2+}-dependent protease,[60,61] the gpIb-IX-cytoskeletal interaction was dissociated by incubating platelets with dibucaine, an agent that activates the Ca^{2+}-dependent protease without causing activation of platelets. In addition to dissociating the complex from the cytoskeleton, dibucaine also decreased the ability of the platelets to bind von Willebrand factor.[62] There was a correlation between the amount of gpIb-IX dissociated from the cytoskeleton and the decreased ability of the platelets to bind ligand. Unfortunately, dibucaine has the additional effect of inducing cleavage of the $gpIb_\alpha$ subunit at an extracellular site close to the plasma membrane. Because $gpIb_\alpha$ is the subunit that contains the von Willebrand factor-binding domain, it is possible that the decreased binding of von Willebrand factor resulted from decreased $gpIb_\alpha$ on the platelet surface rather than from decreased association of the complex with the cytoskeleton. To more directly test the hypothesis that the gpIb-IX cytoskeletal interaction regulates binding of adhesive ligand, it will be important to express gpIb-IX in heterologous cells and to determine the effect of truncations of the cytoplasmic domain on the association of the complex with cytoplasmic elements and on the ability of the complex to bind von Willebrand factor.

B. MECHANISM OF ASSOCIATION OF THE GLYCOPROTEIN IB-IX COMPLEX WITH THE CYTOSKELETON

The experiments that elucidated the mechanism by which the gpIb-IX complex associates with the cytoskeleton provide a good example of the critical effects of different lysis buffers. When platelets are lysed with buffers commonly used for cell extraction, many of the actin filaments remain sufficiently cross-linked that they can be sedimented at low g forces. However, the membrane skeleton dissociates from the underlying cytoplasmic actin filaments and fragments.[41,42] Although conditions have now been established that improve the retention of the membrane skeleton with the cytoplasmic actin filaments,[42] the fragmentation of the skeleton in early experiments using traditional extraction procedures was useful in that it allowed interactions of the gpIb-IX complex to be characterized.[41,60] Thus, when the supernatant remaining after centrifugation of cytoplasmic actin filaments was sedimented through sucrose gradients, much of the gpIb-IX complex sedimented in fractions that also contained a small amount of actin and actin-binding protein (otherwise known as filamin).[41] Under conditions in which the Ca^{2+}-dependent protease was active, actin-binding protein was hydrolyzed, and the gpIb-IX complex was recovered higher in the gradient.[41] If the gpIb-IX-containing fractions were passed over an anti-gpIb affinity column, actin-binding protein was retained along with the gpIb-IX complex.[63] Further actin-binding protein was recovered on crossed immunoelectrophoresis in the same position as gpIb.[64] Finally, when actin filaments remaining in the $15{,}000 \times g$ supernatant were depolymerized with DNase I and any remaining filaments were removed by high-speed centrifugation, actin-binding protein remained in association with the gpIb-IX complex.[60]

Although these results suggested that actin-binding protein mediated the attachment of the gpIb-IX complex to the cytoskeleton, the possibility could not be excluded that additional unidentified proteins were involved. Thus, to determine whether actin-binding protein bound directly to the cytoplasmic domain of the gpIb-IX complex, both the gpIb-IX complex and actin-binding protein were purified to homogeneity. When the gpIb-IX complex was coupled via its extracellular domain to monoclonal antibody-coated beads, it was able to bind purified actin-binding protein in a specific and saturable manner.[61] These results are considered definitive evidence that the interaction of the gpIb-IX complex with the platelet membrane skeleton is mediated by actin-binding protein.

Preliminary experiments have begun to define the domains on the gpIb-IX complex that are involved in the interaction.[65] Synthetic peptides encompassing the entire cytoplasmic domains of $gpIb_\alpha$ and $gpIb_\beta$ have been used in two *in vitro* assays. One assay used peptides coupled to affinity columns. The other used peptides coated to microtiter wells. In both assays, two overlapping peptides encompassing residues of the cytoplasmic domain of $gpIb_\alpha$ bound to actin-binding protein in a specific and saturable manner. A peptide encompassing the cytoplasmic domain of $gpIb_\beta$ or peptides with sequences from the more upstream or downstream regions of the $gpIb_\alpha$ cytoplasmic domain showed much less binding to purified actin-binding protein. Although negative results in an *in vitro* assay do not necessarily mean that these peptides are not involved in the interaction in an intact system (especially one in which synthetic peptides are immobilized), it was of interest that antibodies against one of the overlapping active sequences of gpIb$_\alpha$ were able to inhibit the binding of intact gpIb-IX complex to actin-binding protein. Furthermore, the affinity of the interactions between the overlapping active peptides and actin-binding protein were in the same range ($\sim 1 \times 10^{-7}M$) as that of the interaction between the intact gpIb-IX complex and actin-binding protein. It should be noted, however, that this affinity is considerably lower than that expected for the interaction of the gpIb-IX complex with actin-binding protein.[61] The affinity of this interaction between the proteins isolated from platelet extracts is so high that the complex cannot be dissociated by conventional treatments such as high- or low-salt concentrations. Only incubation with sodium dodecyl sulfate or treatment with *N*-ethylmaleimide has successfully dissociated the complex. Although the low affinity of interaction between purified proteins may result from irreversible modification to one or the other proteins induced by *N*-ethylmaleimide present in the purification schemes, it is also possible that other purification-induced changes are responsible for decreased affinities, or that in the intact cell, domains in addition to those identified in the *in vitro* assays are involved.

One approach that should prove useful in assessing contributions of individual domains and amino acids is to express the complex in cultured cells. The cDNAs for all three subunits are available[66-68] and have been expressed in several different cell types. The three subunits associate and are incorporated into the plasma membrane in a functional form.[69] As in platelets, the gpIb-IX complex is retained with detergent-insoluble material and can be coprecipitated with actin-binding protein. Furthermore, immunofluorescence images have shown that the complex colocalizes with actin-binding protein. Thus, the complex appears to associate with actin-binding protein in these cells in the same way that it does in platelets. Future work with different mutated and truncated forms should provide considerable information on domains of the gpIb-IX complex that are involved in the interaction.

Some information is also available on the domain of actin-binding protein to which the gpIb-IX complex binds. Actin-binding protein is a dimer consisting of two identical elongated molecules of M_r ~250,000. These subunits associate noncovalently through their carboxy-terminal domains, whereas the amino-terminal ends associate with actin filaments.[70-72] In this way actin-binding protein, together with other proteins such as spectrin, is thought to cross-link short filaments of the membrane skeleton into a network that coats the lipid bilayer.[73] If the Ca^{2+}-dependent protease within platelets is activated, for example by addition of dibucaine or the Ca^{2+}-ionophore A23173, actin-binding protein is hydrolyzed.[60,74] The Ca^{2+}-dependent protease cleaves actin-binding protein at two sites.[32] The first cleavage occurs toward the carboxy-terminal end of the molecule, yielding a fragment of M_r ~200,000 and one of M_r ~10,000 that remains as a dimer. A second cleavage subsequently occurs in the M_r ~100,000 fragment. This site is close to the dimerization site, as indicated by the generation of a M_r ~91,000 fragment that no longer dimerizes.[75] Ezzell and co-workers[76] showed that when ionophore A23187-treated platelets are lysed, the gpIb-IX complex coimmunoprecipitates with the M_r ~100,000 fragment, not with the M_r ~200,000 fragment of actin-binding protein. Following more extensive cleavage of actin-binding protein so that the M_r ~91,000 fragment is generated, the gpIb-IX complex is no longer found in association with actin-binding protein. Ezzell and co-workers[76] interpreted these findings as an indication that the gpIb-IX complex bound to the M_r ~100,000 fragment of actin-binding protein close to the site at which calpain cleavage causes generation of the M_r = 91,000 fragment.

Further evidence for this conclusion came from analysis of actin-binding protein cDNA.[72] Sequence analysis revealed that 90% of the sequence of actin-binding protein is made up of 24 repeats. Each repeat consists of approximately 96 amino acid residues that are predicted to exist as β-sheets. Additional sequences occur before repeats 16 and 24 and represent the sites at which calpain cleaves. Hartwig and co-workers[72] suggested that the additional sequence immediately before repeat number 24 may provide an appropriate binding site for the gpIb-IX complex. Figure 4 shows a schematic representation of the association of the gpIb-IX complex with actin-binding protein based on the available information on the domains on gpIb-IX and actin-binding protein that are involved. The availability of an *in vitro* assay for association of the gpIb-IX complex with actin-binding protein[61] and recombinant fragments of actin-binding protein[72] should allow the validity of this model to be tested.

Thus, there is considerable information available on the mechanism by which the gpIb-IX complex associates with the cytoskeleton. Most importantly, the interaction that has been well characterized in *in vitro* assays is the one that is known to occur in intact cells. Future experiments utilizing cultured cells expressing recombinant gpIb-IX complex should prove useful in testing the idea directly that the interaction regulates adhesion and gpIb-IX-mediated adhesion of platelets.

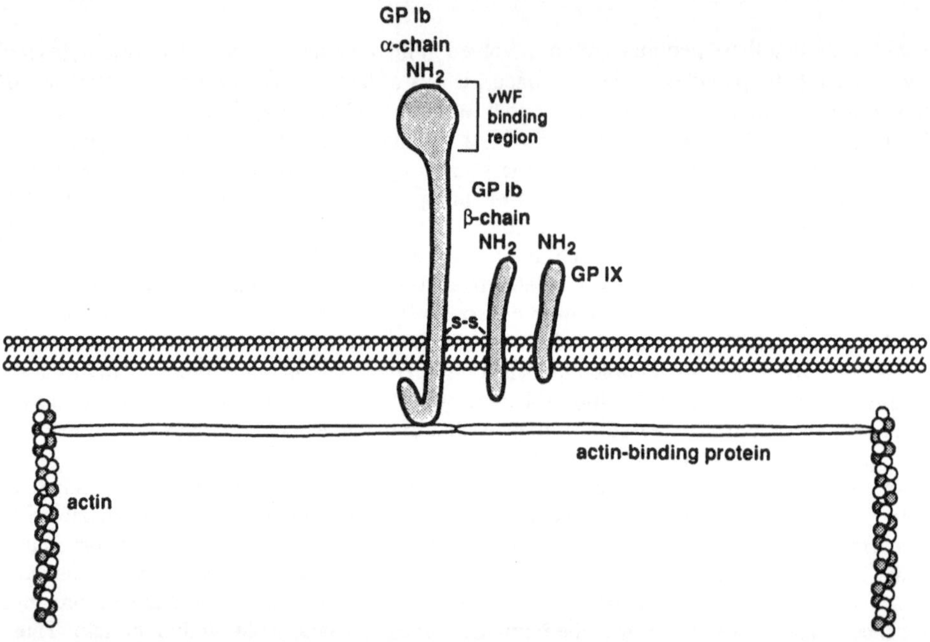

Figure 4 Schematic representation of the association of the gpIb-IX complex with actin-binding protein in the membrane skeleton.

V. SUMMARY

There is now considerable evidence that cadherins, integrins, and the platelet gpIb-IX complex associate with the cytoskeleton. This association may regulate the ability of the receptors to bind ligands that may, in turn, regulate the organization of the cytoskeleton. In the past, the interaction of adhesion receptors with cytoskeletal elements has often gone undetected because cytoskeletons are traditionally defined as the detergent-insoluble material that can be recovered by low-speed centrifugation. Work with the platelet has provided a clear demonstration that some of the traditional lysis buffers do not retain the cytoskeleton in a form that can be sedimented at these g forces; it has shown that the membrane skeleton fragments and requires high g forces to be sedimented. It is now well established that several platelet adhesive receptors are associated with the membrane skeleton. Work with Madin-Darby canine epithelial cells also has shown that adhesive receptors are associated with membrane skeletal structures in these cells. As in the platelet, these skeletal complexes are not recovered in the traditional detergent-insoluble fraction (i.e., low-speed pellet from detergent-solubilized cells) but are driven into these structures as a consequence of binding of adhesive ligands. Based on these observations, it will be interesting to reevaluate cytoskeletal interactions in other cell types to determine whether some of the receptors previously described as being detergent-soluble may in fact be associated with elements of a membrane skeleton.

Considerable evidence from a variety of cell types has suggested that the cytoskeleton may regulate the functional activity of cadherins, integrins, and the gpIb-IX complex. Future work will be needed to test these conclusions directly. The mechanism by which the gpIb-IX complex associates with the cytoskeleton in the intact cell has been identified and characterized in detail. Rapid advances are being made in our understanding of cytoskeletal interactions of the cytoplasmic domains of cadherins and integrins. Future experiments will undoubtedly identify which of these interactions can occur within intact cells. Once these interactions have been identified, it should be possible (for example, by transfection of mutant receptors) to determine directly the importance of each of these interactions in regulating adhesive interactions of cells.

REFERENCES

1. **Edelman, G. M. and Crossin, K. L.,** Cell adhesion molecules: implications for a molecular histology, *Annu. Rev. Biochem.*, 60, 155, 1991.
2. **Boller, K., Vestweber, D., and Kemler, R.,** Cell-adhesion molecule uvomorulin is localized in the intermediate junctions of adult intestinal epithelial cells, *J. Cell Biol.*, 100, 327, 1985.

3. **Volk, T. and Geiger, B.,** A-CAM: a 135-kD receptor of intercellular adherens junctions. I. Immunoelectron microscopic localization and biochemical studies, *J. Cell Biol.,* 103, 1441, 1986.
4. **Hirano, S., Nose, A., Hatta, K., Kawakami, A., and Takeichi, M.,** Calcium-dependent cell-cell adhesion molecules (cadherins): subclass specificities and possible involvement of actin bundles, *J. Cell Biol.,* 105, 2501, 1987.
5. **Volk, T. and Geiger, B.,** A-CAM: a 135-kD receptor of intercellular adherens junctions. II. Antibody-mediated modulation of junction formation, *J. Cell Biol.,* 103, 1451, 1986.
6. **Meza, I., Ibarra, G., Sabanero, M., Martnez-Palomo, A., and Cereijido, M.,** Occluding junctions and cytoskeletal components in a cultured transporting epithelium, *J. Cell Biol.,* 87, 746, 1980.
7. **McNeill, H., Ozawa, M., Kemler, R., and Nelson, W. J.,** Novel function of the cell adhesion molecule uvomorulin as an inducer of cell surface polarity, *Cell,* 62, 309, 1990.
8. **Nagafuchi, A. and Takeichi, M.,** Cell binding function of E-cadherin is regulated by the cytoplasmic domain, *EMBO J.,* 7, 3679, 1988.
9. **Ozawa, M., Baribault, H., and Kemler, R.,** The cytoplasmic domain of the cell adhesion molecule uvomorulin associates with three independent proteins structurally related in different species, *EMBO J.,* 8, 1711, 1989.
10. **Volberg, T., Geiger, B., Kartenbeck, J., and Franke, W. W.,** Changes in membrane-microfilament interaction in intercellular adherens junctions upon removal of extracellular Ca^{2+} ions, *J. Cell Biol.,* 102, 1832, 1986.
11. **Jaffe, S. H., Friedlander, D. R., Matsuzaki, F., Crossin, K. L., Cunningham, B. A., and Edelman, G. M.,** Differential effects of the cytoplasmic domains of cell adhesion molecules on cell aggregation and sorting-out, *Proc. Natl. Acad. Sci. U.S.A.,* 87, 3589, 1990.
12. **Bennett, V.,** The membrane skeleton of human erythrocytes and its implications for more complex cells, *Annu. Rev. Biochem.,* 54, 273, 1985.
13. **Vestweber, D. and Kemler, R.,** Some structural and functional aspects of the cell adhesion molecule uvomorulin, *Cell Differ.,* 15, 269, 1984.
14. **Peyriéras, N., Louvard, D., and Jacob, F.,** Characterization of antigens recognized by monoclonal and polyclonal antibodies directed against uvomorulin, *Proc. Natl. Acad. Sci. U.S.A.,* 82, 8067, 1985.
15. **Vestweber, D., Gossler, A., Boller, K., and Kemler, R.,** Expression and distribution of cell adhesion molecule uvomorulin in mouse preimplantation embryos, *Dev. Biol.,* 124, 451, 1987.
16. **Herrenknecht, K., Ozawa, M., Eckerskorn, C., Lottspeich, F., Lenter, M., and Kemler, R.,** The uvomorulin-anchorage protein α catenin is a vinculin homologue, *Proc. Natl. Acad. Sci. U.S.A.,* 88, 9156, 1991.
17. **Nagafuchi, A. and Takeichi, M.,** Transmembrane control of cadherin-mediated cell adhesion: a 94 kDa protein functionally associated with a specific region of the cytoplasmic domain of E-cadherin, *Cell Regul.,* 1, 37, 1989.
18. **Ozawa, M., Ringwald, M. and Kemler, R.,** Uvomorulin-catenin complex formation is regulated by a specific domain in the cytoplasmic region of the cell adhesion molecule, *Proc. Natl. Acad. Sci. U.S.A.,* 87, 4246, 1990.
19. **Nelson, W. J. and Veshnock, P. J.,** Ankyrin binding to $(Na^{+}+K^{+})$ATPase and implications for the organization of membrane domains in polarized cells, *Nature,* 328, 533, 1987.
20. **Morrow, J. S., Cianci, C. D., Ardito, T., Mann, A. S., and Kashgarian, M.,** Ankyrin links fodrin to the alpha subunit of Na,K-ATPase in Madin-Darby canine kidney cells and in intact renal tubule cells, *J. Cell Biol.,* 108, 455, 1989.
21. **Nelson, W. J. and Hammerton, R. W.,** A membrane-cytoskeletal complex containing Na^{+},K^{+}-ATPase, ankyrin, and fodrin in Madin-Darby canine kidney (MDCK) cells: implications for the biogenesis of epithelial cell polarity, *J. Cell Biol.,* 108, 893, 1989.
22. **Nelson, W. J.,** Development and maintenance of epithelial polarity: a role for the submembranous cytoskeleton, *Mod. Cell Biol.,* 8, 3, 1989.
23. **McCrea, P. D. and Gumbiner, B. M.,** Purification of a 92-kDa cytoplasmic protein tightly associated with the cell-cell adhesion molecule E-cadherin (uvomorulin). Characterization and extractability of the protein complex from the cell cytoskeleton, *J. Biol. Chem.,* 266, 4514, 1991.
24. **Ruoslahti, E.,** Integrins, *J. Clin. Invest.,* 87, 1, 1991.
25. **Phillips, D. R., Charo, I. F., Parise, L. V., and Fitzgerald, L. A.,** The platelet membrane glycoprotein IIb-IIIa complex, *Blood,* 71, 831, 1988.
26. **Shattil, S. J., Cunningham, M., Wiedmer, T., Zhao, J., Sims, P. J., and Brass, L. F.,** Regulation of glycoprotein IIb-IIIa receptor function studied with platelets permeabilized by the pore-forming complement proteins C5b-9, *J. Biol. Chem.,* 267, 18424, 1992.

27. **Kouns, W. C., Fox, C. F., Lamoreaux, W. J., Coons, L. B., and Jennings, L. K.,** The effect of glycoprotein IIb-IIIa receptor occupancy on the cytoskeleton of resting and activated platelets, *J. Biol. Chem.*, 266, 13891, 1991.
28. **Phillips, D. R., Jennings, L. K., and Edwards, H. H.,** Identification of membrane proteins mediating the interaction of human platelets, *J. Cell Biol.*, 86, 77, 1980.
29. **Burridge, K., Fath, K., Kelly, T., Nuckolls, G., and Turner, C.,** Focal adhesions: transmembrane junctions between the extracellular matrix and the cytoskeleton, *Annu. Rev. Cell Biol.*, 4, 487, 1988.
30. **Singer, I. I., Scott, S., Kawka, D. W., Kazazis, D. M., Gailit, J., and Ruoslahti, E.,** Cell surface distribution of fibronectin and vitronectin receptors depends on substrate composition and extracellular matrix accumulation, *J. Cell Biol.*, 106, 2171, 1988.
31. **Fox, J. E. B., Reynolds, C. C., and Austin, C. D.,** The role of GP IIb-IIIa, an integrin, in regulating activation of calpain in platelets, *J. Cell Biol.*, 115, 5a, 1991.
32. **Fox, J. E. B., Goll, D. E., Reynolds, C. C., and Phillips, D. R.,** Identification of two proteins (actin-binding protein and P235) that are hydrolyzed by endogenous Ca^{2+}-dependent protease during platelet aggregation, *J. Biol. Chem.*, 260, 1060, 1985.
33. **Fox, J. E. B., Reynolds, C. C., Morrow, J. S., and Phillips, D. R.,** Spectrin is associated with membrane-bound actin filaments in platelets and is hydrolyzed by the Ca^{2+}-dependent protease during platelet activation, *Blood,* 69, 537, 1987.
34. **Ferrell, J. E., Jr. and Martin, G. S.,** Tyrosine-specific protein phosphorylation is regulated by glycoprotein IIb-IIIa in platelets, *Proc. Natl. Acad. Sci. U.S.A.*, 86, 2234, 1989.
35. **Golden, A., Brugge, J. S., and Shattil, S. J.,** Role of platelet membrane glycoprotein IIb-IIIa in agonist-induced tyrosine phosphorylation of platelet proteins, *J. Cell Biol.*, 111, 3117, 1990.
36. **Rothlein, R., Dustin, M. L., Marlin, S. D., and Springer, T. A.,** A human intercellular adhesion molecule (ICAM-1) distinct from LFA-1, *J. Immunol.*, 137, 1270, 1986.
37. **Rothlein, R. and Springer, T. A.,** The requirement for lymphocyte function-associated antigen 1 in homotypic leukocyte adhesion stimulated by phorbol ester, *J. Exp. Med.*, 163, 1132, 1986.
38. **Hibbs, M. L., Xu, H., Stacker, S. A., and Springer, T. A.,** Regulation of adhesion to ICAM-1 by the cytoplasmic domain of LFA-1 integrin β subunit, *Science,* 251, 1611, 1991.
39. **O'Toole, T. E., Mandelman, D., Forsyth, J., Shattil, S. J., Plow, E. F., and Ginsberg, M. H.,** Modulation of the affinity of integrin $\alpha_{IIb}\beta_3$ (GPIIb-IIIa) by the cytoplasmic domain of α_{IIb}, *Science,* 254, 845, 1991.
40. **Ali, I. U. and Hynes, R. O.,** Effects of cytochalasin B and colchicine on attachment of a major surface protein of fibroblasts, *Biochim. Biophys. Acta,* 471, 16, 1977.
41. **Fox, J. E. B.,** Linkage of a membrane skeleton to integral membrane glycoproteins in human platelets. Identification of one of the glycoproteins as glycoprotein Ib, *J. Clin. Invest.*, 76, 1673, 1985.
42. **Fox, J. E. B., Boyles, J. K., Berndt, M. C., Steffen, P. K., and Anderson, L. K.,** Identification of a membrane skeleton in platelets, *J. Cell Biol.*, 106, 1525, 1988.
43. **Fox, J. E. B., Lipfert, L., Clark, E. A., Reynolds, C. C., Austin, C. D., and Brugge, J. S.,** On the role of the platelet membrane skeleton in mediating signal transduction. Association of GP IIb-IIIa, $pp60^{c\text{-}src}$, $pp62^{c\text{-}yes}$, and the $p21^{ras}$ GTPase-activating protein with the membrane skeleton, *J. Biol. Chem.*, 268, 25973, 1993.
44. **Zhang, J., Fry, M. J., Waterfield, M. D., Jaken, S., Liao, L., Fox, J. E. B., and Rittenhouse, S. E.,** Activated phosphoinositide 3-kinase associates with membrane skeleton in thrombin-exposed platelets, *J. Biol. Chem.*, 267, 4686, 1992.
45. **Smyth, S. S., Hillery, C. A., and Parise, L. V.,** Phosphatidic and lysophosphatidic acid modulate the fibrinogen binding activity of purified platelet glycoprotein IIb-IIIa, *Blood,* 78, 278a, 1991.
46. **Fitzgerald, L. A., Steiner, B., Rall, S. C., Jr., Lo, S. S., and Phillips, D. R.,** Protein sequence of endothelial glycoprotein IIIa derived from a cDNA clone. Identity with platelet glycoprotein IIIa and similarity to "integrin", *J. Biol. Chem.*, 262, 3936, 1987.
47. **Parise, L. V., Criss, A. B., Nannizzi, L., and Wardell, M. R.,** Glycoprotein IIIa is phosphorylated in intact human platelets, *Blood,* 75, 2363, 1990.
48. **Hillery, C. A., Smyth, S. S., and Parise, L. V.,** Phosphorylation of human platelet glycoprotein IIIa (GPIIIa). Dissociation from fibrinogen receptor activation and phosphorylation of GP IIIa *in vitro, J. Biol. Chem.*, 266, 14663, 1991.
49. **Horwitz, A., Duggan, K., Buck, C., Beckerle, M. C., and Burridge, K.,** Interaction of plasma membrane fibronectin receptor with talin — a transmembrane linkage, *Nature,* 320, 531, 1986.

50. **Otey, C. A., Pavalko, F. M., and Burridge, K.,** An interaction between α-actinin and the β_1 integrin subunit in vitro, *J. Cell Biol.,* 111, 721, 1990.
51. **Otey, C. A., Parr, T., Blanchard, A. D., Critchley, D., and Burridge, K.,** Mapping of interactive sites on alpha-actinin and the beta-1 integrin cytoplasmic domain, *J. Cell Biol.,* 115, 166a, 1991.
52. **Painter, R. G., Prodouz, K. N., and Gaarde, W.,** Isolation of a subpopulation of glycoprotein IIb-III from platelet membranes that is bound to membrane actin, *J. Cell Biol.,* 100, 652, 1985.
53. **Hayashi, Y., Haimovich, B., Reszka, A., Boettiger, D., and Horwitz, A.,** Expression and function of chicken integrin β1 subunit and its cytoplasmic domain mutants in mouse NIH 3T3 cells, *J. Cell Biol.,* 110, 175, 1990.
54. **Marcantonio, E. E., Guan, J. L., Trevithick, J. E., and Hynes, R. O.,** Mapping of the functional determinants of the integrin β_1 cytoplasmic domain by site-directed mutagenesis, *Cell Regul.,* 1, 597, 1990.
55. **Solowska, J., Guan, J. L., Marcantonio, E. E., Trevithick, J. E., Buck, C. A., and Hynes, R. O.,** Expression of normal and mutant avian integrin subunits in rodent cells, *J. Cell Biol.,* 109, 853, 1989.
56. **Tapley, P., Horwitz, A., Buck, C., Duggan, K., and Rohrschneider, L.,** Integrins isolated from Rous sarcoma virus-transformed chicken embryo fibroblasts, *Oncogene,* 4, 325, 1989.
57. **Clemetson, K. J.,** Glycoproteins of the platelet plasma membrane, in *Platelet Membrane Glycoproteins,* George, J. N., Nurden, A. T., and Phillips, D. R., Eds., Plenum Press, New York, 1985, 51.
58. **Moake, J. L., Turner, N. A., Stathopoulos, N. A., Nolasco, L., and Hellums, J. D.,** Shear-induced platelet aggregation can be mediated by vWF released from platelets, as well as by exogenous large or unusually large vWF multimers, requires adenosine diphosphate, and is resistant to aspirin, *Blood,* 71, 1366, 1988.
59. **Kroll, M. H., Harris, T. S., Moake, J. L., Handin, R. I., and Schafer, A. I.,** von Willebrand factor binding to platelet GpIb initiates signals for platelet activation, *J. Clin. Invest.,* 88, 1568, 1991.
60. **Fox, J. E. B.,** Identification of actin-binding protein as the protein linking the membrane skeleton to glycoproteins on platelet plasma membranes, *J. Biol. Chem.,* 260, 11970, 1985.
61. **Andrews, R. K. and Fox, J. E. B.,** Interaction of purified actin-binding protein with the platelet membrane glycoprotein Ib-IX complex, *J. Biol. Chem.,* 266, 7144, 1991.
62. **Fox, J. E. B. and Boyles, J. K.,** Platelets contain a skeleton of actin filaments that regulates the shape of the cell and the activity of GP Ib, *Blood,* 66, 304a, 1985.
63. **Okita, J. R., Pidard, D., Newman, P. J., Montgomery, R. R., and Kunicki, T. J.,** On the association of glycoprotein Ib and actin-binding protein in human platelets, *J. Cell Biol.,* 100, 317, 1985.
64. **Solum, N. O., Olsen, T. M., Gogstad, G. O., Hagen, I., and Brosstad, F.,** Demonstration of a new glycoprotein Ib-related component in platelet extracts prepared in the presence of leupeptin, *Biochim. Biophys. Acta,* 729, 53, 1983.
65. **Andrews, R. K. and Fox, J. E. B.,** Identification of a sequence in the cytoplasmic domain of the platelet membrane glycoprotein Ib-IX complex that binds to purified actin-binding protein, *J. Biol. Chem.,* 267, 18605, 1992.
66. **Lopez, J. A., Chung, D. W., Fujikawa, K., Hagen, F. S., Papayannopoulou, T., and Roth, G. J.,** Cloning of the α chain of human platelet glycoprotein Ib: a transmembrane protein with homology to leucine-rich α_2-glycoprotein, *Proc. Natl. Acad. Sci. U.S.A.,* 84, 5615, 1987.
67. **Lopez, J. A., Chung, D. W., Fujikawa, K., Hagen, F. S., Davie, E. W., and Roth, G. J.,** The α and β chains of human platelet glycoprotein Ib are both transmembrane proteins containing a leucine-rich amino acid sequence, *Proc. Natl. Acad. Sci. U.S.A.,* 85, 2135, 1988.
68. **Hickey, M. J., Williams, S. A., and Roth, G. J.,** Human platelet glycoprotein IX: an adhesive prototype of leucine-rich glycoproteins with flank-center-flank structures, *Proc. Natl. Acad. Sci. U.S.A.,* 86, 6773, 1989.
69. **López, J. A., Leung, B., Reynolds, C. C., Li, C. Q., and Fox, J. E. B.,** Efficient plasma membrane expression of a functional platelet glycoprotein Ib-IX complex requires the presence of its three subunits, *J. Biol. Chem.,* 267, 12851, 1992.
70. **Hartwig, J. H., Tyler, J., and Stossel, T. P.,** Actin-binding protein promotes bipolar and perpendicular branching of actin filaments, *J. Cell Biol.,* 87, 841, 1980.
71. **Hartwig, J. H. and Stossel, T. P.,** Structure of macrophage actin-binding protein molecules in solution and interacting with actin filaments, *J. Mol. Biol.,* 145, 563, 1981.
72. **Gorlin, J. B., Yamin, R., Egan, S., Stewart, M., Stossel, T. P., Kwiatkowski, D. J., and Hartwig, J. H.,** Human endothelial actin-binding protein (ABP-280, nonmuscle filamin): a molecular leaf spring, *J. Cell Biol.,* 111, 1089, 1990.

73. **Fox, J. E. B. and Boyles, J. K.,** The membrane skeleton — a distinct structure that regulates the function of cells, *BioEssays,* 8, 14, 1988.
74. **Fox, J. E. B., Austin, C. D., Boyles, J. K., and Steffen, P. K.,** Role of the membrane skeleton in preventing the shedding of procoagulant-rich microvesicles from the platelet plasma membrane, *J. Cell Biol.,* 111, 483, 1990.
75. **Hock, R. S., Davis, G., and Speicher, D. W.,** Purification of human smooth muscle filamin and characterization of structural domains and functional sites, *Biochemistry,* 29, 9441, 1990.
76. **Ezzell, R. M., Kenney, D. M., Egan, S., Stossel, T. P., and Hartwig, J. H.,** Localization of the domain of actin-binding protein that binds to membrane glycoprotein Ib and actin in human platelets, *J. Biol. Chem.,* 263, 13303, 1988.

Section III

Physiology of Adhesion

Chapter 13

Homing of Cells: Stem Cell as a Model

Mehdi Tavassoli (Posthumous)

CONTENTS

I. Introduction and Definition 271
II. Evolutionary Considerations: Nomadic vs. Settled Hemopoiesis 272
III. Fetal and Embryonic Homing 272
  A. The Concept of Directive Endocrinization 272
  B. Microenvironmental Organization 273
IV. Postnatal Cell Homing 273
  A. Recognition vs. Adhesion 273
  B. Molecular Probes 273
  C. *In vivo* Studies 274
  D. *In vitro* Studies 274
  E. Binding Studies 275
V. Molecular Nature of Recognition 277
  A. Homing Receptor (HR) 277
  B. Homing Ligand 279
  C. Membrane-Associated and Extracellular Matrix 280
VI. Anatomical Considerations 280
  A. The Structure of Hemopoietic Marrow 280
  B. A Two-Step Homing 282
VII. Future Directions 282
References 282

Home is the place where, when you
  have to go there,
They have to take you in.

— Robert Frost, *The Death of the Hired Man* (lines 118–119)

I. INTRODUCTION AND DEFINITION

In any organism, the cells of one kind or another tend to segregate into tissues with distinct organization and microstructure (organotypic formations). This selective segregation, although long known as a common sense observation, was first demonstrated experimentally by Moscona.[1] He began to mix different cell types in suspension, such as cartilage and kidney cells, and found that they sorted out according to type to form solid blocks of kidney tissue or cartilage tissue but never ended up as a random mixture of both.

Embryologist Paul Weiss[2] followed up these observations by studying microcinematographically the cells of chick embryo. He found that the cells move around, perfectly at random. However, when they collide with each other, their reaction differs, depending upon whether the colliding cells are homotypic (of the same kind) or heterotypic (of different kind). He noted that on collision, cells of the same kind tend to paralyze each other along the contacting surfaces, so that the mode of locomotion of cells changes. The cells thus associate and, after a period of time, a cluster of homotypic cells is formed, although apparently no permanent mutual cementing occurs. If colliding cells are not of the same kind, they do not recognize each other and do not associate. They simply snap apart and keep moving. They bypass each other, as if they were inert objects. When Weiss subsequently transplanted these reaggregated clusters of cells in the organism, they revascularized and formed organotypic formations of a remarkable degree of ordered complexity.[3]

It was by mere coincidence that Edelman,[4,5] trying to raise monoclonal antibodies reacting with neural tissue, discovered an antibody that reacted with cell adhesion molecule (CAM). It was responsible for the

0-8493-4559-6/95/$0.00+$.50

paralyzing reaction of homotypic cells. Owing to the availability of new methodologies, we have since learned much about the molecular basis of this segregation of homotypic cells into tissues.

Association of cells into organotypic formations makes physiological sense: cells of similar functions can carry out those functions in a microenvironment that may, in the course of evolution, evolve into an optimal one. Thus, we can define "homing" as a recognition system that permits association of cells into organotypic formations to permit the optimal microenvironment for a certain function.

An example may illuminate the matter. In the course of evolution, the first organotypic formations of blood-forming cells are blood islands in segmented worms.[6-9] They form in the wall of the intestinal tract, particularly in the submucosa. What makes this environment particularly suitable is its proximity to absorptive surfaces, where necessary nutrients for a proliferative cell system can be obtained readily. As the system evolves, it becomes increasingly more complex and develops other means of receiving nutrients.[10,11] Nutritional requirement is no longer preeminent. On the other hand, a system of microcirculation offering a very slow and pulsatile flow rate facilitates the function.[6,12] Evolutionary pressure acts to move the site of hemopoiesis into the bone, where the fixed volume offered by its rigid confines permits this essential microenvironmental feature.

In this chapter, the molecular basis of recognition and adherence of hemopoietic cells is used as a model system for homing of cells.

II. EVOLUTIONARY CONSIDERATIONS: NOMADIC VS. SETTLED HEMOPOIESIS

Hemopoiesis offers a particularly suitable model for the study of homing. It offers information in the course of its evolutionary aspects, its fetal and embryonic life, and its postnatal course. Postnatal hemopoiesis as a continuous phenomenon does not exist in the least-developed invertebrates. The life span of these organisms is short. They are born and die with the same blood cells (hemocytes).[7] There is no need for their *de novo* regeneration in the course of life. Nor is there a need for cellular association to organotypic formation: there is no "homing". In certain worms and insects, blood cells may undergo a few divisions within the circulatory space.[13] These divisions may be limited to the nucleus, but as the evolution proceeds, both the nucleus and the cytoplasm are involved. Hemopoiesis may be here considered as "nomadic" or migratory. The most primitive "settlement" of hemopoiesis occurs as organotypic formations in segmented worms.[14-20] These crude "blood islands" consist of a dense association of blood-forming cells in the organism's lateral vascular channels. The cells may or may not be sessile, but certain elements of recognition and adhesiveness certainly exist. Hence, "homing", although in a crude way, begins here.

In the line of evolution, these crude blood islands move from one site to the other and, additionally, their microstructure evolves. They further undergo increasingly more compaction; thus, "homing" becomes increasingly more efficient.[6-8] The difficulties inherent in the study of such vast and varied species have not permitted us to learn more of the molecular events underlying this evolutionary "homing".

III. FETAL AND EMBRYONIC HOMING

A. THE CONCEPT OF DIRECTIVE ENDOCRINIZATION

Embryonic and fetal development is associated with much cell movement and programmed cell death (apoptosis) to achieve the final organization of the fetus. Until recently, little was known of those mechanisms governing these changes that led to more stable organotypic formations. The emergence of such new technologies as monoclonal antibodies and molecular biology techniques has led to the realization that the genetic expression of many differentiation antigens is evident long before differentiation can be observed by other means.[21-23] Many of these differentiation antigens are growth factors, hormone-like substances, their membrane receptors, and their effectual post-receptor machinery. These elements, working in concert, can provide a directive milieu in various parts of the fetus, supervising cell movements and cell death until stable organotypic formations are achieved. Such findings form the basis of the emerging concept of early and directive endocrinization of the embryo.[24] In the same vein, a family of cell adhesion receptors has recently come into focus, called cadherins (for review see Reference 25). Cadherins (so-called because of their Ca^{2+} dependence) are crucial for the mutual association of vertebrate cells, conferring adhesion specificities on cells and regulating morphogenesis.

Again, the movement of hemopoietic stem cells during the embryonic and fetal life and their differentiation potential in different sites provide a relatively well-studied example.

B. MICROENVIRONMENTAL ORGANIZATION

Much of what we know of embryonic and fetal development is in mammalian systems. Here, too, hemopoiesis shows a migratory pattern. It begins as blood islands in the yolk sac, thus being extraembryonic. It then moves to intraembryonic sites, first to the liver and spleen and finally to the marrow. In the yolk sac, cells destined to form blood islands are known as hemangioblasts. They migrate from the primitive streak region of the blastoderm, invaginating and moving laterally.[6-8] The more peripheral of these cells differentiate into angioblasts, while the more central ones are hemopoietic and are known as hemoblasts.[26,27] Yolk sac hemopoiesis is restrictive to the erythroid lineage. Although multipotential stem cells are present in these blood islands and their differentiation potential can be tested by *in vitro* techniques, the milieu of the yolk sac is impermissive to nonerythroid differentiation.

Critical evaluation of data has led to the current consensus that the yolk sac is the origin of stem cells in all subsequently hemopoietic tissues (liver, spleen, marrow), reaching them via the blood stream.[6,7,28] Their presence in the blood stream can be substantiated coincident with the dwindling of their number in one tissue and their rise in the next hemopoietic tissue. That a tissue loses the capacity to retain stem cells while another one gains this potential is a classical example of "cell homing". What governs this phenomenon at a molecular level is not well known, however. The fetal and embryonic system is not quite accessible to experimental manipulation. The directive milieu of a tissue can explain the differing patterns of stem cell differentiation (primarily erythroid in the liver, while mostly granulocytic in the bone marrow). However, the homing of these cells is probably related to the development and loss of a membrane system for recognition and adhesion.

IV. POSTNATAL CELL HOMING

A. RECOGNITION VS. ADHESION

An eminently clear instance of cell homing occurs in bone marrow transplantation (for review see Reference 29). Introduce a source of stem cells anywhere in the blood stream and they selectively and infallibly find their way to the bone marrow, where they home and begin to function. The clarity of the situation matches the clarity of microcinematographic films by Weiss on homotypic cell sorting (vide supra). It may serve as a superior model of "cell homing", so that in its light, other systems of cell homing can be perceived. We must remember that cells of other organs, such as kidney or heart, cannot be so transplanted: they would end up phagocytosed by macrophages.

How do stem cells know their way home? How do they recognize bone marrow as "home"? Before approaching the question, we must define the two terms, "recognition" and "adhesion". "Recognition" implies an element of specificity that is lacking in what the term "adhesion" connotes. Hence, in a molecular sense, recognition is what is implied in the matching of two complementary molecules, such as antigen-antibody, enzyme-substrate, bases of DNA, or lectin-sugar. There is no inference here regarding the affinity and strength of bonding of these molecules. Adhesion, on the other hand, has no inference to specificity and complementarity, but there is implied a notion of affinity and the strength of bonding of the molecules involved.

These definitions are essential to our discussion, for in stem cell homing, an intricate molecular system provides both recognition and adherence. It is essential to distinguish those parts of the system responsible for the recognition from those responsible for the adherence and strengthening of the bonds.

B. MOLECULAR PROBES

Investigation into the molecular mechanism of stem cells homing after their transplantation into the blood stream was precluded by the lack of necessary molecular probes. Early evidence suggested that the molecular basis of recognition and homing specificity involved a lectin-sugar interaction.[30-33] A group of molecular probes were therefore synthesized[34-38] by covalently linking p-aminophenyl derivatives of various biologically active monosaccharides in pyranose form to bovine serum albumin (BSA). The pyranose form is necessary for lectin interaction, and BSA not only renders the molecules large enough not to be diffusible but also permits labeling of the molecule with ^{125}I. Biologically active monosaccharides, involved in the structure of cellular glycoconjugates, are few in number,[39,40] and this makes it possible to synthesize these probes for screening purposes. Since each probe contains monosaccharides of only one type, they can competitively inhibit and thus identify the biologically active residues in membrane glycoconjugates. The biological consequences of these sugar interactions are obviated as a result. The probes are absolutely nontoxic.[41-43]

C. *IN VIVO* STUDIES

To investigate the molecular basis of homing of intravenously transplanted stem cells to the marrow, transplantation was done in the presence or absence of various synthetic neoglycoproteins in inhibiting concentrations.[41] Concentrations of stem cells (CFU-S) and progenitor cells (CFU-GM) were then measured in the tibia at 2 and 24 h and then weekly after transplantation. It was found that among the biologically active sugars, galactose- and mannose-specific probes (galactosyl-BSA, G-BSA and mannosyl-BSA, M-BSA) inhibited homing of stem cells to the marrow. Consequently, the seeding efficiency of stem cells (defined as the concentration of stem cells in the bone marrow 2 and 24 h after intravenous transplantation of isologous stem cells into lethally irradiated mice) declined. Reconstitution of hemopoiesis (as defined by the cellularity of marrow and concentration of CFU-S and CFU-GM after a week) was prevented, and the dose-dependent survival of animals was thus compromised. This type of observation then supported previous suggestions that the recognition and selective homing of stem cells to the marrow are mediated by a lectin-carbohydrate interaction and further assigned the specificity of carbohydrate to galactosyl and mannosyl moieties. Neoglycoproteins with other specificities affected neither homing nor long-term reconstitution of hemopoiesis.

D. *IN VITRO* STUDIES

Similar results were obtained *in vitro* as well, using long-term marrow cultures.[42-44] In this system, an adherent stromal layer is first established. Three weeks later, a second inoculum of fresh marrow is added as the source of hemopoietic progenitors, which attach to the stromal layer. Proliferation and differentiation of hemopoietic cells occur within this layer, in a steady state of equilibrium, and mostly in the granulocytic direction. Mature cells as well as CFU-S and CFU-GM are then released into the supernate, where they may be measured weekly. This system lends itself to the exploration of homing because cultures can be grown in the presence of neoglycoproteins that inhibit the binding of progenitor cells to stromal cells.

In this system, it was found that not only total cell production but also the production of CFU-S and CFU-GM declined in the presence of synthetic neoglycoproteins with galactosyl and mannosyl specificities (Figure 1). The system remained unaffected by neoglycoproteins of other specificities. The decline in the concentration of these progenitors was noted in both the adherent layer and the supernate, indicating that this was not merely a question of compartmentalization. Removal of the inhibiting reagents readily re-established the proliferative capacity of the culture, with the slope of cumulative cell production being parallel to the initial slope (Figure 2), indicating reversibility of the process and further confirming the lack of toxicity of reagents.

The use of combined probes (both galactosyl and mannosyl) in equal concentrations necessary for inhibition did not afford any additive or synergistic effect. This finding, in the context of overall studies, suggested that a configuration of carbohydrate, probably the glycan chain of a glycoconjugate, was involved in the binding and that both galactosyl and mannosyl residues were necessary for this interaction.[29] Therefore, if either of the two was inhibited, the binding was entirely inhibited. Moreover, obtaining similar results by using combined probes strongly suggested that the same lectin (and not two different ones) was involved in the binding of both carbohydrate residues. This conclusion could be confirmed by ligand binding to cloned hematopoietic cells and by purification of the lectin (vide infra).

In long-term marrow culture, proliferation and differentiation of hemopoietic cells led to the formation of morphologically distinct colonies, known as "cobblestones". In the presence of the inhibiting reagents, the absence of adherence was evident: these cobblestones disappeared, leaving only the bare stroma visible (Figure 3). Quantification of these cobblestones (Figure 4) verified the morphological impression.[41-49]

The *in vitro* system has been simplified further by the use of cloned progenitor cells binding to cloned stromal cells.[45] The adherent stromal cell clone, D2X, was grown to near confluence. The progenitor cell FDCP-1, a bipotential clone analogous to CFU-GM, was then used for adherence to the stroma. Binding was quantitated by labeling the progenitor cells with ^{51}Cr and measuring the amount of radioactivity firmly bound to the dish after a 2-h period of contact between stem cells and stromal cells in the presence or absence of neoglycoproteins. Here again, the system reproduced the results of previous observations

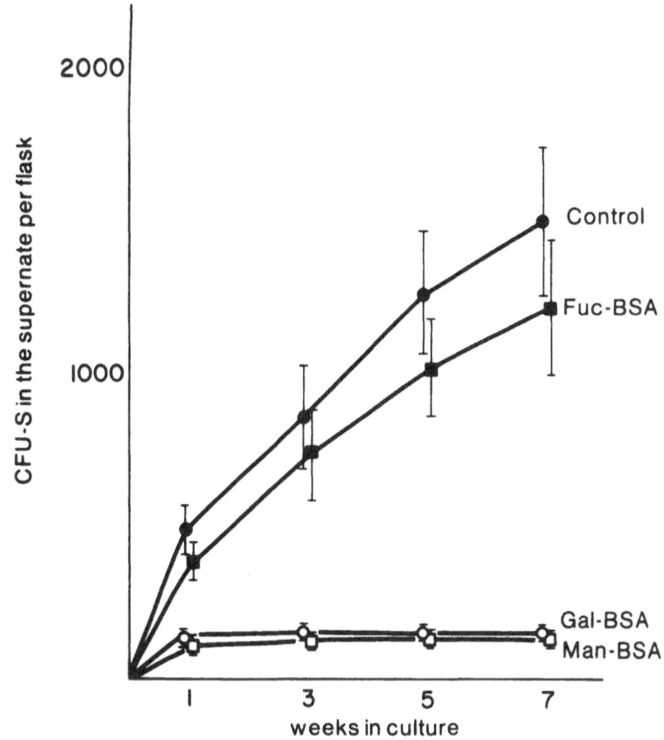

Figure 1 Cumulative CFU-S production in the supernate of long-term marrow cultures in the presence and absence of neoglycoprotein probes. In control cultures, cumulative CFU-S production is a linear function of the time in culture, while in the presence of galactosyl and mannosyl but not fucosyl probes, CFU-S production is inhibited. (Reproduced from *Trans. Assoc. Am. Physicians,* 100, 294, 1987, with permission.)

with consistency: synthetic neoglycoproteins of galactosyl and mannosyl specificities, but not those of other specificities, inhibit the binding of cloned progenitor cells to cloned stroma.

E. BINDING STUDIES

At this point, it was clear that an interaction between a membrane lectin on one side of the equation and a membrane glycoconjugate on the other side was responsible for the selective homing and adherence of progenitor cells to the stroma. It was not clear, however, if the membrane lectins were located on the side of the progenitor cells or the stromal cells. To study this question, selective agglutination of progenitor cells was attempted.[46] The theoretical basis of this experiment exploits the presence of membrane lectins (homing receptors), which are able to cross-link the cells that possess them through the use of neoglycoproteins. Cross-linking was done in whole-marrow cell suspensions, containing all types of progenitor cells. Cross-linked cells could then be agglutinated by centrifugation onto a layer of BSA, providing differential concentrations of progenitor cells in the agglutinated and non-agglutinated fractions. In the absence of homing receptors, agglutination would be random, and cross-linking by neoglycoproteins would not occur. Thus, selective agglutination would not occur, and the concentration of progenitor cells would remain similar throughout the centrifuge tube.

These experiments allowed the assignment of homing receptors to the side of progenitor cells, since both CFU-S and CFU-GM could be selectively agglutinated by neoglycoproteins of galactosyl and mannosyl specificities.[46] This conclusion was subsequently confirmed in cloned progenitor cells (B6Sut, a multipotential, and FDCP-1, a bipotential) through the use of ^{125}I-labeled G-BSA and M-BSA in standard binding assays.[47] Membrane lectins with these specificities, but not those of other sugars, were found on the surface of these cells. In the cell line B6Sut, Scatchard analyses indicated kD of 2.3×10^{-7} *M* and 1.0×10^{-7} *M*, respectively, for G-BSA and M-BSA, with receptor numbers being respectively 10^6 and 3.7×10^5 per cell (Figure 5). Comparable data were also obtained for FDCP-1. In both situations, the

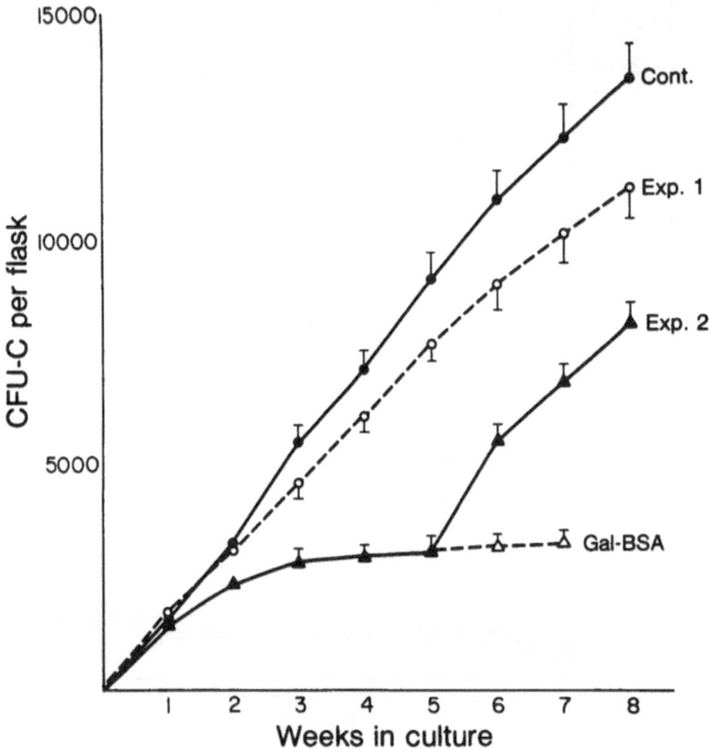

Figure 2 Cumulative CFU-C production in the absence (Cont.) and presence of galactosyl probe (Gal-BSA). In experiment 1 (Exp. 1), the cells in the supernate from cultures exposed to the probe for 1 week were washed and grafted onto control hemopoietic cell-free stroma. In experiment 2 (Exp. 2), the stroma exposed to the probe for 5 weeks was washed free of the probe, and the supernate from control cultures was grafted onto it. In both experiments, CFU-C production continues at a rate parallel to that of control cultures, indicating no functional alteration upon exposure to the probe. (Reproduced from Aizawa, S. and Tavassoli, M., *J. Clin. Invest.*, 80, 1698, 1987, with permission.)

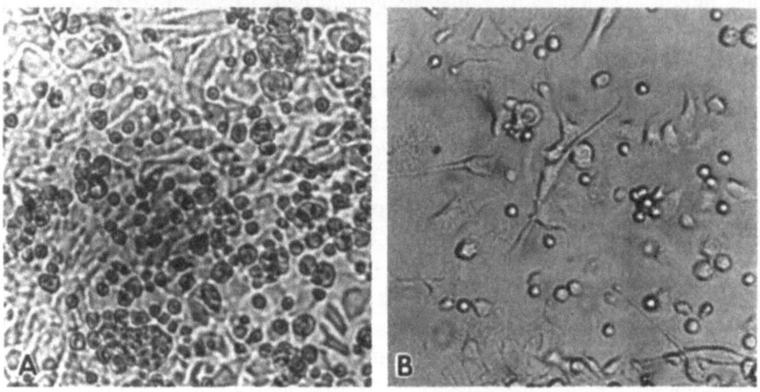

Figure 3 (A) Phase micrograph of a cobblestone in the control culture; (B) culture grown in the presence of galactosyl-BSA. Only stromal cells are seen. (Magnification ×100). (Reproduced from *Trans. Assoc. Am. Physicians,* 100, 294, 1987, with permission.)

labeled synthetic ligand could be displaced readily with cold synthetic ligand of either homologous or heterologous type. In both situations Ca^{2+} was necessary for the binding. The ligand was not internalizable, an expected finding since the natural ligand is membrane-bound (stromal cell membrane), and the entire stromal cell would not be expected to be internalizable. ^{125}I-labeling and binding assays also indicated that stromal cells lack this lectin. The absence of membrane lectin on stromal cells was confirmed, not only in the adherent stroma of long-term marrow culture, which may be heterogenous in nature, but also in

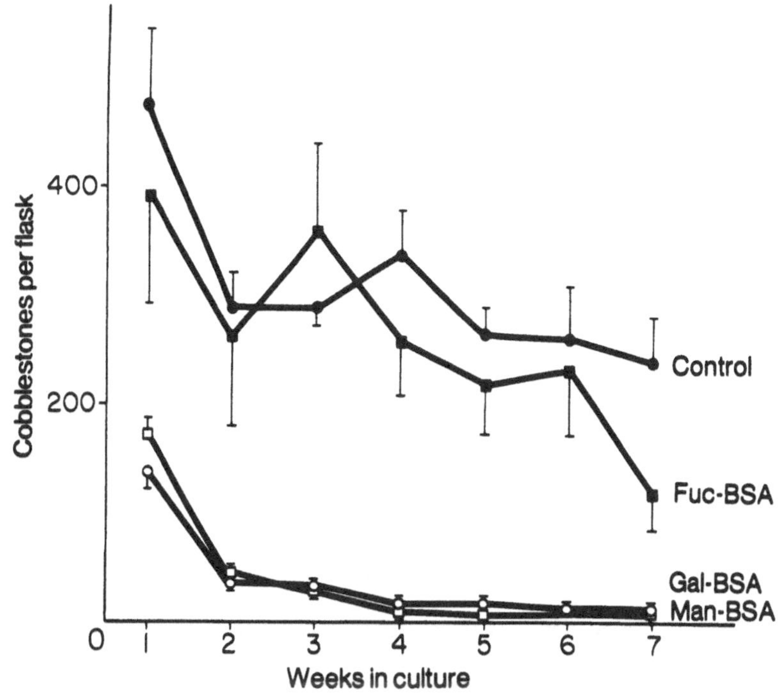

Figure 4 Quantitation of cobblestone areas in long-term marrow cultures. In control cultures and in the presence of fucosyl probe, the number of cobblestones decreases during the first week and then reaches a plateau. In the presence of galactosyl and mannosyl probes, the number of cobblestones is much lower after 1 week and then approaches zero. (Reproduced from Aizawa, S. and Tavassoli, M., *J. Clin. Invest.*, 80, 1698, 1987, with permission.)

cloned stromal cell line D2X.[48] Nor was there any specific binding by mature blood cells, indicating that the homing mechanism is lost in the course of differentiation and maturation (Figure 5).

Consequently, we may assign membrane lectin (homing receptors) to progenitor cells. This lectin is capable of interacting with a certain configuration of a glycoconjugate containing galactosyl and mannosyl groups on the membrane of stromal cells. It is not yet known what the nature of this glycoconjugate is. Its structural analyses require purification, which may also necessitate purification of the lectin to serve as the affinity vehicle. There is, however, evidence that its presence is necessary for the binding of progenitor cells to the stroma: sequential treatment of stromal cell membrane with galactosidase and mannosidase abolishes the binding (homing) of progenitor cells, although it has not been shown that the restitution of these carbohydrate moieties restores homing.[47]

V. MOLECULAR NATURE OF RECOGNITION

Studies on the molecular nature of recognition are all too naturally aimed toward the identification of the molecules involved. In hemopoiesis, this was precluded by the heterogeneity of HPC and their paucity in the marrow cell suspensions. Availability of cloned progenitor cells permitted work on purification of homing receptors.

A. HOMING RECEPTOR (HR)

The technique used for purification of HR is tedious, and the yield is low, but it permits certain characterization.[49] To achieve this, cell membrane fraction was obtained from the cloned progenitor cells, B6Sut or FDCP-1 cells; it was fractionated and solubilized in Triton X-100. Membrane proteins were then labeled with ^{125}I. An affinity column was constructed with either galactosyl or mannosyl groups covalently bound to CNBr-activated Sepharose 4B. Elution of cell membrane proteins was undertaken with Ca^{2+}-containing buffer only until all the unbound radioactivity was eluted. This required a long elution time of about 4 to 5 d at a low elution rate. Subsequent competitive elution with buffer containing either galactosyl or mannosyl sugars in correct form (but not other sugars) led to the elution of a sharp peak of radioactivity. The sharpness of this peak suggested that only a single molecular species was involved.

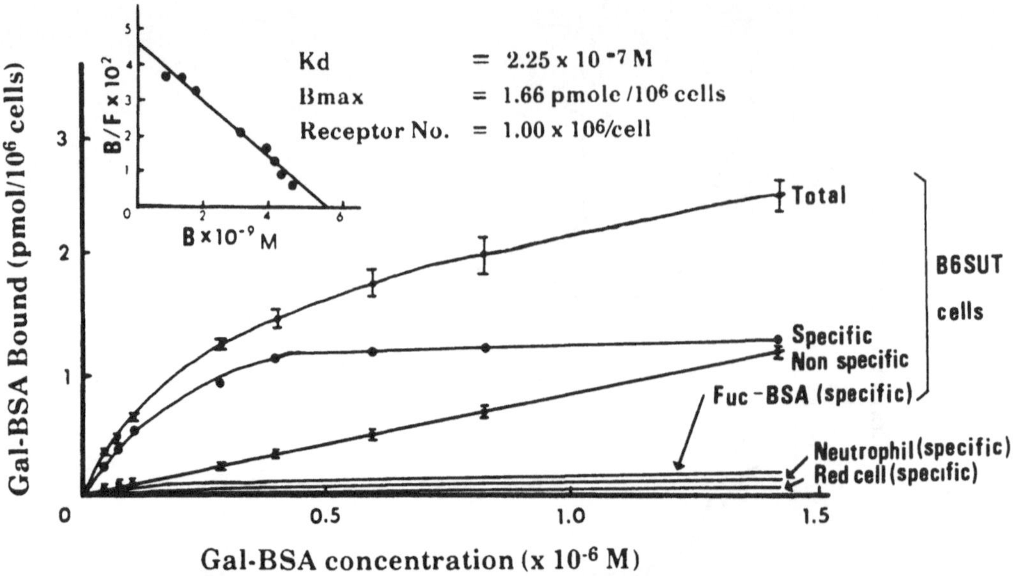

Figure 5 Binding of ^{125}I-Gal-BSA to B6SUT as a function of ligand concentration. Various concentrations of ^{125}I-Gal-BSA were incubated with 10^6 cells at 4°C for 30 min, in a total volume of 0.3 ml. The cells were then washed, and cell-associated radioactivity was counted. This is shown as total binding. Parallel experiments were done in the presence of 40-fold cold ligand. Cell-associated radioactivity is shown as nonspecific binding, which is linear. The latter was subtracted from total binding to obtain specific binding, which is saturable at the ligand concentration of about 0.4×10^{-6} *M*. (Inset) Scatchard plotting of specific binding data shows a single receptor population (kD, 2.25×10^{-7} *M*; B_{max}, 1.66 pmol/10^6 cells; and receptor number, 1.00×10^6/cell). This figure also shows the absence of specific binding for B6SUT cells with Fuc-BSA as ligand, and for mature murine neutrophils and red cells with Gal-BSA as ligand. (Reproduced from Matsuoka, T., Hardy, C., and Tavassoli, M., *J. Clin. Invest.*, 83, 904, 1989, with permission.)

Elution of this peak could be obtained not only with homologous sugar, but also with heterologous sugar (galactosyl-containing buffer on mannosyl column), suggesting that the glycoconjugate ligand also involved a single molecular species and contains both galactosyl and mannosyl grouping. It is the configuration of these two sugars in the molecule that is being recognized by the lectin. The finding is consistent with the binding studies (vide supra), which indicate that inhibition or displacement or radiolabeled ligand can be achieved with heterologous as well as homologous ligand. This information may be helpful in purifying the glycoconjugate. "Stripping" of the protein can also be achieved by the use of 0.5 *M* NaCl, Ca^{2+}-free buffer, or in low or high pH. However, protein degradation may occur.

The peak obtained by affinity chromatography was subjected to polyacrylamide gel electrophoresis followed by autoradiography. Under nonreducing conditions, a major band with a M_r of 110,000 and two lesser bands with a M_r of 87,000 and 23,000 were seen. Only the latter two bands were seen under reducing conditions. Experiments with endoglycosidase F treatment indicated a 5% carbohydrate content in the form of an N-linked glycan chain. Thus, it can be concluded that the HR is a glycoprotein heterodimer, disulfide-bonded with two chains of 87 kD and 23 kD, for a total mass of 110 kD. This is shown diagrammatically in Figure 6. This finding is quite consistent with the wealth of recent information on membrane lectins with a biological recognition function.[50-63] A major group of these lectins is calcium-dependent in their interaction with carbohydrates (as is the hemopoietic homing receptor). These are known as C-lectins (reviewed in References 64 and 65). They are disulfide-bonded, and their recognition function depends on this bonding. They are usually components of the cell membrane and are glycosylated. In these characteristics, homing receptor protein is in the mainstream of C-lectins. Similarities also exist between this homing receptor in hemopoietic progenitors and the homing receptor in lymphocytes.[66-68] In particular, a monoclonal antibody MEL 14 identifies a lectin of 80 to 100 kD, similar to hemopoietic HR. This antibody inhibits homing of lymphocytes. Although carbohydrate specificity is somewhat different, preliminary work in our laboratory suggests that both of these lectins may belong to a single group of peptides that have evolved with the evolution of blood cells.[69]

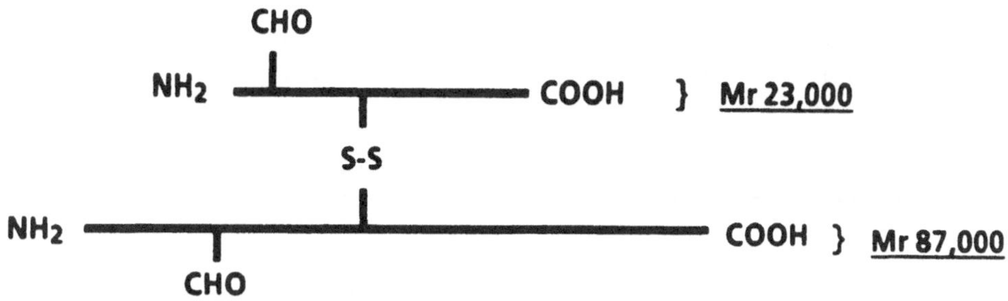

Figure 6 Diagram of the homing protein. The heterodimer is composed of two chains (M_r = 23,000 and 87,000) with a total mass of 110,000. They are disulfide-bonded, although the number of disulfides is not known. The location of CHO group(s) is also unknown, although their presence can be ascertained. (Reproduced from Tavassoli, M. and Hardy, C., *Blood,* 76, 1059, 1990, with permission.)

Sequencing and cloning of hemopoietic HR is underway. It would be of interest in this regard to find out if there exist similarities and differences in the polypeptide sequences between this protein and those involved in the recognition of lymphocytes, or in general with other C-type lectins.

B. HOMING LIGAND

While the nature of homing ligand is still obscure, some recent work has opened the way for its partial characterization. Again, this work (Tavassoli et al., unpublished data) was permitted by the availability of cloned hemopoietic progenitors and stromal cells. To explore the nature of the ligand, a hemopoietic stromal cell line GBl/6 was used.

This cell line has been cloned from mouse long-term bone marrow culture.[70] It binds and supports the growth of several types of hemopoietic progenitor cells. Cells were removed from the culture flask by a teflon spatula and washed twice in phosphate-buffered saline (PBS). Cell membrane was then fractionated, and samples were examined by electron microscopy to ensure the purity of preparations.[49] After centrifugation, the pellets of the cell membrane were solubilized, and its proteins were labeled with ^{125}I. The proteins were then adsorbed to cloned progenitor cells. Subsequently, these cells were washed of unadsorbed radioactivity. Progenitor cells were then counted for adsorbed radioactivity and solubilized with 1% Triton X-100 and subjected to SDS-PAGE under reducing and nonreducing conditions. Gels were then subjected to autoradiography.

Of all labeled GBl/6 membrane proteins, only one was adsorbed to the progenitor cell membrane. On occasions this was a doublet, depending on its state of glycosylation. This protein had M_r of 37,000, and the band obtained was similar under reducing and nonreducing conditions, indicating a single polypeptide chain. This protein was adsorbed to all hemopoietic progenitor cells thus far tested. However, the density of the band was much greater in less differentiated clones compared to more differentiated ones, suggesting the greater expression of homing receptor on the surface of less differentiated cell clones. Adsorption to murine mature erythrocytes did not give a band. This is consistent with our previous observations[46] that homing receptors are present only on the surface of progenitor cells, and they are diluted or degraded as the cells differentiate, so that they are absent on the surface of fully differentiated cells.

Since homing receptor is a membrane lectin that binds to the glycan chain of its stromal cell ligand, by definition the glycan chain is necessary for this interaction. To demonstrate that the glycan chain is requisite for adsorption, we incubated the preparations of labeled proteins from stromal cells with endoglycosidase F and N-glyconase. This treatment totally eliminated the adsorption of the ligand, indicating that the adsorption is via the glycan chain and, therefore, consistent with lectin-carbohydrate molecular interaction needed to define the interaction of homing receptor and its ligand. Also consistent with lectin-carbohydrate interaction was the Ca^{2+}-dependency of this interaction, as EDTA (5 m*M*) prevented the adsorption of this band, while 0.5 or 1 m*M* $CaCl_2$ restored the adsorption.

Thus, it is highly probable that the single 37 K glycoprotein obtained from this hemopoietic stromal cell clone is the ligand for the homing receptor, which characteristically recognizes and interacts with the carbohydrate part of a ligand. It is intriguing that the molecular mass of this ligand is very close to that of the recently discovered c-kit ligand (also known as stem cell factor), which serves as a membrane-associated growth factor (reviewed in Reference 71). Stem cell factor is also heavily glycosylated, but the glycosylation is not necessary for its growth factor function. It is entirely possible that the two molecules are the same or similar; the glycan part serves as the ligand for the homing receptor and is involved in the binding of the progenitor cells, for which the protein moiety serves a trophic faction as the stem cell factor (a membrane-associated growth factor).

C. MEMBRANE-ASSOCIATED AND EXTRACELLULAR MATRIX

The binding of homing receptor to its ligand is of relatively low affinity, in the context of other ligand-receptor interactions (vide supra). It may provide high specificity in recognition, so that intravenously transported cells recognize only marrow stroma as "home". But, recognition, as discussed above, is a somewhat different concept from adherence. That the adherence of hemopoietic cells should be of low affinity is understandable, since in mature form blood cells must be released. However, this low affinity binding as well as work from other laboratories attracted attention to the extracellular matrix and, in particular, to proteoglycans (PG) as an additional mechanism of homing.[72-90] In this role, proteoglycans may provide, if not high specificity as is the case with homing receptors, as least strengthening of the bond between progenitor cells and stroma.

Proteoglycans are a group of matrix-associated molecules consisting of a core protein to which a repeating sequence of usually sulfated glycan structures, called glycosaminoglycan (GAG), are attached.[91-94] They are classified according to their glycan group. Proteoglycans are synthesized by cells of hemopoietic tissues, mostly by stromal cells, and released into the extracellular space. As such, they are part of the extracellular matrix. However, before being released into the extracellular space, they reside for a period of time on the cell membrane, where they are subsequently cleaved by phospholipases. They have recently received considerable attention because emerging evidence indicates that some are involved in regulation of hemopoiesis.[72-90] One line of evidence indicates that they selectively extract and bind hemopoietic growth factors.[89] Other proteoglycans may be instrumental in selective binding of granulocyte-macrophage progenitors,[95-97] while yet another class may be related to the developmental regulation of erythroid cells.[98-102] This latter class is preferentially associated with the cell membrane, rather than with the extracellular matrix.

As regulatory macromolecules, this class of membrane-associated proteoglycans has received relatively little attention, perhaps because the bulk of proteoglycans are not membrane-associated but rather reside in the extracellular space. However, in the course of studying proteoglycans produced by hemopoietic progenitors, it was found that these cells synthesize a considerable amount of PG of one specific class, chondroitin sulfate (CS), that is first associated with the membrane but subsequently is released into the extracellular space.[72] When these progenitor cells are layered on top of stromal cells so that they bind (home) to stromal cells, their homing, which is associated with stabilization of membrane, is also prolonged.[72] Recent work in our laboratory has indicated that membrane-associated CS can also mediate the binding of progenitor cells to stromal cells. At the molecular level, this binding occurs via the interaction between the GAG part of the membrane-associated CS on progenitor cells and the heparin-binding domain of membrane-associated fibronectin (FN) on stromal cells. FN contains two heparin-binding domains, one near the C-terminus with higher affinity for heparin and the other near the cell-binding domain. It is to the latter that the GAG part of CS binds. This interaction subsequently stabilizes membrane-associated CS. The presence of FN on the stromal cell membrane has been well documented.[90,91,96,97] That membrane-associated CS may be involved in the binding of progenitor cells is derived from studies where enzymatic removal of CS abolishes the binding.

Binding of FN to progenitor cell membrane may also occur via the cell-binding domain of fibronectin, which has as one essential structural feature a repeating sequence motif RDG (Arg-Gly-Asp). Here, integrin-like receptors are involved on the progenitor cell membrane. Evidence for the involvement of this particular domain is concluded from experiments in which inhibition of binding can be obtained by the synthetic pentapeptide GRGDS (Gly-Arg-Gly-Asp-Ser). This pentapeptide competitively inhibits the binding of progenitor cells to the tripeptide sequence motif RDG. Control pentapeptide GRGES does not have a similar inhibitory effect (Tavassoli et al., unpublished data).

Thus, at least two domains on the FN molecule can interact with membrane-associated molecules on the surface of progenitor cells to stabilize and strengthen the recognition and binding that is brought about initially by the homing receptor. Figure 7 summarizes our current concept of homing.

VI. ANATOMICAL CONSIDERATIONS

A. THE STRUCTURE OF HEMOPOIETIC MARROW

Anatomical considerations of bone marrow structure indicate that homing of progenitor cells to the marrow may be somewhat more complex: bone marrow is a highly compartmentalized organ,[103,104] with

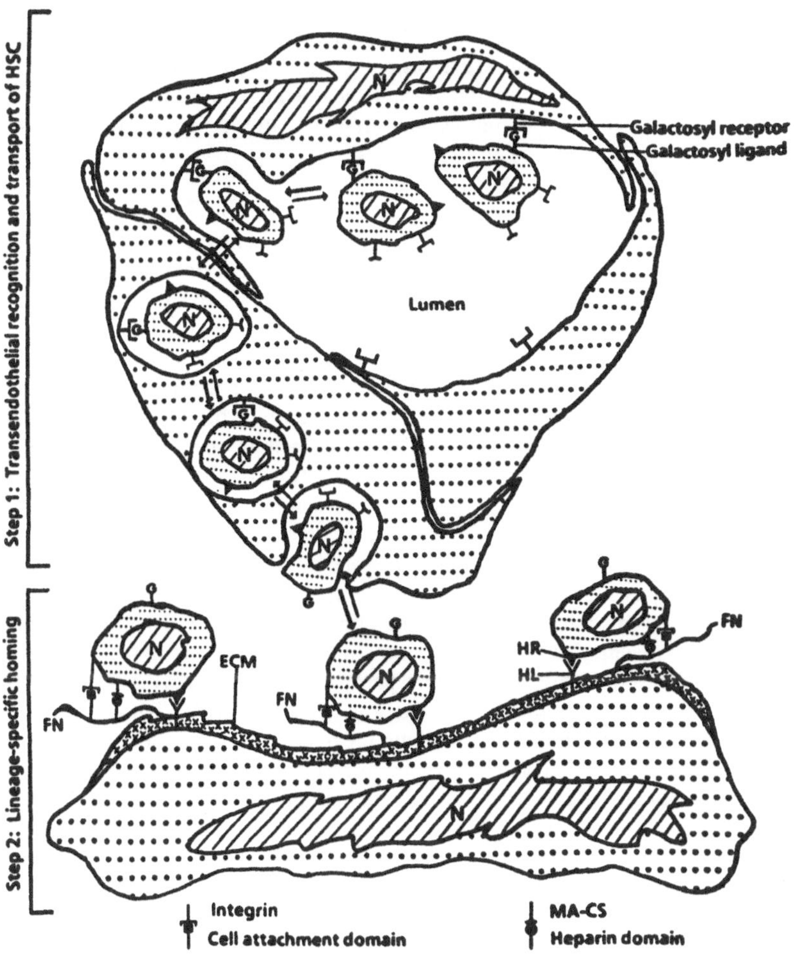

Figure 7 Diagrammatic representation of our proposed two-step homing of stem cells. The first step is its recognition by and transport across the endothelium of marrow sinuses. On the top, a sinus is shown with three endothelial cells. Here, a lectin on the luminal surface of endothelium (┌┴┐) recognizes and binds a galactosyl (G) residue on the membrane of the stem cell. Binding results in internalization of the whole cell and its transport to the hemopoietic compartment, where the lectin becomes dissociated from galactosyl residues. Here, the cell can become associated with lineage-specific stromal cells. This constitutes the second step in homing. Recognition occurs via another lectin, this time on the surface of stem cell (homing receptor or HR). It recognizes and binds galactosyl and mannosyl residues of a glycoconjugate (homing ligand or HL) on the surface of the stromal cell. Binding does not lead to internalization. Other molecular interactions come into play to strengthen the bond between the two cells. Membrane-associated chondroitin sulfate (MA-CS) on stem cells binds to the heparin domain of fibronectin, whose cell attachment domain, containing repeated tripeptide motif RDG, binds to integrin-like molecules on the surface of the stem cell. It is proposed that the homing is in a steady state of equilibrium, with a gradient of stem cell concentration in favor of marrow, but the movement occurs in both directions. This hypothesis has not been tested yet. Moreover, the actual transendothelial (as opposed to interendothelial) passage of the stem cell has not been documented yet. Here, we may extrapolate from the mode of migration of mature cells into the lumen.

its two major compartments being vascular and extravascular ones. At least in mammalian systems, hematopoiesis occurs entirely in the extravascular compartment. Cellular exchange between the two compartments occurs at the level of specialized vessels, known as "sinusoids". The wall of these vessels forms a barrier between the two compartments (bone marrow-blood barrier) and consists of several components, all with well-defined functions. Its most consistent component is a layer of endothelium that controls the nature and magnitude of cellular and molecular exchange between vascular and extravascular compartments. Such substances as iron-transferrin complex do not enter the marrow at the interendothelial junction.[10,11] They are taken up by endothelium on the luminal side through a receptor-mediated mechanism, traverse the endothelial cytoplasm via a vesicular transport system, and are externalized to the abluminal side. Similarly, cellular migration occurs by opening fenestrations through the body of endothelial cells.[6] Fenestrations are then immediately repaired after migration is completed.

B. A TWO-STEP HOMING

Homing of progenitor cells can be conceived as a two-step phenomenon.[29] Progenitor cells arriving in the marrow must first be recognized by and interact with the luminal surface of endothelium. They must then pass through the endothelial layer to enter the hematopoietic compartment. Here the second step of homing occurs. Progenitor cells are recognized by and bind to lineage-specific stromal cells that support their proliferation and maturation. Studies thus far reviewed only pertain to this second step in homing. Less is known of the first step, the recognition and interaction of progenitor cells with endothelium. Available evidence suggests that here, too, a membrane lectin-glycoconjugate interaction may be involved.[105,106] However, here the process is reversed: luminal surface of endothelium provides the lectin and progenitor cells, the carbohydrate moiety. Also, carbohydrate specificity of recognition may be different and be limited to the galactosyl moiety. Evidence for this postulate is derived from the observation that after intravenous infusion of galactosyl-containing neoglycoproteins, the highest uptake per gram of tissue occurs in the bone marrow,[105,106] not, as expected, in the liver, whose hepatocytes are known to possess galactosyl receptors. Similarly desialated glycoproteins, of which the penultimate galactosyl residues are exposed, are rapidly removed by bone marrow. Since bone marrow circulation is closed (i.e., endothelium forms a barrier between the circulatory space and the hematopoietic compartment), it is reasonable to assume that the uptake occurs in the endothelium. In fact, electron microscopic observations, in which labeled neoglycoproteins are perfused in the regional circulation of bone marrow, have confirmed this assumption.[107] Endothelium takes up galactosyl probes but not mannosyl or fucosyl probes. In contrast to the homing protein that mediates the binding of progenitor cells to stromal cells and is not internalizable, the binding of galactosyl probe to endothelium is followed by internalization and subsequent externalization on the abluminal side of endothelium. Of course, this is expected, because the endothelium must mediate the transport from circulation into the hematopoietic compartment.

Certainly a more systematic approach to endothelial recognition and transport of progenitor cells is needed. To provide some coherence to the fragmentary evidence reviewed here, the following conclusions may be helpful.

1. Endothelial recognition of progenitor cells appears to be mediated by a lectin system with galactosyl specificity.
2. It is in the reverse of stromal-progenitor cell recognition, with endothelium providing the lectin and progenitor, the carbohydrate moiety.
3. The adhesion is followed by internalization and then externalization on the abluminal side (transport).

These are summarized in Figure 7 in the context of our overall view of stem cell homing.

VII. FUTURE DIRECTIONS

Certainly, future direction of research into this area must take into consideration the complexity of the homing phenomenon. In both steps of homing, endothelial and stromal levels, the interacting molecules should be purified, sequenced, and cloned. By site-directed mutagenesis, the involved molecular domains should be identified. Monoclonal antibodies reacting with these domains should be raised to serve as tools of investigation. Similarities and differences between this and other homing systems should be explored.

The simplistic view of homing is rapidly changing. It is now being realized that here, too, we are facing a complex interacting system where the investigative horizon is ever expanding. Are we facing a kaleidoscope? Or, perhaps, do we impose complexity upon everything we touch? In the last analysis, science is the game of elusion if not illusion: as we begin to feel that we have caught it and have it in our hands, we look up to see it sitting in a tree laughing at us.

REFERENCES

1. **Moscona, A.,** Development in vitro of chimeric aggregates of dissociated embryonic chick and mouse cells, *Proc. Natl. Acad. Sci. U.S.A.*, 43, 184, 1957.
2. **Weiss, P.,** Cell contact, *Int. Rev. Cytol.*, 7, 391, 1958.

3. **Weiss, P. and Taylor, A. C.,** Reconstitution of complete-organs from single-cell suspensions of chick embryos in advanced stages of differentiation, *Proc. Natl. Acad. Sci. U.S.A.*, 46, 1177, 1960.
4. **Edelman, G. E.,** Cell adhesion molecules, *Science*, 219, 450, 1983.
5. **Rutishanser, U.,** Developmental biology of a neural cell adhesion molecule, *Nature*, 310, 549, 1984.
6. **Tavassoli, M., and Yoffey, J. M.,** *Bone Marrow: Structure and Function*, Alan R. Liss, New York, 1983.
7. **Tavassoli, M.,** Ontogency of hemopoiesis, in *Handbook of Human Growth and Developmental Biology*, Vol. 3, Part A, Messami, E. and Timiras, P. S., Eds., CRC Press, Boca Raton, FL, 1990, 101.
8. **Tavassoli, M.,** Embryonic and fetal hemopoiesis: an overview, *Blood Cell*, 17, 269, 1991.
9. **Charmanier, M.,** Physiologie des invertebres — effets de la regeneration intensive on de la presence d'une Sacculine sur la leucopoies de Pachygrapus marmoratus, *C.R. Acad. Sci. Ser. D*, 276, 2553, 1973.
10. **Irie, S. and Tavassoli, M.,** Transferrin-mediated cellular iron uptake, *Am. J. Med. Sci.*, 292, 103, 1987.
11. **Soda, R. and Tavassoli, M.,** Transendothelial transport of iron-transferrin complex in the bone marrow, *J. Ultrastruct. Res.*, 88, 18, 1984.
12. **Tavassoli, M.,** Bone marrow in boneless fish: lessons of evolution, *Med. Hypoth.*, 20, 9, 1986.
13. **Siminia, T.,** Haematopoiesis in the freshwater snail Lymnaea stagnalis studied by electron microscopy and autoradiography, *Cell Tissue Res.*, 150, 443, 1974.
14. **Hoffmann, J. A.,** The hemopoietic organs of the two orthopterans, Locusta migratoria and Gryllus bimaculatus, *Z. Zellforsch Mikrosk Anat.*, 106, 451, 1972.
15. **Hoffmann, J. A.,** Blood-forming tissues in orthopteran insects: an analogue to vertebrate hemopoietic organs, *Experientia*, 29, 50, 1973.
16. **Nutting, W. L.,** A comparative anatomical study of the heart and accessory structures of the orthopteroid insects, *J. Morphol.*, 89, 501, 1951.
17. **Zachary, D. and Hoffman, J. A.,** The haemocytes of Calliphora erythrocephala, *Z. Zellforsch Mikrosk Anat.*, 141, 55, 1973.
18. **Brehelin, M.,** Presence d'un tissue hematopoietique chez le Coleoptere melolontha, *Experientia*, 29, 1539, 1973.
19. **Akai, H. and Sato, S.,** An ultrastructural study of the haemopoietic organ of the silk worm, Bombyx mori, *J. Invest. Physiol.*, 17, 1665, 1971.
20. **Monpeyssin, M. and Beaulaton, J.,** Hemocytopoiesis in the oak silkworm Antheroae permyi and some other Lepidoptera, *J. Ultrastruct. Res.*, 64, 35, 1978.
21. **Rosenthal, P., Rimm, J. J., Umiel, T., Griffin, J. D., Osathanondh, R., Schlossman, S. F., and Nadler, L. M.,** Ontogeny of human hematopoietic cells: analysis utilizing monoclonal antibodies, *J. Immunol.*, 131, 232, 1983.
22. **Peschle, C.,** Human ontogenic development: studies on the hemopoietic system and the expression of homeo box genes, *Ann. N.Y. Acad. Sci.*, 511, 101, 1987.
23. **Murray, R., Lee, F., and Chiu, C. P.,** The genes for leukemia inhibitory factor and interleukin-6 are expressed in mouse blastocysts prior to the onset of hemopoiesis, *Mol. Cell Biol.*, 10, 4953-1990.
24. **De Pablo, F. and Roth, J.,** Endocrinization of the early embryo, *Trends Biochem. Sci.*, 15, 339, 1990.
25. **Takeichi, M.,** Cadherin cell adhesion receptors as a morphogenetic regulator, *Science*, 251, 1451, 1991.
26. **Aizawa, S. and Tavassoli, M.,** Interaction of murine granulocyte-macrophage progenitors and supporting stroma involves a recognition mechanism with galactosyl and mannosyl specificities, *J. Clin. Invest.*, 80, 1698, 1987.
27. **Aizawa, S. and Tavassoli, M.,** Molecular basis of the recognition of intravenously transplanted hemopoietic cells by bone marrow, *Proc. Natl. Acad. Sci. U.S.A.*, 85, 3180, 1988.
28. **Toles, J. F., Chui, D. H. K., Belbeck, L. W., Starr, E., and Barker, J. E.,** Hemopoietic stem cells in murine embryonic yolk sac and peripheral blood, *Proc. Natl. Acad. Sci. U.S.A.*, 86, 7456, 1989.
29. **Tavassoli, M. and Hardy, C.,** Molecular basis of homing of intravenously transplanted cells to the marrow, *Blood*, 76, 1059, 1990.
30. **Samlowski, W. and Daynes, R. A.,** Bone marrow engraftment efficiency is enhanced by competitive inhibition of the hepatic asialoglycoprotein receptor, *Proc. Natl. Acad. Sci. U.S.A.*, 82, 2508, 1985.
31. **Reisner, Y., Itzicovitch, L., Meshorer, A., and Sharon, N.,** Hemopoietic stem cell transplantation using mouse bone marrow and spleen cells fractionated by lectins, *Proc. Natl. Acad. Sci. U.S.A.*, 75, 2933, 1978.
32. **Baenziger, J. U.,** The role of glycosylation in protein recognition, *Am. J. Pathol.*, 121, 382, 1985.
33. **Greig, R. G. and Jones, M. N.,** Mechanisms of intercellular adhesion, *Biosystems*, 9, 43, 1977.

34. **Monsigny, M., Kieda, C., and Roche, A. C.,** Membrane glycoproteins, glycolipids and membrane lectins as recognition signals in normal and malignant cells, *J. Cell Biol.*, 47, 95, 1983.
35. **Kataoka, M. and Tavassoli, M.,** Synthetic neoglycoproteins: a class of reagents for detection of sugar-recognizing substances, *J. Histochem. Cytochem.*, 32, 1091, 1984.
36. **McBroom, C. R., Samenen, C. H., and Goldstein, I. J.,** Carbohydrate antigens: coupling of carbohydrates to proteins by diazonium and phenylisothiocyanate reactions, *Meth. Enzymol.*, 28, 212, 1972.
37. **Simpson, D. L., Thorne, D. R., and Loh, H. H.,** Lectins: endogeneous carbohydrate-binding proteins from vertebrate tissues: functional rate in recognition process, *Life Sci.*, 22, 727, 1978.
38. **Stowell, C. P. and Lee, Y. C.,** The binding of d-glucosyl-neoglycoproteins to the hepatic asialoglycoprotein receptor, *J. Biol. Chem.*, 253, 6107, 1978.
39. **Sharon, W. and Lis, H.,** Glycoproteins: research booming on long ignored ubiquitous compounds, *Mol. Cell Biochem.*, 42, 167, 1982.
40. **Sharon, N. and Lis, H.,** Lectins as cell recognition molecules, *Science*, 246, 227, 1987.
41. **Aizawa, S. and Tavassoli, M.,** Molecular basis of the recognition of intravenously transplanted hemopoietic cells by bone marrow, *Proc. Natl. Acad. Sci. U.S.A.*, 85, 3180, 1988.
42. **Aizawa, S. and Tavassoli, M.,** Interaction of murine granulocyte-macrophage progenitors and supporting stroma involves a recognition mechanism with galactosyl and mannosyl specificities, *J. Clin. Invest.*, 80, 1698, 1987.
43. **Aizawa, S. and Tavassoli, M.,** In vitro homing of hemopoietic stem cells is mediated by a recognition mechanism with galactosyl and mannosyl specificity, *Proc. Natl. Acad. Sci. U.S.A.*, 84, 4485, 1987.
44. **Tavassoli, M., Aizawa, S., Matsuoka, T., and Hardy, C.,** Molecular basis of the recognition of progenitor cells by marrow stroma, in *Hematopoiesis*, UCLA Symposia on Molecular and Cellular Biology, Vol. 120, Golde, D. and Clark, S., Eds., Wiley-Liss, New York, 1990, 145.
45. **Hardy, C. L. and Tavassoli, M.,** Homing of hemopoietic stem cells to hemopoietic stroma, *Adv. Exp. Med. Biol.*, 241, 129, 1988.
46. **Aizawa, S. and Tavassoli, M.,** Detection of membrane lectins on the surface of hemopoietic progenitor cells and their changing pattern during differentiation, *Exp. Hematol.*, 16, 325, 1988.
47. **Matsuoka, T., Hardy, C., and Tavassoli, M.,** Characterization of membrane homing receptors in two cloned murine hemopoietic progenitor cell lines, *J. Clin. Invest.*, 83, 904, 1989.
48. **Hardy, C. L., Matsuoka, T., and Tavassoli, M.,** Distribution of homing protein on hemopoietic stromal and progenitor cells, *Exp. Hematol.*, 19, 968, 1991.
49. **Matsuoka, T. and Tavassoli, M.,** Purification and partial characterization of membrane-homing receptors in two cloned murine hemopoietic progenitor cell lines, *J. Biol. Chem.*, 264, 20193, 1989.
50. **Chadee, K., Petri, W. A., Innes, D. J., and Ravdin, J. I.,** Rat and human colonic mucins bind to and inhibit adherence lectin of Entamoeba histolytica, *J. Clin. Invest.*, 80, 1238, 1987.
51. **Tavassoli, M., Kishimoto, T., and Kataoka, M.,** Liver endothelium mediates the hepatocyte uptake of ceruloplasmin, *J. Cell Biol.*, 102, 1298, 1986.
52. **Irie, S., Kishimoto, T., and Tavassoli, M.,** Desialation of transferrin by liver endothelium, *J. Clin. Invest.*, 82, 508, 1988.
53. **Kataoka, M. and Tavassoli, M.,** Development of specific surface receptors recognizing D-mannose in cultured monocytes: a possible early marker for differentiation of monocytes into macrophage, *Exp. Hematol.*, 13, 44, 1985.
54. **Prieels, J. P., Pizzo, S. V., Glasgow, L. R., Paulson, J. C., and Hill, R. L.,** Hepatic receptor that specifically binds oligosaccharides containing fucosyl alpha-1 leads to N-acetyl glucosamine linkages, *Proc. Natl. Acad. Sci. U.S.A.*, 75, 2215, 1978.
55. **Lehrman, M. A. and Hill, R. L.,** The binding of fucose-containing glycoproteins by hepatic lectins. Purification of a fucose-binding lectin from rat liver, *J. Biol. Chem.*, 261, 7419, 1986.
56. **Haltiwanger, R. S., Lehrman, M. A., Eckhardt, A. E., and Hill, R. L.,** The distribution and localization of the fucose-binding lectin in rat tissues and the identification of a high affinity form of the mannose N-acetylglucosamine-binding lectin in rat liver, *J. Biol. Chem.*, 261, 7433, 1986.
57. **Stahl, P. D., Rodman, J. S., Miller, M. J., and Schlesinger, P. H.,** Evidence for receptor mediated binding of glycoproteins, glycoconjugate, and lysosomal glycosidases by alveolar macrophages, *Proc. Natl. Acad. Sci. U.S.A.*, 75, 1399, 1978.
58. **Kawasaki, T., Etoh, R., and Yamashina, I.,** Isolation and characterization of a mannan-binding protein from rabbit liver, *Biochem. Biophys. Res. Commun.*, 81, 1018, 1978.
59. **Summerfield, J. A., Vergalla, J., Jones, E. A.,** Modulation of a glycoprotein recognition system on rat hepatic endothelial cells by glucose and diabetes mellitus, *J. Clin. Invest.*, 69, 1337, 1982.

60. **Thornburg, R. W., Day, J. F., Baynes, J. W., and Thorp, S. R.,** Carbohydrate-mediated clearance of immune complexes from the circulation. A role for galactose residues in the hepatic uptake of IgG-antigen complexes, *J. Biol. Chem.*, 255, 6820, 1980.
61. **Day, J. F., Thornburg, R. W., Thorp, S. R., and Baynes, J. W.,** Carbohydrate-mediated clearance of antibody-antigen complexes from the circulation. The role of high mannose oligosaccharides in the hepatic uptake of IgM of antigen complexes, *J. Biol. Chem.*, 255, 2360, 1980.
62. **Hoppe, C. A. and Lee, Y. C.,** Stimulation of mannose-binding activity in the rabbit alveolar macrophage by simple sugars, *J. Biol. Chem.*, 257, 112831, 1982.
63. **Imber, M. J., Pizzo, S. V., Johnson, W. F., and Adams, D. O.,** Selective diminution of the binding of mannose by murine macrophages in the late stages of activation, *J. Biol. Chem.*, 257, 5129, 1982.
64. **Drickamer, K.,** Two distinct classes of carbohydrate-recognition domains in animal lectins, *J. Biol. Chem.*, 257, 5129, 1982.
65. **Drickamer, K. and McCreary, V.,** Exon structure of a mannose-binding protein gene reflects its evolutionary relationship to the asialoglycoprotein receptor and nonfibrillar collagens, *J. Biol. Chem.*, 262, 2582, 1987.
66. **Stoolman, L. M., Tenforde, T. S., and Rosen, S. D.,** Phospho-mannosyl receptors may participate in the adhesive interaction between lymphocytes and high endothelial venules, *J. Cell Biol.*, 99, 1535, 1984.
67. **Yednock, T. A. and Rosen, S. D.,** Lymphocyte homing, *Adv. Immunol.*, 44, 313, 1989.
68. **Siegelman, M. H. and Weissman, I. L.,** Human homologue of mouse lymph node homing receptor: evolutionary conservation at tandem cell interaction domains, *Proc. Natl. Acad. Sci. U.S.A.*, 86, 5562, 1989.
69. **Butcher, E. C.,** Cellular and molecular mechanisms that direct leukocyte traffic, *Am. J. Pathol.*, 136, 3, 1990.
70. **Anklesaria, P., Kase, K., Glowacki, J., Holland, C. A., Sakakeeny, M. A., Wright, J. A., Fitzgerald, T. J., Lee C.-Y., and Greenberger, J. S.,** Engraftment of a clonal bone marrow stromal cell line in vivo stimulates hematopoietic recovery from total body irradiation, *Proc. Natl. Acad. Sci. U.S.A.*, 84, 7681, 1987.
71. **Witte, O. N.,** Steel locus defines new multipotent growth factor, *Cell*, 63, 6, 1990.
72. **Minguell, J. J. and Tavassoli, M.,** Proteoglycan synthesis by hemopoietic progenitor cells, *Blood*, 73, 1821, 1989.
73. **Minguell, J. J., Fernandez, M., Tetas, M., and Lopez, M. C.,** Extracellular matrix production by human bone marrow cells, *Exp. Hematol.*, 14 (Abstr.), 513, 1986.
74. **Tavassoli, M., Eastlund, D. T., Yam, L. T., Neiman, R. S., and Finkel, H.,** Gelatinous-transformation of bone marrow in prolonged self-induced starvation, *Scand. J. Haematol.*, 16, 311, 1976.
75. **Kirby, S. L. and Bentley, S. A.,** Proteoglycan synthesis in two murine bone marrow stromal cell lines, *Blood*, 70, 1777, 1987.
76. **Bentley, S. A., Kirby, S. L., Anklesaria, P., and Greenberger, J. S.,** Bone marrow stromal proteoglycan heterogeneity: phenotypic variability between cell lines and the effects of glucocorticoid, *J. Cell Physiol.*, 136, 182, 1988.
77. **Giancotti, F. G., Comoglio, P. M., and Tarone, G.,** Fibronectin-plasma interaction in the adhesion of hemopoietic cells, *J. Cell Biol.*, 103, 429, 1986.
78. **Coulombel, L., Vuillet, M. H., Leroy, C. and Tchernia, G.,** Lineage- and stage-specific adhesion of human hematopoietic progenitor cells to extracellular matrices from marrow fibroblasts, *Blood*, 71, 329, 1988.
79. **Gospodarowicz, D.,** Extracellular matrix and control of proliferation of vascular endothelial cells, *J. Clin. Invest.*, 65, 1351, 1980.
80. **Gospodarowicz, D., Delgado, D., and Vlodavsky, I.,** Permissive effect of the extracellular matrix on cell proliferation in vitro, *Proc. Natl. Acad. Sci. U.S.A.*, 77, 4094, 1980.
81. **Zuckerman, K. S. and Wicha, M. S.,** Extracellular matrix production by the adherent cells of long-term murine bone marrow cultures, *Blood*, 61, 540, 1983.
82. **Zuckerman, K. S., Rhodes, R. K., Goodrum, D. D., Patel, U. R., Sparks, B., Wells, J., Wicha, M. S., and May, L.,** Inhibition of collagen deposition in the extracellular matrix prevents the establishment of a stroma supportive of hematopoiesis in long-term murine bone marrow culture, *J. Clin. Invest.*, 75, 970, 1985.
83. **Spooncer, E., Gallagher, J. T., Kriza, F., and Dexter, T. M.,** Regulation of hemopoiesis in long term bone marrow cultures. IV. Glycosaminoglycan synthesis and stimulation of hemopoiesis by β-D-xylosides, *J. Cell Biol.*, 96, 510, 1983.

84. **Tavassoli, M. and Minguell, J. J.,** Homing of hemopoietic progenitor cells to the marrow, *Proc. Soc. Exp. Biol. Med.*, 196, 367, 1991.
85. **Pleomacher, R. E., Van Hull, E., and Van Goest, P. L.,** Studies of the hemopoietic microenvironments: effects of acid mucopolysaccharides and dextran sulfate on erythroid and granuloid differentiation in vitro, *Exp. Hematol.*, 6, 311, 1978.
86. **Wight, T. N., Kinsella, M. G., Keating, A., and Singer, J. W.,** Proteoglycans in human long-term bone marrow cultures: biochemical and ultrastructural analysis, *Blood*, 67, 1333, 1986.
87. **Okayama, E., Oguri, K., Kondo, T., and Okayama, M.,** Isolation and characterization of chondroitan 6-sulfate proteoglycans present in the extracellular matrix of rabbit bone marrow, *Blood*, 72, 745, 1988.
88. **Gordon, M. Y., Riley, G. P., Watt, S. M., and Greaves, M. F.,** Compartmentalization of a haemopoietic growth factor (GM-CSF) by glycosaminoglycans in the bone marrow microenvironment, *Nature*, 362, 403, 1987.
89. **Roberts, R., Gallagher, J., Spooncer, E., Allen, T., Bloomfield, F., and Dexter, T. M.,** Heparan sulfate bound growth factors: a mechanism for stromal cell mediated haemopoiesis, *Nature*, 332, 376, 1988.
90. **Del Rosso, M., Cappelletti, R., Dina, G., Fibbi, G., Vannuchi, S., Chiarugi, V., and Guazzelli, C.,** Involvement of glycosaminoglycans in detachment of early myeloid precursors from bone marrow stromal cells, *Biochim. Biophys. Acta*, 676, 129, 1981.
91. **Ruoslahti, E.,** Protoeglycans in cell regulation, *J. Biol. Chem.*, 264, 13369, 1989.
92. **Gospodarowicz, D., Vlodavsky, I., and Savion, N.,** The extracellular matrix and the control of proliferation of vascular endothelial and vascular smooth muscle cells, *J. Supramol. Struct.*, 13, 339, 1980.
93. **Buck, C. A. and Horwitz, A. F.,** Cell surface receptors for extracellular matrix molecules, *Annu. Rev. Cell Biol.*, 3, 179, 1987.
94. **Torok-Storb, B.,** Cellular interaction, *Blood*, 72, 373, 1988.
95. **Campbell, A. D., Long, M. W., and Wicha, M. S.,** Haemonectin, a bone marrow adhesion protein specific for cells of granulocyte lineage, *Nature*, 329, 744, 1987.
96. **Campbell, A., Wicha, M. S., and Long, M.,** Extracellular matrix promotes the growth and differentiation of murine hematopoietic cells in vitro, *J. Clin. Invest.*, 75, 2085, 1985.
97. **Peters, C., O'Shea, K. S., Campbell, A. D., Wicha, M. S., and Long, M. W.,** Fetal expression of hemonectin: an extracellular matrix hematopoietic cytoadhesion molecule, *Blood*, 75, 357, 1990.
98. **Patel, V. P. and Lodish, H. F.,** Loss of adhesion of murine erythroleukemic cells to fibronectin during erythroid differentiation, *Science*, 224, 996, 1984.
99. **Tsai, S., Patel, V., Beaumont, E., Lodish, H. F., Nathan, D. G., and Sieff, C. A.,** Differential binding of erythroid and myeloid progenitors to fibroblasts and fibronectin, *Blood*, 69, 1587, 1987.
100. **Virtanen, I., Ylanne, J., and Vartio, T.,** Human erythroleukemia cells adhere to fibronectin: evidence for a Mr 190,000-receptor protein, *Blood*, 69, 578, 1987.
101. **Weinstein, R., Riodan, M. A., Wenc, K., Kreczko, S., Zhou, M., and Dainiak, N.,** Dual role of fibronectin hematopoietic differentiation, *Blood*, 73, 111, 1989.
102. **Vuillet-Gaugler, M. H., Breton-Gorius, J., Vainchenker, W., Guichard, J., Leroy, C., Tchernia, G., and Coulombel, L.,** Loss of attachment to fibronectin with terminal human erythroid differentiation, *Blood*, 75, 865, 1990.
103. **Tavassoli, M.,** The marrow-blood barrier, *Br. J. Haematol.*, 41, 297, 1979.
104. **Tavassoli, M.,** Red cell delivery and the function of marrow-blood barrier, *Exp. Hematol.*, 6, 257, 1978.
105. **Aizawa, S. and Tavassoli, M.,** Marrow uptake of galactosyl-containing neoglycoprotein: implication in stem cell homing, *Exp. Hematol.*, 16, 811, 1988.
106. **Regoeczi, E., Chindemi, P. A., Hatton, M. W. C., and Berry, L. R.,** Galactose-specific elimination of human asialotransferrin by the bone marrow in the rabbit, *Arch. Biochem. Biophys.*, 205, 76, 1980.
107. **Kataoka, M. and Tavassoli, M.,** Identification of lectin-like substances recognizing galactosyl residues of glycoconjugates on the plasma membrane of marrow sinus endothelium, *Blood*, 65, 1163, 1985.

Chapter 14

Platelet Aggregation and Inhibitors

Kailash C. Agarwal

CONTENTS

I. Introduction 287
II. Measurement of Platelet Aggregation 288
A. Turbidometric or Optical Method 289
B. The Impedance Method 289
C. The Luminescent Method 289
III. Platelet Aggregating Agents 289
A. Adenosine 5′-Diphosphate (ADP) 290
B. Epinephrine 291
C. Thromboxane A_2 (TXA_2) 291
D. Platelet-Activating Factor (PAF) 291
E. Arginine[8]-Vasopressin (AVP) 291
IV. Species Differences in Platelet Aggregation 292
V. Platelet Aggregation — Disorders 292
A. Disorders with Increased Platelet Aggregability 292
1. Diabetes Mellitus 292
2. Myocardial Infarction and Ischemic Heart Disease 293
3. Hypertension 293
B. Disorders with Decreased Platelet Aggregability 293
VI. Platelet Aggregation Inhibitors 293
A. Drugs that Inhibit Arachidonic Acid Metabolism 294
1. Cyclooxygenase (Prostaglandin Synthetase) Inhibitors 294
2. Thromboxane Synthetase Inhibitors and Thromboxane Receptor Antagonists 294
3. Effects of Omega-3 Fatty Acids (Eicosapentaenoic and Docosahexaenoic Acids 294
B. Drugs that Increase Platelet Intracellular cAMP Levels 294
1. PGE_1, PGD_2, PGI_2, and Adenosine 294
2. Dipyridamole 295
C. Drugs that Act by Other Mechanisms 295
1. Ticlopidine 295
2. Nitric Oxide (NO) 296
3. Calcium Channel Blockers (Calcium Antagonists) 296
VII. Acknowledgment 296
References 296

I. INTRODUCTION

Platelets from most mammalian species are similar in appearance. They are disc-shaped, are on average 3 μm in diameter, and are formed by cytoplasmic fragmentation from the megakaryocytes in bone marrow. Each megakaryocyte can form 1000 to 1500 platelets.[1,2] The platelet count in human peripheral blood is 250,000 ± 100,000 per μl, and they circulate within the vascular beds with a lifespan of 7 to 10 d. As platelets age, they undergo a progressive reduction in membrane glycoproteins, phospholipids, cytoplasmic proteins, and adenine nucleotides, causing a decrease in functional activity.[2] Young platelets generally show greater activity and are heavier than older platelets.[2,3] The platelet's main physiological contribution to hemostasis is to initiate the formation of the hemostatic plug, which depends on platelet adhesion, aggregation, and secretion.[4-8] Upon stimulation, platelets undergo a shape change, with the formation of numerous pseudopodia, adhesion, and aggregation, resulting in the release of the granular

0-8493-4559-6/95/$0.00+$.50

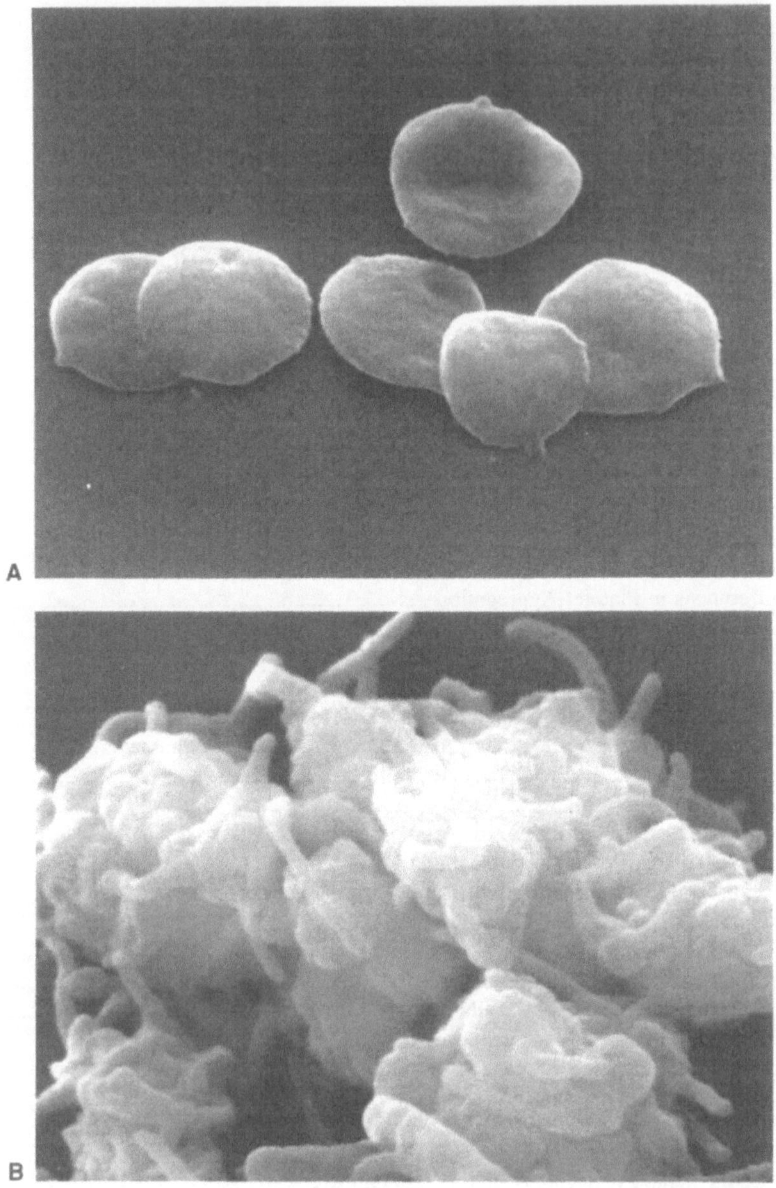

Figure 1 Scanning electron micrographs of human platelets: (A) normal; (B) aggregated platelets. (Kindly provided by Dr. James G. White, University of Minnesota with permission.)

contents, which include ADP, serotonin, calcium, fibrinogen, etc.[4-8] Although these released substances participate primarily in the formation of a hemostatic plug, they also mediate many biological processes, such as vascular permeability, cell proliferation, immune reactions, and many more.[8,9] Thrombosis is one of the major problems of modern medicine, and current concepts of hemostasis and thrombosis universally ascribe a primary role to the platelet and, in particular, to the phenomenon of platelet aggregation. Figures 1a and 1b represent scanning micrographs of human normal and aggregated platelets, respectively.

This review includes sections on aggregation measurement, aggregating agents, species differences, platelet disorders, and aggregation inhibitors.

II. MEASUREMENT OF PLATELET AGGREGATION

Platelet aggregation is perhaps the only platelet function that can be measured *in vitro* with a degree of reliability and validity. Several methods have been developed to measure platelet aggregation that employ platelet-rich plasma (PRP), platelet suspension in an artificial medium, or whole blood. The following are the laboratory methods commonly employed for measuring platelet aggregation.

A. TURBIDOMETRIC OR OPTICAL METHOD

In 1962, Born[10] employed the light scattering principle to measure the coalescence of platelet aggregates. The light transmission through a suspension of platelets is dependent on the number and size of platelets. Induction of platelet shape change and aggregation, therefore, can be recorded easily. In order to quantitate platelet aggregation, lag time to maximum turbidity (shape change), rates of primary and secondary aggregation, and maximum extent of aggregation (minimum turbidity) have been measured. This method is used mostly to measure platelet aggregation in platelet-rich plasma (PRP) of humans and laboratory animals. Generally, the recorded responses are qualitative rather than quantitative in nature. Therefore, to obtain significant information on the activity of an agonist, studies of dose-response effects are more helpful.[11] Computerized data acquisition has further advanced the standardization and quantitation of aggregation tracings.[12]

B. THE IMPEDANCE METHOD

Electronic aggregometry using impedance measurements of whole blood was introduced by Cardinal and Flower in 1980.[13] With this system, blood may be analyzed immediately after sampling without centrifugation. This method allows the measurement of platelet aggregation in a more physiological environment with less manipulation of the sample. Changes in the hematocrit, however, can produce significant effects on impedance. An optimal hematocrit for this method seems to be 30%, achieved by dilution with isotonic saline or platelet-poor plasma.[14] Furthermore, the incubation of whole blood at 37°C for about 5 min is important for stable results. The whole blood aggregometer measures the deposition of platelets on electrodes. The method is suitable for several aggregating agonists, which include ADP, collagen, arachidonic acid, cyclic endoperoxide (PGH_2), and thrombin. Aggregation response with epinephrine is extremely poor, perhaps due to an uptake mechanism or competition by receptors on other cells. Measurement of shape change, disaggregation, and biphasic aggregation are undetectable in whole blood aggregation. On the other hand, measurement of platelet aggregation is not affected by the plasma turbidity of hyperlipidemic blood.

C. THE LUMINESCENT METHOD

With the help of a lumi-aggregometer, one can measure simultaneously platelet aggregation and dense granule secretion from a single sample of whole blood or PRP. The basic principle of the lumi-aggregometer is to monitor continuously the release of ATP from the dense granules by a sensitive luminescent assay (firefly luciferin-luciferase), with the simultaneous measurement of platelet aggregation using the optical light transmission or impedance methods.[15,16] The lumi-aggregation techniques have been useful to assess platelet function in platelet disorders where the release reaction is greatly affected (e.g., storage pool defect, thrombasthenia, thrombopathia). In addition, agents that inhibit cyclooxygenase or thromboxane synthetase or that increase platelet intracellular cAMP levels can be examined for their effects on secondary aggregation (release reaction). The other method that was introduced by these investigators was the employment of a calcium electrode to monitor secretion of calcium from the storage granules.[17] This can be done in the absence of calcium in the medium and, therefore, excludes simultaneous measurement of platelet aggregation.

III. PLATELET AGGREGATING AGENTS

Platelets can be stimulated to form aggregates by a wide variety of biological and nonbiological agents (Table 1). In addition, platelets can be stimulated by immune complexes and complement components.[18] The process of platelet aggregation requires platelet-platelet collision and extracellular cofactors, such as calcium and fibrinogen, to form intracellular bridges between the platelets.[19] Activating agonists express fibrinogen receptors (gpIIb-IIIa) on the platelet membrane.[20,21] A normal human platelet contains approximately 50,000 gpIIb-IIIa complexes, which represent 1 to 2% of total platelet protein.[22] In general, the interaction of an aggregating agent with its specific receptor is followed by the expression of fibrinogen receptors and intracellular mobilization of calcium. This increase in cytoplasmic calcium levels has profound effects on regulatory systems of platelet function.[23,24] A select group of biologically important platelet aggregating agents will be discussed in greater detail.

Table 1 **Platelet aggregating agents**

Biological	Nonbiological
ADP	Calcium ionophores
Arachidonic acid	Kaolin
Calcium	Ristocetin
Collagen	Thimerosal
Epinephrine	
Immune complexes	
Lipoproteins (LDL, VLDL)	
Platelet-activating factor (PAF)	
Prostaglandins G_2 and H_2	
Serotonin	
Vasopressin	
Thrombin	
Thromboxane A_2	
Trypsin	

A. ADENOSINE 5′-DIPHOSPHATE (ADP)

A heat-stable platelet aggregating factor was shown to be present in erythrocytes and platelets.[25] The factor was soon identified as ADP by Gaarder et al.[26] The characteristics of ADP-induced platelet aggregation in PRP have been studied extensively.[27] Platelet aggregation with ADP concentrations (0.1 to 1.0 μ*M*) is usually reversible.[28] With higher ADP concentrations, a second phase of aggregation occurs with the release of endogenous granular contents (Table 2). This secondary aggregation can be blocked by inhibitors of prostaglandin cyclooxygenase (aspirin, indomethacin)[29-31] and thromboxane synthetase (imidazole and analogues).[32] Human platelets contain approximately 9 to 11 m*M* of adenine nucleotides,[33,34] of which ADP is about 40%. They are distributed between the storage (85% in dense granules) and metabolic (15% in cytoplasm) pools. The release of the storage-pool ADP during the second phase of aggregation further potentiates the aggregation phenomenon.[35] Specific receptors for ADP have been characterized on platelet membranes.[36-38] ADP binds slowly to platelet membranes. A Scatchard plot of the binding characteristics shows linearity, with an association constant (K_a) of 6.5×10^{-6} *M*, and total binding of 1200 pmol/mg membrane proteins, suggesting 100,000 molecules/platelet.[38] ADP binding to membranes required a divalent cation and was inhibited by sulfhydryl reagents and 2-chloroadenosine but not by PGE_1.[38] As far as structure-activity relationships are concerned, only 2-substituted ADP analogs (e.g., 2-chloro-ADP and 2-methylthio ADP) showed similar or greater agonistic activity.[39] Other substitutions, either on the purine or ribose moiety of ADP, significantly decreased the activity.[39] Binding studies using a synthetic analogue of ADP, 5′-p-fluorosulfonylbenzoyl adenosine, identified a surface membrane protein, aggregin (mol wt 100,000-Da)[40] that is distinct from the receptor coupled to adenylate cyclase[41] and reportedly is involved in ADP-induced shape change, aggregation, and fibrinogen binding. Platelet aggregation by epinephrine, thromboxane A_2, and collagen also requires ADP binding to aggregin.[41] Reports that ATP can cause platelet aggregation[42] may reflect a substantial contamination of commercial ATP by ADP, or a degradation of ATP to ADP as seen in rat PRP.[43]

Table 2 **Contents of platelet granules**

Lysosomes	Alpha granules	Dense bodies
Acid phosphatases	Fibronectin	Serotonin (5-HT)
(β-glycerophosphatase,	Factor VIII-related antigen	ATP and ADP
p-nitrophenylphosphatase)	Fibrinogen	Pyrophosphate
Arylsulfatase	Platelet-derived growth factor	Calcium
β-glucuronidase	Thrombospondin	
Cathepsin D,E	Factor V	
β-galactosidase	Platelet Factor IV	
β-N-acetylglucosaminidase		
β-N-acetylgalactosominidase		
α-arabinosidase		

B. EPINEPHRINE

Human platelets contain primarily alpha-2 adrenergic receptors.[44] Their number has been estimated to be between 100 and 460 per human platelet.[44] Epinephrine inhibits platelet membrane adenylate cyclase activity[44,45] and thus affects the intracellular cAMP levels of platelets.[46] Epinephrine induces a primary phase of aggregation, which can be followed by secondary aggregation and activation of the arachidonate pathway, leading to the formation of thromboxane A_2 and secretion of granular contents, which include ADP and serotonin. Epinephrine-induced aggregation occurs without a shape change[47,48] and is usually associated with uptake of external calcium.[49] Platelet response to epinephrine in a nonplasma medium containing ionized calcium has not been extensively studied, is variable, and is not readily reproducible.[50] At lower levels ($<1\ \mu M$), epinephrine induces only the primary phase aggregation; however, at these concentrations, epinephrine can potentiate platelet aggregation caused by ADP,[51,52] thrombin, collagen,[53] PAF,[54] or arachidonic acid.[55] There is considerable variability in the responsiveness of platelets to epinephrine among normal subjects.[56] An inherited decrease or absence of response has been reported.[57]

C. THROMBOXANE A_2 (TXA_2)

In 1975, Hamberg et al.[58] demonstrated an aorta-contracting substance, a chemically labile metabolite derived from platelet prostaglandin endoperoxides, which they named TXA_2. It has a half-life of 30 s and is an exceptionally potent inducer of platelet aggregation,[58,59] with specific membrane receptors that may also interact with cyclic endoperoxides (TXA_2/PGH_2 receptors). Scatchard analysis revealed 1500 to 2500 single class binding sites per platelet with a Kd of 10 to 30 n*M*.[59] Most studies have used arachidonic acid, the precursor of TXA_2, to induce platelet aggregation. Two major pathways of oxidative arachidonic acid conversion have been defined in human platelets, the cyclooxygenase and lipoxygenase pathways. The former converts arachidonic acid into PGG_2 and PGH_2,[58] which are rapidly metabolized into TXA_2 by thromboxane synthetase.[60] The other pathway synthesizes 12-HPETE and 12-HETE by 12-lipoxygenase.[58,61] A recent study of 53 healthy women and 65 healthy men showed that the platelet proaggregatory capacity is greater in males than in females due to higher platelet TXB_2 production from arachidonic acid.[62]

D. PLATELET-ACTIVATING FACTOR (PAF)

PAF was discovered in the early 1970s.[63,64] Its structure was elucidated later in the same decade, thus for the first time attributing a biological activity to an ether phospholipid.[65] Both rabbit and human platelets form PAF upon stimulation by calcium ionophore (A 23187) and thrombin.[66] This discovery of another aggregating agent formed by platelets themselves led to the postulate that PAF could be a physiological mediator of platelet activation.[67] The mechanism by which PAF causes platelet aggregation is unclear but may be similar to that of thrombin in that calcium mobilization appears to play a central role.[68] PAF binds to specific receptors on the platelet membrane and causes an influx of calcium.[69] Scatchard analysis revealed the presence of two binding sites for PAF: a high affinity, saturable binding site and a low affinity binding site with an apparently infinite capacity.[70,71] The Kd for the high affinity binding sites for PAF is 37 ± 13 n*M*, and the maximal binding capacity is 1399 ± 498 molecules of PAF per platelet.[71] In contrast, other investigators have reported one type of binding site with a Kd of 16.7 n*M* and 280 sites per platelet.[72] PAF stimulates polyphosphoinositide turnover and induces changes in the gpIIb-IIIa complex that enable fibrinogen binding and aggregation.[68] PAF may also inhibit adenylate cyclase[73] and decrease intracellular cAMP levels. PAF-induced platelet aggregation differs greatly between normal human PRP and plasma-free platelet suspension with EC_{50} values of 80 n*M* and 0.4 n*M*, respectively.[74] The response range for PAF in human PRP varies from 40 to 760 n*M*.[75,76] The large variability in the doses of PAF required to produce irreversible platelet aggregation may reflect several factors, including method of blood collection, PRP preparation, PRP storage time, aggregometer used, factors relating to blood donors, and purity of the PAF.[77] Plasma adenosine may act as a negative modulator of PAF-induced platelet aggregation in human PRP.[78]

E. ARGININE^8-VASOPRESSIN (AVP)

AVP, a biologically active hormone produced by the posterior pituitary gland, has a wide variety of pharmacological actions in the body.[79-82] AVP causes platelet aggregation, with only a partial granular release of serotonin (5-HT).[82] Most of the AVP actions are mediated by two types of AVP-specific membrane receptors, V_1 and V_2.[79,80] Interaction with V_1 receptors causes phosphoinositide hydrolysis and elevation of cytosolic free calcium levels.[83-85] In contrast, binding to V_2 receptors, which predominate in the kidney medulla, stimulates adenylate cyclase.[86] Platelet AVP receptors are of V_1 type,[87,88] similar to

Table 3 **Species differences**

	ADP	Arachidonic acid	Epinephrine	PAF
Human	++	++	++	++
Monkey	++	+	+	+
Rat	+	++	–	–
Rabbit	+	++	–	+++
Sheep	+	++	–	–
Dog	++	++	+	–

Notes: (–): no aggregation; (+): aggregation with no release reaction; (++): aggregation and release reaction; (+++): strong response of aggregation and release reaction.

vascular receptors.[89] They have been shown to have a kD of 5.6 n*M* and B_{max} of 115 fmol/mg protein.[87] AVP can potentiate the aggregating activity of ADP and epinephrine,[90] apparently by augmenting the concentrations of second messengers within the platelet rather than by direct interaction of the membrane receptors or their transmembrane coupling mechanisms.[90] There is considerable individual variation in the platelet response to AVP.[82,91,92] In a recent study, we have demonstrated that platelets of most females are more sensitive to AVP than platelets of most male donors.[92] In addition, plasma adenosine negatively modulates the effect of AVP on platelet aggregation. On the other hand, adenosine antagonists, theophylline and 3,7-dimethyl-1-propargylxanthine (DMPX, a selective A_2 antagonist) potentiate the effect of AVP on this platelet function. Although the biochemical cause for this gender difference is not clear, we speculate that differences in the plasma levels of adenosine or AVP may up- or down-regulate the platelet receptors.[92]

IV. SPECIES DIFFERENCE IN PLATELET AGGREGATION

Large variations in response to various aggregating agonists have been reported in different animal species.[93-97] Aggregation responses to ADP, arachidonic acid, epinephrine, and PAF in humans and several laboratory animals are shown in Table 3. ADP and arachidonic acid induce platelet aggregation in all the species listed. However, the agonist-induced release of granular contents varies greatly among these species. Epinephrine is effective only in humans, monkeys, and dogs and does not cause aggregation in rabbits, rats, and sheep. It induces biphasic aggregation in most humans, whereas in cats, epinephrine produces only monophasic but irreversible platelet aggregation with a long lag period (8 to 12 min).[95] Monkey and dog platelets also respond to epinephrine, but the response is weaker than in human platelets,[93,95] and rat, rabbit, and sheep platelets are completely unresponsive.[94-96] Similarly, PAF induces platelet aggregation only in humans, monkeys, and rabbits, whereas rat, sheep, and dog platelets do not respond to PAF.[97] Rabbit platelets have been reported as being the most sensitive to PAF, about 100-fold more than human platelets.[66,68,97] Collagen and thrombin are strong inducers of platelet aggregation and release in most species.[94,96]

V. PLATELET AGGREGATION — DISORDERS

Blood platelets circulate in a potentially hostile environment, where they are exposed to a number of biologically active agents that may activate or inhibit platelet function. In certain diseases or pathological conditions, the plasma levels of such biological agents are greatly altered and thus may produce abnormal platelet aggregation. The biochemical changes that are responsible for abnormal platelet aggregation in some of the more important diseases are discussed below.

A. DISORDERS WITH INCREASED PLATELET AGGREGABILITY

1. Diabetes Mellitus

The platelets of diabetic subjects generally have increased activity. Studies have shown enhanced second-phase platelet aggregation in response to ADP, epinephrine, collagen, arachidonic acid, or thrombin.[97-99] The effects of diabetes on platelet function have also been examined in animal models. The results, however, are not always in agreement with human studies.[100,101] The hypersensitivity of platelets in diabetic subjects may be due to an increase in the formation of arachidonic acid metabolites. Increased levels of serum PGE_2 and PGF_2,[102] prostaglandin endoperoxides,[103] and TXA_2 levels[98,103-105] have been found in diabetic subjects, and TXA_2 synthesis from exogenous arachidonic acid is enhanced in diabetic

Table 4 **Disorders with decreased platelet aggregation**

Disorders	Abnormal function (decreased activity or absent)	References
	Congenital	
Wiskott-Aldrich syndrome	GP Ia, Ib, dense granules	128,134,135
Bernard-Soulier syndrome	GP Ib	136,137
Glanzmann's thrombasthemia	GP IIb, IIIa	137
Storage pool deficiency	Dense granule contents	138,139
Gray platelet syndrome	Alpha granules	140
Defect in prostaglandin metabolism	Cyclooxygenase or thromboxane synthetase	141,142
	Acquired	
Disseminated intravascular coagulation (DIC)	Dense granules contents	143
Myeloproliferative disorders	Receptors, prostaglandins	144,145
Drug-induced abnormalities	Interaction with receptors	146,147
Acquired immunodeficiency syndrome (AIDS)	Immune system	148,149

patients with or without vascular complications.[98,104,105] The reason for the increased production of arachidonic acid metabolites remains unclear. Takeda et al.[106] have reported increased platelet phospholipase activity in diabetic subjects as compared to normal controls. The enzyme activity was decreased by insulin *in vitro* and by chronic insulin therapy in diabetic patients. More recent findings indicate that increased platelet aggregation may be due to abnormalities of guanylate cyclase function, especially in platelets from subjects with type II diabetes mellitus.[107]

2. Myocardial Infarction and Ischemic Heart Disease

It is generally believed that impairment of coronary artery blood flow is a major cause of angina pectoris, myocardial infarction, and sudden death. Platelets may play a significant role in these phenomena. Enhanced platelet aggregation in patients with myocardial infarction and ischemic heart disease has been reported by several investigators.[108-112] The heightened platelet response to various aggregating agents is seen for 4 to 6 d after myocardial infarction and then gradually decreases over the next 2 to 3 weeks.[113] Platelets from patients with classic angina[114] or cardiac ischemia[115] may generate more TXA_2. Increased platelet aggregability has also been found in patients with thromboembolic disorders (deep vein thrombosis and pulmonary embolism) and acute cerebrovascular disease.[116-118]

3. Hypertension

Poplawski et al.[119] observed significantly increased platelet response to ADP in severely hypertensive patients. More recently, a number of other investigators studied the platelet function in hypertensive patients and in a spontaneously hypertensive rat model.[120-124] The literature is not unanimous as to hypertension-related platelet hyperreactivity. When patients are grouped by age, increased ADP-induced aggregation response is seen in patients over the age of 65. In contrast, there is no abnormal platelet aggregation in response to ADP or epinephrine for slightly younger patients (mean age 50 years) with essential hypertension.[124]

Hyperactive platelets have also been reported in several other common disorders including migraine,[125] gout, and acute renal failure.[99,126]

B. DISORDERS WITH DECREASED PLATELET AGGREGABILITY

A list of such disorders is given in Table 4. Congenital as well as acquired abnormalities can be found. Several excellent reviews are available.[127-133]

VI. PLATELET AGGREGATION INHIBITORS

Antiplatelet drugs that block platelet aggregation and thrombus formation can be grouped in three broad categories: (1) drugs that inhibit the arachidonate metabolism; (2) drugs that increase intracellular cAMP levels; and (3) drugs that inhibit platelet function by other mechanisms.

A. DRUGS THAT INHIBIT ARACHIDONIC ACID METABOLISM

1. Cyclooxygenase (Prostaglandin Synthetase) Inhibitors

These include nonsteroidal antiinflammatory drugs.[147,150,151] Aspirin (acetyl salicylate) irreversibly inactivates cyclooxygenase by acetylating the NH_2- terminal of serine residue, forming an N-acetyl-L serine moiety.[152,153] Since human platelets do not synthesize new proteins, the cyclooxygenase enzyme remains inhibited for the life of platelet (7 to 10 d).[154] Aspirin primarily inhibits the second phase of platelet aggregation. Studies *ex vivo* in humans after 25 to 125 mg/d for 10 d show greater inhibition of aggregation in whole blood than in PRP.[155] Administration of 1 mg/kg of aspirin once a day effectively inhibits platelet cyclooxygenase and TXA_2 production.[150,151] In most of the clinical trials of aspirin as prophylaxis for vaso-occlusive disease, dosages varying from 300 to 1500 mg daily were used.[150,156] The results were contradictory and inconclusive as to the most effective aspirin dosage in preventing arterial thromboses. Recently, in a multicenter study, Dutch investigators compared the effect of aspirin, in a daily dose of 30 mg or 283 mg, in 3131 patients with transient ischemic attacks or minor ischemic strokes.[157] The frequency of death from all vascular causes was similar, 14.7% in the 30-mg group compared to 15.2% in the group receiving 283 mg. However, there were fewer gastrointestinal complaints and less bleeding in the 30-mg group. Other nonsteroidal antiinflammatory drugs (e.g., indomethacin, ibuprofen, sulfinpyrazone, etc.) also inhibit platelet aggregation, primarily the release reaction, by inhibition of cyclooxygenase, but different from aspirin, their effects on the enzyme are reversible.[147,150,151] In addition, they can affect platelet function by other mechanisms; e.g., indomethacin inhibits phospholipase activity,[158] and sulfinpyrazone, an uricosuric agent, blocks platelet adhesion to collagen.[159]

2. Thromboxane Synthetase Inhibitors and Thromboxane Receptor Antagonists

A number of imidazole analogues have been synthesized as potential inhibitors of thromboxane synthetase,[150,160,161] with the expectation that the inhibitors will enhance the production of cyclic endoperoxides (PGG_2, PGH_2) and thus increase the synthesis of PGI_2.[162] In human platelets, thromboxane synthetase inhibitors cause an increase in the synthesis of PGD_2, a potent antiplatelet agent that stimulates adenylate cyclase.[163] The following studies support these findings:

1. The antiplatelet actions of the thromboxane synthetase inhibitor, dazoxiben (UK-37248) can be reversed by SQ 22536, an inhibitor of adenylate cyclase.[164]
2. Synergistic antiplatelet effects are seen when a thromboxane synthetase inhibitor (UK-37,248, OKY-1581, UK-38485, or SC-38249) is combined with a cAMP phosphodiesterase (PDE) inhibitor.[165-167]

These studies suggested that combined use of inhibitors of thromboxane synthetase and cAMP PDE may provide an effective antithrombotic therapy. On the other hand, drugs that antagonize the actions of TXA_2 by blocking its receptors may have greater clinical potential than those that block TXA_2 synthesis,[168] although this has yet to be proven clinically. Several TXA_2 receptor antagonists have been synthesized, and their effects have been examined on human platelets. These antagonists include AH 19437,[169] EP 045,[170] BM 13,177,[171] and GR32191.[168,172] Combining a thromboxane synthetase inhibitor with a thromboxane receptor antagonist produces synergistic inhibition of collagen-induced platelet aggregation in whole blood and may constitute effective antithrombotic therapy.[173]

3. Effects of Omega-3 Fatty Acids (Eicosapentaenoic and Docosahexaenoic Acids)

Several studies show that certain populations (e.g., Greenland Eskimos) consuming fish or fish oils containing eicosapentaenoic acid (EPA) and docosahexaenoic acid (DHA) exhibit prolonged bleeding times and diminished platelet function.[174-176] This suggested that omega-3 fatty acids may be protective against arterial thrombosis and atherosclerotic cardiovascular disorders.[175] Both EPA and DHA lower serum triacylglycerides and reduce platelet aggregation.[175] EPA decreases platelet TXA_2 production with the formation of PGH_3 and TXA_3.[175,177] Neither PGH_3 nor TXA_3 has significant platelet-aggregating activity.[177] In human platelets, PGH_3 produces PGD_3, a potent inhibitor of platelet aggregation.[177] DHA, on the other hand, inhibits conversion of arachidonic acid into TXA_2 and thus blocks platelet aggregation.[178] Healthy subjects receiving EPA demonstrate decreased production of TXA_2 and inhibition of platelet aggregation.[179] The EPA metabolite PGH_3 in vascular endothelium is converted into a prostacyclin analogue, PGI_3, which has potent antiplatelet and vasodilatory activity similar to PGI_2.[180]

B. DRUGS THAT INCREASE PLATELET INTRACELLULAR CYCLIC cAMP LEVELS

1. PGE_1, PGD_2, PGI_2, and Adenosine

Platelet intracellular cAMP levels play a key role in modulating platelet functions.[181-184] An increase in cAMP levels cause inactivation of platelet function.[181,183,184] These biologically produced prostaglandins and adenosine act on their specific membrane receptors that are coupled to adenylate cyclase, causing increases in platelet cAMP levels,[96,181-183] and thereby inhibit platelet aggregation. The IC_{50} values for ADP-induced platelet aggregation range from 1.4 to 60 n*M* for the prostaglandins with the order potency, $PGI_2 > PGD_2 > PGE_1$.[185] PGI_2 is about 20- and 50-fold more inhibitory than PGD_2 and PGE_1, respectively. However, if arachidonic acid is used to induce platelet aggregation, PGD_2 and PGE_1 are almost equipotent with PGI_2.[185,186] PGI_2 and PGE_1 share a common receptor site, which is distinct from that of PGD_2.[182,187] PGI_2 and PGE_1 strongly inhibit platelet aggregation in blood samples of most laboratory animals, whereas PGD_2 inhibits human, sheep, and horse platelets but only weakly rabbit platelets and shows no inhibition on rat platelets.[181,185] Adenosine is also a potent inhibitor of platelet aggregation (IC_{50}, 2 μ*M*),[96] but in comparison to PGI_2, it is 500-fold less inhibitory. However, plasma adenosine levels (100 to 300 n*M*)[188,189] are 400- to 1000-fold greater than PGI_2 levels (0.51 ± 0.20 n*M*, measured as 6-keto PGF_{1a}).[190] Both PGI_2 and adenosine are constantly produced by vascular endothelium[191-193] and play a significant role in keeping platelets in a nonaggregatory state while they circulate in the vascular system.[183,184,194,195] During the past decade, there has been large interest in developing effective adenosine receptor agonists and antagonists as potential therapeutic agents in modulating cell function.[196,197] Platelet aggregation can also be blocked by cAMP PDE inhibitors, which increase intracellular cAMP concentrations.[198-200] Cyclic AMP PDE inhibitors can also potentiate the antiplatelet activities of endogenous stimulators of adenylate cyclase, which include the above prostaglandins and adenosine.[198,201] A large number of cAMP PDE inhibitors have been synthesized and are being studied as potential antithrombotic agents.[198-200]

2. Dipyridamole

Dipyridamole (persantine) is a vasodilatory and antiplatelet agent that has been widely used clinically for many years.[202] Dipyridamole is a weak inhibitor of cAMP PDE (Ki, 20 μ*M*),[203] but a potent inhibitor of the nucleoside transport system in a variety of cells (Ki, 0.05 μ*M*, 400-fold lower compared to the Ki for cAMP PDE).[204] Dipyridamole binds extensively (>99%) to plasma proteins (alpha$_1$-acid glycoprotein and albumin).[205,206] Peak therapeutic values of dipyridamole are usually in the range of 1 to 4 μ*M* (total, bound plus free).[207] Therefore, it appears that clinically relevant blood levels are too low to inhibit cAMP PDE. Because of its potent inhibition of nucleoside transport, dipyridamole causes an elevation in plasma adenosine levels.[188,207,208] Adenosine, which is produced by many tissues,[189,209,210] is rapidly transported into cells (erythrocytes and other cells) and metabolized primarily by adenosine deaminase and adenosine kinase,[211] resulting in low levels of plasma adenosine (100 to 300 n*M*).[188] *In vitro* studies have shown that adenosine, on addition to human whole blood (200 nm or 10 μ*M*), disappears with a $t_{1/2}$ of <30 s.[208,211] However, in the presence of 10 μ*M* dipyridamole, the $t_{1/2}$ of adenosine increases to about 5 min.[208] Clinically, dipyridamole increases plasma adenosine levels.[188] ADP- or collagen-induced platelet aggregation has been found to be inhibited in whole blood samples collected 2 h after an oral intake of 200 mg dipyridamole.[212] This suggests that dipyridamole inhibits platelet aggregation in whole blood by blocking the reuptake of adenosine. Similarly, we were able to show that adenosine, which was noninhibitory to platelet aggregation in whole blood, strongly inhibited platelet aggregation in the presence of 10 μ*M* dipyridamole (IC_{50}, 1.7 μ*M*).[213] Dipyridamole has also been reported to inhibit TXA_2 synthesis;[214] however, the concentrations required to exert this effect are much higher than those that block adenosine uptake.[214] In addition, dipyridamole can inhibit cGMP PDE (IC_{50}, 0.3 μ*M*),[215] and thus potentiate nitric oxide (NO)-mediated inhibition of platelet aggregation.[216]

C. DRUGS THAT ACT BY OTHER MECHANISMS

1. Ticlopidine

This compound inhibits platelet aggregation and release reaction induced by collagen, ADP, arachidonic acid, and PAF, but the mechanism of its antiplatelet action is not well understood.[217,218] It has been suggested that ticlopidine blocks fibrinogen binding to the gpIIb-IIIa complex, the fibrinogen receptor of platelet membrane[219] but more than one mechanism may mediate its antiplatelet action. Both *in vitro* and *ex vivo* studies showed that ticlopidine inhibits especially PAF-induced platelet aggregation.[220,221] The ticlopidine metabolite 2-ketoticlopidine was reported to have about eightfold greater inhibitory activity than ticlopidine.[222] Inhibitors of thromboxane synthetase (UK-38485, dazmegrel) and cAMP PDE (RA-233, a close structural analogue of dipyridamole) strongly potentiate the antiplatelet activity of ticlopidine.[220] Ticlopidine is currently recommended for stroke prevention in patients with symptomatic atherothrombotic cerebrovascular disease.[223] It reduces the stroke rate more effectively than aspirin in patients who have had a transient ischemic attack or minor stroke.[224] Compared with placebo, ticlopidine reduces the rate of recurrence in patients who had a major stroke by 33% at 1 year.[223] In humans, ticlopidine (250 to 1000 mg/d) prolongs the bleeding time and blocks platelet aggregation.[150,225,226]

2. Nitric Oxide (NO)

The first report that endothelium-derived relaxing factor (EDRF) inhibits platelet function was that of Azuma et al. in 1986.[227] Subsequently, two independent groups of investigators demonstrated that EDRF is chemically identical to NO.[228,229] In platelets, NO induces activation of cytoplasmic guanylate cyclase, thereby increasing intracellular cGMP levels, which block platelet aggregation.[230,231] Several studies demonstrate that vascular endothelium[228] and other tissues,[232] including human platelets,[233] synthesize NO from the terminal guanidino nitrogen atom(s) of L-arginine by the biochemical L-arginine/NO pathway. L-Arginine inhibits collagen-induced platelet aggregation both in washed human platelet suspension and whole blood with IC_{50} values of 8.9 ± 1.0 μ*M* and 234 ± 43 μ*M*, respectively. The inhibitory effects can be greatly potentiated by M&B22948, a selective inhibitor of cGMP PDE.[233] These studies suggest that the L-arginine/NO pathway in human platelets may have a significant role in platelet physiology.

3. Calcium Channel Blockers (Calcium Antagonists)

Several laboratories have reported that calcium antagonists inhibit aggregation *in vitro;* however, the antiplatelet effects occur at 5- to 50-fold greater concentrations than those achieved clinically.[234-236] In *ex vivo* studies, a single oral dose of diltiazem (60 mg), nifedipine (10 mg), or verapamil (80 mg) significantly inhibited platelet aggregation induced by ADP and epinephrine[237] but not by collagen. Greater PGI_2 sensitivity of platelets was also found after nifedipine administration.[237] Of the calcium channel blockers tested, diltiazem had the most potent effect on platelet function.[237] A decrease in thromboxane synthesis, enhanced vascular PGI_2 synthesis, and synergistic effects with PGI_2 and PGE_1 have been suggested as underlying mechanisms of action for calcium antagonists.[236-238]

VII. ACKNOWLEDGMENTS

I thank Professor Robert E. Parks, Jr. for his interest and support for many years. The platelet studies have been supported by grants from PHS CA 07340 and the American Heart Association (Rhode Island).

REFERENCES

1. **White, J. G.,** Current concepts of platelet structure, *Am. J. Clin. Pathol.,* 71, 363, 1976.
2. **Karpatkin, S.,** Human platelet senescence, *Annu. Rev. Med.,* 23, 101, 1972.
3. **Harker, L.,** Platelet survival time: its measurement and use, *Prog. Haemost. Thromb.,* 4, 321, 1978.
4. **Stenberg, P. E. and Bainton, D. F.,** Storage organelles in platelets and megakaryocytes, in *Biochemistry of Platelets,* Phillips, D. R. and Shuman, M. A., Eds., Academic Press, New York, 1986, 257.
5. **Zucker, M. B. and Borelli, J.,** Relationship of some blood clotting factors to serotonin release from washed platelets, *J. Appl. Physiol.,* 7, 432, 1955.
6. **White, J. G.,** The dense bodies of human platelets. Origin of serotonin storage particles from platelet granules, *Am. J. Pathol.,* 53, 791, 1968.
7. **Holmsen, H., Day, H. J., and Stormorken, H.,** The blood platelet release reaction, *Scand. J. Haematol.,* 8 (Suppl.), 1, 1969.
8. **Holmsen, H. and Weiss, H. J.,** Secretable storage pools in platelets, *Annu. Rev. Med.,* 30, 119, 1979.
9. **Huang, E. M. and Detwiler, T. C.,** Stimulus-response coupling mechanisms, in *Biochemistry of Platelets,* Phillips, D. R. and Shuman, M. A., Eds., Academic Press, New York, 1986, 1.
10. **Born, G. V. R.,** Aggregation of blood platelets by adenosine diphosphate and its reversal, *Nature,* 194, 927, 1962.
11. **Jancinova, V., Nosal, R., and Petrikova, M.,** Dose-response aggregometry: contribution to the precise platelet function evaluation, *Thromb. Res.,* 65, 1, 1992.
12. **Huzoor-Akbar, Ronstedt, K., and Manhine, B.,** Computerized aggregation instruments: a highly efficient and versatile system for acquisition, quantitation, presentation and management of platelet aggregation data, *Thromb. Res.,* 32, 335, 1983.
13. **Cardinal, D. C. and Flower, R. J.,** The electronic aggregometer: a novel device for assessing platelet behavior in blood, *J. Pharm. Meth.,* 3, 135, 1980.
14. **Mackie, I. J. and Machin, S. J.,** Platelet impedance aggregation in whole blood and its inhibition by antiplatelet drugs, *J. Clin. Pathol.,* 37, 874, 1984.
15. **Feinman, R. D., Lubowsky, J., Charo, I. F., and Zabinski, M. P.,** The lumi-aggregometer: a new instrument for simultaneous measurement of secretion and aggregation by platelets, *J. Lab. Clin. Med.,* 90, 125, 1977.

16. **Charo, I. F., Feinman, R. D., and Detwiler, T. C.,** Interrelations of platelet aggregation and secretion, *J. Clin. Invest.*, 60, 866, 1977.
17. **Feinman, R. D. and Detwiler, T. C.,** Absence of a requirement for extracellular calcium for secretion from platelets, *Thromb. Res.*, 7, 677, 1975.
18. **Polley, M. J. and Nachman, R. L.,** The human complement system in thrombin-mediated platelet function, in *Platelets in Biology and Pathology 2*, Dingle, J. T. and Gordon, J. L., Eds., Elsevier/North-Holland Biomedical Press, Oxford, 1981, 309.
19. **Mustard, J. F., Packham, M. A., Kinlough-Rathbone, R. L., Perry, D. W., and Regoeczi, E.,** Fibrinogen and ADP-induced platelet aggregation, *Blood*, 52, 453, 1978.
20. **Bennett, J. S. and Vilaire, G.,** Exposure of platelet fibrinogen receptors by ADP and epinephrine, *J. Clin. Invest.*, 64, 1393, 1979.
21. **Marguerie, G. A., Plow, E. F., and Edgington, T. S.,** Human platelets possess an inducible and saturable receptor specific for fibrinogen, *J. Biol. Chem.*, 254, 5357, 1979.
22. **Jennings, L. K. and Phillips, D. R.,** Purification of glycoproteins IIb and IIIa from human platelet membranes and characterization of a calcium-dependent glycoprotein IIb-IIIa complex, *J. Biol. Chem.*, 257, 10458, 1982.
23. **Gerrard, J. M., Peterson, D. A., and White, J. G.,** Calcium mobilization, in *Platelets in Biology and Pathology 2*, Dingle, J. T. and Gordon, J. L., Eds., Elsevier/North-Holland Biomedical Press, Oxford, 1981, 407.
24. **Feinstein, M. B., Rodan, G. A., and Cutler, L. S.,** Cyclic AMP and calcium in platelet function. in *Platelets in Biology and Pathology 2*, Dingle, J. T. and Gordon, J. L., Eds., Elsevier/North-Holland Biomedical Press, Oxford, 1981, 437.
25. **Hellem, A. J.,** The adhesiveness of human platelets in vitro, *Scand. J. Clin. Lab. Invest.*, 12 (Suppl. 51), 1, 1960.
26. **Gaarder, A., Jonsen, J., Laland, S., Hellem, A., and Owren, P. A.,** Adenosine diphosphate in red cells as a factor in the adhesiveness of human platelets, *Nature*, 192, 531, 1961.
27. **Haslam, R. J. and Cusack, N. J.,** Blood platelet receptors for ADP and adenosine, in *Purinergic Receptors*, Vol. 12, Receptors and Recognition, Series B, Burnstock, G., Ed., Chapman and Hall, London, 1981, 223.
28. **Mustard, J. F. and Packham, M. A.,** Factors influencing platelet function: adhesion, release and aggregation, *Pharmacol. Rev.*, 22, 97, 1970.
29. **Mills, D. C. B., Robb, J. A., and Roberts, G. C. K.,** The release of nucleotides, 5-hydroxytryptamine and enzymes from human platelet, *J. Physiol.*, 195, 715, 1968.
30. **Zucker, M. B. and Peterson, J.,** Inhibition of adenosine diphosphate-induced secondary aggregation and other platelet functions by acetylsalicylic acid ingestion, *Proc. Soc. Exp. Biol. Med.*, 127, 547, 1968.
31. **Smith, J. B. and Willis, A. L.,** Aspirin selectively inhibits prostaglandin production in human platelets, *Nature*, 231, 235, 1971.
32. **Lasslo, A. and Quintana, R. P.,** Interaction dynamics of blood platelets with medicinal agents and other chemical entities, in *Blood Platelet Function and Medicinal Chemistry*, Lasslo, A., Ed., Elsevier Biomedical, New York, 1984, 229.
33. **Reimers, H.-J.,** Adenine nucleotides in blood platelets, in *Platelets: Physiology and Pharmacology*, Longenecker, G. L., Ed., Academic Press, Orlando, 1985, 85.
34. **Agarwal, K. C. and Parks, R. E., Jr.,** Adenosine analogs and human platelets: effects on adenine nucleotide pools and the aggregation phenomenon, *Biochem. Pharmacol.*, 24, 2239, 1975.
35. **Haslam, R. J.,** Mechanisms of platelet aggregation, in *Physiology of Hemostasis and Thrombosis*, Johnson, S. A. and Seegers, W. H., Eds. Charles C. Thomas, Springfield, IL, 1967, 88.
36. **Haslam, R. J. and Cusack, N. J.,** Blood platelet receptors for ADP and adenosine, in *Purinergic Receptors*, Burnstock, G., Ed., Chapman and Hill, London, 1981, 223.
37. **Mills, D. C. B. and Macfarlane, D. C.** Platelet receptors, in *Platelets in Biology and Medicine*, Gordon, J., Ed., Elsevier/North Holland Biomedical Press, New York, 1976, 159.
38. **Nachman, R. L. and Ferris, B.,** Binding of adenosine diphosphate by isolated membranes from human platelets, *J. Biol. Chem.*, 249, 704, 1974.
39. **Gough, G., Maguire, M. H., and Penglis, F.,** Analogues of adenosine 5′-diphosphate: new platelet aggregators. Influence of purine ring and phosphate chain substitutions on the platelet-aggregating potency of adenosine 5′-diphosphate, *Mol. Pharmacol.*, 8, 170, 1972.
40. **Bennett, J. S., Colman, R. F., and Colman, R. W.,** Identification of adenine nucleotide binding proteins in human platelet membranes by affinity labeling with 5′-p-fluorosulfonylbenzoyl adenosine, *J. Biol. Chem.*, 253, 7346, 1978.

41. **Colman, R. W.,** Aggregin: a platelet ADP receptor that mediates activation, *FASEB J.*, 4, 1425, 1990.
42. **Ts'ao, C.,** Rat platelet aggregation by ATP, *Am. J. Pathol.*, 85, 581, 1976.
43. **Haskel, E. J., Agarwal, K. C., and Parks, R. E., Jr.,** ATP- and ADP-induced rat platelet aggregation: significance of plasma in ATP-induced aggregation, *Thromb. Haemost.*, 42, 1580, 1979.
44. **Scrutton, M. C. and Wallis, R. B.,** Catecholamine receptors, in *Platelets in Biology and Pathology*, Vol. 2, Gordon, J. L., Ed., Elsevier/North-Holland Biomedical Press, Oxford, 1981, 179.
45. **Haslam, R. J., Davidson, M. M. L., and Desjardins, J. V.,** Inhibition of adenylate cyclase by adenosine analogues in preparations of broken and intact human platelets, *Biochem. J.*, 176, 83, 1978.
46. **Steer, M. L. and Wood, A.** Regulation of human platelet adenylate cyclase by epinephrine, prostaglandin E_1, and guanine nucleotides, *J. Biol. Chem.*, 254, 10791, 1979.
47. **O'Brien, J. R.,** Some effects of adrenaline and antiadrenaline compounds on platelets in vitro and in vivo, *Nature*, 200, 763, 1963.
48. **Mills, D. C. B. and Robert, G. C. K.,** Effects of adrenaline on human blood platelets, *J. Physiol. (London)*, 193, 443, 1967.
49. **Owen, N. E., Feinberg, H., and Le Breton, G. C.,** Epinephrine induces Ca^{2+} uptake in human blood platelets, *Am. J. Physiol.*, 239, H483, 1980.
50. **Kinlough-Rathbone, R. L., Mustard, J. F., Packham, M. A., Perry, D. W., Reimbers, H,-J., and Cazenave, J.-P.,** Properties of washed human platelets, *Thromb. Haemost.*, 37, 291, 1977.
51. **Ardlie, N. G., Glew, G., and Schwartz, C. J.,** Influence of catecholamines on nucleotide-induced platelet aggregation, *Nature*, 212, 415, 1966.
52. **Grant, J. A. and Scrutton, M. C.,** Novel alpha-2 adrenoreceptors primarily responsible for inducing platelet aggregation, *Nature*, 277, 659, 1979.
53. **Thomas, D. P.,** The role of platelet catecholamines in the aggregation of platelets by collagen and thrombin, *Exp. Biol. Med.*, 3, 129, 1968.
54. **Vargaftig, B. B., Fouque, F., Benveniste, J., and Odiot, J.,** Adrenaline and PAF-acether synergize to trigger cyclooxygenase-independent activation of plasma-free human platelets, *Thromb. Res.*, 28, 557, 1982.
55. **Rao, G. H. R. and White, J. G.,** Epinephrine potentiation of arachidonate-induced aggregation of cyclooxygenase-deficient platelets, *Am. J. Hematol.*, 11, 355, 1981.
56. **O'Brien, J. R.,** Variability in the aggregation of human platelets by adrenaline, *Nature*, 202, 1188, 1964.
57. **Scrutton, M. C., Clare, K. A., Hutton, R. A., and Bruckdorfer, K. R.,** Depressed responsiveness to adrenaline in platelets from apparently normal human donors: a familial trait, *Br. J. Haematol.*, 49, 303, 1981.
58. **Hamberg, M., Svensson, J., and Samuelsson, B.,** Thromboxanes: a new group of biologically active compounds derived from prostaglandin endoperoxides, *Proc. Natl. Acad. Sci. U.S.A.*, 72, 2994, 1975.
59. **Saussy, D. L., Mais, D. E., Burch, R. M., and Halushka, P. V.,** Identification of a putative TxA_2/ PGH_2 receptor in human platelet membranes, *J. Biol. Chem.*, 261, 3025, 1986.
60. **Needleman, P. M., Moncada, S., Bunting, S., Vane, J., Hamberg, M., and Samuelsson, B.,** Identification of an enzyme in platelet microsomes which generates thromboxane A_2 from prostaglandin endoperoxides, *Nature*, 261, 558, 1976.
61. **Nugteren, D. H.,** Arachidonate lipoxygenase in blood platelets, *Biochim. Biophys. Acta*, 380, 299, 1975.
62. **Pinto, S., Coppo, M., Paniccia, R., Gori, A. M., Attanasio, M., and Abbate, R.,** Sex related differences in platelet TXA_2 generation, *Prostgl. Leukotr. Essentl. Fatty Acid*, 40, 217, 1990.
63. **Henson, P. M.,** Release of vasoactive amines from rabbit platelets induced by sensitized mononuclear leukocytes and antigen, *J. Exp. Med.*, 131, 287, 1970.
64. **Henson, P. M. and Benveniste, J.,** Antibody-leukocyte-platelet interactions, in *Biochemistry of the Acute Allergic Reactions*, Austen, K. F. and Becker, E. L., Eds., Blackwell Scientific, Oxford, 1971, 111.
65. **Benveniste, J., Le Couedic, J. P., Polonsky, J., and Tence, M.,** Structural analysis of purified platelet-activating factor by lipase, *Nature*, 269, 170, 1977.
66. **Chignard, M., Le Couedic, J. P., Vargaftig, B. B., and Benveniste, J.,** The role of platelet-activating factor in platelet aggregation, *Br. J. Haematol.*, 46, 455, 1980.
67. **Vargaftig, B. B., Chignard, M., and Benveniste, J.,** Present concepts on the mechanisms of platelet aggregation, *Biochem. Pharmacol.*, 30, 263, 1981.
68. **Mustard, J. F. Kinlough-Rathbone, R. L., and Packham, M. A.,** Platelet activation: an overview, *Agents Actions*, 21 (Suppl.), 23, 1987.

69. **Valone, F. H.,** Inhibition of platelet activating factor binding to human platelets by calcium channel blockers, in *New Horizons in Platelet Activating Factor Research,* Winslow, C. M. and Lee, M. L., Eds., John Wiley & Sons, 215, 1987.
70. **Valone, F. H., Coles, E., Reinhold, V. R., and Goetzl, E. J.,** Specific binding of phospholipid platelet-activating factor by human platelets, *J. Immunol.,* 129, 1637, 1982.
71. **Valone, F. H.,** Specific binding of AGEPC by human platelets and polymorphonuclear leukocytes, in *Platelet-Activating Factor and Structurally Related Ether-Lipids,* Benveniste, J. and Arnoux, B., Eds., Elsevier Science Publishers, Amsterdam, 1983, 161.
72. **Kloprogge, E. and Akkerman, J. W. N.,** Binding aspects of PAF-acether to intact human platelets, in *Platelet-Activating Factor and Structurally Related Ether-Lipids,* Benveniste, J. and Arnoux, B., Eds., Elsevier Science Publishers, Amsterdam, 1983, 153.
73. **Haslam, R. J. and Vanderwel, M.,** Inhibition of platelet adenylate cyclase by 1-O-alkyl-2-O-acetyl-sn-glyceryl-3-phosphorylcholine (platelet activating factor), *J. Biol. Chem.,* 257, 6879, 1982.
74. **Keraly, C. L. and Delautier, D.,** Aggregation of human platelets by PAF-acether (platelet-activating factor): evidence for its independence from ADP release and thromboxane synthesis, in *Platelet-Activating Factor and Structurally Related Ether-Lipids,* Benveniste, J. and Arnoux, B., Eds., Elsevier Science Publishers, Amsterdam, 1983, 145.
75. **McManus, L. M., Hanaham, D. J., and Pinckard, R. N.,** Human platelet stimulation by acetyl glyceryl ether phosphorylcholine, *J. Clin. Invest.,* 67, 903, 1981.
76. **Marcus, A. J., Safier, L. B., Ullman, H. L. Wong, K. T. H., Broekman, M. J., Weksler, B. B., and Kaplan, K. L.,** Effects of acetyl glyceryl phosphorylcholine on human platelet function in vitro, *Blood,* 58, 1027, 1981.
77. **Misso, N. L. A. and Thompson, P. J.,** The human platelet aggregation model for assessing antiasthma drugs and PAF antagonists, in *Platelet-Activating Factor in Endotoxin and Immune Diseases,* Handley, D. A., Saunders, R. N., Houlihan, W. J., and Tomesch, J. C., Eds., Marcel Dekker, New York, 1990, 245.
78. **Agarwal, K. C., Zhao, Q., and Parks, R. E., Jr.,** Modulation of platelet-activating factor (PAF) actions on human platelets, *FASEB J.,* 6, A1313 (1992).
79. **Jard, S.,** Vasopressin: mechanisms of receptor activation, in *The Neurohypophysis: Structure, Function and Control. Progress in Brain Research,* Vol. 60, Cross, B. A. and Leng. G., Eds., Elsevier, New York, 1983, 383.
80. **Cornett, L. E.,** Vasopressin receptors, in *Peptide Hormone Receptors,* Walter de Gruyter & Co., New York, 1987, 437.
81. **Berecek, K. H. and Swords, B. H.,** Central role for vasopressin in cardiovascular regulation and the pathogenesis of hypertension, *Hypertension,* 16, 213, 1990.
82. **Haslam, R. J. and Rosson, G. M.,** Aggregation of human platelets by vasopressin, *Am. J. Physiol.,* 223, 958, 1972.
83. **Siess, W., Stifel, M., Binder, H., and Weber, P. C.,** Activation of V_1-receptors by vasopressin stimulates inositol phospholipid hydrolysis and arachidonate metabolism in human platelets, *Biochem. J.,* 233, 83, 1986.
84. **Hallam, T. J., Thompson, N. T., Scrutton, M. C., and Rink, T. J.,** The role of cytoplasmic free calcium in the responses of quin2-loaded human platelets to vasopressin, *Biochem. J.,* 221, 897, 1984.
85. **Rink, T. J. and Hallam, T. J.,** What turns platelets on?, *Trends Biochem. Sci.,* 9, 215, 1984.
86. **Michell, R. H., Kirk, C. J., and Billah, M. M.,** Hormonal stimulation of phosphatidylinositol breakdown, with particular reference to the hepatic effects of vasopressin, *Biochem. Soc. Trans.,* 7, 861, 1979.
87. **Berrettini, W. H., Post, R. M., Worthington, E. K., and Casper, J. B.,** Human platelet vasopressin receptors, *Life Sci.,* 30, 425, 1982.
88. **Vanderwel, M., Lum, D. S., and Haslam, R. J.,** Vasopressin inhibits the adenylate cyclase activity of human platelet particulate fraction through V1-receptors, *FEBS Lett.,* 164, 340, 1983.
89. **Fox, A. W.,** Vascular vasopressin receptors, *Gen. Pharmacol.,* 19, 639, 1988.
90. **Grant, J. A. and Scrutton, M. C.,** Positive interaction between agonists in the aggregation of human platelets: interaction between ADP, adrenaline and vasopressin, *Br. J. Haematol.,* 44, 109, 1980.
91. **Vittet, D., Mathieu, M.-N., Cantau, B., and Chevillard, C.,** Vasopressin inhibition of human platelet adenylate cyclase: variable responsiveness between donors and involvement of a G-protein different from Gi, *Eur. J. Pharmacol.,* 150, 367, 1988.
92. **Agarwal, K. C.,** Modulation of vasopressin actions on human platelets by plasma adenosine and theophylline: gender differences, *J. Cardiovasc. Pharmacol.,* 21, 1012, 1993.

93. **Addonizio, V. P., Jr., Edmunds, H. L., and Coleman, R. W.,** The function of monkey (M. mulatta) platelets compared to platelets of pig, sheep, and man, *J. Lab. Clin. Med.*, 91, 989, 1978.
94. **Dodds, W. J.,** Platelet functions in animals: species specificities, in *Platelets: A Multidisciplinary Approach,* de Gaetano, G. and Garattini, S., Eds., Raven Press, New York, 1978, 45.
95. **Hwang, D. H.,** Species variation in platelet aggregation, in *Platelets: Physiology and Pharmacology,* Longenecker, G. L., Ed., Academic Press, Orlando, 1985, 289.
96. **Agarwal, K. C.,** Adenosine and platelet function, in *Role of Adenosine in Cerebral Metabolism and Blood flow,* Stefanovich, V. and Okyayuz-Baklouti, I., Eds., VNU Science Press, Utrecht, The Netherlands, 1988, 107.
97. **Packham, M. A. and Mustard, J. F.** Normal and abnormal platelet activity, in *Blood Platelet Function and Medicinal Chemistry,* Lasslo, A., Ed., Elsevier Biomedical, New York, 1984, 61.
98. **Kwaan, H. C., Colwell, J. A., Cruz, S., Suwanwela, N., and Dobbie, J. G.,** Increased aggregation in diabetes mellitus, *J. Lab. Clin. Med.*, 80, 236, 1972.
99. **Packham, M. A.,** Methods for detection of hypersensitive platelets, *Thromb. Haemost.*, 40, 175, 1978.
100. **Johnson, M., Harrisson, H. E., Hawker, R., and Hawker, L.,** Platelet abnormalities in experimental diabetes, *Thromb. Haemost.*, 42, 333, 1979.
101. **Rosenblum, W. I., El-Sabban, F., and Loria, R. M.,** Platelet aggregation in cerebral and mesenteric microcirculation of mice with genetically determined diabetes, *Diabetes,* 30, 89, 1981.
102. **Chase, H. P., Williams, R. L., and Dupont, J.,** Increased prostaglandin synthesis in childhood diabetes mellitus, *J. Pediatr.*, 94, 185, 1979.
103. **Stuart, M. J., Elrad, H., Graeber, J. E., Hakanson, D. O., Sunderji, S. G., and Barvinchak, M. K.,** Increased synthesis of prostaglandin endoperoxides and platelet hyperfunction in infants of mothers with diabetes mellitus, *J. Lab. Clin. Med.*, 94, 12, 1979.
104. **Butkus, A., Skrinska, V. A., and Schumacher, O. P.,** Thromboxane production and platelet aggregation in diabetic subjects with clinical complications, *Thromb. Res.*, 19, 211, 1980.
105. **Halushka, P. V., Rogers, C., Loadholt, J. A., and Colwell, J.,** Increased platelet thromboxane synthesis in diabetes mellitus, *J. Lab. Clin. Med.*, 97, 87, 1981.
106. **Takeda, H., Maeda, H., Fukushima, H., Nakamura, N., and Uzawa, H.,** Increased platelet phospholipase activity in diabetic subjects, *Thromb. Res.*, 24, 131, 1981.
107. **Chirkov, Y. Y., Tyschuk, I. A., and Severina, I. S.** Guanylate cyclase in human platelets with different aggregability, *Experientia,* 46, 697, 1990.
108. **Dreyfuss, F. and Zahavi, J.,** Adenosine diphosphate induced platelet aggregation in myocardial infarction and ischemic heart disease, *Atherosclerosis,* 17, 107, 1973.
109. **Salky, M. and Dugdale, M.,** Platelet abnormalities in ischemic heart disease, *Am. J. Cardiol.*, 32, 612, 1973.
110. **Guyton, J. P. and Willerson, J. T.,** Peripheral venous platelet aggregates in patients with unstable angina pectoris and acute myocardial infarction, *Angiology,* 28, 695, 1977.
111. **Zahavi, J.,** The role of platelets in myocardial infarction, ischemic heart disease, cerebrovascular disease, thromboembolic disorders and acute idiopathic pericarditis, *Thromb. Haemost.*, 38, 1073, 1977.
112. **Mehta, P. and Mehta, J.,** Platelet function studies in coronary artery disease. V. Evidence for enhanced platelet microthrombus formation activity in acute myocardial infarction, *Am. J. Cardiol.*, 43, 757, 1979.
113. **Chen, Y.-C. and Wu, K. K.,** A comparison of methods for the study of platelet hyperfunction in thromboembolic disorders, *Br. J. Haematol.*, 46, 263, 1981.
114. **Wu, K. K., Barnes, R. W., and Hoak, J. C.,** Platelet hyperaggregability in idiopathic recurrent deep vein thrombosis, *Circulation,* 53, 687, 1976.
115. **Dougherty, J. H., Jr., Levy, D. E., and Weksler, B. B.,** Platelet activation in acute cerebral ischemia, *Lancet,* 1, 821, 1977.
116. **Szczeklik, A. and Musical, J.,** Platelets and ischemic heart disease, in *Platelets: Physiology and Pharmacology,* Longenecker, G. L., Ed., Academic Press, Orlando, 1985, 407.
117. **Mehta, J. and Mehta, P.,** Effects of propranolol therapy on platelet release and prostaglandin generation in patients with coronary heart disease, *Circulation,* 66, 1294, 1982.
118. **Hirsh, P. D., Hillis, L. D., Campbell, W. B., Firth, B. G., and Willerson, J. T.,** Release of prostaglandins and thromboxane into the coronary circulation in patients with ischemic heart disease, *N. Engl. J. Med.*, 304, 685, 1979.
119. **Poplawski, A., Skorulska, M., and Niewiarowski, S.,** Increased platelet adhesiveness in hypertensive cardiovascular disease, *J. Athetoscler. Res.*, 8, 721, 1968.

120. **Hamet, P., Tremblay, J., and Sugimoto, H.** Platelets in hypertension and peripheral vascular disease, in *Platelets: Physiology and Pharmacology*, Longenecker, G. L., Ed., Academic Press, Orlando, 1985, 367.
121. **Saunders, R. N., Burns, T. S., and Rozek, L. F.**, Platelet hyperactivity precedes hypertension in spontaneous hypertensive rats, *Pharmacologist*, 19, 212, 1977.
122. **Olcott, C. and Wylie, E. J.**, Platelet aggregation in patients with severe atherosclerosis, *J. Surg. Res.*, 24, 343, 1978.
123. **Vlachakis, N. D. and Aledort, L.**, Hypertension and propranolol therapy: effect on blood pressure, plasma catecholamines and platelet aggregation, *Am. J. Cardiol.*, 45, 321, 1980.
124. **Mehta, J. and Mehta, P.**, Platelet function in hypertension and effect of therapy, *Am. J. Cardiol.*, 47, 331, 1981.
125. **Kalendovsky, Z. and Austen, J. H.**, Complicated migraine: its association with increased platelet aggregability and abnormal plasma coagulability factors, *Headache*, 15, 18, 1975.
126. **Kuhlmann, U., Steurer, J., Rhyner, K., von Felten, A., Briner, J., and Seigenthaler, W.**, Platelet aggregation and β-thromboglobulin levels in nephrotic patients with and without thrombosis, *J. Lab. Clin. Med.*, 95, 679, 1981.
127. **Stuart, M. J.**, Inherited defects of platelet function, *Semin. Hematol.*, 12, 233, 1975.
128. **Lusher, J. M. and Barnhart, M. I.**, Congenital disorders affecting platelets, *Semin. Thromb. Haemost.*, 4(2), 123, 1977.
129. **Weiss, H. J.**, Congenital disorders of platelet functions, *Semin. Hematol.*, 17, 228, 1980.
130. **Triplett, D.**, Platelet disorders, in *Basic Concepts of Hemostasis and Thrombosis*, Murano, G. and Bick, R. L., Eds., CRC Press, Boca Raton, FL, 1980, 95.
131. **Hardisty, R. M.**, Disorders of platelet function, *Br. Med. Bull.*, 33, 207, 1977.
132. **Malpass, T. W. and Harker, L. A.**, Acquired disorders of platelet function, *Semin. Haematol.*, 17, 242, 1980.
133. **Cowan, D. H.**, Acquired disorders of platelet function, in *Hemostasis and Thrombosis. Basic Principles and Clinical Practice*, Colman, R. W., Hirsh, J., Marder, V. J., and Salzman, E. W., Eds., J. B. Lippincott, Philadelphia, 1982, 516.
134. **Grottum, K. A., Holmsen, H., Abrahamsen, M., Jeremic, M., and Seip, M.**, Wiskott-Aldrich syndrome: qualitative platelet defects and short platelet survival, *Br. J. Haematol.*, 17, 373, 1969.
135. **Baldini, M. G.**, Nature of the platelet defect in the Wiskott-Aldrich syndrome, *Ann. N.Y. Acad. Sci.*, 201, 437, 1972.
136. **Bernard, J. and Soulier, J.-P.**, Sur une nouvelle variete de dystrophie thrombocytaire hemorragipare congenitale, *Sem. Hop. Paris*, 24, 3217, 1948.
137. **George, J. N. and Reimann, T. A.**, Inherited disorders of the platelet membrane: Glanzmann's thrombasthenia and Bernard-Soulier syndrome, in *Hemostasis and Thrombosis. Basic Principles and Clinical Practice*, Colman, R. W., Hirsh, J., Marder, V. J., and Salzman, E. W., Eds., J. B. Lippincott, Philadelphia, 1982, 496.
138. **Holmsen, H. and Weiss, H. J.**, Further evidence for a deficient storage pool of adenine nucleotides in platelets from some patients with thrombocytopathia — "storage pool disease", *Blood*, 39, 197, 1972.
139. **Lages, B., Holmsen, H., Danglemaier, C., and Weiss, H. J.**, Platelet secretion in storage pool deficiency (SPD): response to thrombin and A23187, *Thromb. Haemost.*, 42, 194, 1979.
140. **Weiss, H. J., Witte, L. D., Kaplan, K. L., Lages, B. A., Chernoff, A., Nossel, H. L., Goodman, D. S., and Baumgartner, H. R.**, Heterogeneity in storage pool deficiency: studies on granule-bound substances in 18 patients including variants deficient in α-granules, platelet factor 4, β-thromboglobulin, and platelet-derived growth factor, *Blood*, 54, 1296, 1979.
141. **Malmsten, C., Hamberg, M., Svensson, J., and Samuelsson, B.**, Physiological role of an endoperoxide in human platelets: hemostatic defect due to platelet cyclo-oxygenase deficiency, *Proc. Natl. Acad. Sci. U.S.A.*, 72, 1446, 1975.
142. **Weiss, H. J. and Lages, B.**, Platelet malonadialdehyde production and aggregation responses induced by arachidonate, prostaglandin G_2, collagen, and epinephrine in 12 patients with storage pool deficiency, *Blood*, 58, 27, 1981.
143. **Ulutin, O. N. and Ulutin, S. B.**, Acquired storage pool deficiency of platelets in chronic disseminated intravascular coagulation, Excepta Medica Internat. Cong. Ser. (357), Proc. Intl. Symp. Blood Platelets, Istanbul, Turkey, 1974, 329.
144. **Weinfield, A., Branehog, I., and Kutti, J.**, Platelets in the myeloproliferative syndrome, *Clin. Haematol.*, 4, 373, 1975.

145. **Pareti, F. I., Gugliotta, L., Mannucci, L., Guarini, A., and Mannucci, P. M.,** Biochemical and metabolic aspects of platelet dysfunction in chronic myeloproliferative disorders, *Thromb. Haemost.*, 47, 84, 1982.
146. **Didisheim, P. and Fuster, V.,** Actions of clinical status of platelet-suppressive agents, *Semin. Haematol.*, 15, 55, 1978.
147. **Packham, M. A. and Mustard, J. F.,** Pharmacology of platelet-affecting drugs, *Circulation* 62 (Suppl. 5), 26, 1980.
148. **Morris, L., Distenfeld, A., Amorosi, E., and Karpatkin, S.,** Autoimmune thrombocytopenic purpura in homosexual men, *Ann. Intern. Med.*, 96, 714, 1982.
149. **Leaf, A. N., Laubenstein, L. J., Raphael, B., Hochster, H., Baez, L., and Karpatkin, S.,** Thrombotic thrombocytopenic purpura associated with human immunodeficiency virus type 1 (HIV-1) infection, *Ann. Intern. Med.*, 109, 194, 1987.
150. **Weiss, H. J.,** Antiplatelet drugs: pharmacologic aspects, in *Platelets Pathophysiology and Antiplatelet Drug Therapy*, Alan R. Liss, New York, 1982, 45.
151. **Fuster, V., Badimon, L., Badimon, J., Adams, P. C., Turitto, V., and Chesebro, J. H.,** Drugs interfering with platelet functions: mechanisms and clinical relevance, in *Thrombosis and Haemostasis*, Verstraete, M., Vermylen, J., Lijnen, R., and Arnout, J., Eds., Leuven University Press, Leuven (Belgium), 1987, 349.
152. **Vane, J. R.,** Inhibition of prostaglandin synthesis as a mechanism of action by aspirin-like drugs, *Nat. N. Biol.*, 231, 232, 1971.
153. **Roth, G. S. and Siok, C. J.,** Acetylation of the NH_2-terminal serine of prostaglandin synthetase by aspirin, *J. Biol. Chem.*, 253, 3782, 1978.
154. **Roth, G. L. and Majerus, P. W.,** The mechanism of the effect of aspirin on human platelets. I. Acetylation of particulate fraction protein, *J. Clin. Invest.*, 56, 624, 1975.
155. **De La Cruz, J. P., Camara, S., Bellido, I., Carrasco, T., Sanchez, F., and De La Cuesta, S.,** Platelet aggregation in human whole blood after chronic administration of aspirin, *Thromb. Res.*, 46, 133, 1987.
156. **Lekstrom, J. A. and Bell, W. R.,** Aspirin in the prevention of thrombosis, *Medicine*, 70, 161, 1991.
157. **The Dutch TIA Trial Study Group,** A comparison of two doses of aspirin (30 mg vs. 283 mg a day) in patients after a transient ischemic attack or minor ischemic stroke, *N. Engl. J. Med.*, 325, 1261, 1991.
158. **Minkes, M., Standord, N., Chi, M. M.-Y., Roth, G. J., Raz, A., Needleman, P., and Majerus, P. W.,** Cyclic adenosine 3′,5′-monophosphate inhibits the availability of arachidonate to prostaglandin synthetase in human platelet suspensions, *J. Clin. Invest.*, 59, 449, 1977.
159. **Essien, E. M. and Mustard, J. F.,** Inhibition of platelet adhesion to rabbit aorta by sulphinpyrazone and acetylsalicylic acid, *Atherosclerosis*, 27, 89, 1977.
160. **Moncada, S., Bunting, S., Mullane, K., Throgood, P., Vane, J. R., Raz, A., and Needleman, P.,** Imidazole: a selective inhibitor of thromboxane synthetase, *Prostaglandins*, 13, 611, 1977.
161. **Tai, H. H. and Yuan, B.,** On the inhibitory potency of imidazole and its derivatives on thromboxane synthetase, *Biochem. Biophys. Res. Commun.*, 80, 236, 1978.
162. **FitzGerald, G. A., Reilly, I. A., and Pederson, A. K.,** The biochemical pharmacology of thromboxane synthetase inhibition in man, *Circulation*, 72, 1194, 1985.
163. **Mills, D. C. B. and Macfarlane, D. E.,** Stimulation of human platelet adenylcyclase by prostaglandin D_2, *Thromb. Res.*, 5, 401, 1974.
164. **Bertele, V., Falanga, A., Tomasiak, M., Cerletti, C., and de Gaetano, G.,** SQ 22536, an adenylate cyclase inhibitor, prevents the antiplatelet effect of dazoxiben, a thromboxane synthetase inhibitor, *Thromb. Haemost.*, 51, 125, 1984.
165. **Smith, J. B.,** Effect of thromboxane synthetase inhibitors on platelet function: enhancement by inhibition of phosphodiesterase, *Thromb. Res.*, 28, 477, 1982.
166. **Rajtar, G., Cerletti, C., Castagnoli, M. N., Bertele, V., and De Gaetano, G.,** Prostaglandins and human platelet aggregation: implications for the anti-aggregating activity of thromboxane synthetase inhibitors, *Biochem. Pharmacol.*, 34, 307, 1985.
167. **Agarwal, K. C., Kay, K., Erickson, B. R., and Parks, R. E., Jr.,** Potentiation of forskolin inhibition on platelet aggregation by the inhibitors of cAMP phosphodiesterase and thromboxane synthetase, *Fed. Proc.*, 44, 1665, 1985.
168. **Thomas, M. and Lumley, P.,** Preliminary assessment of a novel thromboxane synthetase A_2 receptor-blocking drug, GR32191, in healthy subjects, *Circulation*, 81, I-53, 1990.
169. **Coleman, R. A., Collington, E. W., Geisow, H. P., Hornby, E. J., Humphery, P. P. A., Kennedy, I., Levy, G. P., Lumley, P., McCabe, P. J., and Wallis, C. J.,** AH19437, a specific thromboxane receptor blocking drug?, *Br. J. Pharmacol.*, 72, 524P, 1972.

170. **Armstrong, R. A., Jones, R. L., Peesapati, V., Will, S. G., and Wilson, N. H.** Competitive antagonism at thromboxane receptors in human platelets, *Br. J. Pharmacol.*, 84, 595, 1985.
171. **Stegmeier, K., Pill, J., Muller-Beckmann, B., Schmidt, F. H., Witte, E. C., Wolff, H.-P., and Patscheke, H.**, The pharmacological profile of thromboxane A_2 antagonist BM 13,177, *Thromb. Res.*, 35, 379, 1984.
172. **Lumley, P., White, B. P., and Humphrey, PPA.** GR32191, a highly potent and specific thromboxane A_2 receptor blocking drug on platelets, vascular and airway smooth muscle in *in vitro, Br. J. Pharmacol.*, 97, 783, 1989.
173. **Gresele, P., Houtte, E. V., Arnout, J., and Vermylen, J.**, Thromboxane synthetase inhibition combined with thromboxane receptor blockade: a step forward in antithrombotic strategy, *Thromb. Haemost.*, 52, 364, 1984.
174. **Dyerberg, J., Bang, H. O., Stofferson, E., Moncada, S., and Vane, J. R.**, Eicosapentaenoic acid and prevention of thrombosis and atherosclerosis, *Lancet*, II, 117, 1978.
175. **Schacky, C. V. and Weber, P. C.**, Metabolism and effects on platelet function of the purified eicosapentaenoic and docosahexaenoic acids in humans, *J. Clin. Invest.*, 76, 2446, 1985.
176. **Knapp, H. R., Reilly, I. A. G., Alessandrini, P., and FitzGerald, G. A.**, In vivo indexes of platelet and vascular function during fish-oil administration in patients with atherosclerosis, *N. Engl. J. Med.*, 314, 937, 1986.
177. **Whitaker, M. O., Wyche, A., Fitzpatrick, F., Sprecher, H., and Needleman, P.**, Triene prostaglandins: prostaglandin D_3 and icosapentaenoic acid as potential antithrombotic substances, *Proc. Natl. Acad. Sci. U.S.A.*, 76, 5919, 1979.
178. **Rao, G. H. R., Radha, E., and White, J. G.**, Effect of docosahexaenoic acid (DHA) on arachidonic acid metabolism and platelet function, *Biochem. Biophys. Res. Commun.*, 117, 549, 1983.
179. **Lands, W. E. M., Culp, B. R., Hirai, A., and Gorman, R.**, Relationship of thromboxane generation to the aggregation of platelets from humans: effects of eicosapentaenoic acid, *Prostaglandins*, 30, 819, 1985.
180. **Needleman, P., Raz, A., Minkes, M. S., Ferrendelli, J. A., and Sprecher, H.**, Triene prostaglandins: prostacyclin and thromboxane biosynthesis and unique biological properties, *Proc. Natl. Acad. Sci., U.S.A.*, 76, 944, 1979.
181. **Mills, D. C. B. and Macfarlane, D. E.**, Platelet receptors, in *Platelets in Biology and Pathology 1*, Gordon, J. L., Ed., Elsevier/North-Holland Biomedical Press, New York, 1976, 159.
182. **Macintyre, D. E.**, Platelet prostaglandin receptors, in *Platelets in Biology and Pathology 2*, Gordon, J. L., Ed., Elsevier/North-Holland Biomedical Press, New York, 1981, 211.
183. **Haslam, R. J. and Cusack, N. J.**, Blood platelet receptors for ADP and for adenosine, in *Purinergic Receptors*, Burnstock, G., Ed., Chapman Hall, New York, 1981, 223.
184. **Mills, D. C. B.**, Platelet aggregation and the adenylate cyclase system, in *Platelets and Thrombosis*, Mills, D. C. B. and Pareti, F. I., Eds., Academic Press, New York, 1977, 63.
185. **Whittle, B. J. R., Moncada, S., and Vane, J. R.**, Comparison of the effects of prostacyclin (PGI_2), prostaglandin E_1 and D_2 on platelet aggregation in different species, *Prostaglandins*, 16, 373, 1978.
186. **DiMinno, G., Silver, M. J., and DeGaetano, G.**, Prostaglandins as inhibitors of human platelet aggregation, *Br. J. Haematol.*, 43, 637, 1979.
187. **Gorman, R. R., Bunting, S., and Miller, O. V.** Modulation of human platelet adenylate cyclase by prostacyclin (PGX), *Prostaglandins*, 13, 377, 1977.
188. **Sollevi, A., Torssell, L., Owall, A., Edlund, A., and Lagerkranser, M.**, Levels and cardiovascular effects of adenosine in humans, in *Topics and Perspectives in Adenosine Research*, Gerlach, E. and Baker, B. F., Eds. Springer-Verlag, Heidelberg, 1987, 559.
189. **Agarwal, K. C.**, Modulation of platelet functions by plasma adenosine levels, in *Role of Adenosine and Adenine Nucleotides in the Biological System*, Imai, S. and Nakazawa, M., Eds. Elsevier Science Publishers, Amsterdam, 1991, 457.
190. **Machin, S. J., Chamone, D. A. F., Defreyn, G., and Vermylen, J.**, The effect of clinical prostacyclin infusions in advanced arterial disease on platelet function and plasma 6-keto PGF_{1a} levels, *Br. J. Haematol.*, 47, 413, 1981.
191. **Nees, S. and Gerlach, E.**, Adenine nucleotides and adenosine metabolism in cultured coronary endothelial cells: formation and release of adenine compounds and possible functional implications, in *Regulatory Functions of Adenosine*, Berne, R. M., Rall, T. W., and Rubio, R., Eds., Martinus Nijhoff, The Hague, 1983, 347.
192. **Moncada, S. and Vane, J. R.**, Discovery, biological significance, and therapeutic potential of prostacyclin, in *Clinical Pharmacology of Prostacyclin*, Lewis, P. J. and O'Grady, J., Eds., Raven Press, New York, 1981, 1.

193. **Weksler, B. B., Marcus, A. J., and Jaffe, E. A.,** Synthesis of prostaglandin I_2 (prostacyclin) by cultured human and bovine endothelial cells, *Proc. Natl. Acad. Sci. U.S.A.*, 74, 3922, 1977.
194. **Cliveden, P. B. and Salzman, E. W.,** Platelet metabolism and the effects of drugs, in *Hemostasis and Thrombosis*, Bowie, E. J. W. and Sharp, A. A., Eds., Butterworths, London, 1985, 1.
195. **Moncada, S. and Vane, J. R.,** Prostacyclin, platelet aggregation and thrombosis, in *Platelets: A Multidisciplinary Approach*, de Gaetano, G. and Garattini, S., Eds., Raven Press, New York, 1978, 239.
196. **Jacobson, K. A.,** Adenosine (P1) and (P2) receptors, in *Comprehensive Medical Chemistry*, Vol. 3, Emmett, J. C., Ed., Pergamon Press, New York, 1990, 601.
197. **Williams, M.,** Adenosine receptors as drug targets: fulfilling the promise?, in *Purines in Cellular Signaling: Targets for New Drugs*, Jacobson, K. A., Daly, J. W., and Manganiello, V., Eds., Springer-Verlag, New York, 1990, 174.
198. **Cucuianu, M. P., Nishizawa, E. E., and Mustard, J. F.,** Effect of pyrimidopyrimidine compounds on platelet function, *J. Lab. Clin. Med.*, 77, 958, 1971.
199. **Meanwell, A.,** Inhibitors of platelet phosphodiesterase: potential antithrombotic agents, *Drug News Perspect.*, 4, 400, 1991.
200. **Murray, K. J., England, P. J., Hallam, T. J., Maguire, J., Moores, K., Reeves, M. L., Simpson, A. W. M., and Rink, T. J.,** The effect of siguazodan, a selective phosphodiesterase inhibitor, on platelet function, *Br. J. Pharmacol.*, 99, 612, 1990.
201. **Agarwal, K. C., Buckley, R. S., and Parks, R. E., Jr.,** Role of plasma adenosine in the antiplatelet action of HL 725, a potent inhibitor of cAMP phosphodiesterase: species differences, *Thromb. Res.*, 47, 191, 1987.
202. **Stafford, A.,** Potentiation of adenosine and adenine nucleotides by dipyridamole, *Br. J. Pharmacol.*, 28, 218, 1966.
203. **Asano, T., Ochiai, Y., and Hidaka, H.,** Selective inhibition of separated forms of human platelet cyclic nucleotide phosphodiesterase by platelet aggregation inhibitors, *Mol. Pharmacol.*, 13, 400, 1977.
204. **Paterson, A. R. P., Jakobs, B. S., Harley, E. R., Fu, N. U., Robins, M. J., and Cass, C. E.,** Inhibition of nucleoside transport, in *Regulatory Functions of Adenosine*, Berne, R. M., Rall, T. W., and Rubio, R., Eds., Martinus Nijhoff, The Hague, 1983, 203.
205. **El-Gamel, S., Wollert, U., and Muller, W. E.,** Optical studies on the specific interaction of dipyridamole with α-acid glycoprotein (orosomucoid), *J. Pharm. Pharmacol.*, 34, 152, 1981.
206. **Agarwal, K. C. and Parks, R. E., Jr.,** Adenosine analogs and human platelets. II. Inhibition of ADP-induced aggregation by carbocyclic adenosine and imidazole ring modified analogs: significance of alterations of the nucleotide pools, *Biochem. Pharmacol.*, 28, 501, 1979.
207. **Klabunde, R. E.,** Dipyridamole inhibition of adenosine metabolism in human blood, *Eur. J. Pharmacol.*, 93, 21, 1983.
208. **Dawicki, D. D., Agarwal, K. C., and Parks, R. E., Jr.,** Adenosine metabolism in human whole blood. Effects of nucleoside transport inhibitors and phosphate concentration, *Biochem. Pharmacol.*, 37, 621, 1988.
209. **Nees, S. and Gerlach, E.,** Adenine nucleotide and adenosine metabolism in cultured coronary endothelial cells, in *Regulatory Functions of Adenosine*, Berne, R. M., Rall, T. W., and Rubio, R., Eds., Martinus Nijhoff, The Hague, 1983, 347.
210. **Berne, R. M., Winn, H. R., Knabb, R. M., Ely, S. W., and Rubio, R.,** Blood flow regulation by adenosine in heart, brain and skeletal muscle, in *Regulatory Functions of Adenosine*, Berne, R. M., Rall, T. W., and Rubio, R., Eds., Martinus Nijhoff, The Hague, 1983, 293.
211. **Dawicki, D. D., Agarwal, K. C., and Parks, R. E., Jr.,** Role of adenosine uptake and metabolism by blood cells in the antiplatelet actions of dipyridamole, dilazep and nitrobenzylthioinosine, *Biochem. Pharmacol.*, 34, 3965, 1985.
212. **Gresele, P., Arnout, J., Deckmyn, H., and Vermylen, J.,** Mechanism of the antiplatelet action of the dipyridamole in whole blood: modulation of adenosine concentration and activity, *Thomb. Haemost.*, 55, 12, 1986.
213. **Dawicki, D. D., Agarwal, K. C., and Parks, R. E., Jr.,** Potentiation of the antiplatelet action of adenosine in whole blood by dipyridamole or dilazep and the cAMP phosphodiesterase inhibitor, RA233, *Thromb. Res.*, 43, 161, 1986.
214. **Best, L. C., McGuire, M. B., Jones, P. B. B., Holland, T. K., Martin, T. J., Preston, F. E., Segal, D. S., and Russel, R. G. G.,** Mode of action of dipyridamole on human platelets, *Thromb. Res.*, 16, 367, 1979.

215. **Lugnier, C., Schoeffter, P., Bec, A. L., Strouthou, E., and Stoclet, J. C.,** Selective inhibition cyclic nucleotide phosphodiesterases of human, bovine and rat aorta, *Biochem. Pharmacol.*, 35, 1743, 1986.
216. **Bult, H., Fret, H. R. L., Jordaens, F. H., and Herman, A. G.,** Dipyridamole potentiates platelet inhibition by nitric oxide, *Thromb. Haemost.*, 66, 343, 1991.
217. **Ashida, S. I. and Abiko, Y.,** Inhibition of platelet aggregation by a new agent, ticlopidine, *Thromb. Haemost.*, 40, 542, 1978.
218. **Keraly, C. L., Delautier, D., Delabassee, D., Chignard, M., and Benveniste, J.,** Inhibition by ticlopidine of PAF-acether-induced in vitro aggregation of rabbit and human platelets, *Thromb. Res.*, 34, 463, 1984.
219. **Saltiel, E. and Ward, A.,** Ticlopidine: a review of its pharmacodynamic and pharmacokinetic properties, and therapeutic efficacy in platelet-dependent disease states, *Drugs*, 34, 222, 1987.
220. **Barquet, G. J., Agarwal, K. C., and Parks, R. E., Jr.,** Antiplatelet activity of ticlopidine: new insights into the mechanism of action, *Pharmacologist*, 34, 141, 1992.
221. **Chignard, M., Keraly, C. L., Delautier, D., Sebag, C., Motte, G., and Benveniste, J.,** Reduced sensitivity of human platelets to PAF-acether following ticlopidine intake, *Haemostasis*, 19, 213, 1989.
222. **Lecrubier, C., Conard, J., Samama, M., and Bousser, M. G.,** Essai randomise d'um nouval agent antiagregant: la ticlopidine, *Therapie*, 32, 189, 1977.
223. **Gent, M., Blakely, J. A., Easton, J. D., Ellis, D. J., Hachinski, V. C., Harbison, J. W., Panak, E., Roberts, R. S., Sicurella, J., and Turpie, A. G.,** The Canadian American Ticlopidine Study (CATS) in thromboembolic stroke, *Lancet*, 1, 1215, 1989.
224. **Hass, W. K., Easton, J. D., Adams, H. P., Jr., Pryse-Phillips, W., Molony, B. A., Anderson, S., and Kamm, B.,** A randomized trial comparing ticlopidine hydrochloride with aspirin for prevention of stroke in high-risk patients, *N. Engl. J. Med.*, 321, 501, 1989.
225. **Thebault, J. J., Blatrix, C. E., Blanchard, J. F., and Panak, E. A.,** Effects of ticlopidine, a new platelet aggregation inhibitor in man, *Clin. Pharmacol. Ther.*, 18, 486, 1975.
226. **David, J.-L., Monfort, F., Herion, F., and Raskinet, R.,** Compared effects of three dose-levels of ticlopidine on platelet function in normal subjects, *Thromb. Res.*, 16, 663, 1979.
227. **Azuma, H., Ishikawa, M., and Sekizaka, S.,** Endothelium dependent inhibition of platelet aggregation, *Br. J. Pharmacol.*, 88, 411, 1986.
228. **Palmer, R. M. J., Ashton, D. S., and Moncada, S.,** Vascular endothelial cells synthesize nitric oxide from L-arginine, *Nature*, 333, 664, 1988.
229. **Ignarro, L. J., Buga, G. M., Wood, K. S., Byrns, R. E., and Chaudhuri, G.,** Endothelium-derived relaxing factor produced and released from artery and vein is nitric oxide, *Proc. Natl. Acad. Sci. U.S.A.*, 84, 9265, 1987.
230. **Mellion, B. T., Ignarro, L. J., Ohlstein, E. H., Pontecorvo, E. G., Hyman, A. L., and Kadowitz, P. J.,** Evidence for the inhibitory role of guanosine 3′, 5′-monophosphate in ADP-induced human platelet aggregation in the presence of nitric oxide and related vasodilators, *Blood*, 57, 946, 1981.
231. **Loscalzo, J.,** Antiplatelet and antithrombotic effects of organic nitrates, *Am. J. Cardiol.*, 70, 18B, 1992.
232. **Moncada, S., Palmer, R. M. J., and Higgs, E. A.,** Nitric oxide: physiology, pathology and pharmacology, *Pharmacol. Rev.*, 43, 109, 1991.
233. **Radomski, M. W., Palmer, R. M. J., and Moncada, S.,** An L-arginine/nitric oxide pathway present in human platelets regulates aggregation, *Proc. Natl. Acad. Sci. U.S.A.*, 87, 5193, 1990.
234. **Addonizio, V. P., Fisher, C., Wachtfogel, Y., Colman, R. W., Edmunds, L. H., and Josephson, M.,** Verapamil and diltiazem as antiplatelet agents: spectrum and mechanisms of activity, *J. Clin. Res.*, 31, 307, 1983.
235. **Johnsson, H.** Effects of nifedipine on platelet function *in vitro* and *in vivo*, *Thromb. Res.*, 21, 523, 1981.
236. **Mehta, J. L., Mehta, P., and Ostrowski, N.,** Calcium blocker diltiazem inhibits platelet activation and stimulates vascular prostacyclin synthesis, *Am. J. Med. Sci.*, 291, 20, 1986.
237. **Weiss, K., Fitscha, P., Virgolini, I., O'Grady, J., and Sinzinger, H.,** Influence of single dose of calcium channel blockers on platelet function, *Platelets*, 2, 41, 1991.
238. **Khan, S. N., Oppong, S., Conroy, D. M., Smith, A. D., and Zilkha, K. J.,** Synergistic effect of PGE_1 and calcium antagonists verapamilin disaggregation of aggregated platelets and inhibition of platelet aggregation induced by PAF-acether or arachidonic acid, *Prog. Lipid Res.*, 25, 295, 1986.

Chapter 15

Platelet Adhesion

Manfred Steiner

CONTENTS

I. Introduction 307
II. The Platelet Shape Change: The First Step on the Road to a Platelet Plug 307
III. Activation of Platelet Glycoprotein IIb/IIIa: The Aggregation Protein Complex But Also an Important Mediator of Adhesion 308
IV. Leucine-Rich Glycoproteins: Important Mediators of Platelet Adhesion 310
V. Platelet Adhesion Without Stimulation: Some Receptors are Always Ready to Interact with Adhesive Surfaces 311
VI. Platelet Adhesion: The Effect of Flow Dynamics 312
VII. Conclusions 312
References 312

I. INTRODUCTION

Adhesion is the primary functional manifestation of platelets. These cellular elements of the hemostatic system are the first to become activated when this complex system of cells, clotting factors, and their inhibitors are called upon to stem the flow of blood from a vessel that has lost its continuity. The interaction of platelets with specific elements of the vessel wall represents the first event that in time leads to formation of a firm clot composed of a fibrin meshwork intimately linked to platelets and intermingled to varying degrees with red cells. Platelets normally circulate in blood as small, oblate ellipsoids that shun interaction with each other and are shielded from adhesive surfaces by the continuum of the endothelial surface lining the blood vessels. Contact with certain agonists, however, induces a profound morphological and ultrastructural reorganization, which also renders the surface of the platelet sticky and ready to interact with adhesive proteins. This altered reactivity of platelet membranes, as well as their change in shape from a discoid structure to a spiny sphere, can also be induced by shear forces of varying strength. The latter represents an exceedingly important mechanism of platelet activation, especially under pathological circumstances where flow alterations due to inhomogeneities in the wall may be a prominent feature.

Platelet adhesion can be studied in a system that allows direct observation of their interaction with adhesive surfaces under varying flow conditions. The availability of such analytic methods facilitated not only the identification of significant physiological factors but also permitted evaluation of certain inhibitors of this platelet function.

This chapter is organized so that the reader will be able to follow the sequential steps of platelet activation up to the point of interaction of specific membrane receptors with their adhesive ligands. Because details of the physical and biochemical structure of some of these receptors have been given in other chapters of this book, we shall limit our remarks to those features that are uniquely important to platelet adhesion. Thus, integrins and LECAMs will not be discussed to the same extent as receptors such as the gpIb/IX complex.

II. THE PLATELET SHAPE CHANGE: THE FIRST STEP ON THE ROAD TO A PLATELET PLUG

The interaction of specific agonists with their respective receptor sites on the platelet surface as well as shear stress transform the inactive discoid platelet into a spherical body from which long, thin pseudopodia protrude. This morphological alteration takes place in less than 1 s and is associated with profound changes in the platelet cytoskeleton, the distribution of subcellular granules, as well as the conformation of certain platelet glycoproteins. Stopped-flow experiments[1-4] have resolved this rapid activation event into at least one intermediate stage prior to its final activation stage. Gear has demonstrated that the shape

0-8493-4559-6/95/$0.00+$.50

change of the spiny sphere character is preceded by protrusion of short, broad-based pseudopodia that extend only fractions of the distance that the long spiny ones of the completed shape alteration do.[1]

The extrusion of pseudopodial processes is probably preceded by an increase in intracellular pH, which apparently is due to an activation of the Na^+/H^+ exchange mechanism.[5] A number of investigators have pointed out the importance of this pump mechanism for normal platelet activation, including shape change, indomethacin-sensitive secretion, and calcium mobilization.[6-8] The studies of Siffert et al., of Leven et al., and others have clearly shown the importance of Na^+ and pH_i for the platelet shape change. Nevertheless, none of the studies correlating changes in pH_i and Na flux to the various parameters of platelet activation were performed by the use of a stopped-flow approach. As the time scale for these events is extremely short, uncertainty about the exact role of the Na^+/H^+ exchange mechanism for platelet activation will remain. Slowing the reaction sequence by reducing the temperature has been tried by Zavoico and Cragoe[9] who found that Ca^{2+} is mobilized from intracellular sources before alkalinization of the cytoplasm occurs. As Ca^{2+} is not a prerequisite for the platelet shape change but is essential for the subsequent activation process, their findings cast doubt on the essential role of Na^+/H^+ exchange in the production of platelet shape change. In view of the cold-induced disassembly of microtubules, which by itself is capable of inducing a platelet "shape change", the significance of their findings remains debatable. As the monovalent cation ionophores monensin and nigericin can cause a platelet shape change[5] and amiloride, a potent sodium transport blocker, is capable of inhibiting such shape change without major alteration of calcium availability,[9] a strong case can be made that the morphological alteration of platelets in response to agonists is at least in part triggered by changes in intracellular pH and Na^+ influx.

III. ACTIVATION OF PLATELET GLYCOPROTEIN IIB/IIIA: THE AGGREGATION PROTEIN COMPLEX BUT ALSO AN IMPORTANT MEDIATOR OF ADHESION

This Ca^{2+}-dependent heterodimer glycoprotein is represented with approximately 50,000 copies in the platelet membrane.[10] As described in other parts of this volume, the glycoprotein complex belongs to a family of cell surface receptors, integrins, that are important mediators of cell-to-cell adhesion in many different cell types. gpIIb, a sulfhydryl-cleavable protein of 140,000 mol wt constitutes the α subunit and the 105,000 mol wt gpIIIa, represents the β subunit of this integrin. Upon reduction, the former readily dissociates into a smaller subunit, $gpIIb_\beta$, of 25,000 mol wt, and a larger portion of the molecule, $gpIIb_\alpha$, of 125,000 mol wt In the presence of Ca^{2+}, it exists as a heterodimer in the platelet surface, but chelation of this cation disassembles the glycoprotein complex into its subunits.[11] gpIIb, the Ca^{2+}-binding portion of the heterodimer, changes its conformation when Ca^{2+} is removed and thus loses its function.[12]

Conformational changes in the gpIIb/IIIa heterodimer occur in response to activation of platelets by a variety of stimuli, such as thrombin, collagen, ADP, and epinephrine.[13] A change in the conformation of this protein complex uncovers cryptic fibrinogen receptors, which thus become available for interaction with this important coagulation factor. There is good evidence that fibrinogen may form the "glue" that holds platelets together once they have been stimulated by agonists. Not only fibrinogen but also fibronectin[14] and von Willebrand factor[15] can bind to this activated gpIIb/IIIa complex. The spreading of platelets upon adherence to a surface may be mediated by fibronectin,[16] which has also been reported to be necessary for platelet thrombus growth under flow conditions.[17] Von Willebrand factor is of eminent importance in the interaction of platelets with subendothelial matrix collagen.[18,19]

Congenital abnormalities of gpIIb/IIIa, known as Glanzman's thrombasthenia,[20] are not only associated with quantitative reductions of this glycoprotein complex,[21] but also functional alterations of normally abundant complexes have been described.[22] The primary abnormality is reduced or even absent aggregability of platelets in response to most agonists, but in addition, platelet adherence has been found to be reduced at high shear rates ($1000\ s^{-1}$).[18] These findings establish the gpIIb/IIIa complex as the major aggregation protein of platelets and in addition show that it functions as an adhesion receptor, which can be induced by shear stress.

The characteristics of integrins have been described elsewhere in this book (see Chapter 8), but suffice it to say that the β subunit of this platelet integrin carries a single transmembrane domain, a large number of cysteines with a major portion of them clustered into four segments tandemly repeated, and at least six potential N-linked glycosylation sites in the extracellular domain. All of the cysteine residues appear to exist as disulfide cysteines. Six areas of high homology have been identified, in which glycoprotein IIIa,[23] the fibronectin receptor β subunit,[24] and the leukocyte β subunit[25] differ only minimally. A potential

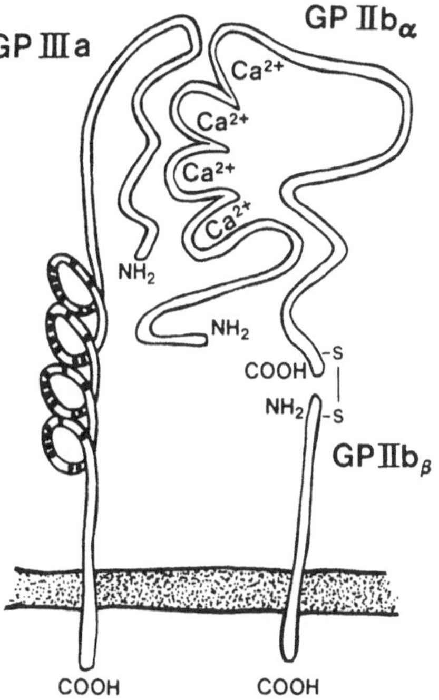

Figure 1 Model of the gpIIb/IIIa complex of platelets. (Reprinted from Phillips, D.R. et al., *Blood,* 71, 831, 1988, with permission.)

tyrosine phosphorylation site exists in the cytoplasmic portion of the molecule in a sequence that shows homology to epidermal growth factor and insulin receptors.

The α subunit, α_{4b} or platelet gpIIb,[26] also has great similarities to other α chains that have been examined in detail.[27,28] They all exhibit one transmembrane domain, multiple calcium binding repeats that have Asp, Asn, and Gly in a spacing characteristic of calcium-binding proteins. Similar to the β chain of this complex, the α chain carries a large number of cysteines, but, different from the conserved nature of these residues that is observed throughout the β chains of the integrins, differences exist among the α chain cysteines. The small chain of gpIIb, gpIIb-β, also carries multiple cysteines, all of which appear to be linked via disulfide bonds. Multiple potential N-linked glycosylation sites are present on the α_2b subunit, although not as many as in other α chains of the integrin family.

Electron microscopic studies[29] have revealed that the gpIIb-IIIa complex consists of a globular domain with two filamentous domains extending from one of its sides, which appear to represent gpIIbβ and the C-terminal end of gpIIIa (Figure 1). The complex associates by itself into rosettes held together by its filamentous regions. In the resting cell, the gpIIb-IIIa complex appears to be randomly distributed, but upon stimulation of the platelet, clustering of receptors takes place.[30] Ligand binding is probably taking place in the globular domain where α and β subunits of the integrin interface. Stereo-electron microscopy has shown the location of fibrinogen and its receptors at sites where platelets interact.[30] In adherent platelets, fibrinogen was found to be distributed on the plasma membrane, primarily over the granulomere and at platelet-platelet contact regions. In addition, free, unoccupied gpIIb-IIIa receptor was localized over the hyalomere where fibrinogen did not bind to any extent. There are a number of possible explanations for this finding. Without going into the detail of such speculations, it is important to realize that the gpIIb-IIIa receptor may exist on the platelet surface in forms that react differently with ligands, suggesting one or more subgroups of this integrin.

Hillery et al.[31] have shown a limited phosphorylation of gpIIIa in intact platelets and a slightly increased phosphorylation in activated platelets. On a mole basis, however, only 1 to 3% of gpIIIa was phosphorylated. Thus, only a small subgroup of this platelet integrin exhibits this posttranslational modification. No functional significance can be ascribed presently to this phosphorylated species. Selective interaction of some subpopulation of this integrin with cytoskeletal structures (for more details see chapter 8) and palmitoylation of a fraction of IIIa[32] further underscores the concept of the existence of discreet populations of gpIIb/IIIa species in the platelet membrane.

gpIIb/IIIa is known to express its fibrinogen-binding site only after activation by agonists such as ADP, epinephrine, or thrombin,[33] which seem to act via a G protein-dependent mechanism that can be activated by the products of phospholipase C or of the arachidonic acid pathway.[34] Other mechanisms

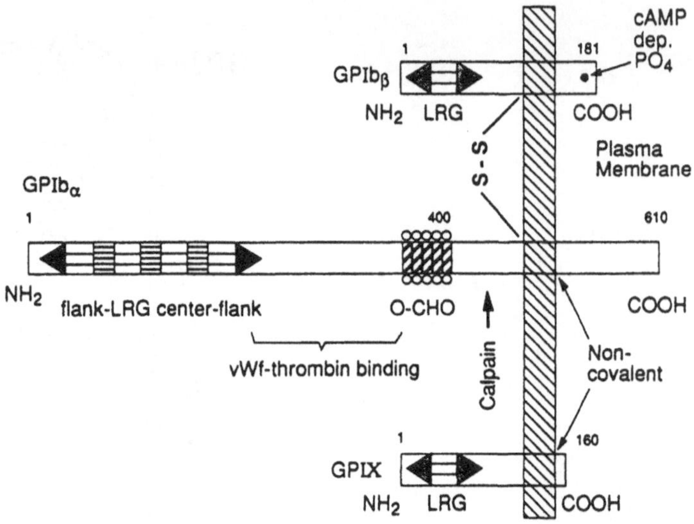

Figure 2 Sketch of the primary structure of the gpIb/IX complex. Proteins are depicted as open bars (▭) with superscripts referring to amino acid number and subscripts to NH_2– and COOH– termini. Transmembrane domains demarcate larger extracellular from smaller intracellular regions extending on either side of the plasma membrane (▨). Disulfide bonds, S–S, link the gpIb α and β chains, while gpIX is noncovalently associated with Ib. Flank-LRG center-flank structures include conserved, 22-amino acid flanking sequences (◀ ▶) on either side of central, 24-amino acid LRG sequences present as one segment (▭, Ib_β, and IX) or as seven tandem repeats (▭, Ib_α). Calpain releases the extracellular portion of $gpIb^{\alpha\prime}$ glycocalicin, with its macroglycopeptide domain marked by five, nine-amino tandem repeats and associated O-linked carbohydrate (▨). The hinge region between the LRG structure and the O-carbohydrate region of Ib_α mediates binding to von Willebrand factor (vWF) and thrombin. $gpIb_\beta$ contains a cAMP-dependent phosphorylation (PO_4) site. (Reprinted from Roth, G.J., *Blood,* 77, 5, 1991, with permission.)

have been suggested but are not clearly defined as yet. In addition to this mechanism of integrin interaction with its respective ligands, the interaction of unstimulated platelets with fibrinogen adsorbed onto glass[35] or polymer beads[36] has been known for many years. A non-RGDS-containing C-terminal peptide of γ-fibrinogen has been shown to bind to unstimulated platelets[37] and could be responsible for this observation. Partial adherence of gpIIb/IIIa to fibronectin and von Willebrand factor has also been reported.[37] Thus, the concept that this platelet integrin is not constitutively active in platelets may have to be revised.

IV. LEUCINE-RICH GLYCOPROTEINS: IMPORTANT MEDIATORS OF PLATELET ADHESION

The glycoprotein complex Ib/IX is the primary receptor for von Willebrand factor (vWF),[38] which mediates the adhesion to collagenous tissue.[39,40] The adhesion mediated by gpIb/von Willebrand factor is shear rate-dependent, requiring shear rates >650 s^{-1}.[41,42]

gpIb is in fact composed of two dissimilar subunits, gpIbα, a large peptide of 143,000 mol wt, and a genomically distinct gpIbβ, a peptide of 22,000 mol wt, which is linked to the larger one via disulfide bridge(s) (Figure 2). Each of the polypeptides participating in the gpIb/IX complex has two hydrophobic domains at the two ends of the molecule, the C- and the N-terminus.[43-45] The most characteristic feature of these peptides is leucine-rich regions consisting of 24 amino acid sequences that are unique among the members of the leucine-rich glycoprotein family. As in other proteins containing such leucine-rich sequences, they are found in a series of tandem repeats in gpIbα and are located near the N-terminal part of the protein. gpIbβ and gpIX carry only one such leucine-rich segment. There is large conservation in the flanking sequences that surround the leucine-rich core repeats among the three participants of the gpIb/IX complex.[38] The intracellular anchors for gpIbβ and gpIX are very limited. An interesting feature is the presence of an unpaired cysteine in the transmembrane domain of gpIX, which is probably the site of fatty acylation binding this protein to the lipids of the membrane. gpIb$_\beta$ is known to have two phosphorylation sites subject to cAMP-dependent phosphorylation, an event that appears to inhibit polymerization of actin and influences the link between the gpIb/IX complex and the platelet cytoskeleton.[46] Calpain, a calcium-dependent protease, releases the extracellular portion of gpIbα.[47] This portion of the molecule, glycocalicin, contains the carbohydrate-binding region of gpIb$_\alpha$ as well as the binding site for vWF, which is located in the hinge region between the leucine-rich repeats and the carbohydrate domain.[48]

The function of the leucine-rich segments is not clear. Since its first discovery in leucine-rich α_2 glycoprotein of human plasma,[49] a wide variety of proteins has been found that carry leucine-rich sequences in a series of tandem repeats.[50-54] Most of these proteins have nothing to do with adhesive events. Whether these segments are in some way involved in the shear-dependent conformational change of gpIb$_\alpha$ that allows it to interact with vWF as suggested by Roth[38] remains to be seen.

A glycoprotein closely associated with the gpIb/IX complex is gpV.[55,56] Similar to the other leucine-rich glycoproteins, at least six leucine-rich repeats have been identified in gpV.[56] Further documentation for the close association of gpV with the gpIb/IX complex can be seen from the concomitant deficiency of all these proteins in Bernard-Soulier's syndrome.[57] This congenital abnormality of platelets results in severe bleeding and defective adhesion and vWF binding. There is, however, no direct physical linkage between this protein and the Ib/IX complex. gpV is a membrane protein that can be released, probably by calpain proteolysis.[55] It consists of a single chain of approximately $M_r = 82{,}000$. Thrombin cleaves gpV and releases a soluble glycoprotein of 69,000 mol wt.[55,56] gpV contains as much as one third of its weight in carbohydrate.[58]

V. PLATELET ADHESION WITHOUT STIMULATION: SOME RECEPTORS ARE ALWAYS READY TO INTERACT WITH ADHESIVE SURFACES

In contrast to the activation-dependent interaction of gpIIb/IIIa or gpIb/IX with their specific ligands, some platelet receptors for adhesive proteins are always ready to interact with their ligands. Such constitutively expressed adhesion receptors are far less numerous than the aforementioned but are essential for the normal interaction of platelets with the exposed subendothelium.

A number of integrins are present on the platelet surface, which have similarity to the VLA (very late activation) receptors of lymphocytes. Among these, gpIc/IIa ($\alpha_F\beta 1$)[59] represents a fibronectin receptor present on the platelet in low abundance. It can react with its ligand in the absence of platelet activation.[60] gpIa/IIa ($\alpha_c\beta 1$),[61,62] a cation-dependent (Mg^{2+}) collagen receptor, and $\alpha_V\beta 3$,[63] the vitronectin receptor of platelets, are all capable of interacting with their ligands without prior platelet activation.

gpIV, a glycoprotein of 88,000 mol wt, is a major platelet membrane glycoprotein,[64] which has been proposed as the receptor for thrombospondin[65] and collagen.[66] The interaction of gp IV with thrombospondin, however, has recently been questioned.[67] Little is known about the biochemical and molecular structure of this protein, but it is known to be identical to the differentiation antigen CD36. It is present in a wide variety of cells. Its primary sequence has been determined from placental cDNA.[68] Absence of this glycoprotein has been reported in certain populations, e.g., Japanese, who have a high percentage of Nak^a-negative phenotype, although they do harbor the corresponding mRNA.[69] Individuals with this abnormality appear to be healthy without obvious hematological abnormalities. They do, however, risk the development of isoantibodies when Nak^{a+} cells are transfused.

Finally, gpVI, a glycoprotein of 62,000 mol wt, has been shown to be necessary to bind platelets to type I and III collagen fibrils.[70] It is interesting that the patient described by Moroi et al.[70] who lacked this protein showed absent collagen-induced aggregation and adhesion. As normal levels of the glycoproteins Ia, Ib, IIa, IIb, IIIa, and IV were found in this patient, all of which have previously been claimed in one form or another to be mediators of collagen adhesion (see above), questions arise about their *in vivo* function as collagen receptors. An antibody in a patient with immune thrombocytopenia was found to react with gpVI and to induce aggregation in normal control but not in gpVI-deficient platelets.[71] Although this provides some support for the concept of gpVI being the collagen receptor of platelets, it does not completely exclude the other glycoproteins as potential collagen receptors, as autoimmune sera often contain more than one antibody. Therefore, definitive proof will have to await immunological studies with a specific gpVI-neutralizing antibody.

VI. PLATELET ADHESION: THE EFFECT OF FLOW DYNAMICS

In vivo, shear stress represents probably the most important activator of platelets. It is, therefore, not surprising that in recent years a considerable effort has been made to examine this effect on platelet aggregation and adhesion. Such studies have been facilitated by the development of appropriate instruments for measuring shear stress-induced aggregation and adhesion.[72] Shear-induced changes in the conformation of gpIIb/IIIa that allow them to bind fibrinogen and lead to platelet aggregation appear to be operative at a relatively low shear force.[73,74] Shear force plays a crucial role in the interaction of vWF with gpIb. At shear forces >80 dyn/cm^2, shear-induced aggregation is dependent on the presence of von Willebrand factor as well as gpIb/IX and gpIIb/IIIa.[73,74] On the other hand, various platelet receptors have been reported to interact with adhesive proteins in the absence of shear or platelet activation (see above). Laminin, fibronectin, and various collagens have all been shown to bind to nonstimulated platelets. This suggests that the reactive sites of the receptors are always exposed and ready to interact with their respective ligands. Such interaction, however, does not preclude an additional effect of shear stress as more reactive sites may become exposed under such conditions.

Considerable importance has been ascribed to the adhesive behavior of platelets under different shear stresses. This is important from a pathophysiological standpoint as platelets are subjected to different shear stresses in different areas of the circulation. Whether platelets adhere to an adhesive surface depends, however, on the relative contribution of two opposing factors, the number of sites that can interact with the adhesive surface and the strength of the force that dislodges platelets from their sites of adhesion. Platelets play a major role in the initiation of thrombotic events in the arterial circulation. This is an area in which high shear rates generally prevail; thus, platelet adhesion has been examined especially under such conditions. This, however, ignores the fact that most of the thromboses in the cerebral and coronary circulation, the two areas most extensively studied, begin in fissures of arteriosclerotic plaques, sites at which flow is almost stagnant and shear rates are consequently low. Therefore, far closer attention should be paid to the adhesiveness of platelets at low than at high shear rates.

VII. CONCLUSIONS

Platelets are unusual in their ability to express adhesiveness upon stimulation by specific agonists. This characteristic of being able to express extreme stickiness so quickly has made them preferred subjects of investigation. Different classes of adhesion receptors are present on the platelet surface, some of which are constitutively in a state where they can interact with ligands, whereas others have to undergo conformational changes before they are able to interact with adhesive surfaces. Some adhesion receptors are expressed only after the platelets undergo major morphological changes as membranes of subcellular granules fuse with the platelet surface. Another important aspect of platelet adhesiveness is the shear stress-induced expression of some adhesion receptors. This is of great pathophysiological importance, as platelets are the initiators of hemostasis. As platelets are readily available and can be obtained in large amounts for functional as well as biochemical study, many adhesion phenomena have been examined in platelets. However, platelets are anuclear cells; therefore, the molecular aspect of synthesis and regulation of adhesion receptors cannot be investigated in them. Nevertheless, we believe that the platelet will remain a favorite model for investigation of adhesion phenomena and testing potential inhibitors. Correlations between *in vitro* and *in vivo* observation relating to platelets are easily made, which is of great advantage in relating cellular functions to clinical situations.

REFERENCES

1. **Deranleau, D. A., Dubler, D., Rothen, C. and Luscher, E. F.,** Transient kinetics of the rapid shape change of unstirred human blood platelets stimulated with ADP, *Proc. Natl. Acad. Sci. U.S.A.*, 79, 7297, 1982.
2. **Gear, A. R.** Rapid reactions of platelets studied by a quenched flow approach: aggregation kinetics, *J. Lab. Clin. Med.*, 100, 866, 1982.
3. **Gear, A. R.** Rapid platelet morphological changes visualized by scanning electron microscopy: kinetics derived from a quenched flow approach, *Br. J. Haematol.*, 56, 387, 1984.
4. **Hantgan, R. R.** A study of the kinetics of ADP-triggered platelet shape change, *Blood,* 64, 896, 1984.

5. **Leven, R. M., Gonella, P. A., Reeber, M. J., and Nachmias, V. T.,** Platelet shape change and cytoskeletal assembly: effects of pH and monovalent cation ionophores, *Thromb. Haemost.*, 49, 230, 1983.
6. **Siffert, W. and Akkerman, J. W. N.,** Activation of sodium-proton exchange is a prerequisite for Ca^{2+} mobilization in human platelets, *Nature*, 325, 456, 1987.
7. **Siffert, W., Muckenhoff, K., and Scheid, P.,** Evidence for a role of Na^+/H^+ exchange in platelets activated with calcium ionophore, *Biochem. Biophys. Res. Commun.*, 125, 1123, 1984.
8. **Connolly, T. M. and Limbird, L. E.,** Removal of extracellular Na^+ indomethacin-sensitive secretion from human platelets stimulated by epinephrine, ADP and thrombin, *Proc. Natl. Acad. Sci. U.S.A.*, 80, 5320, 1983.
9. **Zavoica, G. B. and Cragoe, E. J.,** Ca^{2+} mobilization can occur independent of acceleration of Na^+/H^+ exchange in thrombin stimulated human platelets, *J. Biol. Chem.*, 263, 9635, 1988.
10. **Jennings, L. K. and Phillips, D. R.,** Purification of glycoprotein IIb and IIIa from human platelet membranes and characterization of a calcium-dependent glycoprotein IIb-IIIa complex, *J. Biol. Chem.*, 257, 10458, 1982.
11. **Kunicki, T. J., Pidard, D., Rosa, J. P., and Nurden, A. T.,** The formation of calcium-dependent complexes of platelet membrane glycoproteins IIb-IIIa in solution as determined by crossed immunoelectrophoresis, *Blood*, 58, 268, 1981.
12. **McEver, R. P. and Martin, M. N.,** A monoclonal antibody to a membrane glycoprotein binds only to activated platelets, *J. Biol. Chem.*, 259, 9799, 1984.
13. **Bennett, J. S. and Vilaire, G.,** Exposure of platelet fibrinogen receptors by ADP and epinephrine, *J. Clin. Invest.*, 64, 1393, 1979.
14. **Plow, E. F. and Ginsberg, M. H.,** Specific and saturable binding of plasma fibronectin to thrombin-stimulated platelets, *J. Biol. Chem.*, 256, 9477, 1981.
15. **Plow, E. F., Ginsberg, M. H., and Marguerie, G. A.,** Expression and function of adhesive proteins on the platelet surface, in *Biochemistry of Platelets*, Phillips, D. R. and Shuman, M. A., Eds., Academic Press, San Diego, 1986, 226.
16. **Ill, C. R., Engvall, E., and Ruoslahti, E.,** Adhesion of platelets to laminin in the absence of activation, *J. Cell Biol.*, 99, 2140, 1984.
17. **Bastida, E., Escolar, G., Ordinas, A., and Sixma, J. J.,** Fibronectin is required for platelet adhesion and for thrombus formation on subendothelium and collagen surfaces, *Blood*, 70, 1437, 1987.
18. **Sakariassen, K. S., Nievelstein, P. F. E. M., Coller, B. S., and Sixma, J. J.,** The role of platelet membrane glycoprotein Ib and IIb/IIIa in platelet adherence to human artery subendothelium, *Br. J. Haematol.*, 63, 681, 1986.
19. **Houdijk, W. P. M., Sakariassen, K. S., Nievelstein, P. F. E. M., and Sixma, J. J.,** Role of factor VIII-von Willebrand factor and fibronectin in the interaction of platelets in flowing blood with monomeric and fibrillar human collagen types I and III, *J. Clin. Invest.*, 75, 531, 1985.
20. **Caen, J. P., Castaldi, P. A., Leclerc, J. C., Inceman, S., Larrieu, M. J., Probst, M., and Bernard, J.,** Congenital bleeding disorders with long bleeding time and normal platelet count. I. Glanzmann's thrombasthenia (report of fifteen patients), *Am. J. Med.*, 41, 4, 1966.
21. **Phillips, D. R. and Agin, P. P.,** Platelet membrane defects in Glanzmann's thrombasthenia. Evidence for decreased amounts of two major glycoproteins, *J. Clin. Invest.*, 60, 535, 1977.
22. **Nurden, A. T., Rosa, J. P., Fournier, D., Legrande, C., Didry, D., Parquet, A., and Pidard, D. A.,** Variant of Glanzmann's thrombasthenia with abnormal glycoprotein IIb-IIIa complexes in the platelet membrane, *J. Clin. Invest.*, 79, 962, 1987.
23. **Fitzgerald, L. A., Steiner, B., Rall, S. C., Lo, S. S., and Phillips, D. R.,** Protein sequence of endothelial glycoprotein IIIa derived from a cDNA clone. Identity with platelet glycoprotein IIIa and similarity to "integrin", *J. Biol. Chem.*, 262, 3936, 1987.
24. **Tamkun, J. W., DeSimone, D. W., Fonda, D., Patel, R. S., Buck, C., Horwitz, A. F., and Hynes, R. O.,** Structure of integrin, a glycoprotein involved in the transmembrane linkage between fibronectin and actin, *Cell*, 46, 271, 1986.
25. **Suzuki, S., Argraves, W. S., Pytela, R., Arai, H., Krusius, T., Pierschbacher, M. D., and Ruoslahti, E.,** cDNA and amino acid sequences of the cell adhesion protein receptor recognizing vitronectin reveal a transmembrane domain and homologies with other adhesion protein receptors, *Proc. Natl. Acad. Sci. U.S.A.*, 83, 8614, 1986.

26. **Poncz, M., Eisman, R., Heidenreich, R., Silver, S. M., Vilaire, G., Surrey, S., Schwartz, E., and Bennett, J. S.,** Structure of the platelet membrane glycoprotein IIb: homology to the alpha subunits of the vitronectin and fibronectin membrane receptors, *J. Biol. Chem.*, 262, 8476, 1987.
27. **Argraves, W. S., Susuki, S., Arai, H., Thompson, K., Pierschbacher, M. D., and Ruoslahti, E.,** Amino acid sequence of the human fibronectin receptor, *J. Cell Biol.*, 105, 1183, 1987.
28. **Suzuki, S., Argraves, W. S., Arai, H., Languino, L. R., Pierschbacher, M. D., and Ruoslahti, E.,** Amino acid sequence of the vitronectin receptor α-subunit and comparative expression of adhesion receptor mRNAs, *J. Biol. Chem.*, 262, 14080, 1987.
29. **Carrell, N. A., Fitzgerald, L. A., Steiner, B., Erikson, H. P., and Phillips, D. R.,** Structure of human platelet membrane glycoprotein IIb and IIIa as determined by electron microscopy, *J. Biol. Chem.*, 260, 1743, 1985.
30. **Lewis, J. C., Hantgan, R. R., Stevenson, S. C., Thornburg, T., Kieffer, N., Guichard, J., and Breton-Gorius, J.,** Fibrinogen and glycoprotein IIb/IIIa localization during platelet adhesion, *Am. J. Pathol.*, 136, 239, 1990.
31. **Hillery, C. A., Smyth, S. S., and Parise, L. V.,** Phosphorylation of human platelet glycoprotein IIIa (GPIIIa), *J. Biol. Chem.*, 266, 14663, 1991.
32. **Cierniewski, C. S., Krzeslowska, J., Pawlowska, Z., Witas, H., and Meyer, M.,** Palmitylation of the glycoprotein IIb-IIIa complex in human blood platelets, *J. Biol. Chem.*, 264, 12158, 1989.
33. **Marguerie, G. A., Plow, E. F., and Edgington, T. S.,** Human platelets possess an inducible and saturable receptor specific for fibrinogen, *J. Biol. Chem.*, 254, 5357, 1979.
34. **Shattil, S. J. and Brass, L. F.,** Induction of fibrinogen receptors on human platelets by intracellular mediators, *J. Biol. Chem.*, 262, 992, 1987.
35. **Packham, M. A., Evans, G., Glynn, M. F., and Mustard, J. F.,** The effect of plasma proteins on the interaction of platelets with glass surfaces, *J. Lab. Clin. Med.*, 73, 686, 1969.
36. **Coller, B. S.,** Interaction of normal, thrombasthenic, and Bernard-Soulier platelets with immobilized fibrinogen: defective platelet-fibrinogen interaction in thrombasthenia, *Blood*, 55, 169, 1980.
37. **Savage, B. and Ruggeri, Z. M.,** Selective recognition of adhesive sites in surface-bound fibrinogen by glycoprotein IIb-IIIa on nonactivated platelets, *J. Biol. Chem.*, 266, 11227, 1991.
38. **Roth, G. J.,** Developing relationships: arterial platelet adhesion, glycoprotein Ib and leucine-rich glycoproteins, *Blood*, 77, 5, 1991.
39. **Tschopp, T. B., Weiss, H. J., and Baumgartner, H. R.,** Decreased adhesion of platelets to subendothelium in von Willebrand's disease, *J. Lab. Clin. Med.*, 83, 296, 1974.
40. **Weiss, H. J., Tschopp, T. B., Baumgartner, H. R., Sussman, I. I., Johnson, M. M., and Egan, J. J.,** Decreased adhesion of giant (Bernard-Soulier) platelets to subendothelium, *Am. J. Med.*, 57, 920, 1974.
41. **Baumgartner, H. R.,** The role of blood flow in platelet adhesion, fibrin deposition and formation of mural thrombi, *Microvasc. Res.*, 5, 167, 1973.
42. **Sakariassen, K. S., Bolhuis, P. A., and Sixma, J. J.,** Human blood platelet adhesion to artery subendothelium is mediated by factor VIII/von Willebrand factor bound to subendothelium, *Nature*, 279, 636, 1979.
43. **Lopez, J. A., Chung, D. W., Fujikawa, K., Hagen, F. S., Papayamopoulta, T., and Roth, G. J.,** Cloning of the alpha chain of human platelet glycoprotein Ib: a transmembrane protein with homology to leucine-rich alpha 2 glycoprotein, *Proc. Natl. Acad. Sci. U.S.A.*, 84, 5615, 1987.
44. **Lopez, J. A., Chung, D. W., Fujikawa, K., Hagen, F. S., Davie, E. W., and Roth, G. J.,** The alpha and beta chains of human platelet glycoprotein Ib are both transmembrane proteins containing a leucine-rich amino acid sequence, *Proc. Natl. Acad. Sci. U.S.A.*, 85, 2135, 1988.
45. **Hickey, M. J., Deaven, L. L., and Roth, G. J.,** Human platelet glycoprotein IX: characterization of a full length DNA and localization to chromosome 3, *FEBS Lett.*, 274, 189, 1990.
46. **Wardell, M. R., Reynolds, C. C., Berndt, M. C., Wallace, R. W., and Fox, J. E. B.,** Platelet glycoprotein Ib^{beta} is phosphorylated on serine 166 by cyclic AMP-dependent protein kinase, *J. Biol. Chem.*, 264, 15656, 1989.
47. **Okumura, T., Lombart, C., and Jamieson, G. A.,** Platelet glycocalicin. II. Purification and characterization, *J. Biol. Chem.*, 251, 5950, 1976.
48. **Vicente, V., Houghton, R. A., and Ruggeri, Z. M.,** Identification of a site in the alpha chain of platelet glycoprotein Ib that participates in von Willebrand factor binding, *J. Biol. Chem.*, 265, 274, 1990.

49. **Takehashi, N., Takehashi, Y., and Putnam, F. W.,** Periodicity of leucine and tandem repetition of a 24-amino acid segment in the primary structure of leucine-rich alpha$_2$-glycoprotein of human serum, *Proc. Natl. Acad. Sci. U.S.A.*, 82, 1906, 1985.
50. **Kataoka, T., Broek, D., and Wigler, M.,** DNA sequence and characterization of the *S. cerevisiae* gene encoding adenylate cyclase, *Cell*, 43, 493, 1985.
51. **McFarland, K. C., Sprengel, R., Phillips, H. S., Kohler, M., Rosemblit, N., Nikolics, K., Segaloff, D. L., and Seeburg, P. H.,** Lutropin-choriogonadotropin receptor: an unusual member of the G protein-coupled receptor family, *Science*, 245, 494, 1989.
52. **Hashimoto, C., Hudson, K. L., and Anderson, K. V.,** The *toll* gene of Drosophila, required for dorsal-ventral embryonic polarity, appears to encode a transmembrane protein, *Cell*, 52, 269, 1988.
53. **Reinke, R., Krantz, D. E., Yen, D., and Zipursky, S. L.,** Chaoptin, a cell surface glycoprotein required for Drosophila photoreceptor cell morphogenesis contains a repeat motif found in yeast and humans, *Cell*, 52, 291, 1988.
54. **Krusius, T. and Ruoslahti, E.,** Primary structure of an extracellular matrix proteoglycan core protein deduced from cloned cDNA, *Proc. Natl. Acad. Sci. U.S.A.*, 83, 7683, 1986.
55. **Bienz, D., Schnippering, W., and Clemetson, K. J.,** Glycoprotein V is not the thrombin activation receptor on human platelets, *Blood*, 68, 720, 1986.
56. **Shimomura, T., Fujimura, K., Maehama, S., Takemoto, M., Oda, K., Fujimoto, T., Oyama, R., Suzuki, M., Ichihara-Tanaka, K., Titani, K., and Kuramoto, A.,** Rapid purification and characterization of human platelet glycoprotein V: the amino acid sequence contains leucine-rich repetitive modules as in glycoprotein Ib, *Blood*, 75, 2349, 1990.
57. **Clemetson, K. J., McGregor, J. L., James, E., Dechavanne, M., and Luscher, E. F.,** Characterization of the platelet membrane glycoprotein abnormalities in Bernard-Soulier syndrome and comparison with normal by surface labeling techniques and high resolution two-dimensional electrophoresis, *J. Clin. Invest.*, 70, 304, 1982.
58. **Zafar, R. S. and Walz, D. A.,** Platelet membrane glycoprotein V: characterization of the thrombin-sensitive glycoprotein from human platelets, *Thromb. Res.*, 53, 31, 1989.
59. **Giancotti, F. G., Languino, L. R., Zanetti, A., Peri, G., Tarone, G., and Dejama, E.,** Platelets express a membrane protein complex immunologically related to the fibroblast fibronectin receptor and distinct from gpIIb/IIIa, *Blood*, 69, 1535, 1987.
60. **Piotrowicz, R. S., Orchekowski, R. P., Nugent, D. J., Yamada, K. Y., and Kunicki, T. J.,** Glycoprotein Ic-IIa functions as an activation independent fibronectin receptor on human platelets, *J. Cell Biol.*, 106, 1359, 1988.
61. **Nieuwenhuis, H. K., Akkerman, J. W. N., Houdijk, W. P. M., and Sixma, J. J.,** Human blood platelets showing no response to collagen fail to express surface glycoprotein Ia, *Nature*, 318, 470, 1985.
62. **Santoro, S. A.,** Identification of 160,000 dalton platelet membrane protein that mediates the initial divalent cation dependent adhesion of platelets to collagen, *Cell*, 46, 913, 1986.
63. **Lam, S. C. T., Plow, E. F., D'Souza, S. E., Cheresh, D. A., Frelinger, A. L., and Ginsberg, M. H.,** Isolation and characterization of a platelet membrane protein related to the vitronectin receptor, *J. Biol. Chem.*, 264, 3742, 1989.
64. **Okumura, T. and Jamieson, G. A.,** Platelet glycocalicin. I. Orientation of glycoproteins of the human platelet surface, *J. Biol. Chem.*, 251, 5944, 1976.
65. **Silverstein, R. L., Asch, A. S., and Nachman, R. L.,** Glycoprotein IV mediates thrombospondin-dependent platelet-monocyte and platelet-U937 cell adhesion, *J. Clin. Invest.*, 84, 546, 1989.
66. **Tandon, N. N., Kralisz, U., and Jamieson, G. A.,** Identification of glycoprotein IV (CD36) as a primary receptor for platelet collagen adhesion, *J. Biol. Chem.*, 264, 7576, 1989.
67. **Tuszynski, G. P. and Kowalska, M. A.,** Thrombospondin-induced adhesion of human platelets, *J. Clin. Invest.*, 87, 1387, 1991.
68. **Oquendo, P., Hundt, E., Lawler, J., and Seed, B.,** CD36 directly mediates cytoadherence of *Plasmodium falciparum* infected erythrocytes, *Cell*, 58, 95, 1989.
69. **Lipsky, R. H., Sobieski, D. A., Tandon, N. N., Herman, J., Ikeda, H., and Jamieson, G. A.,** Detection of gpIV (CD36) mRNA in Nak^{a-} platelets, *Thromb. Haemost.*, 65, 456, 1991.
70. **Moroi, M., Jung, S. M., Okuma, M., and Shinmyozu, K.,** A patient with platelets deficient in glycoprotein VI that lack both collagen-induced aggregation and adhesion, *J. Clin. Invest.*, 84, 1440, 1989.

71. **Sugiyama, T., Okuma, M., Ushikubi, F., Sensaki, S., Kanaji, K., and Uchino, H.,** A novel platelet aggregating factor found in a patient with defective collagen-induced platelet aggregation and autoimmune thrombocytopenia, *Blood,* 69, 1712, 1987.
72. **Fukuyama, M., Sakai, K., Itagaki, I., Kawano, K., Murata, M., Kawai, Y., Watanabe, K., Handa, M., and Ikeda, Y.,** Continuous measurement of shear-induced platelet aggregation, *Thromb. Res.,* 54, 253, 1989.
73. **Hantgan, R. R., Hindriks, G., Taylor, R. G., Sixma, J. J., and DeGroot, P. D.,** Glycoprotein Ib, von Willebrand factor, and glycoprotein IIb:IIIa are all involved in platelet adhesion to fibrin in flowing whole blood, *Blood,* 76, 345, 1990.
74. **Ikeda, Y., Handa, M., Kawano, K., Kamata, T., Murata, M., Araki, Y., Anbo, H., Kawai, Y., Watanabe, K., Itagaki, I., Sakai, K., and Ruggeri, Z. M.,** The role of von Willebrand factor and fibrinogen in platelet aggregation, *J. Clin. Invest.,* 87, 1234, 1991.

Chapter 16

Interactions Among Platelets, Tumor Cells, and the Vessel Wall

Kenneth V. Honn, Yong Q. Chen, and Dean G. Tang

CONTENTS

I. Metastasis 317
II. Hematogenous Dissemination of Tumor Cells: Heterotypic Interactions 317
A. Tumor Cell Induction of Platelet Aggregation 317
1. Soluble Agonists 318
2. Tumor Cell Surface Molecules 319
a. Carbohydrate-Carbohydrate Recognition 319
b. Carbohydrate-Protein Recognition 319
c. Protein-Protein Interactions 320
B. Platelet Action on Tumor Cells 321
C. Tumor Cell Interaction with Endothelium 322
1. "Docking" and "Locking" Hypothesis 324
III. Signal Transduction Pathways Involved in Platelet-Tumor Cell-Vessel Wall Interactions 325
IV. Perspective 327
References 327

I. METASTASIS

Tumor progression is a multistep process. Neoplastic clones are usually generated by a series of somatic and/or inherited mutations. Those genetic changes allow cells to bypass programmed cell death, avoid the need for growth factors (positive regulators), ignore restraining signals (negative regulators), escape from immunological surveillance, stimulate the formation of their own blood supply (angiogenesis), breach surrounding tissues (invasion), and often colonize distant sites (metastasis). Malignant tumor cells generally metastasize via lymphatic and/or vascular systems. During the hematogenous phase of metastasis, tumor cells undergo extensive cell-cell (such as tumor cell-tumor cell, tumor cell-stromal cell, tumor cell-platelet, tumor cell-EC) and cell-ECM interactions.[1-3] These interactions are mediated by a wide spectrum of surface adhesion molecules[4] present on both tumor cells and host cells, in particular target organ microvascular endothelial cells.[5-7] Therefore, adhesions are involved essentially, although at different levels, in all of the tumor cell-host cell interactions. In fact, cell adhesion is the fundamental process of all biological systems. Altered adhesion could result in a loss of cell-cell contact, an increase in cell motility, and a commitment to the first step of invasion. Also, altered adhesion could enhance tumor cell interaction with platelets, augment tumor cell lodgement in the microvasculature and adhesion to endothelium, establish tumor cell linkage with subendothelial matrix, and lead to tumor cell extravasation.

II. HEMATOGENOUS DISSEMINATION OF TUMOR CELLS: HETEROTYPIC INTERACTIONS

During hematogenous dissemination, tumor cells undergo homotypic and heterotypic interactions. Homotypic tumor cell adhesion has been reported to enhance experimental metastasis,[8] probably because cell aggregates have a greater chance of survival in circulation and of lodging in the microvasculature. In this review, we will focus our attention on heterotypic interactions, especially the tumor cell-platelet and tumor cell-endothelium interactions, which play pivotal roles in tumor cell metastasis.

A. TUMOR CELL INDUCTION OF PLATELET AGGREGATION

Numerous experiments using both human and animal models have provided ample evidence that platelet-tumor cell interaction may facilitate cancer metastasis.[3,9-14] Although platelets may not play an

0-8493-4559-6/95/$0.00+$.50

important role in the initial arrest of malignant cells in the microvasculature,[3,15-17] tumor cell-induced platelet aggregation and subsequent activation contribute to tumor cell adhesion to endothelium and the subendothelial matrix[12,13] and to tumor cell extravasation by indirectly inducing EC retraction.[3,18,19] In addition, formation of tumor cell-platelet aggregates may protect tumor cells from killing by immune cells.[17] Tumor cell-platelet interactions can be described by three defined, sequential events, (1) receptor-mediated platelet adhesion to the tumor cell plasma membrane, (2) tumor cell cytoskeletal alterations and the expression of prothrombogenic activity, and (3) platelet release and activation. At the very initial stage, few platelets adhere to the tumor cell plasma membrane.[20] Subsequently, more platelets are recruited to the initial site as the attached platelets respond to tumor cells by shape change, degranulation, and activation.[20,21] Platelets disappear soon after tumor cells establish adhesion to the subendothelial matrix.[16]

In vitro experiments have demonstrated that tumor cells can induce platelet aggregation and that the ability to aggregate platelets, in most cases, correlates with the metastatic potential *in vivo*. Different tumor cells may use different mechanisms to induce platelet aggregation; therefore, multiple factors so far have been implicated in tumor cell-induced platelet aggregation (TCIPA). Generally, these factors can be classified into (1) soluble agonists shed by circulating tumor cells and (2) tumor cell surface molecules.

1. Soluble Agonists

Thrombin is the most potent physiological agonist to aggregate platelets. It is conceivable that many tumor cells can invoke the conversion of plasma prothrombin to thrombin and thus induce platelet aggregation. Pearlstein et al.[22] observed that platelet aggregation triggered by human colon carcinomas could be inhibited by the thrombin inhibitor, dansylarginine N-(3-ethyl-1,5-pentanediyl) amide, implying that thrombin is the aggregating agonist. TCIPA caused by B16a melanoma and 3LL cells could also be blocked by the thrombin inhibitor hirudin but not by ADP scavengers like apyrase.[23] Thrombin generation triggered by tumor cells does not necessarily require platelets.[24] The ability to invoke thrombin generation and subsequent platelet aggregation by these tumor cell lines is correlated with their lung colonization potential. This ability may be partially attributed to enhanced expression on tumor cells of class II MHC and/or tissue factor, which can trigger the conversion of prothrombin to thrombin.[25,26] Antitissue factor antibody has been found to inhibit sarcoma[27] and neuroblastoma cell- induced[28] platelet aggregation, which is also inhibited by hirudin. Thrombin is a multifunctional protein. Injection of thrombin with tumor cells has been shown to enhance pulmonary metastasis.[24] Thrombin binds to the high-affinity thrombin receptor on platelets[29,30] and, via different G proteins, sets in motion several signaling pathways. Activation of phospholipase C pathway via G_P, of phospholipase A pathway via G_A, and of adenylate cyclase pathway via G_i and G_S leads to, among other responses, drastic morphological changes in platelets, increased arachidonate metabolism, and functional activation of integrin αIIbβ3. The concerted action of all of these processes is essential for the phenotypic expression of platelet aggregation. Thus, cytoskeletal rearrangements, arising from activated myosin light chain kinase and depressed PKA, confers on platelets maximal elasticity in cell shape, facilitating rapid homotypic cell contact. Increased arachidonic acid metabolism will give rise to a variety of bioactive products (TXA_2, 12-HETE, etc.), which can further act on both platelets and tumor cells. The exposure of fibrinogen binding site (i.e., the activation of αIIbβ3 integrin) plays a critical role in both the development of tight, irreversible platelet aggregation and formation of tumor cell-platelet aggregates.

Platelet aggregation induced by some tumor cell lines is not mediated by thrombin generation. Instead, ADP produced by tumor cells is involved.[31] Pretreatment of benzopyrene-induced sarcoma cells with apyrase significantly decreased platelet aggregation,[32] suggesting that tumor cell-derived ADP is responsible for TCIPA. In contrast, human rhabdosarcoma cell-induced platelet aggregation appears to involve ADP derived from both tumor cells and platelets, since apyrase treatment of either tumor cells or platelets resulted in the inhibition of TCIPA.[33] Platelets release ADP in response to a host of agonists, including ADP itself. ADP acts on platelets by different molecular mechanisms from thrombin. ADP receptor (aggregin) appears to bear little resemblance to the thrombin receptor, a typical seven-pass receptor that is linked to G-proteins.[33] Exactly how ADP elicits platelet aggregation is at present unclear, but calcium mobilization, TxA_2 synthesis, and release appear to be involved.[33] A controversy has arisen recently as to whether ADP generated by tumor cells and ADP-mediated TCIPA are related to metastatic potential. Human 253J urinary carcinoma cells and variants of the B16 melanoma cell line activate heparinized human platelets by an ADP-dependent mechanism. This TCIPA can be blocked by creatine phosphate/creatine phosphokinase, but the ability to produce ADP and support platelet aggregation appears not to be correlated with metastatic potential.[34] In contrast, other authors[35] have demonstrated that increased

ADP production by tumor cells and subsequent enhancement of platelet-aggregating activity are closely related to metastatic potential. This discrepancy may be explained by the fact that different tumor cell lines have been used and that tumor cell populations are highly heterogeneous. The other alternative could be that, in contrast to thrombin-induced TCIPA, ADP-triggered TCIPA may lack significant tumor cell-platelet interdigitation, since ADP-induced TCIPA has never been verified at the morphological level.

2. Tumor Cell Surface Molecules

Another group of tumor cells induce platelet aggregation, neither by generating thrombin nor by ADP but by endogenous cell membrane molecules. These molecules can be classified generally as adhesion molecules, considering that they mediate the initial tumor cell-platelet interaction and subsequent platelet aggregation. As a matter of fact, TCIPA mediated by soluble agonists or by these adhesion molecules is by no means exclusive but rather closely related. For those tumor cells that release ADP or other materials or trigger thrombin generation, adhesion molecules, in particular activated $\alpha IIb\beta 3$, are clearly required for the eventual cell-cell aggregation. In other words, agonists produced by tumor cells act to modulate platelet adhesion molecules and therefore its general adhesive property. Initial platelet-tumor cell adhesion requires participation of these agonists and many other substances released by locally activated platelets in order to recruit more platelets and tumor cells. As a consequence, many general features of cell adhesion appear to apply to tumor cell-platelet and/or platelet-platelet interactions. Biochemically, cell-cell adhesion involves four general molecular mechanisms, i.e., carbohydrate-carbohydrate, carbohydrate-protein, direct protein-protein, and indirect protein-protein interactions.

a. Carbohydrate-Carbohydrate Recognition

Essentially all cells have an exterior cell coat made up of glycosylated proteins, glycosphingolipids, and other glycoconjugates. It is well known that platelets have an unusually enriched glycocalyx.[36] Various tumor cells also express abundant aberrantly glycosylated glycosphingolipids and glycoproteins or tumor-associated carbohydrate antigens.[37,38] The important role of carbohydrate-based recognition mechanisms (carbohydrate-carbohydrate or carbohydrate-lectin) in mediating cell-cell adhesion and determining the specificity of a variety of physiological (e.g., tissue pattern formation in embryogenesis and morphogenesis) as well as pathological (e.g., inflammation and metastasis) processes[39] has come to our attention only very recently. Depending on different glycoconjugate combinations, carbohydrate-carbohydrate interactions can be strong, moderate, or weak. Carbohydrates (e.g., glycosphingolipids) have been found to exist in patches on the cell membrane, and cell adhesion may be initiated by interactions between these patches and subsequently strengthened by other adhesion mechanisms.[40] Whether carbohydrate-carbohydrate recognition is involved in initial tumor cell-platelet interaction has not been determined. Circumstantial evidence provides some clues. Early work by Gasic et al.[41,42] demonstrated that neuraminidase-rendered thrombocytopenia greatly reduced metastasis, suggesting an important role for platelet membrane sialic acid-containing glycoconjugates in cancer metastasis. Subsequently, the sialic acid content of platelet-aggregating material and the degree of sialylation of surface glycoconjugates in cultured renal sarcoma cells have been found to correlate with TCIPA and tumor metastasis.[43] The platelet-aggregating protein extracted from virus-transformed tumors is also a sialylated glycoprotein, and a good correlation has been obtained between *in vivo* metastatic potential and cell surface sialic acid.[44] In chemically-induced rat fibrosarcoma cells, platelet-aggregating material was found to be a high mol wt substance that contains sialic acid.[45] This substance required Ca^{2+} for its activity.[45] A sialyl-transferase inhibitor demonstrated potent inhibitory effect on the pulmonary metastasis of murine colon carcinoma, an effect mediated by inhibiting tumor cell-induced platelet aggregation.[46] Treatment of tumor cells with neuraminidase, sialic acid, fucose, and sialyllactose also inhibited TCIPA and experimental metastasis.[46] So far, none of these sialoglycoconjugates has been biochemically or molecularly characterized, nor has any possible carbohydrate ligand on platelets been defined. However the presence of such a ligand is supported by the experiment showing that treatment of platelets with neuraminidase inhibits experimental metastasis.[47]

b. Carbohydrate-Protein Recognition

Widespread expression of lectin activity in platelets and tumor cells underlines the possibility of carbohydrate-protein interactions. Recent identification of the ligands of selectins (both P- and E-selectins) to be typical carbohydrates, i.e., sialyl Le^a and sialyl Le^x, strongly supports this possibility.[48-50] Tumor cells, especially metastatic variants, express various aberrant forms of endogenous lectins or sialoglycoconjugates.[37,38,51] Metastatic human rectal carcinoma and lung adenocarcinoma cells have been shown to express increased sialyl-dimeric Le^x antigen.[52,53] Although up-regulated expression of these carbohydrate antigens is most likely to be related to tumor cell targeting to specific organ microvascular EC, the possibility cannot be excluded that they are also involved in tumor cell-platelet interactions, since platelets express abundant P-selectin. On the other hand, many of the sialylated glycoconjugates described above in tumors may well represent selectin-like molecules, which recognize the carbohydrate ligands in platelets.

c. *Protein-Protein Interactions*

As in many other cell adhesion systems, cell-cell interactions based on protein-protein recognition mechanisms are required to establish tight tumor cell-platelet association as well as late-phase platelet aggregation. Direct protein-protein interactions are exemplified by homophilic association of cadherins and some Ig family members (such as N-CAM)[54] or heterophilic integrin Ig associations (such as VCAM-1 and VLA-4 and LFA-1 and ICAM-1).[54,55] Platelets express PECAM-1 molecules, an Ig family members,[56] and some tumor cells express a PECAM-1-like protein,[19] but it is not known whether these molecules participate in platelet-tumor cell interactions. The most significant mechanism involved in stable tumor cell-platelet association is indirect protein-protein interaction using ligands as the bridge. Integrin receptor αIIbβ3 is the key player in this process. It has been known that this integrin receptor is required for agonist-induced second-phase (irreversible) platelet aggregation. It is postulated that, following agonist activation, functional αIIbβ3 binds to a ligand, which further binds to another αIIbβ3 on other platelets, therefore mediating platelet aggregation. Physiologically, fibrinogen is the predominant bridging ligand due to its high concentration. Some other adhesive glycoproteins can also bind to αIIbβ3 and mediate platelet aggregation. For instance, von Willebrand factor can mediate platelet aggregation at high shear rates,[57,58] and thrombospondin,[59] fibronectin, and possibly others may also participate in aggregation when localized concentrations of these proteins are dramatically elevated in the vicinity of activated platelets. The multivalency (i.e., multiple binding sites) of fibrinogen makes the molecule structurally appropriate for this linking function. The crucial importance of αIIbβ3 in mediating platelet aggregation also gains strong support from a genetic disorder, Glanzmann's thrombasthenia, in which mutations on either αIIb or β3 predispose patients to serious bleeding problems.[57,60] Grossi et al.[61] and Bastida et al.[62] independently reported that treatment of platelets by antibodies to GpIb or αIIbβ3 resulted in diminished TCIPA, which has been confirmed since by other laboratories.[44,63] These observations suggest that tumor cells may express functional surface αIIbβ3. We provided the first experimental evidence that tumor cells derived from human solid tumors express functional αIIbβ3 as well as GpIb, since treatment of tumor cells with antibodies against these two glycoproteins inhibited TCIPA.[64] In addition, anti-αIIbβ3 pretreatment of tumor cells also blocked tumor cell adhesion to fibronectin.[64] Functional αIIbβ3 was soon identified on rat[20] and murine[1] tumor cells. These experimental findings were confirmed by others.[65-68] Tumor cells organize the αIIbβ3 receptor to focal adhesions, extending cell filapodia, or to the cell surface.[64,69] Tumor cells possess an apparent intracellular pool of αIIbβ3 which appear to be constantly cycling with cell surface receptors and can be rapidly mobilized to the membrane by stimulation with TPA or 12(S)-HETE.[67,69] Intracellularly, this integrin receptor colocalizes with the intermediate filament protein vimentin,[71] although receptor is also localized to the microfilament endings at focal adhesions.[69,72] Functionally, tumor cell αIIbβ3 receptors are actively involved in TCIPA, tumor cell adhesion to EC, and subendothelial matrix[3,66,68,73,74] and possibly in tumor cell proliferation.[69]

It was a major concern that all of the experiments on αIIbβ3 mentioned above were mainly based on antibody studies, therefore, either the term "αIIbβ3-like protein" or "immunologically related αIIbβ3" (i.e., IRGp IIb/IIIa) was used instead of αIIbβ3. Recently, reverse transcription followed by polymerase chain reaction (RT-PCR) using specific primers detected an αIIb band in B16a cells. Partial DNA sequence and deduced amino acid data demonstrated that B16a αIIb shares an 80 and 90% sequence homology with that of human and rat αIIb, respectively, but only 26% homology with human αv subunit.[75] This result indicates that the αIIbβ3-like integrin is the authentic αIIbβ3. In addition, RT-PCR combined with sequencing was employed to screen the expression of αIIb on a wide spectrum of tumor cells (about 20 lines) from different histological and pathological types. Preliminary results indicate that αIIb is widely expressed in tumor cells (Chen and Honn, unpublished observation). This wide distribution of αIIbβ3 suggests the possibility that this integrin plays a significant role in tumor cell-platelet-EC interactions.

The αIIbβ3-mediated tumor cell-platelet interactions and platelet aggregation can be triggered by tumor cell released agonists and strengthened by platelet activation and release products such as TxA_2 and 12(S)-HETE. As we discussed above, these cell-cell interactions can also result from the activation of αIIbβ3 by ligand occupancy. Tumor cells themselves may readily provide this ligand function. Melanoma cells can quickly form a coating of fibrinogen on their surface.[76] This surface-immobilized fibrinogen can function as ligand bridging to mediate αIIbβ3-dependent homotypic and heterotypic cell-cell interactions.[77] In fact, αIIbβ3 on nonactivated platelets can selectively recognize adhesive sites in surface-bound fibrinogen[78] as long as there is enough Ca^{2+}.[79] This recognition alone greatly modulates and potentiates the receptor ligand-triggered conformational change, independent of any intracellular signaling pathways.[80-83] Another modular protein that functions similarly to fibrinogen is thrombospondin, the most abundant protein component of platelet alpha granules from which it is rapidly secreted upon platelet activation at sites of injury and thrombosis.[59] Secreted thrombospondin binds to the surface of tumor cells through specific thrombospondin receptors, which are distinct from integrin receptors, and mediates tumor cell-platelet interaction through platelet gpIV.[84,85] Thrombospondin has been reported to participate in irreversible platelet aggregation, mediate tumor cell adhesion, and promote cancer metastasis. All of these effects can be inhibited by peptides CSVTCG and CSTSCG present in the type I repeats of the thrombospondin molecule.[86,87] Therefore, this protein-protein interaction system (thrombospondin receptor-thrombospondin-gpIV), like αIIbβ3-fibrinogen-αIIbβ3 binding, may also play an important role in TCIPA.

Many other tumor cell membrane proteins may also be involved in tumor cell-platelet interactions. The 44-kD protein that is responsible for TCIPA has been detected in colon carcinoma cells.[88] Similar platelet-aggregating factors are also present in melanoma cells.[89] The mechanism by which this protein mediates platelet aggregation remains unknown.

B. PLATELET ACTION ON TUMOR CELLS

In a series of *in vivo* studies of tumor cell-vessel wall interactions, Crissman[15,16] and colleagues[11] demonstrated that

1. cancer cell arrest in the microvasculature is independent of platelet activation
2. platelet activation most probably occurs after cancer cell arrest and is localized to specific areas of the cancer cell plasma membrane
3. EC retraction occurs concomitant with the presence of activated platelets
4. platelets are transiently (i.e., 24 to 48 h) associated with intravascular tumor cells
5. cancer cells can remain intravascular-attached to the subendothelial matrix in the absence of platelet activation, suggesting that the factors responsible for cancer cell-induced platelet activation are transiently expressed

These observations and conclusions are supported by others[23,24,90] and also by *in vitro* experiments.[11,20,91] In the vasculature, tumor cells undergo a coordinated sequence of morphological changes in response to platelet adhesion and platelet products, which include a rapid disappearance of microvilli in areas not involved with platelet contact and the formation of cellular processes that interdigitate with a focal platelet aggregate.[11,92] These responsive cell shape changes depend on concerted tumor cell cytoskeleton reorganization, which could result from either αIIbβ3-transduced signals, since αIIbβ3 is most richly localized in the extending cell filapodia in moving cells[69] and possesses a diversity of signaling functions, or cytoskeletal rearrangement induced by released platelet substances such as 12(S)-HETE.[69,72,93,94] Disruption of tumor cell microfilaments and intermediate filaments inhibited TCIPA *in vitro* and platelet association with tumor cells *in vivo* and decreased lung colony formation.[20,21,71]

Platelets can drastically modulate tumor cell properties and TCIPA by their membrane structures or activation metabolites. Stimulated by agonists or tumor cells, platelets are activated and release a large array of bioactive granular substances, including amines, adhesive glycoproteins, growth factors, and arachidonic acid metabolites. These products greatly modify circulating tumor cells, participate in TCIPA, and to a large extent, are the major contributory factors in metastasis-promoting effects of platelets. Activated platelets release 5-HT. A correlation between 5-HT release and tumor cell lodgement was observed when it was demonstrated that animals given tumor cells showed a two-fold increase in 5-HT levels compared to animals injected with saline.[95] Treatment of animals with a 5-HT receptor blocker reduced 5-HT levels to 63% and tumor cell lodgement to 72%.[96] A 5-HT_2 specific antagonist (R56413) demonstrated an even greater reduction in tumor cell lodgement.[96] No clear mechanism of the above actions by this amine is known.

Platelet α-granules contain many adhesive glycoproteins, including thrombospondin, fibronectin, fibrinogen, and von Willebrand factor. After release, these proteins can function as ligands for various adhesion molecules, thus promoting tumor cell-platelet-EC interactions. For example, platelet fibronectin released in response to tumor cell stimulation can quickly bind to the tumor cell surface.[97] Fibrinogen molecules also become associated with the tumor cell surface. Surface-bound fibrinogen can either serve as an αIIbβ3 ligand or protect tumor cells from killing by lymphokine-activated killer cells.[76] The most abundant platelet granule protein thrombospondin, after release, mediates a diversity of interactions, including TCIPA and tumor cell adhesion to ECM.[59,86,87] Von Willebrand factor can also be involved in various adhesive processes. Platelet α-granules also contain many growth factors, the most prominent being PDGF and TGF-β. PDGF dose-dependently stimulated the growth of highly metastatic colon carcinoma cells, and this effect was interpreted as relevant to the intravascular proliferation of tumor cells.[98] The *in vitro* invasive capacity was augmented by pretreating tumor cells with TGF-β or with activated platelets.[99] Purified platelet-derived TGF-β has been shown to be a chemoattractant for bone-metastasizing tumor cells.[100]

It is possible that the most important regulatory mediators from activated platelets are the arachidonic acid metabolites. Although early experiments only indicated the involvement of the cyclooxygenase (COX) pathway, it becomes clear that both COX and lipoxygenase (LOX) pathways are actively involved in TCIPA and cancer metastasis.[101-103] In the COX pathway, thromboxane A_2 (TxA_2) is the most potent and important mediator. Initially, a positive correlation was observed between the TCIPA potential and tumor cell-triggered platelet TxA_2 production.[104] Treatment of platelets but not tumor cells with aspirin or indobufen resulted in diminished TCIPA.[104,105] Walker 256 carcinosarcoma cells induced the aggregation of rat platelets and concomitant production of both TxA_2 and 12(S)-hydroxyeicosatetraenoic acid [12(S)-HETE], a 12-LOX metabolite of arachidonic acid.[101] TCIPA was found to be dependent on the concentration of tumor cells initiating aggregation, as well as COX and LOX products. At a low concentration (8.8×10^4 tumor cells), COX inhibitors but not LOX inhibitors blocked TCIPA. At a high concentration (5×10^5 tumor cells), neither COX nor LOX inhibitors alone affected platelet aggregation. However, the combined inhibition of both pathways resulted in subsequent inhibition of TCIPA, regardless of tumor cell concentrations.[101,102] In these experiments, the inhibitors did not significantly inhibit PKC activity at the doses tested. Thrombin stimulation of washed rat platelets resulted in significantly enhanced surface expression of αIIbβ3, and this effect was not inhibited by a thromboxane synthase inhibitor or a TxA_2 receptor antagonist.[101] However, αIIbβ3 expression was blocked by LOX inhibition and increased by 12(S)-HETE.[101] These results suggest that both the platelet LOX [i.e., 12(S)-HETE] and COX (i.e., TxA_2) pathways are important for TCIPA but that different mechanisms of action are involved. Later experiments established that tumor cell COX and LOX pathways also play an important modulatory role in TCIPA and metastasis.[3,73,74,101,106]

The hydroxy fatty acid, 12-(S)-HETE, can be produced by a variant of normal cells, in particular platelets,[106] as well as tumor cells.[108] Since it was first implicated in TCIPA, a wide diversity of functions have been ascribed to this eicosanoid. 12(S)-HETE enhances TCIPA,[3,70,106] promotes tumor cell adhesion to EC,[18,67,71] and promotes tumor cell spreading on adhesive glycoproteins,[110] induces reversible and nondestructive retraction of both large vessel and microvessel EC,[18,111-113] stimulates tumor cells to release cathepsin B,[114] promotes tumor cell motility by increasing the autocrine-motility factor expression,[115] and augments tumor cell metastatic potential.[3,73,74,103] Many of these effects can be antagonized by 13-HODE (13-hydroxyoctadecaenoic acid).[67,109,116] This 15-LOX metabolite of linoleic acid also regulates the adhesivity of the vascular wall.[117]

C. TUMOR CELL INTERACTION WITH ENDOTHELIUM

Tumor-microvasculature interaction constitutes one of the most essential steps for the successful completion of metastasis. Tumor cells have to survive constant shear force, mechanical damaging, and immune surveillance before they can arrest in an organ microvessel.[1,92,118] In circulation, extensive tumor cell-tumor cell and tumor cell-host cell interactions occur, some of them being deleterious to tumor cells, some being favorable for the survival of tumor cells. Platelets and a variety of other factors can facilitate or modulate this interaction.[3,10,11,74,119-121] Tumor cells use diverse groups of adhesion molecules to adhere to target organ capillary EC, and the specificity of this interaction results from contributions of the surface receptors on both cell types. Conceivably, the molecular mechanisms could be none other than those we discussed in platelet-tumor cell interaction, i.e., carbohydrate-carbohydrate, carbohydrate-protein, indirect protein-protein, or direct protein-protein hydrophobic interactions. In principle, tumor cell-EC interaction can involve all of the above mechanisms individually or in combination.

Blood-borne tumor cells generally take advantage of the adhesion molecules used by blood cells to attach to EC; this is especially true for lymphomas.[6,121-123] For example, lymphoma cell variants deficient in LFA-1 are unable to invade and metastasize *in vivo*.[124] Modification of tumor cell surface oligosaccharides by tunicamycin and swainosine can greatly alter the tumor cell and EC adhesive properties and metastasis,[121] suggesting that carbohydrate-carbohydrate or carbohydrate-protein recognitions are involved. Endothelium is a very heterogeneous organ that demonstrates wide variability in morphology and biochemistry, including different lectin stainability.[6] Tumor cells also differ greatly in the expression of surface glycoconjugates.[39,40,125-128] Expression of a specific type of carbohydrate may be one of the most important determinants for tumor cell "homing" to the target organ endothelium, which bears much resemblance to lymphocyte homing.[129] Thus, it is not surprising that some metastatic tumor cells express higher sialyl dimeric Le^x antigen than nonmetastatic tumor cells.[130] Colon carcinoma cell adhesion to activated vascular EC (by TNF-α and IL-1) can be blocked by anti-E-selectin antibody,[125] suggesting these tumor cells express sLe^x or similar antigens.

Various melanoma cell lines have been widely used as models to investigate solid tumor cell-EC interactions, and numerous relevant molecules have thus been reported. Rice et al.[131] first demonstrated an increased adhesion of human melanoma cells to EC that had been activated by cytokine. The adhesion occurred directly between two cell types, which was not affected by RGD peptide treatment.[131] Subsequent experiments[132] showed that increased EC VCAM-1 expression was responsible for increased melanoma cell adhesion, and up-regulated E-selectin expression in EC supported human colon carcinoma cell adhesion. Recently it was reported that the molecule responsible for mediating melanoma cell adhesion to IL-1-activated EC in VLA-4, and its expression varies widely among different clones.[133] Melanoma cells also express ICAM-1[136,137] and CD44,[133,134] which mediates tumor cell adhesion to EC via surface hyaluronate, and the expression of these two molecules has been directly correlated with the progression of metastasis. Interestingly, EC do not express LFA-1 or Mac-1, the natural ligand for ICAM-1. Therefore, the function of the ICAM-1 on tumor cells was hypothesized to be attributable to ICAM-1-mediated tumor cell aggregate formation.[138] Recently, highly metastatic melanoma cells that constitutively express ICAM-1 were found to secrete large amounts of IL-1,[138] which enhanced tumor cell ICAM-1 expression in an autocrine fashion and increased expression of ICAM-1, E-selectin, and VCAM-1 on EC. However, none of the antibodies to these molecules could inhibit tumor cell adhesion to EC, which was instead inhibited by about 50% by anti-αIIbβ3,[138] strongly suggesting that activated tumor cell adhesion to EC is, at least in part, mediated by tumor cell integrin αIIbβ3. This result is in accord with our experimental findings[3,69,139] and those of others.[66] The importance of widespread αIIbβ3 expression[64,75,140] is further strengthened by observations that this integrin receptor is only expressed in metastatic melanoma cell clones,[65] that the expression of this integrin is highly correlated with the lung-colonizing capacities of melanoma cells,[103,141] and that it not only mediates tumor cell-platelet interaction (see previous section), but also activates tumor cell adhesion to EC and matrix.[64,67-69] Similar to αIIbβ3-mediated tumor cell-platelet interaction, tumor cell-EC interaction may also employ fibrinogen to interlink the αIIbβ3 in tumor cells and $\alpha_v\beta3$ in EC.[77]

Tumor cell-EC interaction can also be mediated by unique EC surface adhesion molecules. Such a molecule (termed Lu-ECAM-1) has been isolated recently from lung matrix-modulated bovine aortic EC.[142] This 90 kD protein, Lu-ECAM-1 (whose ligand is unknown) promotes the selective attachment of lung-metastatic B16-F10 melanoma cells. It is not up-regulated by cytokines and displays extremely selective cell-binding properties, serving as an adhesion molecule for metastatic clones of a specific tumor cell type (lung metastatic B16a cells). These authors conclude that a constitutively expressed EC phenotype, modulated by tissue-specific matrix, could serve as the EC "addressin" for the "homing" of metastasizing tumor cells.[142]

The expression of organ-specific molecules by vascular EC is not an intrinsic feature of each distinct organ but is regulated by the composite of ECM on which the EC resides. Therefore, EC expression of integrins and membrane glycoconjugates can be modulated by organ-specific biomatrices.[7,143] Aortic EC grown on ECM extracted from either lung or liver preferentially bind to tumor cells that selectively colonize one organ or the other.[5] Tumor cell-EC adhesion is also modulated by a variety of bioactive substances. Following TGF-β treatment (7 to 12 h), human osteosarcoma cells underwent a selective loss of cell adhesion to laminin, resulting from decreased α3β1 (a laminin/collagen receptor) transcription and increased α2β1 (a collagen receptor) and α5β1 (fibronectin receptor).[136] This decreased adhesion to laminin induced by TGF-β may partly explain its inhibitory effect on the growth and metastasis of some tumors. In contrast, some breast cancer cells challenged with estrogen and progestins dramatically increased laminin receptor expression.[144] Treatment of neuroblastoma cells with retinoic acid resulted in up-regulated α1β1 expression.[145] Expression of VCAM-1, E-selectin, P-selectin, ICAM-1 on EC is stimulated by cytokines such as IL-1, TNF-α and TNF-β.[49,131,132,138,146,147] Platelets and some tumor cells synthesize TGF-β[148] and many other cytokines. Moreover, we have demonstrated that 12(S)-HETE exerts a variety of modulatory effects (see previous section). Modulation of tumor cell-EC adhesion process by these cytokines, e.g., by tumor cell-released TGF-β, IL-1, or by 12(S)-HETE derived from tumor cells and activated platelets, will promote and/or strengthen the bonding between tumor cells and ECs, leading to enhanced metastasis.[3,72,149]

After tumor cells have established a stable bond with EC, tumor cells may leave the circulation (extravasate), either by inducing EC retraction[18,19] or by damage and penetration of the EC.[150] Once the subendothelial matrix is exposed, tumor cells again face all of the challenges that were overcome during their initial invasion. Therefore, cell motility, secretion of degradative enzymes, adhesion to ECM components, and responses to the target organ autocrine and paracrine growth factors will all determine the eventual fate of the metastasizing tumor cells. Due to the structural and biochemical complexity of the basement membrane (or ECM) itself, tumor cell-matrix interactions are very intricate.[134,150,151] Several general features can be summarized as follows:

1. Tumor cells generally use multiple integrin receptors to attach to ECM, as evidenced by the inability of antibodies against a single integrin receptor to block completely tumor cell adhesion.[69,131,138] Generally, tumor cells express increased levels of laminin receptors[144,152] and diminished levels of fibronectin receptors.[153,154] On examination of the profile of integrin expression by human colon carcinoma cells, Schreiner et al.[155] observed that these cancer cells do not express α5β1 at all, but express large amounts of α6β1, a typical laminin receptor. Other molecules involved in tumor cell-matrix interactions are CD44 and many other nonintegrin receptors.
2. Tumor cells demonstrate a preferential binding affinity for the ECM derived from the target organ.[119-121] Rat mammary carcinoma cells preferentially adhered to syngeneic target (lung) organ-derived subendothelial matrix, and this adhesion correlated with spontaneous metastatic potential.[156] Similarly, neuroblastoma cells selected for increased metastatic potential by *in vivo* passaging demonstrate enhanced attachment to organ-derived ECM.[157]
3. Clearly, organ-specific ECM is functioning as a positive selector for metastasizing tumor cells. This may be related to the fact that great heterogeneity exists in ECM components among different tissues. Pulmonary ECM is rich in type IV collagen, laminin, and elastin, but these molecules are much less abundant in the liver, which instead contains a high concentration of heparan sulfate proteoglycans. Conceivably, tumor cells that express high levels of receptors for collagen/laminin/elastin will preferentially attach to lung ECM and therefore metastasize to the lung, while those tumor cells expressing receptors for proteoglycan (such as CD44) may selectively colonize the liver. In addition, this organ matrix-specified repertoire of receptors may work as the sensors for AMF-directed random motility, since these adhesion molecules are enriched in the extending pseudopodia of mobilizing tumor cells.[158] Also, interactions of tumor cells and host cells (e.g., fibroblasts) may trigger protease release by both cell types,[121] which is supported by a recent experiment showing that challenge of fibroblasts with anti-α5β1 can induce collagenase and stromelysin gene expression.[159] The expression of elastase and type IV collagenases triggered by tumor cell interaction with lung-specific matrix is an important correlate of pulmonary metastasis.[160]
4. Tumor cell-ECM interactions are reciprocal. Therefore, organ-specific matrix may induce tumor cells to express specific integrin receptors. On the other hand, tumor cells induced to express the specific receptors may remodel the ECM. A typical example is given by a recent experiment showing that highly metastatic melanoma cells grown in three-dimensional collagen lattices are induced to express high levels of α2β1 mRNA and protein. By doing so, these tumor cells caused a significant contraction of the collagen fibrils.
5. Finally, the adhesion, motility, protease secretion, and growth of metastasizing cells are under the influence of a vast array of host-derived paracrine and/or autocrine growth factors. Tumor cell responses to these factors have great impact on its survival. Therefore, the expression of tumor cell adhesion molecules (mainly integrins) can be modulated by many cytokines, such as TGF-β, TNF-α, IL-1, IL-6, etc.[161] Cytokines may also alter the existing tumor cell integrin expression pattern.[162]

1. "Docking" and "Locking" Hypothesis

Tumor cell metastasis can be achieved through either a series of relatively specific adhesion events between tumor cells and host cells or by adhesive processes mediated via a few "sticky" molecules present on either or on both cell types. Accumulated data suggest that tumor cells must undergo multistep adhesive interactions before they can colonize a distant organ successfully. Studies on leukocyte recruitment to inflammatory sites and lymphocyte homing to specific organ EC have enlightened this area of cancer metastasis research and provided important clues to tumor cell dissemination.[163] In these two processes, one physiological (lymphocyte recirculation) and the other triggered by inflammatory stimuli, the end result is leukocyte adhesion and extravasation, which is exquisitely regulated by mechanisms of selective leukocyte-endothelial recognition.[163-166] Several distinct steps have been discerned, i.e., reversible rolling of neutrophils on EC mediated by L-selectin and EC addressin, followed by neutrophil activation by chemoattractants or other activating factors, and then followed by activation-dependent, stable binding of neutrophils to EC mediated by integrins such as Mac-1. In light of these distinguishing features of leukocyte adhesion and extravasation and based on experimental findings of our own and others, Chen and Honn[75] proposed a "docking and locking" hypothesis to dissect the tumor cell-EC interactions molecularly. This hypothesis states that specific tumor cell adhesion to target organ EC involves two basic interdependent but distinct steps. An initial recognition ("docking") step occurs between the "rolling" tumor cell and underlying EC, mediated by relatively weak and transient adhesion mechanisms involving carbohydrate-carbohydrate and/or carbohydrate-protein interactions. The candidate molecules can be selectins, sLe^x, or similar molecules. The subsequent (late phase) firm adhesion ("locking") is mediated by activation-dependent integrins. The accomplishment of these two steps to a large extent determines the specificity of tumor cell adhesion and thus the organ selectivity of metastasis and is under the regulation of a number of factors, such as arachidonic acid metabolites. This hypothesis can find support from the following lines of experimental evidence:

1. Adhesion of B16 melanoma cells to EC *in vitro* involves two adhesion systems, one mediated by the GM3 ganglioside and the other by integrin receptors for fibronectin and laminin.[167] Carbohydrate-carbohydrate-mediated adhesion occurs within 20 min, while integrin-mediated adhesion reached a maximum after 60 min.[167]
2. Initial tumor cell adhesion to EC is mediated by carbohydrate-carbohydrate or carbohydrate-protein interactions. The level of the glycosphingolipid GM3 expressed in four B16 melanoma variants is directly correlated with both their ability to adhere to EC and cell metastatic potential. The initial tumor adhesion was mediated by GM3 on tumor cells and LacCer in EC.[128]
3. Treatment of B16a melanoma cells with 12(S)-HETE increased $\alpha IIb\beta 3$ surface expression and induced functional activation of this integrin, leading to enhanced adhesion to EC and fibronectin.[69,139] Antibodies to $\alpha IIb\beta 3$ blocked 12(S)-HETE stimulated adhesion but had little or no effect on basal (unstimulated) B16a cell adhesion, implying that this integrin-mediated tumor cell adhesion needs activation.
4. Similarly, VCAM-1- and E-selectin-mediated tumor cell adhesion only needs EC to be activated, without requirement for tumor cell stimulation,[131-133] suggesting that carbohydrate-protein recognition does not need preactivation. In support of this hypothesis, peptide GRGDSP patially blocked tumor cell adhesion to ECM matrix but had no effect on adhesion to activated EC monolayer.
5. More drastically, a human melanoma cell line that constitutively expresses high levels of ICAM-1 was found to secrete large amounts of IL-1, which constantly stimulates both tumor cell and EC during an *in vitro* adhesion assay.[138] Although secreted IL-1 significantly stimulated tumor cells to express more ICAM-1 and EC to express high levels of VCAM-1, E-selectin, and ICAM-1, antibodies to these molecules did not show any inhibitory effect on tumor cell adhesion to EC. Instead, an antibody to $\alpha IIb\beta 3$ demonstrated 44% inhibition,[138] strongly supporting the concept that stimulated tumor cell adhesion to EC is largely mediated by integrins such as $\alpha IIb\beta 3$.

III. SIGNAL TRANSDUCTION PATHWAYS INVOLVED IN PLATELET-TUMOR CELL-VESSEL WALL INTERACTIONS

During the tumor cell-platelet-endothelium interactions, tumor cells induce platelet aggregation, and platelets respond to and act on tumor cells. Tumor cells and platelets adhere to endothelial cells, and endothelial cells retract in response to platelet-potentiated tumor cell induction. These interactions are dynamic, mutually influential processes involving soluble factors, cell surface receptors, and cell skeleton networks, with constant cell-cell communication and signal transduction. Mounting evidence indicates that most agonists stimulate platelet aggregation by binding to surface receptors and eliciting a cascade of intracellular second messengers through different G-proteins. To date, at least six G-proteins, i.e., G_s, G_z, three variants of G_i (G_A, G_P, and G'_P), and G_q have been identified in platelets.[169] G-proteins couple agonist-induced conformational changes of the receptors to various effector systems, including adenyl cyclase (AC), phospholipase C (PLC), phospholipase A_2 (PLA_2), GMP-selective phosphodiesterase, and ion channels.[170] In platelets, the first three signaling pathways, i.e., AC, PLC, and PLA pathways, are employed by agonists. Thrombin, vasopressin, epinephrine, platelet-activating factor (PAF), and TxA_2 activate PLC by GTP-dependent mechanism. It has been suggested that actually two types of G-proteins mediate the activation of PLC. One type is pertussis toxin-sensitive and belongs to the G_i family. These G proteins interact with receptors for thrombin, PAF, and vasopressin. The other type, which interacts with the TxA_2 receptors, is pertussis-toxin-resistant.[169] Activation of PLC produces two intracellular second messengers, inositol 1,4,5-triphosphate (IP_3) and diacylglycerol (DAG). IP_3 releases Ca^{2+} from the platelet-dense tubular system, raising the cytosolic free Ca^{2+} concentration. Elevated Ca^{2+} level in turn activates PLA_2 localized on both the membrane of the dense tubular system and plasma membrane. PLA_2 releases arachidonic acid from membrane phospholipid, such as phosphatidylcholine. Arachidonic acid can be metabolized to COX products such as TxA_2 and LOX products such as 12(S)-HETE. TxA_2 diffuses out of the cell, binds to its own set of unique receptors, activates G'_p and causes further platelet activation.[170] Increased Ca^{2+} also binds to calmodulin and Ca^{2+}/calmodulin complex activates myosin light chain kinase (MLCK), which subsequently phosphorylates myosin light chain (MLC) and initiates contraction and cell shape change.[36] DAG, together with Ca^{2+}, activates protein kinase C (PKC), triggering granule secretion and the activation of αIIbβ3. Release of platelet granule contents is also dependent on the cytoskeleton and cytosolic free Ca^{2+}, but not TxA_2.[36] In the case of cell adhesion receptors, the molecular mechanism for exposure of fibrinogen binding sites on αIIbβ3 is still very nebulous. It is known that αIIbβ3 on unstimulated platelets needs to be "activated" by agonist-elicited cellular mediators.[171,172] Recently, accumulated data suggest that the functional activation of αIIbβ3 may involve all of the three signaling pathways mentioned above. Both PLA_2 and the PLC systems appear to be essential for inducing fibrinogen binding, since complete inhibition can be achieved by either indomethacin or PKC inhibitors.[172] PKC has been shown to phosphorylate the β3 subunit and the phosphorylation can be blocked by PKC inhibitors.[172-174] In contrast, activation of AC pathway (e.g., by PGI_2), which elevates cAMP levels, prevents the exposure of binding sites, and PKC and cAMP have been shown to regulate reversible affinity modulation of αIIbβ3.[173]

In addition to the PLC pathway, thrombin also activates platelets via G-protein-coupled PLA_2 and AC systems. Thrombin can directly activate PLA_2 through a unique G protein (G_A).[169] Activation of PLA_2 by thrombin will result in a series of events identical to the PLA_2 activation induced by Ca^{2+}. Thrombin can also act on AC via another G-protein, i.e., a typical G_i protein. This leads to a decreased cAMP production and thus deactivation of cAMP-dependent protein kinase (PKA). PKA primarily serves to phosphorylate MLCK. Phosphorylated MLCK loses its ability to catalyze the phosphorylation of MLC; therefore, cytoskeleton-motivated cell contraction and shape change is inhibited. Thrombin also binds to the GpIb-IX complex, which is linked to the membrane skeleton. Binding of thrombin somehow transduces signals into the cell, which activate Ca^{2+}-dependent protease and disrupt the GpIb-IX-membrane skeleton. This series of events has been thought to contribute to the shape change of platelets during aggregation and clot retraction after aggregation.[175] Recently the receptors for thrombin,[30] TxA_2,[176] and PAF[177] have been cloned and characterized. As typical G-protein coupling receptors, they are all characterized by a unique protein structure, in which a single poylpeptide chain crosses the plasma membrane seven times. The mechanism of activation of thrombin receptor is very novel and unique since it involves the proteolysis by thrombin of the extracellular domain of the receptor, exposing a new aminoterminus capable of self-activating.[30]

Evidence also indicates that αIIbβ3 activated by either agonist or ligand occupancy can further trigger a variety of transducing signals. Ligand-receptor interactions induce clustering of αIIbβ3 on the cell surface[178] and the clustering promotes an incorporation of actin-binding protein, α-actinin, talin, and αIIbβ3 into the Triton-insoluble residue.[179] Agonist stimulation causes a rapid association of several enzymes ($PP^{60c\text{-}src}$ tyrosine kinase, PLC, diacylglycerol kinases), responsible for lipid phosphorylation and hydrolysis, with the cytoskeletons.[180] Activated αIIbβ3 has been implicated in protein tyrosine phosphorylation,[181,182] increasing intracellular Ca^{2+} level by working as Ca^{2+} channels,[183,184] promoting TxA_2 production,[183] activating PKC,[185] synthesis of membrane phosphatidylinositol 3,4-bisphosphate (PIP2),[53] and increased N^+/H^+ exchanges.[186] The exact role of these secondary responses invoked by αIIbβ3 is not clear at present, but suffice it to say that they can modulate platelet responses to agonists or tumor cells. We also do not know whether tumor cell αIIbβ3 can participate in these signaling processes. What has become clear is that tumor cell αIIbβ3 is also responsive to platelet products such as 12(S)-HETE (see previous section) in a PKC-dependent manner, although it may not be stimulated by platelet agonists like thrombin.[187]

Many 12(S)-HETE effects on tumor cells and endothelial cells have been discussed earlier. The wide-spectrum activity of 12(S)-HETE may be attributed to its ability to translocate and activate PKC.[19,109,188] Activated PKC subsequently phosphorylates the cytoskeletal proteins in both tumor cells and EC.[94,112] Rearrangement of cytoskeleton, together with enhanced integrin surface expression on both tumor cells[3,67,71] and EC,[19,74] is largely responsible for most of the 12(S)-HETE effects.

IV. PERSPECTIVE

Tumor cell-platelet-endothelium interaction during metastasis is an old problem with exciting prospectives. Although the tumor cell-endothelium interaction has been considered as a critical step in tumor cell hematogenous dissemination and has drawn many scientists' attention, the interaction of platelets and tumor cells has until recently been considered as an epiphenomenon. In the last decade, graced by the studies of receptors for both agonists and antagonists, mediators of signal transduction, especially cell adhesion molecules, our understanding of the molecular mechanism of tumor cell-platelet-endothelium interaction has increased tremendously. In the next decade, one will expect that this area of research will develop in multiple directions:

1. More and more receptors, adhesion molecules, and signal transduction mediators will be discovered, and their roles in homotypic as well as heterotypic cell-cell interactions will be elucidated.
2. The molecular mechanism of the cell-cell interaction will be addressed at two levels: (a) how the receptors, adhesion molecules, and mediators are regulated at the gene level and their transcription and translation are influenced by cell-cell communication signals, and (b) how these molecules and proteins are regulated at the posttranslational level, such as phosphorylation, glycosylation, sialylation, intracellular transportation, cell surface expression, etc.
3. The advances of research on the molecular mechanism of cell interactions will not only enrich fundamental cell biology, but also will open a new vista of opportunity for development of antimetastatic therapies.

REFERENCES

1. **Weiss, L., Orr, F. W., and Honn, K. V.,** Interactions of cancer cells with the microvasculature during metastasis, *FASEB J.*, 2, 12, 1988.
2. **Weiss, L., Orr, F. W., and Honn, K. V.,** Interactions between cancer and the microvasculature: a rate-regulator for metastasis, *Clin. Expl. Metastasis,* 7, 127, 1989.
3. **Honn, K. V., Grossi, I. M., Timar, J., Chopra, H., and Taylor, J. D.,** Platelets and cancer metastasis, in *Microcirculation in Cancer Metastasis,* Orr, F. W., Buchanen, M., and Weiss, L., Eds., CRC Press, Boca Raton, FL, 1991.
4. **Hart, I. R., Goode, N. T., and Wilson, R. E.,** Molecular aspects of the metastatic cascade, *Biochim. Biophys. Acta,* 989, 65, 1989.
5. **Pauli, B. U. and Lee, C. L.,** Organ preference in metastasis: the role of organ-specifically modulated endothelial cells, *Lab. Invest.*, 58, 379, 1988.
6. **Belloni, P. N. and Tressler, R. J.,** Microvascular endothelial cell heterogeneity: interactions with leukocytes and tumor cells, *Cancer Metastasis Rev.*, 8:353, 1990.
7. **Pauli, B. U., Augustin-Voss, H. G., El-Sabban, M. E., Johnson, R. C., and Hammar, D. A.,** Organ preference of metastasis: the role of endothelial cell adhesion molecules, *Cancer Metastasis Rev.*, 9, 175, 1990.
8. **Updyke, T. V. and Nicholson, G. L.,** Malignant melanoma cell lines selected in vitro for increased homotypic adhesion properties have increased experimental metastatic potential, *Clin. Expl. Metastasis,* 4, 273, 1986.
9. **Honn, K. V., Menter, D. G., Steinert, B. W., Taylor, J. D., Onoda, J. M., and Sloane, B. F.,** Analysis of platelet, tumor cell and endothelial cell interactions in vivo and in vitro, in *Prostaglandins in Cancer Research,* Garaci, E., Paoletti, R., and Santoro, M. G., Eds., Springer-Verlag, Berlin, 1987.
10. **Honn, K. V., Menter, D. G., Onoda, J. M., Steinert, B. W., Diglio, C. A., Taylor, J. D., and Sloane, B. F.,** Tumor cell-platelet-endothelial cell interaction and prostaglandin metabolism, *Adv. Prostaglandin Thromboxane Leukotriene Res.*, 17, 981, 1987.
11. **Menter, D. G., Hatfield, J. S., Harkins, C., Sloane, B. F., Taylor, J. D., Crissman, J. D., and Honn, K. V.,** Tumor cell-platelet interactions in vitro and their relationship to in vivo arrest of hematogenously circulating tumor cells, *Clin. Expl. Metastasis,* 5, 65, 1987.
12. **Menter, D. G., Harkins, C., Onoda, J. M., Riorden, W., Sloane, B. F., Taylor, J. D., and Honn, K. V.,** Inhibition of tumor cell induced platelet aggregation by prostacyclin and carbacyclin: an ultrastructural study, *Invasion Metastasis,* 7, 109, 1987.
13. **Menter, D. G., Steinert, B. W., Sloane, B. F., Gundlach, N., O'Gara, C. Y., Marnett, L. J., Diglio, C., Walz, D., Taylor, J. D., and Honn, K. V.,** Role of platelet membrane in enhancement of tumor cell adhesion to endothelial cell and extracellular matrix, *Cancer Res.*, 47, 6751, 1987.

14. **Menter, D. G., Sloane, B. F., Steinert, B. W., Onodo, J., Craig, R., Harkins, C., Taylor, J. D., and Honn, K. V.**, Platelet enhancement of tumor cell adhesion to subendothelial matrix: role of platelet cytoskeleton and platelet membrane, *J. Natl. Cancer Inst.*, 79, 1077, 1987.
15. **Crissman, J. D., Hatfield, J., Schaldenbrand, M., Sloane, B. F., and Honn, K. V.** Arrest and extravasation of B16 amelanotic melanoma in murine lungs: a light and electron microscopic study, *Lab Invest.*, 53, 470, 1985.
16. **Crissman, J. D., Hatfield, J., Menter, D. G., Sloane, B. F., and Honn, K. V.**, Morphological study of the interaction of intravascular tumor cells with endothelial cells and subendothelial matrix, *Cancer Res.*, 48, 4065, 1988.
17. **Mareel, M. M., DeBaestselier, P., and VanRoy, F. M.**, *Mechanisms of Invasion and Metastasis*, CRC Press, Boca Raton, FL, 1991.
18. **Honn, K. V., Grossi, I. M., Diglio, C. A., Wojtukiewicz, M., and Taylor, J. D.**, Enhanced tumor cell adhesion to the subendothelial matrix resulting from 12(S)-HETE-induced endothelial cell retraction, *FASEB J.*, 3, 2285, 1989.
19. **Tang, D. G., Diglio, C. A., and Honn, K. V.**, Tumor cells express PECAM-1-like molecules involved in mediating basal tumor cell adhesion to endothelium, *Proc. Am. Assoc. Cancer Res.*, (Abstr.), 33, 33, 1992.
20. **Chopra, H., Hatfield, J. S., Chang, Y. S., Grossi, I. M., Fitzgerald, L. A., O'Gara, C. Y., Marnett, L. J., Diglio, C. A., Taylor, J. D., and Honn, K. V.**, Role of tumor cell cytoskeleton and membrane glycoprotein IRGpIIb/IIIa in platelet adhesion to tumor cell membrane and tumor cell-induced platelet aggregation, *Cancer Res.*, 48, 3787, 1988.
21. **Chopra, H., Fligiel, S. E. G., Hatfield, J. S., Nelson, K. K., Diglio, C. A., Taylor, J. D., and Honn, K. V.**, An in vivo study of the role of the tumor cell cytoskeleton in tumor cell-platelet-endothelial cell interactions, *Cancer Res.*, 50, 7686, 1990.
22. **Pearlstein, E., Ambrogio, C., Gasic, G., and Karpatkin, S.**, Inhibition of the platelet-aggregating activity of two human adenocarcinomas of the colon and an anaplastic murine tumor with a specific thrombin inhibitor: dansylarginine N-(3-ethyl-1, 5-pentanediyl) amide, *Prog. Clin. Biol. Res.*, 89, 479, 1982.
23. **Tohgo, A., Tanaka, N., Ashida, S., and Ogawa, H.**, Platelet-aggregating activities of metastasizing tumor cells. II. Variety of the aggregation mechanisms, *Invasion Metastasis*, 3, 209, 1984.
24. **Tohgo, A., Tanaka, N., and Ogawa, H.**, Platelet-aggregating activities of metastasizing tumor cells. III. Platelet aggregation as resulting from thrombin generation by tumor cells, *Invasion Metastasis*, 5, 96, 1985.
25. **Chelladurai, M., Honn, K. V., and Walz, D. A.**, HLA-DR is a procoagulant, *Biochem. Biophys. Res. Commun.*, 178, 467, 1991.
26. **Guha, A., Bach, R., Konigsburg, W., and Nemerson, Y.**, Affinity purification of human tissue factor: interaction of factor VII and tissue factor in detergent micelles, *Proc. Natl. Acad. Sci. U.S.A.*, 83, 299, 1986.
27. **Grignani, G. and Jamieson, G. A.**, Tissue factor-dependent activation of platelets by cells and microvessels of SK-05-10 human osteogenic sarcoma cell line, *Invasion Metastasis*, 7, 172, 1987.
28. **Esumi, N., Todo, S., and Imashuku, S.**, Platelet aggregating activity mediated by thrombin generation in the NCG human neuroblastoma cell line, *Cancer Res.*, 47, 2129, 1987.
29. **Nierodzik, M. L., Plotkin, A., Kajumo, F., and Karpatkin, S.**, Thrombin stimulates tumor platelet adhesion in vitro and metastasis in vivo, *J. Clin. Invest.*, 87, 229, 1991.
30. **Vu, T. K., Hung, D. T., Wheaton, V. I., and Coughlin, S. R.**, Molecular cloning of a functional thrombin receptor reveals a novel proteolytic mechanism of receptor activation, *Cell*, 64, 1057, 1991.
31. **Bastida, E., Ordinas, A., and Jamieson, G. A.**, Differing platelet aggregating effects by two tumor cell lines: absence of role for platelet-derived ADP, *Am. J. Hematol.*, 11, 367, 1981.
32. **Grignani, G., Pacchiarini, M., and Pagliarino, M.**, The possible role of blood platelets in tumor growth and dissemination, *Haematologica*, 71, 245, 1986.
33. **Longenecker, G. L., Beyers, B. J., Bowen, R. J., and King, T.**, Human rhabdosarcoma cell-induced aggregation of blood platelets, *Cancer Res.*, 49, 16, 1989.
34. **Grignani, G. and Jamieson, G. A.**, Platelets in tumor metastasis: generation of adenosine diphosphate by tumor cells is specific but unrelated to metastatic potential, *Blood*, 71, 844, 1988.
35. **Mogi, Y., Kogawa, K., Takayama, T., Yoshizaki, N., Bannai, K., Maramatsu, H., Koike, K., Kohgo, Y., Watanabe, N., and Niitsu, Y.**, Platelet aggregation induced by adenosine diphosphate release from cloned murine fibrosarcoma cells is positively correlated with the experimental metastatic potential of the cells, *Jpn. J. Cancer Res.*, 82, 192, 1991.

36. **Mackie, I. J. and Neal, C. R.,** The platelet, in *Platelet-Vessel Wall Interactions*, Pittilo, R. M. and Machin, S. J., Eds., Springer-Verlag, Berlin, 1988.
37. **Raz, A. and Lotan, R.,** Endogenous galactose-binding lectins: a new class of functional tumor cell surface molecules related to metastasis, *Cancer Metastasis Rev.*, 6, 433, 1987.
38. **Dennis, J. W. and Laferle, S.,** Importance of cell surface carbohydrates in tumor cell metastasis, in *Cancer Metastasis*, Schirmacher, V. and Schwartz-Albiez, R., Eds., Springer-Verlag, Berlin, 1989.
39. **Eggens, I., Fenderson, B., Toyokuni, T., Dean, B., Stroud, M., and Hakamori, S.,** Specific interaction between Le^x and Le^x determinants. A possible basis for cell recognition in preimplantation embryos and in embryonal carcinoma cells, *J. Biol. Chem.*, 264, 9476, 1989.
40. **Hakamori, S.,** New directions in cancer therapy based on aberrant expression of glycosphingolipids: anti-adhesion and ortho-signaling therapy, *Cancer Cells*, 3, 461, 1991.
41. **Gasic, G. J. and Gasic, T. B.,** Removal of sialic acid from the cell coat in tumor cells and vascular endothelium and its effects on metastasis, *Proc. Natl. Acad. Sci. U.S.A.*, 48, 1172, 1962.
42. **Gasic, G. J., Gasic, T. B., and Stewart, C. C.,** Antimetastatic effects associated with platelet reduction, *Proc. Natl. Acad. Sci. U.S.A.*, 61, 46, 1968.
43. **Pearlstein, E., Salk, P. L., Yogeeswaran, G., and Karpatkin, S.,** Correlation between spontaneous metastatic potential, platelet-aggregating activity of cell surface extracts, and cell surface sialylation in 10 metastatic variant derivatives of rat renal sarcoma cell line, *Proc. Natl. Acad. Sci. U.S.A.*, 77:4336, 1980.
44. **Karpatkin, S., Pearlstein, E., Salk, P. L., and Yogeeswaren, G.,** Platelet-aggregating material (PAM) of two virally-transformed tumors: SV3T3 mouse fibroblast and PW20 rat renal sarcoma. Role of cell surface sialylation, *Prog. Clin. Biol. Res.*, 89, 445, 1982.
45. **Mohanty, D. and Hilgard, P.,** A new platelet aggregating material (PAM) in an experimentally induced rat fibrosarcoma, *Thromb. Haemost.*, 51, 192, 1984.
46. **Kijima, S. I., Miyazawa, T., Itoh, M., Toyoshima, S., and Osawa, T.,** Possible mechanism of inhibition of experimental pulmonary metastasis of mouse colon adenocarcinoma 26 sublines by a sialic acid conjugate, *Cancer Res.*, 48, 3728, 1988.
47. **Mahaligam, M., Ugen, K. E., Kao, K. J., and Klein, P. A.,** Functional role of platelets in experimental metastasis studies with cloned murine fibrosarcoma cell variants, *Cancer Res.*, 48, 1460, 1988.
48. **Takahashi, N., Takahashi, Y., and Putman, F. W.,** Periodicity of leucine and tandem repetition of a 24 amino acid segment in the primary structure of leucine rich α2-glycoprotein of human serum, *Proc. Natl. Acad. Sci., U.S.A.*, 82, 1906, 1985.
49. **McEver, R. P.,** Selectins: novel receptors that mediate leukocyte adhesion during inflammation, *Thromb. Haemost.*, 65, 223, 1991.
50. **Berg, E. L., Robinson, M. K., Mansson, O., Butcher, E. C., and Magnani, J. L.,** A carbohydrate domain common to both sialyl Le^a and sialyl Le^x is recognized by the endothelial cell leukocyte adhesion molecule ELAM-1, *J. Biol. Chem.*, 266, 14869, 1991.
51. **Bresalier, R. S., Rockwell, R. W., Dahiya, R., Duh, Q. Y., and Kim, Y. S.,** Cell surface sialoprotein alterations in metastatic murine colon cancer cell lines selected in an animal model for colon cancer metastasis, *Cancer Res.*, 50, 1299, 1990.
52. **Hoff, S. D., Matsushita, Y., Ota, D. M., Cleary, K. R., Tamori, T., Hakomori, S., and Irimura, T.,** Increased expression of sialyl-dimeric Le^x antigen in liver metastases of human colorectal carcinoma, *Cancer Res.*, 49, 6883, 1989.
53. **Sultan, C., Plantavid, M., Bachelot, C., Grondin, P., Breton, M., Mauco, G., Lievy-Toledano, S., Cann, J. P., and Chap, H.,** Involvement of platelet glycoprotein IIb-IIIa (αIIbβ3 integrin) in thrombin-induced synthesis of phosphatidylinositol 3,4-biphosphate, *J. Biol. Chem.*, 266, 23554, 1991.
54. **Edelman, G. M. and Crossin, K. V.,** Cell adhesion molecules: implications for a molecular histology, *Annu. Rev. Biochem.*, 60, 155, 1991.
55. **Elices, M. I., Osborn, L., Takada, Y., Crouse, C., Luhowsky, S., Hemler, M. E., and Lobb, R. R.,** VCAM-1 on activated endothelium interacts with the leukocyte VLA-4 at a site distinct from the VLA-4 fibronectin binding site, *Cell*, 60, 577, 1990.
56. **Simmons, D. L., Walker, C., Power, C., and Pigott, R.,** Molecular cloning of CD31, a putative intercellular adhesion molecule closely related to carcinoembryonic antigen, *J. Exp. Med.*, 171, 2147, 1990.
57. **Kieffer, N. and Phillips, D. R.,** Platelet membrane glycoproteins: functions in cellular interactions, *Annu. Rev. Cell. Biol.*, 6, 329, 1990.
58. **Phillips, D. R., Charo, I. F., and Scarborough, R. M.,** Gp-IIb-IIIa: the "responsive" integrin, *Cell*, 65, 359, 1991.

59. **Frazier, W. A.,** Thrombospondins, *Curr. Opinion Cell Biol.*, 3, 792, 1991.
60. **Newman, P. J., Seligsohn, U., Lyman, S., and Coller, B. S.,** The molecular genetic basis of Glanzmann thrombasthenia in the Iraqi-Jewish and Arab populations in Israel, *Proc. Natl. Acad. Sci. U.S.A.*, 88, 3160, 1991.
61. **Grossi, I. M., Fitzgerald, L. A., Kendall, A., Taylor, J. D., Sloane, B. F., and Honn, K. V.,** Inhibition of human tumor cell induced platelet aggregation by antibodies to platelet glycoproteins Ib and IIB/IIIa, *Proc. Soc. Exp. Biol. Med.*, 186, 378, 1987.
62. **Bastida, E., Almiral, L., and Ordinas, A.,** Tumor-cell-induced platelet aggregation is a glycoprotein-dependent and lipoxygenase-associated process, *Int. J. Cancer*, 39, 760, 1987.
63. **Kitagawa, H., Yamamoto, N., Yamamoto, K., Tanoue, K., Kosaki, G., and Yamazaki, H.,** Involvement of platelet membrane glycoprotein Ib and glycoprotein IIb/IIIa complex in thrombin-dependent and independent platelet aggregations induced by tumor cells, *Cancer Res.*, 49, 537, 1989.
64. **Grossi, I. M., Hatfield, J. S., Fitzgerald, L. A., Newcombe, M., Taylor, J. D., and Honn, K. V.,** Role of tumor cell glycoproteins immunologically related to glycoproteins Ib and IIb/IIIa in tumor cell-platelet and tumor cell-matrix interactions, *FASEB J.*, 2, 2385, 1988.
65. **McGregor, B. C., McGregor, J. L., Weiss, L. M., Wood, G. S., Hu, C. H., Boukerche, H., and Warnke, R. A.,** Presence of cytoadhesions (IIb-IIIa-like glycoproteins) on human metastatic melanomas but not on benign melanomas, *Am. J. Clin. Pathol.*, 92, 495, 1989.
66. **Boukerche, H., Berthier-Vergnes, O., Tabone, E., Dore, J. F., Leung, L. L. K., and McGregor, J. L.,** Platelet-melanoma cell interaction is mediated by the glycoprotein IIb-IIIa complex, *Blood*, 74, 658, 1989.
67. **Boukerche, H., Berthier-Vergnes, O., Tabone, E., Dore, J. F., Leung, L. L. K., and McGregor, J. L.,** A monoclonal antibody (LYP18) directed against the blood platelet glycoprotein IIb/IIIA complex inhibits human melanoma growth in vivo, *Blood*, 74, 909, 1989.
68. **Knudsen, K. A., Smith, L., Smith, S., Karczewski, J., and Tuszynski, G. P.,** Role of IIb-IIIa-like glycoproteins in cell-substratum adhesion of human melanoma cells, *J. Cell Physiol.*, 136, 471, 1988.
69. **Chopra, H., Timar, J., Chen, Y. Q., Rong, X. M., Grossi, I. M., Fitzgerald, L. A., Taylor, J. D., and Honn, K. V.,** The lipoxygenase metabolite 12(S)-HETE induces a cytoskeleton-dependent increase in surface expression of integrin αIIbβ3 on melanoma cells, *Int. J. Cancer*, 49, 774, 1991.
70. **Grossi, I. M., Fitzgerald, L. A., Umbarger, L. A., Nelson, K. K., Diglio, C. A., Taylor, J. D., and Honn, K. V.,** Bidirectional control of membrane expression and/or activation of the tumor cell IRGpIIb/IIIa receptor and tumor cell adhesion by lipoxygenase products of arachidonic acid and linoleic acid, *Cancer Res.*, 49, 1029, 1989.
71. **Chopra, H., Timar, J., Rong, X. M., Grossi, I. M., Hatfield, J. S., Fligiel, S. E. G., Finch, C. A., Taylor, J. D., and Honn, K. V.,** Is there a role for the tumor cell integrin αIIbβ3 and cytoskeleton in tumor cell-platelet interaction?, *Clin. Exp. Metastasis*, 10, 125, 1992.
72. **Timar, J., Chopra, H., Grossi, I. M., Taylor, J. D., and Honn, K. V.,** The lipoxygenase metabolite of arachidonic acid, 12(S)-HETE, induces cytoskeleton dependent upregulation of integrin αIIBβ3 in melanoma cells, in *Eicosanoids and Other Bioactive Lipids in Cancer and Radiation Injury*, Honn, K. V., Marnett, L. J., Nigam, S., and Walden T., Eds., Kluwer Academic Publishers, Boston, 1991.
73. **Honn, K. V., Grossi, I. M., Steinert, B. W., Chopra, H., Onoda, J., Nelson, K. K., and Taylor, J. D.,** Lipoxygenase regulation of membrane expression of tumor cell glycoproteins and subsequent metastasis, *Adv. Prostaglandin Thromboxane Leukotriene Res.*, 19, 439, 1989.
74. **Honn, K. V., Grossi, I. M., Diglio, C. A., and Taylor, J. D.,** Role of 12-lipoxygenase metabolites in integrin glycoprotein receptors in metastasis, in *New Concepts in Cancer: Metastasis, Oncogenes, and Growth Factors*, Etievant, C., Cros, J., and Rustum, Y. M., Eds., MacMillan, New York, 1990.
75. **Chen, Y. Q. and Honn, K. V.,** Eicosanoid regulation of tumor cell-platelet and -endothelium interaction during arrest and extravasation, in *Eicosanoids and Other Bioactive Lipids in Cancer, Inflammation and Radiation Injury*, Nigam, S., Honn, K. V., Marnett, L. J., and Walden, T., Eds., Kluwer Academic Publishers, Boston, 1992.
76. **Cardinali, M., Uchino, R., and Chung, S., II,** Interaction of fibrinogen with murine melanoma cells: covalent association with cell membranes and protection against recognition by lymphokine-activated killer cells, *Cancer Res.*, 50, 8010, 1990.
77. **Gawaz, M. P., Loftus, J. C., Bajt, M. L., Frojmovic, M. M., Plow, E. F., and Ginsberg, M. H.,** Ligand bridging mediates integrin αIIbβ3 (Platelet GpIIb-IIIa)-dependent homotypic and heterotypic cell-cell interactions, *J. Clin. Invest.*, 88, 1128, 1991.
78. **Savage, B. and Ruggeri, Z. M.,** Selective recognition of adhesive sites in surface-bound fibrinogen by glycoprotein IIb-IIIa on non-activated platelets, *J. Biol. Chem.*, 266, 11227, 1991.

79. **Guilino, D., Boudignon, C., Zhang, L., Concord, E., Rabiet, M. J., and Marguerie, G.,** Ca^{2+}-binding properties of the platelet glycoprotein IIb ligand-interacting domain, *J. Biol. Chem.,* 267, 1001, 1992.
80. **Frelinger, A. L., Lamm, S. C-T., III, Plow, E. F., Smith, M. A., Loftus, J. C., and Ginsberg, M. H.,** Occupancy of an adhesive glycoprotein receptor modulates expression of an antigenic site involved in cell adhesion, *J. Biol. Chem.,* 263, 12397, 1988.
81. **O'Toole, T. E., Loftus, J. C., Du, X. P., Glass, A. A., Ruggeri, Z. M., Shattil, S. J., Plow, E. F., and Ginsberg, M. H.,** Affinity modulation of the αIIbβ3 integrin (platelet GpIIb-IIIa) is an intrinsic property of the receptor, *Cell Regul.,* 1, 883, 1990.
82. **Kouns, W. C. and Jennings, L. K.,** Activation-dependent exposure of the GpIIb-IIIa fibrinogen receptor, *Thromb. Res.,* 63, 343, 1991.
83. **Du, X. P., Plow, E. F., Freelinger, A. L., III, O'Toole, T. E., Loftus, J. C., and Ginsberg, M. H.,** Ligands "activate" integrin αIIbβ3 (platelet GpIIb-IIIa), *Cell,* 65, 409, 1991.
84. **Yabkowitz, R. and Dixit, V. M.,** Human carcinoma cells express receptors for distinct domains of thrombospondin, *Cancer Res.,* 51, 1645, 1991.
85. **Clezardin, P., Serre, C. M., Trzeeciak, M. C., Drouin, J., and Delmas, P. D.,** Thrombospondin binds to the surface of human osteosarcoma cells and mediates platelet-osteosarcoma cell interaction, *Cancer Res.,* 51, 2621, 1991.
86. **Tuszynski, G. P., Gasic, G. B., Rothman, V. L., Knudsen, K. A., and Gasic, G. J.,** Thrombospondin, a potentiator of tumor cell metastasis, *Cancer Res.,* 47, 4130, 1987.
87. **Tuszynski, G. P., Rothman, V. L., Deutch, A. H., and Hamilton, B. K.,** Biological activities of peptides and peptide analogues derived from common sequences present in thrombospondin, properdin, and malarial proteins, *J. Cell. Biol.,* 116, 209, 1992.
88. **Watanabe, M., Okchini, E., Sugimoto, Y., and Tsuruo, T.,** Identification of a platelet-aggregating factor of murine colon adenocarcinoma 26: M_r 44,000 membrane protein as determined by monoclonal antibodies, *Cancer Res.,* 48, 6411, 1988.
89. **Watanabe, M., Okchini, E., Sugimoto, Y., and Tsuruo, T.,** Expression of a M_r 41,000 glycoprotein associated with thrombin-independent platelet aggregation in high metastatic variants of murine B16 melanoma, *Cancer Res.,* 50, 6657, 1990.
90. **Tanaka, N., Tohgo, A., and Ogawa, H.,** Platelet aggregating activities of metastasizing tumor cells. V. In situ roles of platelets in hematogenous metastases, *Invasion Metastasis,* 6, 209, 1986.
91. **Honn, K. V., Menter, D. G., Onoda, J. M., Taylor, J. D., and Sloane, B. F.,** Role of prostacyclin as a natural deterrent to hematogenous tumor metastasis, in *Cancer Invasion and Metastasis: Biologic and Therapeutic Aspects,* Nicolson, G. L. and Milas, L., Eds., Raven Press, New York, 1984.
92. **Honn, K. V. and Sloane, B. F.,** *Hemostatic Mechanisms and Metastasis,* Martinus Nijhoff, Boston, 1984.
93. **Timar, J., Liu, B., Bazaz, R., Taylor, J. D., and Honn, K. V.,** Fatty acid modulation of cancer cell spreading and cytoskeleton rearrangement, in *Eicosanoids and Other Bioactive Lipids in Cancer, Inflammation and Radiation Injury,* Nigam, S., Honn, K. V., Marnett, L. J., Walden, T., Eds., Kluwer Academic Publishers, Boston, 1992.
94. **Taylor, J. D., Timar, J., Tang, D., Bazaz, R., Chopra, H., Kimler, V., and Honn, K. V.,** 12(S)-HETE induces cytoskeleton phosphorylation and rearrangement in melanoma cells, in *Eicosanoids and Other Bioactive Lipids in Cancer, Inflammation and Radiation Injury,* Nigam, S., Honn, K. V., Marnett, L. J., and Walden, T., Eds., Kluwer Academic Publishers, Boston, 1992.
95. **Skolnik, G., Bagge, V., Dahlstrom, A., and Ahlman, M.,** The importance of 5-HT for tumor cell lodgement in the liver, *Int. J. Cancer,* 33, 519, 1984.
96. **Skolnik, G., Bagge, V., Blomquist, G., Djarv, L., and Ahlman, H.,** The role of calcium channels and serotonin (5-HT_2) receptors for tumor cell lodgement in the liver, *Clin. Exp. Metastasis,* 7, 169, 1989.
97. **Turner, W. A., Szlag, D. C., and Taylor, J. D.,** Platelet fibronectin release induced by Walker 256 rat carcinoma tumor cells, *Clin. Exp. Metastasis,* 3, 209, 1985.
98. **Tsuruo, T., Watanabe, M., and Oh-hara, T.,** Stimulation of the growth of metastatic clones of mouse colon adenocarcinoma 26 in vitro by platelet-derived growth factor, *Jpn. J. Cancer Res.,* 80, 136, 1989.
99. **Akedo, H., Shinkai, K. Mukai, M., and Komatsu, K.,** Potentiation and inhibition of tumor cell invasion by host cells and mediators, *Invasion Metastasis,* 9, 134, 1989.
100. **Orr, F. W., Millar, B. W., and Singh, G.,** Chemotactic activity of bone and platelet-derived TGF-beta for bone-metastasizing rat Walker 256 carcinoma cells, *Invasion Metastasis,* 10, 241, 1990.
101. **Honn, K. V., Steinert, B. W., Moin, K., Onoda, J. M., Taylor, J. D., and Sloane, B. F.,** The role of platelet cyclooxygenase and lipoxygenase pathways in tumor cell induced platelet aggregation, *Biochem. Biophys. Res. Commun.,* 145, 384, 1987.

102. **Steinert, B. W., Tang, D. G., Jones, C. L., Grossi, I. M., Umbarger, L. A., and Honn, K. V.,** Studies on the role of platelet eicosanoid metabolism and integrin αIIbβ3 in tumor cell induced platelet aggregation, *Int. J. Cancer,* 54, 92, 1993.
103. **Honn, K. V. and Chen, Y. Q.,** Prostacyclin, hydroxy fatty acids and cancer metastasis, in *Prostacyclin: New Perspectives in Basic Research and Novel Therapeutic Indications,* Rubaryi, G. M., and Vane, J. R., Eds., Elsevier Science Publishers, Amsterdam, 1992.
104. **Grignani, G., Pacchiarini, L., Almasio, P., Pagliarino, M., and Gamba, G.,** Activation of platelet prostaglandin biosynthesis pathway during neoplastic cell-induced platelet aggregation, *Thromb. Res.,* 34, 147, 1984.
105. **Pacchiarini, L., Zucchella, M., Milanesi, G., Tocconi, F., Bonomi, E., Canevari, A., and Grignani, G.,** Thromboxane production by platelets during tumor cell-induced platelet activation, *Invasion Metastasis,* 12, 102, 1991.
106. **Honn, K. V., Grossi, I. M., Chopra, M., Steinert, B. W., Onoda, J. M., Nelson, K. K., and Taylor, J. D.,** Role of tumor cell eicosanoids and membrane glycoproteins IRGpIIb/IIIa in metastasis, in *Lipid Peroxidation and Cancer,* Nigam, S. and Slater, M. B., Eds., Springer-Verlag, Berlin, 1988.
107. **Spector, A. A., Gordon, J. A., and Moore, S. A.,** Hydroxyeicosatetraenoic acids (HETEs), *Prog. Lipid Res.,* 27, 271, 1988.
108. **Marnett, L. J., Leithauser, M. T., Richards, K. M., Blair, I., Honn, K. V., Yamamoto, S., and Yoshimoto, T.,** Arachidonic acid metabolism of cytosolic fractions of Lewis lung carcinoma cells, *Adv. Prostaglandin Thromboxane Leukotriene Res.,* 21, 895, 1990.
109. **Liu, B., Timar, J., Howlett, J., Diglio, C. A., and Honn, K. V.,** Lipoxygenase metabolites of arachidonic and linoleic acids modulate the adhesion of tumor cells to endothelium via regulation of protein kinase, *Cell Regul.,* 2, 1045, 1991.
110. **Timar, J., Chen, Y. Q., Liu, B., Bazaz, R., Taylor, J. D., and Honn, K. V.,** The lipoxygenase metabolite 12(S)-HETE promotes αIIbβ3 integrin mediated tumor cell spreading on fibronectin, *Int. J. Cancer,* 52, 594, 1992.
111. **Grossi, I. M., Diglio, C. A., and Honn, K. V.,** Control of tumor cell induced endothelial cell retraction by lipoxygenase metabolites, prostacyclin and prostacyclin analogs, in *Eicosanoids and Other Bioactive Lipids in Cancer and Radiation Injury,* Honn, K. V., Marnett, L. J., Nigam, S., and Walden, T., Eds., Kluwer Academic Publishers, Boston, 1991.
112. **Tang, D. G., Diglio, C. A., and Honn, K. V.,** 12(S)-HETE-induced microvascular endothelial cell retraction is mediated by cytoskeletal rearrangement dependent on PKC activation, in *Eicosanoids and Other Bioactive Lipids in Cancer, Inflammation and Radiation Injury,* Nigam, S., Honn, K. V., Marnett, L. J., and Walden, T., Eds., Kluwer Academic Publishers, Boston, 1992.
113. **Honn, K. V., Grossi, I. M., Chang, Y. S., and Chen, Y.,** Lipoxygenases, integrin receptors, and metastasis, in *Eicosanoids and Other Bioactive Lipids in Cancer and Radiation Injury,* Honn, K. V., Nigam, S., Marnett, L. J., and Walden, T., Eds., Kluwer Academic Publishers, Boston, 1991.
114. **Sloane, B. F., Rozhin, J., Gomez, A. P., Grossi, I. M., and Honn, K. V.,** Effects of 12-hydroxyeicosatetraenoic acid on release of cathepsin B and cysteine proteinase inhibitors from malignant melanoma cells, in *Eicosanoids and Other Bioactive Lipids in Cancer and Radiation Injury,* Honn, K. V., Nigam, S., Marnett, L. J., and Walden, T., Eds., Kluwer Academic Publishers, Boston, 1991.
115. **Raz, A., Silletti, S., and Honn, K. V.,** Effect of 12-HETE on the expression of autocrine motility factor-receptor and motility in melanoma cells, in *Eicosanoids and Other Bioactive Lipids in Cancer, Inflammation and Radiation Injury,* Nigam, S., Honn, K. V., Marnett, L. J., and Walden, T., Eds., Kluwer Academic Publishers, Boston, 1992.
116. **Honn, K. V., Nelson, K. K., Renaud, C., Bazaz, R., Diglio, C. A., and Timar, J.,** Fatty acid modulation of tumor cell adhesion to microvessel endothelium and experimental metastasis, *Prostaglandins,* 44, 413, 1992.
117. **Buchanan, M. R., Bertomeu, M. C., Bastida, E., Orr, F. W., and Gallo, S.,** Eicosanoid metabolism and tumor cell endothelial cell adhesion, in *Eicosanoids and Other Bioactive Lipids in Cancer and Radiation Injury,* Honn, K. V., Nigam, S., Marnett, L. J., and Walden, T., Eds., Kluwer Academic Publishers, Boston, 1991.
118. **Filder, I. J. and Hart, I. R.,** Biological diversity in metastatic neoplasms: origins and implication, *Science,* 217, 998, 1982.
119. **Nicolson, G. L.,** Cancer metastasis: organ colonization and cell surface properties of malignant cells, *Biochim. Biophys. Acta,* 695, 113, 1982.
120. **Nicolson, G. L.,** Organ specificity of tumor metastasis: role of preferential adhesion, invasion, and growth of malignant tumor cells at specific secondary sites, *Cancer Metastasis Rev.,* 7, 143, 1988.

121. **Nicolson, G. L.,** Tumor and host molecules important in the organ preference of metastasis, *Semin. Cancer Biol.,* 2, 143, 1991.
122. **Holzmann, B. and Weissmann, I. L.,** Peyer's patch specific lymphocyte homing receptor consists of a VLA4-like alpha chain associated with either of two beta chains, one of which is novel, *EMBO J.,* 8, 1735, 1989.
123. **Dustin, M. L. and Springer, T. A.,** Role of lymphocyte adhesion receptors in transient interactions and cell locomotion, *Annu. Rev. Immunol.,* 9, 27, 1991.
124. **Roosien, F. F., DeRijk, D., Birkker, A., and Roos, E.,** Involvement of LFA-1 in lymphoma invasion and metastasis demonstrated with LFA-1 deficient mutants, *J. Cell. Biol.,* 108, 1979, 1989.
125. **Lauri, D., Needham, L., Martin-Padura, I., and Dejana, E.,** Tumor cell adhesion to endothelium: endothelial leukocyte adhesion molecule-1 as an inducible adhesive receptor for colon carcinoma cells, *J. Natl. Cancer Inst.,* 81, 556, 1991.
126. **Hakamori, S.,** Aberrant glycosylation in tumors and tumor associated carbohydrate antigen, *Adv. Cancer Res.,* 52, 257, 1989.
127. **Hakamori, S.,** Bifunctional role of glycosphingolipids: modulators for transmembrane signaling and mediators for cellular interactions, *J. Biol. Chem.,* 265, 18713, 1990.
128. **Kojima, N. and Hakamori, S.,** Cell adhesion, spreading, and motility of GM3-expressing cells based on glycosphingolipid interactions, *J. Biol. Chem.,* 264, 20159, 1991.
129. **Yednock, T. A. and Rosen, S. D.,** Lymphocyte homing, *Adv. Immunol.,* 44, 313, 1989.
130. **Matsushita, Y., Hoff, S. D., Nudelman, Otaka, M., Hakamori, S., Ota, D. M., Cleary, K. R., and Irimura, T.,** Metastatic behavior and cell surface properties of HT-29 human colon carcinoma variant cells selected for their different expression of sialy dimeric Le^x antigen, *Clin. Expl. Metastasis,* 9, 283, 1991.
131. **Rice, G. E., Gimbrone, M. A., and Bevilaqua, M. P.,** Increased adhesion of human melanoma cells to activated vascular endothelium, *Am. J. Pathol.,* 133, 204, 1988.
132. **Rice, C. E. and Bevilacqua, M. P.,** An inducible endothelial cell glycoprotein mediates melanoma adhesion, *Science,* 246, 1303, 1989.
133. **Martin-Padura, I., Mortraini, R., Lauri, D., Bernasconi, S., Sanchez-Madrid, F., Parmiani, G., Mantovani, A., Anichi, A., and Dejana, E.,** Heterogeneity in human melanoma cell adhesion to cytokine activated endothelium correlates with VLA4 expression, *Cancer Res.,* 51, 2239, 1991.
134. **Haynes, B. F., Telen, M. J., Hale, L. P., and Denning, S. M.,** CD44: a molecule involved in leukocyte adherence and T cell activation, *Immunol. Today,* 10, 423, 1989.
135. **Haynes, B. F., Liao, H. X., and Patton, K. L.,** The transmembrane hyaluronate receptor (CD44): multiple function, multiple forms, *Cancer Cells,* 3, 347, 1991.
136. **Heino, J. and Massague, J.,** Transforming growth factor-β switches the pattern of integrins expressed in MG-63 human osteosarcoma cells and causes a selective loss of cell adhesion to laminin, *J. Biol. Chem.,* 264, 21806, 1989.
137. **Johnson, J. P., Stade, B. G., Holzmann, B., Schwable, W., and Riethmuller, G.,** De novo expression of intercellular adhesion molecule-1 in melanoma correlates with increased risk of metastasis, *Proc. Natl. Acad. Sci. U.S.A.,* 86, 641, 1989.
138. **Burrows, F. J., Haskard, D. O., Hart, I. R., Marshall, J. F., Selkirk, S., Poole, S., and Thorpe, P. E.,** Influence of tumor-derived interleukin-1 on melanoma-endothelial cell interactions in vitro, *Cancer Res.,* 51, 4768, 1991.
139. **Honn, K. V., Chen, Y. Q., Timar, J., Onoda, J. M., Hatfield, J. S., Fligiel, S. E. G., Steinert, B. W., Diglio, C. A., Grossi, I. M., Nelson, K. K., and Taylor, J. D.,** $\alpha IIb\beta 3$ integrin expression and function in subpopulations of murine tumors, *Exp. Cell Res.,* 201, 23, 1992.
140. **Chang, Y. S., Chen, Y. Q., Fitzgerald, L. A., and Honn, K. V.,** Analysis of integrin mRNA in human and rodent tumor cells, *Biochem. Biophys. Res. Commun.,* 176, 108, 1991.
141. **Chang, Y. S., Chen, Y. Q., Timar, J., Nelson, K. K., Grossi, I. M., Fitzgerald, L. A., Diglio, C. A., and Honn, K. V.,** Increased expression of $\alpha IIb\beta 3$ integrin in lung-colonizing subpopulations of murine melanoma cells, *Int. J. Cancer,* 51, 445, 1992.
142. **Zhu, D. Z., Cheng, C. F., and Pauli, B. U.,** Mediation of lung metastasis of murine melanomas by a lung-specific endothelial cell adhesion molecule, *Proc. Natl. Acad. Sci. U.S.A.,* 88, 9568, 1991.
143. **Handa, K. Y., Igarashi, Y., Nisar, M., and Hakamori, S.,** Downregulation of GMP-140 expression on platelets by N,N-dimethyl and N,N,N,-trimethyl derivatives of sphingosine, *Biochemistry,* 30, 11682, 1991.
144. **Castronovo, Y., Taraboletti, G., Liotta, L. A., and Sobel, M. E.,** Modulation of laminin receptor expression by estrogen and progestins in human breast cancer cell lines, *J. Natl. Cancer Inst.,* 8, 781, 1989.

145. **Rossino, P., Defilippi, R., Silengo, L., and Tarone, G.,** Up-regulation of the integrin α1/β1 in human neuroblastoma cells differentiated by retinoic acid: correlation with increased neurite outgrowth responses to laminin, *Cell Regul.*, 2, 1021, 1991.
146. **Bevilacqua, M. P., Pober, J. S., Mendrick, D. L., Cotran, R. S., and Gimbrone, M. A.,** Identification of an inducible endothelial-leukocyte adhesion molecule, *Proc. Natl. Acad. Sci. U.S.A.*, 84, 9238, 1987.
147. **Rosen, S. D.,** The LEC-CAMs: an emerging family of cell-cell adhesion receptors based upon carbohydrate recognition, *Am. J. Respir. Cell Mol. Biol.*, 3, 397, 1990.
148. **Sporn, M. B. and Robert, A. B.,** Transforming growth factor-β: multiple actions and potential clinical applications, *JAMA*, 262, 938, 1989.
149. **Giavazzi, R., Garfalo, A., Bani, M. R., Abbate, M., Ghezzi, P., Boraschi, D., Mantovani, A., and Dejana, E.,** Interleukin-1-induced augmentation of experimental metastasis from a human melanoma in nude mice, *Cancer Res.*, 50, 4771, 1990.
150. **Auerbach, R.,** Pattern of tumor metastasis: organ selectivity in the spread of cancer cells, *Lab Invest.*, 58, 361, 1988.
151. **Liotta, L. A.,** Tumor invasion and metastasis: role of the extracellular matrix, *Cancer, Res.*, 46, 1, 1986.
152. **Cioce, V., Castronovo, V., Shmookler, B. M., Garbisa, S., Grigioni, W. F., Liotta, L. A., and Sobel, M. E.,** Increased expression of the laminin receptor in human colon cancer, *J. Natl. Cancer Inst.*, 83, 29, 1991.
153. **Zutter, M. M., Mazoujian, G., and Santoro, S. A.,** Decreased expression of integrin adhesive protein receptors in adenocarcinoma of the breast, *Am. J. Pathol.*, 137, 863, 1990.
154. **Feldman, L. E., Shin, K. C., Natale, R. B., and Tod R. F., III,** β1 integrin expression on human small cell lung cancers, *Cancer Res.*, 51, 1065, 1991.
155. **Schreiner, C., Bauer, J. Margolis, M., and Juliano, R. L.,** Expression and role of integrins in adhesion of human colonic carcinoma cells to extracellular matrix components, *Clin. Expl. Metastasis*, 2, 163, 1991.
156. **Lichtner, R. B., Belloni, P. N., and Nicolson, G. L.,** Differential adhesion of metastatic rat mammary carcinoma cells to organ-derived microvessel endothelial cells and subendothelial matrix, *Exp. Cell. Biol.*, 57, 146, 1989.
157. **Hutchinson, R., Fligiel, S., Appleyard, J., Varani, J., and Wicha, M.,** Attachment of neuroblastoma cells to extracellular matrix: correlation with metastatic capacity, *J. Lab. Med.*, 113, 561, 1989.
158. **Liotta, L. A., Stracke, M. I., Aznavoorian, S. A., Becker, M. E., and Schiffman, A.,** Tumor cell motility, *Semin. Cancer Biol.*, 2, 111, 1991.
159. **Werb, Z., Tremble, P. M., Behrendtsen, O., Crowley, E., and Damsky, C. H.,** Signal transduction through the fibronectin receptor induces collagenase and stromelysin gene expression, *J. Cell Biol.*, 109, 877, 1989.
160. **Liotta, L. A. and Stetler-Stevenson, W.,** Metalloproteinases and malignant conversion: does correlation imply causality?, *J. Natl. Cancer Inst.*, 81, 556, 1989.
161. **Herlyn, M. and Malkowicz, S. B.,** Regulatory pathways in tumor growth and invasion, *J. Clin. Invest.*, 65, 262, 1991.
162. **Mortarini, R., Anichini, A., and Parmiani, G.,** Heterogeneity for integrin expression and cytokine-mediated VLA modulation can influence the adhesion of human melanoma cells to extracellular matrix proteins, *Int. J. Cancer*, 47, 551, 1991.
163. **Smith, C. W. and Anderson, D. C.,** PMN adhesion and extravasation as a paradigm for tumor cell dissemination, *Cancer Metastasis Rev.*, 10, 61, 1991.
164. **Butcher, E. C.,** Leukocyte-endothelial cell recognition: three (or more) steps to specificity and diversity, *Cell*, 61, 1033, 1991.
165. **Kishimoto, T. K.,** A dynamic model for neutrophil localization to inflammatory site, *J. N.I.H. Res.*, 3, 75, 1991.
166. **von Adrian, U. H., Chambers, J. D., McEvoy, L. M., Bargatze, R. F., Arfors, K. E., and Butcher, E. C.,** Two-step model of leukocyte-endothelial cell interaction in inflammation: distinct roles for LECAM-1 and the leukocyte β2 integrins in vivo, *Proc. Natl. Acad. Sci. U.S.A.*, 88, 7538, 1991.
167. **Manning, D. R. and Brass, L. F.,** The role of GTP-binding proteins in platelet activation, *Thromb. Haemost.*, 66, 393, 1991.
168. **Freissmuth, M., Casey, P. J., and Gilman, A. G.,** G proteins control diverse pathways of transmembrane signaling, *FASEB J.*, 3, 2125, 1989.

169. **Kojima, N. and Hakamori, S.,** Synergistic effect of 2 cell recognition systems: glycosphongolipid-glycosphingolipid interaction and integrin-receptor interaction with pericellular matrix protein, *Glycobiology,* in press, 1992.
170. **Smreka, A. V., Helper, J. R., Brown, K. O., and Sternweiss, P. C.,** Regulation of polyphosphoinositide-specific phospholipase C activity by purified Gq, *Science,* 251, 804, 1991.
171. **Shattil, S. J., Hoxie, J. A., Cunnningham, M., and Brass, L. F.,** Changes in the platelet membrane glycoprotein IIb-IIIa complex during platelet activation, *J. Biol. Chem.,* 260, 11107, 1985.
172. **Shattil, S. J. and Brass, L. F.,** Induction of the fibrinogen receptor on human platelets by intracellular mediators, *J. Biol. Chem.,* 262, 992, 1987.
173. **van Willigen, G. and Akkerman, J.-W. N.,** Protein kinase C and cyclic AMP regulate reversible exposure of binding sites for fibrinogen on the glycoprotein IIB-IIIA complex of human platelets, *Biochem. J.,* 273, 115, 1991.
174. **Hillery, C. A., Smith, S. S., and Parise, L. V.,** Phosphorylation of human platelet glycoprotein IIa (GpIIIa). Dissociation from fibrinogen receptor activation and phosphorylation of GpIIIa in vitro, *J. Biol. Chem.,* 266, 14663, 1991.
175. **Fox, J. E. B. and Boyles, J. K.,** The membrane skeleton — a distinct structure that regulates the function of cells, *Bioessays,* 8, 14, 1988.
176. **Hirata, M., Hayashi, Y., Ushikubi, F., Nakanishi, S., and Narumiya, S.,** Cloning and expression of cDNA for a human thromboxane A_2 receptor, *Nature,* 349, 617, 1991.
177. **Honda, Z., Nakamura, M., Miki, I., Minami, M., Watanabe, T., Seyama, Y., Okado, H., Toh, H., Ito, K., Miyamoto, T., and Shimizu, T.,** Cloning by functional expression of platelet-activating factor receptor from guinea pig lung, *Science,* 349, 342, 1991.
178. **Isenberg, W. M., McEver, R. P., Phillips, D. R., Shuman, M. A., and Bainton, D. F.,** The platelet fibrinogen receptor: an immunogold-surface replica study of agonist-induced ligand binding and receptor clustering, *J. Cell. Biol.,* 104, 1655, 1987.
179. **Kouns, W. C., Fox, C. F., Lamoreaux, W. J., Loons, L. B., and Jennings, L. K.,** The effect of glycoprotein IIb-IIIa receptor occupancy on the cytoskeleton of resting and activated platelets, *J. Biol. Chem.,* 266, 13891, 1991.
180. **Grondin, P., Plantavid, M., Sultan, C., Breton, M., Mauco, G., and Chap, H.,** Interaction of $pp^{60c\text{-}src}$, phospholipase C, inositol-lipid, and diacylglycerol kinases with the cytoskeletons of thrombin-stimulated platelets, *J. Biol. Chem.,* 266, 15705, 1991.
181. **Ferrel, J. E. and Martin, G. S.,** Tyrosine-specific protein phosphorylation is regulated by glycoprotein IIb-IIIa in platelets, *Proc. Natl. Acad. Sci. U.S.A.,* 86, 2234, 1989.
182. **Golden, A., Brugge, J. S., and Shattil, S. J.,** Role of platelet membrane glycoprotein IIb-IIIa in agonist-induced tyrosine phosphorylation of platelet proteins, *J. Cell Biol.,* 111, 3117, 1990.
183. **Yamaguchi, A., Tanoue, K., and Yamazaki, H.,** Secondary signals mediated by GpIIb/IIIa in thrombin-activated platelets, *Biochim. Biophys. Acta,* 1054, 3, 1990.
184. **Fujimoto, T., Fujimura, K., and Kuramoto, A.,** Functional Ca^{2+} channel produced by purified platelet membrane glycoprotein IIb-IIIa complex incorporated into planar phospholipid bilayer, *Thromb. Haemost.,* 66, 598, 1991.
185. **Bachelot, C., Rendu, F., Boucheix, C., Hogg, N., and Levy-Toledano, S.,** Activation of platelets induced by mAb P_{256} specific for glycoprotein IIb-IIIa. Possible evidence for a role for IIb-IIIa in membrane signal transduction, *Eur. J. Biochem.,* 190, 177, 1990.
186. **Banga, H. S., Simmons, E. R., Brass, L. F., and Rittenhouse, S. E.,** Activation of phospholipase A and C in human platelets exposed to epinephrine, *Proc. Natl. Acad. Sci. U.S.A.,* 83, 9197, 1986.
187. **Kieffer, N., Fitzgerald, L. A., Wolf, D., Cheresh, D. A., and Phillips, D. R.,** Adhesive properties of the β3 integrins: comparison of GPIIb-IIIa and the vitronectin receptor individually expressed in human melanoma cells, *J. Cell Biol.,* 113, 451, 1991.
188. **Liu, B., Renaud, C., Kowynia, J., Nelson, K. K., Roudacherski, E., Snyder, D., Timar, J., and Honn, K. V.,** Activation of protein kinase C by 12(S)-HETE: role in tumor cell metastasis, in *Eicosanoids and Other Bioactive Lipids in Cancer, Inflammation and Radiation Injury,* Nigam, S., Honn, K. V., Marnett, L. J., and Walden, T., Eds., Kluwer Academic Publishers, Boston, 1992.

Chapter 17

Pharmacological Intervention of Platelet Adhesion

Doloretta D. Dawicki

CONTENTS

I. Introduction 337
II. Specific Agents 337
A. Aspirin 337
B. Indomethacin 338
C. Sulphinpyrazone 338
D. Dipyridamole 338
E. Prostacyclin and Analogues 339
F. Prostaglandin E_1 339
G. Thromboxane Synthetase Inhibitors 339
H. Ticlopidine 340
I. Peptides 340
J. Antibodies 341
III. Miscellaneous 342
IV. Conclusions/Future Directions 343
References 343

I. INTRODUCTION

The participation of platelets in the progression of atherosclerosis, in blood vessel occlusion, and thus in subsequent manifestations such as myocardial infarction and stroke is clinically and experimentally substantiated. Potential antiplatelet agents are studied initially in the laboratory as antagonists of platelet aggregation. Adhesion to the subendothelium, the first step in hemostasis and thrombosis, is a logical and important process to investigate. However, due to informational and methodological problems, past research has not been directed heavily toward elucidation of the molecular aspects of platelet adhesion. This chapter is a survey of the studies that have examined the effects of various antithrombotic compounds on platelet adhesion. Only reports that employed conditions simulating the *in vivo* environment and that examined adhesion, not aggregation, are cited. Adhesion under static conditions or to glass bead columns is not discussed. Clinical and experimental studies aimed at assessing the efficacy of antiplatelet agents or anticoagulants in preventing myocardial infarction, stroke, restenosis, or vessel or graft occlusion are not mentioned, as these studies do not evaluate adhesion per se. Detailed descriptions of the techniques used to measure adhesion (e.g., Baumgartner system, parallel-plate assembly) are given in Chapters 1 and 3. Earlier reviews on platelet adhesion and the effects of antiaggregatory and antiadhesive agents are available.[1,2]

II. SPECIFIC AGENTS

A. ASPIRIN

Aspirin exerts an antithrombotic effect via irreversibly inhibiting cyclooxygenase, thus blocking the formation of the most potent aggregating agonist, thromboxane A_2. Currently, aspirin is the most widely used drug in the prevention of thrombosis. However, experimental findings have indicated that aspirin is not active as an antiadhesive agent. In fact most reports show that aspirin enhances the contact and spreading of platelets on the subendothelium. *In vitro* and *ex vivo* studies employing the Baumgartner technique (see Chapter 1) have demonstrated that high doses of aspirin (*ex vivo* — ≃1 g/d for humans or 18 to 30 mg/kg for rabbits; *in vitro* — 0.1 to 1 m*M*) do not significantly decrease the percentage of deendothelialized rabbit aorta covered by contact and/or spread rabbit or human platelets at shear rates of 800 to 2600 s^{-1} in citrated or native (i.e., nonanticoagulated) blood.[3-7] Casenave and colleagues,[8,9] utilizing a technique in which two rods coated with denuded aorta or collagen are rotated (200 rpm, 40 s^{-1}) through a suspension of ^{51}Cr-labeled

0-8493-4559-6/95/$0.00+$.50

rabbit platelets and red blood cells, confirmed these observations. Furthermore, more recent investigations with lower doses of aspirin (*ex vivo* — 0.05 to 0.15 g/d; *in vitro* — 0.1 m*M*) have corroborated the earlier findings.[10-12]

In vivo studies have given conflicting results. Dejana et al.[9] and Baumgartner and Muggli[3] have found that aspirin administration (25 to 100 mg/kg, I.V.) to rabbits does not inhibit platelet adherence to aortae damaged *in situ* with a balloon catheter. However, Silver et al.[13] observed that aspirin (8 mg/kg, I.P.) decreases adhesion to rabbit ear arteries damaged *in situ* with forceps. As pointed out by Silver et al.,[13] the discrepancy in results may be due to the greater extent of vessel damage caused by balloon catheterization. The artery and the shear rate will also affect results.

B. INDOMETHACIN

Unlike aspirin, indomethacin, a reversible inhibitor of cyclooxygenase, has been shown to interfere with platelet adhesion.[8] Adhesion was assessed using ^{51}Cr-labeled rabbit platelets resuspended in modified Tyrodes buffer in the presence of red blood cells (40% hematocrit; Hct). Cylinders coated with collagen (type I) or everted denuded rabbit aorta were attached to probes, which were rotated through the platelet/rbc suspension at 200 rpm (40 s^{-1}) for 10 min at 37°C. The probes were then rotated in Tyrodes buffer containing 10 m*M* EDTA to remove aggregates and loosely adherent cells. Indomethacin (1 to 100 μ*M*) significantly inhibited adhesion, whereas aspirin (≤10 m*M*) was ineffective. A later report by Menys and Davies[14] corroborated these findings and provided additional results, which suggest that the cyclooxygenase and lipoxygenase pathways are not involved in platelet adhesion.

Rao[12] has shown, using human platelets and the Baumgartner system (800 s^{-1}), that ibuprofen (200 μ*M*), another nonsteroidal antiinflammatory agent, does not affect platelet interaction with rabbit abdominal aortic subendothelium.

C. SULPHINPYRAZONE

Sulphinpyrazone, a phenylbutazone derivative, antagonizes platelet aggregation (cyclooxygenase inhibitor) and protects the vascular endothelium from injury. Davies et al.[15] have studied the effect of sulphinpyrazone on the adhesion of ^{51}Cr-labeled rabbit platelets to rabbit deendothelialized thoracic aorta, using a rotating probe device similar to that described by Casenave et al.[8] They found that incubation of platelets for 10 min with 100 μ*M* sulphinpyrazone did not affect adhesion, while 250 μ*M* sulphinpyrazone significantly reduced the number of adherent platelets. Under the same experimental conditions, aspirin (100 or 250 μ*M*) was ineffective.

Ex vivo studies seem contradictory. The adhesion of platelets at a shear rate of 1300 s^{-1} from rabbits orally administered megadoses of sulphinpyrazone (100 μmol/kg, b.i.d. × 1 d) was greater than controls in studies by Baumgartner et al.[6,16] However, Davies and Menys,[17] employing the system of Casenave et al.,[8] observed inhibition by sulphinpyrazone (20 mg/kg/d × 5 d, p.o.) but only when the platelets were resuspended in plasma as opposed to buffer. These data suggest that multiple doses are necessary for the production of a therapeutic level of a metabolite of sulphinpyrazone, which is responsible for impeding platelet adhesion.

D. DIPYRIDAMOLE

Hundreds of studies have been conducted to elucidate the antiplatelet mechanism of action of dipyridamole. Presently, the antiaggregatory activity of dipyridamole may result from one or more of the following activities: (1) inhibition of platelet cAMP phosphodiesterase with resulting potentiation of the effects of endogenous components such as prostacyclin, nitric oxide, and/or adenosine, (2) inhibition of nucleoside, specifically adenosine, transport by endothelial cells and erythrocytes, thereby elevating the plasma adenosine concentration, or (3) stimulation of endothelial prostacyclin production. Reports on the antiadhesive nature of dipyridamole are not as numerous and just as inconclusive.

In vitro, Baumgartner et al.[16] have observed inhibition of rabbit platelet adherence to rabbit subendothelium by an extremely high concentration (1 m*M*) of dipyridamole. Casenave et al.[18] and Groves et al.[19] have also reported antagonistic effects of dipyridamole (100 μ*M*) on platelet adherence to a collagen-coated surface or rabbit subendothelium. Unfortunately, these investigators did not examine physiological doses (2 to 5 μ*M*) of dipyridamole.

Ex vivo adhesion studies employing the Baumgartner system have shown that clinical doses of dipyridamole do not affect the adherence of human or rabbit platelets to de-endothelialized aorta.[6,7,10] More recently, Muller and colleagues[20] also have noted a lack of effect of dipyridamole administration on platelet adhesion to coverslips coated with subendothelial matrix produced by cultured endothelial cells. Of potential importance are the findings of Aznar-Salatti et al.[21] These investigators found that the extracellular matrix from endothelial cells treated in culture for 2 d with dipyridamole (10 μM) was less adhesive than control matrix. This effect was shear rate-dependent with significant inhibition obtained at 1300 s^{-1} but not at 300 s^{-1}.

As research goes, reports have been published that contradict the aforementioned *ex vivo* results. Employing a rotating probe system similar to Casenave et al.,[8] Lauri and co-workers[11] and Groves and associates[19] have demonstrated that platelets from humans or rabbits administered dipyridamole do not adhere as well as controls. Silver et al.[13] performed an *in vivo* study in which rabbits were treated with dipyridamole (1.5 mg/kg, i.p.) five times over 3 consecutive d, after which the ear arteries were damaged with forceps. After 30 min of flow, the arteries were fixed *in situ* and adhesion assessed morphometrically. In this system, dipyridamole treatment attenuated adhesion.

E. PROSTACYCLIN AND ANALOGUES

Prostacyclin (PGI_2), an adenylate cyclase stimulator that is the most potent inhibitor of platelet aggregation, has been studied as an antiadhesive agent. Concentrations much greater than those that inhibit platelet aggregation are required to antagonize adhesion. Higgs et al.[22] found that addition of at least 56 nM PGI_2 to citrated human blood just prior to perfusion at 800 s^{-1} in a Baumgartner assembly was necessary to lower the number of platelets spread on deendothelialized rabbit abdominal aorta significantly. The activity of PGI_2 may be shear rate-dependent. Weiss and Turitto[23] observed that 100 nM PGI_2 was noninhibitory at a shear rate of 800 s^{-1}, but at 2600 s^{-1}, PGI_2 (10 to 1000 nM) dose-dependently reduced adhesion. As PGI_2 is rapidly degraded in whole blood at 37°C ($t_{1/2}$ = 3 min) the different incubation times in the cited studies may account for the variation in PGI_2 activity at a shear rate of 800 s^{-1}. Casenave et al.[24] and Groves et al.[19] have made use of a rotating probe device to examine the effect of PGI_2 on adhesion. In this system also, PGI_2 (≥100 nM) inhibits rabbit or human platelet adherence to collagen or damaged aorta. However, even at concentrations as high as 10 μM, only 50% inhibition was obtained.[24]

Adelman et al.[25] performed an *in vivo* study in which rabbit aorta was de-endothelialized *in situ* by balloon catheterization. PGI_2 infused at 50 to 100 ng/kg/min 10 min prior to de-endothelialization resulted in about 20% less surface coverage by adherent platelets, while a dose of 650 to 850 ng/kg/min caused approximately 80% reduction in surface coverage and 63% decrease in the number of attached platelets.

Analogues of greater stability than PGI_2 have been assessed as antiaggregatory agents. The effect of ZK 36374 (Iloprost), a carbocyclin derivative, on platelet adhesion has also been examined. This analogue, which is equipotent to PGI_2 as an inhibitor of human platelet aggregation, has been found to be equally active as PGI_2 in blocking adhesion.[26]

F. PROSTAGLANDIN E_1

Prostaglandin E_1 (PGE_1), another stimulator of adenylate cyclase, has been examined briefly for its effect on platelet adhesion. In 1976 Baumgartner et al.[16] reported that PGE_1 (1 μM) strongly inhibits the initial contact and spreading of rabbit platelets on rabbit aortic subendothelium. Using a rotating collagen-coated probe system, Casenave et al.[18] later duplicated these results. Unexplainably, Rao[12] did not observe an inhibitory effect by 1 μM PGE_1 in the Baumgartner system. Rao[12] utilized human rather than rabbit PRP and a long incubation time (30 min) with PGE_1 prior to perfusion. Importantly, in another study by Casenave et al.,[24] 1 μM PGE_1 significantly reduced the attachment of rabbit platelets to a collagen-coated (or subendothelium-coated) glass probe, whereas 10 μM PGE_1 was required to block human platelet adherence. Thus, results may be affected by the animal source of the platelets.

G. THROMBOXANE SYNTHETASE INHIBITORS

Using the rotating probe system described by Casenave et al.,[8] Menys and Davies[27] have investigated the effect of dazoxiben, a specific inhibitor of thromboxane synthetase, on platelet adhesion. A 3 min preincubation with 1 or 10 μM dazoxiben reduced the number of rabbit platelets adhering to damaged rabbit aorta by about 40% but did not hinder adhesion to collagen. The inhibitory effect of dazoxiben was abolished by pretreating the aorta with 15-HPETE (Hydroperoxyeicosatetraenoic acid), an antagonist of PGI_2 synthetase. These data imply that thromboxane synthetase inhibitors block platelet adhesion via redirecting platelet endoperoxides to PGI_2 synthesis by the endothelium.

Other thromboxane synthetase inhibitors, designated OKY-046 and OKY-1581, have been used for *in vitro* and *in vivo* studies by Rao[12] and Silver et al.,[13] respectively. At a dose of 100 μg/ml, OKY-046 did not significantly reduce human platelet adherence to de-endothelialized rabbit aorta in the Baumgartner model.[12] Administration of OKY-1581 (15 mg/kg, i.p.) to rabbits at a dose that caused 90% inhibition of rabbit platelet thromboxane synthesis *ex vivo* did not decrease platelet adhesion to damaged arteries.[13]

H. TICLOPIDINE

Ticlopidine, a thienopyridine derivative, has been shown to have antiaggregating activity in *in vitro, in vivo,* and *ex vivo* studies. *In vivo* activity is greater than *in vitro* efficacy, suggesting that a metabolite or an *in vivo* factor is involved. The exact mechanism of action is unknown. In a study by Cattaneo et al.,[28] rats and rabbits were given ticlopidine (100 mg/kg or 400 mg/kg) orally 48 h prior to experiments and daily during experiments. Ticlopidine inhibited ADP-, collagen-, thrombin-, and arachidonic acid-induced aggregation in *ex vivo* experiments and prolonged platelet survival. However, adhesion to indwelling catheters and aortae damaged by the catheters was not reduced. Additionally, in rabbits in which aortae were de-endothelialized *in situ,* this drug did not impede adhesion. In a study conducted by Escolar et al.,[29] human volunteers were given ticlopidine 250 mg b.i.d. for 7 d. Adhesion was evaluated in a Baumgartner perfusion system using citrated blood at a shear rate of 800 s^{-1}. Drug treatment did not alter the area of surface covered by contact and spread platelets.

Orlando et al.[30] have conducted a cross-over *ex vivo* adhesion study with insulin-dependent diabetics. Adhesion of [^{3}H]adenine-labeled platelets to everted air-damaged saphena veins in a rotating probe device system (570 s^{-1}) was examined. Patients receiving ticlopidine (250 mg, b.i.d. for 7 d) presented a consistently significant decrease in adhesion.

I. PEPTIDES

Peptides containing the RGD-sequence, or the AGDV-sequence of the carboxy terminal of the γ chain of fibrinogen, recognized by the platelet gpIIb-IIIa integrin, have been examined for their effects on platelet adhesion. Under static conditions these peptides have been found to inhibit adhesion of activated platelets to surfaces coated with fibrinogen, fibronectin, or von Willebrand Factor (vWF).[31]

Nievelstein and Sixma[32] attempted to define the role of the RGD-sequence in platelet adhesion to fibronectin under flow conditions. Adhesion to the matrix of endothelial cells (previously shown by these investigators to be fibronectin-dependent) in a flat perfusion chamber was not blocked by the hexapeptide GRGDSP (1 m*M*) at shear rates of 300, 800, or 1500 s^{-1}. Adhesion to nonfibrillar collagen type I (also fibronectin-dependent) was significantly inhibited only with 1 m*M* GRGDSP and only at a shear rate of 1500 s^{-1}.

The effects of RGDS-containing peptides derived from the sequence of vWF and the dodecapeptide HHLGGAKQAGDV, corresponding to the carboxyl terminal of the γ chain of fibrinogen, on human platelet adhesion to collagen have also been assessed. As shown by Fressinaud et al.,[33] using a parallel-plate perfusion chamber, these peptides (50 to 300 μ*M*) dose-dependently antagonize platelet adhesion to collagen (equine or human type III) at a shear rate of 2600 s^{-1}. RGDS-peptides were more effective than the AGDV-containing dodecapeptide, although complete inhibition of adhesion was not obtained at the highest concentration tested (500 μ*M*). These peptides were inactive at lower shear rates of 100 and 650 s^{-1}. Additionally, RGD-containing peptides and the dodecapeptide carboxy terminus of the γ chain of fibrinogen have been documented to hinder platelet adhesion to fibrinogen- and fibrin-coated surfaces at shear rates of 300 or 1300 s^{-1}.[34] Of potential importance, this study indicated that the peptide D-RGDW, containing the proteolytically resistant D-stereoisomer of arginine as well as the hydrophobic residue tryptophan, is 50-fold more effective than RGDS. Unexplainably, the RGDS peptide exhibited biphasic responses, inhibition at concentrations of 100 μ*M* (fibrin surface) or 300 μ*M* (fibrinogen surface) and no effect at concentrations of 300 to 600 μ*M* (fibrin) or 600 μ*M* (fibrinogen). The dodecapeptide fragment of fibrinogen was the least active in these studies.

Employing a complex surface, Weiss et al.[35] have shown that the tetrapeptide RGDS (78 to 500 μ*M*) and the dodecapeptide HHLGGAKQAGDV (25 or 500 μ*M*) dose-dependently inhibit adhesion of human platelets to de-endothelialized rabbit aorta at a shear rate of 2600 s^{-1}, although complete inhibition was not obtained. The RGDS tetrapeptide was less effective at a shear rate of 800 s^{-1} than at 2600 s^{-1}.

Of interest, although perplexing, are the findings of Lawrence et al.[36] Peptides (50 μ*M*) containing the RGD sequence in native or transposed configuration or with conservative substitution, inhibited human platelet adherence to human umbilical artery subendothelium at a shear rate of 2600 s^{-1} (not at 800 s^{-1}). More surprisingly, these peptides did not inhibit thrombus growth.

RGD-containing cysteine-rich peptides from the venoms of viper snakes are much more potent (n*M* vs. μ*M*) in blocking platelet aggregation and adhesion than short RGD-peptides.[37] Consequently, these peptides have been named "disintegrins". Musial et al.[38] have shown that disintegrins protect platelets during simulated extracorporeal circulation. The tetrapeptide RGDS was much less effective. Jen and Lin have recorded platelet adhesion continuously over a particular time period by using acridine-labeled platelets and an epifluorescence video microscope.[39] Fluorescence images on video were analyzed by digital image processing techniques. These investigations found that halysin (5 μg/ml), an RGD-containing peptide from the venom of *Agkistrodon halys,* completely prevented platelet adhesion to fibrinogen and fibrin at a shear rate of 445 s^{-1}. Under the same conditions, the RGDS tetrapeptide (0.1 m*M*) was ineffective in blocking adhesion to fibrin and only somewhat inhibitory with fibrinogen.

J. ANTIBODIES

Antibodies to adhesion-related proteins have been studied for their effects on platelet interaction with the subendothelium or individual subendothelial components. These studies, along with results from patients with von Willebrand's disease, Glanzmann's thrombasthenia, or Bernard-Soulier syndrome, have identified the major platelet glycoproteins and plasma or subendothelial proteins involved in platelet adhesion.

von Willebrand Factor (vWF) has received a great deal of attention, as this molecule is involved in platelet adhesion, platelet aggregation, and coagulation. Reports by several groups have indicated that antibodies to vWF inhibit human platelet adherence to rabbit aorta and human umbilical artery subendothelium in a shear rate-dependent manner, with little effect at shear rates <650 s^{-1}.[40-42] Inhibition is also dose-dependent, with complete inhibition of adhesion often occurring. These results are similar to findings obtained with blood from von Willebrand's patients. Investigations conducted by Bellinger and associates[43] failed to show an antiadhesive effect of an antibody to porcine vWF *in vitro,* while an antithrombotic state was induced *in vivo.* This potential for selectivity was initially indicated in the reports of Myer et al.[41] and Stel et al.,[44] in which several monoclonal antibodies to vWF were found to exhibit differential potencies against platelet adhesion and Ristocetin-induced platelet agglutination.

The studies discussed above were conducted using the Baumgartner system. The adhesive role of vWF has also been examined with parallel-plate perfusion chambers containing protein-coated slides. Baruch et al.[45] have shown that an anti-vWF antibody, which interferes with the binding of vWF to platelet gpIb, dose-dependently reduces the area of endothelial extracellular matrix-coated coverslips covered with platelets at a shear rate of 1600 s^{-1}. Anti-vWF antibodies, which block the binding of vWF to gpIb, gpIIb-IIIa, or collagen, reduce adhesion to collagen (equine or human type III)-coated surfaces at high shear rate (2600 s^{-1}).[46] Fressinaud et al.[47] extended the investigations of Sakariassen et al.[46] by showing that inhibition of adhesion to collagen by these antibodies was dose- (2.5 to 60 μg/ml) and shear rate- (200 to 2600 s^{-1}) dependent. Inhibition by any single antibody was statistically significant only at a shear rate greater than 650 sec^{-1}. At 2600 s^{-1}, complete inhibition was obtained by a mixture of these monoclonal antibodies (7 μg of each antibody per ml). These findings were supported by Jorieux et al.,[48] who utilized a different anti-vWF antibody that antagonizes ristocetin-induced platelet aggregation. Furthermore, an antibody to vWF that interferes with the binding of vWF to gpIb has been shown to hinder platelet adhesion to a fibrin-coated surface.[49] At a shear rate of 1300 s^{-1}, inhibition of adhesion was statistically significant and dose-dependent.

Complementary to the above studies, investigations have been performed with antibodies to platelet gpIb and gpIIb-IIIa.[46,50,51] The three cited reports utilized the same antibodies. The anti-gpIIb-IIIa antibody inhibits vWF, fibronectin, and fibrinogen binding to platelets. The anti-gpIb antibody inhibits ristocetin-induced platelet agglutination and vWF binding. These antibodies significantly reduced platelet adhesion to human umbilical artery subendothelium at a shear rate of 2600 s^{-1}.[50] Sakariassen et al.[46,51] showed that these antibodies were effective only at shear rates ≥500 s^{-1}. At 1800 s^{-1} the antibody to gpIb was more inhibitory than the antibodies to gpIIb-IIIa. Antibodies against gpIIb-IIIa seem to inhibit platelet spreading, while antibodies to gpIb block the initial contact. The shear rate-dependent effect of anti-gpIIb-IIIa antibodies has also been shown by Weiss et al.,[35,52] who examined human platelet adhesion to rabbit aorta subendothelium. Platelet adherence to an endothelial extracellular matrix-coated surface has been reported to be inhibited by an antibody to gpIIb-IIIa at shear rates ≥800 s^{-1}.[32] Antibodies to gpIIb-IIIa and gpIb have been shown to decrease adhesion to a fibrin-coated surface.[49] Interestingly, in this study, the anti-gpIIb-IIIa antibody was more effective than the anti-gpIb antibody, and the anti-gpIIb-IIIa antibody produced a significant reduction at a shear rate of 300 s^{-1}. Jen and Lin[39] also found that an antibody against gpIIb-IIIa dramatically decreased platelet adhesion to fibrin- and fibrinogen-coated slides at 445 s^{-1}, while anti gpIb antibody was inactive. Likewise, platelet deposition on artificial grafts (polytetrafluroethylene and polyethylene) at a shear rate of 312 s^{-1} was greatly reduced by antibodies to gpIIb-IIIa (or gpIb).[53,54] In these systems (i.e., fibrin(ogen) surfaces), initial adhesion appears to be mediated by gpIIb-IIIa, while further platelet accrual is dependent on gpIb.

The importance of subendothelial fibronectin in platelet adhesion was illustrated by Houdjik and Sixma.[55] These investigators obtained significant inhibition of adhesion at shear rates of 800 and 1800 s^{-1} (not at 400 s^{-1}), when the human umbilical artery was treated with a rabbit antibody to human fibronectin prior to perfusion. Similarly, Nievelstein and Sixma[32] have reported that preincubation of endothelial cell matrix with antifibronectin $F(ab')_2$ fragments reduces platelet adhesion.

III. MISCELLANEOUS

The necessity of calcium ions for platelet aggregation is well known. Likewise, extracellular calcium appears to be necessary for platelet adhesion. Studies by Baumgartner et al.,[16] Cazenave et al.,[56] and Badimon et al.[57] have shown that EDTA (3 to 9 m*M*) inhibits platelet adhesion to subendothelium at shear rates of 800, 40, and 848 to 1690 s^{-1}, respectively. Jen and Lin[39] reported that EGTA (5 m*M*) completely blocks adhesion to fibrin- and fibrinogen-coated surfaces. The effect of sodium citrate on adhesion is not as apparent. This chelator has been reported to enhance platelet adhesion (compared to native blood) as a consequence of decreased aggregation and thrombus formation (i.e., more available surface area for contact and spreading).[6,58,59] This effect was observed when the plasma citrate concentration was no greater than ≃20 m*M* and the shear rate was ≤1300 s^{-1}. At higher concentrations of citrate (>20 m*M*) and higher shear rates (≥1300 s^{-1}), inhibition of adhesion resulted.[6,7,59] Thus it seems that citrate is antagonistic when present at a high concentration and under high shear rate. However, Badimon et al.[60] have shown that platelet adhesion to porcine aortic subendothelium or a collagen I-coated slide in blood anticoagulated with 15 m*M* citrate (≃plasma concentration) is significantly less than adhesion in heparinized blood at a shear rate of 1690 s^{-1}. (The authors, however, state that there is no difference.) Furthermore, in the system of Cazenave et al.,[56] 13 m*M* citrate significantly reduced the number of platelets adhering to subendothelium- or collagen-coated probes at a shear rate of only 40 s^{-1}. An influx of Ca^{2+} into the platelet is not necessary as calcium antagonists (e.g., verapamil, diltiazem, and nifedipine) do not affect platelet adhesion.[12,61]

Heparin is often used as an anticoagulant for *in vitro* studies, as well as clinically. The investigators who examined the effect of heparin on platelet adhesion and aggregation did not observe modulation of platelet interaction with subendothelium by heparin *in vitro, ex vivo,* or *in vivo.*[6,57,60,62,63] As heparin is a heterogeneous mixture of sulfated polysaccharide chains, a specific fraction of heparin may be more effective than unfractionated heparin. In support of this hypothesis, Kupinski et al.[64] have shown that a low mol wt heparin inhibits human platelet adhesion to bovine subendothelial extracellular matrix in a static system, whereas unfractionated heparin is inactive. Heparin bound to prosthetic surfaces appears to antagonize adhesion.[65-67] The thrombin inhibitor hirudin is inactive as an antiadhesive agent.[57,60,62]

Nitric oxide (endothelium-derived relaxing factor), produced by endothelial cells, macrophages, platelets, and neurons, modulates a variety of physiological functions, including vasodilation, inflammation, neurotransmission, and platelet aggregation. Most of these effects are produced by stimulation of guanylate cyclase with a concomitant increase in cGMP. Inhibitors of nitric oxide synthesis have been shown to increase platelet adhesion to endothelial cell matrices *in vitro* and to damaged endothelium *in vivo.*[68,69] Furthermore, exogenous authentic nitric oxide inhibits platelet adhesion to endothelial matrix at shear rates of 300 to 1800 s^{-1}.[69] As expected, SIN-1 (3-morpholino-syndonimine), an exogenous donor of nitric oxide, administered (10 μg/kg/min) to pigs 1 h prior to and for the duration of the experiment, has been found to significantly reduce platelet adhesion to carotid arteries superficially damaged by balloon angioplasty.[70,71] Whether pharmacological intervention of platelet adhesion by nitric oxide or an analogue is feasible remains to be determined.

The effects of natural products such as vitamins and fatty acids on platelet adhesion have been examined somewhat. Vitamin E, an antioxidant, has been shown in *ex vivo* studies to reduce platelet adhesion dramatically. Jandak, Steiner, and Richardson,[72] using platelet-rich plasma and a laminar flow chamber coated with collagen I, fibrinogen, or fibronectin, have documented antiadhesive activity with doses of vitamin E as low as 200 IU/d (× 2 weeks) at a shear rate of 20 to 25 s^{-1}. An interesting aside resulting from these studies is that platelets preferentially adhered to sites that were previously occupied by other platelets. Possibly, detached platelets leave behind some remnant that is conducive to further adhesion. The same laboratory[73] has shown that fish oil, a good source of the omega-3 polyunsaturated fatty acid eicosapentaenoic acid (EPA), is also an inhibitor of platelet adherence. Volunteers consumed a large dose (≃25 capsules/d, ≃6 g EPA/d) of MaxEPA for 14 to 25 d prior to adhesion studies. Adhesion to both fibrinogen and collagen was significantly reduced within 14 d after initiation of fish oil ingestion, whereas the control vegetable oil was without effect. As with vitamin E, EPA also lowered the reutilization of previously occupied sites. Subsequently, Li and Steiner[74] reported that a dose of 3 g EPA/d (× 3 weeks) was sufficient to reduce platelet adherence at a low (20 to 25 s^{-1}) or high (>760 s^{-1}) shear rate. In contrast, Owens and Cave[75] have found that human consumption of fish oil (15 capsules MaxEPA/d, ≃3 g EPA/d × 4 weeks) does not modify platelet adhesion to rabbit aortic subendothelium at a shear rate of 850 s^{-1}. The discrepancy between Owen and Cave's and Li and Steiner's results may reflect the different adhesion techniques and/or adhesive surfaces employed.

IV. CONCLUSIONS/FUTURE DIRECTIONS

The occlusion of blood vessels, vascular prosthetic grafts, and extracorporeal circuits by platelets warrants the (re)search for effective selective antiplatelet therapy. An anticoagulant or dipyridamole and aspirin comprise the current treatment modalities, treatments that have been employed for decades. While these agents are undoubtedly beneficial in the prevention of thrombosis, they are not potent antiadhesive drugs. In researching the current antiplatelet agents for their effects on the interaction of platelets with the subendothelium, it was sometimes difficult to come to a definite conclusion. Adhesion is affected by many factors, such as the shear rate, hematocrit, amount and type of anticoagulant, vascular bed, extent of vascular damage, experimental animal, and adhesion technique. All factors considered, the present antiplatelet drugs are not antiadhesive. For some agents (e.g., prostacyclin), megadoses are active but of course would not be tolerated in the individual. Many of these compounds are also not platelet-specific, limiting their use. Therefore, the need for active and selective antiadhesive (as well as antiaggregatory) antiplatelet chemotherapy exists. Unfortunately the development of novel, more efficacious regimens has not materialized.

Much information has accrued over the past several years on the cellular receptors and ligands involved in adhesion of cells to extracellular matrices and to each other. This is highly relevant for platelet physiology, in which the initial event in hemostasis and thrombosis is adhesion. Certain glycoproteins on the platelet surface have been identified as the receptors for several adhesive ligands (see Chapters 8, 10 to 12, and 15). The two receptors that have been studied extensively are gpIb and gpIIb-IIIa. The receptors and the adhesive ligands for these receptors, i.e., vWF and fibrinogen/fibronectin, have been dissected to pinpoint the regions involved in receptor-ligand interaction. Molecular studies of this nature have opened the door to the design of novel, selective antiadhesive (as well as antiaggregatory) agents. *In vivo* and *ex vivo* studies employing antibodies against the gpIIb-IIIa complex or RGD-containing peptides in nonhuman hosts have illustrated the greater antithrombotic effects of these agents as compared to conventional therapy, such as aspirin, dipyridamole, and/or heparin.[76-79] Early clinical testing with the $F(ab')_2$ fragment of an antibody to gpIIb-IIIa has given promising results in settings of unstable angina and coronary angioplasty.[80,81] These studies are preliminary, and complications of toxicity, immunogenicity, and nonspecific effects on other cell types need to be carefully investigated. However, they show the potential for the development of antiplatelet agents, which interfere specifically with adhesion. The upcoming decade should be exciting and productive in this area.

REFERENCES

1. **Packham, M. A. and Mustard, J. F.,** Platelet adhesion, in *Progress in Hemostasis and Thrombosis,* Vol. 7, Spaet, T. H., Ed., Grune & Stratton, New York, 1984, 211.
2. **Hawiger, J.,** Platelet-vessel wall interactions; platelet adhesion and aggregation, in *Atherosclerosis Reviews,* Vol. 21, Leaf, A. and Weber, P. C., Eds., Raven Press, New York, 1990, 165.
3. **Baumgartner, H. R. and Muggli, R.,** Effect of acetylsalicylic acid on platelet adhesion to subendothelium and on the formation of mural platelet thrombi, *Thromb. Diath. Haemorrh. Suppl.,* 60, 345, 1974.
4. **Weiss, H. J., Tschopp, T. B., and Baumgartner, H. R.,** Impaired interaction (adhesion-aggregation) of platelets with the subendothelium in storage-pool disease and after aspirin ingestion, *N. Engl. J. Med.,* 293, 619, 1977.
5. **Tschopp, T. B.,** Aspirin inhibits platelet aggregation on, but not adhesion to, collagen fibrils: an assessment of platelet adhesion and deposited platelet mass by morphometry and ^{51}Cr-labeling, *Thromb. Res.,* 11, 619, 1977.

6. **Baumgartner, H. R.,** Effects of acetylsalicylic acid, sulfinpyrazone and dipyridamole on platelet adhesion and aggregation in flowing native and anticoagulated blood, *Haemostasis,* 8, 340, 1979.
7. **Weiss, H. J., Turitto, V. T., Vicic, W. J., and Baumgartner, H. R.,** Effect of aspirin and dipyridamole on the interaction of human platelets with subendothelium: studies using citrated and native blood, *Thromb. Haemost.,* 45, 136, 1981.
8. **Cazenave, J.-P., Kinlough-Rathbone, R. L., Packham, M. A., and Mustard, J. F.,** The effect of acetylsalicylic acid and indomethacin on rabbit platelet adherence to collagen and the subendothelium in the presence of a low or high hematocrit, *Thromb. Res.,* 13, 971, 1978.
9. **Dejana, E., Cazenave, J.-P., Groves, H. M., Kinlough-Rathbone, R. L., Richardson, M., Packham, M. A., and Mustard, J. F.,** The effect of aspirin inhibition of PGI_2 on platelet adherence to normal and damaged rabbit aortae, *Thromb. Res.,* 17, 453, 1980.
10. **Escolar, G., Bastida, E., Ordinas, A., and Castillo, R.,** Interaction of platelets with subendothelium in humans treated with aspirin and dipyridamole alone or in combination, *Thromb. Res.,* 40, 419, 1985.
11. **Lauri, D., Zanetti, A., Dejana, E., and de Gaetano, G.,** Effects of dipyridamole and low-dose aspirin therapy on platelet adhesion to vascular subendothelium, *Am. J. Cardiol.,* 58, 1261, 1986.
12. **Rao, G. H. R.,** Influence of anti-platelet drugs on platelet-vessel wall interactions, *Prostaglandins, Leukotrienes Med.,* 30, 133, 1987.
13. **Silver, M. J., Ingerman-Wojenski, C. M., Sedar, A. W., and Smith, M.,** Model system to study interaction of platelets with damaged arterial wall. I. Inhibition of platelet adhesion to subendothelium by aspirin and dipyridamole, *Exp. Mol. Pathol.,* 41, 141, 1984.
14. **Menys, V. C. and Davies, J. A.,** Inhibition of platelet adhesion to aortic subendothelium by indomethacin — an effect unrelated to inhibition of arachidonic acid metabolism, *Thromb. Res.,* 37, 225, 1985.
15. **Davies, J. A., Essien, E., Cazenave, J.-P., Kinlough-Rathbone, R. L., Gent, M., and Mustard, J. F.,** The influence of red blood cells on the effects of aspirin and sulphinpyrazone on platelet adherence to damaged rabbit aorta, *Br. J. Haematol.,* 42, 283, 1979.
16. **Baumgartner, H. R., Muggli, R., Tschopp, T. B., and Turitto, V. T.,** Platelet adhesion, release and aggregation in flowing blood: effects of surface properties and platelet function, *Thromb. Haemost.,* 35, 124, 1976.
17. **Davies, J. A. and Menys, V. C.,** Effect of sulphinpyrazone and aspirin on platelet adhesion to subendothelium following oral administration to rabbits, *Thromb. Res.,* 21, 329, 1981.
18. **Cazenave, J.-P., Packham, M. A., Kinlough-Rathbone, R. L., and Mustard, J. F.,** Platelet adherence to the vessel wall and to collagen-coated surfaces, in *Adv. Exp. Med. Biol.,* Vol. 102, Day, H. J., Molony, B. A., Nishizawa, E. E., and Rynbrandt, R. H., Eds., Plenum Press, New York, 1978, 31.
19. **Groves, H. M., Kinlough-Rathbone, R. L., Casenave, J.-P., Dejana, E., Richardson, M., and Mustard, J. F.,** Effect of dipyridamole and prostacyclin on rabbit platelet adherence in vitro and in vivo, *J. Lab. Clin. Med.,* 99, 548, 1982.
20. **Muller, T. H., Su., C. A. P. F., Weisenberger, H., Brickl, R., Nehmiz, G., and Eisert, W. G.,** Dipyridamole alone or combined with low-dose acetyl salicylic acid inhibits platelet aggregation in human whole blood ex vivo, *Br. J. Clin. Pharmac.,* 30, 179, 1990.
21. **Aznar-Salatti, J., Bastida, E., Escolar, G., Almirall, L., Diaz-Ricart, M., Anton, P., Castillo, R., and Ordinas, A.,** Dipyridamole induces changes in the thrombogenic properties of extracellular matrix generated by endothelial cells in culture, *Thromb. Res.,* 64, 341, 1991.
22. **Higgs, E. A., Moncada, S., Vane, J. R., Caen, J. P., Michel, H., and Tobelem, G.,** Effect of prostacyclin (PGI_2) on platelet adhesion to rabbit arterial subendothelium, *Prostaglandins,* 16, 17, 1978.
23. **Weiss, H. J. and Turitto, V. T.,** Prostacyclin (Prostaglandin I_2, PGI_2) inhibits platelet adhesion and thrombus formation on subendothelium, *Blood,* 53, 244, 1979.
24. **Cazenave, J.-P., Dejana, E., Kinlough-Rathbone, R. L., Richardson, M., Packham, M. A., and Mustard, J. F.,** Prostaglandins I_2 and E_1 reduce rabbit and human platelet adherence without inhibiting serotonin release from adherent platelets, *Thromb. Res.,* 15, 273, 1979.
25. **Adelman, B., Stemerman, M. B., Mennell, D., and Handin, R. I.,** The interaction of platelets with aortic subendothelium: inhibition of adhesion and secretion by prostaglandin I_2, *Blood,* 58, 198, 1981.
26. **Menys, V. C. and Davies, J. A.,** Inhibitory effects of ZK36374, a stable prostacyclin analogue, on adhesion of rabbit platelets to damaged aorta and serotonin release by adherent platelets, *Clin. Sci.,* 65, 149, 1983.

27. **Menys, V. C. and Davies, J. A.,** Selective inhibition of thromboxane synthetase with dazoxiben-basis of its inhibitory effect on platelet adhesion, *Thromb. Haemost.*, 49, 96, 1983.
28. **Cattaneo, M., Winocour, P. D., Somers, D. A., Groves, H. M., Kinlough-Rathbone, R. L., Packham, M. A., and Mustard, J. F.,** Effect of ticlopidine on platelet aggregation, adherence to damaged vessels, thrombus formation and platelet survival, *Thromb. Res.*, 37, 29, 1985.
29. **Escolar, G., Bastida, E., Castillo, R., and Ordinas, A.,** Ticlopidine inhibits platelet thrombus formation studied in a flowing system, *Thromb. Res.*, 45, 561, 1987.
30. **Orlando, E., Cortelazzo, S., Nosari, I., Lepore, G., Pagani, G., de Gaetano, G., and Barbui, T.,** Inhibition by ticlopidine of platelet adhesion to human venous subendothelium in patients with diabetes, *J. Lab. Clin. Med.*, 112, 583, 1988.
31. **Haverstick, D. M., Cowan, J. F., Yamada, K. M., and Santoro, S. A.,** Inhibition of platelet adhesion to fibronectin, fibrinogen and von Willebrand factor substrates by a synthetic tetrapeptide derived from the cell-binding domain of fibronectin, *Blood*, 66, 946, 1985.
32. **Nievelstein, P. F. E. M. and Sixma, J. J.,** Glycoprotein IIb-IIIa and RGD(S) are not important for fibronectin-dependent platelet adhesion under flow conditions, *Blood*, 72, 82, 1988.
33. **Fressinaud, E., Girma, J. P., Sadler, J. E., Baumgartner, H. R., and Meyer, D.,** Synthetic RGDS-containing peptides of von Willebrand factor inhibit platelet adhesion to collagen, *Thromb. Haemost.*, 64, 589, 1990.
34. **Hantgan, R. R., Endenburg, S. C., Cavero, I., Marguerie, G., Uzan, A., Sixma, J. J., and DeGroot, P. G.,** Inhibition of platelet adhesion to fibrin(ogen) in flowing whole blood by Arg-Gly-Asp and fibrinogen γ chain carboxy terminal peptides, *Thromb. Haemost.*, 68, 694, 1992.
35. **Weiss, H. J., Hawiger, J., Ruggeri, Z. M., Turitto, V. T., Thiagarajan, P., and Hoffmann, T.,** Fibrinogen-independent platelet adhesion and thrombus formation on subendothelium mediated by glycoprotein IIb-IIIa complex at high shear rate, *J. Clin. Invest.*, 83, 288, 1989.
36. **Lawrence, J. B., Kramer, W. S., McKeown, L. P., Williams, S. B., and Gralnick, H. R.,** Arginine-glycine-aspartic acid and fibrinogen γ-chain carboxy-terminal peptides inhibit platelet adherence to arterial subendothelium at high wall shear rates; an effect dissociable from interference with adhesive protein binding, *J. Clin. Invest.*, 86, 1715, 1990.
37. **Gould, R. J., Polokoff, M. A., Friedman, P. A., Huang, T. F., Holt, J. C., Cook, J. J., and Niewiarowski, S.,** Disintegrins: a family of integrin inhibitory proteins from viper venoms, *Publ. Soc. Exp. Biol. Med.*, 195, 168, 1990.
38. **Musial, J., Niewiarowski, S., Rucinski, B., Stewart, G. J., Cook, J. J., Williams, J. A., and Edmunds, L. H.,** Inhibition of platelet adhesion to surfaces of extracorporeal circuits by disintegrins, RGD-containing peptides from viper venoms, *Circulation*, 82, 261, 1990.
39. **Jen, C. J. and Lin, J. S.,** Direct observation of platelet adhesion to fibrinogen- and fibrin-coated surfaces, *Am. J. Physiol.*, 261, H1457, 1991.
40. **Baumgartner, H. R., Tschopp, T. B., and Meyer, D.,** Shear rate dependent inhibition of platelet adhesion and aggregation on collagenous surfaces by antibodies to human factor VIII/von Willebrand factor, *Br. J. Haematol.*, 44, 127, 1980.
41. **Meyer, D., Baumgartner, H. R., and Edgington, T. S.,** Hybridoma antibodies to human von Willebrand factor. II. Relative role of intramolecular loci in mediation of platelet adhesion to subendothelium, *Br. J. Haematol.*, 57, 609, 1984.
42. **Stel, H. V., Sakariassen, K. S., de Groot, P. G., van Mourik, J. A., and Sixma, J. J.,** Von Willebrand factor in the vessel wall mediates platelet adherence, *Blood*, 65, 85, 1985.
43. **Bellinger, D. A., Nichols, T. C., Read, M. S., Reddick, R. L., Lamb, M. A., Brinkhous, K. M., Evatt, B. L., and Griggs, T. R.,** Prevention of occlusive coronary artery thrombosis by a murine monoclonal antibody to porcine von Willebrand factor, *Proc. Natl. Acad. Sci. U.S.A.*, 84, 8100, 1987.
44. **Stel, H. V., Sakariassen, K. S., Scholte, B. J., Veerman, E. C. I., van der Kwast, Th. H., de Groot, Ph. G., Sixma, J. J., and Mourik, J. A.,** Characterization of 25 monoclonal antibodies to factor VIII-von Willebrand factor: relationship between ristocetin-induced platelet aggregation and platelet adherence to subendothelium, *Blood*, 63, 1408, 1984.
45. **Baruch, D., Denis, C., Marteaux, C., Schoevaert, D., Coulombel, L., and Meyer, D.,** Role of von Willebrand factor associated to extracellular matrices in platelet adhesion, *Blood*, 77, 519, 1991.
46. **Sakariassen, K. S., Fressinaud, E., Girma, J.-P., Meyer, D., and Baumgartner, H. R.,** Role of platelet membrane glycoproteins and von Willebrand factor in adhesion of platelets to subendothelium and collagen, *Ann. N.Y. Acad. Sci.*, 516, 52, 1987.

47. **Fressinaud, E., Baruch, D., Girma, J.-P., Sakariassen, K. S., Baumgartner, H. R., and Meyer, D.,** Von Willebrand factor-mediated platelet adhesion to collagen involves platelet membrane glycoprotein IIb-IIIa as well as glycoprotein Ib, *J. Lab. Clin. Med.,* 112, 58, 1988.
48. **Jorieux, S., de Romeuf, C., Samor, B., Goudemand, M., and Mazurier, C.,** Characterization of a monoclonal antibody to von Willebrand factor as a potent inhibitor of ristocetin-mediated platelet interaction and platelet adhesion, *Thromb. Haemost.,* 57, 278, 1987.
49. **Hantgan, R. R., Hindriks, G., Taylor, R. G., Sixma, J. J., and de Groot, P. G.,** Glycoprotein Ib, von Willebrand factor, and glycoprotein IIb:IIIa are all involved in platelet adhesion to fibrin in flowing whole blood, *Blood,* 76, 345, 1990.
50. **Lawrence, J. B. and Gralnick, H. R.,** Monoclonal antibodies to the glycoprotein IIb-IIIa epitopes involved in adhesive protein binding: effects on platelet spreading and ultrastructure on human arterial subendothelium, *J. Lab. Clin. Med.,* 109, 495, 1987.
51. **Sakariassen, K. S., Nievelstein, P. F. E. M., Coller, B. S., and Sixma, J. J.,** The role of platelet membrane glycoproteins Ib and IIb-IIIa in platelet adherence to human artery subendothelium, *Br. J. Haematol.,* 63, 681, 1986.
52. **Weiss, H. J., Turitto, V. T., and Baumgartner, H. R.,** Platelet adhesion and thrombus formation on subendothelium in platelets deficient in glycoproteins IIb-IIIa, Ib, and storage granules, *Blood,* 67, 322, 1986.
53. **Johnson, P. C., Sheppeck, R. A., Hribar, S. R., Bentz, M. L., Janosky, J., and Dickson, C. S.,** Inhibition of platelet retention on artificial microvascular grafts with monoclonal antibodies and a high-affinity peptide directed against platelet membrane glycoproteins, *Arterioscl. Thromb.,* 11, 552, 1991.
54. **Sheppeck, R. A., Bentz, M., Dickson, C., Hribar, S., White, J., Janosky, J., Berceli, S. A., Borovetz, H. S., and Johnson, P. C.,** Examination of the roles of glycoprotein Ib and glycoprotein IIb/IIIa in platelet deposition on an artificial surface using clinical antiplatelet agents and monoclonal antibody blockade, *Blood,* 78, 673, 1991.
55. **Houdijk, W. P. M. and Sixma, J. J.,** Fibronectin in artery subendothelium is important for platelet adhesion, *Blood,* 65, 598, 1985.
56. **Cazenave, J. -P., Blondowska, D., Richardson, M., Kinlough-Rathbone, R. L., Packham, M., and Mustard, J. F.,** Quantitative radioisotopic measurement and scanning electron microscopic study of platelet adherence to a collagen-coated surface and to subendothelium with a rotating probe device, *J. Lab. Clin. Med.,* 93, 60, 1979.
57. **Badimon, L., Badimon, J. J., Turitto, V. T., and Fuster, V.,** Role of von Willebrand factor in mediating platelet-vessel wall interaction at low shear rate; the importance of perfusion conditions, *Blood,* 73, 961, 1989.
58. **Weiss, H. J., Turitto, V. T., and Baumgartner, H. R.,** Effect of shear rate on platelet interaction with subendothelium in citrated and native blood. I. Shear rate-dependent decrease of adhesion in von Willebrand's disease and the Bernard-Soulier syndrome, *J. Lab. Clin. Med.,* 92, 750, 1978.
59. **Baumgartner, H. R., Turitto, V., and Weiss, H. J.,** Effect of shear rate on platelet interaction with subendothelium in citrated and native blood. II. Relationships among platelet adhesion, thrombus dimension and fibrin formation, *J. Lab. Clin. Med.,* 95, 208, 1980.
60. **Badimon, L., Badimon, J. J., Lassila, R., Heras, M., Chesebro, J. H., and Fuster, V.,** Thrombin regulation of platelet interaction with damaged vessel wall and isolated collagen type I at arterial flow conditions in a porcine model: effects of hirudins, heparin and calcium chelation, *Blood,* 78, 423, 1991.
61. **Rao, G. H. R., Smith, C. M., and White, J. G.,** Influence of calcium antagonists on thrombin-induced calcium mobilization and platelet-vessel wall interactions, *Biochem. Med. Metab. Biol.,* 47, 226, 1992.
62. **Lam, J. Y. T. Chesebro, J. H., Steele, P. M., Heras, M., Webster, M. W. I., Badimon, L., and Fuster, V.,** Antithrombotic therapy for deep arterial injury by angioplasty. Efficacy of common platelet inhibition compared with thrombin inhibition in pigs, *Circulation,* 84, 814, 1991.
63. **Inauen, W., Baumgartner, H. R., Bombeli, T., Haeberli, A., and Straub, P. W.,** Dose- and shear rate-dependent effects of heparin on thrombogenesis induced by rabbit aorta subendothelium exposed to flowing human blood, *Arteriosclerosis,* 10, 607, 1990.
64. **Krupinski, K., Basic-Micic, M., Lindhoff, E., and Breddin, H. K.,** Inhibition of coagulation and platelet adhesion to extracellular matrix by unfractionated heparin and a low mol wt heparin, *Blut,* 61, 289, 1990.
65. **Venkataramani, E. S., Senatore, F., Feola, M., and Tran, R. C.,** Nonthrombogenic small-caliber human umbilical vein vascular prosthesis, *Surgery,* 99, 735, 1986.

66. **Engbers, G. H. M., Dost, L., Hennink, W. E., Aarts, P. A. M. M., Sixma, J. J., and Feijen, J.,** An in vitro study of adhesion of blood platelets onto vascular catheters. Part I., *J. Biomed. Mater. Res.*, 21, 613, 1987.
67. **Nojiri, C. N., Park, K. D., Grainger, D. W., Jacobs, H. A., Okano, T., Koyanagi, H., and Kim, S. W.,** In vivo nonthrombogenicity of heparin immobilized polymer surfaces, *ASAIO Trans.*, 36, M168, 1990.
68. **Herbaczynska-Cedro, K., Lembowicz, K., and Pytel, B.,** N^6-Monomethyl-L-Arginine increases platelet deposition on damaged endothelium in vivo. A scanning electron microscopic study, *Thromb. Res.* 64, 1, 1991.
69. **DeGraaf, J. C., Banga, J. D., Moncada, S., Palmer, R. M. J., DeGroot, P. G., and Sixma, J. J.,** Nitric oxide functions as an inhibitor of platelet adhesion under flow conditions, *Circulation*, 85, 2284, 1992.
70. **Groves, P. H., Penny, W. J., Cheadle, H. A., and Lewis, M. J.,** Exogenous nitric oxide inhibits in vivo platelet adhesion following balloon angioplasty, *Cardiovasc. Res.*, 26, 615, 1992.
71. **Groves, P. H., Lewis, M. J., Cheadle, H. A., and Penny, W. J.,** SIN-1 reduces platelet adhesion and platelet thrombus formation in a porcine model of balloon angioplasty, *Circulation*, 87, 590, 1993.
72. **Jandak, J., Steiner, M., and Richardson, P.,** α-tocopherol, an effective inhibitor of platelet adhesion, *Blood*, 73, 141, 1989.
73. **Li, X. and Steiner, M.,** Fish oil: a potent inhibitor of platelet adhesiveness, *Blood*, 76, 938, 1990.
74. **Li, X. and Steiner, M.,** Dose response of dietary fish oil supplementation on platelet adhesion, *Arteriosci. Thromb.*, 11, 39, 1991.
75. **Owens, M. R. and Cave, W. T., Jr.,** Dietary fish lipids do not diminish platelet adhesion to subendothelium, *Br. J. Haematol.*, 75, 82, 1990.
76. **Shebuski, R. J., Ramjit, D. R., Sitko, G. R., Lumma, P. K., and Garsky, V. M.,** Prevention of canine coronary artery thrombosis with echistatin, a potent inhibitor of platelet aggregation from the venom of the viper, Echis carinatus, *Thromb. Haemost.*, 64, 576, 1990.
77. **Yasuda, T., Gold, H. K., Leinbach, R. C., Yaoita, H., Fallon, J. T., Guerrero, L., Napier, M. A., Bunting, S., and Collen, D.,** Kistrin, a polypeptide platelet GPIIb/IIIa receptor antagonist, enhances and sustains coronary arterial thrombolysis with recombinant tissue-type plasminogen activator in a canine preparation, *Circulation*, 83, 1038, 1991.
78. **Hanson, S. R., Pareti, F. I., Ruggeri, Z. M., Marzec, U. M., Kumicki, T. J., Montgomery, R. R., Zimmerman, T. S., and Harker, L. A.,** Effects of monoclonal antibodies against the platelet glycoprotein IIb/IIIa complex on thrombosis and hemostasis in the baboon, *J. Clin. Invest.*, 81, 149, 1988.
79. **Bates, E. R., McGillem, M. J., Mickelson, J. K., Pitt, B., and Mancini, G. B. J.,** A monoclonal antibody against the platelet glycoprotein IIb/IIIa receptor complex prevents platelet aggregation and thrombosis in a canine model of coronary angioplasty, *Circulation*, 84, 2463, 1991.
80. **Gold, H. K., Gimple, L. W., Yasuda, T., Leinbach, R. C., Werner, W., Holt, R., Jordan, R., Berger, H., Collen, D., and Coller, B. S.,** Pharmacodynamic study of $F(ab')_2$ fragments of murine monoclonal antibody 7E3 directed against human platelet glycoprotein IIb/IIIa in patients with unstable angina pectoris, *J. Clin. Invest.*, 86, 651, 1990.
81. **Ellis, S. G., Navetta, F. L., Tcheng, J. T., Weisman, H. F., Wang, A. L., Pitt, B., and Topol, C. J.,** Antiplatelet gpIIb/IIIa (7E3) antibody in elective PTCA; safety and inhibition of platelet function, *Circulation*, 82, III-191, 1990.

Chapter 18

Platelet Adhesion to Damaged Vessel Wall

S. Fazal Mohammad

CONTENTS

I. Introduction 349
II. Blood Platelets: Activation, Adhesion, Aggregation, and Hemostatic Plug Formation 350
III. The Endothelium 352
A. Structure and Function 352
B. Thromboresistance 354
IV. Endothelial Injury 355
V. Reaction of Endothelium to Injury 356
VI. Platelet Adhesion to Damaged Vessel Wall 356
VII. Concluding Remarks 361
VIII. Acknowledgments 362
References 362

I. INTRODUCTION

Although it is quite well established that blood platelets play an important role in a number of pathophysiological reactions, their ability to adhere to one another (aggregation) and to the subendothelial structures following intimal injury is certainly one of their most important functions.[1-3] Platelets are endowed with powerful biochemical stimuli, which are released on site when they adhere. These stimuli include (1) vasoconstricting agents that help reduce the size of the opening following vascular injury, (2) activating agents that recruit other platelets, activate the coagulation pathway, and help consolidate the hemostatic plug, and (3) growth factors that promote the proliferation of smooth muscle cells to initiate the processes that help repair the damaged vessel wall.[4-6] These reactions manifest the intricacies of a complex series of responses that proceed in an orderly sequence to achieve effective hemostasis.

Platelet adhesion to damaged endothelium/subendothelium is an important step in hemostasis.[1,3,7] If platelets do not adhere normally or if adherent platelets do not undergo the necessary sequence of changes, the hemostatic plug will either not form or it will not have the required strength to withstand the force exerted by the circulating blood.[8] On the other hand, excessive adhesion of platelets or excessive release of potent bioactive agents from adherent platelets may result in the formation of large aggregates of platelets at the site of injury. Uncontrolled adhesion of platelets and deposition of fibrin may occlude the blood vessel. Rapid accumulation of a large number of platelets increases the risk of embolization, with serious clinical consequences when emboli traveling downstream reach critical organs.[9,10] Therefore, neither hyporeactive nor hyperreactive platelets are desirable, and a critical balance and optimal response to appropriate stimuli are necessary to achieve (normal) hemostasis and the repair of the damaged vessel wall.

The ability of platelets to adhere to one another and to other substrates is probably unparalleled, and the ease with which platelets quickly adhere and release various chemical stimuli is truly remarkable. Platelets not only adhere to the subendothelium or injured endothelium as a part of a physiological response to pathological conditions, but they also adhere to "nonbiological" (foreign) surfaces.[11-13] This has presented a major challenge in the development and use of blood-contacting prosthetic devices.[14] Somewhat analogous to their participation in the hemostatic process, platelets adherent to artificial surfaces release potent bioactive agents and thus recruit a large number of activated platelets at the blood-material interface. In conjunction with the ensuing activation of the coagulation pathway, this results in adherent platelet-fibrin mass at the interface.[15-17] Deposition of a large amount of blood components at the surface entails the risk of thromboembolization and eventual dysfunction of the device. Thus, the expression of an otherwise normal, physiological response of platelets to a nonendothelial surface becomes a major concern when prosthetic devices contact blood. These problems have rendered the development of blood-contacting devices frustratingly slow.

0-8493-4559-6/95/$0.00+$.50

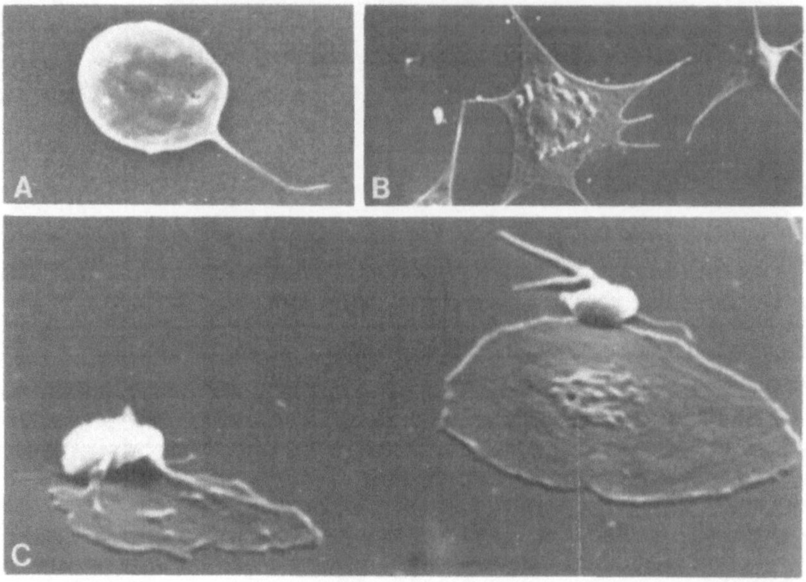

Figure 1A–C

It should be evident from the preceding chapters that considerable progress has been made during the last decade in elucidating the mechanism by which platelets adhere,[18] spread, and undergo release reaction.[19-21] Several glycoprotein receptors that participate in platelet adhesion have been identified.[21-23] The use of specific antibodies against these receptors or their ligands has provided important clues as to how platelets interact with one another or with the subendothelium. It is now well established that the adhesion of platelets and the release of cytoplasmic constituents in the surrounding medium is important for platelets to discharge their physiological functions. Why platelets neither adhere to one another nor to the intact endothelium under normal conditions has been a subject of considerable discussion.[24-26] A proper understanding of the underlying mechanisms by which adhesion of platelets to the intact endothelium is prevented but rapid activation and adhesion occurs when the vessel wall is damaged is of fundamental interest in understanding how cells interact with one another and how they respond to the changes in their environment. In addition, a better understanding of the interplay between platelets, vessel wall, and coagulation pathway may pave the way for a better clinical management of patients with hemostatic disorders or patients with cardiovascular prostheses.

II. BLOOD PLATELETS: ACTIVATION, ADHESION, AGGREGATION, AND HEMOSTATIC PLUG FORMATION

As fragments of megakaryocyte cytoplasm, platelets can be considered as circulating bundles of highly potent bioactive agents.[26-28] Since this discussion is primarily focused on the adhesion of platelets to the damaged endothelium, only a brief review of the adhesive properties of platelets may be in order. For a detailed discussion on structure and function of platelets, readers are referred to other chapters in this book or several excellent reviews on the subject elsewhere.[28,29]

Platelets in circulating blood maintain a discoid shape with the help of microtubules and other contractile elements in their cytoplasm. However, in the presence of minute amounts of certain chemical stimuli, platelets lose their discoid shape (activation, Figure 1A) and rapidly form pseudopods.[28,30] This could be considered the first step in their activation. In the presence of flow (or stirring *in vitro* to enhance platelet-platelet encounter) activated platelets (with extended pseudopodia) intertwine to form an aggregate of platelets.[31,32] Aggregated platelets release a number of highly potent stimuli into the surrounding medium, including ADP, serotonin, thromboxane A_2, and platelet-activating factor.[33] Upon contact with a surface (i.e., injured vessel wall) under appropriate conditions, platelets adhere, spread, and release their cytoplasmic constituents (Figure 1B, 1C).[34-36] The agonists released by adherent or aggregated platelets activate other platelets in the vicinity, thus facilitating their participation in the hemostatic process (Figure 1D, 1E). Platelets that are normally approximately 3 μm in diameter will cover 10 μm or more in diameter when they are fully spread.

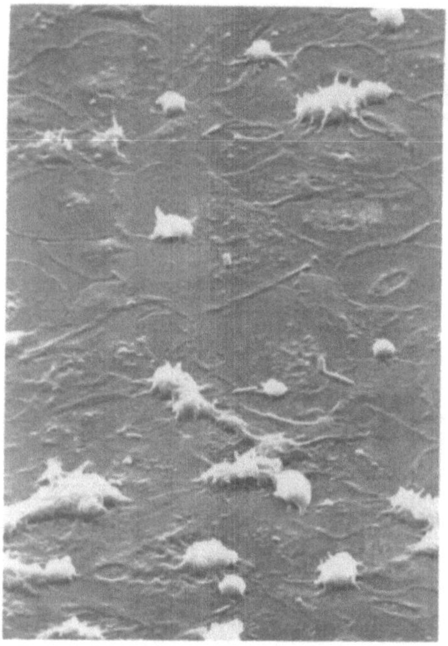

Figure 1D

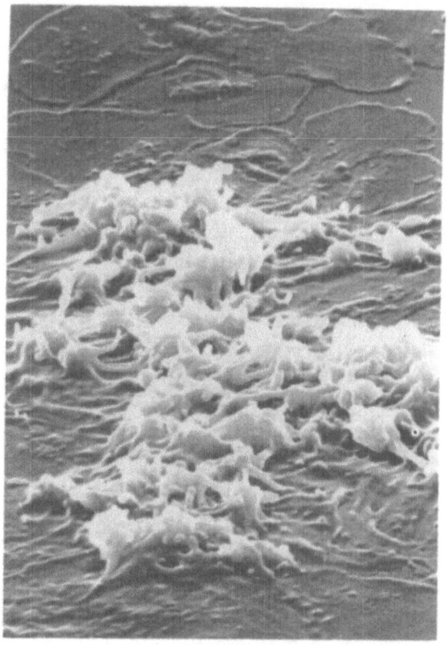

Figure 1E

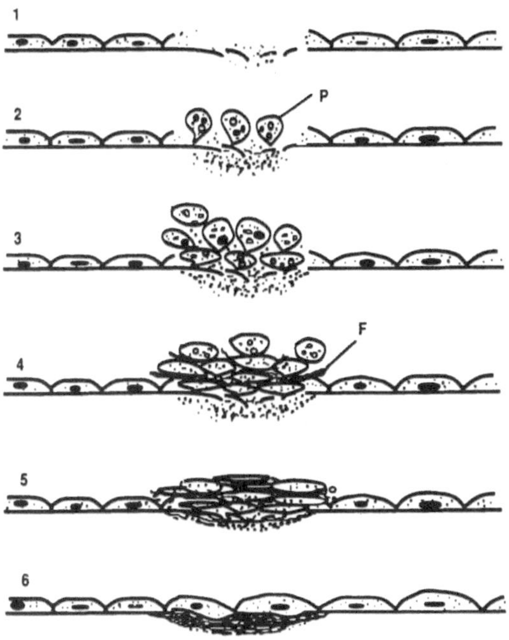

Figure 1F

Figure 1 Adhesion of blood platelets proceeds through a sequence of steps that include activation and formation of pseudopods (A). The tips of pseudopodia move away from the body of the platelets (B); the space between the pseudopods is filled in by stretching platelet plasma membrane (C) to achieve complete spreading of the adherent platelet. Additional platelets then adhere to the spread platelets (D) and form aggregates (E). Please note in Figure 1E that the original surface (glass) is completely covered with spread platelets. (F) Schematic representation of hemostatic plug formation. Following vascular injury (1), platelets rapidly adhere to the damaged endothelium/subendothelium (2). Adherent platelets spread and release agonists that recruit additional platelets (P) to form an aggregate (3). While platelet aggregate formation is in progress, expression of thromboplastic activity by activated platelets and damaged endothelium results in the formation of fibrin (F, 4), which then cements the platelets and consolidates the hemostatic plug (5). Activation of fibrinolysis helps in the dissolution of adherent mass of platelets and fibrin, smooth muscle cells repair the damaged vessel wall, and the luminal surface is covered with endothelial cells to complete the reparative process (6).

As discussed later, subendothelial structures that are exposed when the vessel wall is damaged attract platelets. Since the damaged endothelial cells and the subendothelium are unable to prevent platelets from adhering (whereas normal endothelium does), platelets in the vicinity of injured endothelium can adhere and form aggregates. Besides the factors present in the surrounding medium that activate platelets in the area of injury, platelets are endowed with specific glycoprotein receptors on their surface with which they can interact with certain components of subendothelium. Aided by disturbed flow conditions, these factors together help platelets adhere and accumulate in large numbers at the site of injury. It is important to note that, in these reactions, only the first layer of platelets attaches to the subendothelium. All subsequent layers of platelets adhere by platelet-platelet interaction (Figure 1D, 1E); this latter process is aided by the agonists released by adherent platelets. Platelets participating in the aggregate formation in turn also release their granular content to support the process of platelet plug formation. Activation of the coagulation cascade both by the injured vessel wall and by the activated, adherent platelets results in the formation of fibrin, which cements the aggregated platelets and allows for the formation of a consolidated hemostatic plug (Figure 1F). This hemostatic plug offers the desired tenacity to withstand the pressures exerted by the circulating blood and thus prevent further bleeding. Following intimal injury, cessation of blood loss is the immediate goal of the hemostatic process. Once this task has been accomplished, reparative processes are initiated. Activation of the fibrinolytic pathway helps remove deposited blood components, and regeneration of endothelium and smooth muscle cells is aided by the growth factors that are released from activated platelets.

From this brief discussion, it is obvious that platelets armed with a number of highly potent chemical stimuli play a critical role in achieving effective hemostasis. Inadequate response on the part of platelets leads to abnormal hemostasis. Bleeding is the consequence when platelets fail to adhere normally or when they are unable to release normal constituents. Excessive adhesion or release reaction entails the risk of thrombosis or thromboembolism. Therefore, in order to achieve an effective hemostasis, platelets must mount a carefully orchestrated response in which platelet adhesion, platelet release reaction, and generation of fibrin proceed rapidly in an orderly sequence following vascular injury. It is also necessary that these reactions cease quickly once the bleeding has been arrested.

III. THE ENDOTHELIUM

A. STRUCTURE AND FUNCTION

The endothelium is a continuous sheet of flat, mononuclear cells that line the entire vascular system and thus separate the blood from a very thrombogenic subendothelium. Based on scanning electron microscopic assessment of fixed segments of blood vessels, the endothelial cells appear elongated, with cells aligned longitudinally in the direction of blood flow (Figure 2). However, since all blood vessels undergo longitudinal as well as radial contraction, the morphological appearance of endothelial cells in excised blood vessels is altered considerably. Published studies have indicated that severed blood vessels may contract as much as 60%; this forces the bulging and elongation of firmly anchored endothelial cells to accommodate the changing conditions.[37-39] When blood vessels are fixed *in situ* with minimal longitudinal and radial contraction, the endothelium appears very flat, an almost glass-like surface with neither bulging nor elongation of the endothelial cells.[40] Under these conditions, as shown in Figure 3, the endothelium is so thin that subendothelial structures can be resolved easily and individual endothelial cells cannot be distinguished because of the tight cell junctions and invisible cell boundaries. When the blood vessel is allowed a limited contraction at the time of fixation, distinct cell boundaries showing the contours of individual cells appear, confirming the presence of endothelial cells (Figure 4).

Individual endothelial cells, which form tight junctions with adjacent cells to form a monolayer that withstands the stresses imposed by circulating blood, are endowed with important structural and functional properties to facilitate uninterrupted blood flow through the vessels. Each endothelial cell has two surfaces, the luminal surface, which faces the blood; and the abluminal surface, which is anchored to the basal lamina. The luminal surface faces a constant rhythmic current of plasma and formed elements of blood, with continuous rubbing, traction, and shearing effects. In contrast, the abluminal side of the endothelial cell appears to remain firmly fixed to the subendothelium and never comes in contact with blood. Although it has been known for some time that normal blood vessels with intact endothelium are characteristically nonthrombogenic,[41-43] how endothelium achieves this hemocompatibility is poorly understood.

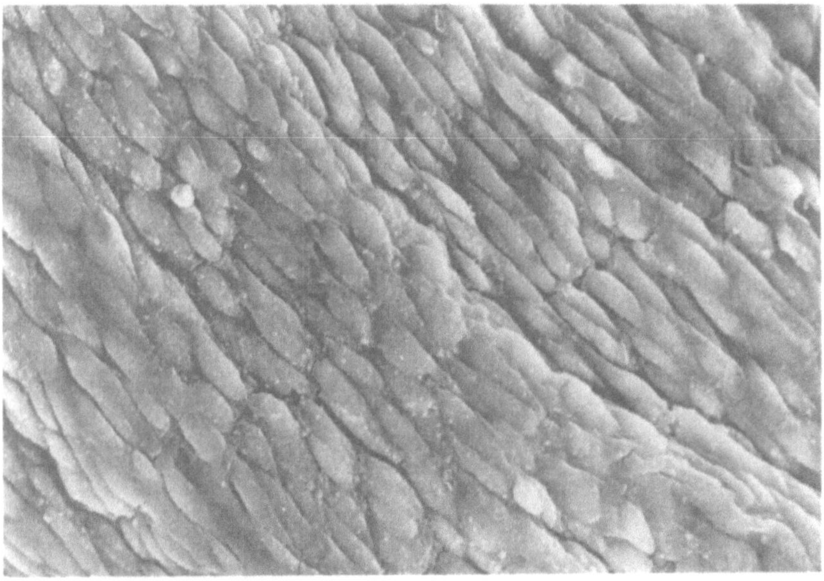

Figure 2 Morphological appearance of human umbilical vein endothelium. A segment of the blood vessel was removed after the umbilical cord was severed. The vein lumen was rinsed with 0.15 *M* NaCl, fixed with buffered glutaraldehyde, and processed for scanning electron microscopy. Note that endothelial cells are aligned longitudinally in the direction of blood flow. (Magnification ×450.)

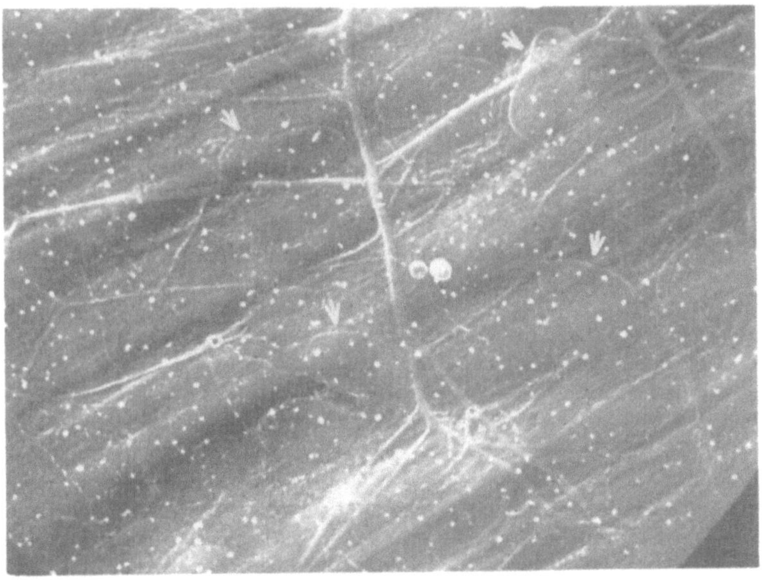

Figure 3 Morphological appearance of dog saphenous vein endothelium fixed *in situ*. The blood vessel was isolated and immediately ligated at two ends approximately 5 cm apart. The lumen of the ligated segment was rinsed with Ca^{2+} and Mg^{2+} free Tyrode's solution containing papaverine to remove blood components and to prevent vessel contraction. The lumen was then filled with buffered glutaraldehyde, while the pressure inside the lumen was maintained at 100 mmHg throughout the fixation period (1 h). The ligated points were attached to a clamp to prevent longitudinal contraction of the blood vessel when it was severed for further processing and remained in place until the fixed segment was dehydrated. Please note a smooth, glass-like appearance of the endothelium in which the cell boundaries are not visible. The outline of the endothelial cell nuclei (arrows) suggest that the fibrous structures are underneath the endothelium. (Magnification ×2000.)

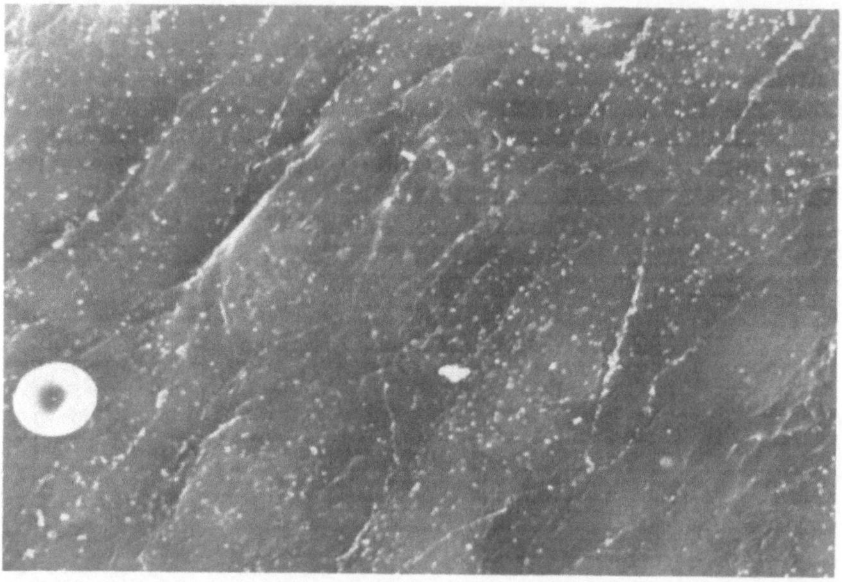

Figure 4 The blood vessel shown in this micrograph was processed as indicated in Figure 3, with the exception that during the fixation, the vessel was allowed to undergo a slight (approximately 10%) longitudinal contraction. Please note that individual cells are still flat, but now their cell borders are visible and some bulging of the cell in areas where nuclei are located is evident. When the blood vessel undergoes maximal radial and longitudinal contraction, each endothelial cell becomes rounded and longitudinally apposed in the direction of flow, as seen in Figure 2. (Magnification ×1500.)

B. THROMBORESISTANCE

The nonthrombogenic nature of endothelium on the luminal side is believed to be due to a variety of factors, including the presentation of a hemocompatible glycocalyx and production and release of powerful chemical agents that prevent activation of platelets and coagulation. The factors that may contribute to the thromboresistance of endothelium include the following:

1. Presence of heparin and heparin-like molecules on the surface of endothelial cells may impart a net negative charge and also prevent contact activation of the coagulation pathway.[44,45]
2. Alpha 2 macroglobulin present on the surface of endothelial cells has been implicated in endothelial thromboresistance.
3. Since endothelial cells have been shown to synthesize a number of plasminogen activators[47-50], they have been implicated to play an active role in fibrinolysis.
4. However, to assure that the fibrinolysis is limited to the affected areas, activation of plasminogen to plasmin is carefully regulated by plasminogen activator inhibitor also present in endothelial cells.[51-53]
5. Endothelial cells synthesize and release prostacyclin, a potent platelet inhibitor. As discussed in other sections of this book, prostacyclin is of considerable significance in vascular pathophysiology.[54-57]
6. Endothelial cells can also bind and thus neutralize thrombin. A receptor for thrombin on endothelial cell surface has been identified. Published studies suggest that attachment of thrombin to the endothelial cell surface stimulates the arachidonate metabolism, which in turn results in enhanced production of prostacyclin. Thrombin may also have mitogenic or growth-stimulating effect on endothelial cells. Thus, binding of thrombin to endothelial cells appears to serve two functions: (a) endothelium removes thrombin from blood. This is critical given the fact that thrombin is a potent enzyme that rapidly converts fibrinogen to fibrin. The ability of endothelium to bind thrombin and neutralize its enzyme activity may be important in vessel wall physiology. (b) Moreover, since thrombin generation signals rapid activation of the coagulation pathway, endothelium may respond to the presence of thrombin by up-regulating prostacyclin production and activation of fibrinolysis,[58-60] which provide an important protection. Thrombin-binding activity at the endothelial cell surface may be due to the association of heparin-like glycosaminoglycans, which are present at the endothelial cell surface. In addition, thrombomodulin present at the endothelial cell surface may also bind thrombin.[61] Thrombin bound in this manner activates protein C, a potent inhibitor of factors Va and VIIIa.[62,63A] In a published study,

Shanberge et al.[63B] found that thrombin alone did not affect platelet adhesion to endothelial cells. However, adhesion of thrombin-stimulated platelets to endothelial cells was considerably greater when antiheparin agents (protamine, polybrene) were used, suggesting that heparin-like molecules on the endothelial cells may be responsible for inhibiting platelet adhesion.

7. Endothelial cells bind antithrombin III, a naturally occurring anticoagulant presumably through surface-associated glycosaminoglycans.[64] When bound to the surface of endothelium, AT III may also inhibit activation of blood coagulation factors.
8. Endothelial cells may serve as scavengers. They bind hormones[65,66] and fibrin monomers[67] and remove adenine nucleotides,[68] stable prostaglandin metabolites,[69] bradykinin, and many other substances.[70]

Through these activities, intact endothelium prevents platelet adhesion and activation of coagulation factors and facilitates dissolution of deposited blood components once they are formed; thus, it helps maintain a delicate balance necessary for uninterrupted flow of blood through the blood vessels. Some of these functions of endothelial cells may be compromised when endothelium is damaged. This may shift the balance in favor of platelet adhesion and activation of coagulation pathway in the area of vascular injury.

Perhaps one of the most important properties of the endothelial cells is their ability to prevent platelets from adhering to endothelium *in vivo* under normal conditions.[60] This is noteworthy, since platelets generally adhere to almost any other surface, regardless of whether it is an artificial surface or a surface presented by other cell types. The mechanisms by which endothelium prevents platelet adhesion are complex and may involve release of antiplatelet agents, as well as presenting a nonattractive surface to the circulating platelets. Prostacyclin remains one of the most studied inhibitors of platelet function.[61-65] Prostacyclin also down-regulates the expression of glycoprotein receptors on platelet membrane.[71] In the absence of an adequate number of these receptors, ligand binding is reduced or diminished, and platelets fail to adhere to the vessel wall. Endothelial-cell-derived relaxing factor (EDRF), adenosine diphosphatase, glycosaminoglycans, and associated AT III also inhibit platelet adhesion to the endothelial cells.[72,73] EDRF also down-regulates the expression of glycoproteins on platelet membrane.[74] It is likely that any one of the above factors, acting alone, may not render endothelium hemocompatible. However, acting in concert, these factors collectively make the endothelium a most hemocompatible interface.

It is also interesting to note that while the luminal surface of the endothelial cell is hemocompatible, the abluminal surface of these cells, as well as subendothelium may be the most thrombogenic interface. Therefore, when a blood vessel is damaged, and the abluminal side of the endothelial cells and the subendothelium is exposed to blood, platelets adhere to the damaged vessel wall within seconds and initiate the process of hemostatic plug formation. Under normal conditions, platelet deposition on damaged vessel wall proceeds in a balanced manner, and the hemostatic plug formation is generally restricted to the area of vascular injury (Figure 1F). As pointed out in the preceding section, a shift in this balance may have disastrous consequences: a shift toward increased coagulation or platelet deposition may lead to thrombosis, whereas a shift toward suppressed coagulation or inadequate participation by platelets may result in serious (even fatal) hemorrhage.

IV. ENDOTHELIAL INJURY

Under normal conditions, endothelial cells adherent to the basement membrane on the abluminal surface and to the neighboring endothelial cells through tight intercellular junctions have sufficient strength and tenacity to withstand the stresses normally exerted by the circulating blood. However, the endothelium can be damaged easily under a variety of conditions. Increased shear stresses, particularly in areas of turbulent or vortical flow, cause endothelial injury[75,76] and a shorter lifespan of endothelial cells in areas with abnormal flow has been observed.[77] Enhanced vascular permeability and subendothelial edema in areas with abnormal flow have also been noted.[78] Endothelial cells are susceptible to injury when deprived of oxygen, which happens when blood vessels are occluded by thromboemboli. Hempel et al.[79] found that endothelial cells subjected to anoxia followed by reoxygenation produce less prostacyclin when challenged by thrombin or other agonists and exhibit increased platelet adhesion. Endothelium can be injured by circulating immune complexes or antibodies directed against specific antigens on the endothelial cell surface that will elicit the antibody response.[80,81] Adhesion of platelets and leukocytes to endothelial cells damaged by antibodies has been documented.[82,83] Endothelium may be injured also by bacteria and bacterial products. Severe injury, including disruption of intercellular junctions, occurs when endothelial cells are exposed to cholera toxin.[84] Endothelial injury leading to thrombosis followed anthrax and meningococcal septicemia[85] and endotoxins have been noted to cause desquamation of endothelial cells.[86] Endothelium may be damaged by elevated levels of vasoactive substances,[87] hypertension, and a variety of chemical agents.[88] Balloon angioplasty can produce severe endothelial injury, resulting in deposition of platelets at the injured site.[89] However, heat-damaged endothelial cells do not appear to attract platelets, suggesting that the risk of thrombogenicity following laser or thermal angioplasty may be less than those methods that involve mechanical angioplasty.[90] Iatrogenic vascular injury has also become a concern in recent years because of the increasing number of surgeries, catherization procedures, implantation of prosthetic devices, and procedures that involve extracorporeal circulation, all of which cause endothelial damage.

V. REACTION OF ENDOTHELIUM TO INJURY

Depending upon the nature of the injurious agent, endothelial cells may suffer sublethal or lethal injury. Sublethal injury is frequently manifested by swelling of the cells, formation of protrusions in the endothelial cell plasma membrane (Figure 5), and development of craters (Figure 6). Severe endothelial injury may lead to cell death and desquamation (Figure 7).[91] Sublethal injury results in increased vascular permeability; however, depending upon the strength of the injury, cells may recover and restore their normal function.[92] Cell death and desquamation result in exposure of the subendothelium to blood components, thus exposing a very thrombogenic, subendothelial surface. Platelets quickly adhere to the exposed subendothelium as well as the damaged endothelial cells (Figures 7A to 7C)[93] and go through the sequence of steps shown in Figure 1. As pointed out in the preceding sections, adherent platelets release potent chemical stimuli and recruit additional platelets to form an aggregate; this process continues until the entire exposed subendothelium is covered with platelets.[94] Figure 8 gives an idea of the extent of damage to the intima and media following venipuncture with a 22-gauge hypodermic needle.

VI. PLATELET ADHESION TO DAMAGED VESSEL WALL

Adhesion of platelets to the damaged vessel wall may be considered as one of the first steps of the hemostatic process. As shown in Figure 1, initial platelet contact with subendothelium or damaged endothelial cells results in activation of platelets. Activated platelets form pseudopods, which can stretch several micrometers away from the body of the platelets. Formation of pseudopods also signals changes in the platelet cytoplasm; these changes include activation of a contractile system that may help platelets release their granule constituents. The sensitivity of platelets to the presence of various agonists and their ability to quickly effect the release reaction are facilitated by a unique surface canalicular system (SCS). The SCS is formed presumably by infolding of platelet plasma membrane that extends deep into the cytoplasm. The presence of SCS permits the platelets to transport their granule constituents to the outside rapidly, rather than relying on diffusion, and platelets use this system efficiently to respond to the changes in their environment.

Platelet activation is followed by adhesion, spreading, and release of contents stored in platelet granules. Agonists released by adherent platelets recruit additional platelets to form a platelet aggregate. Fibrin then forms to consolidate the aggregated platelets (Figure 1F). The initial events proceed through two key steps: (1) adhesion of platelets to the components of the damaged vessel wall, and (2) adhesion of platelets to one another.

A number of factors contribute to facilitate platelet adhesion. This includes rheological factors, which help bring circulating platelets to the proximity of the injured area, participation of plasma proteins that provide key components needed for adhesion, and unique glycoprotein receptors on platelet membranes. Although it has been known for some time that the damaged endothelium/subendothelium offers a substratum for platelets to adhere and initiate the hemostatic process, the underlying mechanisms by which platelets adhere and the role of glycoprotein receptors on the platelet surface and appropriate ligands that interact with these receptors has been elucidated only in recent years, and this subject has been covered in detail in other chapters in this book. The following is a brief review of our current understanding of the mechanism of platelet adhesion.

As mentioned previously, endothelial cells synthesize and release von Willebrand factor (vWF), a large multimeric glycoprotein, into the circulating blood.[95-99] The vWF in conjunction with coagulation factor VIII provides the circulating vWF:VIII complex, a critical component of the coagulation pathway.[100,101] Endothelial cells release vWF on the abluminal as well as luminal side.[102,103] The vWF released from the luminal surface circulates in blood as a normal constituent of plasma/cells. The vWF released on the abluminal side is deposited in the form of extracellular matrix and is of critical importance as a

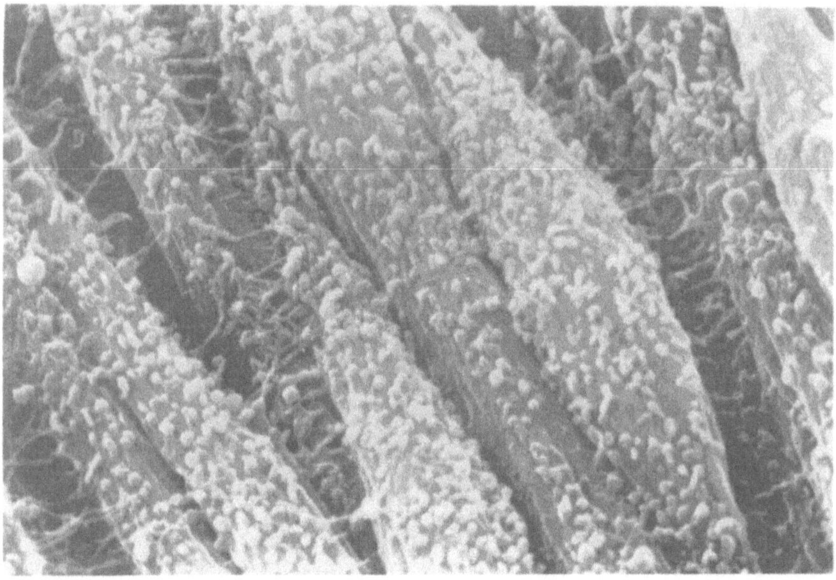

Figure 5 One of the early responses of endothelium to mechanical injury may be the formation of cytoplasmic protrusions. (Magnification ×3500.)

primary site for platelet adhesion when the endothelium is removed or when the vessel wall is damaged.[104-106] vWF has functional binding domains for factor VIII, heparin, gpIb, collagen, and gpIIb-IIIa.[107] The gpIb binding domain on vWF appears to be cryptic in normal circulation and becomes expressed only when the vWF is associated with either extracellular matrix or with fibrin.[108,109] This explains why platelets do not bind vWF in circulating blood and do not undergo spontaneous aggregation. Using a parallel plate perfusion chamber and a monoclonal antibody to vWF that specifically blocked vWF binding to platelet gpIb, Baruch et al.[106] observed that by inhibiting the vWF-gpIb interaction, platelet adhesion was inhibited by approximately 45%. This suggested that the binding of platelet gpIb to vWF

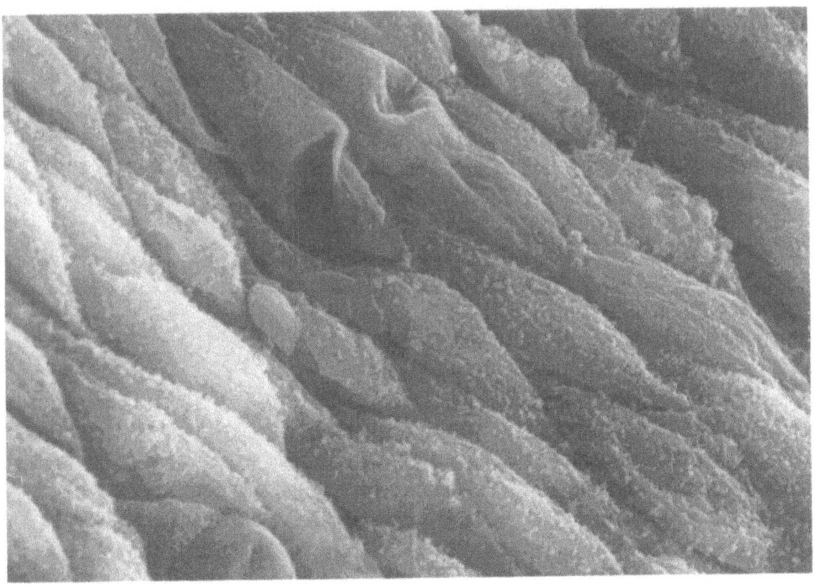

Figure 6 Dog saphenous vein endothelium following relatively severe mechanical injury (the blood vessel was lightly clamped for 15 s). Please note the appearance of crater-like structures. (Magnification ×2600.)

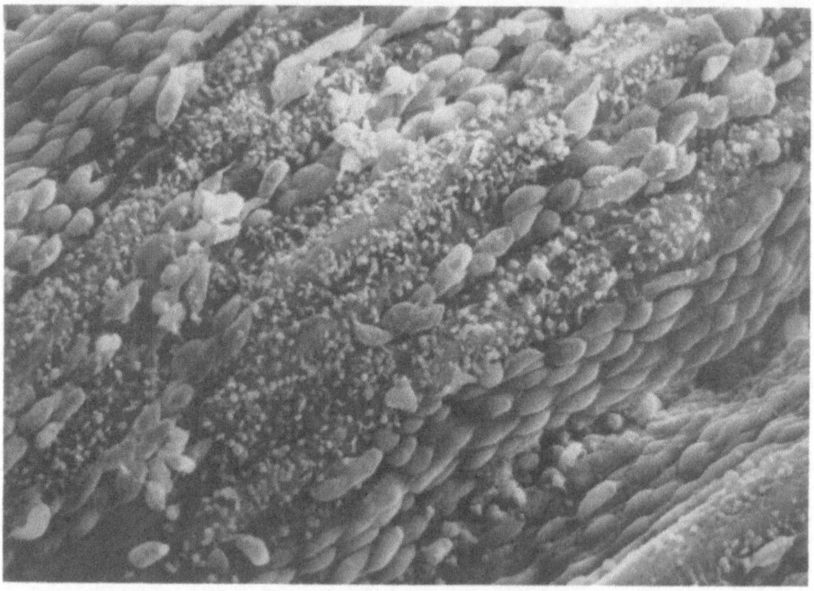

Figure 7A

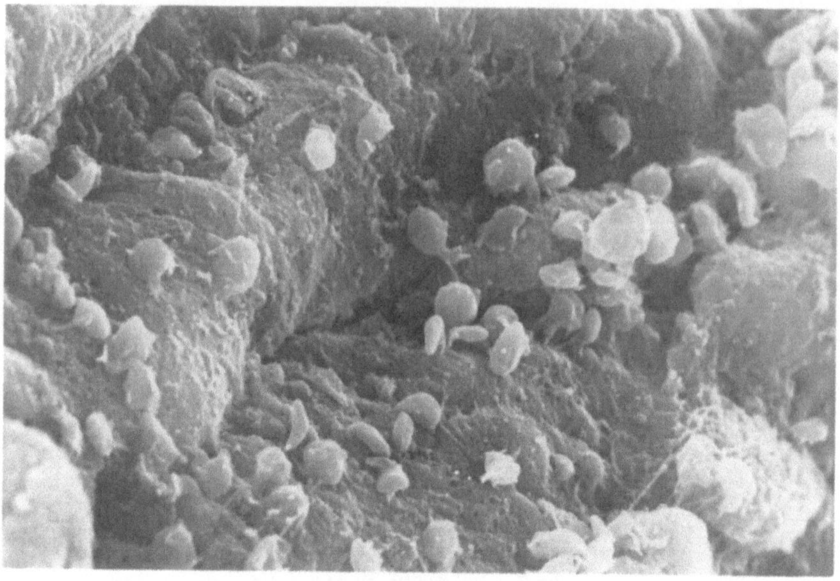

Figure 7B

may be a major contributing factor in platelet adhesion to subendothelium. Results reported from other laboratories support this conclusion[109,110] and further suggest that vWF promotes platelet adhesion at higher shear stress.[109-112] This is interesting, since high shear stresses are encountered in microcirculation or when blood vessels are partially occluded. While platelet gpIb is the primary binding site for vWF, gpIIb-IIIa and vitronectin receptors also bind vWF.[113-115] Interaction of vWF localized on extracellular matrix with platelet gpIb or gpIIb-IIIa appears to be the primary mechanism for anchoring the platelets to the exposed subendothelium. Factors that inhibit or suppress the synthesis of vWF by the endothelial cells or prevent its deposition on the abluminal side may adversely affect the hemostatic process. Patients with von Willebrands disease have prolonged bleeding time primarily because of the inability of their platelets to form an effective hemostatic plug in the absence of vWF.[116] That the presence of vWF in the

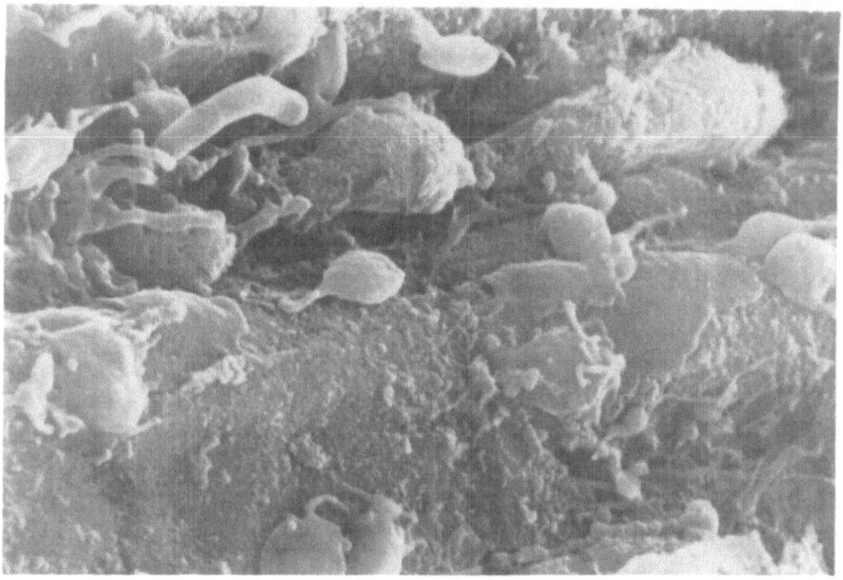

Figure 7C

Figure 7 (A) Severe mechanical injury leads to endothelial desquamation. Depending on the nature of injury, this may occur in patches or complete denudation of large areas of the intima. The following micrographs show a dog saphenous vein that was clamped for 5 min to completely prevent the blood flow through the vessel. Immediately after removing the clamp, the vessel segment was removed, rinsed with 0.15 *M* NaCl, fixed with buffered glutaraldehyde, and processed for scanning electron microscopy. This micrograph shows a view of the intima at lower magnification to document that: (1) clamping led to patchy loss of endothelium, and (2) platelets adhered only in areas where endothelium was damaged. Note that platelets did not adhere in areas where endothelium was still intact. (Magnification ×450.) (B) An area of Figure 7A at higher magnification to show that despite less than 1 min exposure, platelets have started adhering to the exposed subendothelium. (Magnification ×2800); (C) A view of the exposed subendothelium showing that adherent platelets have begun the process of spreading (arrows). (Magnification ×4000.)

extracellular matrix is critical for platelet adhesion is further supported by the results reported by de Groot and co-workers.[117] They noted that the treatment of endothelial cells with interleukin-1 resulted in decreased synthesis of vWF. One of the consequences of this reduction in vWF synthesis was a reduction in the concentration of vWF in the extracellular matrix produced by endothelial cells treated with interleukin-1 and a corresponding reduction in platelet adhesion to this matrix.[117] These observations suggest that vWF is critical for platelet adhesion to subendothelium. Given the significance of vWF in platelet adhesion, it is not surprising that vWF is also stored in the α granules of platelets. Published studies document that the concentration of vWF in platelets may be 60 times greater than the vWF concentration in circulating blood.[118] Stored vWF is released when platelets are activated and facilitates the adhesion as well as spreading of the adherent platelets. This may explain why the patients with von Willebrands disease type I, who are deficient in circulating vWF but have normal vWF content in the platelet alpha granule, have a relatively shorter bleeding time.[119]

In addition to the gpIb-vWF interaction, glycoprotein receptors on platelet membrane for other components of the extracellular matrix have been identified. Platelet membrane gpIa-IIa serves as a receptor for collagen,[120] gpIc′-IIa is a receptor for laminin,[121] and gpIc-IIa has been shown to be a receptor for fibronectin.[122] Collectively, these glycoprotein receptors bind to their ligands present on the damaged vessel wall. As pointed out earlier, adherent platelets release their cytoplasmic constituents and recruit more platelets to sustain the process of hemostatic plug formation. However, it should be noted that after the first layer of platelets has adhered to the damaged surface, newly arriving platelets see only the membrane of the adherent platelets; ECM-associated vWF remains covered under the first layer of

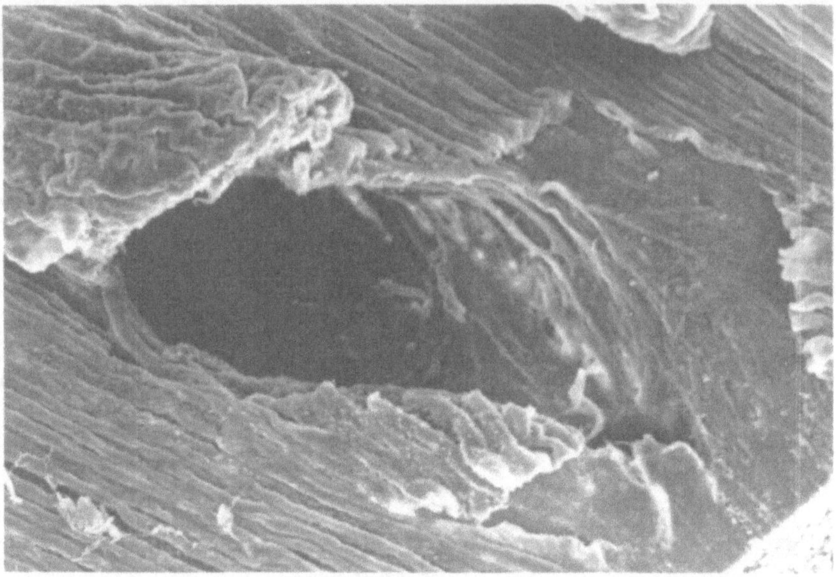

Figure 8 Scanning electron microscopic view of the lumen of a blood vessel (mesenteric artery in a calf), which was isolated, ligated at two ends, and rinsed free of blood components with 0.15 *M* NaCl. A 22-gauge hypodermic needle was then inserted and allowed to remain in place for 1 min. After gently withdrawing the needle, the blood vessel segment was removed, fixed in buffered glutaraldehyde, and processed for scanning electron microscopy. (Magnification ×100.)

adherent platelets. Therefore, to sustain the hemostatic process, platelets arriving on the scene after the adhesion of first layer of platelets must interact with those platelets that are already adherent. This is accomplished by platelet membrane gpIIb-IIIa, a promiscuous receptor and one of the most widely studied platelet membrane glycoproteins.[123-125] When activated, platelets express this receptor.[126,127] Fibrinogen present in the plasma or released from the platelet α granule serves as a ligand for this receptor

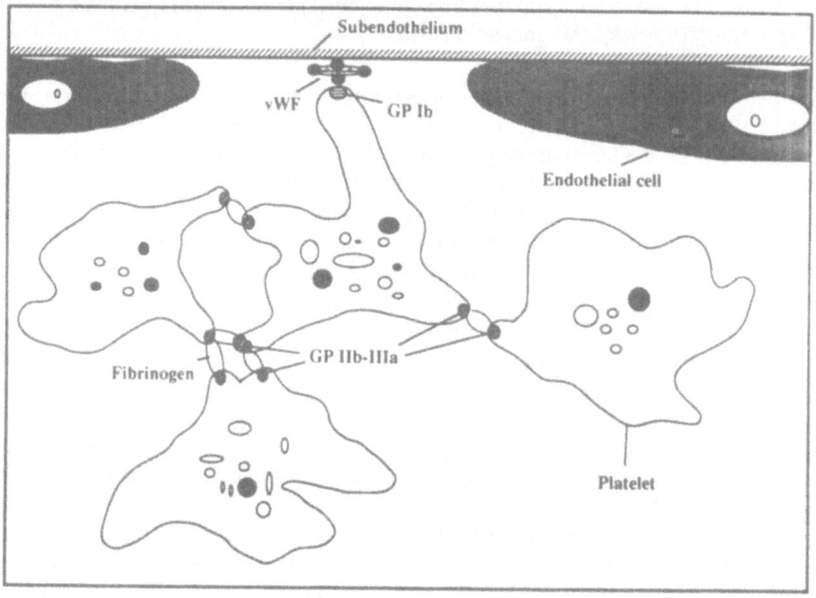

Figure 9 Schematic illustration of glycoprotein receptor mediated adhesion of platelets. vWF: von Willebrand factor; gplb, gpllb-Illa: gplb and llb-Illa. vWF also binds to gpllb-Illa and thus participates in the formation of platelet aggregates.[135,136]

(Figure 9). In addition, gpIIb-IIIa also binds vWF, fibronectin, collagen, and thrombospondin.[123] Fibrinogen also readily binds to the gpIIb-IIIa receptor expressed on platelets adherent to the vessel wall via gp1b-vWF. By forming a bridge between two platelets, fibrinogen helps anchor nonadherent platelets to the surface. Aggregated platelets may also adhere to the subendothelium by either extracellular matrix-associated vWF binding to gpIb on the platelets or fibrinogen bridging gpIIb-IIIa on two separate platelets.[128]

While gpIb and gpIIb-IIIa are major platelet membrane receptors, other glycoprotein receptors have also been identified. gpIV serves as a receptor for thrombospondin or collagen, gpIb-IX as a receptor for vWF and thrombin, vitronectin as a receptor that interacts with vitronectin, vWF, fibronectin, fibrinogen, and thrombospondin.[113]

VII. CONCLUDING REMARKS

Armed with a number of receptors on their surface, vWF, fibrinogen, and other ligands in their granules, and equipped with a number of agonists, growth factors, and other proteins/peptides, platelets have the arsenal needed for adhesion on surfaces and the means to sustain the hemostatic process until their mission is accomplished.

Many of the platelet, leukocyte, and endothelial cell membrane glycoprotein receptors are currently under intense scrutiny, and it is expected that our understanding of the precise mechanism of action of each of the glycoproteins and their relationship in the overall homeostatic processes will improve in the coming years. Endothelial cell adhesion molecules have attracted considerable attention in recent years and have provided the basis for understanding the trafficking of leukocytes[129] and the interaction of malignant cells with endothelium.[129,130] Endothelial cells express specific cell adhesion molecules at the appropriate time to facilitate leukocyte migration across the endothelium. Intercellular adhesion molecules ICAM-1, ICAM-2, ELAM-1, VCAM-1, PAF receptors, etc. are part of the arsenal that endothelial cells utilize to mediate the adhesion and extravasation of leukocytes.[129] Cytokines, IL-1, interferon-γ, and tissue necrosis factor, increase the expression of cell adhesion molecules in cultured endothelial cells.[131,132] Recent studies suggest an increase in the expression of adhesion molecules in various malignancies.[133,134]

Although our understanding of platelet-endothelial cell interaction is limited, the possible involvement of platelet cell adhesion molecules on endothelial cells following injury has not been explored with the same intensity as leukocyte-endothelial cell interaction. It is likely that, in addition to exposing subendothelium, injured endothelial cells support hemostatic plug formation by expressing specific molecules targeted for platelets or other components of the hemostatic plug.[135-137] At least two such adhesive molecules, granule membrane protein-140 (GMP-140, also called PADGEM) and platelet-endothelial cell adhesion molecule-1 (PECAM-1), have been identified. GMP-140 is present in platelet granules and is expressed on the surface of platelets when release occurs. GMP-140 is also present in endothelial cell Weibel-Palade bodies and becomes expressed on the surface of endothelial cells following the release of these organelles.[113] This protein facilitates platelet-leukocyte or endothelial cell-leukocyte interaction, and its primary function may be to encourage the macrophage to remove activated, consumed platelets. The function of PECAM-1 is not known, but its possible involvement in intercellular junctions in endothelial cells has been considered.[113,138]

Based on our current understanding, it is obvious that the interplay of the glycoprotein receptors on various cells and their interactions with the damaged vessel wall are complex. The hemostatic process proceeds through many steps to assure that the rate of hemostatic plug formation will be adequate and that this process is limited to arresting the bleeding and not occluding unaffected blood vessels. The balance in these various processes is of profound importance in assuring that each reaction proceeds to the desired level. A shift in this balance on either side results in either ineffective hemostasis that leads to bleeding, or thrombosis and thromboembolism; both are serious life-threatening consequences requiring immediate intervention.

Evolving as a lining that separates blood from a very thrombogenic environment, the endothelium is perhaps the only truly hemocompatible interface. Endothelium keeps circulating blood from activating on the luminal side and facilitates rapid activation on the abluminal side; both of these properties are necessary for the endothelium to perform its assigned tasks as a barrier between blood and vessel wall. Limiting the activation of hemostatic and complement pathways is perhaps one of the most challenging tasks that intact endothelial cells face. Recent technological advances and improvements in surgical techniques, when coupled with our ability to substitute malfunctioning organs with prosthetic devices, are placing an ever-increasing demand on the hemostatic system. Direct contact of blood with materials on which proteins can be adsorbed and activated, and platelets can readily adhere, present a clinical challenge requiring the use of anticoagulants and antiplatelet agents. Angioplasty, indwelling catheters, extracorporeal oxygenation of blood, hemodialysis, implantation of mechanical heart valves, vascular grafts, ventricular assist devices, and total artificial hearts serve as examples of clinical procedures that are affected by the hemostatic system. Therefore, it is imperative that the delicate roles of both platelets and endothelium/subendothelium are well understood in order to best utilize what these systems offer in keeping the blood fluid as it circulates through the body. While advances made in the recent past have raised our expectations, there are lessons to be learned as to how platelets and damaged vessel wall interact when injuries challenge the integrity of the vascular system.

VIII. ACKNOWLEDGMENTS

The author gratefully acknowledges the help of Drs. Olga Sehnalova and Ingrid de Bruijn for help in preparing the scanning electron micrographs and of my associates for allowing me the time to write this chapter.

REFERENCES

1. **Packham, M. A. and Mustard, J. F.,** Normal and abnormal platelet activity, in *Blood Platelet Function and Medicinal Chemistry,* Lasslo, A., Ed., Elsevier Biomedical, New York, 1984, 61.
2. **Coller, B. S.,** Blood elements at surfaces: platelets, *Ann. N.Y. Acad. Sci.,* 516, 362, 1987.
3. **Packham, M. A. and Mustard, J. F.,** Platelet adhesion, *Prog. Hemost. Thromb.,* 7, 211, 1984.
4. **Hawiger, J., Steer, M. L., and Salzman, E. W.,** Intercellular regulatory processes in platelets, in *Hemostasis and Thrombosis: Basic Principles and Clinical Practice,* 2nd ed., Colman, R. W., Marder, V. J., Salzman, E. W., and Hirsh, J., Eds., J. B. Lippincott, Philadelphia, 1987, 710.
5. **Mason, R. G., Mohammad, S. F., and Chuang, H. Y. et al.,** The adhesion of platelets to subendothelium, collagen, and artificial surfaces, *Semin. Thromb. Hemost.,* 3, 98, 1976.
6. **Weksler, B. B.,** Platelet interactions with the blood vessel wall, in *Hemostasis and Thrombosis: Basic Principles and Clinical Practice,* 2nd ed., Colman, R. W., Marder, V. J., Salzman, E. W., and Hirsh, J., Eds., J. B. Lippincott, Philadelphia, 1987, 804.
7. **Baumgartner, H. R., Muggli, R., Tschopp, P. B., and Turitto, V. T.,** Platelet adhesion, release and aggregation in flowing blood: effect of surface properties and platelet function. *Thromb. Hemost.,* 35, 124, 1976.
8. **Sixma, J. J. and Bouma, B. N.,** Physiology and biochemistry of the hemostatic system, in *The Thromboembolic Disorders,* van de Loo, J., Prentice, C. R. M., and Beller, F. K., Eds., Schattaur Verlag, New York, 1983, 3.
9. **Goldsmith, H. L. and Karino, T.,** Interaction of human blood cells with vascular endothelium, *Ann. N.Y. Acad. Sci.,* 516, 468, 1987.
10. **Grabowski, E. F.,** Platelet aggregation in flowing blood at the site of injury to an endothelial cell monolayer: quantitation and real time imaging with TAB monoclone antibody, *Blood,* 75, 390, 1990.
11. **Mason, R. G., Chuang, H. Y. K., Mohammad, S. F., and Saba, H. I.,** Thrombosis on artificial surfaces, in *The Thromboembolic Disorders,* Van de Loo, J., Prentice, C. R. M., and Beller, F. K., Eds., Schattaur Verlag, New York, 1983, 533.
12. **Mohammad, S. F., and Mason, R. G.,** A human model for the study of blood vascular wall interactions: effect of enzymatic treatment of intima, *Arch. Pathol. Lab. Med.,* 105, 2, 1981.
13. **Magari, K., Fukushima, H., and Ishimaru, S. et al.,** Platelet adhesion to a biomimetic prosthesis prepared from canine arterial wall, *Artif. Org.,* 15, 492, 1991.
14. **Mohammad, S. F.,** Extracorpreal thrombogeneis, in *Replacement of Renal Function by Dialysis,* 3rd ed., Maher, J., Ed., Kluwer Academic Boston, 1989, 229.
15. **Packham, M. A. and Mustard, J. F.,** Platelet adhesion, *Prog. Hemost. Thromb.,* 7, 211, 1984.
16. **Colman, R. W., Scott, C. F., Schmaier, A. H. et al.** Initiation of blood coagulation at artificial surfaces, *Ann. N.Y. Acad. Sci.,* 516, 253, 1987.
17. **Galletti, P. M.,** Thrombosis in extracorporeal devices, *Ann. N.Y. Acad. Sci.,* 516, 679, 1987.
18. **White, J. G.,** Ultrastructure and regulatory mechanisms, in *Blood Platelet Function and Medicinal Chemistry,* Lasslo, A., Ed., Elsevier Biomedical, New York, 1984, 15.
19. **Day, H. J. and Holmsen, H.,** Concepts of the blood platelet release reaction, *Ser. Hematol.,* 4, 3, 1971.
20. **Baumgartner, H. R., Muggli, R., Tschopp, T. B., and Turitto, V. T.,** Platelet surface properties and platelet function, *Thromb. Haemost.,* 35, 124, 1976.

21. **Fitzgerald, L. A. and Phillips, D. R.,** The molecular biology of platelets and endothelial cell adhesion receptors, *Prog. Clin. Biol. Res.,* 283, 387, 1988.
22. **Fressinaud, E. and Meyer, D.,** von Willibrand factor and platelet interactions with the vessel wall. Blood coagulation, *Fibrinolysis,* 2, 333, 1991.
23. **Plow, E. F. and Ginsberg, M. H.,** Cellular adhesion: GpIIb-IIIa as a prototypic adhesion receptor, *Prog. Hemost. Thromb.,* 9, 17, 1989.
24. **Mustard, J. F., Groves, H. M., Kinlough-Rathbone, R. L., and Packham, M. A.,** Thrombogenic and nonthrombogenic biological surfaces, *Ann. N.Y. Acad. Sci.,* 516, 12, 1987.
25. **Mason, R. G., Mohammad, S. F., and Chuang, H. Y. et al.,** The adhesion of platelets to subendothelium, collagen and artificial surfaces, *Semin. Thromb. Hemost.,* 3, 98, 1976.
26. **Colman, R. W., Marder, V. J., Salzman, E. W., Hirsh, J.,** Overview of hemostasis, in *Hemostasis and Thrombosis: Basic Principles and Clinical Practice,* 2nd ed., Colman, R. W., Marder, V. J., Salzman, E. W., and Hirsh, J., Eds., J. B. Lippincott, Philadelphia, 1987, 3.
27. **Pennington, D. G.,** Formation of platelets, in *Platelets in Biology and Pathology,* Vol. 2, Gordon, J. L., Ed., Elsevier North Holland, New York, 1981, 19.
28. **White, J. G.,** Anatomy and structural organization of the platelet, in *Hemostasis and Thrombosis: Basic Principles and Clinical Practice,* 2nd ed., Colman, R. W., Marder, V. J., Salzman, E. W., Hirsh, J., Eds., J. B. Lippincott, Philadelphia, 1987, 537.
29. **White, J. G.,** Arrangements of actin filaments in the cytoskeleton of human platelets, *Am. J. Pathol.,* 117, 207, 1984.
30. **Turitto, V. T. and Baumgartner, H. R.** Platelet surface interactions, in *Hemostasis and Thrombosis: Basic Principles and Clinical Practice,* 2nd ed., Colman, R. W., Marder, V. J., Salzman, E. W., and Hirsh, J., Eds., J. B. Lippincott, Philadelphia, 1987, 555.
31. **Grabowski, E. F.,** The roles of blood flow in platelet adhesion and aggregation, *Adv. Exp. Med. Biol.,* 102, 73, 1978.
32. **Colman, R. W. and Walsh, P. N.,** Mechanism of platelet aggregation, in *Hemostasis and Thrombosis: Basic Principles and Clinical Practice,* 2nd ed., Colman, R. W., Marder, V. J., Salzman, E. W., Hirsh, J., Eds., J.B. Lippincott, Philadelphia, 1987, 594.
33. **Holmsen, H., Kaplan, K. L., and Dangelmaier, C. A.,** Differential requirements for platelet responses: a simultaneous study of dense granule, alpha granule and acid hydrolase secretion, arachidonate liberation, phosphatidyl inositol turn over and phosphatidase formation, *Biochem. J.,* 208, 9, 1982.
34. **Muggli, R., Baumgartner, H. R., Tschopp, P. B., and Keller, H.,** Automated microdensitometry and protein assays as a measure for platelet adhesion and aggregation on collagen coated slides under controlled flow conditions, *J. Lab. Clin. Med.,* 95, 195, 1980.
35. **Sakariassen, K. S., Aarts, P. A. M. M., and de Groot, P. G. et al.,** A perfusion chamber developed to investigate platelet interaction in flowing blood with human vessel wall cells, the extracellular matrix and purified components, *J. Lab. Clin. Med.,* 102, 522, 1983.
36. **Baumgartner, H. R., Muggli, R., and Tschopp, T. B. et al.,** Platelet adhesion, release and aggregation in flowing blood: effects of surface properties and platelet function, *Thromb. Hemost.,* 35, 124, 1976.
37. **Baumann, H. G., Katinella, F. B., Cunningham, J. N., and Spencer, F. C.,** Vein contraction and smooth muscle cell extensions as causes of endothelial damage during graft preparation, *Ann. Surg.,* 194, 199, 1981.
38. **Haudenschild, T. C., Gould, K. E., Quist, W. C., and LoGerfo, F. W.,** Protection of endothelium in vessel segments excised for grafting, *Circulation,* 64 (Suppl. 2), 101, 1981.
39. **Mohammad, S. F.,** unpublished observations.
40. **LoGerfo, F. W., Sidaway, A. N., and Quist, W. C.,** A technique for prevention of a spasm in in situ vein grafts, *Contemp. Surg.,* 28, 75, 1986.
41. **Mason, R. G., Mohammad, S. F., Saba, H. I., and Chuang, H. Y. K. et al.,** Functions of endothelium, *Pathobiol. Annu.,* 9, 1, 1979.
42. **Danon, D. and Skutelsky, E.,** Endothelial cells' charge and its possible relationship to thrombogenesis, *Ann. N.Y. Acad. Sci.,* 275, 47, 1976.
43. **Vasiliev, J. M. and Gelsand, I. N.,** Mechanism of non-adhesiveness of endothelial and epithelial surfaces, *Nature,* 274, 710, 1978.
44. **Bounassisi, V. and Root, N.,** Enzymatic degradation of heparin related mucopolysaccharides from the surface of endothelial cell cultures, *Biochim. Biophys. Acta,* 385, 1, 1975.
45. **Glimelius, B., Busch, C., and Hook, M.,** Binding of heparin on the surface of cultured endothelial cells, *Thromb. Res.,* 12, 773, 1978.

46. **Becker, C. G. and Harpel, B. C.,** Alpha 2 macroglobulin on human vascular endothelium, *J. Exp. Med.,* 144, 9, 1976.
47. **Laug, W. E., Tokes, Z. E., Benedict, W. F., and Sorgente, N.,** Anchorage independent growth and plasminogen activated production by bovine endothelial cells, *J. Cell Biol.,* 84, 281, 1980.
48. **Loskutoff, D. J.,** Fibrinolytic components as specified by vascular endothelial cells, *Fed. Proc.,* 36, 676, 1977.
49. **Emeis, J. J.,** The vascular wall and fibrinolysis, *Haemostasis,* 8, 33, 1979.
50. **Laug, W. E.,** Secretion of plasminogen activators by cultured bovine endothelial cells: partial purification, characterization, and evidence for multiple forms, *Thromb. Hemost.,* 45, 219, 1981.
51. **Loskutoff, D. J. and Edgington, T. S.,** Synthesis of fibrinolytic activator and inhibitor by endothelial cells, *Proc. Natl. Acad. Sci. U.S.A.,* 74, 3903, 1977.
52. **Dosne, A. M., Dupuy, E., and Badevin, E.,** Production of a fibrinolytic inhibitor by cultured endothelial cells derived from human umbilical vein, *Thromb. Res.,* 12, 377, 1978.
53. **Levin, E. G. and Loskutoff, D. J.,** Serum mediated suppression of cell-associated plasminogen activator activity in cultured endothelial cells, *Cell,* 22, 701, 1980.
54. **Moncada, S. and Vane, J. R.,** Prostacyclin, in *Perspective in Prostacyclin,* Vane, J. R., Bergstrom, S., Eds., Raven Press, New York, 1979, 5.
55. **Vane, J. R. and Moncada, S.,** The antithrombotic effects of prostacyclin, *Acta Med. Scand.* 642, 11, 1980.
56. **Hook, J. C., Brotherton, A. A., Czervionke, R. L., and Fry, G. L.,** Role of prostacyclin in inhibiting platelet adherence to cells of the vessel wall, in *Interaction of Platelets with the Vessel Wall,* Oates, J. A., Hawiger, J., and Ross, R., Eds., American Physiological Society, Bethesda, MD 1985, 117.
57. **Brash, A. R., Jackson, E. K., Caggese, C., Lawson, and J. A., et al.,** The metabolic disposition of prostacyclin in man, *J. Pharmacol. Exp. Ther.,* 26, 78, 1983.
58. **Awbrey, B. J., Hook, J. C., and Owen, W. G.,** Binding of human thrombin to cultured human endothelial cells, *J. Biochem.,* 254, 4092, 1979.
59. **Stein, M. M. and Hook, J. C.,** Thrombin receptors on human endothelial cells: a morphologic study, *J. Ultrastruct. Res.,* 74, 136, 1981.
60. **Kaplan, J. E., Moon, D. G., and Westin, L. K. et al.,** Platelets adhere to thrombin treated endothelial cells in vitro, *Am. J. Physiol.,* 57, 423, 1989.
61. **Esmon, C. T. and Owen, W. G.,** Identification of an endothelial cell cofactor for thrombin catalyzed activation of protein C, *Proc. Natl. Acad. Sci. U.S.A.,* 78, 2249, 1981.
62. **Owen, W. G. and Esmon, C. T.,** Functional properties of an endothelial cell cofactor for thrombin catalyzed activation of protein C, *J. Biochem.,* 256, 5532, 1981.
63. **Owen, W. G.,** Protein C, in *Hemostasis and Thrombosis: Basic Principles and Clinical Practice,* 2nd ed., Colman, R. W., Marder, V. J., Salzman, E. W., and Hirsh, J., Eds., J.B. Lippincott, Philadelphia, 1987, 235.
64. **Hiebert, L. M. and Jaques, L. B.,** The observation of heparin on endothelium after injection, *Thromb. Res.,* 8, 195, 1976.
65. **Colburn, P. and Buonassisi, V.,** Estrogen binding sites in endothelial cells, *Science,* 201, 8173, 1978.
66. **Buonassisi, V. and Colburn, P.,** Hormone and surface receptors in vascular endothelium, *Adv. Microcirc.,* 9, 76, 1980.
67. **Kisker, C. T., Hook, J. C., Taylor, B. J., and Fry, G. L.,** Binding of fibrin monomer to vascular endothelium, *Blood,* 50, 273, 1977.
68. **Smith, U. and Ryan, J. W.,** Pulmonary endothelial cells and metabolism of adenine nucleotides, guanines, and angiotensin-1, *Adv. Exp. Med. Biol.,* 4, 267, 1972.
69. **Ryan, J. W., Niemeyer, R. S., and Ryan, U.,** Metabolism of prostaglandin F_1 alpha in the pulmonary circulation, *Prostaglandins,* 10, 101, 1975.
70. **Thorgeirsson, G. and Robertson, A. L.,** The vascular endothelium XXX biologic significance, *Am. J. Pathol.,* 93, 803, 1978.
71. **Hawiger, J., Grabber, S. E., and Timmons, S.,** Prostacyclin regulation of platelet receptors for adhesive macromolecules, in *Interaction of Platelets with the Vessel Wall,* Oates, J. A., Hawiger, J., and Ross, R., Ed., American Physiologic Society, Bethesda, MD, 1985, 89.
72. **Sneddon, J. M. and Vane, J. R.,** Endothelium derived relaxing factor reduces platelet adhesion to bovine endothelial cells, *Proc. Natl. Acad. Sci. U.S.A.,* 85, 2800, 1988.

73. **Venturini, C. M., Del-Vecchio, P. J., and Kaplan, J. E.,** Thrombin induced platelet adhesion to endothelium is modified by endothelial derived relaxing factor. *Biochem. Biophys. Res. Commun.*, 159, 349, 1989.
74. **Bassenge, E.,** Antiplatelet effect of endothelium derived relaxing factor and nitric oxide donors, *Eur. Heart J.*, 12 (Suppl. E), 12, 1991.
75. **Reidy, M. A. and Bowyer, D. E.,** Scanning electron microscopy of arteries: the morphology of aortic endothelium in hemodynamically stressed areas associated with branches, *Atherosclerosis,* 26, 181, 1977.
76. **Flaherty, J. T., Pierce, J. E., and Ferrans, V. J. et al.,** Endothelial nuclear patterns in the canine arterial tree with particular reference to hemodynamic events, *Circ. Res.*, 30, 23, 1972.
77. **Hertzer, N. R., Beven, E. G., and Benjamin, S. P.,** Ultramicroscopic ulcerations and thrombi of the carotid bifurcation, *Arch. Surg.*, 112, 1394, 1977.
78. **Fonklsrud, E. W., Sanches, M., Zerubavel, R., and Mahoney, A.,** Serial changes in arterial endothelium following ischemia and perfusion, *Surgery,* 81, 527, 1977.
79. **Hempel, S. L., Haycraft, D. L., Hoak, J. C., and Spector, A. A.,** Reduced prostacyclin formation after reoxygenation of anoxic endothelium, *Am. J. Physiol.*, 259, 738, 1990.
80. **Moraes, J. R. and Stastny, P.,** Eight groups of human endothelial cell alloantigens, *Tissue Antigens,* 8, 273, 1976.
81. **Moraes, J. R. and Stastny P.,** A new antigen system expressed in human endothelial cells, *J. Clin. Invest.*, 60, 449, 1977.
82. **Hibbs, J. B.,** Platelets and the renal vascular endothelium, *Perspect. Nephrol. Hypertension,* 1, 907, 1973.
83. **Reidy, M. A. and Bowyer, D. E.,** Scanning electron microscopic studies of endothelium of aortic allografts in the rabbit. Effect of azathiorprine, prednisolone and promethazine on early cellular invasion, *J. Pathol.*, 124, 1, 1978.
84. **Hashimoto, P. H., Takaesu, S., Chazono, M., and Amano, T.,** Vascular leakage through intraendothelial channels induced by cholera toxin in the skin of guinea pigs, *Am. J. Pathol.*, 75, 171, 1974.
85. **Dalldorf, F. G. and Jennette, J. C.,** Fatal menningococcal septicemia, *Arch. Pathol. Lab. Med.*, 101, 6, 1977.
86. **Balis, J. U., Rappaport, E. S., Gergel, L., and Fareed, J. et al.,** A primate model for prolonged endotoxin shock, Lab. Invest., 38, 511, 1978.
87. **Mason, R. G. and Balis, J. U.,** Pathology of the endothelium, in *Pathobiology of Cell Membranes II,* Trump, B. F. and Arstilla, A. U., Eds., Academic Press, New York, 1980, 425.
88. **Bertomeu, M. C., Gallo, S., Lauri, D., and Levine, M. N. et al.,** Chemotherapy enhances endothelial cell reactivity to platelets, *Clin. Exp. Metast.*, 8, 511, 1990.
89. **Ryan, T. J., Saxon, D., Gunnar, R., and Kennedy, J. W., et al.,** Guidelines for PTCA: a report of the American College of Cardiology-American Heart Association task force on assessment of diagnostic and therapeutic cardiovascular procedures, *J. Am. Coll. Cardiol.*, 12, 529, 1988.
90. **Borst, C., Bos, A. N., and Zwaginga, J. J. et al.,** Loss of blood platelet adhesion after heating native and cultured human endothelium to 100°C. *Cardiovasc. Res.*, 24, 665, 1990.
91. **Dries, D., Mohammad, S. F., Woodward, S. C., and Nelson, R. M.,** The influence of harvesting technique on endothelial preservation in saphenous veins, *J. Surg. Res.*, 52, 219, 1992.
92. **Mason, R. G., Mohammad, S. F., and Saba, H. I. et al.,** Functions of endothelium, *Pathobiol. Annu.*, 9, 1, 1979.
93. **Huang, T. W. and Benditt, T. P.,** Mechanism of platelet adhesion to basal lamina, *Am. J. Pathol.*, 92, 99, 1979.
94. **Bull, H. A. and Machin, H.,** The hemostatic function of the vascular endothelial cell, *Blut,* 55, 71, 1987.
95. **Sixma, J. J. and de Groot, P. G.,** von Willebrand factor and the blood vessel wall, *Mayo Clin. Proc.* 66, 628, 1991.
96. **Stel, H. V., Sakariassen, K. S., and de Groot, P. G. et al.,** von Willebrand factor in the vessel wall mediates platelet adherence, *Blood,* 65, 85, 1985.
97. **Sadler, J. E.,** The molecular biology of von Willebrand factor, in *Thrombosis and Hemostasis,* Verstraete, M., Vermylen, J., Lijnen, R., and Aernout, J., Eds., Leuven University Press, 1987, 61.
98. **Wagner, D. D. and Bonfanti, R.,** von Willebrand factor and the endothelium, *Mayo Clin. Proc.*, 66, 621, 1991.

99. **Wagner, D. D. and Marder, V. J.,** Biosynthesis of von Willebrand protein by human endothelial cells: processing steps and the intracellular localization, *J. Cell Biol.,* 99, 2193, 1984.
100. **Coller, B. S.,** von Willebrands disease, in *Hemostasis and Thrombosis: Basic Principles and Clinical Practice,* 2nd ed., Colman, R. W., Marder, V. J., Salzman, E. W., and Hirsh, J., Eds., J.B. Lippincott, Philadelphia, 1987, 60.
101. **Ruggeri, Z. M. and Zimmerman, T. S.,** Variant von Willebrand disease: characterization of two subtypes by analysis of multimeric composition of VIII-vWF in plasma and platelets, *J. Clin. Invest.,* 65, 1318, 1980.
102. **Lynch, D. C., Williams, R., and Zimmerman, T. S. et al.,** Biosynthesis of the subunit of VIIIR:Ag by bovine aortic endothelial cells, *Proc. Natl. Acad. Sci. U.S.A.,* 80, 2738, 1983.
103. **Nievelstein, P. F., de Groot, P. G., and D'Alessio, P. et al.,** Platelet adhesion to vascular cells. The role of exogenous vWF in platelet adhesion, *Arteriosclerosis,* 10, 462, 1990.
104. **Sakariassen, K. S., Muggli, R., and Baumgartner, H. R.,** Measurements of platelet interaction with components of the vessel wall in flowing blood, *Meth. Enzymol.,* 169, 37, 1989.
105. **Eldor, A., Suks, Z., Levine, R., and Vlodavsky, L.,** Measurement of platelet and megakaryocyte interaction with subendothelial extracellular matrix, *Meth. Enzymol.,* 169, 76, 1989.
106. **Baruch, D., Denis, C., Marteaux, C., and Schoevaert, D. et al.,** Role of vWF associated to extracellular matrices in platelet adhesion, *Blood,* 77, 519, 1991.
107. **Coller, B. S., Peerschke, E. I., Scudder, L. E., and Sullivan, C. A.,** Studies with a murine monoclonal antibody that abolishes ristocetin induced binding of vWF to platelets: additional evidence in support of GP lb as a platelet receptor for vWF, *Blood,* 61, 99, 1983.
108. **Fujimura, Y., Fukui, H., Usami, Y., and Titani, K.,** Domain structure of human vWF, and its modulators involved in platelet adhesion process in vitro, *Rinsho-Ketsueki* (Japan), 32, 475, 1991.
109. **Wagner, D. D., Fay, P. J., and Sporn, L. A. et al.,** Divergent fates of vWF and its polypeptide (von Willebrand antigen II) after secretion from endothelial cells, *Proc. Natl. Acad. Sci. U.S.A.,* 84, 1987.
110. **Sakariassen, K. S., Fessinaud, E., and Girma, J. P. et al.,** Role of platelet membrane glycoproteins and vWF in adhesion of platelets to subendothelium and collagen, *Ann. N.Y. Acad. Sci.,* 516, 52, 1987.
111. **de Groot, P. G. and Sixma, J. J.,** Role of vWF in the vessel wall, *Semin. Hemost. Thromb.,* 13, 416, 1987.
112. **Lawrence, J. B., Kramer, W. S., McKeown, L. P. et al.,** Arginine-glycine-aspartic acid and fibrinogen gamma chain carboxy terminal peptides inhibit platelet adherence to arterial subendothelium at high wall shear rate. An effect dissociable from interference with adhesive protein binding, *J. Clin. Invest.,* 86, 1715, 1990.
113. **Kieffer, N. and Phillips, D. R.,** Platelet membrane glycoproteins: functions in cellular interactions, *Annu. Rev. Cell Biol.,* 6, 329, 1990.
114. **Clemetson, K. J.,** Glycoproteins of the platelet plasma membrane, in *Platelet Membrane Glycoproteins,* George, J. N., Nurden, A. T., and Phillips, D. R., Eds, Plenum Press, New York, 1985, 51.
115. **Cheresh, D. A. and Spiro, R. C.,** Biosynthetic and functional properties of an Arg-Gly-Asp-directed receptor involved in human melanoma cell attachment to vitronectin, fibrinogen and vWF, *J. Biol. Chem.,* 262, 17703, 1987.
116. **Gralnick, H. R., Williams, S. B., and McKeown, L. P. et al.,** In vitro correction of the abnormal multimeric structure of vWF in type IIa von Willebrands disease. *Proc. Natl. Acad. Sci. U.S.A.,* 82, 5968, 1985.
117. **de Groot, P. G., Verweij, C. L., and Nawroth, P. P. et al.,** Interleukin-1 inhibits the synthesis of vWF in endothelial cells, which results in a decreased reactivity of their matrix toward platelets, Arteriosclerosis, 7, 605, 1987.
118. **Bouma, B. N., Hordik-Hos, J. M., de Graaf, S., and Sixma, J. J. et al.,** Presence of VIII related antigen in blood platelets of patients with von Willebrands disease, *Nature,* 257, 510, 1975.
119. **Gralnick, H. R., Rick, M. E., and McKeown, L. P. et al.,** Platelet vWF: an important determinant of the bleeding time in type I von Willebrands disease, *Blood,* 68, 58, 1986.
120. **Pischel, K. D., Bluestein, H. G., and Woods, V. L.,** Platelet glycoprotein Ia, Ic and IIa are physicochemically indistinguishable from the very late activation antigen adhesion-related proteins of lymphocytes and other cell types, *J. Clin. Invest.,* 81, 505, 1988.
121. **Ill, C. R., Engvall, E., and Ruoslahti, E.,** Adhesion of platelets to laminin in the absence of activation, *J. Cell Biol.,* 99, 2140, 1984.
122. **Piotrowicz, R. S., Orchekowski, R. P., and Nugent, D. J. et al.,** Glycoprotein Ic-IIa functions as an activation independent fibronectin receptor on human platelets, *J. Cell Biol.,* 106, 1359, 1988.

123. **Fitzgerald, L. A. and Phillips, D. R.,** Platelet membrane glycoprotein, in *Hemostasis and Thrombosis: Basic Principles and Clinical Practice,* 2nd ed., Colman, R. W., Marder, V. J., Salzman, E. W., and Hirsh, J., Eds., J.B. Lippincott, Philadelphia, 1987, 572.
124. **Fitzgerald, L. A. and Phillips, D. R.,** Ca^{2+} regulation of platelet membrane glycoprotein IIb-IIIa complex, *J. Biol. Chem.,* 260, 11366, 1985.
125. **Brass, L. F.,** Ca^{2+} transport across the platelet plasma membrane: a role for membrane glycoproteins IIb-IIIa, *J. Biol. Chem.,* 260, 2231, 1985.
126. **Phillips, D. R., Jennings, L. K., and Edwards, H. H.,** Identification of membrane proteins mediating the interaction of human platelets, *J. Cell Biol.,* 86, 77, 1980.
127. **George, J. N., Nurden, A. T., and Phillips, D. R.,** Molecular defects in interaction of platelets with the vessel wall. *N. Engl. J. Med.,* 311, 1084, 1984.
128. **Roth, G. J.,** Platelets and blood vessels: the adhesion event, *Immunol. Today,* 13, 100, 1992.
129. **Zimmerman, G. A., Prescott, S. M., and McIntyre, T. M.,** Endothelial cell interactions with granulocytes: tethering and signaling molecules, *Immunol. Today,* 13, 93, 1992.
130. **Butcher, E. C.,** Cellular and molecular mechanisms that direct leukocyte traffic, *Am. J. Pathol.,* 136, 1, 1990.
131. **Osborn, L., Hession, C., and Tizard, R. et al.,** Direct expression cloning of vascular adhesion molecule-1, a cytokine-induced endothelial protein that binds to lymphocytes, *Cell,* 59, 1203, 1989.
132. **Ruco, L. P., Pomponi, D., and Pigott, R. et al.,** Expression and cell distribution of the intercellular adhesion molecule, vascular cell adhesion molecule, and endothelial cell adhesion molecule (CD 31) in reactive human lymph nodes in Hodgkin's disease, *Am. J. Pathol.* 140, 1337, 1992.
133. **Sher, B. T., Bargatze, R., and Holzmann, B. et al.,** Homing receptors and metastasis, *Adv. Cancer Res.,* 51, 361, 1988.
134. **Ronkonen, R., Paavonen, T., Nortame, P., and Gahmberg, C. G.,** Expression of endothelial adhesion molecules in vivo. Increased endothelial ICAM-2 expression in lymphoid malignancies, *Am. J. Pathol.,* 140, 763, 1992.
135. **Meyer, D. and Girma, J.-P.,** von Willebrand factor: structure and function, *Thromb. Haemost.,* 70, 99, 1993.
136. **Ruggeri, Z. M. and Ware, J.,** The structure and function of von Willebrand factor, *Thromb. Haemost.,* 67, 594, 1992.
137. **Ruggeri, Z. M.,** Mechanism of sheer-induced platelet adhesion and aggregation, *Thromb. Haemost.,* 70, 119, 1993.
138. **Wagner, D. D.,** The Weibel-Palade Body: the storage granule for von Willebrand factor and P-selectin, *Thromb. Haemost.,* 70, 105, 1993.

INDEX

A

Abelson proto-oncogene, 194
N-Acetylgalactosamine, 93
N-Acetylglucosamine, 90, 93, 207
N-Acetylneuraminic acid, 93
Acetylsalicylate, 294
Acquired immune deficiency syndrome (AIDS), 293
Actin
  E-cadherin colocalization, 254, 256
  gpIb-IX interactions, 261–264
  integrin interactions, 167, 260
  platelet, 257
Actin-binding protein, 262–264
α-Actinin-integrin interactions, 167, 259, 260–261
Actinomyces, 91
  naeslundii, 90
  viscosus, 90
Addressin, 325
  CD34, 147
  mucosal lymph node vascular addressin (MAd), 209
  peripheral lymph node vascular addressin (PNAd), 209
  tumor cell homing, 323
Adenine nucleotides, endothelial cells and, 355
Adenosine
  dipyridamole and, 338
  plasma, dipyridamole and, 295
  platelet, AVP and, 292
  and platelet-activating factor, 291
  and platelet aggregation, 295
  platelet release of, 288
Adenosine 5′-diphosphate (ADP), 171, 350, 355
  and gpIIb/IIIa conformation, 308, 309
  platelet, 290
    AVP and, 292
    epinephrine and, 291
    in hypertension, 293
    species differences, 292
  platelet activation by tumor cells, 318–319
  and platelet aggregation, 290
    calcium channel blockers and, 296
    epinephrine and, 291
    ticlopidine and, 340
Adenylate cyclase, 318, 325, 326
Adherence index, 23
Adherens junction, 254
Adhesins, see Lectins
Adhesion, see also specific cells
  vs. aggregation, 8
  bacterial, mechanisms of, 92
  ECM proteins inhibiting, 239
  mechanical strength of, 23
  stages of, 8
  tumor cell, 177, 324
Adhesion
  efficiency, 9
  energy, 24–28
  energy density, 23–25, 28–32, 35
  plaques, 167–169
Adhesive strength, 23
Adhesive stress, 24, 25
Affective disorders, 195
Agglutination, bacterial, 93
Aggregation, see also specific cells
  vs. adhesion, 8
  bacterial, 93
  Dictyostelium discoideum, see also *Dictyostelium discoideum*
Agkistrodon halys, 341
Agrin, 229–230
AH 19437, 294
Albumin-carbohydrate complexes, 207
α, subunits and heterodimers, 164
α_1
  DNA sequences, 166
  structure, 168
  vascular smooth muscle, 171
$\alpha_1\beta_1$
  in inflammatory bowel disease, 177
  neuroblastoma cells, retinoic acid and, 323
α_2
  DNA sequences, 166
  structure, 168
  vascular smooth muscle, 171
$\alpha_2\beta_1$
  epidermal cell localization, 172
  epithelial cell localization, 171
  Glanzmann's thrombasthenia and, 174
  in inflammation, 176–177
  melanoma cells, 324
  osteosarcoma cell binding, TGFβ and, 323
  regulation of, cytokines and, 172
  tumor cells, 323, 324
  type I collagen interactions, 168
α_{IIb}
  β_3 subunit interaction, 170
  calcium-binding domain, 169
  DNA sequences, 166
  mutations, 165
  structure, 168
$\alpha_{IIb}\beta_3$, 163, 164
  agonists activating, 290
  antibodies to, 341
  β_3 subunit blockade of, 169

carcinoma cell lines, 171
cytoskeleton interactions, 257–261
regulation of, 257–259
discrete populations, 309
fibrinogen binding force, 149
in inflammation, 177
MCAs, therapeutic use of, 177
melanoma cells, 325
platelet-activating factor and, 291
platelet activation, by tumor cells, 318–321
and platelet adhesion, 308–310
and platelet aggregation, shear-induced, 312
platelet-tumor cell-vessel wall interactions, 326
regulation, 171
shear stress and, 312, 326
structure, 165, 166
structure-function relationships, 169
ticlopidine and, 295
tumor cells, 171, 325
von Willebrand factor binding domains, 357
α_3
DNA sequences, 166
structure, 168
vascular smooth muscle, 171
$\alpha_3\beta_1$
distribution of, 170
epidermal cell localization, 172
epithelial cell localization, 171
osteosarcoma cell binding, TGFβ and, 323
pathogen interactions, 175
regulation of, cytokines and, 172
α_4
DNA sequences, 166
number of subunits and heterodimers, 164
α_{4b} (gpIIb), 309
$\alpha_4\beta_1$ (VLA-4)
CS-1 peptide binding, 168
epitope mapping, 170
ICAM-1 and, 146
leukocyte adhesion deficiencies and, 174
leukocyte-endothelial cell interactions, 172
ligands and cell distribution, 144
lymphocyte homing, 152, 153
lymphoid organs, 171
monocyte adhesion in atherosclerosis, 153
monocyte-macrophage differentiation and, 145
pathogen interactions, 175
structure, 143
T-cell conversion from naive to memory cells and, 145
T-cell receptor crosslinking and, 172
tumor cell-platelet-endothelial cell interactions, 320, 323
VCAM-1 binding, 168
$\alpha_4\beta_7$
ligands and cell distribution, 144
lymphocyte homing, 152
MAdCAM-1 and, 147
α_5
DNA sequences, 166
structure, 168
$\alpha_5\beta_1$
cell distribution, 171
cytoskeleton proteins and, 259
endocytosis of in CHO cells, 173
in inflammation, 176–177
pathogen interactions, 175
phorbol esters and, 172
regulation of, cytokines and, 172
T-cell receptor crosslinking and, 172
tumor cell expression, 324
in wound healing, 177
α_6
DNA sequences, 166
number of subunits and heterodimers, 164
structure, 168
$\alpha_6\beta_1$
epithelial cell localization, 171
in inflammation, 176–177
macrophage laminin binding, 172
T-cell receptor crosslinking and, 172
tumor cell expression, 324
$\alpha_6\beta_4$
extracellular matrix binding, 170
epithelial cell localization, 171
α_7 structure, 168
$\alpha_C\beta_1$ (gpIa/IIa), 311
$\alpha_F\beta_1$ (gpIc/IIa), 311
α_L
DNA sequences, 166
structure, 168
$\alpha_L\beta_2$, 144, 320
activation by lipid, 258
ICAM-1, cytotoxic T cell-target cell adhesion, 29, 30
ICAM-1 and ICAM-2 binding, 168–169
ICAM-1 avidity, 172
ICAM-1 induction and, 146
leukocyte-endothelial cell interactions, 172
leukocytes, 171
ligands and cell distribution, 144
lymphocyte homing, 152
lymphoma cells, 322–323
neutrophil adherence in inflammation, 152
and neutrophil adhesion, 211
phorbol esters and, 257
regulation, 171
T lymphocyte conversion from naive to memory cells and, 145
α_M
DNA sequences, 166
structure, 168
$\alpha_M\beta_2$, 144, 325

ICAM-1 and ICAM-2 binding, 168
ICAM-1 induction and, 146
leukocyte–endothelial cell interactions, 172
leukocytes, 171
ligands and cell distribution, 144
LPS binding, 169
monocyte-macrophage differentiation and, 145
neutrophil adherence in inflammation, 152
and neutrophil adhesion, 211
number of receptors on stimulated neutrophil, 149
pathogen interactions, 175
structure, 143
α subunits, 164
heterodimer formation, 169
structure, 165, 166
tissue distribution, 170
α_v
in Alzheimer's disease, 177
DNA sequences, 166
extracellular matrix binding, 170
integrin cell–substratum interactions, classification of, 170
number of subunits and heterodimers, 164
structure, 168
vascular smooth muscle, 171
$\alpha_v\beta_3$, 311
manganese and, 167
platelet ($\alpha v\beta 3$), 311
structure–function relationships, 169
α_X
DNA sequences, 166
structure, 168
$\alpha_X\beta_2$
leukocytes, 171
ligands and cell distribution, 144
monocyte-macrophage differentiation and, 145
pathogen interactions, 175
Alpha 2 macroglobulin, 354
Alternative splicing
integrin subunits, 165
N-CAM, 190, 191
Alzheimer's disease, 177
American Society for Testing and Materials, 5
Amiloride, 308
Amines, platelet, 321, see also Epinephrine
Amino acid analysis, P-selectin, 207
Amino acid composition
biomineralization matrices, 132
DOPA proteins, 113–115
Amino acid sequences
integrin homologies, 165
RGD recognition, 167
Amyotrophic lateral sclerosis, 195
Anaphylatoxins, 233
Ankryn, 255, 256
Annelids, 108, 110, 117
Antibodies, see also Monoclonal antibodies
bacterial, and bacterial attachment, 92, 93
and bacterial attachment, 93
cross-linking by, 9
endothelial injury, 355
and platelet adhesion, 340
Antibody-dependent cytotoxicity, LAD and, 174
Antibody techniques, 4
Anticoagulants, biological, 9
Antithrombin III, 354–355
AP-1, 207
Aplysia, 194
Appendix, 171
Apyrase, 318
Arachidonic acid pathway, 354
platelet
in diabetes, 292–203
epinephrine and, 291
species differences, 292
thromboxane A_2 and, 291
and tumor cells, 321
platelet activation, 309, 318, 320–321
and platelet adhesion/aggregation, 290, 294–295
calcium channel blockers and, 296
epinephrine and, 291
indomethacin and, 340
ticlopidine and, 340
tumor cell-induced, 322
platelet-tumor cell-vessel wall interactions, 318, 320–322, 326
Arg-Gly-Asp (RGD) domain, see RGD sequence
Arginine^8-vasopressin, 209
L-Arginine-NO pathway, 296
ASA, and platelet adhesion, 17
Aspirin, 294, 337–338
Atherosclerosis, 153–154, 213
Atomic mechanisms, 3–4, 6–7
ATP, platelet, 290
Atrina serrata, 114–115
Aulacomya ater, 114–115, 117, 129
Autocrine motility factor, 322
Avidity, 23
Axisymmetric forward stagnation point flow, 12, 13

B

Bacteria, 8–9
Dictyostelium discoideum adhesion, 65–66
endothelial injury by, 355
lectins, cell–cell recognition and infectious diseases, 88–98
biology, 88
carbohydrate receptors and recognition, 90–91
epithelial cells, adhesion to, 94–95
experimental evidence, 94–98
fimbriae, 88–90

phagocytic cells, adhesion to, 95–97
phase variation, population shift in fimbrial phase, 91–94
polymers, adhesion to, 97–98
LPS, see Lipopolysaccharide
nonspecific factors affecting adherence, 98–100
bacterial cell surface molecules, 99
bacteriophoresis, 100
extracellular matrix, 99–100
lipoteichoic acids, 99
mechanical and physicochemical forces on mucosae, 98–99
terminology, 87–88
Bacterial proteases, 91
Bacteriophoresis, 100
Bamin, 230
Basement membrane
integrins
distribution of, 170
malignant cells, 175–176
as natural substrate for cell adhesion, 219–221
protein interactions, 234–237
collagen IV self-interaction, 235–236
laminin self-interaction, 235
mixtures, 236–237
protein structure, 221–230
basement membrane proteoglycans, 228–229
collagen type IV, 227–228
laminin, 221–225
laminin-nidogen complex, 227
miscellaneous proteins, 229–230
nidogen/enactin, 225–226
tumor cells, 175–176, 324
Basic fibroblast growth factor, 230
B cells
integrins and ligands, 144, 171
L-selectin, 205
Bdelloura candida, 110, 114–115, 122, 123
Bernard-Soulier syndrome, 293, 311, 341
β, number of subunits and heterodimers, 164
β_1 (CD29), 144
chimeric subunits, 169
congenital absence of, 174
cytoplasmic domain, deletional analysis, 169
cytoskeletal interactions, 167
DNA sequences, 166
extracellular matrix binding, 170
glycosylation, 165, 166
integrin cell–substratum interactions, classification of, 170
number of subunits and heterodimers, 164
phosphorylation, 172
vascular smooth muscle, 171
β_1 integrins
cytoskeleton interactions, 259–261
during development, 174
distribution, 164
β_2 (CD18), 144
DNA sequences, 166
leukocytes, 171
number of subunits and heterodimers, 164
pathogen interactions, 175
β_2 integrins
in inflammation and immunity, 152, 211
monocyte adhesion in atherosclerosis, 153
nomenclature, 164
β_3 (CD61)
α_{IIb} subunit interaction with, 170
chimeric subunits, 169
cytoskeletal interactions, 167
DNA sequences, 166
Glanzmann's thrombasthenia, 174
glycosylation, 166
structure, 165, 166
structure–function relationships, 169
vascular smooth muscle, 171
β_4
DNA sequences, 166
integrin cell–substratum interactions, classification of, 170
structure, 165, 166
β_5
DNA sequences, 166
vascular smooth muscle, 171
β_6
DNA sequences, 166
vascular smooth muscle, 171
β_7, DNA sequences, 166
β subunit, 164
chimeric, 169
gpIIb/IIIa, 308–309
heterodimer formation, 169
structure, 165, 166
Bioadhesives, polymeric
biology of sclerotized matrices, 109–112
composites, 111–112
structural adhesives, 110–111
varnishes, 112
composition and chemical composition of phenolics, 113
DOPA protein amino acid composition, 114–115
essential properties, 112–113
phenolic protein consensus motif sequences, 116–118
phenolic protein structure and function, 113–134
curing, 133
DOPA protein diversity, 123–127
evolutionary implications of DOPA protein diversity, 125–127
Fasciola eggcase DOPA protein precursors, 123
motif structure, 113–119
mussel adhesive DOPA protein precursors, 119–122
permeability, 133

polypeptide resin matrices, 131–133
schistosome eggcase phenolic protein precursors, 122–123
sclerotized structure biosynthesis, 130–131
secondary structure, 127–130
sclerotized adhesives and matrices, 108–109
Biophysical approaches, 5–7, see also Mathematical modeling
Biopolymers, see Bioadhesives, polymeric
Bioreactors, see Substrate surface chemical factors
Blood cells, see Homing of hemopoietic cells
Blood group antigens, 88
Blood oxygenation, extracorporeal, 9
BM 13, 177, 294
BM-600/nicein, 222
BM-90, 233
BM-90/fibulin, 226
Bond density, equilibrium properties of adherent cell system, 34
Bonding energy, 23
Bone marrow
cell culture, 274, 275
hempoiesis, 273
homing of hemopoietic cells, 273
integrins, $\alpha_4\beta_1$, 171
structure, 280–282
Bone marrow metastasis, N-CAM and, 195
Bordetella pertussis, 175
Born aggregometer, 15
Boundary conditions of adhesion regions, 31–32, 34
Brachidontes exustus, 114–116, 125, 127, 129
Brachionectin, 233–234
Bradykinin, 151
Brain, N-CAM, 188, 191
Breast, collagen/laminin receptors, 171
Breast adenocarcinoma integrins, 176
Breast cancer, 176, 323
Brownian motion, 9
Byssus, 109, 112, 113

C

Cadherin E, 254
Cadherins
cytoskeleton interactions, 253–257
directive endocrinization, 272–273
platelet activation by tumor cells, 320
Caenorhabditis elegans, 227
Caffeine, 77
Calcium
and *Discyostelium discoideum* adhesion, 80
and integrin distribution in cultured keratinocytes, 172
integrin interactions with ligands, 167, 173
platelet, 288, 290
AVP and, 291
and intracellular alkalinization, 308
and platelet activation by tumor cells, 321
and platelet aggregation, 290, 291, 342
platelet-tumor cell-vessel wall interactions, 326
red cell aggregation, 9
selectin binding, 205, 208
Calcium binding sites
α_{IIb}, 169
integrins, 165
Calcium channel blockers, and platelet aggregation, 296
Calcium electrode, 289
Calcium fluxes, N-CAM, 194
Calcium ionophores
endothelial activation, and P-selectin redistribution, 207
and platelet aggregation, 290, 219
Calpain, 257, 311
cAMP
Dictyostelium discoideum, 64, 75–79
dipyridamole and, 338
platelet
epinephrine and, 291
drugs affecting, 294–295
platelet-tumor cell-vessel wall interactions, 326
Carbohydrates, 207, see also Lectins; specific carbohydrates
bacterial attachment, 90–92
tumor cell induction of platelet aggregation, 319–320
Cardiac bypass surgery, 10–11
Catch bond, 31, 35
Catechol oxidases, 109
Catenins, 255, 256
Cathepsin B, 322
Cation ionophores, and platelet shape change, 308
Cations, integrin interactions with ligands, 167
CD antigens, integrin nomenclature, 164
CD2/LFA-3, 29, 30
CD8, 29, 30
CD11a/CD18, see $\alpha_L\beta_2$
CD11b/CD18, see $\alpha_M\beta_2$
CD11c/CD18, see $\alpha_X\beta_2$
CD15, 147, 210, 213, see also Lewis^x
CD18 (β_2 subunit), 144, 148, see also β_2 subunit
CD29 (β_1 subunit), 144, see also β_1 subunit
CD31 (platelet endothelial adhesion molecule-1), 146
CD34, 152
CD36 (gpIV), 311, 360
CD44, 147, 324
CD49d/CD29, see $\alpha_4\beta_1$
CD54, see ICAM-1
CD61 (β_3 subunit), 144
CD62, see P-Selectin
Cell attachment assays, 238
Cell culture, see also Tumor cells; specific cell lines
integrins, 170–172, 175
marrow cells, 274–276
N-CAM and neurite outgrowth, 192
Cell cycle synchronization, 42

Cell motility, see Motility
Cell sorting, *Dictyostelium discoideum*
  interspecies, 74–75
  prespore and prestalk cells, 79–80
Cell–substrate interactions, see Integrins; specific integrins
Cell surface complex, 99
Centrifugal force, 23
Centrifugation, George Test, 10
CFU-C, 276
CFU-GM, 274, 275
CFU-S, 274, 275
Chemical cross-linking studies, β_3 integrins, 169
Chemoattractants, 4
Chemotactic factors
  Dictyostelium discoideum, cAMP and, 64
  monocyte-macrophage, and Mac-1 and p150,95 expression, 145
  and L-selectin shedding, 146
Childhood malignancies, N-CAM in, 195
Chimeras
  β_1 and β_3, 169
  E-cadherin-N-CAM, 254
  P- and L-selectin, 210
Chinese hamster ovary cells, see CHO cells
Chitin, 108, 111
CHO cells
  cell retention on synthetic solid supports, see Substrate surface chemical factors
  endocytosis of $\alpha_5\beta_1$, 173
  integrins
    and cell motility, 173–174
    pathogens and, 175
    phorbol esters and, 172
Chondroitin sulfate, 229, 280
Chordata, 110
Choromytilus chorus, 114–115
Chromosomal localization
  N-CAM, 192
  selectins, 204
Circulating immune complexes, 355
CLA (cutaneous lymphocyte antigen), 153
Classical mechanics, 3–7
C-lectins, hemopoietic cell homing receptor, 278
Cloning, molecular, 6
Cluster 1 antigen, 195
Cnidaria, 110
Coagulase-negative staphylococci, 93
Coagulation pathway, 11, 354, 355
  integrins and, 177
  selectins in, 212–213
Coefficient of restitution, 3
Coelenterates, fibronectin, 230
Collagen
  and bacterial attachment, 87, 100
  fibroblast binding, anti-β_1 MAB and, 172
  gpIV as receptor, 311, 361
  integrin binding, 167, 170, 171
  and platelet aggregation, 290
    gpIIb/IIIa binding, 360
    gpIIb/IIIa conformation, 308
    P-selectin and, 213
    ticlopidine and, 340
    vitamin E and, 342
    von Willebrand factor binding domains, 356
  platelet binding, unstimulated platelets, 312
Collagenase, anti-$\alpha_5\beta_1$ and, 324
Collagen I
  $\alpha_2\beta_1$ interactions, 168
  integrin recognition of RGD sequence, 168
Collagen II, 100
Collagen IV, 220
  basement membrane binding sites, 226
  collagen VII NC domains and, 230
  integrin expression, 170
  integrin localization in keratinocytes, 172
  interactions with other basmement membrane proteins, 237, 238
  pulmonary ECM, 324
  self-interaction, 235–236
  SPARC binding, 233
  structure, 227–228
  tumor cell interactions, 324
Collagen IV-laminin interaction, 236
Collagen/laminin integrins, tissue distribution, 171
Collagen receptor, osteosarcoma cell binding, TGFβ and, 323
Collagen VII, 220, 230
Colon carcinoma
  integrins, 176, 177, 324
  platelet activation by, 321
  E-selectin, 213
Colorectal carcinoma, 319–320
Complement C5b-9 complex, 207
Complement component anaphylatoxins, 233
Complement component iC3b, 144, 167
Concanavalin A, 73, 74
Conditioning film, substrate surface properties, 49–50, 57–58
Consensus sequences
  DOPA protein precursors, 119–120
  phenolic proteins, 116–119
Conservation of energy equation, 24, 25
Contact angle analysis, 44, 46, 47, 49
Contact potential data, 44, 57
Continuum mechanics analysis, 24–28
Continuum models, 5
Contractility, integrins and, 172
Convective transport theory, 12–13
Co-translational processing, integrin subunits, 165
CR3 receptor, see $\alpha_M\beta_2$
Crassostrea virginica, 132
Critical surface tension, substrate surface properties, 46

Cross-linking, 4–5
by antibodies, 9
equilibrium properties of adherent cell system, 34
CS-1, $\alpha_4\beta_1$ binding, 168
CS-5, $\alpha_4\beta_1$ binding, 168
CSIII, $\alpha_4\beta_1$ binding, 168
Cutaneous lymphocyte antigen (CLA), 147, 153
Cuticle, 111
Cyclooxygenase pathway
and platelet adhesion, 340
platelet aggregation, tumor cell-induced, 322
inhibitors, 294
CYP21, 234
Cytoadhesins (β_3 integrins), 144
Cytochalasins, 254, 261–262
Cytokines, 361, see also IL-1; TGFβ; TNFα
and endothelial cell adhesion molecules, 323
in hemostasis, 213
ICAM-1 regulation, 146, 171, 323, see also ICAM-1
inflammation
integrins in, 177
neutrophil adherence in, 151
and integrins, 172, 177
in vivo flow assay system, 151
and monocyte adhesion in atherosclerosis, 153, 154
neutrophil activation
in inflammation, 151
and selectin expression, 206
selectins, 206, 212, 213
and tumor cell adhesion molecules, 324
Cytoskeleton, 4, 6
adhesion receptor families, 253
cadherins and, 254–257
gpIb-IX interactions, 261–264, 311
mechanism of association, 262–264
role of cytoskeleton, 261–262
gpIIIa interactions, 309
integrins and, 167, 257–261
adhesive interactions, 257–259
cell motility effects, 172, 173
extracellular matrix linkages, 22
gpIb-IX, 261–264, 311
gpIIIa interactions, 309
phosphorylation and, 172
proteins mediating, 259–261
regulation of, 172
structure–function relationships, 169
transmembrane signaling, 257
N-CAM linkage, 189
platelet
and adhesion, 307–308
gpIb-IX, 261–264, 311
gpIIIa interactions, 309
tumor cell interactions, 318, 321
Cytotactin, 228, 233–234, 239
Cytotoxic T lymphocyte-target cell pairs, 27–30

D

Dansylarginine, 318
Dazmegrel, 295
Dazoxiben, 339
Debye length, 6, 29
Deletional analysis, β_1 subunit cytoplasmic domain, 169
Dental plaque, 8–9
Development, see also Homing of hemopoietic cells
cadherins in, 253
integrin expression, 170, 174
N-CAM and, 188, 194
Dextrans, 9, 36
Diabetes mellitus, 292–293
Dibucaine, 262
Dictyostelium discoideum
adherence to bacteria and substrate, 65–66
cAMP relay during development, 64
cohesion after aggregation, 73–74
cohesion prior to aggregation, 66–73
contact site A identification, 67
contact site B, identification of, 71–73
contact sites A and B, definition of, 66–67
gp80, structure–function analysis, 67–70
gp80-mediated, 67
gp80-mediated, developmental relevance of, 70–71
functional effects of cohesion, 74–80
cAMP relay, 76–79
gene expression, 75–76
interspecies sorting, 84–75
prespore and prestalk cell sorting, 79–80
life cycle, 63–64
Dictyostelium
mucoroides, 75
purpureum, 75
Differentiation antigens, 272
3,4–Dihydroxyphenylalanine, 107
Diltiazem, 296
Dimer formation, 5, 9
Dipyridamole, 295, 338–339
Directive endocrinization, 272–273
Discoidin I, 66, 75
Disintegrins, 340
Disseminated intravascular coagulation, 293
DLVO approach, 7
DNA sequences, integrins, 165, 166
Docosapentaenoic acid, 294
Dogfish (*Scyliorhinus canicula*), 133–134
DOPA proteins, 108–112
amino acid composition, 114–115
biomineralization matrices, 132, 133
curing and crosslinking, 133
diversity of, 123–127
precursors, mussel, 119–122
sclerotized structure biosynthesis, 131
secondary structure, 127–130
Doublets, cell, 5, 9

DREG.56, see L-Selectin
Dreissena polymorpha, 113, 123
Drosophila
collagen type IV, 227–228
fibronectin, 230
laminin A chain, 224
position-specific antigen-1, 174–175
D2X, 274, 277
Dynamic models, 35–36

E

E-cadherin, 254–256
EDTA, 78–80
EHS tumor, 235
bamin, 230
basement membrane proteoglycans, 228
BM-40, 231–233
proteoglycans, 229
Eicosapentaenoic acid, 294, 342–343
ELAM-1, 361, see also E-Selectin
Elastase, 324
Elastic model, 26
Elastic strain energy, 26
Elastin, pulmonary ECM, 324
Electric fields, 3
Electron microscopy, 4
Electrophoretic mobility, bacteriopheresis, 100
Electrostatic interactions, 6, 7, 28–29
and bacterial attachment, 87, 93
Dictyostelium discoideum, 65
phenolic bioadhesives, 113
Ellipsometry, 44, 49
Embolization, tests of, 11
Embolus formation, see Platelet aggregation
Embryonic development
hemopoietic cell homing, 272–273
integrin regulation, 174
N-CAM in, 188, 195
neural crest cells, 173
Enactin, structure, 225–227
Endocrine factors, and tumor cell receptors, 323, 355
Endocrine tissue and tumors, N-CAM, 194
Endocrinization, directive, 272–273
Endocytosis, 9, 173
Endothelial cells
flow and, 18
ICAM-1, see ICAM-1
ICAM-1 up-regulation, 146
integrins
cell–cell adhesion, 17
in vitro, *in vivo*, and *in situ*, 170
leukocyte adhesion deficiencies and, 174
leukocyte-endothelial cell interactions, 172
localization of, 172
tissue distribution, 171
mathematical modeling, 36
and neutrophil adhesion, 211–212
leukocyte adhesion deficiencies and, 174
leukocyte interactions, 172
selectins, 208
E-selectin, 206, 207, 212, 213
P-selectin, 147, 207
viscoelastic standard solid model, 26
Endothelium, see also specific cells
basement membrane, see Basement membrane
cell interactions, see Platelets; specific cells
bone marrow structure, 280–282
extracellular matrix, see also Extracellular matrix
hemopoietic cell homing, 282
in inflammation, 206, see also Inflammation
organ-specific biomatrices, 323
Endothelium-derived relaxing factor (NO), 296, 342, 355
Endothelium-leukocyte interactions, see Leukocyte-endothelial cell adhesion, flow effects
Endothelium-platelet interactions, see Platelet adhesion; Platelet aggregation
Endothelium-tumor cell interactions, see also Tumor cell-platelet-vessel wall interactions
Endotoxins, 177, 355
Energy, adhesion, 24–28
Energy density, surface adhesion, 23–25, 28–32, 35
Englebreth-Holm-Swarm tumor, 222, 224–225
Enterobacter aerogenes, 76, 77
Eosinophils, 144, 146
EP-045, 294
Epidermal cells, integrin localization, 172
Epidermal growth factor
ECM sequence motifs, 223
gpIIIa tyrosine phosphorylation sites and, 308–309
Epinephrine, platelet
AVP and, 292
calcium channel blockers and, 296
and gpIIb/IIIa conformation, 308
and gpIIb/IIIa fibrinogen-binding site expression, 309
in hypertension, 293
and platelet aggregation, 290, 291, 296
platelet-tumor cell-vessel wall interactions, 325, 326
species differences, 292
Epithelial cells/epithelium
bacterial adhesion, 94–98
mucosal cleansing and, 98–99
specificity, 93–94
basement membrane, see Basement membrane
extracellular matrix, see Extracellular matrix
integrin localization, 171
Epitope mapping, $\alpha_4\beta_1$, 170
Equilibrium models, 35
Equilibrium properties of adherent cell system, 32–35
ESCA, see X-ray photoelectron spectroscopy
Escherichia coli
bacterial surface agglutinin specificity, 90

carbohydrate receptors, 91
extracellular matrix and, 100
integrins and, 175
pap gene cluster, 90
phagocytosis, 97
synthetic substrates, 42
N-(3–Ethy-1,5–pentanediyl)amide, 318
Evolution
hempoietic cells, 272
laminin/nidogen complex, 227
Ewing's sarcoma, 195
Exocytosis, 9
Extensins, 118
Extracellular ligands, integrins and
phosphorylation and, 172
regulation of, 172
structure–function relationships, 169
Extracellular matrix
and bacterial attachment, 99–100
cell adhesion to, 237–239
fibroblast binding, anti-β_1 MAB and, 172
hemopoietic cell homing, 280
integrins
binding, 170
cell motility effects, 172
distribution of, 170
embryogenesis and morphogenesis, 174
localization in keratinocytes, 172
tissue distribution, 170
lung, tumor cell preferential binding, 324
organ-specific, 323
platelet interactions, 308
platelet receptors, 359
protein structure, 230–234
fibronectin, 230–231
SPARC/BM-40/osteonectin, 231–233
tenascin/cytotactin/hexabrachion, 233–234
Vollagen VII, 230
tumor adhesion, 325
tumor cell extravasation, 324
tumor cell interactions, 317
von Willebrand factor binding, 356
Extracorporeal oxygenation, 9
Extravasation
integrins and, 211
leukocyte, 361
flow effects, 147–148
E-selectin and, 207
tumor cell, 324
Ex vivo test, platelet aggregation, 11–12

F

Factor V, 290
Factor VIII, 356
Factor VIII-related antigen, 290
Factor X, 144
Fasciola hepatica, 108, 114–115, 117, 122, 123, 126, 133, 134, 141
Fatty acids
12(S)-HETE, 321, 322
omega-3, 294
and platelet adhesion, 342–343
and platelet aggregation, 342
FDCP-1, 274, 275
Ferreascidin, 128
Fertilization, 9
Fetal development
hemopoietic cell homing, 272–273
N-CAM and, 195
skin, $\alpha_5\beta_1$, 171
Fibrillae, 87
Fibrin, 352, 354, 356
Fibrinogen, 7
ECM sequence motifs, 223
Glanzmann's thrombasthenia and, 174
integrin binding, 144, see also $\alpha_{IIb}\beta_3$
$\alpha_{IIb}\beta_{III}$ binding β subunit mutations and, 169
epithelial cells, 171
RGD sequence, 16
integrin expression, 144, 170
platelet, 290
and tumor cells, 321, 322
release of, 288
platelet-activating factor and, 291
platelet activation
cytoskeleton rearrangements, 257
by tumor cells, 321
platelet aggregation, vitamin E and, 342
platelet binding by unstimulated platelets, 310
red cell aggregation, 9
tenascin homology, 233–234
vitronectin and, 361
Fibrinogen, soluble, 171
Fibrinogen receptors, platelet, see also $\alpha_{IIb}\beta_3$
gpIb-IX complex-cytoskeleton interactions, 261
platelet gpIIb/IIIa conformation and, 308
Fibrinolysis, 351
Fibroblasts
adhesion, flow and, 10–18
E-cadherin transfection, 256
integrins and, in wound healing, 177
Fibroin, 128
Fibronectin, 5, 8, 221
antibodies to, 342
and bacterial attachment, 87, 100
cell attachment, 239
ECM sequence motifs, 223
gpIc-IIa as receptor, 359
gpIIb/IIIa binding, 359
hemopoietic cell homing, 280
integrins and, 167

$\alpha_4\beta_1$-mediated adhesion, inhibition by monoclonal ABs, 170
binding and cell distribution, 144
cell motility effects, 172
during development, 174
in epithelial cells, 170–172
gpIIb/IIIa adherence, 169, 310
in inflammation, 176–177
in lymphoid tissue, 171
RGD sequence recognition, 167–168
tissue distribution of, 170
in wound healing, 177
perlecan vs., 228
platelet adhesion, 308
and platelet aggregation, vitamin E and, 342
platelet binding, unstimulated platelets, 312
platelet granule contents, 290
platelet-tumor cell interactions, 321, 322
structure, 230–231
vitronectin and, 361
Fibronectin-fibrinogen integrins, tissue distribution, 171
Fibronectin receptors, see also $\alpha_4\beta_1$; $\alpha_5\beta_1$
$\alpha_5\beta_1$, cell distribution, 171
platelet, gpIc/IIa, 311
tumor cell, 323, 324
Fibronectin type III, tenascin-homologue protein, 234
Fibulin, 226, 233
Filamentous hemagglutinin, 175
Filamin, 262–264
Fimbriae, 87
fimbrillin subunit, 89
lectins, 88–90
opsonization, 97
population shifts, 91–94
Fimbrillin, 89, 90
Fimbrin, 259
Flow chambers
leukocyte–endothelial cell adhesion, 150–151
platelet adhesion and aggregation studies, 18
platelet aggregation and adhesion studies, 13–15
substrate surface properties, 44, 52, 54
Flow dynamics, 7, see also Shear stress
and endothelial cell adhesion, 18
and fibroblast adhesion, 18–19
and leukocyte–endothelial cell adhesion
model systems, 150–151
physiological applications, 151–154
receptor function, 147–150
receptor properties, 144–147
and platelet adhesion, 10–18, 307, 312, 358
Flow streamlines, 12–14
Fluid dynamics, 7
Fluid stress, see Shear stress
fMLP, 146, 149
Focal contacts, 167, 168, 257, 260
Fodrin/ankryn, 255, 256
Follistatin, 232
Footprints, platelet adherence and, 15–17
Formyl-Met-Leu-Phe, 146, 149
Fracture stress, 23
Fucoidan, 207, 211
Fucoidin, 205, 209
Fucose, 90, 207
Fucose-1–phosphate, 205, 208–209
Fusion, membrane, 9

G

Galactose, 88, 90
Galactose Galα1–4Gal moiety, 88, 89
Galactose Galα4Galβ, 90
Galactose Galα4GlcNAc, 90, 95
β-D-Galactosido(-1,3–)*N*-D-galactosamine residues, 90
γ, see Adhesion energy density
Gas plasma treatment, see Substrate surface chemical factors
G-CSF, 206
George Test, 10
Geukensia demissa, 111, 113–116, 124, 125, 127, 129
Glanzmann's thrombasthenia, 169, 174, 293, 308, 341
GlcNAc, 95
GlcNAcβ3Gal, 90
Glioma cells, 5
Glioma mesenchymal extracellular matrix antigen (GMEM), 233–234
Glomerular basement membranes, 230
GlyCAM-1, 147, 152
Glycine-rich proteins, 108, 128, 130
Glycocalin, 311
Glycocalyx
electrostatic charge, 28–29
sialyl-Lewis^x, 147
Glycoconjugates
bacterial, and bacterial attachment, 92, 93
homing of hemopoietic cells, 273, 274
organ-specific biomatrices regulating, 323
platelet activation by tumor cells, 319
silalyation, 319, see also Sialyl-Lewis^x
Glycolipids, fimbriae recognizing, 88
Glycoproteins
charge of, 6
Dictyostelium discoideum
gp24, 72–73, 79, 80
gp80, 67–71
gp95, 73
gp126, 71–72
gp150, 74, 79, 80
gpIa/IIa ($\alpha_C\beta_1$), 311
gpIb, 360
antibodies to, 341

platelet activation by tumor cells, 320
von Willebrand factor binding domains, 357
gpIb/IIIa, 360
gpIb/IX complex, 253, 312, 326
cytoskeleton interactions, 261–264
and platelet adhesion, 310–311
gpIc′-IIa, 359
gpIc/IIa ($\alpha_F\beta_1$), 311
gpIIb/IIIa, 6, 360–361, see also $\alpha_{IIb}\beta_3$
gpIV (CD36), 311, 361
gpIV-vWF, 359
gpV, 311
gpVI, 311
gp63, 175
gp90, see L-Selectin
gp250, 233–234
integrin nomenclature, 164
laminin, 222, 224–225
platelet, 358–361, see also $\alpha_{IIb}\beta_3$
and adhesion, 310–311
and shape change, 307–308
Glycosaminoglycans, 355
Glycosphingolipid GM3, 325
Glycosylaminoglycans, 280
Glycosylation
collagen type IV, 227
homing ligand and stem cell factor, 279–280
integrin subunits, 165–166
laminin, 225
N-CAM, 188, 190
platelet integrin, 308–309
selectins, 205–207
GM3 ganglioside, 325
GM-CSF, L-selectin expression, 206
GMP-140, see P-Selectin
Goodpasture antigen, 228
Gott Ring Test, 11
G-proteins
lymphocyte homing, 153
platelet activation, 309
platelet aggregation, 318
platelet-tumor cell-vessel wall interactions, 325–326
GR32191, 294
Grafts, 8
Granulocytes
endothelial cell adhesion, see Neutrophils
hempoiesis, 273
Gray platelet syndrome, 293
GRGD sequence, *Dictyostelium discoideum* discoidin I, 66
Griffith crack tip, 27
Group A streptococci, 99, 100
Group C streptococci, 100
Group G streptococci, 100
Growth factors, 321, 323, 324, see also specific growth factors

H

Halocynthia roretzi, 113
Halysin, 341
Hamster cheek pouch assay system, 151
Harrick RFGP device, see Substrate surface chemical factors
HCAM (CD44), 147, 324
Healing processes, integrins in, 176–177
Hele-Shaw chambers, 13
Helmholtz free energy, 32
Hemagglutination, lectins causing, 88
Hemagglutinin, filamentous, 175
Hematogenous dissemination of tumor cells, 317–325
Hemidesmosomes, 222
Hemopoietic cell homing, see Homing of hemopoietic cells
Hemostasis, 212–213, 352, 357, see also Platelet adhesion; Platelet aggregation
Heparan sulfate, 221, 229, 237
Heparan sulfate proteoglycan
interactions with other basmement membrane proteins, 237, 238
liver ECM composition, 324
Heparin, 9, 237, 354
and platelet aggregation, 342
von Willebrand factor binding domains, 357
Heparin-binding domains, hemopoietic cell homing, 280
Heparin-like molecules, N-CAM and, 190
Hermes antigen, see CD44
12(S)-HETE, 323, 324
and melanoma integrins, 325
platelet-tumor cell-vessel wall interactions, 326
Hexabrachion, 228, 233–234, 239
Hiditron, see Substrate surface chemical factors
High-density proteoglycan, 221
High endothelial venules
lymphocyte homing, 147, 152
selectins and, 205
Hirudin, 318
Histamine, 151, 177, 207
Histoplasma capsulatum, 175
HL60 cells, 206–208, 210
13–HODE, 322
Homing
lymphoid cells, 147, 152–153
selectins and, 206
tumor cell, 323
Homing ligand, 279–280
Homing of hemopoietic cells
anatomical considerations, 280–282
definition, 271–272
evolutionary considerations, 272
fetal and embryonic, 272–273
molecular nature of recognition, 277–280

postnatal, 273–277
binding studies, 275–277
in vitro studies, 274–275
in vivo studies, 274
molecular probes, 273–274
recognition vs. adhesion, 273
Homing receptor
hemopoietic cell, 280
hemopoietic progenitor cells, 277–279
integrins, in lymphoid tissue, 171
Homotypic cell segregation, 272
Hormones, 323, 355
HT-29, 213
Human serum albumin-carbohydrate complexes, 207
Hyaluronic acid, CD44 and, 147
Hydration pressure, 7
Hydrocephalus, 195
Hydrophobic interactions, *Dictyostelium discoideum*, 65
Hydrophobicity, 7
and bacterial attachment, 87, 93, 98
bacterial fimbriae, 88–89
Hypertension, 293, 356

I

Ibuprofen, 294
iC3b, 167, 169
ICAM-1, 146, 361
cytokine stimulation and, 151, 323
cytoskeleton interactions, 253n
cytotoxic T cell-target cell adhesion, 29, 30
in inflammation, 151
and integrins, 144, 168, 169, 172
in inflammation, 177
regulation of, 171
subunit associations and ligand specificities, 167
lymphocyte homing, 152
melanoma cell expression, 323
neutrophil adherence and adhesion
in inflammation, 152
selectins vs. integrins and, 211–212
platelet activation by tumor cells, 320
tumor cells and, 152, 320, 325
ICAM-2, 361
cytotoxic T cell-target cell adhesion, 29, 30
integrins binding, 144, 167–169
LFA-1 and, 146
ICAM-3, 144, 146
ICAM-R, 146
IL-1, 361
and endothelial cells, 323
and E-selectin, 146, 206
von Willebrand factor production, 359
inflammation, 151, 177
and integrins, 177
tumor cell-interactions, 324, 325
IL-1β, 212
IL-4, 146
IL-6, 324
IL-8, 151, 154
Immune complexes, and platelet aggregation, 290
Immune thrombocytopenia, 311
Immunity, selectin roles in, 211–212
Immunoglobulin-like domain
Dictyostelium discoideum gp80 homology, 68
ECM sequence motifs, 223
N-CAM, 191
Immunoglobulin superfamily molecules
cytoskeleton interactions, 253n
leukocyte–endothelial interactions, 144–146
platelet activation by tumor cells, 320
Impedance method, platelet aggregation measurement, 289
Indomethacin, 294, 338
Infections, see also Bacteria
integrins in, 175
leukocyte adhesion deficiencies and, 174
Inflammation, 324
ICAM-1 induction and, 146
integrins in, 170, 172, 176–177
leukocyte-endothelial cell interactions, 172
lymphocyte homing, 153
neutrophil adherence under flow, 151–152
selectin roles in, 146–147, 206, 207, 211–212
Inflammatory bowel disease, 177
Infrared spectroscopy, 44–45, 49
Insulin receptors, 309
Integrins, 325, see also specific α-β heterodimers
cytoskeleton interactions, 221, 253, 257–261
cytoskeleton linkage with extracellular matrix, 221
in disease, 174–177
congenital dysfunction, 174–175
infectious agents, 175
inflammatory and healing processes, 176–177
malignancy, 175–176
distribution, 164
function, 166–169
in inflammation and immunity, 211
initial description of, 162–164
leukocyte–endothelial cell interactions
flow effects on receptor function, 149
receptor properties, 144–145
nomenclature, 164–165
normal function, 170–174
cell–cell adhesion, 172
cell motility, 172–174
developmental regulation, 174
expression, 170–171
regulation, 171–172
organ-specific biomatrices regulating, 323
platelet
and adhesion, 308–311
and adhesion without stimulation, 311

platelet activation by tumor cells, 318
structure, 165–166
structure and function of, 165–170
therapeutic manipulation of function, 177
tumor cells
adhesion, 325
ECM interactions, 325
platelet activation, 318
cytokines and, 324
Intercellular adhesion molecules, see ICAM-1; ICAM-2
Interfacial tension, 7
Interferon-γ, 361
Interleukin-1, see IL-1
Intermolecular forces, mathematical modeling, 28–32
Intestine
collagen/laminin receptors, 171
integrins, in inflammatory bowel disease, 177
Intgegrin-like receptors, hemopoietic cell homing, 280
Intracellular pH, platelet, 308
Invasin, 175
Invertebrates, see Bioadhesives, polymeric
In vitro testing, 8
In vivo systems, leukocyte–endothelial cell adhesion, 151
Ischemia-reperfusion injury, 177
Ischemic heart disease, 293

J

J1 glycoprotein, 233–234
Jump conditions, boundary conditions of adhesion regions, 31–32, 34
Junctional epidermolysis bullosa, 222
"Juxtacrine" activation mechanisms, 151

K

Kalinin, 222, 230
Kalinin/nicein, 222
Kaolin, 290
K88 cluster, 90
Keratinocytes, 172
Kinetic theory of aggregation, 9–10
K-laminin, 230
Klebsiella, 93, 95, 97
pneumoniae, 90, 91

L

LacCer, 325
Lacto-N-fucopentose III (LNF III), 210
LAD, 169
LAD-1, 152
LAD-2, 152
LAM-1, see L-Selectin
LAM-1 antigen, 205
Lamina densa, 220, 230
Lamina fibroreticularis, 220
Lamina lucida, 220, 221
Laminates, see Bioadhesives, polymeric
Laminin, 220
and bacterial attachment, 87, 100
cell attachment, 239
collagen VII NC domains and, 230
fibroblast binding, anti-β_1 MAB and, 172
gpIc′-IIa as receptor, 359
integrin binding, 170, 171
integrins and, 167, 173
cell/tissue distribution, 170
cytokine regulation, 172
cytokines and, 172
during development, 174
in inflammation, 176–177
localization in keratinocytes, 172
RGD sequence recognition, 168
osteosarcoma cell binding, TGFβ and, 323
perlecan vs., 228
platelet binding, unstimulated platelets, 312
pulmonary ECM, 324
self-interaction, 235
sequence motifs, 223
Laminin-collagen type IV interaction, 236
Laminin/nidogen complex
interactions with other basmement membrane proteins, 237, 238
structure, 227
Laminin receptors, tumor cells, 323, 324
Langmuir-Blodgett films, 59
LDL, and platelet aggregation, 290
LDL-receptor, ECM sequence motifs, 223
Lead poisoning, 195
LECAM-1, see L-Selectin
LECAM-2, see E-Selectin
LECAM-3, see P-Selectin
LEC-CAMs, see Selectins
Lectins, see also Selectins
bacterial
biology, 88
carbohydrate receptors and recognition, 90–91
epithelial cells, adhesion to, 94–95
experimental evidence, 94–98
fimbriae, 88–90
forms of attachment, 92–93
phagocytic cells, adhesion to, 95–97
phase variation, population shift in fimbrial phase, 91–94
polymers, adhesion to, 97–98
Dictyostelium discoideum, discoidin I, 66
Dictyostelium discoideum gp95 staining, 73
Dictyostelium discoideum receptors, 65
hemopoietic cell homing, 273, 274, 278, 282
platelet activation by tumor cells, 319
Lectin-type interactions, *Dictyostelium discoideum*, 65

Leishmania mexicana, 175
Leu-8, see L-Selectin
LeuCAMs, 145, see also $\alpha_4\beta_1$
Leukemia cell lines, 206–208, 210
Leukocyte adhesion, see Neutrophils
Leukocyte adhesion deficiency, 174
Leukocyte adhesion deficiency-2 (LAD-2), 147
Leukocyte adhesion deficiency-1 (LAD-1), 144
Leukocyte antigens, 145, see also $\alpha_4\beta_1$
Leukocyte-endothelial cell adhesion, flow effects, 143–154
  adhesion receptor function, 147–150
    extravasation process, 147–148
    integrin-peptide-mediated adhesion, 149
    leukocyte recruitment, 149–150
    selectin-carbohydrate-mediated adhesion, 148–149
  model systems for study, 150–151
  physiological applications, 151–154
    inflammation, 151–152
    lymphocyte homing, 152–153
    monocyte adhesion in atherosclerosis, 153–154
  receptor properties, 144–147
    Ig superfamily, 145–146
    integrins, 144–145
    miscellaneous adhesion molecules, 147
    selectins, 146–147
  regulation of integrins, 171
Leukocyte homing, see Lymphocyte homing; Homing of hemopoietic cells
Leukocyte migration
  integrins and
    cell–cell adhesion, 172
    in inflammation, 177
  E-selectin and, 207
Leukocyte motility, integrins and, 172
Leukocyte-platelet interactions, P-selectin and, 212
Leukocyte recruitment, 324
  flow effects, 149–150
  ICAM-1 induction and, 146
Leukocytes, 6, see also Lymphocytes; Neutrophils
  bacterial phagocytosis, 95–97
  hempoiesis, 273
  integrins, 171
    in inflammation, 177
    nomenclature, 164
  selectins, 212–213, see also L-selectin
Lewis^x, 147, 210, 213
  platelet activation by tumor cells, 319–320
  tumor cell adhesion to endothelium, 323
  tumor expression, 213
LFA-1, 145, see also $\alpha_L\beta_2$
LFA-3, 29, 30
LHR, see L-Selectin
Ligand, 87
Ligand-receptor complexes, 4
Linear elastic fracture mechanics, 27
Lipid phosphorylation, platelet-tumor cell-vessel wall interactions, 326
Lipophilicity, and bacterial attachment, 87, 93
Lipopolysaccharide (LPS)
  $\alpha_M\beta_2$ binding, 169
  inflammation, 151
  and integrins, 175
  E-selectin induction, 146, 212
Lipoproteins, and platelet aggregation, 290
Lipoteichoic acid, 87, 99
Lipoxygenase pathway, 338
Liver
  ECM composition, 324
  hemopoiesis, 273
Local stress, 23
Local velocity gradient, 11–12
Loricrin, 118, 128
Lu-ECAM-1, 323
Lumi-aggregometry, 289
Lung
  collagen/laminin receptors, 171
  melanoma cell-endothelial cell interactions, 323
Lung carcinoma
  N-CAM in, 195
  platelet activation, 319–320
Lung metastasis, 318, 324
  platelets and, 321
  tumor cell homing, 323
Luteinizing hormone-releasing neurons, 194
Lymph nodes, collagen/laminin receptors, 171
Lymphocyte homing, 324, see also L-Selectin
  adhesion under flow, 152–153
  leukocyte adhesion deficiencies and, 174
  selectins and, 206
Lymphocyte recirculation, 325
Lymphocytes, 6
  CD44 and, 147
  leukocyte adhesion deficiencies and, 174
  selectin, 208–209, 212, see also L-Selectin
  surface adhesion energy density, cytotoxic T cell-target cell pairs, 27–28
  VAP-1 and, 147
  viscoelastic standard solid model, 25
Lymphoid organs
  integrins, 171
  selectins, 205
Lymphomas, 322
Lymphotoxin, 151

M

Mac-1, see $\alpha_M\beta_2$
Macrophages, 361
  integrins, 144, 177, see also $\alpha_L\beta_2$
    differentiation, Mac-1, VLA-4 and p150,95

expression, 145
Mac-1 expression, 145
pathogens and, 175
laminin binding, 172
Macroscopic mechanisms, 3–7
Magnesium
and *Discyostelium discoideum* adhesion, 80
integrin interactions with ligands, 167
Malignancy, see Tumor cell-platelet-vessel wall interactions; Tumor cells
Mammary carcinoma cells, 324
Manganese
integrin interactions with ligands, 167
integrin receptor ligand affinity, 173
Mannose, 207
bacterial surface agglutinin specificity, 90
fimbriae recognizing, 88
and phagocytosis, 96
Mannose-6–phosphate, 207
and selectin-dependent binding, 205
and L-selectin-mediated binding, 208
Mannose-resistant (MR) fimbriae, 88, 89
Mannosyl residues, bacterial lectins and, 91
Marine cellular adhesion molecules, 9, see also Bioadhesives, polymeric
Marine invertebrate bioadhesives, see Bioadhesives, polymeric
Mathematical modeling, 5, 23–38
continuum mechanics analysis, 24–28
dynamic models, 35–36
equilibrium models, 35
mechanical property model, 37
molecular considerations, 28–32
molecular level, 37
platelet adhesion in flow chamber, 15–17
surface roughness and, 36–37
thermodynamics formulation, 32–35
M&B22948, 296
MECA 79, 209
Mechanical strength of adhesiveness, 23
Mechanoreceptor, endothelial, 18
Megakaryocytes, 207
MEL-14, 205, 206, 209, 211, see also L-Selectin
Melanoma cells
adhesion systems, 325
endothelial cell interactions, 323
integrins, 171, 175–176
platelet activation by, 321
platelet aggregation, TICPA, 318
Membrane anchoring, *Discyostelium discoideum* gp80, 70
Membrane assembly, 9
Membrane filters, see Substrate surface chemical factors
Membrane fusion, 9
Membranes
Discyostelium discoideum gp80, 70
equilibrium properties of adherent cell system, 34
and red cell adhesion film, 8
Membranes, synthetic, 4, see also Substrate surface chemical factors
Memory T cells, 145
lymphocyte homing, 152
L-selectin expression, 146
Meningitis, 177
Mercenaria mercenaria, 132
Merosin, 222
Metalloproteins, integrins as, 167
Metastasis, see also Tumor cell-platelet-vessel wall interactions; Tumor cells
integrins and, 175–177
N-CAM and, 195
platelet-tumor cell-vessel wall interactions, 317
selectins and, 213
MHC-I, cytotoxic T cell-target cell adhesion, 29, 30
MHC II, tumor cell expression, 318
MHC III, 234
Microglia, 177
Mobility, cell, see Motility, cell
Mobility, membrane
Discyostelium discoideum gp80, 70
equilibrium properties of adherent cell system, 34
Modiolus modiolus squamosus, 114–115, 117, 125
Molecular mechanisms, 3–4, 6–7
mathematical models, 28–32, 37
thermodynamics formulation, 33–35
Mollusks, 109–110, 112–115
phenolic protein consensus sequences, 116–118
polypeptide resins as matrices for biomineralization, 132–133
Monensin, 308
Monoclonal antibodies
$\alpha_4\beta_1$ epitope mapping, 170
α-specific, immunolocalization in cultured endothelial and skin cells, 172
β_1, fibroblast binding effects, 172
integrin, therapeutic use of, 177
Monocyte chemotactic protein-1, 153, 154
Monocytes, see also Macrophages
in atherosclerosis, 153–154
bacterial phagocytosis, 95–97
differentiation of, Mac-1, VLA-4 and p150,95 expression, 145
integrins and ligands, 144
Mac-1 expression, 145
and E-selectin expression, 209
L-selectin, 146, 205, 206, 208, 212
P-selectin and
in hemostasis, 213
platelet and endothelial cell binding, 207
as receptor for, 208
Morphogenesis, see Development

Motility, cell, see also Cytoskeleton
integrins and, 170, 172–174
tumor cell, 322, 324
Mucosa
bacterial attachment, 88–89, 98–99
lymphocyte homing, 152
Mucosal addressin cell adhesion molecule (MAdCAM-1), 146
lymphocyte homing, 152
as L-selectin and $\alpha_4\beta_7$ ligand, 147
integrins binding and cell distribution, 144
Mucosal lymph node vascular addressin, 209
Multiple-attenuated internal reflection IR spectroscopy, 44–45
Multiple myeloma, 194
Muscle
disorders of, N-CAM in, 195
integrins and, 171, 173
during development, 174
Drosophila myospheroid phenotype, 175
Mussels, 114–115
DOPA protein precursors, 119–122
DOPA protein secondary structures, 129
phenolic protein consensus sequences, 116–119
polypeptide resins as matrices for biomineralization, 132–133
Mycoplasma
gallisepticum, 90, 91
pneumoniae, 90, 91
Myeloproliferative disorders, 293
Myocardial infarction, 293
Myosin light chain kinase, 318
Myospheroid phenotype *Drosophila*, 174–175
Mytella guyanensis, 114–116
Mytilus
californianus, 114–116, 127
chilensis, 114–115
edulis, 113–115, 124, 127
phenolic protein consensus sequences, 116, 119
polypeptide resins as matrices for biomineralization, 132–133

N

Natural killer cells, 144, 194
Nerve cells, see also Neural cell adhesion molecule
integrins and
during development, 174
in Alzheimer's disease, 177
N-CAM and neurite outgrowth, 192
NeuAc, bacterial adhesion, 95
NeuNAcα2–3Galβ3GalNAc, 90
NeuNAcα2–3Galβ4Glc, 90
Neural cell adhesion molecule (N-CAM), 187–195
biological studies, 194–195
Dictyostelium discoideum gp80 sequence homology, 68
in disease, 195
E-cadherin chimeras, 254
ECM sequence motifs, 223
isoforms, 189, 190
molecular biology, 190–194
platelet activation by tumor cells, 320
structural studies, 187–190
Neural cell adhesion molecule-related motifs, perlecan, 228–229
Neural crest cells
endocytosis of $\alpha_5\beta_1$, 173
integrins, 171, 174
N-CAM and, 194
Neural extracellular matrix, tenascin-homologue protein, 234
Neuraminidase, 205, 209
Neurite outgrowth, N-CAM and, 192
Neuroblastoma cells, 195, 323, 324
Neurological disorders, N-CAM in, 195
Neutrophils
adherence to endothelium
endothelial injury and, 355
E-selectin and, 206
flow studies, 151–152
adhesion studies, 151–152, 325
bacterial phagocytosis, 95–97
extravasation of, 147–148
vs. lymphocyte homing, 153
pathogens and, 175
integrins and, 172
leukocyte adhesion deficiencies and, 174
platelet interactions, 212, 213
E-selectin, 206–209
L-selectin, 146, 148, 205, 206, 208, 212
P-selectin, 206, 208
in hemostasis, 213
in inflammation, 211–212
platelet interactions, 153, 213
as receptor for, 208
sialyl-Lewis^x expression, 147
viscoelastic standard solid model, 25
Newtonian fluids, 12, 13
NF-κB, 207
NF-κB-like factor, 147
Nicein, 222
Nidogen, 228, 235, 236, 239
Nidogen/entactin, 220, 225–227
Nifepidine, 296
Nigericin, 308
Nitric oxide
dipyridamole and, 338
and platelet aggregation, 296, 342
Nitrous oxide, 355
Nonsteroidal anti-inflammatory drugs, 294

Nude mouse skin assay systems, 151

O

OKY-046, 339
OKY-1581, 294, 339, 340
Omega-3 fatty acids, 294, 342–343
Oncofetal antigen, N-CAM as, 195
Opsonization, and $\alpha_M\beta_2$ binding, 169
Opsonophagocytosis, 95–97
Optical methods, platelet aggregation measurement, 289
Optical microscopy, 4
Orbital mechanics, 4
Organic composites, see Bioadhesives, polymeric
Organotypic formations, blood-forming organs, 272
Organ-specific matrix components, 324
Osteonectin, 231–233
Osteosarcoma cells, 172
Ovomucoid, 232
Oxygenation, extracorporeal, 9

P

p150,95, see $\alpha_X\beta_2$
PADGEM, 361, see also P-Selectin
Palmitate, *Discyostelium discoideum* gp80, 70
Pancreatic secretory trypsin inhibitor, 232
Pap gene cluster, 89, 90
Parallel-plate flow chamber, 150
Parasites, see Bioadhesives, polymeric
Patch clamp technique, 6
P-blood group, 88
PECAM-1, 361
  cytoskeleton interactions, 253n
  platelet activation by tumor cells, 320
Peptides, 340
Peripheral lymph nodes
  homing receptor, see L-Selectin
  lymphocyte homing, 152
Peripheral lymph node vascular addressin (PNAd), 209
Peritoneal macrophages, 172
Peritonitis, 177
Perlecan, 226, 228, 237
Persantine, 295
Pertactin, 175
Peyer's patches, 152
P-fimbriae, 88, 89
pH, intracellular, platelet, 308
Phagocytes, bacterial adhesion to, 95–97
Phagocytosis, 174, 177
Phase contrast microscopy, substrate surface properties, 44–45
Phaseolus limensis, 88
Phenolases, 109
Phenolic protein, 108, see also Bioadhesives, polymeric
Phenoloxidases, 109
Phorbol esters
  $\alpha_L\beta_2$-I-CAM binding, 257
  integrin receptor ligand affinity, 173
  macrophage laminin binding, 172
Phosphatidyl inositol, see Signal transduction
Phospholipase A, 318, 325, 326
Phospholipase C, 309, 325, 326
Phosphorylation
  gpIb/IX, 310, 311
  gpIIb/IIIa, 308–309
  gpIIIa, 258–259, 309
  integrin-cytoskeleton interactions, 260
  integrin receptor ligand affinity, 173
  platelet activation, 318
  platelet integrin-cytoskeletal interactions, 257, 258, 259
  platelet-tumor cell-vessel wall interactions, 326
Photoaffinity studies, β_3 integrins, 169
Phragmatopoma, 110
  californica, 114–115, 117, 123
Physicochemical forces, 5–7
  and bacterial attachment, 98–99
  mathematical modeling, see Mathematical modeling
Pi region, N-CAM, 191, 192
Plane-cone viscometer, 9
Plant hydroxyproline-rich proteins, 118
Plants
  glycine-rich proteins, 128, 130
  phenolic protein consensus sequences, 118, 119
Plaque, dental, 8–9
Plasminogen activators, 354
Platelet-activating factor, 350, 361
  and neutrophils in inflammation, 151
  and platelet aggregation, 290, 291
  platelet-tumor cell-vessel wall interactions, 325, 326
  species differences, 292
Platelet activation, 289–292
  and gpIIB/IIIa fibrinogen-binding site expression, 309
  and P-selectin, 207–208
Platelet adhesion, 10, 307–312
  vs. aggregation, 8
  to damaged vessel walls, 349–362
    adhesion process, 356–361
    endothelial injury, 355–356
    endothelial reaction to injury, 356
    endothelium, 352–355
    platelet factors, 350–352
  endothelial injury and, 355
  flow and, 10–18
  flow dynamics and, 312
  integrins and
    gpIIb/IIIa activation, 308–310
    in inflammation, 176–177
  leucine-rich glycoproteins, 310–311

pharmacological intervention, 337–343
antibodies, 341–342
aspirin, 337–338
dipyridamole, 338–339
indomethacin, 338
miscellaneous, 342–343
peptides, 340–341
prostacyclin analogues, 339
prostaglandin E_1, 339
sulphinpyrazine, 338
thromboxane synthetase inhibitors, 339–340
ticlopidine, 340
shape changes, 307–308
without stimulation, 311
Platelet-aggregating factor, tumor cell, 321
Platelet aggregation, 8, 10
vs. adhesion, 8
agents inducing, 289–292
ADP, 290
arginine^8-vasopression, 291–292
epinephrine, 291
platelet activating factor, 291
thromboxane A_2, 291
disorders of, 292–293
Glanzmann's thrombasthenia, 174
inhibitors of, 293–296
agents affecting arachidonic acid metabolism, 294–295
agents affecting cAMP, 294–295
calcium antagonists, 296
nitric oxide, 296
ticlopidine, 295–296
integrins and, see also $\alpha_{IIb}\beta_3$
β_3 subunit blockade of $\alpha_{IIb}\beta_{III}$ binding, 169
cytoskeleton interactions, 257–261
gpIIb/IIIa and, 308–310
regulation, 171
measurement methods, 288–289
species differences, 292
tumor cell induction of, 317–321
Platelet aggregation protein complex, see gpIIb-IIIa; $\alpha_{IIb}\beta_3$
Platelet-derived growth factor, 290
Platelet endothelial adhesion molecule-1 (CD31), 146
Platelet factor IV, 290
Platelet gpIIb-IIIa, see gpIIb-IIIa
Platelet integrin (gpIIb/IIIa), see $\alpha_{IIb}\beta_3$
Platelets, 6
gpIb-IX complex-cytoskeleton interactions, 253, 261–264
integrins, pathogens and, 175
selectins, see also P-Selectin
characteristics, 208
expression of, 207
in hemostasis, 212–213
in inflammation, 212
in tumor metastasis, 213
thrombin receptor, 318
tumor cell interactions, 213
platelet action on tumor cells, 321–322
signal transduction pathways, 325–327
tumor cell induction of platelet aggregation, 317–321
Platyhelminths, 110, 117
Pneumonitis, 177
Point contact, 36
Polybrene, 355
Polyethylene, see Substrate surface chemical factors
Polylysine, 93
Polymers, see also Substrate surface chemical factors
bacterial adhesion to, 97–98
biological, see Bioadhesives, polymeric
synthetic, see Substrate surface chemical factors
Polymorphonuclear leukocytes, see also Neutrophils
bacterial phagocytosis, 95–97
integrins, pathogens and, 175
Polyphosphomannan ester (PPME), 205, 211
Polysialic acid, N-CAM, 188, 194, 195
Polysphondylium
pallidum, 74, 75
violaceum, 74
Polystyrene, see Substrate surface chemical factors
Population balance method, 9
Porifera, 110, 230
Position-specific antigen-1, 174–175
pp60src, 258, 326
Prostacyclin and analogues, 339, 354, 355
dipyridamole and, 338
endothelial cell production, 18
Prostaglandins, 339
endothelial cells and, 355
in diabetes, 292
and platelet aggregation, 290, 295
calcium channel blockers and, 296
omega-3 fatty acids and, 294
Prostaglandin synthetase inhibitors, 294
Prostheses, medical, 8
Protamine, 354
Proteases
bacterial, 91
and *Discyostelium discoideum* gp150, 80
and P-selectin binding, 210
tumor cells, 324
Proteinaceous conditioning film, substrate surface properties, 44, 49–50, 57–58
Protein C, 354
Protein kinases, see also Phosphorylation
gpIIIa phosphorylation, 258
integrin-cytoskeleton interactions, 260
integrin regulation, 172
Proteoglycans, 220, 237
hemopoietic cell homing, 280

high-density, 221
interactions with other basmement membrane proteins, 237, 238
liver ECM composition, 324
structure, 228–229
Proteolytic shedding of L-selectin, 149, 206
Pseudomonas aeruginosa, 91, 93, 95
bacterial surface agglutinin specificity, 90
phagocytosis, 97
polymer adhesion, 98
Pseudoplasmodium (slug) stage, *Dictyostelium discoideum*, 64
Pseudopodia, 5
Pyura
stolonifera, 113
stolonifera, 114–115

Q

Quantification methods, 7–9
Quantum viewpoint, 4
Quinone-tanned sclerotin, 108

R

R56413, 321
RA-233, 295
Radiofrequency-gas-plasma treatment (RFGPT), see Substrate surface chemical factors
Reaction kinetics, 4
Reaction rate constants, red cell aggregation, 9–10
Receptor binding, 4
Receptor cross-linking, 4–5
Receptor cloning studies, 6
Receptor destroying enzyme (RDE), 90
Rectal carcinoma, 319–320
Red cells
adhesion, 7, 9–10
vs. aggregation, 8
types of films, 8
aggregation, 9–10
doublets, 5, 7
hempoiesis, 273
lectins causing hemagglutination, 88
rouleau formation, 36
Resins, see Bioadhesives, polymeric
Reticular lamina, 220
Retinoic acid, 323
Reuse of sites index (RUSI), 17
RGD (Arg-Gly-Asp) sequence, 144
cell adhesion and binding to ECM, 239
fibronectin, 240
hemopoietic cell homing, 280
and integrins, 167, 171, 309
malignant cells, 175–176
pathogen interactions, 175
therapeutic use of monoclonal antibodies, 177
nidogen/enactin, 225
and platelet adhesion, 340–341
platelet integrin α subunit, 309
RGDS peptides, 340
Rhabdomyosarcoma, 175, 195
Rheology, see also Flow dynamics; Shear stress
Ricin, 88
Ricinus communis, 88
Ristocetin, 290
Rodent ear flow assays, 151
Rouleaux formation, 9–10

S

Salmonella, 90, 91
Sambucus nigra lectin, 210
Sarcoma cells, platelet activation, 318
SC-38249, 294
Scanning electron microscopy, substrate surface properties, 44, 48–49
Schistosome eggcase bioadhesives, 108, 122–123
Schistosomes, 113–115
motif substitution, 127
phenolic protein consensus sequences, 117
phenolic protein precursors, 122–123
polymer secondary structures, 129
sclerotized structure biosynthesis, 131
Schizophrenia, 195
Sclerotin, quinone-tanned, 108
*Scyliorhinus canicul*a (dogfish), 133–134
Sea snail (*Aplysia*), 194
Second messenger systems, see also cAMP; G-proteins; Signal transduction
N-CAM, 194
Secreted Protein Acidic and Rich in Cysteine (SPARC), 231–233
Sedimentation rate, 10
E-Selectin
carbohydrates, 209–210
characteristics and functions, 206–208
cytokine stimulation and, 151, 323
induction of, 146–147
in inflammation, 151, 212
lymphocyte homing, 153
melanoma cell-endothelial cell interactions, 323
and monocyte adhesion in atherosclerosis, 153
platelet activation by tumor cells, 319–320
relative size of molecule, 148
sialyl-Lewis^x binding, 147
structure, 204
terminology, 204
tumor adhesion, 323, 325
tumor cell platelet activation, 319–320
tumor cell secretion of IL-1 and, 325
L-Selectin, 146, 325

carbohydrates, 208–209
characteristics and functions, 205–206, 208
expression on lymphocytes, 212
in inflammation, 151, 212
lectin domain, mapping to, 209
localization on neutrophils, 148
lymphocyte homing, 152
relative size of molecule, 148
terminology, 204
P-Selectin, 361
carbohydrates, 210
characteristics and functions, 207–208
endothelium, cytokine stimulation, 323
in hemostasis and coagulation, 212–213
induction of, 147, 151
in inflammation and immunity, 151, 211–212
and neutrophil adhesion, 211–212
platelet activation by tumor cells, 319–320
relative size of molecule, 148
sialyl-Lewis^x binding, 147
structure, 204
terminology, 204
Selectins, 253
carbohydrate ligands, 208–210
characteristics and functions, 205–208
leukocyte–endothelial cell interactions
flow effects on receptor function, 148–149
receptor properties, 146–147
leukocyte–endothelium interactions, 146–147
molecular structure, 203–205
and monocyte adhesion in atherosclerosis, 153
number of receptors on stimulated neutrophil, 149
roles of in health and disease, 211–213
structure–function relationships, 210–211
terminology, 204
tumor cell
Lex interactions, 213
platelet activation, 319–320
Septifer bifurcatus, 114–116, 127
Sericin, 128
Serotonin, 350
platelet, 288, 290
AVP and, 291
and tumor cells, 321
and platelet aggregation, 290
Serratia marcescens, 90
Sgp, 209
Sgp90, see CD34
Shear stress, 7–9, 23, see also Flow dynamics
local velocity gradient and shear stress in, 11–12
and monocyte adhesion in atherosclerosis, 154
and platelet adhesion, 18–19, 307, 312
red cell disaggregation, 9
and shear flow, 12–14
substrate surface properties and, 59
tumor cells, 322
Shigella flexneri, 90
Sialyl-Lewis^x (SLe^x), 147
CLA and, 153
E-selectin ligand, 209–210
neutrophil adherence in inflammation, 152
platelet activation by tumor cells, 319–320
P-selectin expression, 210
tumor cell adhesion to endothelium, 323
Signal transduction, 4
N-CAM, 189, 190, 194
neutrophil adherence in inflammation, 151–152
platelet aggregation, 318
platelet integrin–cytoskeletal interactions, 257–259
platelet-tumor cell-vessel wall interactions, 325–327
Silks, 128, 129
Skeletal muscle, integrins and, 173
Skin
collagen/laminin receptors, 171
integrin localization, 172
Skin cancer, integrins, 176
Skin inflammation, lymphocyte homing, 153
S-laminin, 222
SLe^x, see Leukocyte adhesion deficiency-2; Sialyl-Lewis^x
Slip bonds, 31
Slug (pseudoplasmodium) stage, 64
Small cell lung carcinoma, 195
Smooth muscle integrins, 171
Sodium fluxes, and platelet shape change, 308
Sodium pump, 256
Soluble forms of L-selectin, 204, 206
Sorting, see Cell sorting
Sorus, *Dictyostelium discoideum*, 64
SPARC/BM/osteonectin structure, 231–233
Spectrin
E-cadherin-cytoskeletal interactions, 255
N-CAM and, 189
Spectroscopy, 4, 44–45
Spleen
hempoiesis, 273
a_4b_1, 171
lymphocyte homing, 152
Spores, *Dictyostelium discoideum*, 64, 75
SQ 22536, 294
Stalk, *Dictyostelium discoideum*, 64
Stamper-Woodruff assay, 150
Staphylococci, 87, 93, 100
Staphylococcus
aureus, 97, 99, 100
saprophyticus, 90, 93, 95
cell surface complex and, 99
lipoteichoic acid, 99
phagocytosis, 96, 99
polymer adhesion, 98
Static assay systems, leukocyte–endothelial cell adhesion, 150

Stem cell differentiation, see Homing of hemopoietic cells
Stem cell factor, 279–280
Stem cell homing, see Homing of hemopoietic cells
Stereochemistry, 6, 7
Steric hindrance, selectin–carbohydrate interactions, 148
Stokes flow theory, 7
Storage pool deficiency, 293
Streptococci, 87
  extracellular matrix and, 100
  lipoteichoic acid, 99
Streptococcus
  mitis, 90
  pneumoniae, 90, 93, 95
  sangui, 90
Stress, see also Shear stress
  adhesive, 24, 25
  continuum mechanics analysis, 24–28
Stress fibers, 257
Stroke prevention, 295
Stromal cells
  marrow culture, 274–277
  tumor cell interactions, 317
Stromelysin, 324
Stylus profilometry, 44
Subbasal dense plaque, 220
Substance P, 151
Substrate interactions, see also Integrins; specific integrins
  basement membrane, see Basement membrane; Extracellular matrix
  Dictyostelium discoideum, 65–66
Substrate surface chemical factors
  analytical methods, 44–45
  attachment strength, 52–55
  bacterial cells, 42–43
  cast films, 49
  cell assays, 45
  cell attachment/retention and morphology, 58–59
  cell attachment studies, 50–52
  cell processing of conditioning films, 57–58
  conditioning film, 49–50
  growth rate studies, 52
  MAIR-IR spectroscopy, 45–48
  masking studies, 56
  polyethyene and polystyrene RFGPT treatment, 43–44
  SEM, 48–49
  surface contact potential, 57
  UV-treated polyethylene and polystyrene, 49, 56
  XPS analysis, 48, 56–57
Sugars
  bacterial attachment, 90, 92, 93
  homing of hemopoietic cells, 273, 274
  and selectin-dependent binding, 205
Sulfated glycolipid, 209
Sulfatide, 209
Sulfinpyrazone, 294, 338
Surface adhesion energy density, 23–25
Surface charge, see Electrostatic interactions
Surface charge density, 6
Surface chemical factors, see Substrate surface chemical factors
Surface properties, physical
  and bacterial attachment, 99
  mathematical modeling, 28–32
  and platelet aggregation, 11
  substrate properties, 46, 48–50
  and bacterial attachment, 99
Surface potential, 44, 57
Swainosine, 323
Synchronization of cells, 42
Synthetic membranes, 4
Synthetic substrates, see Substrate surface chemical factors

T

Talin
  β subunit cytoplasmic domain and, 169
  cell motility effects, 172
  integrin interactions, 167, 172, 259, 260
  T-cell receptor and, 172
T cell receptor, 144
  and $\alpha_L\beta_2$-ICAM-1 binding, 172
  cytotoxic T cell-target cell adhesion, 29, 30
  and L-selectin expression, 206
T cells
  CD31 and, 146
  conversion from naive to memory cells, 145
    $\alpha_4\beta_1$, 171
    regulation of, 172
  cytotoxic T cell-target cell adhesion, 27–30
  integrins and ligands, 144
  lymphocyte homing, 152
  L-selectin expression, 146
  selectins and, 206
Tenascin
  cell attachment, 239
  perlecan vs., 228
  structure, 233–234
Tensin, 259
TGFβ
  in inflammation, 177
  integrin regulation, 172, 177
  osteosarcoma cell binding to laminin, 323
  platelet synthesis, 323
  tumor cells, 323, 324
Theophylline, 292
Thermodynamics, 4–5
Thermodynamics formulation, 32–35
Thimerosal, 290
Thrombin, 171
  gpIb-IX as receptor, 361
  and gpV, 311

inflammation, 151, 177
and integrins
gpIIb/IIIa conformation, 308
gpIIB/IIIa fibrinogen-binding site expression, 309
in inflammation, 177
and platelet aggregation, 290
epinephrine and, 291
P-selectin and, 207, 213
ticlopidine and, 340
platelet-tumor cell-vessel wall interactions, 325, 326
tumor cell factors, 318
Thromboembolic disorders, see also Platelet adhesion; Thrombus formation
bacterial infections and, 355
integrins and, 176–177
platelet activation in, 293
selectins in, 212–213
Thrombomodulin, 213, 354
Thrombospondin, 230, 232
cell attachment, 239
gpIV as receptor, 311, 361
integrin binding, 359
integrin recognition of RGD sequence, 168
integrin subunit associations and ligand specificities, 167
platelet granule contents, 290
tumor cell-platelet interactions, 321, 322
vitronectin and, 361
Thromboxane A_2, 350
in cardiac ischemia, 293
in diabetes, 292
omega-3 fatty acids and, 294
and platelet aggregation, 290, 291, 296, 322
platelet-tumor cell-vessel wall interactions, 318, 320–322, 325, 326
Thromboxane receptor antagonists, 294
Thromboxane synthetase, 296
Thromboxane synthetase inhibitors, 294, 295, 339–340
Thrombus formation, see also Thromboembolic disorders
in flow chamber wall, 18
tests of, 11
Thymocytes, N-CAM, 194
Thymus, integrins, 171
Thyroglobulin-like motif, ECM, 223
Ticlopidine, 295–296, 340
Tissue factor, 318
Tissue necrosis factor, 361
Tissue-specific isoforms, laminin, 222
Tissue specificity
bacterial adhesion, 93–94
integrins, 170
laminin, 222
N-CAM, 194
TNF-α
and endothelial cell adhesion molecules, 146, 206, 212, 323
in inflammation, 151, 177
E-selectin induction, 146, 212
L-selectin expression, 206
and tumor cell adhesion molecules, 323, 324
TNF-β, 323
TQ1, see L-Selectin
Transgenic animals, 6
Trichomya hirsuta, 114–115, 117, 125
Trihydroxyphenylalanine, 119
Trimers, red cell, 9–10
Trypsin, 290
Tumor cell-induced platelet aggregation (TICPA), 318–325
Tumor cell-platelet-vessel wall interactions, 317–327
hematogenous dissemination, 317–325
platelet action on tumor cells, 321–322
platelet aggregation, tumor cell induction of, 317–321
tumor cell-endothelium interaction, 322–325
metastasis, 317
signal transduction pathways, 325–327
Tumor cells
integrins and, 175–176
N-CAM, 194–195
selectins and, 213
Tumor typing, N-CAM and, 195
Tunicamycin, 323
Tunicates, 110, 112
Tyrosinases, 109
Tyrosine, see DOPA proteins
Tyrosine hydroxylation, 109
Tyrosine phosphorylation
gpIIIa, 258
integrin α and β subunits, 172
platelet integrin–cytoskeletal interactions, 257
platelet-tumor cell-vessel wall interactions, 326

U

U937, 206, 208, 210
UK-37248, 294
UK-38485, 294, 295
Uroepithelial cells, bacterial adhesion, 95
Uvomorulin (cadherin E), 254

V

van der Waals interactions, 7, 28, 113
VAP-1, 147
Varnishes, 112
Vascular activation, see Endothelial cells; Endothelium; specific blood cells
Vascular injury, 8, 349–361
and E-selectin expression, 206
and P-selectin redistribution, 207
Vascular smooth muscle integrins, 171

VASE segment, N-CAM, 191, 192
Vasoactive substances, endothelial injury by, 355
Vasoconstriction, in hemostasis, 212–213
Vasopressin
and platelet aggregation, 290
platelet-tumor cell-vessel wall interactions, 325, 326
VCAM-1, 146, 361
$\alpha_4\beta_1$ binding, 168, 170
cytokine stimulation, 323
IL-4 and, 146
integrins and
$\alpha_4\beta_1$, 172
binding and cell distribution, 144
in inflammation, 177
leukocyte adhesion deficiencies and, 174
subunit associations and ligand specificities, 167
lymphocyte homing, 152
melanoma cell-endothelial cell interactions, 323
monocyte adhesion in atherosclerosis, 153
platelet activation by tumor cells, 320
tumor adhesion, 325
tumor cell secretion of IL-1 and, 325
VCAM-1 receptor, see $\alpha_4\beta$
Verapamil, 296
Vessel walls
platelet adhesion, 349–361
tumor cells and, see Tumor cell-platelet-vessel wall interactions
Vibrio cholerae, 90
Vinculin
β subunit cytoplasmic domain and, 169
integrin–cytoskeleton interactions, 259
integrin interactions, 167
Virus infection, 9
Viscoelastic standard solid model, 26
Viscometry, plane-cone, 9
Vitamins, and platelet aggregation, 342
Vitronectin
and bacterial attachment, 87
fibroblast binding, anti-β_1 MAB and, 172
integrins and
$\alpha_{IIb}\beta_{III}$ binding β subunit mutations and, 169
distribution of, 170
during development, 174
epithelial cell binding, 171
fibroblast binding, anti-β_1 MAB and, 172
RGD sequence recognition, 168
platelet receptors, 169, 360
von Willebrand factor binding domains, 357
Vitronectin receptor, see $\alpha_v\beta_3$
VLA-4, see $\alpha_4\beta_1$
VLA-5, 145
VLA-6, 145
von Willebrand factor, 356
Bernard-Soulier syndrome, 311
gpIb/IX as receptor, 310–311, 360
gpIb/IX-cytoskeleton interactions, 261–262
gpIIb/IIIa binding, 169, 310, 360
integrin subunit associations and ligand specificities, 167
platelet adhesion, 308
and platelet aggregation, shear-induced, 312
P-selectin and, 207
vitronectin and, 361
and tumor cells, 321, 322
von Willebrand's disease, 341, 359
VpB protein, *Fasciola hepatica*, 126

W

Weibel-Palade bodies, 147, 207
Wheat germ agglutinin, 65, 73
White cells, see Leukocytes
Wilms' tumor, 195
Wiskott-Aldrich syndrome, 293
Wound healing, 170, 177

X

X-ray photoelectron spectroscopy, 48, 56–57

Y

Yeast, 65
Yersinia pseudotuberculosis, 175
Yolk sac, 273
Young-Dupre equation, 25, 27–28

Z

Zebra mussel, 113, 123
Zisman plots, 46, 47
ZK 36374, 339